AF670412

GENETIC ASPECTS OF PLANT MINERAL NUTRITION

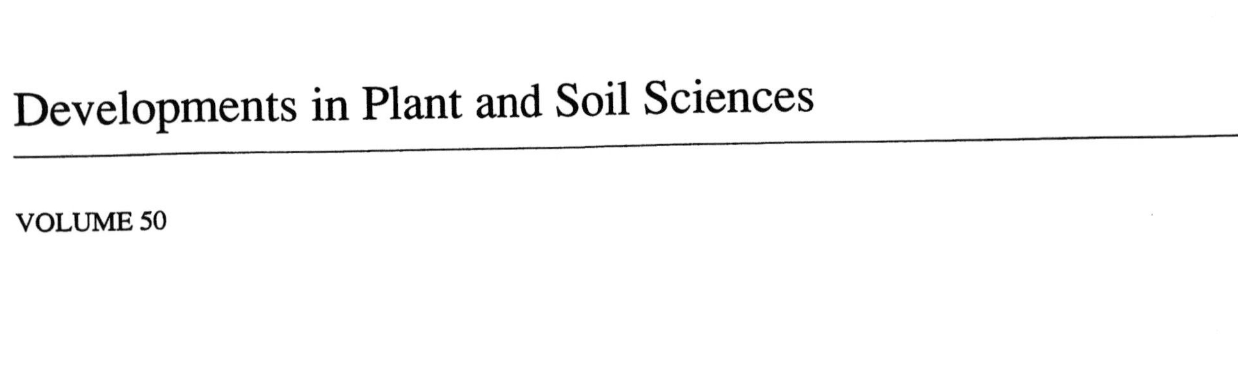

Developments in Plant and Soil Sciences

VOLUME 50

The titles published in this series are listed at the end of this volume.

Genetic Aspects of Plant Mineral Nutrition

The Fourth International Symposium on Genetic Aspects of Plant Mineral Nutrition, 30 September – 4 October 1991, Canberra, Australia

Edited by
P. J. RANDALL
E. DELHAIZE
R. A. RICHARDS
R. MUNNS
Division of Plant Industry
Commonwealth Scientific and Industrial Research Organization (CSIRO)
Canberra, Australia

Contributions with an asterisk in the table of contents were first published in *Plant and Soil*, Volume 146 (1992)

Springer Science+Business Media, B.V.

Library of Congress Cataloging in Publication Data

International Symposium on Genetic Aspects of Plant Mineral Nutrition (4th : 1991 : Canberra, A.C.T.)
Genetic aspects of plant mineral nutrition / the Fourth International Symposium on Genetic Aspects of Plant Mineral Nutrition, Canberra, Australia, 30 September - 4 October 1991 ; edited by P.J. Randall ... [et al.].
p. cm. -- (Developments in plant and soil sciences ; v. 50)
ISBN 0-7923-2118-9 (acid free paper)
1. Plants--Nutrition--Genetic aspects--Congresses. 2. Plants, Effect of minerals on--Congresses. 3. Crops--Nutrition--Genetic aspects--Congresses. 4. Crops--Effect of minerals on--Congresses.
I. Randall, P. J. (Peter J.) II. Title. III. Series.
QK867.I424 1991
581.1'335--dc20 92-43811

ISBN 978-94-010-4721-0 ISBN 978-94-011-1650-3 (eBook)
DOI 10.1007/978-94-011-1650-3

Contents

* Contributions indicated with an asterisk were first published in *Plant and Soil*, Volume 146 (1992).

The Fourth International Symposium on Genetic Aspects of Plant Mineral Nutrition was held in Canberra, Australia from 30 September to 4 October 1991

The Organizing Committee

E Delhaize[1]
D DeMarco[1]
T J V Higgins[1]
N Marcar[2]
Rana Munns[1]
P J Randall[1] (Chairman)
R A Richards[1]

The following provided financial support:

The Australian Wheat Research Council
The International Conference Support Scheme of the Department of Industry, Technology and Commerce.
The Australian Tourist Commission.
The Sulphur Institute, Washington DC.
The Australian International Development Assistance Bureau.

[1]Division of Plant Industry and [2]Division of Forestry Commonwealth Scientific and Industrial Research Organization.

Preface

This volume contains papers presented at the Fourth International Symposium on Genetic Aspects of Plant Mineral Nutrition, held in Canberra, Australia in spring 1991. The manuscripts were reviewed by at least two independent reviewers. The Editors thank Professor H. Lambers and A. Houwers of the Editorial Office of Plant and Soil for their assistance in this process.

There is a need to optimize the productivity of infertile and problem soils in order to meet increasing world-wide demand for agricultural and forestry products. The Symposium and its predecessors (Belgrade 1982, Madison 1985 and Braunschweig 1988 – published as Developments in Plant and Soil Sciences numbers 8, 27 and 42 respectively) recognize the increasingly important role of selection and breeding of plants specifically for such soils. Plant breeding solutions will complement agronomic methods to achieve these objectives in a manner which is both economically sound and ecologically responsible.

Considerable progress has been made in plant improvement for problem soils. Many cultivars and provenances have been developed for their ability to overcome edaphic constraints. New cultivars now allow economic production on soils where yields of standard varieties are normally limited by mineral toxicity of nutrient deficiency. There is a strong emphasis in this volume on studies relating to plant breeding, the strategies to be adopted, the development of selection criteria and the testing and evaluation of promising genotypes. Several contributions draw upon data from already successful plant breeding programs. These include papers on tolerance to acid soils in wheat by Fisher and Scott and in phalaris by Oram, Culvenor and Ridley, a paper on iron efficiency of soybeans growing in calcareous soils by Cianzio and the papers by Paull and Moody and their colleagues on tolerance to high boron in wheat. It is worth noting in this work the value of comparisons using closely related genotypes or near isogenic lines for demonstrating the yield advantage of such characters. There was recognition at the Symposium of the value of developing isogenic lines for physiological and molecular studies as well as for agronomic evaluation of nutrition-related characters.

Future progress will depend on identifying agronomically useful genes and understanding how they operate at the physiological, biochemical and genetic levels. The techniques of molecular biology will play an increasingly important part in this. One fifth of the papers presented in the oral sessions of the Symposium were in the "molecular" area and some of these are represented in this volume. The next Symposium in the series will be held in Davis, California, USA in 1994 with the slightly altered title "Fifth International Symposium on the Genetics and Molecular Biology of Plant Nutrition".

Peter Randall

P. J. Randall et al. (Eds.), Genetic aspects of plant mineral nutrition, 1–8.

PLSO SV01

Are we justified in breeding wheat for tolerance to acid soils in southern New South Wales?

JOHN A. FISHER and BRENDAN J. SCOTT
Agricultural Research Institute, Wagga Wagga 2650, NSW, Australia

Key words: acid soils, aluminium tolerance, breeding, manganese tolerance, selection, *Triticum aestivum*, wheat

Abstract

Acid soils in southern New South Wales (NSW) were recognised as a problem in the late 1960's. Initial research determined the range of tolerance to Aluminium (Al) and Manganese (Mn), the benefits of these tolerances and the interaction with liming. Screening methods were evaluated and the utility of the haematoxylin root tip stain test for Al tolerance confirmed. A nutrient solution system for determining Mn tolerance was developed.

In the southern NSW wheat belt, where the soils are slowly acidifying, selecting for Al tolerance was estimated to increase grain yields by 1.4% at present, increasing to 3.2% in ten years time. This is a worthwhile improvement compared to other objectives of the breeding programme. Selecting for Mn tolerance was not justified.

Varieties in the breeding programme are evaluated with the haematoxylin stain test after the second year of trials on non-acid soils. Subsequent testing is on acid and non-acid soils. This scheme is not designed to maximise Al tolerance, but gives a reasonable balance between the many different objectives which the breeding programme has to consider. It is important that varieties recommended for acid soils also perform well on non-acid soil.

Introduction

Accurate estimation of the regional extent and severity of constraints is essential for rational decisions on resource allocation and research direction (Munns and Scott, 1987). The benefit obtained from breeding for tolerance to mineral stress is a function of the area and severity of the problem, the improvement that can be obtained from breeding, and the benefit of this tolerance compared to other potential breeding objectives.

The acid soils problem was first recognised in southern New South Wales, Australia (NSW) in the early 1960's, initially in the wetter fringes of the wheat belt. By the late 1970's formerly mildly acid soils had become strongly acid and plant production was affected (Osborne et al., 1978). Soil acidity decreases plant growth in many ways, but toxicities of aluminium manganese and hydrogen ions have been recognised as one of the most common causes of reduced yields (Richie, 1989). The mechanisms leading to this acidification have been discussed by Helyar and Porter (1989). In the last 30–40 years the red earths in southern NSW have acidified by approximately 0.5 pH units, these soils may have toxic levels of Mn or Al (Chartres et al., 1990). The current gross value of cereal production lost due to acid soils in the seven shires around Wagga Wagga was estimated to be $14.2 million, this was predicted to increase to $17.3 million in ten years time if soils were not limed (Fraser and Geeves 1990).

Tolerance of soil acidity can play an important role as a component of acid soils management (Cregan et al., 1989; Scott and Fisher, 1989). In severely acid soils, cultivar tolerance is important in addition to liming because of the problem of acidity below the depth of lime incorporation, the variability of acidity across a site and problems with uneven spreading or incorporation of lime. In sites with moderate acidity, cultivar tolerance may provide a yield benefit, even though the yield response is not sufficient to make liming economical.

There are two possible approaches to selecting varieties which tolerate a stress; to select for tolerance directly in the presence of the stress or to select for tolerance indirectly (Atlin and Frey, 1989) by selecting for a character, or in another environment, which correlates with stress tolerance. The benefit of indirect selection compared to direct selection is the ratio of the expected responses (Falconer, 1981). This ratio can be determined by the equation;

$$CR_X/R_X = r_G * i_Y/i_X * h_Y/h_X$$

where R_X is the response to direct selection for character X, CR_X is the correlated response in X when Y is selected for, r_G is the genetic correlation between X and Y, i_Y and i_X are the selection intensities and h_Y and h_X are the square root of the heritabilities. Indirect selection is likely to be more successful than direct selection if the genetic correlation is high and either a large population can be used ($i_Y > i_X$) or there is an improvement in heritability ($h_Y > h_X$).

The option of selecting for yield per se, rather than to selecting for stress tolerance also needs to be considered (Richards, 1983). This option is preferable if the stress is very variable across the field, and the response to selection for stress tolerance is likely to be lower than the response to selection for yield in the absence of stress.

Because acid soils in southern NSW are very variable (Cregan et al., 1989), direct selection for tolerance in the field, based on single plants or unreplicated plots is ineffective. However indirect selection for acid soils tolerance, by selecting for tolerance to Al or Mn is feasible. There is an adequate range of tolerance of both Al and Mn available in wheat, both tolerances are simply inherited and can be selected for efficiently (Foy et al., 1988; Scott and Fisher, 1989). A rapid test for Al tolerance is available (Polle et al., 1978). In this test young seedlings (4 days old) are exposed to Al (0.18, 0.36 and 0.72 m*M*) overnight and the proportion of root staining, after exposure to haematoxylin, is assessed. This test identifies major genes on chromosomes 2D and 4D (Takagi et al., 1982). Our current procedure to identify Mn tolerance is to grow plants in a solution culture for four weeks and to measure the dry weight of the plant tops (Scott, unpublished). The haematoxylin stain test is suitable for large populations, but the Mn testing procedure is too slow. These procedures were used in a pilot breeding programme (Fisher and Scott 1983).

An acceptable cultivar must satisfy a range of requirements. These include, in addition to grain yield, grain value (quality aspects) and reliability/cost of production (disease resistance and agronomic characteristics). The breeding programme needs to integrate all these different objectives. Failure to do this could result in a new cultivar being unacceptable to the industry, even though the narrow objective of improving acid soils tolerance may have been met.

In this paper we determined the benefit which could be obtained, in southern NSW, from selecting wheat with Mn or Al tolerance. These benefits were compared with the other main breeding objectives. We then describe how breeding for acid soils in southern NSW was integrated into the overall wheat breeding programme.

Materials and methods

Estimating the benefit of Mn tolerance

Two series of trials were conducted. In series A four pairs of closely related lines differing in Mn tolerance, but tolerant of Al, were selected from a backcrossing programme designed to transfer acid soils tolerance from Carazinho to Egret (Fisher and Scott, 1987). These lines were grown in a trial at Borambola on a yellow podzolic soil in 1985. The acidity problem in this soil is confined to the top 30 cm, the pH_{Ca} (1:2 soil:0.01 *M* $CaCl_2$ method) at the surface is 3.8 increasing to 4.4 at 50 cm. In series B twelve other lines from the same crossing programme, six susceptible to Mn and six tolerant, were sown in five trials over two years. In 1986 at Wagga, and in 1985 and 1986 at Mendooran and Borambola.

Estimating the benefit of Al tolerance

The results from 65 wheat variety trials in southern NSW between 1986 and 1990 were used. These trials covered the range of soil types and environments within the wheat belt; they were not intentionally sown on acid sites. Sowing dates were early May to mid June. Soil pH_{Ca} of the surface 0–10 was determined for each site after the trial was sown. The Al tolerance of the varieties was determined using the haematoxylin root tip stain method of Polle et al. (1978). Four susceptible varieties (Banks, Sunbird, Sunstar and Vulcan) were present in all trials. The tolerant variety Dollarbird was also present in all trials. The tolerant crossbred K1939 was present from 1988 to 1990. In 1986 and 1987 a similar tolerant crossbred K1056 was present. The benefit of tolerance (ΔY) was calculated as the mean yield of the tolerant varieties minus the mean yield of the susceptible varieties. The actual yield difference, rather than the ratio of tolerant to susceptible yield, was used since the ratio obscures the economic implications of tolerance (Scott and Fisher, 1989).

The Gompertz curve (Payne et al., 1987) was fitted to the data set. The equation for this curve is:

$$\Delta Y = A + C * \exp(-\exp(B * (pH - M))$$

where A is the lower asymptote, A + C is the upper asymptote, B is a slope parameter and M is the point of inflexion. The maximum benefit of tolerance is C and the benefit of tolerance at other pHs is calculated as $\Delta Y - A$.

Soil pHs from Chartres et al. (1990) and Fraser and Geeves (1990) were used to calculate the current regional benefit of tolerance in southern NSW. The benefit of tolerance in ten years time was calculated using the soil pHs predicted by Fraser and Geeves (1990). This prediction was obtained from the 'Lime-it' model (Hochman et al. 1989) and assumes that no lime is applied.

Results and discussion

Estimating the benefit of Mn tolerance in southern NSW

In the series A trial with the closely related pairs of lines, there was no advantage from Mn tolerance. The Mn tolerant lines had a mean yield of 2.15 t ha^{-1} and the Mn susceptible lines a yield of 2.13 t ha^{-1}, the LSD5% was 0.21 t ha^{-1}. In the five series B trials with lines differing in Mn tolerance, the Mn tolerant lines had a mean yield of 2.04 t ha^{-1} and the Mn susceptible lines 2.08 t ha^{-1}, the LSD5% was 0.23 t ha^{-1}. The interaction between Mn tolerance and sites was not significant.

The lack of evidence for any yield benefit of Mn tolerance in southern NSW and the lack of a quick test for tolerance led us to conclude that we should not include selection for Mn tolerance as an objective of the breeding programme at present.

However a number of successful Australian wheat varieties are tolerant of Mn e.g. Egret, Corella, Grebe, Warigal, Sunstar (Scott, unpublished). Since this tolerance was not specifically selected, it implies that Mn tolerance may be of benefit in some situations.

Estimating the benefit of Al tolerance in southern NSW

A model relating the yield of tolerant and susceptible wheats to soil pH was presented by Scott and Fisher (1989), and is reproduced here as Figure 1. The relationship between the points A, B, C, D and pH will depend on the relationship between pH and Al in the particular soil. Point D is where the concentration of Al is sufficient to start to affect plant growth. The yields of both tolerant and susceptible lines decline with decreasing soil pH, as the availability of Al increases. At point A there is a serious Al toxicity which restricts the growth of both the tolerant and sensitive cultivar. From this model we expect ΔY to be constant above a critical pH (point D), this constant will be zero if there is no inherent yield difference between the varieties. Below this critical pH, as pH declines ΔY should

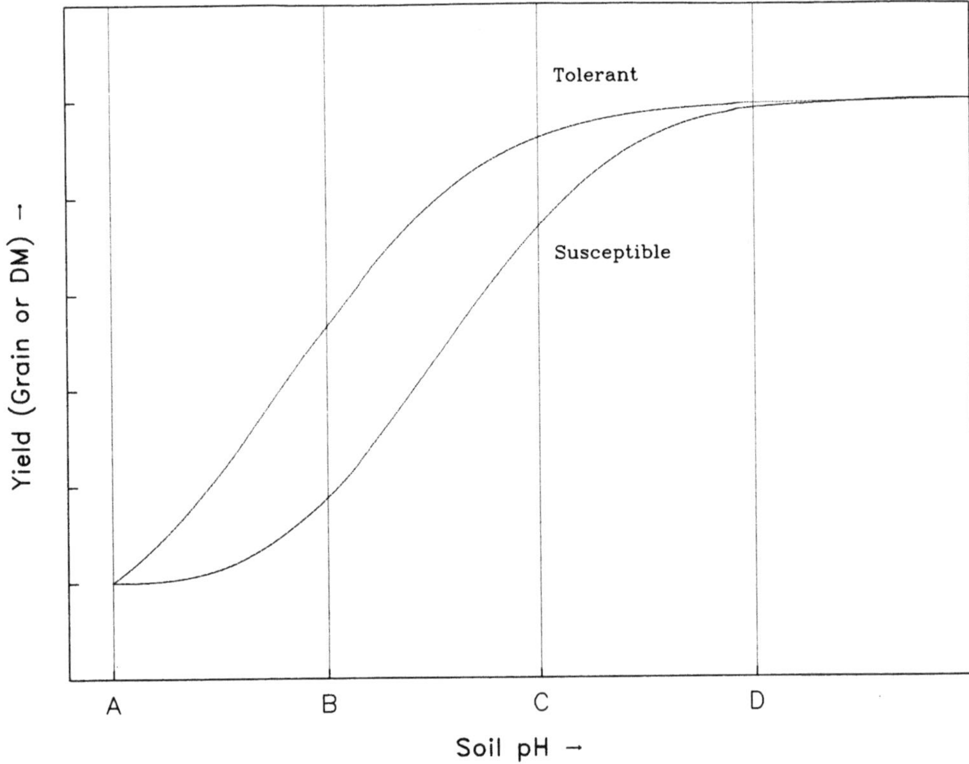

Fig. 1. Idealised response by Al tolerant and susceptible wheat cultivars to soil pH.

increase until point C, then remain constant until point B is reached. If extremely acid soils are included, ΔY will decline from point B and become zero at point A.

Severely acid sites

Closely related pairs of lines differing in a gene for Al tolerance were used in a previous experiment to determine the benefit of Al tolerance on two severely acid sites (Fisher and Scott, 1987). At Mendooran, on a deep acid sand pH 4.1, the tolerant lines yielded 0.88 t ha^{-1} and the susceptible lines 0.44 t ha^{-1}. In the other trial at Borambola, pH 3.8 at the surface, the tolerant lines yielded 2.09 t ha^{-1} and the susceptible lines 1.22 t ha^{-1}. At sites like these, a combination of tolerance and lime is desirable for optimum production.

Since the area of the wheat crop in this category is very small, the only breeding approach which could be justified would be to backcross a major gene for Al tolerance into an existing cultivar.

Less acid sites

If tolerance of Al gives a small yield benefit across a larger area of the wheat belt, then a more extensive breeding programme may be justified. The relationship between ΔY and pH over 65 variety trials in southern NSW is shown in Figure 2; the relationship between pH and exchangeable Al in these trials is shown in Figure 3. The mean yield in these trials was 3.3 t ha^{-1}, this is more than the shire mean yield (1.78 t ha^{-1}) but less than the mean potential yield of 3.69 t ha^{-1} (Cornish and Murray, 1989).

The equation $\Delta Y = 0.0928 + 1.012 * \exp(-\exp(7.19 * (\text{pH-}4.3160)))$ accounted for 52% of the variance in ΔY. The inherent yield difference between the tolerant and susceptible varieties was 0.0928 t ha^{-1}. As the tolerant varieties were developed more recently than the susceptible varieties, this yield difference is probably due to breeding progress in recent years. The yield benefit from tolerance to Al is $\Delta Y - 0.0928$. Above a pH of 4.4 there was no benefit from tolerance.

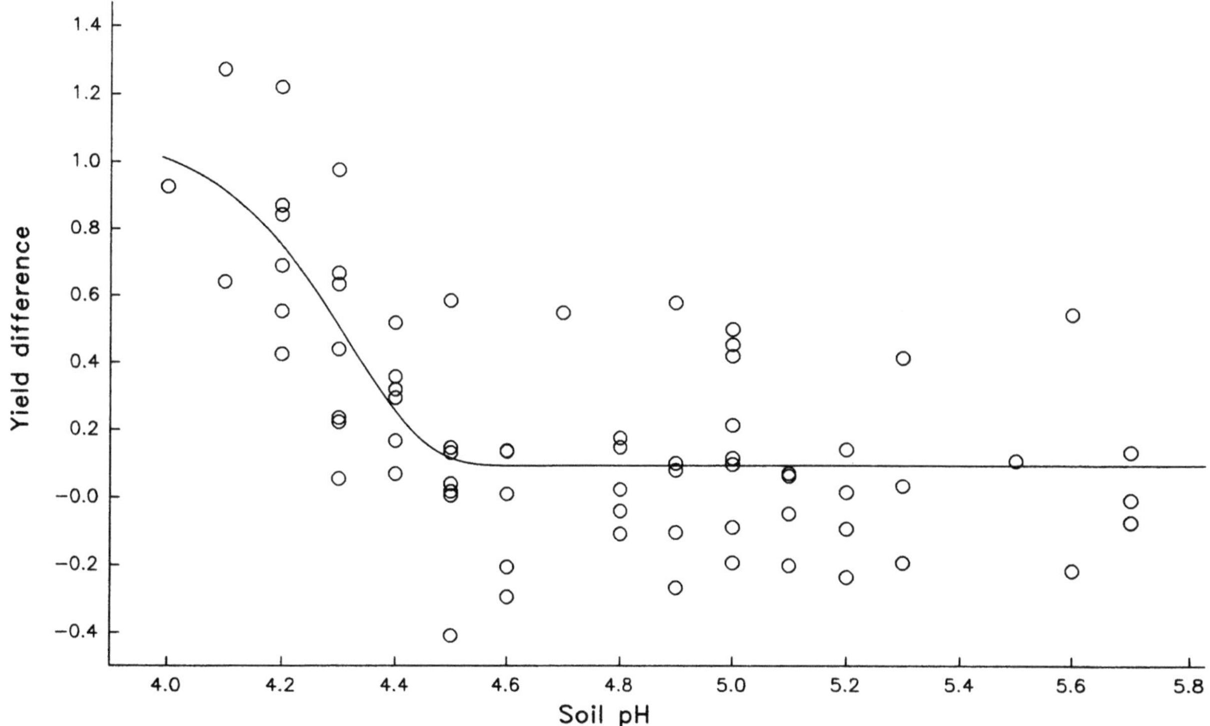

Fig. 2. Relationship between the yield difference (ΔY = Tolerant-susceptible) and surface soil pH_{Ca} in variety trials in southern NSW, 1986–1990.

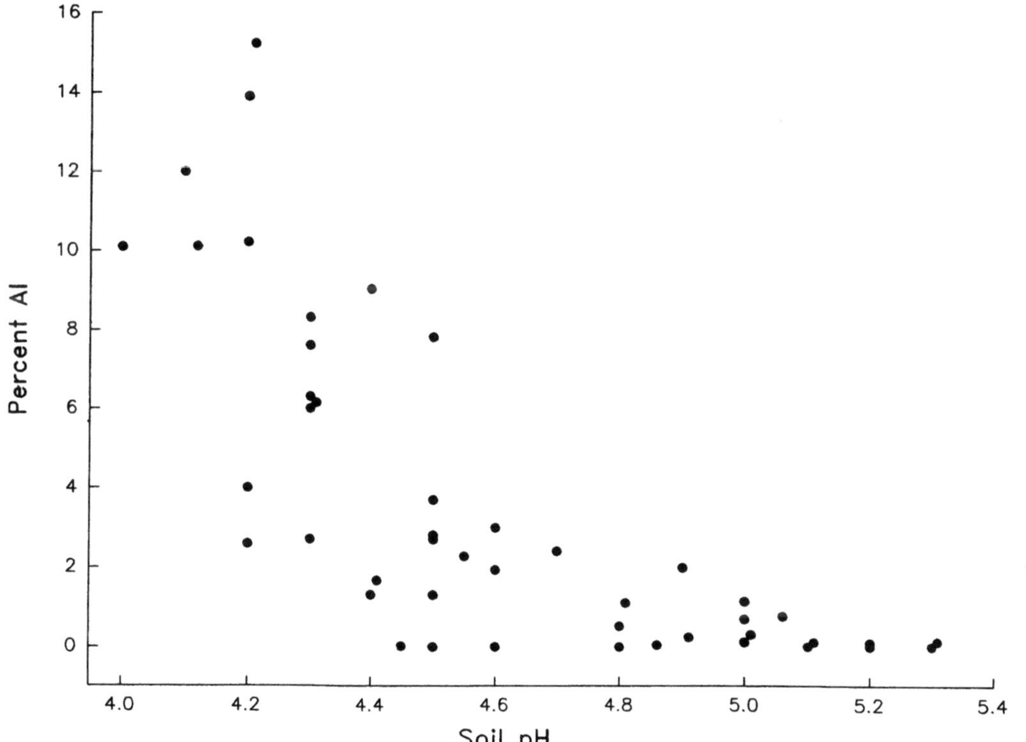

Fig. 3. Relationship between calcium chloride extractable Al (percent exchangeable cations) and pH_{Ca} for sites with $pH_{Ca} < 5.4$.

Using the soil pH data of Chartres et al. (1990) for a transect through the shires of Coolamon and Junee, the estimated benefit of selecting for tolerance is a 3.9% improvement in grain yield. The potential yield benefit from tolerance was also calculated using the current and predicted change in surface soil pH for different soil groups, and the area of each soil group in the seven shires around Wagga Wagga (Fraser and Geeves, 1990). The present estimated increase in grain yield, in these shires, from selecting for tolerance is 4.2%. This benefit is expected to increase to 9.3% in ten years time as soil pH declines. Most of this increase is due to the decline of the pH in the red earth soils, Fraser and Geeves group 6, from a mean pH of 4.38 to 4.26. These seven shires planted 204,000 hectares of wheat in 1987/88 which was 16.6% of the crop in southern NSW. Since soils in the shires to the west of this area generally have a higher pH, the results of Helyar et al. (1990) were used to extrapolate from these shires to the southern NSW wheat belt. The overall current benefit of tolerance is 1.4%, increasing to 3.2% in ten years. Based on an average wheat production in southern NSW of 3.06 million tonnes and a wheat price of $139.83/tonne (Brennan and Murray, 1988), the current value of this tolerance to the community, should be $5.99 million rising to $13.69 million in ten years time.

Comparison with other breeding objectives

Disease resistance

The economic benefit of breeding for resistance to the major diseases in southern NSW is given in Table 1 (Brennan and Murray 1988). Breeding for resistance needs to be a continuing process, since diseases can mutate to become virulent on resistant varieties. It is also necessary to ensure that disease resistance is maintained when breeding for other objectives. Selection for disease resistance is usually carried out during the early stages of selection, and the objective is to achieve an adequate level of resistance.

Table 1. Current and potential benefits of selecting for disease resistance in southern NSW

Disease	Benefit expressed as % Yield	
	Current	Potential
Septoria tritici blotch	8.7	8.0
Stem rust	10.0	0.2
Stripe rust	14.7	2.0
Flagsmut	2.0	0.0
Leaf rust	1.2	0.2

Derived from Brennan and Murray (1988). Current is the percentage yield loss that has been saved by breeding, Potential is the additional saving which could be achieved.

Table 2. Benefits of selecting for grain yield and some quality attributes

Character	Potential improvement	Expressed as % Yield	
		Total	Marginal
Grain yield	5%	5.0	0.12
Flour extraction	2%	3.2	0.07
Protein content	1%	2.4	0.06
Bake score	2 units	6.3	0.14

Potential and total are the improvement expected in the current breeding population, Marginal is the improvement due to the last 20% (400 to 500) of the population.

Quantitative characters

When allocating resources to the improvement of characters controlled by many genes, it is necessary to consider the marginal improvement with each increase in population size. The mean response to selection is $i\sigma_P h^2$ where i is selection intensity, σ_P is phenotypic standard deviation, and h^2 is heritability (Falconer, 1981). If only population size is varied then the marginal change in response to selection is proportional to the change in selection intensity.

Brennan (1990) has assessed the economic value of breeding for various characters in wheat. Table 2 was derived from Brennan's figures, but with value expressed as the equivalent percentage yield improvement. The values given for potential improvement are derived from the population in the current breeding programme at Wagga Wagga. In this programme after selecting for disease resistance and other simply inherited characters, about 500 new lines are available for yield and quality evaluation each year. The marginal value is the predicted improvement due to the last 20% of population size (from 400 to 500 lines).

Selecting for Al tolerance

We conclude that breeding for Al tolerance would be a worthwhile objective for the main wheat belt in southern NSW. The predicted benefit is greater than the predicted marginal benefits of selecting for grain yield per se or various quality attributes, so that we are justified in diverting resources from them to Al tolerance. Aluminium tolerance is less important than the current major diseases.

Incorporation of Al tolerance into the wheat breeding programme

In our current programme, some crosses specifically include parents with tolerance to Al, mainly of CIMMYT origin. Selection during the inbreeding generations is for disease resistance and agronomic characteristics. Lines are evaluated initially for their yield and grain quality on non-acid soils. Yield ability per se is important because of the variable nature of soil acidity. On most farms with acid soils, there is still a significant area of land which is unaffected. Aluminium tolerance is determined using the haematoxylin stain test and lines which are tolerant or moderately tolerant in this test are subsequently included in trials on acid soils. Final yield testing includes trials on acid soils because the haematoxylin stain test may identify only one mechanism of tolerance, which is consistent with the exclusion of Al from the growing point (Wallace et al., 1982). This test accounted for 48% of the variation in Al tolerance in a nutrient solution experiment (Scott and Fisher, 1992). In addition there are likely to be other environmental factors at acid sites which affect the relative performance of varieties, e.g. the Podzolic soils in the eastern part of the wheat belt in southern NSW, are often waterlogged during winter. This scheme is not designed to maximise Al tolerance, but provides a reasonable balance between the different objectives.

References

Atlin G N and Frey K J 1989 Predicting the relative effectiveness of direct versus indirect selection for oat yield in three types of stress environments. Euphytica 44, 137–142.

Brennan J P 1990 Valuing the breeding characteristics of wheat. Agricultural Economics Bulletin 7. NSW Agriculture and Fisheries, Yanco, Australia.

Brennan J P and Murray G M 1988 Australian Wheat Diseases: Assessing their economic importance. Agric. Sci. 2, 26–35.

Chartres C J, Cumming R W, Beattie J A, Bowman G M and Wood J T 1991 Acidification of soils on a transect from plains to slopes, south-western New South Wales. Aust. J. Soil Res. 28, 539–548.

Cornish P S and Murray G M 1989 Low rainfall rarely limits wheat yields in southern New South Wales. Aust. J. Exp. Agric. 29, 77–83.

Cregan P D, Hirth J R and Conyers M K 1989 Amelioration of soil acidity by liming and other amendments. *In* Acidity and Plant Growth. Ed A D Robson. pp 205–264. Academic Press, Sydney, Australia.

Falconer D S 1981 Introduction to quantitative Genetics. 2nd Ed. Longman Group Ltd., London.

Fisher J A and Scott B J 1987 Response to selection for aluminium tolerance. *In* Priorities in Soil/Plant Relations Research. Eds. P G E Searle and B G Davey. pp 135–137. School of Crop Sciences, University of Sydney, Australia.

Fisher J A and Scott B J 1983 Breeding wheats for tolerance to acid soils. *In* Proceedings of the Australian Plant Breeding Conference, Adelaide, February 1983, p 333.

Foy C D, Scott B J and Fisher J A 1988 Genetic differences in plant tolerance to manganese toxicity. *In* Manganese in Soils and Plants. Eds. R D Graham, R J Hannam and N C Uren, pp 293–307. Kluwer Academic Publishers, Dordrecht, The Netherlands.

Fraser K and Geeves G 1990 The economics of acid soils. Australian Agricultural Economics conference, 13–15 February 1990, University of Queensland.

Helyar K R, Cregan P D and Godyn D L 1991 Soil acidity in New South Wales: Current pH values and estimates of acidification rates. Aust. J. Soil Res. 28, 523–537.

Helyar K R and Porter W M 1989 Soil acidification, its measurement and the processes involved. *In* Acidity and Plant Growth. Ed. A D Robson. pp 167–203. Academic Press, Sydney, Australia.

Hochman Z, Godyn D L and Scott B J 1989 The integration of data on lime use by modelling. *In* Acidity and Plant Growth. Ed A D Robson. pp 265–301. Academic Press, Sydney, Australia.

Munns D N and Scott B J 1987 Biological aspects of mineral toxicities. *In* Priorities for soil/plant research for plant production. Eds. P G E Searle and B G Davey. pp 91–109. School of Crop Science, University of Sydney, Australia.

Osborne G J, Wright W A, and Sykes J A 1978 Increasing soil acidity threatens farming system. Agr. Gazette NSW 89, 21.

Payne R W, Lane P W, Ainsley A E, Bicknell K E, Digby P G N, Harding S A, Leech P K, Simpson H R, Todd A D, Verrier P J, White R P 1987 Genstat 5 Reference Manual. pp 366–368. Oxford University Press, Oxford, UK.

Polle E, Konzak A F and Kittrick J A 1978 Visual detection of aluminium tolerance levels in wheat by hematoxylin staining of seedling roots. Crop Science 18, 823–827.

Richards R A 1983 Should selection for yield in saline regions be made on saline or non-saline soils? Euphytica 32, 431–438.

Ritchie G S P 1989 The chemical behaviour of aluminium, hydrogen and manganese in acid soils. *In* Acidity and Plant

Growth. Ed. A D Robson. pp 1–60. Academic Press, Sydney, Australia.

Scott B J and Fisher J A 1989 Selection of genotypes tolerant of aluminium and manganese. *In* Acidity and Plant Growth. Ed. A D Robson. pp 167–203. Academic Press, Sydney, Australia.

Scott B J and Fisher J A 1992 Tolerance of Australian wheat varieties to aluminium toxicity. Commun. Soil Sci. Plant Anal. 23, 509–526.

Takagi H, Namai N and Murakami K 1982 Location of genetic factors of aluminium tolerance in common wheat cv. Chinese Spring. Jap. J. Breeding 32 (Suppl. 1), 28–29.

Wallace S U, Henning S L and Anderson IC 1982 Elongation, Al concentration, and haematoxylin staining of aluminium-treated wheat roots. Iowa State J. of Res. 57, 97–106.

P. J. Randall et al. (Eds.), Genetic aspects of plant mineral nutrition, 9–16.

PLSO SV02

Comparison of techniques for determining the effect of aluminium on the growth of, and the inheritance of aluminium tolerance in wheat

D.M. WHEELER[1], D.C. EDMEADES[1], R.A. CHRISTIE[1] and R. GARDNER[2]
[1]*Ruakura Agricultural Centre, Ministry of Agriculture and Fisheries, Private Bag, Hamilton, New Zealand and* [2]*Department of Molecular Biology, Auckland University, Private Bag, Auckland*

Key words: aluminium, genetics, inheritance, toxicity, wheat

Abstract

The effect of Al on the growth of plants derived from the F3 generation of a cross between Al tolerant (Waalt) and Al sensitive (Warigal) wheat cultivars, grown in low ionic strength nutrient solutions, were assessed by a number of methods viz; root length and haematoxylin stain after 3 days exposure to Al and plant top and root yields, and root length and visual assessment for Al damage after 4 weeks growth.

Of these methods haematoxylin stain (3 days) and visual assessment at 4 weeks identified the same plants as being sensitive or tolerant to Al and clearly segregated the 2 populations. Consequently these 2 methods were used as 'standard' techniques to determine the ability of the other methods to distinguish between tolerant and sensitive plants.

The ratio of plant top:root yields clearly segregated the 2 populations. The 2 populations could not be clearly distinguished based on plant top or root yields, or on root length either after 3 days or 4 weeks exposure to Al.

Within the population of tolerant plants, root length was significantly correlated with root weight ($r^2 = 0.86$) and top weight ($r^2 = 0.71$). None of these relationships were significant for the population of sensitive plants.

These techniques were applied in a number of separate experiments on the F2 and F3 populations from a Waalt × Warigal cross. The results indicate that Al tolerance in wheat is inherited by a single gene and that this gene has incomplete dominance.

Introduction

Although Al tolerance in a number of cultivars of wheat has been reported (see review Foy, 1988), there are conflicting reports on the inheritance of Al tolerance in wheat. Kerridge and Kronstad (1968) and Larkin (1987) reported that Al tolerance was inherited by a single gene. In contrast, Campbell and Lafever (1978) concluded that the inheritance of Al tolerance was more complex than an incomplete dominant gene, while Aniol (1990) concluded that Al tolerance was controlled by several genes, minor modifying genes and suppression genes.

It is possible that these inconsistencies may be due in part to the different techniques which have been used to study the inheritance of Al tolerance in wheat. For instance, Larkin (1987) used a staining technique, Aniol (1990) used a technique involving the measurement of root length in short term experiments and Campbell and Lafever (1978) measured root length over a longer time period. There is little information on whether different techniques could result in different conclusions on the inheritance of Al tolerance.

The yield of the more Al tolerant cultivars of wheat have been shown to be reduced by 50%

by 2–3 μM Al^{3+} activity (10–15 μM nominal Al concentration) in solution (Wheeler et al., 1991). However, many of the techniques used to identify Al tolerant and sensitive plants have used high nominal Al concentrations (>37 μM Al). The toxic effects of the low Al rates used by Wheeler et al. (1991) were attributed to the low ionic strength of the nutrient solution used. The importance of using a low ionic strength nutrient solution in terms of ranking plants for tolerance to Al are discussed by Blamey et al. (1991). Therefore, a range of techniques were selected such that the effects of Al were measured over different time periods and modified if necessary to ensure that low Al concentrations were used.

Two wheat cultivars, Waalt (Al tolerant) and Warigal (Al sensitive), have been used in experiments to investigate factors which influence Al toxicity (Edmeades et al., 1990) and to isolate the gene(s) controlling the expression of Al tolerance in these cultivars (Putterill, 1991). Al tolerance in these cultivars was reported by Larkin (1987) to be inherited by a single dominant gene.

This paper reports the results of a series of experiments in which five techniques to measure Al tolerance were assessed. These techniques were applied to a F2 and F3 population from a Waalt × Warigal cross to confirm or otherwise the results of Larkin (1987).

Methods and materials

Plants from the F1 generation from a cross between Waalt and Warigal were selfed and seeds collected. This gave F2 seeds which were used in subsequent experiments. During the course of one experiment (experiment 3 listed below) a range in plant root density was observed. Within the population of tolerant plants, 2 categories were visually identified; those with a high density of fine roots (HDFR) and those with a low density of fine roots (LDFR). All plants in the sensitive population were similar with a low density of short stubby roots (SSR). Individual plants from each category (14/category) were on-grown and selfed. F3 seed from this material was used in subsequent experiments. Also, F3 plants that had been identified as homozygous dominant (Larkin pers. comm.) were selfed and seeds collected (F4 seeds).

In all experiments, seeds were germinated on moistened filter paper. Six methods were examined to assess Al tolerance based on either short term (3 days) or long term (4 weeks) exposure to Al:

a. Plant yield (4 weeks). Seedlings were transplanted into plastic tubs (15 plants per tub) containing 40 litres of low ionic strength nutrient solution and grown for 4 weeks in a temperature controlled (12°C min, 25°C max) glasshouse. The technique, analytical procedures and nutrient composition are outlined elsewhere (Edmeades et al., 1991). Al was added as $Al_2SO_46H_2O$ 2–3 days after transplanting. After 4 weeks plants were harvested, dried and top and root yields determined.
b. Visual assessment (4 weeks). The roots of plants grown in nutrient solution (as above) were visually assessed for Al damage 4 weeks after transplanting. Sensitive plants were classified as those that had short, thick roots with little or no lateral root development.
c. Root length and average root width (4 weeks). These were determined using a Delta-10 root length meter system on plants grown in the glasshouse (technique (a) above). Resolution was approximately 0.015 mm.
d. Root elongation (3 days). The average change in root length of the 2 longest roots was determined after 3 days growth in a low ionic strength nutrient solution of the same composition to that used in solution culture but containing 10 μM Al. The method is similar to that proposed by Kinraide et al. (1985) except that a low ionic strength nutrient solution was used instead of 0.02 M $CaSO_4$ and plants were grown for 3 days instead of 2.
e. Stain (3 days). Plants were grown using technique (d) as above and then stained with haematoxylin using the procedure outlined by Polle et al. (1978). At least two of the plant roots had to stain to be classified as sensitive. If the change in root length was also to be determined, this was measured before staining.

Assessment of techniques for measuring Al tolerance (Experiment 1)

To assess the ability of the techniques to identify Al tolerant and sensitive plants, 215 F3 seedlings were assessed using the stain and root elongation (3 days) techniques. Of these, 120 plants, including 16 assessed as sensitive using the stain technique, were transferred to nutrient solutions in the glasshouse. The roots were trimmed to 2 cm after 3 days and 10 μM Al added. Four weeks after transferring to the nutrient solution, yield and visual assessment of damage were determined on all plants and root length was determined on 60 plants including 14 assessed as sensitive using the stain technique.

Assessment of inheritance of Al tolerance

Nine more experiments were conducted over a 2 year period using the techniques described previously. The first 3 experiments were designed to test whether populations of the F2 segregated in a 3:1 ratio. The yield and segregation ratios in the F3 seed from the LDFR and HDFR plants were assessed in experiments 5, 6, 7 and 8. Experiment 9 was designed to compare the yields from the homozygous tolerant F3 and F4 populations with the original tolerant parent (Waalt). Experiment 10 was undertaken to determine the Al tolerance of plant material selected from previous experiments based on improved vigour in the presence of Al.

The details for each experiment are:

- Experiment 2. 102 F2 seedlings were assessed using the stain technique.
- Experiment 3. 577 F2 seedlings were grown in solution culture at 5 (285 seedlings) and 10 (292 seedlings) μM Al and measurements of yield and visual assessment were made after 4 weeks.
- Experiment 4. 480 F2 seedlings were grown in solution culture at 10 μM Al and measurements of yield and visual assessment were made after 4 weeks. Root length was determined on 57 randomly selected plants before drying and weighing.
- Experiment 5. F3 seedlings from 5 of the SSR plants were grown in solution culture at 5 and 10 μM Al using 3 plants per line per tub with 3 replicates. Visual assessments were made after 4 weeks.
- Experiment 6. F3 seedlings from twelve (plants 1–10, 12, 14) of the LDFR plants were grown in solution culture at 10 μM Al using 6 lines per tub, 2 tubs/replicate and 8 replicates. Visual assessments were made after 4 weeks.
- Experiment 7. F3 seedlings from 4 (plants 1, 2, 4, 5) HDFR and 5 (plants 1, 3, 8, 11, 13) LDFR plants were grown in solution culture at 5, 10 and 20 μM Al (2 replicates) and visually assessments were made after 4 weeks.
- Experiment 8. F3 seedlings from all the HDFR and LDFR plants (1098 seedlings total) were assessed by staining.
- Experiment 9. Homozygous tolerant F3 and F4, and Waalt seedlings (6, 6, 3 plants per treatment respectively) were grown in solution culture at 0, 10, 20 and 30 μM Al (2 replicates) and top and root yields of each plant were determined after 4 weeks. Yield data was transformed ($\log_e$) prior to analysis by ANOVA.
- Experiment 10. Seedlings from seven plants selected from previous experiments due to better growth and colour in the presence of Al were grown in solution culture at 0, 5, 10 and 20 μM Al in one experiment (2 replicates), at 0, 5 and 10 μM Al (1 replicate) in a second and at 10 μM Al (8 replicates) in a third. In each experiment, top and root yields of each plant were determined after 4 weeks. Yield data was transformed ($\log_e$) prior to analysis by ANOVA.

Results

Assessment of techniques

Only the haematoxylin stain (3 days) and visual assessment (4 weeks) techniques unambiguously identified sensitive and tolerant plants in a population of 215 F3 seedlings. Furthermore, plants assessed as sensitive or tolerant using the stain technique were also assessed as sensitive or tolerant visually. These 2 methods were therefore used as the 'standards' to identify and label plants as either sensitive or tolerant to Al in subsequent experiments. The frequency distribu-

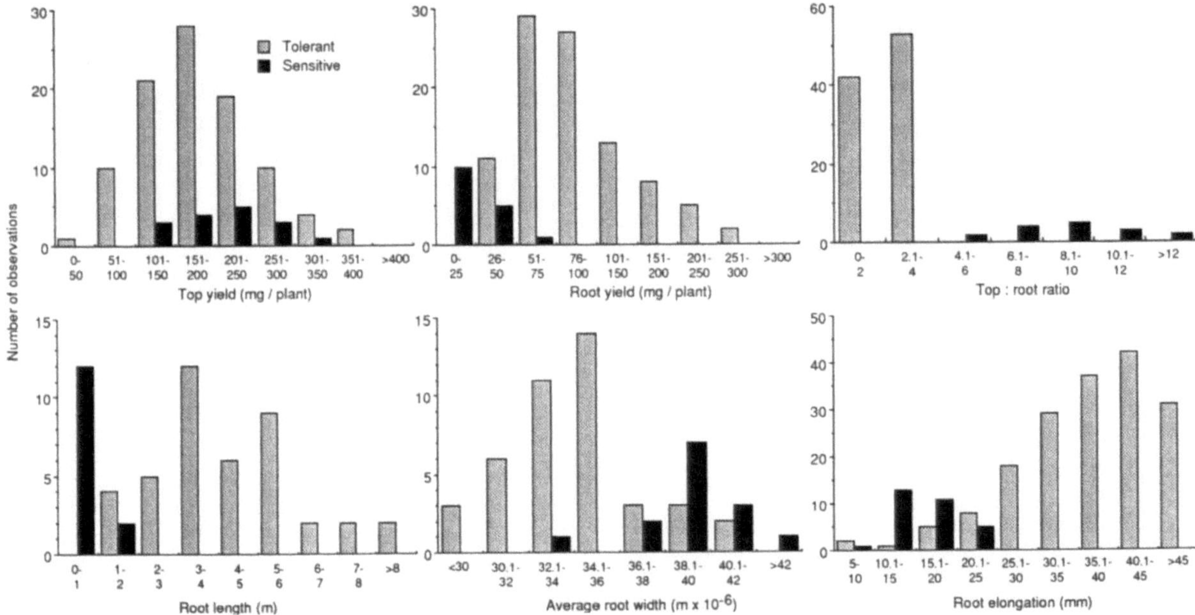

Fig. 1. Distribution of top and root yields, top:root ratio, root length, root width after 4 weeks growth and root elongation after 3 days growth in 10 μM aluminium (Al) for Al tolerant and sensitive plants from a F3 population from a Waalt (Al tolerant) × Warigal (Al sensitive) cross.

tions of the other measured plant attributes are shown in Fig. 1 in relation to these so-labelled plants.

The ratio of plant top:root yields clearly distinguished between sensitive and tolerant plants assessed by the 'standard' techniques (Fig. 1). However, plant top or root yields singularly were unable to distinguish between the 2 populations (Fig. 1). Other root attributes such as root length or root diameter were more discriminating than root yield but were also unable to clearly distinguish between the 2 populations. Similarly, root elongation (3 days) did not separate the two populations.

For the population of tolerant plants, root weight was positively related to root length ($r^2 = 0.86$) and top weight ($r^2 = 0.71$) (Fig. 2). However root length was not related to root elongation over 3 days ($r^2 = 0.07$) (Fig. 2). None of these relationships were statistically significant for the population of sensitive plants.

Assessment of genetic inheritance

The results from experiments 2, 3 and 4 are summarised in Table 1 and show that for the F2 population from the Waalt × Warigal cross, the ratio of sensitive to tolerant plants determined by staining (3 days) or visually (4 weeks) is not significantly different from the expected ratio of

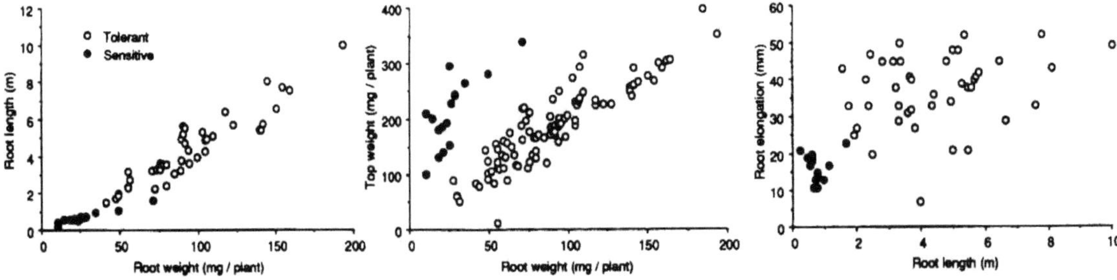

Fig. 2. Relationship between root weight and root length, root weight and top weight, and root length and root elongation for Al tolerant and sensitive plants from a F3 population from a Waalt (Al tolerant) × Warigal (Al sensitive) cross.

Table 1. The number and ratio of aluminium (Al) tolerant and sensitive plants from the F2 population from the Waalt (Al tolerant) × Warigal (Al sensitive) cross

Experiment	Technique	Number of tolerant	Number of sensitive	Ratio Sensitive: Tolerant	
2	Stain (3 days)	80	22	1:3.64	n.s.[a]
3	Visual (4 weeks)	424	153	1:2.77	n.s.
4	Visual (4 weeks)	374	106	1:3.53	n.s.
	Total	878	280	1:3.12	n.s.

[a] Significance of the hypothesis that the number of sensitive and tolerant plants was different from the 1:3 ratio expected for a single gene using a Chi-square test (ns = not significant).

1:3, assuming a single dominant gene for Al tolerance. This confirms the results of Larkin (1987).

The tolerant and sensitive populations in the F2 (Experiment 3) could not be clearly distinguished when top or root yields (Fig. 3), or the ratio of top to root yield (not shown) were measured.

The results from either the stain technique or visual assessment for the F3 seedlings from the LDFR and HDFR plants are summarised in Table 2. Thirteen out of 14 of the tolerant F2 plants selected for low root density (LDFR) were heterozygous dominant (sensitive:tolerant 1:3) for Al tolerance (Table 2). In contrast, only 5 of the HDFR plants were heterozygous. Of the other 9 plants, 3 were homozygous dominant for Al tolerance. The other 6 plants had sensitive progeny, although the ratio of sensitive to tolerant indicate that these plants were not heterozygous for a single gene. All the progeny from the sensitive F2 plants (SSR) were visually assessed as sensitive (Experiment 5).

There were no significant differences in yield between plants from the original tolerant parent (Waalt) and plants from the homozygous tolerant F3 and F4 populations (Table 3; Experiment 9). Similarly, there were no significant differences in yield between Waalt and the seven selected lines (Experiment 10; data not shown).

After 4 weeks growth in nutrient solutions containing Al, the Al tolerant plants still showed symptoms similar to Mg and/or Ca deficiency in the tops. The visual symptoms were associated with low plant Mg concentrations (<0.12% Mg).

Discussion

A range of techniques were used to assess the

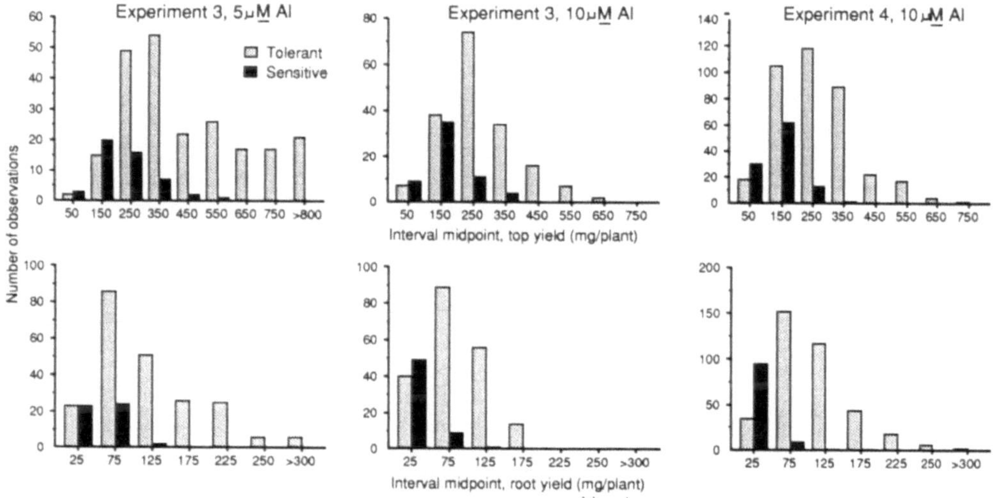

Fig. 3. Distribution of top and root yields (mg/plant) for aluminium (Al) tolerant and sensitive plants from a F2 population from a Waalt (Al tolerant) × Warigal (Al sensitive) cross determined in 2 experiments at the aluminium concentrations shown.

Table 2. The number of aluminium (Al) tolerant (T) and sensitive (S) plants determined by visual assessment (experiment 6 and 7) or by stain (experiment 1 and 8) and over all experiments (Total) over all Al concentrations for progeny (F3) from the LDFR and HDFR (see text) plants selected from the F2 of a Waalt (Al-tolerant) × Warigal (Al-sensitive) cross

Line	Expt. 6, 7		Expt. 1		Expt. 8		Total			
	T	S	T	S	T	S	T	S	Ratio	
LDFR										
1	26	8	5	4	48	23	79	35	1:2.3	n.s.[a]
2	10	3	7	1	89	29	106	33	1:3.2	n.s.
3	49	4	14	3	24	13	87	20	1:4.4	n.s.
4	11	4	–	–	–	–	11	4	1:2.8	n.s.
5	14	2	6	1	44	15	64	18	1:3.6	n.s.
6	13	3	5	1	25	15	43	19	1:2.3	n.s.
7	12	0	–	–	–	–	12	0	–	*
8	29	9	7	3	–	–	36	12	1:3.0	n.s.
9	13	3	–	–	–	–	13	3	1:4.3	n.s.
10	12	4	7	3	48	18	67	25	1:2.7	n.s.
11	18	6	7	1	27	8	52	15	1:3.5	n.s.
12	11	5	4	2	30	12	45	19	1:2.4	n.s.
13	18	6	8	2	7	6	33	14	1:2.4	n.s.
14	4	3	4	3	–	–	8	6	1:1.3	n.s.
HDFR										
1	24	0	7	0	140	1	171	1	1:171	***
2	19	5	4	1	50	1	73	6	1:12.2	***
3	–	–	9	0	98	0	107	0	–	***
4	48	0	–	–	–	–	48	0	–	***
5	19	5	10	0	60	6	89	11	1:8.1	**
6	–	–	10	0	8	2	18	2	1:9.0	n.s.
7	–	–	10	0	91	2	101	2	1:50.5	***
8	–	–	8	2	25	2	27	4	1:6.8	n.s.
9	–	–	8	0	44	2	52	2	1:26	***
10	–	–	9	1	14	6	23	7	1:3.3	n.s.
11	–	–	10	0	2	2	12	2	1:6.0	n.s.
12	–	–	9	0	22	2	31	2	1:5.5	*
13	–	–	7	2	19	5	26	7	1:3.7	n.s.
14	–	–	10	0	13	0	23	0	–	**

[a] significance of the hypothesis that the number of sensitive and tolerant plants was different from the 1:3 ratio expected for a single gene using a Chi-square test (ns = not significant, * = $p < 0.5$, ** $p < 0.1$, *** $p < 0.001$).

Table 3. The effect of solution aluminium concentration (μM Al) on the top and root yields (mg/plant, $\log_e$ transformed) of Waalt and the F3 and F4 homozygous tolerant progeny from a Waalt (Al-tolerant) × Warigal (Al-sensitive) cross

Al (μM)	Tops			Roots		
	Waalt	F3	F4	Waalt	F3	F4
0	7.304	7.389	7.322	6.005	6.072	5.934
10	5.983	5.976	5.966	5.159	4.885	4.812
20	4.959	5.133	5.030	4.260	4.208	3.881
30	4.823	4.891	5.030	4.082	4.011	4.027
SED						
min rep[a]		0.1201			0.1869	
min-max rep[b]		0.1040			0.1619	
max rep[c]		0.0849			0.1322	

[a] for comparisons within Waalt.
[b] for comparisons between Waalt and F3 or F4.
[c] for comparisons within F3 and F4.

inheritance of Al tolerance from the wheat cultivar Waalt. The stain and visual assessment techniques indicated that Al tolerance was inherited by a single dominant gene. However, this gene could not alleviate all the symptoms of Al toxicity as indicated by the Al-induced Mg deficiency in the tops of the Al tolerant plants after 4 weeks growth in solution culture.

The ability to select a predominantly heterozygous population from the F2 plants with a low density of roots and a population with a low number of heterozygotes from the F2 plants with a high density of roots indicates that the single gene identified has incomplete dominance. However, there were 6 F2 plants with a high root density which were not heterozygous for a single gene but had a low proportion of Al-sensitive progeny (Table 2). The reason for the low proportion of sensitive plants is unclear but could be due to misclassification, pollen contamination or due to a more complex genetic control of Al tolerance. If the low proportion of sensitive plants observed in the HDFR are due to a more complex genetic control such as reported by Aniol (1990) or Campbell and Lafever (1978), then the results indicate that the inheritance of Al tolerance in wheat can be approximated by a single gene. It is noted that the selection of plants with either a low or high density of fine roots is subjective and thus the categories are not necessarily mutually exclusive, and that the two populations of root density were only observed when plants were grown for 4 weeks in solution culture.

The ability to assess a large number of plants in a short time period indicates that the stain technique would be the preferred method to determine the inheritance of Al tolerance in wheat. Although this technique clearly identified sensitive and tolerant plants and thus the presence of a single gene, it could not identify the incomplete dominance, nor the Al-induce Mg deficiency. The identification of these aspects required longer term experiments (visual assessment after at least 3 weeks growth in solution culture).

Techniques based on plant yields (tops and roots), root length and width (4 weeks) or root elongation (3 days) could not clearly identify sensitive and tolerant plants in the F2 population. Thus it was not possible, based on these techniques alone, to conclude that a single gene controlled Al tolerance or to identify the incomplete dominance of the Al tolerant gene. Although the top to root ratio clearly segregated sensitive and tolerant plants in the first set of experiments, the results were not conclusive in the second set of experiments. This suggests that measurements of plant yield and root length may not be very useful when determining the inheritance of Al tolerance in wheat.

The poor segregation between the tolerant and sensitive plants using techniques based on plant weight or length measurements may be due to the incomplete dominance of the Al tolerant gene. These results could also suggest that other genes are modifying the effect of a single gene. However, the lack of significant differences between the F3, F4 and Waalt and the inability to select plants that were more tolerant than Waalt indicate that it is unlikely that other genes are modifying the Al tolerant gene identified in Waalt.

Although a single incomplete dominant gene was identified in these experiments, the symptoms of Al toxicity in the tops of Waalt (Al-induced Mg deficiency) indicate that factors other than those controlled by the Al tolerant gene identified in these experiments are involved in the expression of Al toxicity. As shown elsewhere (Wheeler et al., 1991), there is a range in the relative tolerance to Al in wheat cultivars and Waalt is only intermediate within this range. It is possible that genes expressed in other cultivars of wheat may overcome these secondary effects of Al toxicity.

Acknowledgements

We wish to thank P J Larkin for performing the initial cross between Waalt and Warigal and for the supply of the F1 and F3 homozygous dominant seed.

References

Aniol A 1990 Genetics of tolerance to aluminium in wheat (*Triticum aestivum* L. Thell). Plant and Soil 123, 223–227.

Blamey F P C, Edmeades D C, Asher C J, Edwards D G and Wheeler D M 1990 Evaluation of solution culture techniques for studying aluminum toxicity in plants. *In* Plant-Soil Interactions at Low pH. Eds. R J Wright, V C Baligar and R P Murrmann. pp 905–912. Kluwer Academic Publishers, Dordrecht, The Netherlands.

Campbell L G and Lafever H N 1978 Heritability and gene effects for aluminium tolerance in wheat. Proc. Fifth Int. Wheat Genet. Symp. Ed. S Ramarujan. Indian Society of Genetics and Plant Breeding. New Delhi, India. pp 963–977.

Edmeades D C, Wheeler D M and Christie R A 1991 The effect of aluminium and pH on the growth of a range of temperate grass species and cultivars. *In* Plant-Soil Interactions at low pH. Eds. R J Wright, V C Baligar and R P Murrmann. pp 913–924. Kluwer Academic Publishers, Dordrecht, The Netherlands.

Foy C D 1988 Plant adaptation to acid, aluminium-toxic soils. Commun. Soil Sci. Plant Anal. 19, 959–987.

Kerridge P C and Kronstad W E 1968 Evidence of genetic resistance to aluminium toxicity in wheat (*Triticum aestivum* Vill. Host.). Agron. J. 60, 710–711.

Kinraide B K, Arnold R C and Baligar V C 1985 A rapid assay for aluminium phytotoxicity at submicromolar concentrations. Physiol. Plant. 65, 245–250.

Larkin P J 1987 Calmodulin levels are not responsible for aluminium tolerance in wheat. Aust. J. Plant Physiol. 14, 377–385.

Polle E, Konzak C F and Kittrick J A 1978 Visual detection of aluminium tolerance levels in wheat by haematoxylin staining of seedling roots. Crop Sci. 18, 823–827.

Putterill J J, Richards K D, Boyd L, Konigstorfer A, Richardson T E and Gardner R 1991 Molecular approaches to aluminium tolerance in plants. Curr. Top. Plant Biochem. Physiol. 10, 142–147.

Wheeler D M, Edmeades C D and Christie R A 1991 Effect of aluminium on yield and nutrient concentrations on 18 cultivars of cereals grown in a low ionic strength solution. J. Plant Nutr. 15, 403–418.

P. J. Randall et al. (Eds.), Genetic aspects of plant mineral nutrition, 17–22.

PLSO SV04

Breeding the perennial pasture grass *Phalaris aquatica* for acid soils

R.N. ORAM[1], R.A. CULVENOR[1] and A.M. RIDLEY[2]
[1]*CSIRO Division of Plant Industry, G.P.O. Box 1600, Canberra, A.C.T. 2601, Australia and*
[2]*Department of Agriculture, Rutherglen Research Institute, Rutherglen, Victoria, 3685, Australia*

Key words: acidity, aluminium, genes, grass, hybrids, introgression, manganese, palatability, perennial, phalaris, resistance

Abstract

Previous studies on the reactions of a range of cultivars and Mediterranean geographic races of *P. aquatica* in nutrient solutions containing varying levels of manganese and aluminium, and in acid soils in situ and in pot experiments in a glasshouse, are reviewed. All populations tested were comparatively resistant to manganese; they tolerated tissue manganese concentrations of 700–1000 $\mu g\ g^{-1}$ without significant yield loss. By contrast, phalaris populations contained a highly aluminium sensitive component, which was severely damaged by 2 mg Al L^{-1} in nutrient solution, and a moderately tolerant component, which required up to 6 mg Al L^{-1} for 50% growth reduction. Segregations suggested that the two classes differed mainly by two complementary dominant genes for tolerance. Polygenic variation appeared to be extensive in the moderately tolerant class; both kinds of genetic variation were used to develop cv. Holdfast.

Some *P. arundinacea* accessions exhibited more tolerance than *P. aquatica* in Al solutions and acid soils. F_1 hybrids and some plants in first and second backcrosses to *P. aquatica* also were acid soil tolerant. Data from a new experiment are presented to show that three out of four backcross derivatives were higher yielding than 13 phalaris and cocksfoot control cultivars on an unlimed, acid soil, but all backcross lines were lower yielding than the phalaris cultivars Australian and Sirosa on this soil after liming. The development of new acid tolerant cultivars from these backcross populations, and from selected *P. aquatica* populations, is discussed.

Introduction

Soil acidity is increasing inexorably in the poorly buffered soils used for growing sheep and cattle pastures in the higher rainfall zones of Australia. Typically, the pH of soils under subterranean clover pasture has declined by one pH unit over 50 years (Bromfield et al., 1983; Williams, 1980). In one 73-year old experiment with unfertilized, and limed and unlimed fertilized annual pastures, Ridley et al. (1990b) calculated a rate of acidification in soils under pastures improved by sowing subterranean clover and adding superphosphate of 1.42 kmol H-ion ha^{-1} $year^{-1}$ compared with a rate of 0.16 kmol H-ion ha^{-1} $year^{-1}$ in soils under unfertilized pastures. About two-thirds of the acidification in the fertilized soil was due to carbon cycle processes, particularly those which increased, firstly, the content of organic matter, where most additional exchange sites were occupied by H-ions, and secondly, removal of an excess of bases over acids in animal products and hay. The remaining acidification was due to nitrogen cycle processes, i.e. the fixation of atmospheric nitrogen and the leaching of nitrate ions and accompanying cations from the root zone. This is particularly likely to occur in Mediterranean and modified Mediterranean climates in southern and south-eastern Australia after the first autumn rains, when annual pasture seedlings have shallow root-systems (Helyar and Porter, 1989; Ridley et al., 1990a). Deeper-rooted perennial pasture species such as *Phalaris aquatica* L. (phalaris) and *Medicago sativa* L. (lucerne) are sufficiently drought hardy to survive the long dry summers and sufficiently deep rooted to capture subsoil nitrate and cations, thereby reducing the rate of acidification. Ridley et al. (1990c) found that nitrate leaching was substantial under both annual grass and phalaris pastures in north-eastern Victoria, but acidification of the upper 60 cm of soil was apparently reduced by 1.01 kmol H-ion ha^{-1} $year^{-1}$ under phalaris pasture. However, alkali addition probably is necessary on both kinds of pastures to allow aluminium-sensitive species to be grown.

Both lucerne and phalaris are sensitive to soil aluminium (Al), but some genetic variation for greater tolerance exists, particularly in phalaris and its close relative, *P. arundinacea* L. Here, we review our earlier studies of this variability, and the development of a new moderately acid tolerant cultivar of phalaris, Holdfast. Finally, unreported results will be presented from a recent field experiment aimed at the construction of an even more tolerant cultivar by introgression of genes from *P. arundinacea* into *P. aquatica*.

Review

Responses of phalaris genotypes to manganese and aluminium

Manganese (Mn)

Sixteen cultivars and Mediterranean geographic races were grown by Culvenor (1985) in nutrient solutions containing 0.5 to 210 mg Mn L^{-1}. The pH was adjusted to 5.0 daily and the solutions were changed twice weekly. About half of the populations responded to increasing Mn concentrations, giving maximal growth at 40 mg kg^{-1}. In contrast, a concentration of 0.5 mg Mn L^{-1} was optimal for wheat, barley and subterranean clover in the same experiment. Relative to these control species, phalaris was highly resistant to manganese. The accessions tolerated, without significant yield loss, manganese concentrations of at least 700–1000 $\mu g\,g^{-1}$ in the older leaf tissue. The presence of aluminium in the solution at pH 4.5 strongly reduced manganese uptake.

The effects on animal production of high dietary intakes of Mn were reviewed briefly by Culvenor et al. (1986c). The evidence is somewhat conflicting, but overall suggests that excessive levels depress appetite and growth rate of young ruminants. Therefore, Culvenor et al. (1986c) investigated the extent of genetic variability in the uptake of Mn by phalaris from acid, manganiferous soils. The Mn concentration in the second-lowest leaf on spaced plants from a breeding population at the early heading stage was found to represent adequately the concentrations of Mn in the whole shoots. Significant family differences were observed. A sample of plants with high or low leaf Mn was transferred to sand in a glasshouse, and watered daily with nutrient solutions containing varying Mn levels. Five high Mn genotypes contained 2942–3231 μg Mn g^{-1} in their shoot tissue, whereas five low Mn genotypes contained 2179–2531 $\mu g\,g^{-1}$ (LSD between plants 327 $\mu g\,g^{-1}$), when the nutrient solution contained 200 μg Mn mL^{-1}.

Much lower Mn concentrations of 250–500 $\mu g\,g^{-1}$ were found in the leaf of plants grown in the field on a soil of pH ($CaCl_2$) 4.0 containing 30 $\mu g\,g^{-1}$ of Mn extractable in $CaCl_2$ when measured in late winter (all $CaCl_2$ extractant solutions in this paper were 0.01 *M*). Fifty-two half-sib families and their parents from a breeding population were grown as spaced plants on this soil, and young leaves sampled at the vegetative stage in July of the second year. The parental mean Mn concentrations differed by more than three-fold, and the offspring means by almost two-fold. The narrow-sense heritability of Mn concentration was 0.72 ± 0.30 in the half-sib family analysis, and 0.40 ± 0.28 from the parent-offspring regression. Thus selection for lower tissue manganese would be effective, if cultivars must be developed for very high-Mn soils.

Aluminium

Thirty-two accessions and seven cultivars of *P. aquatica* and five accessions of *P. arundinacea* were grown in nutrient solutions containing 0–8 mg Al L^{-1} at pH 4.1 (Culvenor et al. 1986a). Growth was reduced 50% by Al concentrations ranging from 2 to 6 mg l^{-1} among the *P. aquatica* lines. Although the most sensitive character was root extension, the ranking of lines according to yield in solution gave better agreement ($r = 0.81$, $p < 0.01$) with ranking according to relative yield in unlimed vs. limed Al-toxic soil. The ionic strength of the nutrient solution was considerably higher than those of soil solutions; Edmeades et al. (1991) have criticised the use of such nutrient solutions in studies of Al toxicity, but offered no evidence on the effects of ionic strength on genotype or species rankings in tolerance to Al. The purpose of genotype comparisons in nutrient solutions is the prediction of performance in Al-toxic soils; the system used by Culvenor et al. (1986a) gave quite good correlations with rankings in one unlimed vs. limed soil. Whether better correlations could be obtained with lower ionic strengths remains to be investigated.

Most of the *P. aquatica* lines tested by Culvenor et al. (1986a) contained varying ratios of highly Al sensitive and moderately tolerant plants. The root extension or dry weight growth of the extremely sensitive class was reduced to 50% by 2 mg Al L^{-1} whereas the moderately tolerant class required 4–6 mg L^{-1} for 50% root growth reduction. This heterogeneity had been noted previously by plant nutritionists, e.g. in cv. Australian by S.M. Bromfield (pers. comm.), but not taken into account in the analysis or interpretation of experiments on aluminium, phosphate or liming effects on phalaris. Culvenor et al. (1986a) found that the Al tolerance of 39 lines was correlated with the proportion of sensitive plants, the coefficient r ranging from 0.57 for shoot dry weight to 0.90 for root extension. The proportion of sensitive plants ranged from 0 to 100% in this representative sample of phalaris populations. Therefore, all studies of genetic differences in Al tolerance within *P. aquatica*, or comparisons of particular phalaris populations with other species must determine and report the relative frequencies of the highly sensitive and the moderately tolerant classes.

P. arundinacea was more tolerant to Al than *P. aquatica*: CPI 69103 *P. arundinacea* was as tolerant as Demeter tall fescue (*Festuca arundinacea* Schreb.), but less tolerant than Currie cocksfoot (*Dactylis glomerata* L.).

The inheritance of the difference between the extremely sensitive and moderately tolerant classes of *P. aquatica* was investigated by Culvenor et al. (1986b). A dominant allele at each of two independent loci appears to be necessary for moderate tolerance. Modifiers may also be involved. There was extensive variation within the moderately tolerant class, variation which appeared to be polygenic. Analysis of variation within and between half-sib families gave narrow-sense heritability estimates of 0.48 to 0.75 for different growth periods in a phalaris breeding population grown in nutrient solution. By contrast, family effects on Al tolerance were not significant in the same set of half-sib families grown in the field. Variation in soil Al concentration appeared to be the main cause of this lack of significance in the field. Therefore, selection aimed at developing a new Al-tolerant cultivar was carried out in a variable breeding population in nutrient solution containing 8 or 10 mg Al L^{-1} for two generations. Selection was imposed on the same population in the same two generations for high herbage and seed yields, good retention of ripe seeds in the panicles and high survival over three years in a marginal environment. This selection program resulted in the development of the new cultivar, Holdfast (Anon., 1990), which performs better on some acid soil sites than the similar cultivar, Sirosa, which was not selected for acid tolerance.

The finding that some *P. arundinacea* accessions were more tolerant to Al than *P. aquatica* led to the testing of first and second backcross populations in acid soil in situ and in pots, as well as in nutrient solutions containing Al (Oram et al., 1990). Also, the tolerance to soil acidity of six other *Phalaris* species with which *P. aquatica* can be hybridised was determined but none except one accession of *P. minor* was more tolerant than *P. aquatica*. The seedling weights of first-backcross individuals grown on an acid soil in a glasshouse were unimodally distributed, indicating that more than a few genes or chromosome segments condition the additional Al tolerance of *P. arundinacea*. Nevertheless a few percent of seedlings grew as well or better than a highly tolerant *P. arundinacea* accession, showing that the number of tolerance loci or segments is not very large. Therefore, it should be possible to transfer these to *P. aquatica* by backcrossing.

First and second backcross progeny were selected for the ability to produce roots in nutrient solutions containing 10 mg Al L^{-1}. Fifty tolerant individuals were clonally multiplied and transplanted to four acid sites. Only half of these genotypes grew better than the sensitive *P. aquatica* controls, indicating that factor(s) other than Al affect the tolerance of the backcross plants to soil acidity. These factor(s) have not been identified, so further selection should be practised primarily under realistic conditions in acid soils.

Experimental

Four second backcross individuals which performed well in the 4-site spaced plant trial (Oram et al., 1990) were interpollinated at random to produce BC_2F_2 families. To engender more recombination between the *P. aquatica* chromosomes and the *P. arundinacea* segments in the backcross population, the whole set of first and second backcross genotypes was interpollinated at random, and two progeny of each genotype were again interpollinated. The growth of one of these latter open-pollinated families and three of the four BC_2F_2 families was then compared on a limed and unlimed acid soil with the growth of a range of phalaris and cocksfoot (*Dactylis glomerata* L.) cultivars differing in acid soil tolerance. These comparisons are then used to plan certain aspects of the development of future acid tolerant cultivars of phalaris.

Methods and materials

Seeds of the four second backcross families and 14 control cultivars of phalaris and cocksfoot were sown on an acid soil at Swanpool, Victoria, on April 24, 1990. Most of the controls shown in Table 1 are described in Oram (1990); descriptions of the others follow. AT 88 is a phalaris breeding line developed by interpollinating the 20 most Al-tolerant genotypes among the 45 parents of cv. Holdfast, together with four additional genotypes. Wana and Kara cocksfoots are recent New Zealand cultivars, and Siro 1146 is the F_1 hybrid between Seedmaster *P. aquatica* and CPI 10446 *P. arundinacea*. There were also four cocksfoot breeding lines from New Zealand which are not shown in Table 1.

The Swanpool soil had pH($CaCl_2$) values in the A_1, A_2 and B horizons of 4.2, 4.4 and 4.4, respectively. The aluminium concentrations extractable with 0.01 *M* $CaCl_2$ ranged from 13.1 to 21.1 $\mu g\,g^{-1}$ (mean 15.8) in the top 10 cm. Extractable manganese concentrations in this soil were low, averaging 15.3 $\mu g\,g^{-1}$. Four replicates each contained limed and unlimed main plots. Liming involved the incorporation to 10 cm depth of 2.5 t fine $CaCO_3$ ha^{-1} six weeks before sowing. Seeds were sown at 4.5 kg ha^{-1} in open furrows with 60 kg ha^{-1} each of $CaCO_3$ and single superphosphate. The furrows were lightly harrowed after sowing. These subplots in the split plot design consisted of single rows 4 m long and 45 cm apart.

The plots were scored on a 1–5 (increasing) scale on November 1, 1990, when most growth had ceased. One fall of rain during summer induced some further growth, and the total amount of dry matter was scored again on March 28, 1991.

Results

The 18 lines differed significantly ($p < 0.01$) and the lime response was highly significant ($p < 0.001$) on both scoring occasions. The lime × entry interaction was significant only on the second occasion, and the lime response was calculated as the ratio of the unlimed score/limed score for each replicate.

Table 1 shows that there was considerable variation within phalaris and cocksfoot for estimated yield at both times of scoring. The ratio of growth on the limed and unlimed plots fell within the range of 0.32–0.38 for the phalaris cultivars Australian, Sirosa and Sirolan, which are generally regarded as sensitive to Al. The breeding line AT 88 tended to be less affected by acidity, but the difference from the sensitive group was not significant. Holdfast performed poorly on this particular acid soil. The Australian cultivars of cocksfoot, Currie and Porto, were more tolerant to acidity than Wana and Kara.

The backcross phalaris entries were less sensi-

Table 1. Mean scores in late spring and early autumn (1–5 increasing) of dry matter produced by control cultivars and breeding lines in the first growing season in a limed and unlimed acid soil at Swanpool, Victoria

Species	Entry	Dates of scoring				Ratio $\frac{U}{L}$
		Nov. 1, 1990		Mar. 28, 1991		
		Limed	Unlimed	Limed	Unlimed	
Phalaris	Australian	3.75	1.88	3.61	1.25	0.38
	Sirosa	4.13	2.00	3.75	0.97	0.33
	Sirolan	2.75	1.75	2.22	0.56	0.32
	Holdfast	2.75	1.25	2.08	0.42	0.36
	AT 88	3.00	2.38	2.22	1.67	0.77
Cocksfoot	Currie	2.13	1.75	0.56	0.56	1.00
	Porto	4.25	2.13	1.81	1.25	0.71
	Wana	2.50	1.75	1.11	0.28	0.21
	Kara	3.25	1.88	1.25	0.28	0.38
P. aquatica × *P. arundinacea*	Siro 1146	2.50	1.88	2.64	1.39	0.57
Backcrosses	BC2-5-26-1-9-1	2.75	2.50	1.94	1.81	0.88
	BC2-6-4	2.63	2.25	2.08	2.22	1.06
	BC2-10-20	2.63	1.38	2.08	1.11	0.69
	BC2-11-36	2.88	3.25	2.64	2.50	1.32
	LSD ($p = 0.05$)	1.04	1.04	1.26	1.26	0.76

tive to soil acidity than the sensitive *P. aquatica* cultivars, and tended to equal or surpass AT 88 and the entries which are recognised as Al-tolerant viz. Porto and Currie cocksfoot, and Siro 1146 hybrid phalaris (Culvenor et al., 1986a).

General discussion

Recent results are consistent with the earlier conclusion that sensitivity to Al, but not to Mn, is the primary cause of the poor growth of phalaris on acid soils. However, there is further evidence that Al-tolerance is not the only factor to be improved by selection for the development of new cultivars for acid soils. For example, Holdfast performs better than Sirosa on some acid soils, such as a yellow podzol on which a grazing trial is being conducted near Canberra, whereas both Holdfast and Sirosa perform poorly on unlimed soil at Swanpool. These other factors affecting phalaris performance remain to be identified. Meanwhile, empirical testing on representative target soils, either in situ or in glasshouse tests, appears to be the best strategy for further cultivar development.

The AT 88 breeding population within *P. aquatica* and some selected *P. aquatica* × *P. arundinacea* × *P. aquatica* backcross genotypes and progenies were more productive than the phalaris and cocksfoot controls on unlimed soil at Swanpool (Table 1), but less productive than Sirosa phalaris and Porto cocksfoot on limed soil. The extent to which acid soil tolerance can be combined with high yields on limed soils remains to be determined.

There is extensive genetic variation remaining in Holdfast, AT 88 and other seed-retaining *P. aquatica* populations, and also in the backcross population, for survival and herbage yield on acid soils such as that at Swanpool. This variation can be exploited by selection under realistic field conditions, but not by seedling tests on acid topsoils in the glasshouse. However, the seedling tests are rapid, and permit one generation of selection and inter-pollination to be completed each year, and therefore are useful in the early generations of selections in sensitive × tolerant populations. Field tests require two or three years per generation. In view of the urgent need for acid tolerant cultivars, the best strategy appears to be to develop an interim cultivar by one additional generation of selection within the seed-retaining *P. aquatica* populations. However, two or more generations of selection in the backcross population should produce an even more productive, acid-tolerant cultivar.

Acknowledgements

The financial support of the Wool Research and Development Corporation and the National Soil Conservation Program (Federal and Victorian branches) is greatly appreciated. Mr John Hunter and Ms Sue Windsor provided the yield scores made on November 11, 1990, and have helped in many ways to maintain the Swanpool experiment.

References

Anonymous 1990 Phalaris (*Phalaris aquatica* L.) variety 'Holdfast'. Aust. Plant Varieties J. 3, 12–13.

Bromfield S M, Cumming R W, David D J and Williams C H 1983 Change in soil pH, manganese and aluminium under subterranean clover pasture. Aust. J. Exp. Agric. Anim. Husb. 23, 181–191.

Culvenor R A 1985 Tolerance of *Phalaris aquatica* L. populations and some other agricultural species to excess manganese, and the effect of aluminium on manganese tolerance in *P. aquatica*. Aust. J. Agric. Res. 36, 695–708.

Culvenor R A, Oram R N and Fazekas de St. Groth C 1986a Variation in tolerance in *Phalaris aquatica* L. and a related species to aluminium in nutrient solution and soil. Aust. J. Agric. Res. 37, 383–395.

Culvenor R A, Oram R N and Wood J T 1986b Inheritance of aluminium tolerance in *Phalaris aquatica* L. Aust. J. Agric. Res. 37, 397–408.

Culvenor R A, Oram R N and David D J 1986c Genetic variability for manganese concentration in *Phalaris aquatica* growing in acid soil. Aust. J. Agric. Res. 37, 409–416.

Edmeades D C, Blamey F P C, Asher C J and Edwards D G 1991 Effects of pH and aluminium on the growth of temperate pasture species. I. Temperate grasses and legumes supplied with inorganic nitrogen. Aust. J. Agric. Res. 42, 559–569.

Helyar K R and Porter W M 1989 Soil acidification, its measurement and the processes involved. *In* Soil Acidity and Plant Growth. Ed. A D Robson. pp 61–101. Academic Press, Sydney, Australia.

Oram R N (Compiler) 1990 Register of Australian Herbage Plant Cultivars. 3rd edition. CSIRO, Melbourne, Australia. 304 p.

Oram R N, Ridley A M, Hill M J, Hunter J, Hedges D A, Standen R L and Bennison L 1990 Improving the tolerance of *Phalaris aquatica* L. to soil acidity by introgression of genes from *P. arundinacea* L. Aust. J. Agric. Res. 41, 657–668.

Ridley A M, Helyar K R and Slattery W J 1990a Soil acidification under subterranean clover (*Trifolium subterranean* L.) pastures in north-eastern Victoria. Aust. J. Exp. Agric. 30, 195–201.

Ridley A M, Slattery W J, Helyar K R and Cowling A 1990b The importance of the carbon cycle to acidification of a grazed annual pasture. Aust. J. Exp. Agric. 30, 529–537.

Ridley A M, Slattery W J, Helyar K R and Cowling A 1990c Acidification under grazed annual and perennial grass based pastures. Aust. J. Exp. Agric. 30, 539–544.

Williams C H 1980 Soil acidification under clover pasture. Aust. J. Exp. Agric. Anim. Husb. 20, 561–567.

P. J. Randall et al. (Eds.), Genetic aspects of plant mineral nutrition, 23–33.

PLSO SV05

Screening perennial rye-grass from New Zealand for aluminium tolerance

D.M. WHEELER[1], D.C. EDMEADES[1], D.R. SMITH[2] and M.E. WEDDERBURN[3]
[1]*Ruakura Agriculture Centre, Ministry of Agriculture and Fisheries, Private Bag, Hamilton, NZ,*
[2]*Hastings Research Centre, Ministry of Agriculture and Fisheries, Box 85, Hastings, NZ and*
[3]*Whatawhata Research Centre, Ministry of Agriculture and Fisheries, Private Bag, Hamilton, NZ*

Key words: aluminium tolerance, breeding, heritability, *Lolium perenne* L., perennial rye-grass

Abstract

Approximately 11,500 seedlings from 510 lines of perennial rye-grass (*Lolium perenne* L.) were screened for tolerance to aluminium (Al) using a low ionic strength 'still' solution culture technique. Although none of the individual lines were consistently more tolerant than any other line, 23 individual plants were selected from 13 lines for superior vigour and colour in the presence of Al.

The growth of three of these elite plants was examined on a reconstructed acid soil profile protected from prevailing weather conditions allowing control of the moisture status of the soil. The plants selected for Al tolerance in solution culture had significantly higher yields before drought and after recovery from drought than the rye-grass cultivars Ariki, Ellett and Droughtmaster and 4 other hill country lines which were previously selected for high yields in the presence and absence of nitrogen, and for drought and grassgrub resistance. Of the total number of plants tested from all cultivars and lines, <2% had yields that were greater than one third of the yields of the 3 Al tolerant plants. The better performance of the Al tolerant plants is attributed to better root growth in the acid soil.

Three polycrosses were made from the 23 Al tolerant plants selected in solution culture. When tested in solution culture, the yields of the half-sib families in the presence of Al averaged approximately twice that of Grasslands Nui in one experiment, but were similar to Grasslands Nui in another. Heritability of total yield and relative yield in the presence of Al, calculated from half-sib measurements on a single replicate basis, averaged 0.33 and 0.24 respectively. Individual plants from the half-sib families from two polycrosses were grown in a nursery and heading date and vigour recorded. There were no significant differences in heading data between the polycross lines and either of the cultivars Grasslands Nui or Yatsyn. Although there were significant differences in spring vigour between lines, they were not significantly different from either Grasslands Nui or Yatsyn. Twelve of the polycross lines showed decreased vigour in summer and autumn. This decline in vigour was attributed to damage from Argentine stem weevil (*Listronotus bonariensis*) as a consequence of low levels of lolium endophyte (*Acremonoim lolii*).

Introduction

Aluminium toxicity is a limiting factor for pasture and crop production on many acid soils. In New Zealand (NZ), the hill country topsoils are generally acid ($pH(H_2O) < 5.5$) and the liming of many of these soils has become uneconomic (Edmeades et al., 1985). Similarly, the liming of acidic subsoils ($pH(H_2O) < 5.0$) is either impractical or uneconomic. In Australia, increasing soil acidification has been highlighted as a major problem limiting plant production (Heylar and Porter, 1989).

Perennial rye-grass (*Lolium perenne* L.) is an

important component of many permanent pastures and is preferred because of its good cool season growth, high forage quality and ease of management. However, Edmeades et al. (1991b) reported that perennial rye-grass was moderately sensitive to Al (yield reduced by 50% by 1 – 4 μM Al^{3+} activity) and that the NZ derived cultivars were more sensitive to Al than the Australian and European derived cultivars tested. Earlier work (Edmeades et al., 1991a) had established that NZ derived cultivars of perennial rye-grass were more sensitive to Al than white clover (*Trifolium repens*), also a important component of many permanent pastures.

It was therefore decided to screen a large perennial rye-grass population to identify individuals or lines which were tolerant to Al. This paper reports the results of this screening and the initial agronomic experiments on the resultant elite material.

Methods

Initial screenings and plant selection

Perennial rye-grass (*Lolium perenne* L.) seedlings from 510 lines were screened for Al tolerance using a solution culture technique (Edmeades et al., 1991b). Each line consisted of the progeny from individual plants from a polycross which was part of a program to breed high producing cultivars. The source of the original plant material used for this program was from New Zealand, including some lines (not identified) that were selected from the same ecotype as the cultivar Grasslands Nui was selected from. Within these lines, 270 lines were early flowering and 240 lines were late flowering.

Plants were screened for tolerance to Al in a temperature controlled glasshouse (min 14°C, max 28°C). After germination on filter paper, seedlings were placed in pots suspended over a plastic tub (15 pots per tub) containing 40 litres of nutrient solution. The technique, analytical methods and solution composition are described elsewhere (Edmeades et al., 1991b). In general, nutrient concentrations were maintained at (μM) 500 Ca, 100 Mg, 300 K, 600 N (150 NH_4, 450 NO_3), 600 SO_4, 2.5 P, 5 Fe, 6 Mn, 1 Zn, 1 BO_3 and 0.1 Cu and the solution pH maintained at 4.7 throughout the course of the experiment. Aluminium was added as $Al_2(SO_4)_3$ and maintained at the nominal rates. The Al rates 10, 20, 30, 40 and 60 μM Al used in subsequent experiments contained 2.2, 4.4, 6.5, 8.7 and 13 μM Al^{3+} activity respectivity as calculated by Geochem (Sposito and Mattigod, 1980) using the modifications of Parker et al. (1987).

The rye-grass lines were screened at 3 Al concentrations (10, 20 and 30 μM Al) in 15 experiments (5 experiments per Al concentration) between August and November, 1987. Each experiment consisted of 8 tubs all containing nutrient solution at one Al concentration, with 14 lines plus a standard cultivar (Grasslands Nui) per tub and with 5 plants per pot. After 4 weeks growth, plants were assessed for colour (the degree of yellowing) and tops were harvested and weighed. From these initial experiments, 3 plants were identified which had good vigour and were considerably ‘greener’ than other plants in the presence of Al (labelled D1-D3).

From the above initial screening, 112 lines were selected based on the following criteria: the yields of plants from the selected lines were greater than Grasslands Nui at all levels of solution Al, the yields at 10 and 20 μM were similar, or the yield and colour rating at 20 and 30 μM were better than Grasslands Nui. The selected lines were further screened between January and February, 1988, in 5 separate experiments at four Al concentrations (0, 10, 20 and 30 μM Al), with 14 lines plus Grasslands Nui per tub and 4 plants per pot. One additional elite plant was identified (labelled D5).

The results for the standard cultivar Grasslands Nui from these 2 screenings showed that the relative yield (tops) was 100, 50, 30 and 15% for the Al rates 0, 10, 20 and 30 μM Al respectively. Therefore, a second selection of 84 lines from the 112 lines previously selected was made on the basis that relative yields were >50% and >40% at 10 and 20 μM Al respectively. These 84 selected lines were subjected to a third screening in March 1988 (Al concentrations of 0, 10, 20 and 30 μM Al) with 14 lines plus grasslands Nui per tub, and 4 plants per pot. The plants were harvested after 4 weeks growth and the tops weighed.

Thirteen lines from this third screening had relative yields greater than 80% and 50% at 10 and 20 μM Al respectively. Also, a further thirteen lines had a number of individual plants that appeared slightly greener and/or bigger than other plants when under Al stress. Seedlings from these 26 lines (approximately 1500) were screened at 10 μM Al. From this final screening, 19 individual plants (labelled D6-D24) were selected based on vigour and colour in the presence of Al. These 19 plants, together with the 4 individual plants previously identified in earlier screening were increased in number using tiller isolates and on-grown in potting mix for further experiments.

To confirm that the plants selected for vigour and colour were in fact Al tolerant, the tillers of these selected plants were tested by the solution culture technique. Tillers from five plants of the cultivar Grasslands Nui were also included. Individual tillers from the parent plants were removed and trimmed (roots to 2 cm and tops to 5 cm) before transplanting into the nutrient solution. In preliminary experiments using a standard cultivar (Grasslands Nui), it was established that tillers required a higher concentration of Al than seedlings to give the same reduction in yield. Therefore, the Al concentrations used to test tillers were higher than when seedlings were used (up to 60 μM Al for tillers compared to up to 20 μM Al for seedlings). The tillers were screened at 0, 20, 40 and 60 μM Al in November, 1988 using 2 replicates, 2 tubs per replicate and 15 lines per tub, with 3 plants per pot. The plants were harvested after 4 weeks growth and the tops and roots weighed.

Evaluation in reconstructed soil profiles

The first 3 elite Al tolerant plants identified and set aside in the initial screening (plants labelled D1, D2 and D3) were included in a larger experiment to determine their response to nitrogen and their recovery after drought and grass grub (*Costelytra zealandica*) infestation. This experiment was conducted in soil bins in which an acid soil profile had been reconstructed (Wedderburn et al., 1989). The reconstructed soil had a topsoil with a pH of 5 in water and a subsoil with 288 mg g^{-1} soil KCl-extractable Al (Wedderburn et al., 1989).

The experiment included 15 lines or cultivars of perennial rye-grass, each line consisting of 15 individual plants. The lines reported in this paper included the established cultivars Ariki, Ellett and Droughtmaster, and 4 lines selected from local hill country ecotypes for high dry matter production in the presence of N (HY + N), the absence of N (NY − N), drought resistance (DR) and grass grub resistance (GG). The hill country selections were based on results from previous screening experiments (Wedderburn et al., 1990). The 3 Al tolerant plants were included in a line with 12 other plants which were initial hill country selections.

The experimental design consisted of 2 bins, with each bin split into 6 replicates (see Wedderburn et al. (1990) for details). In each replicate there were 15 rows, each row representing a line. Within each row, tiller isolates (2 tillers) from each of the 15 plants in a line were planted. Rows and plants within rows were randomised.

Nitrogen in the form of $1(NH_4NO_3):2(Ca(NO_3)_2)$ at a rate equivalent to 3.5 kg N ha^{-1} $week^{-1}$ was applied to bin 2 (6 replicates) every fortnight from August 23 to December 20, 1988. A single harvest was taken in December 19, 1988 from the 2 bins.

A rain shelter was placed over both bins to simulate drought on January 4 to May 10, 1989. An index of the number of live tillers (TSC, 0 = no live tillers, 1 = 25% tillers alive, 2 = 50% tillers alive, 3 = 75% tillers alive and 4 = 100% tillers alive) was made on January 10. As soon as plant wilt commenced, wilt scores (1 = all brown leaves, 2 = yellowing leaves, 3 = faded leaf colour, 4 = limp green leaves, 5 = turgid green leaves) were recorded weekly for 7 weeks, and a total wilt score (TWS) derived. Grass grubs were placed in the soil of 3 replicates of each bin on March 31. On May 11, the grass grubs were killed using DASANIT insecticide.

On May 11, the soils in the bins were rewetted. Individual plants were harvested 20 days after rewetting and dried and weighed. The yield data (DM) was analysed using the log transformation $\text{Log}_e((DM/0.02) + 1)$ to stabilise the variance.

Estimation of heritability

Heritability of yield and relative yield at various solution Al concentrations was estimated from the progeny of 3 polycrosses using the 23 individual plants selected for Al tolerance in the initial screenings. The polycrosses were: 1) a polycross between the 23 plants selected for Al tolerance, 2) a polycross of 5 elite plants (labelled D5, D7, D9, D14, D21) selected for either high absolute dry weights or high relative dry weights in the presence of Al as determined from tiller testing (Table 1) and 3) a polycross of 11 plants (labelled D1, D2, D3, D6, D8, D12, D13, D17, D20, D22, D23) which had better absolute or relative yields than Grasslands Nui in the presence of Al as determined by tiller testing (Table 1). The range of responses between the plants used in this polycross and in the second polycross overlap slightly. This assignment of plants was made to ensure adequate phenotypic variation in each population.

Table 1. Effect of solution aluminium (μM) on the top yield (mg pot^{-1}) of perennial rye-grass (*Lolium perenne*) grown form tiller isolates from 23 selected aluminium tolerant plants and the standard cultivar Grasslands Nui

Line	Aluminium concentration (μM)			
	0	20	40	60
1	260	129	107	49
2	328	249	92	69
3	508	199	112	59
5	717	323	183	119
6	404	236	130	76
7	452	412	211	120
8	416	285	117	106
9	626	223	174	94
10	417	125	89	76
11	371	107	76	73
12	548	255	158	97
13	560	195	109	79
14	385	254	256	189
15	330	123	95	51
16	643	269	165	109
17	403	195	92	78
18	274	105	81	44
19	507	171	72	58
20	308	242	127	82
21	496	292	216	138
22	518	308	142	93
23	351	201	246	36
24	541	174	94	53
Grasslands Nui	709	176	118	86
SED	209[a]	60	47	28
	181[b]	52	41	24

[a] for comparing half-sib families.
[b] for comparing half-sib families with Grasslands Nui.

Seedlings from the half-sib families in each polycross were assessed for Al tolerance in 2 separate experiments using the solution culture technique. In the first experiment, seedlings from the 23 half-sib families from the first polycross were assessed at 0, 10, 20 and 30 μM Al in April, 1990. The experimental design consisted of 5 blocks with 2 replicates. Within each block, 4 lines and the cultivar Grasslands Nui were assessed using 3 pots per line per tub and 4 plants per pot. In the second experiment (June, 1990) seedlings from 18 of the 23 half-sib families from the first polycross, the half-sib families from the second 2 polycrosses and the cultivar Grasslands Nui were assessed at 0, 10 and 20 μM Al. There were 2 replicates of the 0 Al treatment, and 3 replicates of the 10 and 20 μM Al treatments. The lines were randomly allocated among 4 tubs, with 2 plants per pot. In all experiments, tops and roots were harvested, dried and weighed after 4 weeks growth.

Yield data was transformed using natural logarithms. Heritability of yield (top, root and total) in solution culture was estimated separately for each Al level as:

$$\text{Variance(line)}/(\text{Variance(line)} + \text{Variance(error)})\,.$$

Heritability was calculated on a single unit basis in order to facilitate comparisons between experiments (Hanson, 1963). Variance components were estimated by maximum likelihood using the Varcomp procedure of SAS. The reference populations for these estimates were considered to be the theoretical populations synthesized from the plants selected for inclusion in the polycross. The experimental unit was defined as the single 'pot' containing either 3 or 4 half-sib plants. Thus these estimates of heritability relate to selections amongst unreplicated plots of half-sib families under similar experimental conditions.

Relative tolerance was defined as the ratio of the yield at either 10, 20 or 30 μM Al to the yield in the absence of Al. On a log scale these ratios equate to simple differences i.e.

$$\log(\text{dry weight } Al_{20}/\text{dry weight } Al_0) = \log(\text{dry weight } Al_{20}) - \log(\text{dry weight } Al_0)$$

Analysis of variance of log dry weight at two levels of Al provides the mean square for this difference over lines as the Al × line interaction term. Thus heritability of relative tolerance to Al was determined as:

$$\text{Variance}(Al \times \text{line})/(\text{Variance}(Al \times \text{line}) + \text{Variance}(\text{error}))$$

where variance components were calculated from analysis at two levels of Al (0 and either 10, 20 or 30 μM).

Field nursery evaluation

Seed of the 16 half-sib families from polycross 2 and 3, and 2 control cultivars (Grasslands Nui and Yatsyn) were sown into 1 metre nursery rows at Poukawa Research Station, Hawkes Bay, New Zealand on a fertile soil on 3 April, 1990. The experiment was a randomised complete block design with 2 replicates. Sowing rate was 0.5 g seed per 1 metre row.

Rows were scored for general vigour of growth at approximately 20 day intervals from late winter 1990 to autumn 1991. The nursery was mown to a height of 2.5 cm whenever the control cultivars attained an average height of 18–20 cm. Adult Argentine stem weevil (*Listronotus bonariensis*) feeding damage and rust reaction were scored in February.

In an adjacent block, 25 seedlings of each of the polycross half-sib families were transplanted into a spaced plant nursery to observe plant morphology. Plant spacing was 25 cm between plants within lines and 1 metre between lines. During October/early November, individual plants were scored at 2–3 day intervals for heading (exsertion of the first spike from its sheath) and median heading date estimated.

Results

Selection of plants

The initial screening of 510 lines was designed to identify perennial rye-grass lines in which all individual plants had considerably better yield or growth characteristics such as colour or vigour in the presence of Al than the standard rye-grass cultivar (Grasslands Nui). None of the 510 lines screened met this criteria, although 3 individual plants were identified initially that had considerably better vigour than Grasslands Nui in the presence of Al. Although lines were identified with greater yield and better colour than Grasslands Nui in the presence of Al and included in the second and third screenings, the individual plants in these selected lines did not consistently show the same vigour or colour in the presence of Al as the 3 individual plants initially selected. However, twenty-six lines were identified that consistently produced individual plants that had superior colour, vigour or yield relative to Grasslands Nui in the presence of Al. Further screening of these 26 lines identified 19 individual plants that performed well in the presence of Al.

The 23 plants selected for Al tolerance came from only 14 lines (4 lines gave 3 plants each, 1 line gave 2 plants each and 9 lines gave 1 plant each). Of these 13 lines, 8 were from the early flowering and 6 were from the late flowering selection lines.

Plant top yields of tillers taken from the 23 selected plants are shown in Table 1. Sixteen of the selected plants had yields or relative yields greater than the standard cultivar, Grasslands Nui. The rankings of the plants were similar when root yields were measured (data not shown).

Assessment in reconstructed profile

There were no significant interaction between the nitrogen (presence or absence of N), drought (before or after) or grass grub infestation (present or absence) treatments. Therefore, only the results for the line effects are presented.

The 3 individual plants, identified as superior under Al stress in the initial screening had higher yields prior to drought, higher yields on recovery from drought and grew longer into the drought as indicated by the total wilt score than the commercial perennial rye-grass cultivars Ariki, Ellett and Droughtmaster, or the selected hill country ecotypes (Table 2). The higher yields of these plants were partially attributed to higher

Table 2. Dry matter yield (mg plant^{-1}) before (DMY 1) and after drought (DMY 2), tiller score (TSC) before drought and the total wilt score (TWS) during drought for 3 plants selected for aluminium tolerance (Al Selections), the cultivars Ariki, Ellett and Droughtmaster and 4 hill country selections (see text) of perennial rye-grass (*Lolium perenne*)

Source	Line	DMY 1[a]	DMY 2[a]	TSC	TWS
Al selections	1	4.21	3.02	2.00	7.00
	2	4.15	3.59	2.08	7.00
	3	3.98	3.34	1.92	7.00
	Mean[b]	4.11	3.32	2.00	7.00
Cultivars	Ariki	1.74	0.61	1.04	4.86
	Ellet	1.97	1.18	1.08	6.06
	Droughtmaster	1.65	0.86	1.30	5.66
Hill-country selections	HY − N	2.23	1.24	1.21	6.31
	HY + N	1.96	1.10	1.06	5.92
	DR	2.53	1.46	1.37	6.43
	GG	1.91	1.17	1.08	6.03
SED[c]		0.153	0.158	0.075	0.34

[a] Transformed by $\log_e((DM/0.02)+1)$, each unit increase represents an increase in yield of approximately 2.7 times.
[b] SED for comparisons between the Al tolerant lines were 0.25, 0.28, 0.16 and 0.67 for DMY 1, DMY 2, TSC and TWS respectively.
[c] SED is for comparisons between the mean Al Selections and the other cultivars and selections tested.

tiller survival as indicated by the tiller score in the Al lines compared to the other cultivars and lines (Table 2). There were no significant differences between the 3 Al tolerant plants in yield prior to or on recovery from drought, tiller score or total wilt score (Table 2).

Table 3 shows the distribution of the 218 plants tested (other than the 3 Al tolerant selections) for each parameter measured. Less than 2% of the other plants had yields prior to or on recovery from drought that were greater than a third of the yield of the mean of the Al tolerant plants (difference in transformed yields > 1). However, 9% of the other perennial rye-grass plants grew as long into the drought as the Al tolerant selections, as indicated by the total wilt score.

Heritability of Al tolerance in solution culture

The overall mean total yield for the progeny of each half-sib family from the three polycrosses is

Table 3. The distribution of 218 perennial rye-grass (*Lolium perenne*) lines (see text) for the parameters dry matter yield (mg plant^{-1}) before (DMY 1) and after drought (DMY 2), tiller score before drought (TSC) and the total wilt score during drought (TWS). The mean of 3 aluminium tolerant plants (Al Selections) is also shown for comparison

Parameter	Percent in each category			Al selections
DMY 1[a]	>2.5	>3	>3.5	4.11
	15%	1.8%	0%	
DMY 2[a]	>2	>2.5	>3.0	3.32
	2.3%	1.1%	0%	
TSC	>1.5	>1.75	>1.8	2.00
	7%	2%	0%	
TWS	>6.5	7.0	>7.0	7.00
	29%	9%	0%	

[a] Transformed by $\log_e((DM/0.02)+1)$, each unit increase represents an increase in yield of approximately 2.7 times.

Table 4. Comparison of the mean total yield (mg plant^{-1}) of the half-sib families of perennial rye-grass (*Lolium perenne*) from 3 polycrosses (1, 2, and 3, see text) and the total yield of Grasslands Nui (Nui) at 4 solution aluminium concentrations for 2 separate experiments. Relative yield (expressed as a percentage of yield in the absence of aluminium) is shown in parenthesis

Experiment	Polycross	Aluminium concentration (μM)			
		0	10	20	30
1	1	2020	1120 (55)	690 (34)	380 (19)
		ns[a]	ns (ns)	ns (ns)	ns (ns)
	Nui	2050	1120 (55)	740 (36)	480 (23)
2	1	501	460 (92)	284 (57)	
		ns	*** ***	*** ***	
	2	562	469 (83)	287 (51)	
		ns	*** ***	*** ***	
	3	635	562 (89)	338 (53)	
		ns	*** ***	*** ***	
	Nui	554	251 (45)	167 (30)	

[a] ns = no significant difference between the polycross mean and the standard cultivar Grassland Nui. *** = significantly different at $p = 0.001$.

shown in Table 4. In the first experiment, there were no significant differences between the overall mean of the half-sib families in polycross 1 and Grasslands Nui in total or relative yield for all Al treatments (Table 4). In comparison, in Experiment 2 (using progeny from the half-sib families from polycrosses 1, 2 and 3), the mean total and relative yield for each polycross were approximately twice those for Grasslands Nui at 10 and 20 μM Al (Table 4).

There were significant differences in total or relative yield at each Al concentration between half-sib families within each polycross in each experiment (data not shown). In Experiment 1 (polycross 1), the differences between families in total yield and relative yield (as indicated by var(line) or var(Al × line) variance component) was most pronounced at 20 μM Al (Table 6). In Experiment 2 (polycross 1–3), the magnitude of the genetic variation depended upon the population (polycross) as well as upon the level of Al. In the population generated from polycross 1, variation was greater at 0 and 10 μM Al than at 20 μM Al. In the narrow based second polycross (derived from parents selected for high yields in both the presence and absence of Al), genetic variation was observed only at 20 μM Al. In the more broadly based third polycross population, the genetic variance of yield and relative yield were similar at 10 and 20 μM.

Over all polycrosses, the mean heritability of total yield in the absence and presence of Al average 0.25 and 0.33 respectively and averaged 0.24 for relative yield (Table 5). The pattern of genetic variance and the heritability of total and relative yields for top and root yields were similar to that of total yield (data not shown).

The rankings of the families in polycross 1 was not consistent between Experiments 1 and 2 (data not shown). This was further confirmed by the non significant correlations between measurements made in Experiment 1 and in Experiment 2 (Table 6).

Performance in the field nursery

In spring, there were no significant differences in vigour between the half-sib families of polycross 2 and 3 and the control cultivars Grasslands Nui and Yatsyn (Table 7). Several families maintained this vigour into summer and autumn with the family D2 showing the highest vigour over this period. However, vigour in many of the families declined dramatically during late summer/early autumn (Table 7). These lines also had high Argentine stem weevil damage scores. Examination of a sample of tillers from the lines with high weevil damage scores showed that lolium endophyte (*Acremonoim lolii*) infection levels were low. There were no significant differences in mean heading date between the polycross families and the control cultivars.

Table 5. Estimates of the variance components and heritability (h^2) of the log of total yield at 0, 10, and 20 μM aluminium (Al) and the log of relative yield (expressed as a fraction of yield in the absence of Al) for 3 polycrosses of perennial rye-grass (*Lolium perenne*)

Parameter	Al	Polycross	Var (line)[b]	Var (error)	h[2c]
Total yield	0	1[a]	0.030	0.051	0.37 (0.10)
		1	0.046	0.063	0.43 (0.20)
		2	0.000	0.096	0.00 (0.32)
		3	0.000	0.025	0.00 (0.40)
	10	1[a]	0.092	0.183	0.33 (0.10)
		1	0.032	0.059	0.35 (0.15)
		2	0.046	0.072	0.39 (0.20)
		3	0.000	0.082	0.00 (0.29)
	20	1[a]	0.213	0.150	0.59 (0.09)
		1	0.010	0.113	0.08 (0.15)
		2	0.036	0.074	0.32 (0.20)
		3	0.060	0.033	0.60 (0.23)
	30	1[a]	0.146	0.170	0.46 (0.10)
Relative yield	0, 10	1[a]	0.013	0.117	0.10 (0.08)
		1	0.036	0.061	0.37 (0.18)
		2	0.036	0.080	0.31 (0.25)
		3	0.000	0.061	0.00 (0.35)
	0, 20	1[a]	0.065	0.101	0.39 (0.10)
		1	0.001	0.098	0.01 (0.20)
		2	0.022	0.081	0.22 (0.26)
		3	0.034	0.033	0.51 (0.31)
	0, 30	1[a]	0.038	0.110	0.26 (0.10)

[a] Results from experiment 1 (see text).
[b] Var(line) for total yield and var(Al × line) for relative yield.
[c] Heritability of half-sib family measurements, calculated on a single replicate basis (total weight of 4 plants per pot). Standard error of estimate given in parenthesis.

Table 6. Correlation coefficients for the relationship between total yield of half-sib families of perennial rye-grass (*Lolium perenne*) from polycross 1 (see text) measured in 2 separate experiments for 4 aluminium (Al) concentration. Correlation coefficients for relative yield are shown in parenthesis

Experiment 3 Al concentration (μM)	Experiment 2 Al concentration (μM)			
	0	10	20	30
0	0.15[a]	0.12	0.00	−0.11
10	0.25	0.19 (0.01)	0.20 (0.09)	0.31 (0.33)
20	0.11	−0.16 (−0.32)	−0.10 (−0.18)	−0.09 (0.03)

[a] Correlation coefficients were not significantly differed from zero at $p = 0.05$.

The correlations coefficients between Al tolerance as measured by total yield or relative yield in solution culture and attributes measured in the field (heading date and vigour) were not significant (Table 8).

Discussion

Approximately 11,500 seedlings from 510 perennial rye-grass lines were screened for Al tolerance. The 23 plants selected for Al tolerance

Table 7. Median heading date, vigour scores and Argentine stem weevil (ASW) damage in perennial rye-grass (*Lolium perenne*) for the half-sib families from polycross 2 and 3 (see text) and the standard cultivars Grasslands Nui and Yatsyn

Line	Median heading date[a]	Vigour scores[b]			ASW Damage[c]
		Spring	Summer	Autumn	
D1	−1	7.7	6.0	5.8	2.0
D2	0	8.2	8.8	8.3	1.0
D3	2	7.7	7.4	7.8	1.0
D5	1	7.8	5.7	5.2	2.5
D6	5	7.8	5.9	5.7	2.5
D7	0	7.5	5.7	5.5	2.5
D8	3	8.2	6.3	6.0	2.0
D9	2	6.8	5.9	5.5	2.0
D12	3	7.2	5.2	5.7	2.0
D13	3	7.2	6.0	6.0	2.5
D14	0	7.5	7.3	6.8	1.0
D17	1	7.8	5.8	5.7	2.0
D20	2	7.2	7.0	6.0	2.0
D21	−1	7.2	5.6	4.7	2.0
D22	3	7.3	5.7	5.8	2.0
D23	2	8.0	7.5	7.0	1.0
Grasslands Nui	1	7.7	7.1	6.5	1.0
Yatsyn	0	7.5	6.8	6.8	1.0
LSD	ns	1.3	0.9	1.4	0.8

[a] Estimated date of 50% heading, measured from 22.10.90.
[b] Scored at 20 day intervals. Scale 1 (poor) to 10 (excellent).
[c] Scored on 19.01.91 for adult Argentine stem weevil feeding damage. Scale: 1 = less than 20% of tillers damage; 2 = 20–40% damage; 3 = 40–60% damage.

Table 8. Correlation coefficients for the relationship between total yield and the field attributes median heading date and vigour scores for spring, summer and autumn in perennial rye-grass (*Lolium perenne*) from 16 half-sib families from 2 polycrosses (see text). The correlation coefficients for relative yields are shown in parenthesis

		Aluminium concentration (μM)				
		0	10		20	
Median heading date[a]		−0.20[c]	−0.23	(−0.15)	0.26	(0.07)
Vigour score[b]	Spring	−0.43	−0.30	(−0.21)	0.17	(0.25)
	Summer	−0.21	−0.48	(−0.12)	0.40	(−0.07)
	Autumn	−0.26	−0.39	(−0.01)	0.10	(−0.08)

[a] Estimated date of 50% heading, measured from 22.10.90.
[b] Scored at 20 day intervals.
[c] Correlation coefficients were not significantly differed from zero at $p = 0.05$.

came from only 13 of the original perennial rye-grass lines.

The reason for the superior performance of 3 of the selected Al tolerant plants in the reconstructed acid soil profiles is not known. However, given that these plants were selected for Al tolerance and given that they were tested in an acid soil, it is probable that these plants could explore a larger soil volume in an acid Al toxic soil. This ability to explore a larger soil volume (and thus a larger soil nutrient and/or soil moisture pool) could explain the higher yields before and after recovery from drought.

The heritability of total yield and relative yield in the presence of Al (averaged 0.33 and 0.24 respectively) was moderate and is similar to that reported for tiller numbers and plant yields before and after drought (Wedderburn pers. comm.). Thus, although further improvements in Al tolerance is perennial rye-grass could be made, progress is likely to be slow. However, the fact that the yields of the half-sib families were twice that of Grasslands Nui in one experiment and that three of the plants preformed exceptionally well in the reconstructed soil profiles indicate that the plants selected for Al tolerance and their progeny could be significantly better than Grasslands Nui in some acid soils.

It is unclear why the rankings of half-sib families in polycross 1 were inconsistent between the 2 experiments. Given that the experiments were conducted at different times of the year, there is the possibility that the mechanism of Al tolerance may interact with unknown environmental factors to modify the response of plants to Al. The different Al concentrations that are toxic to seedlings and tillers indicates that plant age may influence the expression of Al tolerance, although the mechanism of such an effect is unknown. A better understanding of the mechanism(s) of Al tolerance and of the environmental factors modifying the response of plants to Al, could improve the ability to select Al tolerant material.

With New Zealand, many of the previously selected perennial rye-grass cultivars have been selected on fertile (i.e. no soil pH limitation to growth) soils for traits such as high yield and vigour. The vigour scores reported for the plants selected for Al tolerance (Table 7) indicates that on fertile soils, the agronomic performance of these plants is as good as those cultivars previously selected for high yield and vigour. The good agronomic potential in fertile soils of these selected plants is attributed to the fact that the original 510 lines were taken from a program to breed high producing rye-grass. Although screening rye-grass material from a more acid environment might increase the number of plants selected, it is noted that 3 of the plants selected for Al tolerance performed better in the reconstructed acid soil profiles before and after drought than those plants selected from the hill country, some of which would have been grown on an acid soil profile similar to that reconstructed in the bins. Also, the fact that Al tolerance was not related to vigour (Table 8) and that the established cultivars were selected for high vigour could explain why many of the established cultivars were reported as being moderately sensitive to Al by Edmeades et al. (1991b).

The present results indicate that the perennial rye-grass plants selected for Al tolerance have good agronomic potential in acid and fertile soils. However, lolium endophyte (*Acremonoim lolii*) will need to be reintroduced to these plants to prevent damage and thus a lack of persistence due to Argentine stem weevil infestation (Prestidge et al., 1982). The next stage of testing these plants in acid soils under field conditions has commenced.

References

Edmeades D C, Pringle R M, Mansell G P, Shannon P W, Ritchie J and Stewart K W 1985 Effects of lime on pasture production on soil in the North Island of New Zealand. 5. Description of a lime recommendation scheme. N.Z. J. Exp. Agric. Res. 13, 47–58.

Edmeades D C, Blamey F P C, Asher C J and Edwards D G 1991a Effects of pH and aluminium on the growth of temperate pasture species. 1. Temperate grasses and legumes supplied with inorganic nitrogen. Aust. J. Agric. Res. 42, 559–69.

Edmeades D C, Wheeler D M and Christie R A 1991b The effect of Aluminium and pH on the Growth of a Range of Temperate Grass Species and Cultivars. *In* Plant-Soil Interactions at Low pH. Eds. R J Wright, V C Baligar and R P Murrmann. pp 913–924. Kluwer Academic Publishers, Dordrecht, The Netherlands.

Hanson W D 1963 Heritability. *In* Statistical Genetics and Plant Breeding. Eds. W D Hanson and H F Robson. Publication 962, National Academy of Science/National Research Council, Washington DC.

Helyar K R and Porter W M 1989 Soil acidification, its measurement and the processes involved. *In* Soil Acidity and Plant Growth. Ed. A D Robson. pp 61–101. Academic Press, Australia.

Parker D R, Zelazny L W and Kinraide T B 1987 Improvements to the program GEOCHEM. Soil Sci. Soc. Am. J. 51, 488–491.

Prestidge R A, Pottinger R P and Barker G M 1982 An association of Lolium endophyte with ryegrass resistance to Argentine Stem Weevil. Proc. N.Z. Weed and Pest Control Conference 35, 119–22.

Sposito G and Mattigod S V 1980 GEOCHEM: A Computer Program for the Calculation of Chemical Equilibra in Soil Solution and other Natural Water Systems. Kearney Foundation of Soil Science, University of California, Riverside, CA.

Wedderburn M E, Pengelly W J and Tucker M A 1989 Simulated soil profiles as a plant screening technique. Proc. of the N.Z. Grass. Assoc. 50, 249–252.

Wedderburn M E, Tucker M A, Pengelly W J and Ledgard S F 1990 Responses of a New Zealand North Island hill perennial ryegrass collection to nitrogen, moisture stress, and grass grub (*Costelytra zealandica*) infestation. N.Z. J. Agric. Res. 33, 405–411.

P. J. Randall et al. (Eds.), Genetic aspects of plant mineral nutrition, 35–44.

PLSO SV07

Somaclonal variation in plant adaptation to acid soil in the tropical forage legume *Stylosanthes guianensis*

I.M. RAO, W.M. ROCA, M.A. AYARZA, E. TABARES and R. GARCIA
Centro Internacional de Agricultura Tropical (CIAT), Apartado Aéreo 6713, Cali, Colombia

Key words: biomass partitioning, nutrient uptake, plant adaptation, soil acidity, somaclonal variation, *Stylosanthes guianensis*

Abstract

Somaclonal variation offers the possibility to obtain changes in one or a few characters of an otherwise outstanding cultivar without altering the remaining, and often unique, part of the genotype. It has been shown to be heritable for some species. A check line of *Stylosanthes guianensis* (Aubl.) Sw., CIAT 2243 and 14 somaclones in the R_4 generation, selected after three generations from the original 114 plants regenerated from callus cultures, were used in a glasshouse trial. The main objective of the study was to evaluate the physiological basis of the differences in agronomic performance of certain somaclones over the check genotype when grown in a sandy loam acid soil at low or high fertility level. Measurements at the time of harvest (170 days of plant age) included dry matter distribution between shoot and roots, leaf area production, nutrient levels in soil and plant parts, and uptake of nutrients from soil. Somaclones differed with the check genotype in terms of (i) partitioning of fixed carbon between the shoot and roots; (ii) root biomass production and (iii) uptake of nitrogen and phosphorus. Positive relationships were found between total nitrogen uptake and total biomass, and total phosphorus uptake and total biomass, and total phosphorus uptake and total nitrogen uptake. The results of this study provide an insight into the potential use of somaclonal variation for the improvement of plant adaptation to acid soil conditions.

Introduction

Somaclonal variation is described as the variation arising from the culture of plant cells, tissues, organs and explants and their subsequent regeneration into whole plants (Larkin, 1987; Larkin and Scowcroft, 1981). Many mechanisms for such variation have either been identified or proposed (Karp, 1990; Larkin, 1987; Larkin and Scowcroft, 1981; Phillips et al., 1990). These mechanisms include changes in chromosome structure and/or number, apparent point mutations, changes in gene expression as a result of chromosome structural changes (heterochromatin and position effects) or activation of transportable elements, chromatin loss, DNA amplification, somatic crossing-over, somatic reduction, and changes in the DNA of cytoplasmic organelles. Somaclonal variation offers the possibility to obtain changes in one or a few characters of an otherwise outstanding cultivar without altering the remaining, and often unique, part of the genotype. In some species, somaclonal variation has been shown to be heritable (Larkin, 1987), a necessary condition for somaclonal variation to be useful as a tool for studies on genetic plant adaptation to the environment. However, very few reports exist on the use of somaclonal variation for plant adaptation to abiotic stress (Adkins et al., 1990; Bagley and Taylor, 1987; Conner and Meredith, 1985; McHughen and Swartz, 1984; Miller et al., 1992; Ojima and Ohira, 1988).

Plant growth in many acid soils of the world is

affected by excess aluminum and low availability of phosphorus (Foy, 1988). The management of soil acidity by liming is not always economically feasible, especially in strongly acid soils of tropical America. Therefore, breeding programs directed toward developing cultivars adapted to acid soil conditions have been undertaken recently (Aniol and Gustafson, 1990; Duncan and Baligar, 1990). Newly identified genetic sources of tolerance to soil acidity combined with efficient screening techniques and greater understanding of the physiological and genetic basis of tolerance to acid soil conditions assure rapid progress in developing plants for improved adaptation to nutrient-poor acid soils.

Stylosanthes guianensis (Aubl.) is a tropical forage legume adapted to acid soil conditions of tropical America. It is highly amenable to *in vitro* regeneration using protoplast and cultured callus tissue (Meijer and Broughton, 1981; Szabados and Roca, 1986). Somaclonal variation for morphological and agronomic characters has been reported independently by two research teams for primary regenerants (R_1) and selfed progenies (R_2) of *S. guianensis* (Godwin et al., 1987; Miles et al., 1989). A research team at CIAT (Miles et al., 1989) has analyzed the regenerated plants of *S. guianensis* from callus culture of leaf or hypocotyl explants originating from a single seedling of germplasm accession CIAT 2243 (highly inbred and uniform) for ploidy level and agronomic performance. Approximately 20% of the total of 114 plants regenerated were tetraploids. For most agronomic traits evaluated under acid-soil field conditions, tetraploids and some diploid lines differed to the check genotype, demonstrating the generation of heritable genetic variation through in vitro tissue culture (CIAT, 1989; Miles et al., 1989). The present study aims to determine the physiological basis for the differences in agronomic performance of certain somaclones over the check genotype.

Materials and methods

Fourteen somaclone lines (nos. 4, 5(4×), 9, 15, 18, 22, 26, 36, 39, 40, 45, 52, 57, 67) along with the check line #74 of *Stylosanthes guianensis* (Aubl.) Sw. CIAT 2243, selected after three selfed generations from the original 114 plants regenerated from callus cultures (Szabados and Roca, 1986), were used in a pot culture study conducted in a glasshouse during February–July 1990. The maximum and minimum day/night temperature and relative humidity were 36/19°C and 96/48%, respectively. The maximum photon flux density during the day was 1100 μmol $m^{-2} s^{-1}$. Agronomic performance (based on plant vigor, dry matter production, lateral growth, and seed production potential) evaluated under field conditions at Quilichao, Colombia allowed classifying these somaclones as low (lines #22, 15, 45), medium (lines #39, 5(4×), 36, 57, 18, 74 check line, 4, 67), high (lines #40, 26, 9) and higher (line #52) performers (CIAT, 1989). The characteristic features of certain somaclone lines include dwarf and bushy (#39), tetraploid (#5(4×)), chlorotic (#36) and variable number (1 to 3) of leaflets (#22).

A glasshouse trial was conducted at CIAT headquarters using an Oxisol (Coarse-loamy, kaolinitic, isohyperthermic Typic Haplustox) collected from the field 'Alegria' at Carimagua, Meta Department, Colombia (4.5°N, 71.5°W, 150 m elevation, 2300 mm mean annual rainfall). The soil was a sandy loam with 17% clay, 65% sand, 18% silt and 0.89% organic matter. The chemical characteristics of the soil before fertilization were: pH 5.1; available $N–NH_4$, 0.26 mmol kg^{-1}; total N, 24 mmol kg^{-1}; total carbon, 433 mmol kg^{-1}; C/N ratio, 15.5; Bray II extracted P, 0.06 μmol g^{-1}, Bray II extracted K, 0.03 cmol kg^{-1}; extracted S, 0.17 mmol kg^{-1}; extracted Al, Ca and Mg, 0.7, 0.13 and 0.08 cmol kg^{-1}, respectively. The Al saturation was 77% while the effective cation exchange capacity was 0.94 cmol kg^{-1}. The micronutrient levels in μmol kg^{-1} were 2.44, 2.98, 13.46 and 7.4 for Zn, Cu, Mn and B, respectively. The analytical methods used to determine the soil chemical characteristics were as described in Salinas and Saif (1990).

Two fertility levels, low and high, were selected to represent the recommended fertility levels for pasture establishment and crop-pasture rotations, respectively. The soil was thoroughly mixed with the fertilizers before transferring to plastic pots. Individual rates of fertilizer added

Table 1. Fertilizer details to establish two fertility levels

Element	Compound	Composition (%)	Application rate of element ($kg\ h^{-1}$)[a]	
			Low fertility	High fertility
N	Urea	46	0	40
P	Triple super phosphate	20	20	50
K	KCl	52	20	100
Ca	Dolomitic lime	22.8	33	66
Ca	Triple super phosphate	14	14	35
Mg	Dolomitic lime	9.8	14.2	28.4
S	Elemental Sulfur	86.1	10	20
Zn	$ZnCl_2$	48	0	2
Cu	$CuCl_2$	37.2	0	2
B	H_3BO_3	17.5	0	0.1
Mo	$Na_2MoO_42H_2O$	40	0	0.1

[a] Application rates are on a surface-area basis, $100\ kg\ ha^{-1} = 206.2$ mg/container.

to each pot having 4 kg of dry soil in 2.7 liter volume are shown in Table 1. It is important to note that nitrogen and micronutrients were not added to the low fertility treatment. Seedlings of R_4 somaclones were raised in soil and selected seedlings of similar shoot and root growth were transferred to plastic pots. Each pot had 1 plant and was inoculated with mycorrhizae (isolated from soil with *Brachiaria decumbens* roots) and rhizobium. The trial was arranged in a completely randomized design with 3 replications. The total number of pots was 90 (15 lines × 2 fertility levels × 3 replications). The soil was maintained near field capacity throughout the experiment, and plants were harvested at 170 days after germination.

At harvest, the following measurements were made: plant height, leaf number per plant, internode length, number of main branches, stem diameter, leaf area per plant, dry matter distribution between shoot and roots, nutrient levels in soil and plant parts, and uptake of N and P in shoot and total biomass. Each plant was harvested and separated into different plant parts. For the separation of roots from the soil, each soil block from the pots was soaked overnight in water containing sodium bicarbonate, which helped to disperse the clay. Roots were washed from the soil on a 1-mm sieve and 'live' roots were separated by hand from organic matter. The dry matter distribution into different plant parts (leaves, stems and roots) was determined by drying them for at least 2 days in the oven at 70°C to constant weight. Leaf area was determined using a leaf area meter (LI 3000; LI-COR, Inc., Lincoln, NE). Leaf chlorophyll, soluble protein and inorganic phosphorus for some selected clones were determined as described in Rao and Terry (1989). Nutrient composition of different plant parts was determined according to the methods outlined by Salinas and García (1985).

Nutrient uptake was determined using the values of biomass production (shoot or shoot + root) and nutrient concentration in the biomass. The determination of N uptake includes N_2 fixation by root nodules as well as uptake from the soil. Analysis of variance was calculated, using the SAS computer program (SAS, 1982). A probability level of 0.05 was considered statistically significant.

Results

Variation among somaclonal lines in plant morphological characteristics of *S. guianensis* is shown in Figure 1. Plant height in the dwarf, bushy, somaclone #39 was 24% or 34% less at low or high fertility treatments, respectively, when compared with the check line #74 (Fig. 1A). Maximum plant height among somaclones was observed for somaclonal lines #5(4×) and #9 with low- and high-fertility treatments, respectively. The dwarf somaclone had more than twice leaves as the check under high fertility (Fig. 1B). The number of main stems under low fertility increased by 84% in somaclone #22

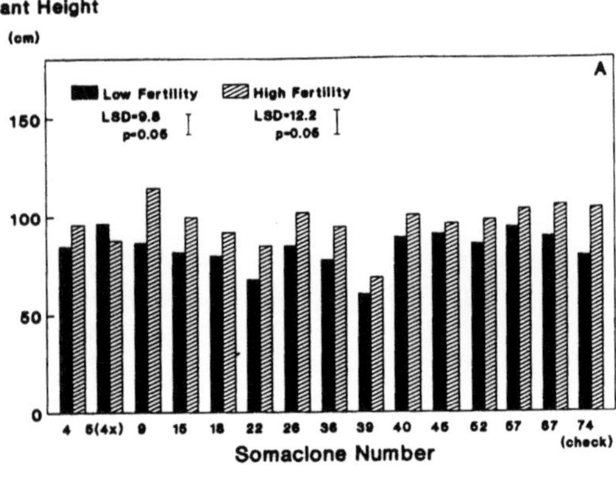

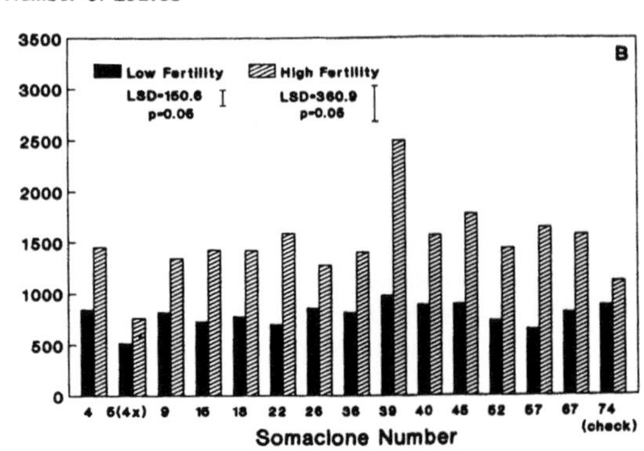

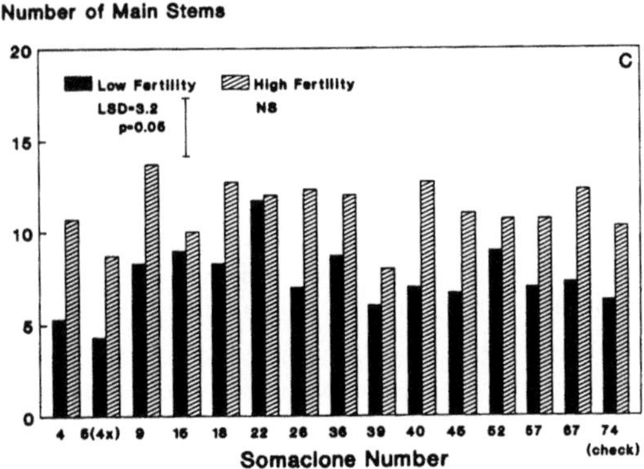

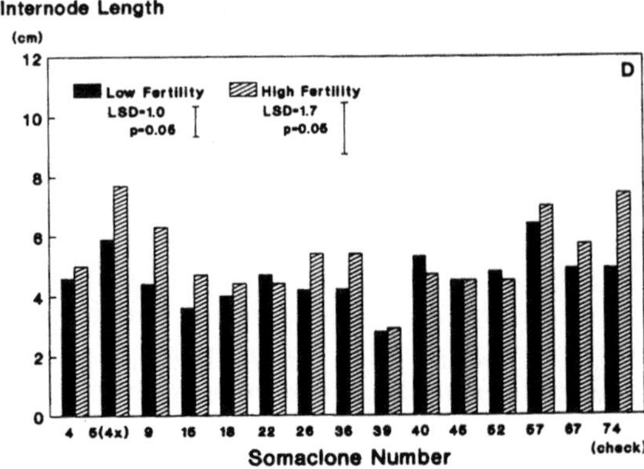

Fig. 1. Somaclonal variation in morphological characteristics of *S. guianensis* grown under low- or high-fertility acid-soil conditions. LSD values are at the 0.05 probability level. NS = not significant.

(Fig. 1C). The length of the internodes decreased by as much as 61% under high fertility in the dwarf somaclone #39 (Fig. 1D).

Somaclones did not differ from the check line (#74) in total biomass (shoot + root) under low fertility (Fig. 2A). An increase in soil fertility improved total biomass production in most of the somaclones compared with that of the check line. Maximum biomass production under high fertility was observed with somaclone #40. Total biomass production in somaclone #40 was 60% higher than that of the check line under high fertility. Somaclones did not differ in shoot production under low fertility (Fig. 2B). But an increase in soil fertility improved shoot production much more in the somaclones than in the check line, except the somaclone #5(4×). Several somaclones under higher fertility had 30% more shoot biomass than the check line.

In contrast to shoot production, root production in some of the somaclones was higher than the check line under both fertility levels (Fig. 2C). The increase in root biomass in somaclone #15 under low fertility was 85%, while in somaclone #40 it was as much as 3-fold under high fertility when compared with the check line. However, high fertility decreased root biomass production in the tetraploid, somaclone (#5). Differences among somaclones in partitioning of dry matter between shoot and roots are shown in Figure 2D. Low fertility decreased the shoot to root ratio in most of the somaclones but the decrease was more distinct with somaclones #52 and 15. In the case of high fertility, there was a dramatic drop in shoot to root ratio with somaclone #40, while the ratio was increased in the tetraploid #5(4×).

Variation of somaclonal lines in leaf biomass,

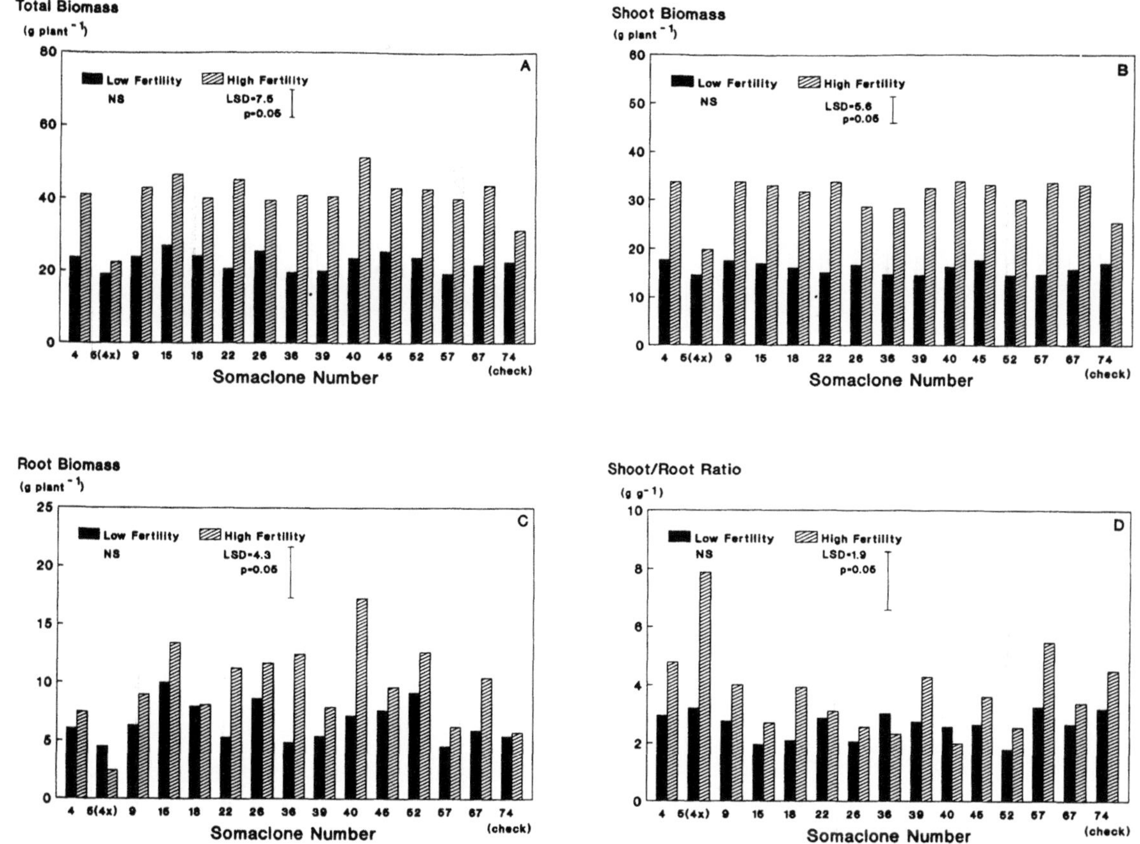

Fig. 2. Somaclonal variation in biomass production and partitioning of dry matter between shoot and roots in *S. guianensis* grown under low- or high-fertility acid-soil conditions. LSD values are at the 0.05 probability level. NS = not significant.

leaf area, specific leaf weight and leaf-stem ratio as influenced by soil fertility is shown in Figure 3. Leaf biomass per plant decreased in the somaclone #5 (tetraploid) which had larger leaves but fewer in number at low and high-fertility level (Fig. 1B and Fig. 3A). Leaf biomass under high fertility in the dwarf, bushy somaclone #39 was 1.8-fold higher than that of the check line, which is correlated with high number of leaves (Fig. 1B). Although differences among somaclones in leaf area production were not significant, leaf area per plant increased by 36% in somaclone #57 under low fertility while the increase was 2.3-fold with somaclone #39 under high fertility when compared with the check line (Fig. 3B). Similar to leaf area production, somaclonal variation in specific leaf weight within the fertility treatment was also not significant (Fig. 3C). The leaf to stem ratio of the dwarf somaclone #39 was markedly higher than that of all other somaclones as well as that of the check line under both fertility conditions (Fig. 3D).

Differences in leaf area ratio, specific leaf area and leaf inorganic phosphorus among 4 selected somaclones and the check line were not significant ($p = 0.05$) under both fertility levels (Table 2). But soluble leaf protein in the tetraploid somaclone (#5), was significantly higher than the check line under both fertility levels. Total chlorophyll content of somaclone #57 under low fertility was lower than the check line. Shoot N and total N (shoot + root) uptake of somaclone #5(4×) under high fertility was markedly lower than that of the check line (Fig. 4A, C). Variation of somaclones in N and P uptake under low fertility was not significant (Fig. 4A to D). Shoot P uptake was a similar in somaclones #26 and 52 when compared with that of the check line under

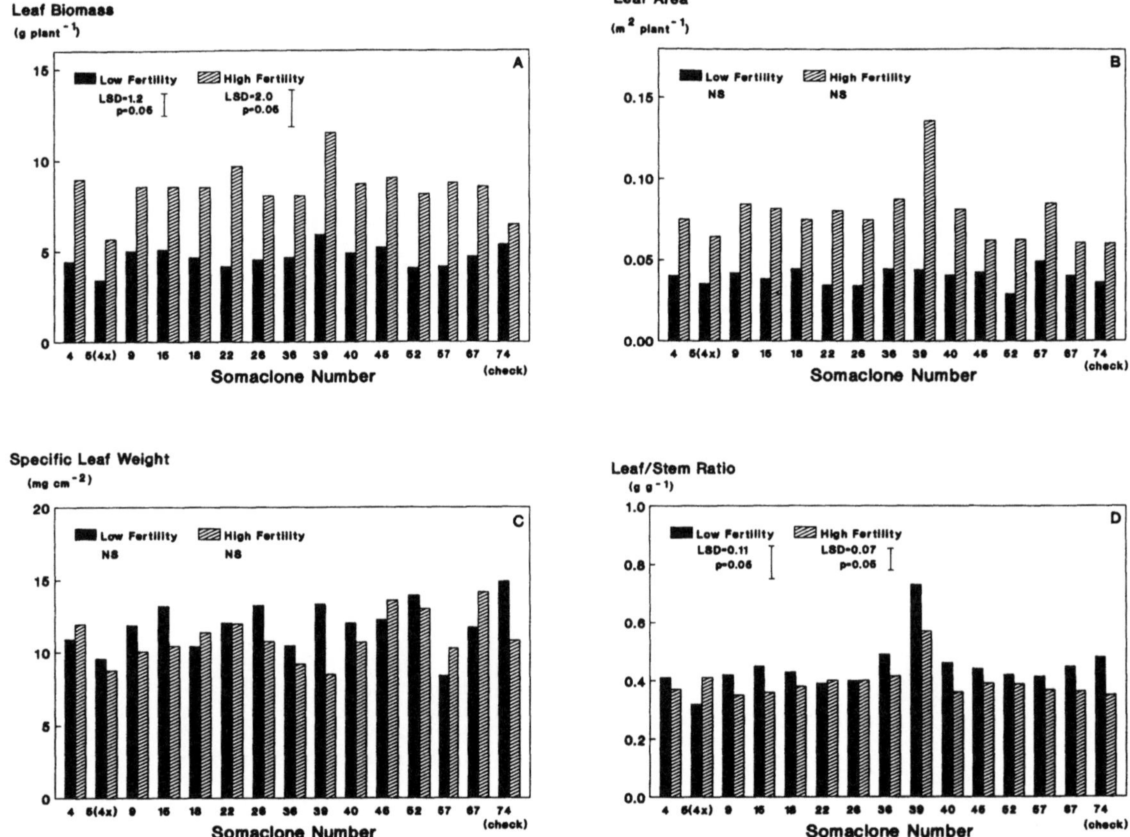

Fig. 3. Somaclonal variation in leaf biomass, leaf area, specific leaf weight and leaf/steam ratio in *S. guianensis* grown under low- or high-fertility acid-soil conditions. LSD values are at the 0.05 probability level. NS = not significant.

Table 2. Variation of some somaclones in leaf physiological characteristics. LSD values are at 0.05 probability level

Characteristics	Fertility	Somaclone number					
		74 (check)	4	5 (tetraploid)	39 (dwarf)	57	LSD ($p = 0.05$)
Total biomass	Low	22.4	23.7	19.1	20.0	19.2	NS
(g plant^{-1})	High	31.8	41.2	22.3	40.4	39.9	10.3
Leaf area ratio	Low	1.6	1.7	1.9	2.2	2.6	NS
(m^2 kg^{-1} plant wt)	High	1.9	1.8	2.5	3.4	2.1	NS
Specific leaf area	Low	6.7	9.1	10.7	7.6	12.2	NS
(m^2 kg^{-1} of leaf wt)	High	9.3	8.4	9.8	12.0	9.8	NS
Leaf weight ratio	Low	0.24	0.19	0.18	0.29	0.21	NS
(g g^{-1} plant wt)	High	0.21	0.22	0.25	0.28	0.22	0.02
Total chlorophyll	Low	652	414	854	479	349	259
(mg m^{-2})	High	590	511	672	500	630	NS
Soluble leaf protein	Low	2.6	2.6	3.6	2.5	3.2	0.7
(g m^{-2})	High	2.5	3.0	4.5	2.7	2.8	1.4
Leaf inorganic	Low	12.8	15.8	20.3	19.5	19.0	NS
phosphorus (mg m^{-2})	High	23.8	17.4	19.8	17.1	16.9	NS

NS = not significant

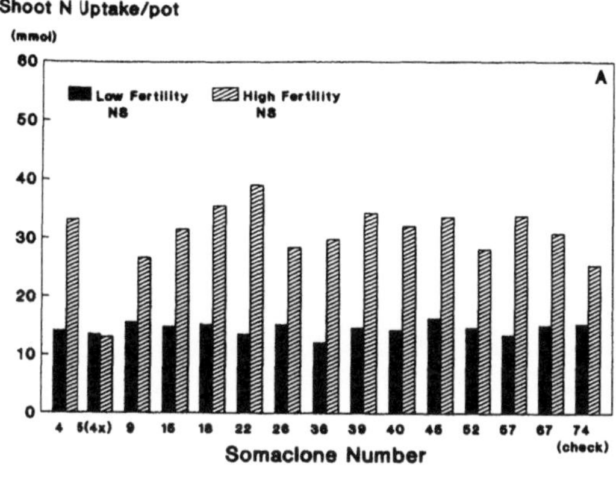

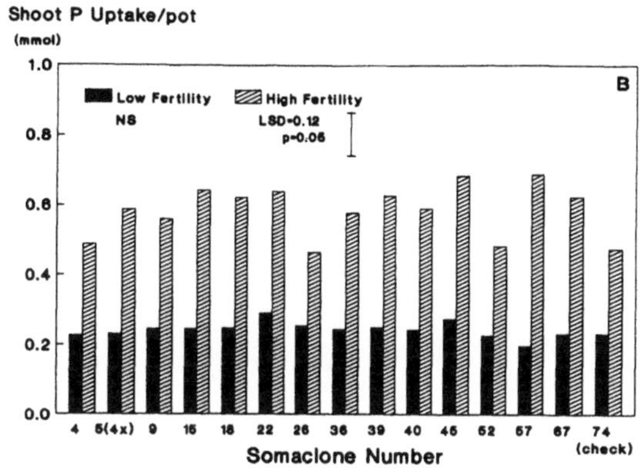

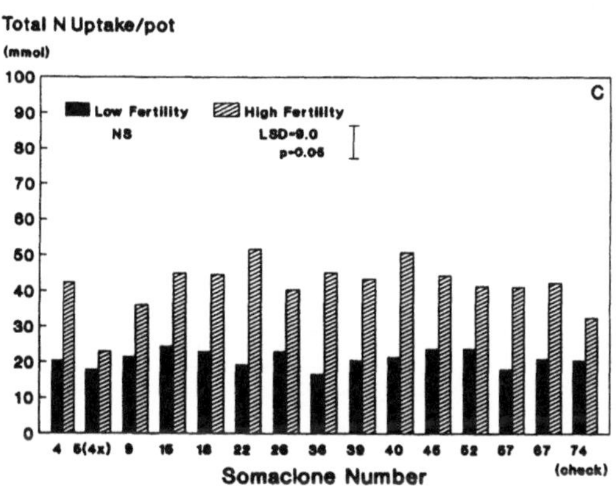

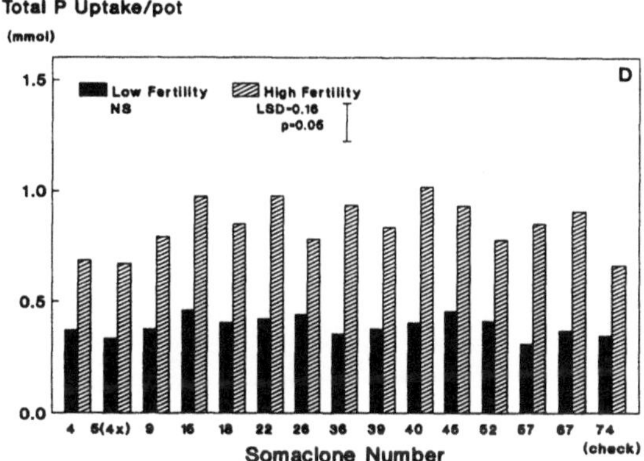

Fig. 4. Somaclonal variation in shoot and total uptake of N and P per pot in *S. guianensis* grown under low- or high-fertility acid-soil conditions. LSD values are at the 0.05 probability level. NS = not significant.

high fertility conditions (Fig. 4B). Total P uptake of somaclone #40 under high fertility was 1.5-fold higher than that of the check line (Fig. 4D).

Total N uptake was positively correlated with total biomass production at both levels of soil fertility (Fig. 5A, D). At high fertility, 13 of the total 14 somaclones had higher total N uptake and total biomass production than the check line. Increased total uptake of N resulted into higher biomass production at both levels of fertility. A similar relationship was also observed between total P uptake and total biomass (Fig. 5B, E). Furthermore, total P uptake was also positively correlated with total N uptake (Fig. 5C, F).

Discussion

Adaptation or tolerance to soil acidity is generally considered a complex trait (Aniol and Gustafson, 1990). An understanding of the physiological and biochemical mechanisms and identification of the specific physiological traits conferring adaptation to acid soil conditions could play a major role in the development of plant improvement strategies. Somaclonal variation in adaptation to acid soil conditions provides a window of opportunity in this direction.

Most of the phenotypic differences among somaclones (CIAT, 1989; Miles et al., 1989) appear to be stable over four sexual generations. Variation for dry matter partitioning between

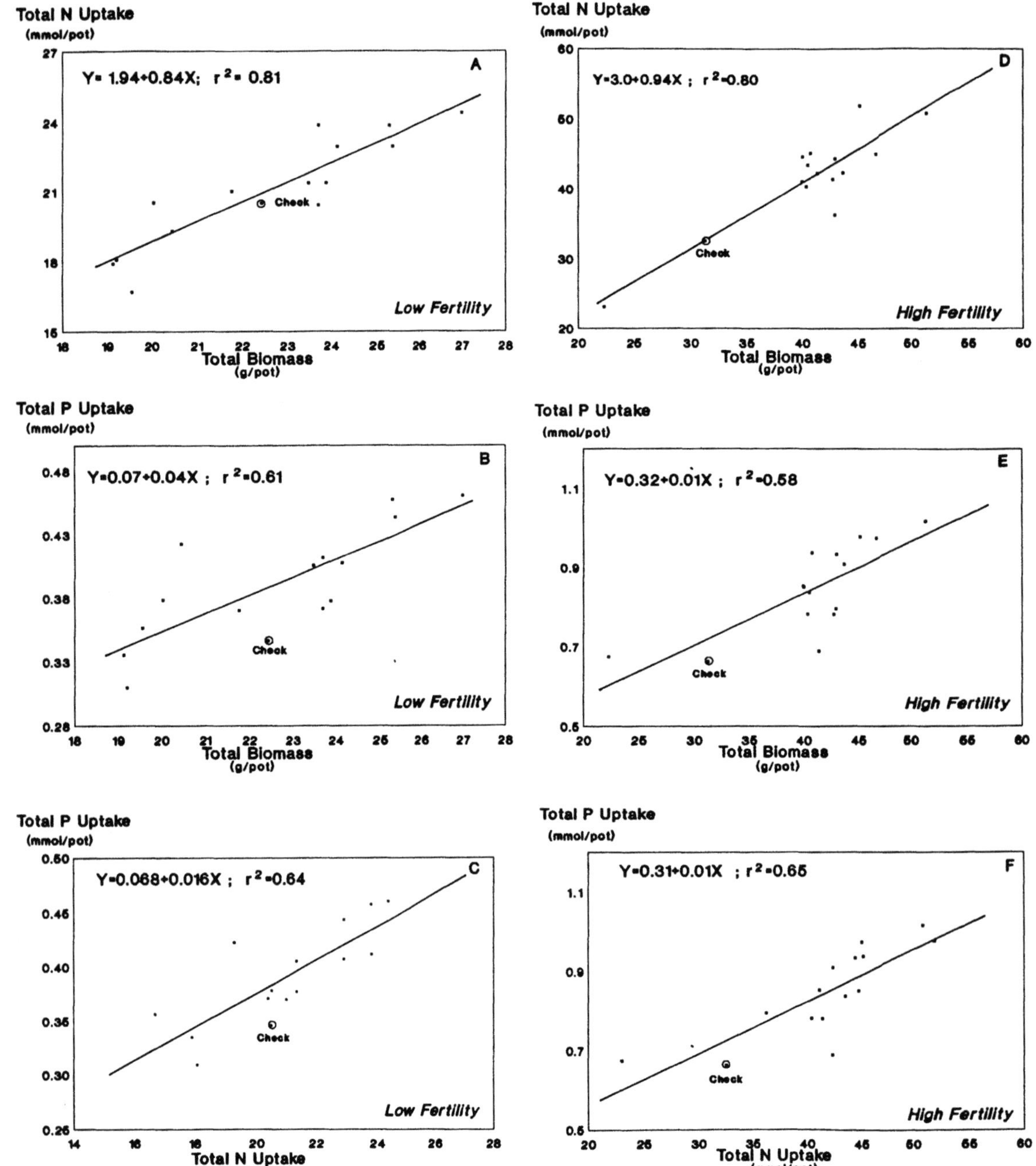

Fig. 5. Relationships between total N uptake and total biomass (**A, D**) total P uptake and total biomass (**B, E**); and total P uptake and total N uptake (**C, F**) of 14 somaclones in comparison to the check line in *S. guianensis* at low or high soil fertility.

shoot and roots was found in the present study among the somaclones of *S. guianensis*. Increased variation for the same traits under high-fertility conditions indicates that an adequate supply of nutrients is essential for the full expression of the genetic potential of these somaclones. However, in somaclone #15 under low fertility conditions, root weight was 85% more than that of the check genotype, while in somaclone #40 under high fertility it was 3-fold higher. This particular observation needs further verification under field conditions. In contrast to somaclones #40 and 15, the tetraploid somaclone (#5) produced fewer roots under high fertility than low fertility conditions. These differences in root biomass compared with the check line indicate that expression of genetic potential for root production in *S. guianensis* is greatly influenced by nutrient supply. Genetic analyses of somaclones #40, #15, and #5, in comparison with the check line, may provide some clues towards understanding the mechanisms of root production in *S. guianensis*. Variation in shoot to root ratio found in somaclones #52, 15 and 5(4×) in relation to the check line suggests the occurrence of genetic diversity in the partitioning of fixed carbon between the shoot and roots in *S. guianensis*.

The higher levels of soluble protein in leaves of the somaclone #5 (tetraploid), under both fertility levels may be due to an increase in size of chloroplasts per unit leaf area. However, total biomass production of this somaclone was the lowest of all somaclones. It is possible that the photosynthetic nitrogen use efficiency (photosynthetic rate per unit leaf nitrogen) was inherently lower for this somaclone than for the others. Also, it is possible that the respiratory carbon losses were significantly higher in this somaclone due to an increased activity of mitochondria per unit biomass.

Field evaluations of somaclones have identified a number of interesting agronomic variants in rice, wheat, potato, sorghum, aromatic grass and *S. guianensis* (Adkins et al., 1990; Bhaskaran et al., 1987; Evans et al., 1986; Godwin et al., 1990; Larkin et al., 1984; Mathur et al., 1988; Miller et al., 1992; Sun et al., 1983). Recently, Miller et al. (1992) reported tissue culture-derived sorghum lines with improved acid soil stress tolerance and other agronomic traits. These examples indicate the potential of somaclonal variation to develop genetic stocks for use in studies to define the mechanisms governing plant adaptation to stressful environments.

Adaptation to acid soils may be a product of several physiological processes that govern the tolerance to, or avoidance of, aluminum, combined with efficient uptake and utilization of certain key nutrients such as phosphorus. Differences among our somaclones and the check line in total biomass production under high fertility, and to a lesser degree under low fertility conditions, were largely associated with changes in total N or P uptake (Fig. 5). A positive correlation of total P uptake with total N uptake may indicate the importance of P to N_2 fixation process of root nodules (Cadisch et al., 1989). It appears that the differences in response of certain somaclones over the check line under high fertility conditions in acid soil may be due to a combination of changes in allocation of biomass among plant parts and changes in physiological characteristics such as uptake of nutrients. Our results showed that somaclones differ with the check genotype in terms of (i) biomass partitioning among plant parts; (ii) root production; and (iii) uptake of nitrogen and phosphorus.

In summary, results of this study indicate that in *S. guianensis* somaclones differ in the partitioning of fixed carbon between the shoot and roots; expression of genetic potential per root production is related to nutrient supply; and biochemical and genetic analysis of somaclones #40, #15 and the tetraploid #5, in comparison with the check line, may provide some clues towards the understanding of the genetic basis of root production and nutrient uptake in *S. guianensis*.

Acknowledgements

We are indebted to Vicente Borrero, Marceliano Calero and Plutarco Alvarez for their valuable assistance.

References

Adkins S W, Shiraishi T, McComb J A, Ratanopol S, Kupkanchanatul T, Armstrong L J and Schultz A L 1990 Somaclonal variation in rice-submergence tolerance and other agronomic characters. Physiol. Plant. 80, 647–654.

Aniol and Gustafson J P 1990 Genetic tolerance in agronomic plants. *In* Heavy Metal Tolerance in Plants: Evolutionary Aspects. Ed. A J Shaw. pp 255–267. CRC Press, Boca Raton, FL.

Bagley P C and N L Taylor 1987 Evaluation of phosphorus efficiency in somaclones of red clover. Iowa State Journal of Research 61, 459–480.

Bhaskaran S, Smith R H, Paliwal S and Schretz K F 1987 Somaclonal variation from *Sorghum bicolor* (L.) Moench cell culture. Plant Cell, Tiss. Org. Cult. 9, 189–196.

Cadisch G, Sylvester-Bradley R and J Nösberger 1989 ^{15}N-based estimation of nitrogen fixation by eight tropical forage-legumes at two levels of P:K supply. Field Crops Res. 22, 181–194.

CIAT (Centro Internacional de Agricultura Tropical) 1989 Biotechnology Research Unit. Annual Report 1988. Working Document No. WD 48, pp. 91–96. Cali, Colombia.

Conner A J and C P Meredith 1985 Large scale selection of aluminum-resistant mutants from plant cell culture: Expression and inheritance in seedlings. Theor. Appl. Genet. 71, 159–165.

Duncan R R and Baligar V C 1990 Genetics, breeding, and physiological mechanisms of nutrient uptake and use efficiently: An overview. *In* Crops as Enhancers of Nutrient Use. Eds. V C Baligar and R R Duncan, pp 3–35. Academic Press, San Diego-London.

Evans N E, Foulger D, Farrer L and Bright S W J 1986 Somaclonal variation in explant-derived potato clones over three tuber generations. Euphytica 35, 353–361.

Foy C D 1988 Plant adaptation to acid, aluminum-toxic soils. Commun. Soil Sci. Plant Anal. 19, 959–987.

Godwin I D, Gordon G H and Cameron D F 1987 Callus culture-derived somaclonal variation in the tropical pasture legume *Stylosanthes guianensis* (Aubl.) Sw. Plant Breeding 98, 220–227.

Godwin I D, Cameron D F and Gordon G H 1990 Variation among somaclonal progenies from three species of *Stylosanthes*. Aust. J. Agric. Res. 41, 645–656.

Karp A 1990 On the current understanding of somaclonal variation. Oxford Surveys Plant Molec. Cell Biol. 7, 1–58.

Larkin P J 1987 Somaclonal variation: history, method, and meaning. Iowa State J. Res. 61, 393–434.

Larkin P J and Scowcroft W R 1981 Somaclonal variation – a novel source of variability from cell culture for plant improvement. Theor. Appl. Genet. 60, 197–214.

Larkin P J, Ryan S A, Brettel R I S and Scowcroft W R 1984 Heritable somaclonal variation in wheat. Theor. Appl. Genet. 67, 443–455.

Mathur A K, Ahuja P S, Pandey B, Kukreja A K and Mandal S 1988 Screening and evaluation of somaclonal variation for qualitative and quantitative traits in an aromatic grass, *Cymbopogon winterianus* Jowitt. Plant Breeding 101, 321–334.

McHughen A and Swartz M 1984 A tissue culture derived salt-tolerant line of flax (*Linum usitatissium*). J. Plant Physiol. 117, 109–117.

Meijer E G M and Broughton W J 1981 Regeneration of whole plants from hypocotyl-, root-, and leaf-derived tissue cultures of the pasture legume *Stylosanthes guianensis*. Physiol. Plant. 52, 280–284.

Miles J W, Roca W M and Tabares E 1989 Assessment of somaclonal variation in *Stylosanthes guianensis*, a tropical forage legume. *In* International Symposium on Genetic Manipulation in Crops. Eds. A Mujeeb-Kazi and L A Sitch. pp 249–257. CIMMYT and IRRI, Mexico City, Mexico, and Manila, Philippines.

Miller D R, Waston R M, Duncan R R, Chapman P L, Brick M A, Hanning G E, Timm D A and Nabors M W 1992 Acid soil stress tolerance in tissue culture-derived sorghum lines. Crop Sci. 32, 324–327.

Ojima K and Ohira K 1988 Aluminum-tolerance and citric acid release from a stress-selected cell line of carrot. Commun. Soil Sci. Plant Anal. 19, 1229–1236.

Phillips R L, Kaeppler S M and Peschke V M 1990 Do we understand somaclonal variation? *In* Progress in Plant Cellular and Molecular biology. Eds. H J J Nijkamp, L H W Van der Plas and J Van Aartrijk. pp 131–141. Kluwer Academic Publishers, Dordrecht-Boston-London.

Rao I M and Terry N 1989 Leaf phosphate status, photosynthesis, and carbon partitioning in sugar beet. I. Changes in growth, gas exchange, and Calvin cycle enzymes. Plant Physiol. 90, 814–819.

Salinas J G and García R 1985 Métodos químicos para el análisis de suelos ácidos y plantas forrajeras. CIAT, Cali, Colombia, 83 p.

Salinas J G and Saif S ur R 1990 Nutritional requirements of *Andropogon gayanus*. *In Andropogon gayanus* Kunth: A Grass for Tropical Acid Soils. Eds. J M Toledo, R Vera, C Lascano and J M Lenné. pp 99–155. Centro Internacional de Agricultura Tropical (CIAT), Cali, Colombia.

SAS Institute, Inc. 1982 SAS Users' Guide: Statistics. SAS Institute, Inc., Cary, NC.

Sun Z X, Chen-Zhang Z, Kang-Le Z, Xiu-Fang Q, and Ya-Ping F 1983 Somaclonal genetics of rice, *Oryza sativa* L. Theor. Appl. Genet. 67, 67–73.

Szabados L and Roca W 1986 Regeneration of isolated mesophyll and cell suspension protoplasts to plants in *Stylosanthes guianensis:* A tropical forage legume. Plant Cell Reports 3, 174–177.

P. J. Randall et al. (Eds.), Genetic aspects of plant mineral nutrition, 45–52.

PLSO SV06

Tolerance to manganese toxicity among cultivars of lucerne (*Medicago sativa* L.)

P.W.G. SALE[1], D.I. COUPER[1], P.L. CACHIA[1] and P.J. LARKIN[2]
[1]*School of Agriculture, La Trobe University, Bundoora 3038, Australia and* [2]*CSIRO Divison of Plant Industry, G.P.O. Box 1600, Canberra 2601, Australia*

Key words: acid-soil, alfalfa, height, lucerne, manganese, symptoms, toxicity

Abstract

Two sand culture experiments were carried out to identify commercial cultivars of lucerne or alfalfa (*Medicago sativa* L.) which contain elite, Mn-tolerant plants for use in a selection programme to increase the acid-soil tolerance of this perennial legume. Differences in Mn tolerance, both within and between cultivars, were observed when a range of cultivars were exposed to regular waterings with dilute nutrient solution containing 20 or 25 mg Mn L^{-1}. Under these moderately toxic regimes, the winter dormant cultivars Cimmaron and WL 318 were found to contain elite plants that had greater dry matter yields than their mean cultivar yield under non-toxic Mn conditions.

Cultivars which contained elite, Mn-tolerant plants could not be identified by phenotypic characteristics such as their height or their toxicity symptom score, nor by their winter dormancy class. Possible reasons for the occurrence of elite plants in these cultivars are discussed. The elite, high yielding Mn-tolerant plants could not be identified from the other plants within their cultivar population by their Mn toxicity symptoms nor by their height.

Introduction

Lucerne (*Medicago sativa* L.), also known as alfalfa, is probably the most widely grown forage legume around the world. However it grows poorly on acid soils because it is intolerant of the high concentrations of aluminium (Al) and manganese (Mn) (Lanyon and Griffith, 1988), that generally increase in the soil solution as soils become acidic (Marschner, 1986). Any breeding programme that aims to increase the acid-soil tolerance of lucerne must achieve a substantial improvement in its tolerance to toxicities of both these elements.

Previous work suggests that differential tolerance to Mn toxicity does occur both between and within lucerne cultivars, and that conventional screening and selection procedures will lead to a significant improvement in Mn tolerance. For example, Dessureaux and Ouellette (1958) screened a range of European and North American cultivars under Mn toxic conditions in sand culture and then crossed the most Mn tolerant plants. Maximum relative yields (% of control yields) under moderate Mn toxicity increased from 48% to 95% in one cycle of selection. These findings indicate that variation for Mn tolerance does exist in these older lucerne cultivars and that the trait for Mn tolerance is heritable. A more recent study, based on differences in the severity of Mn toxicity symptoms, suggests that variation in Mn tolerance also occurs in a range of Australian cultivars (Salisbury and Downes, 1982).

This paper describes experiments that form part of a larger project which aims to increase the acid-soil tolerance of lucerne. The objects of these studies were to identify modern, commercial lucerne cultivars that vary widely in their tolerance to Mn toxicity, and then to determine how best to locate the high-yielding, Mn-tolerant plants from populations of these cultivars. The first of two experiments measured the response of three cultivars to increasing concentrations of Mn in the root medium, while the second screened eleven lucerne cultivars under two Mn toxicity regimes that were selected on the basis of responses in the first experiment.

Materials and methods

Treatments and experimental design

Two sand culture experiments were carried out in a glasshouse at Bundoora in Victoria, Australia (latitude 37°44′S) in the autumn and the spring respectively of 1990. In the first experiment three commercial cultivars, WL 318, PB 581 and Maxidor II were grown in sand that was regularly flushed with nutrient solutions that contained either 0.06 (control), 10, 20, 40 or 60 mg Mn L^{-1}, added as $MnSO_4$. The experiment was laid out as a split plot design, with Mn levels as main plots and cultivars as sub plots. All treatments were replicated twice.

In the second experiment a range of eleven lucerne cultivars that varied in winter dormancy were grown in sand with either 0.06, 12.5 or 25 mg Mn L^{-1} in the nutrient solution. The cultivars included WL 318, PB 545 and Cimmaron (winter dormant), PB 581, Validor, and WL SS (semi-winter dormant), Aurora, Sheffield, and Trifecta (winter active), and PB 577 and Maxidor II (highly-winter active). The design was also a split plot with two replicates. Mn treatments (each applied to four plastic boxes) were the main plots, with cultivars being the sub plots. Approximately 40 individual plants of each cultivar were monitored in each sub plot in the two experiments.

Growing conditions

The lucerne cultivars were grown for 50 days (experiment 1) or for 40 days (experiment 2) in a temperature-controlled glasshouse with a diurnal range of 20 to 30°C (±5°C), in fine-grained river sand that had been repeatedly washed with deionised water and heated at 80°C for 24 hours. Approximately 12.5 kg of sand was then placed in free-draining plastic boxes (350 × 270 × 180 mm) over a thin layer of washed gravel. The surface sand layers of each box were divided into three sections, 90 mm wide, by placing 20 mm deep plastic dividers along the length of the box. Lucerne seed was sown in single rows (experiment 1) or in two rows, 5 cm apart (experiment 2) along the box in the central portion of each section, and then thinned after 8 days to approximately 40 equally spaced, uniform seedlings. Each section was flushed four times a day with approximately 3 L of a dilute nutrient solution; the solution subsequently drained rapidly and was discarded. The solution (pH 6.0) contained the macro-nutrients (in mM) 2.3N as NH_4NO_3, KNO_3 and $Ca(NO_3)_2$, 0.71 K as KNO_3 and KH_2PO_4; 0.01 P as KH_2PO_4; 0.5 Ca as $Ca(NO_3)_2$; 0.3 Mg as $MgSO_4$; and 0.3 S as $MgSO_4$; and the micronutrients (in μM) 4 Fe as Fe-EDTA; 1 Mn as $MnSO_4$; 1 Zn as $ZnSO_4$; 0.5 Cu as $CuSO_4$; 2 B as H_3BO_3; and 0.07 Mo as $(NH_4)_6Mo_7O_{24}$. The nitrate to ammonium ratio was 6.5 to 1. Manganese additions were made to this solution 14 and 11 days after sowing in experiments 1 and 2 respectively.

Measurements

Manganese toxicity symptoms consisted of initial chlorosis on the distal portions of the expanding leaves, followed by more complete chlorosis and, in the high Mn treatments, eventual necrosis of the younger leaves. The severity of the symptoms of individual plants in the first experiment was assessed and given a score from 1 to 10 with increasing severity of the symptoms. This scoring was made at 8 and then 36 days after the plants were exposed to the Mn treatments. In the second experiment each group of 40 plants from each cultivar was given a symptom score at 6, 13 and 20 after exposure to the Mn treatments.

At harvest, the height from the sand surface to the stem apex of each plant was measured, as was the length of any major lateral branches (experiment 1). Lateral branching was less apparent in the second experiment possibly because the plants were harvested 10 days earlier. Individual plants were harvested at the sand surface and oven dried at 80°C for 48 hours and separately weighed. Individual plants from the moderate Mn toxicity treatment (20 or 25 mg Mn L^{-1} in experiments 1 and 2 respectively) were arbitrarily considered to be 'elite' and Mn-tolerant if the dry weight of their tops exceeded their cultivar mean dry weight for the control (0.06 mg Mn L^{-1}) treatment.

The extent to which a cultivar contained elite plants under moderate Mn toxicity was indicated by the parameter 'Δ% yield'. This was the difference in dry weight between the highest yielding individual plant and the cultivar mean dry weight expressed as a percentage of the cultivar mean dry weight (all under moderate Mn toxicity). This effectively removed possible confounding effects of the growth patterns, associated with winter dormancy, on yields under the optimum temperatures within the glasshouse. Winter dormant cultivars tend to have higher growth rates under these conditions.

Results

Yield differences under Mn toxicity: cultivar populations and elite plants

Increasing the Mn concentration in the nutrient solution resulted in a marked reduction in the dry weight of the tops of the three lucerne cultivars (Fig. 1). However, there were differences in magnitude of yield reductions between the three cultivars. At the moderate Mn toxicity regime of 20 mg L^{-1} cultivar WL 318 had a significantly higher ($p < 0.05$) absolute as well as relative yield than the other cultivars, while Maxidor II had the smallest mean yields, at the lower Mn levels and also at the 0.06 mg Mn L^{-1} level (Fig. 1). Under severe Mn toxicity (60 mg Mn L^{-1}) all cultivars produced plants with similar dry matter yields which were approximately 20 per cent of the yield of their control plants.

The frequency distribution of plant dry weight categories for the three cultivars were examined to determine which cultivars contained high-yielding plants under Mn toxic regimes. The distribution of categories for the control (0.06) and 20 mg Mn L^{-1} treatments are shown in Figure 2. The dry weights of plants from all three cultivars grown in the 0.06 mg Mn L^{-1} treatment varied considerably. In the moderate Mn toxicity treatment (20 mg Mn L^{-1}) the range in plant dry weights diminished considerably (Fig. 2). However there was one high yielding, elite plant (see definition in materials and methods) of cultivar WL 318 in the 110 mg dry weight category. In contrast, the heaviest yielding plants of cultivars PB 581 and Maxidor II under moderate Mn toxicity were in the 70 mg dry weight category. Under the more severe Mn stress at 40 mg Mn

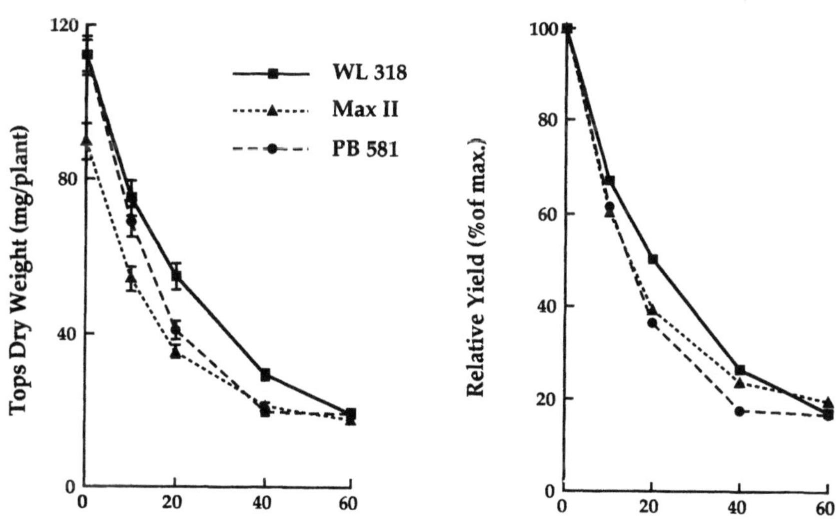

Fig. 1. The effect of Mn concentration in the nutrient solution on the absolute and relative dry matter yields of cultivars WL 318, Maxidor II and PB 581 in experiment 1. Data are means of approximately 40 plants per cultivar per treatment.

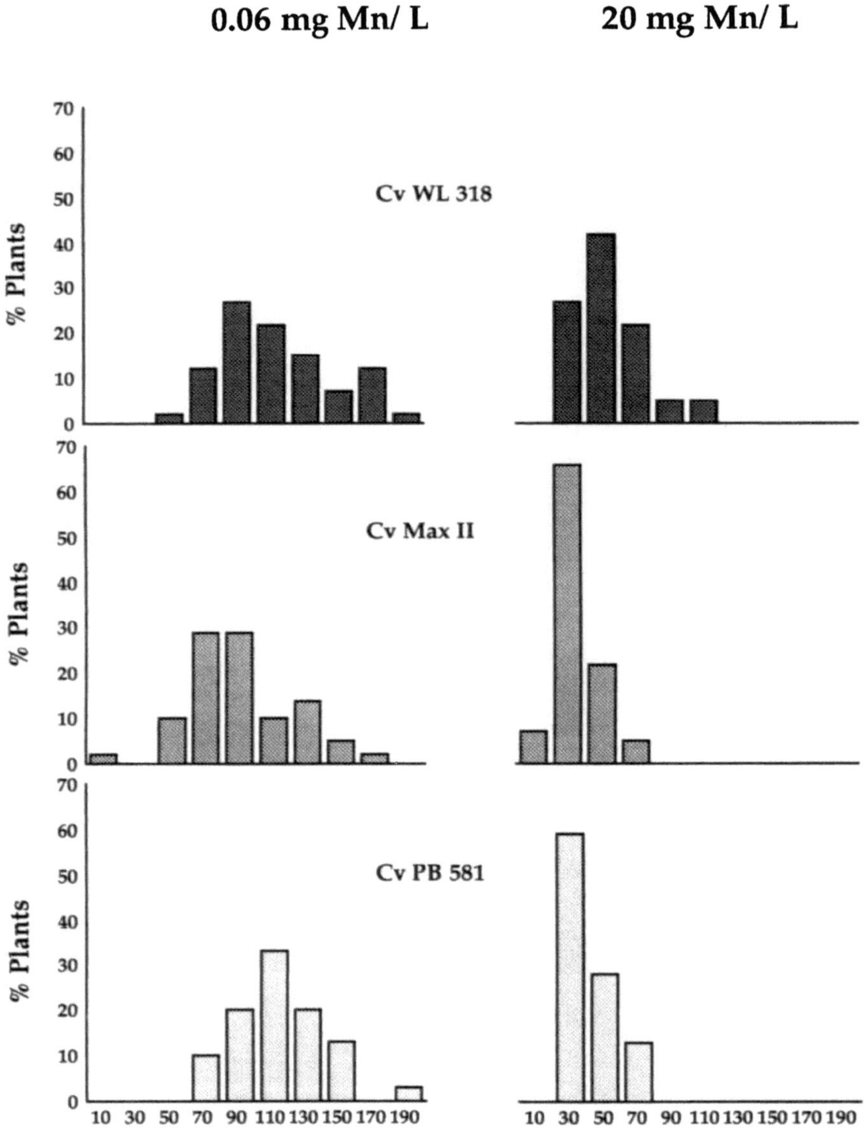

Fig. 2. Frequency distributions of dry weight categories of plant tops for cutlivars WL 318, Maxidor II and PB 581 grown under control and moderate Mn toxicity regimes (0.06 and 20 mg Mn L^{-1} respectively) in experiment 1.

L^{-1} (data not shown) cultivar WL 318 again contained the highest yielding individual plant (61 mg) compared to the highest yielding plants in Maxidor II and PB 581 (40 and 28 mg respectively).

In the second experiment with Mn concentrations of 25 mg L^{-1} in the nutrient solution, there were marked differences in the absolute and relative yields among the 11 lucerne cultivars (Table 1). The more winter dormant cultivars tended to have the highest mean yields while the more winter active cultivars had the lowest yields. Cultivar rankings for absolute yield did not closely correlate with rankings for relative yield: the highest yielding cultivar, Validor ranked 5th for relative yield, while the 8th highest yielding cultivar Aurora had the 3rd highest relative yield in the group. In general the rela-

Table 1. Absolute and relative mean dry matter yield of lucerne cultivars, grown at 25 mg Mn L^{-1}, and the dry weight of the highest yielding plant within that cultivar population, together with rankings for these measurements

Cultivar	Winter[a] dormancy	Population yields				Yield of top plant	
		Absolute (mg $plant^{-1}$)	Rank	Relative[c] (%)	Rank	Yield (mg $plant^{-1}$)	Rank
Validor	SWD	28.7 (±1.5)[b]	1	45	5	51	3
WL SS	SWD	26.3 (±1.3)	2	64	1	43	7
WL 318	WD	25.6 (±1.9)	3	58	2	83	2
PB 581	SWD	24.6 (±1.2)	4	38	9	50	4
Sheffield	WA	24.1 (±1.1)	5	44	6	46	5
Cimmaron	WD	23.8 (±2.0)	6	50	4	90	1
Trifecta	WA	22.5 (±1.2)	7	37	11	46	5
Aurora	WA	22.4 (±1.1)	8	56	3	39	8
PB 545	WD	20.7 (±1.0)	9	43	7	39	8
PB 577	HWA	19.7 (±1.0)	10	38	9	38	10
Maxidor II	HWA	14.9 (±0.9)	11	39	8	31	11

[a]SWD, WD, WA, and HWA indicate the winter dormancy classes of the lucerne cultivars were slightly winter dormant, winter dormant, winter active and highly winter active respectively.
[b]±Standard error of means
[c]Relative yield is the yield (at 25 mg Mn L^{-1}) expressed as a % of the maximum yield.

tionship was poor between the rankings for either absolute or relative yields and the ranking for the yield of the highest yielding individual plant in the cultivar population. The cultivar with the highest yielding plant (90 mg) was Cimmaron which had a population mean yield of 23.8 mg that ranked 6th, and a mean relative yield of 50% that ranked 4th in the group. Cimmaron contained two elite plants while WL 318 contained one elite high yielding plant (83 mg) with the 25 mg Mn L^{-1} treatment in this second experiment. Under the less severe Mn toxicity regime (12.5 mg Mn L^{-1}) Cimmaron and WL 318 again had high yielding plants with dry matter yields of 85 and 86 mg respectively, as did the cultivars PB 581 and Trifecta, with yields of 88 and 100 mg, respectively (data not shown).

Identification of cultivars containing elite Mn tolerant plants

Phenotypic characters such as the height, or the toxicity symptom score of cultivars exposed to moderate Mn toxicity were poor indicators of

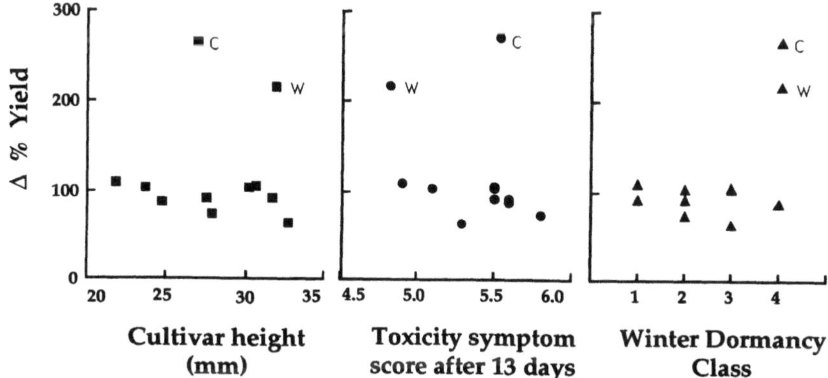

Fig. 3. The relationship between the 'Δ% yield' (see materials and methods for definition) of each cultivar and the mean cultivar height (■), the cultivar symptom score after 13 days of exposure to Mn (●), and the winter dormancy class (▲) of the cultivar under moderate Mn toxicity (25 mg Mn L^{-1}) in experiment 2. Cultivars that are highly winter active, winter active, slightly winter dormant and winter dormant are in winter dormancy classes 1 to 4 respectively. Cimmaron (c) and WL 318 (w) were cultivars that contained elite Mn-tolerant plants.

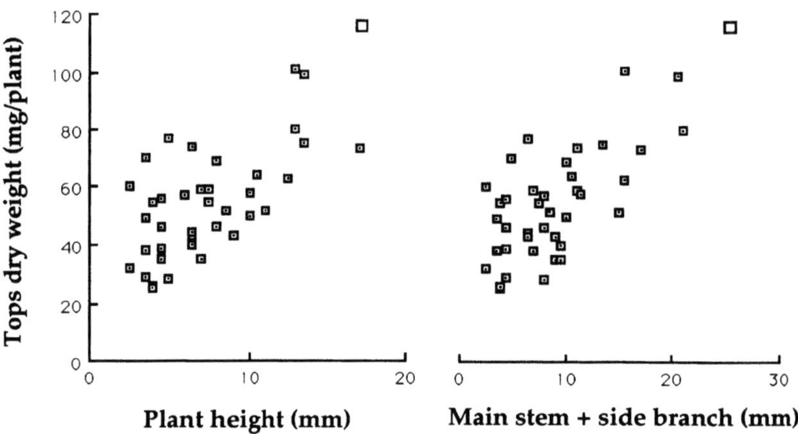

Fig. 4. The relationship between the dry weight and height or total stem length (length of main stem plus lateral branch) of individual plants of WL 318, including the elite plant (□) grown under moderate Mn toxicity (20 mg Mn L^{-1}) in experiment 1.

whether a cultivar contained elite, Mn-tolerant plants in their populations. The parameter used to indicate whether a cultivar did contain elite plants under moderate Mn toxicity, the 'Δ% yield' (see definition in materials and methods) bore no relationship (R^2 = zero) to the mean cultivar height (Fig. 3). Cimmaron had the 4th lowest mean height while WL 318 was the second tallest cultivar in the group.

The severity of mn toxicity symptoms displayed by plants 13 days after exposure to the 25 mg Mn L^{-1} treatment was poorly related ($R^2 = 0.08$) to the 'Δ% yield' of cultivars (Fig. 3). Cimmaron had a similar symptom score to those of cultivars that ranked in the middle of the cultivar group. The toxicity symptom scores 6 days after Mn exposure were even less useful with Cimmaron displaying the most severe toxicity symptoms in the group (data not shown).

The winter dormancy score of cultivars was also poorly related ($R^2 = 0.29$) to their 'Δ% yield'. This was partly due to the low percentage yield increase of the third winter dormant cultivar PB545 which was less than half of the increases for Cimmaron and WL 318 (Fig. 3).

Identification of elite, Mn-tolerant plants within cultivar populations

The tallest plant in a population exposed to Mn toxicity was not always the plant with the highest dry matter yield in that population. For example there were two equally tall plants of WL 318 of which one was the elite plant while the other ranked 8th in tops dry weight (Fig. 4). The ability to identify the elite plant in the WL 318 population was improved by adding the length of the major side branch to the height of the plant. When this was done the plant with the longest combined stem length (main stem plus side branch) was in fact the elite plant of that population (Fig. 4).

The usefulness of Mn toxicity symptom scores to identify the elite plant in the WL 318 population exposed to a moderate Mn toxicity in the first experiment, was very limited. After 8 days the eventual elite plant had the lowest symptom score along with 15 other plants, out of a population of 42 plants. However, after 35 days this plant had the second lowest score along with 20 other plants, while five other plants had lower symptom scores.

Discussion

The key finding in this study is that substantial differential tolerance to Mn toxicity exists both between (Fig. 1) and within (Fig. 2) a selection of commercial lucerne cultivars that are currently being grown by farmers in southern Australia. The presence of this variation in a limited selection of lucerne genotypes would suggest that the systematic screening of a larger range of genetic material would provide sufficient genetic variation on which to base a selection programme for Mn tolerance. This confirms that the differential Mn tolerance that occurred in European lucerne cultivars (Dessureaux and Ouellette, 1958) is also present in the cultivars currently grown in Australia.

The occurrence of elite plants that displayed marked tolerance to Mn toxicity was restricted to a limited number of cultivars (Table 1). There were no obvious phenotypic measurements that could be used to identify these cultivars, apart from measuring dry weights of a relatively large number of individuals from each cultivar. Neither the mean height of the cultivar, nor the extent to which the cultivar displayed Mn toxicity symptoms gave any useful indication of cultivars which contained elite, Mn-tolerant plants (Fig. 3). The winter dormancy rating of the cultivar does seem to be involved in that the two cultivars that contained elite plants, Cimmaron and WL 318, were winter dormant lines. However, the third winter dormant cultivar in the second experiment, PB 545, contained minimal variation for Mn tolerance (Table 1). An understanding of why elite, Mn-tolerant plants occur in some winter dormant cultivars and not in others would be useful in order to identify other sources of Mn tolerant lucerne germplasm.

One common feature of the two cultivars that contained elite, Mn-tolerant plants is that both were selected from crosses from a range of lines of which a number were of Flemish parentage (I. Kahne, pers comm.). It is interesting to note that one of the most Mn tolerant crosses produced by Dessureaux and Ouellette (1958) had Flandria as a parent, which is an old Flemish variety.

Another common feature in the development of these cultivars was that the selection was carried out in soils of high water content. For WL 318, a major objective was the provision of resistance to Phytophthora root rot (*Phytophthora megasperma*). Selection was therefore carried out at high soil water content either in the field or in sealed tanks containing Phytophthora-infested soil (Beard and Kawaguchi 1978). Such conditions may have produced a poor soil aeration (waterlogging) which would have resulted in increased concentrations of Mn^{2+} in the soil solution (Schlichting and Sparrow, 1988), particularly if the soil contained sufficient quantities of reducible Mn. Cimmaron was also selected for Phytophthora resistance, and has been noted as a more waterlogging-tolerant lucerne cultivar in South Australia (I Kahne, pers comm). Presumably some of its parental lines would have been exposed to waterlogging regimes and potential Mn toxic conditions which may have resulted in inadvertent selection for tolerance to Mn toxicity.

The identification of the elite, Mn-tolerant plant in populations of WL 318 grown under moderate Mn toxicity was best achieved by selecting the plant with the longest combined stem length (Fig. 4). The leaves of the elite plant seemed to gradually adapt to the high concentrations of Mn in the nutrient solution. Expanding leaves were at first chlorotic but gradually became green with time as more leaves developed. This suggests that the young leaves were able to adapt to high internal Mn concentrations in their tissue, perhaps by the compartmentation of the Mn into cell vacuoles (Horst, 1988).

The use of Mn toxicity symptoms per se to identify the elite, Mn-tolerant plant from a population growing under Mn toxicity was not effective. This was because the elite plant also displayed toxicity symptoms on the upper leaves. Any system therefore, that is based on the selection of Mn tolerant plants solely on the basis of their toxicity symptoms, such as that used by Salisbury and Downes (1982), may fail to select elite Mn-tolerant plants.

Acknowledgements

The authors wish to acknowledge the Land and Water Resources Research and Development Corporation and the Salinity Bureau of the Victorian State Government for the provision of research funds for this project.

References

Beard D F and Kawaguchi I I 1978 Registration of WL 311 and WL 318 alfalfa cuttings. Crop Sci. 18, 523.

Dessureaux L and Ouellette G J 1958 Tolerance of alfalfa to manganese toxicity in sand culture. Can. J. Soil Sci. 38, 8–13.

Horst W J 1988 The physiology of manganese toxicity. *In* Manganese in Soils and Plants. Eds. R D Graham, R J Hannam and N C Uren. pp 175–188. Kluwer Academic Publishers, Dordrecht, The Netherlands.

Lanyon L E and Griffith W K 1988 Nutrition and fertilizer use. *In* Alfalfa and Alfalfa Improvement. Eds. A A Hanson, D K Barnes and R R Hill, JR. pp 333–372. American Society of Agronomy, Crop Science Society of America and Soil Science Society of America, Madison, WI.

Marschner H 1986 Mineral Nutrition in Higher Plants. Academic Press, London, 674 p.

Salisbury P A and Downes R W 1982 Breeding lucerne for tolerance to acid soils. *In* Proceedings of the Second Australian Agronomy Conference, Wagga Wagga, July 1982. Ed. M J J Norman. p 339.

Schlichting E and Sparrow L A 1988 Distribution and amelioration of manganese toxic soils. *In* Manganese in Soils and Plants. Eds. R D Graham, R J Hannam and N C Uren. pp 277–292. Kluwer Academic Publishers, Dordrecht, The Netherlands.

P. J. Randall et al. (Eds.), Genetic aspects of plant mineral nutrition, 53–57.

PLSO SV19

Response to soil aluminium of two white clover (*Trifolium repens* L.) genotypes

J.R. CRUSH and J.R. CARADUS
DSIR Grasslands, Private Bag, Palmerston North, New Zealand

Key words: acetylene reduction, aluminium toxicity, chemical composition, genotypic variation, growth, roots, *Trifolium repens* L., white clover

Abstract

An aluminium (Al) tolerant genotype of white clover was compared with an Al susceptible genotype in artificial soil profiles in which exchangeable Al increased with depth. The tolerant genotype had a greater proportion of its root mass deeper in the soil than the susceptible genotype. Nitrogenase activity showed a similar pattern. Shoot Al concentration did not vary between the genotypes but root Al in the susceptible line was twice that in the tolerant genotype. Plant potassium content in the susceptible line was relatively less, probably in response to higher aluminium content.

Introduction

White clover (*Trifolium repens* L.) is rarely found in soils with a pH lower than 4.8 (Snaydon, 1962). This inability to grow in very acid soils is in part due to the sensitivity of white clover to aluminium (Al) (Edmeades et al., 1983). The dissolved aluminium concentration in most soils increases as pH decreases below about 5 (Foy, 1984).

While variation between natural populations (Caradus, 1987) or cultivars (Mackay et al., 1990) of white clover for tolerance to Al has not been demonstrated, tolerant and susceptible genotypes have been identified within cultivars on the basis of shoot growth (Caradus, 1987; Caradus et al., 1991). Aluminium tolerance in white clover appears to be inherited predominantly as a recessive character (Caradus et al., 1991), however in some genotypes Al tolerance was inherited as a dominant character.

The aim of the present study was to compare an Al-tolerant genotype with an Al-sensitive genotype and identify plant characteristics that might explain the differences in Al-tolerance.

Materials and methods

Plant material

Two genotypes selected within *Trifolium repens* L. cv. 'Grasslands Huia' for variation in shoot growth responses to soil available aluminium were examined. Genotype 77 shows more tolerance of aluminium than genotype 129 (Caradus et al., 1991).

Stolon tips from container grown plants were surface sterilised in 3% hypochlorite solution for 3 minutes, rinsed with water and planted in sterilised sand. After 19 days the stolon tips were washed out of the sand, sorted and uniform, rooted cuttings were transplanted into the experimental pots.

Treatments

The experimental pots were 330 mm deep × 108 mm diameter cylinders with soil aluminium concentration increasing in three stages with depth, after premixing soil with aluminium sulphate solution. Wainui silt loam, a Typic Dystrochrept with pH 4.9 that has previously been used in the programme (Mackay et al., 1990) was used in the experiment. Data for soil cation concentrations under aluminium additions are given by MacKay et al., (1990). A peat-based *Rhizobium trifolii* inoculant was also premixed through all the soil and it was packed into the cylinders at the field volume weight ($0.8\,g \times cm^{-3}$). Soil sulphate levels were balanced with $CaSO_4$ and phosphorus and potassium additions restricted to the surface layer (Table 1). The soil was then held at −8°C for five days to reduce soil nematodes (Watson, 1990) while retaining other soil microflora. After thawing, the soil was held moist in the glasshouse for eight days before being planted.

At the end of the experiment exchangeable Al (leached with 1 *N* KCl; Mclean 1965) levels were $59\,mg\,kg^{-1}$ in the 0–100 mm zone and $131\,mg\,kg^{-1}$ and $212\,mg\,kg^{-1}$ in the 101–200 and 201–300 mm zones respectively.

Measurements

After 59, 76 and 93 days growth, nine replicates of each genotype, selected randomly from the fully randomised layout, were harvested. The soil-root columns were divided at 100 and 200 mm for acetylene reduction (AR) assays of nitrogenase activity in the three aluminium levels. General procedures for AR assays followed Crush and Tough (1981). After the AR assays the roots were washed free of soil and root and shoot dry matter (DM) recorded. Shoots and roots from the third harvest were analyzed for nitrogen (Kjeldahl digestion and distillation of ammonia) and other elements (P, K, Ca, Mg, Na, S, Al, B, Co, Cu, Mn, Zn) by dry ashing and plasma emission spectrometry. Shoot samples were also analysed for cyanide content (AOAC 1984). At the end of the experiment the roots systems of some spare replicates were divided into the three zones, carefully washed clean and preserved in formalin-acetic acid-alcohol (Phillips and Hayman, 1970). These were examined microscopically for distal root diameter, root hair occurrence and the condition of root tips. Treatment effects were tested for statistical significance by one- and two-way analysis of variance, depending on the factors involved.

Table 1. Rates ($mg\,kg^{-1}$ dry soil) and sources of Al, P, K and S mixed through the experimental soil profiles

Soil depth (mm)	Al $Al_2(SO_4)_3$	P $CaHPO_4$ K_3PO_4	K K_2SO_4 K_3PO_4	S $CaSO_4$ $Al_2(SO_4)_3$ K_2SO_4
0–100	0	150	190	625
101–200	150	0	0	625
201–300	350	0	0	625

Results

The two genotypes did not differ in plant weight or root/shoot ratio during the experiment but the distribution of roots with depth in the soil did differ significantly after the first harvest (Table 2). Genotype 77 had a greater proportion of its root mass deeper in the soil than did genotype 129.

Acetylene reduction rates increased with plant age and decreased with depth in the soil but there was no difference in activity per plant between genotypes at all three harvests. Acetylene reduction per unit root weight was greater for line 77 than line 129 in the 0–100 mm soil zone at harvest 2 (4.6 vs 2.5 μM $C_2H_4\,g^{-1}$ root $DM\,h^{-1}$, $p < 0.05$), but there were no other significant differences in specific activity. There were, however, significant differences in the distribution of AR activity with depth in the soil, between the two genotypes by the third harvest (Table 3). Genotype 77 had more of its nodule activity in the 101–200 soil zone. There was little nodule activity and no genotype differences in the 201–300 mm zone.

There were no differences between the genotypes in shoot aluminium concentration or any of the trace elements, or total or labile cyanide. Genotype 77 shoots were significantly lower in nitrogen and sodium and higher in potassium but phosphorus, calcium and magnesium did not differ significantly (Table 4).

The aluminium tolerant genotype 77 had half the root aluminium concentration of the susceptible genotype 129 (Table 4). Genotype 77 had significantly higher root concentrations of P, K, S, B and Mn but was lower in Ca.

Table 2. Root distribution (% of total root mass) in each depth zone in the experimental soil profiles for the two clover genotypes over three harvests. Statistical comparisons are between genotypes, within harvests and soil depth

Soil depth (mm)	Days growth								
	59			76			93		
	G77		G129	G77		G129	G77		G129
0–100	61	ns	66	65	ns	72	60	***	74
101–200	33	ns	30	20	ns	21	22	ns	19
201–300	6	ns	4	15	*	7	18	**	7

* Significant difference at $p < 0.05$, $^{**}p < 0.01$, $p^{***} < 0.001$.

Table 3. Distribution of acetylene reduction activity at different soil depths as a percentage of total plant activity for genotypes 77 and 129. Statistical comparisons are between genotypes within harvests

Soil depth (mm)	Days growth								
	59			76			93		
	G77		G129	G77		G129	G77		G129
0–100	93	ns	90	96	ns	97	85	***	94
101–200	6	ns	9	4	ns	3	13	*	5
201–300	1	ns	1	0	ns	0	2	ns	1

Table 4. Genotype differences in shoot and root chemical composition for lines 77 and 129 after 93 days growth. Statistical comparisons are between genotypes for each element

Genotype	Shoots			Roots						
	N	K	Na	P	K	Ca	S	Al	B	Mn
	(% in DM)			(% in DM)					(mg kg^{-1} DM)	
77	1.96	2.01	0.08	0.22	1.30	0.29	0.49	0.34	2.9	208.1
	**	*	***	**	***	***	***	**	*	**
129	2.34	1.63	0.12	0.17	0.84	0.36	0.39	0.63	2.1	155.7

Total plant cations (expressed as milli-equivalent %) did not differ significantly between the genotypes. However genotype 129 contained more aluminium and proportionately less potassium than genotype 77.

There was no difference in distal root diameter between the genotypes or over the soil aluminium range (overall mean diameter 0.22 mm). The proportion of distal roots bearing root hairs exceeded 90% for both genotypes and there was no visible damage to root tips, even at the highest soil aluminium concentrations.

Discussion

The only morphological difference between the two clovers was the ability of the aluminium tolerant genotype to have more roots deeper down the profile in the more Al toxic soil. This change in root distribution was accompanied by a small shift in nitrogenase activity into the intermediate soil depth zone. These differences may be of considerable agronomic importance during periods of moisture stress. The restriction of white clover roots to the surface of acid soils means the plant is prone to premature wilting during dry conditions compared to an acid-tolerant legume such as Lotus (Lowther, 1980). Better access to subsoil moisture would prolong growth of white clover as soils dried. White clover nodules are susceptible to moisture stress (Engin and Sprent, 1973; Foulds, 1978) and nitrogenase activity below 75 mm depth in soil may have survival value to the plant (Hoglund and Brock, 1978). Genotypic variations in rooting habit in white clover may be exploitable to produce a white clover better adapted to acid soils and less susceptible to moisture stress.

The increase in aluminium concentration in roots compared to shoots, observed in the experiment, is normal in forage legumes (Haynes and Ludecke, 1981; Macleod and Jackson, 1965; Munns, 1965). The ability of genotype 77 to maintain low aluminium in its roots rather better than genotype 129 did, is a novel result for white clover. No attempt was made to determine the form of the additional root aluminium in genotype 129. The relatively low root P concentration in genotype 129 tends to rule out accumulation of insoluble metal phosphates. This was suggested as a mechanism for white clover to cope with excess aluminium by Haynes and Ludecke (1981) and Jarvis and Hatch (1987). The cation balances of the two genotypes suggest that most of the plant aluminium was present in solution or in exchangeable form as suggested by Clarkson (1967). The difference in aluminium uptake between the genotypes seems to result from the ability of genotype 77 either to prevent entry of aluminium or maintain high aluminium efflux rates. Immobilisation of aluminium in the rhizosphere through the action of a root exudate is a possible exclusion mechanism although this is thought not to operate in Lotus (Blamey et al., 1990). The increased cation loading from the extra aluminium was balanced by a reduction in potassium. Andrew et al., (1973) show a reduction in milli-equivalents % potassium in white clover herbage as treatment aluminium increased from 1.0 to 2.0 ppm in solution culture. The small increase in me% sodium in genotype 129 is more likely to be associated with reduced potassium content, than any direct effect of aluminium.

Mackay et al., (1990) compared the aluminium responses of 14 clover cultivars but found no cultivar with greater Al-tolerance than 'Grasslands Huia'. From this they concluded that selection for Al tolerance in white clover could be restricted to germplasm with wide agronomic adaptation. The original screening of Huia that produced genotypes 77 and 129 was done on shoot growth, but in the current work, the plant response was restricted to the root system. Improved root growth in acid soils by lucerne selected for shoot vigour under acid conditions was recorded by Devine et al., (1976) and Bouton et al., (1982). Because root architecture has such major implications for persistence and growth of clovers under nutrient-, water- or pest-stress it would seem wise to monitor root and shoot growth in plant screening programmes.

Acknowledgements

J P M Evans for expert technical assistance; staff of DSIR Grasslands Analytical Unit for ICP analyses; staff of DSIR Grasslands who contributed to selection of the genotypes used.

References

AOAC 1984 Official Methods of Analysis, 14th Edn. Association of Official Analytical Chemists, Arlington, VA.

Andrew C S, Johnson A D and Sandland R L 1973 Effect of aluminium on the growth and chemical composition of some tropical and temperate pasture legumes. Aust. J. Agri. Res. 24, 325–339.

Blamey F P C, Wheeler D M, Edmeades D C and Christie R A 1990 Independence of differential aluminium tolerance in Lotus on changes in rhizosphere pH or excretion of organic ligands. J. Plant Nutr. 13, 713–728.

Bouton J H, Hammel J E and Sumner M E 1982 Alfalfa, *Medicago sativa* L., in highly weathered, acid soils. IV. Root growth into acid subsoil of plants selected for acid tolerance. Plant and Soil 65, 187–192.

Caradus J R 1987 Intraspecific variation for tolerance to aluminium toxicity in white clover. J. Plant Nutr. 10, 821–830.

Caradus J R, Mackay A D and Wewala S 1991 Selection for tolerance and susceptibility to aluminium within *Trifolium repens* L. *In* Plant-Soil Interactions at Low pH. Eds. R J Wright et al. pp 1029–1036. Kluwer Academic Publishers, Dordrecht, The Netherlands.

Clarkson D T 1967 Interactions between aluminium and phosphorus on root surface and cell wall material. Plant and Soil 27, 347–356.

Crush J R and Tough H J 1981 Hydrogen evolution from white clover. N.Z.J. Agric. Res. 24, 365–270.

Devine T E, Foy C D, Fleming A L, Hanson C H, Campbell T A, McMurtrey J E and Schwartz J W 1976 Development of alfalfa strains with differential tolerance to aluminium toxicity. Plant and Soil 44, 73–79.

Edmeades D C, Smart C E and Wheeler D M 1983 Aluminium toxicity in New Zealand soils, preliminary results on the development of diagnostic criteria. N.Z.J. Agric. Res. 26, 493–501.

Engin M and Sprent J I 1973 Effects of water stress on

growth and nitrogen fixing activity of *Trifolium repens*. New Phytol. 72, 117–126.

Foulds M E 1978 Response to soil moisture supply in three leguminous species. III. Rate of N_2 (C_2H_2) fixation. New Phytol. 80, 547–555.

Foy C D 1984 Physiological effects of hydrogen, aluminium and manganese toxicities in acid soil. *In* Soil Acidity and Liming. Ed. F Adams. Agronomy 12, 57–97.

Haynes R J and Ludecke T D 1981 Yield, root morphology and chemical composition of two pasture legumes as affected by lime and phosphorus application to acid soil. Plant and Soil 62, 241–254.

Hoglund J H and Brock J L 1978 Regulation of nitrogen fixation in a grazed pasture. N.Z.J. Agric. Res. 21, 73–82.

Jarvis S C and Hatch D J 1987 Differential effect of low concentrations of aluminium on the growth of four genotypes of white clover. Plant and Soil 99, 241–253.

Lowther W L 1980 Establishment and growth of clovers and lotus on acid soils. N.Z.J. Exp. Agric. 8, 131–138.

Mackay A D, Caradus J R and Pritchard M W 1990 Variation for aluminium tolerance in white clover. Plant and Soil 123, 101–105.

Macleod L B and Jackson L P 1965 Effect of concentration of the aluminium ion on root development and establishment of legume seedlings. Can. J. Soil Sci. 45, 221–234.

Mclean E O 1965 Aluminium. *In* Methods of Soil Analysis. Part 2. Chemical and Microbiological Properties. Ed. C A Black, Agronomy 9, 978–998.

Munns D M 1965 Soil acidity and growth of a legume II. Reactions of aluminium and phosphate in solution and effects of aluminium, phosphate, calcium and pH on *Medicago sativa* L. and *Trifolium subterraneum* L. in solution culture. Aust. J. Agric. Res. 16, 743–755.

Phillips J M and Hayman D S 1970 Improved procedures for clearing roots and staining parasitic and vesicular-arbuscular mycorrhizal fungi for rapid assessment of infection. Trans. Br. Mycol. Soc. 55, 158–161.

Snaydon R W 1962 Microdistribution of *Trifolium repens* and its relation to soil factors. J. Ecol. 50, 133–143.

Watson R N 1990 Effects of plant nematodes and Acremonium endophyte on white clover establishment with ryegrass or tall fescue. Proc. N.Z. Weed Pest Control Conf. 43, 347–351.

P. J. Randall et al. (Eds.), Genetic aspects of plant mineral nutrition, 59–68.

PLSO SV09

Differences in calcium efficiency between cowpea (*Vigna unguiculata* (L.) Walp.) cultivars

W. J. HORST, C. CURRLE and A. H. WISSEMEIER
Institute for Plant Nutrition, University of Hannover, Herrenhäuser Str. 2, DW-3000 Hannover 21, Germany

Key words: calcium, cowpea, efficiency, genotypic differences, *Vigna unguiculata*

Abstract

The cowpea (*Vigna unguiculata* (L.) Walp.) cultivars TVu 354 and Solojo were grown in solution culture at 10 to 1000 μM Ca supply. The Ca supply did not vary by more than 10% during the experiment. The pH value was kept constant within 0.1 units at 4.0 by automatic titration. The cultivar TVu 354 proved to be much more Ca-efficient than Solojo. At 10 μM Ca supply Solojo died, whereas TVu 354 was hardly affected in dry matter production. The differences in Ca efficiency were independent of the P supply. They could not be explained by differences in Ca uptake or Ca concentrations in the plant tissue. Short-term studies using ^{45}Ca, both in the dark and in the light, indicated better transport of Ca from the roots to the shoots and within the shoots to the younger leaves in the Ca-efficient cultivar TVu 354. However, the main reason for the differences between the cultivars in sensitivity to low Ca supply were differences in the Ca requirement of the plant tissue to maintain tissue organization and function. Sequential fractionation of the freeze-dried leaf tissue with hot water, 0.5 M NaCl, 1 M CH_3COOH, and 2 M HCl did not reveal cultivar differences in Ca binding state. The results clearly show that considerable genetic potential in tolerance to low Ca supply exists in cowpea. However, a better understanding of the physiological/biochemical reasons for low internal Ca requirement is needed.

Introduction

In large areas of the world crop production is limited by soil acidity. Productivity of acid soils may be increased by amelioration of soil acidity, especially by liming, or by growing plant species or cultivars adapted to acid soils. In acid mineral soils, Al toxicity is an important soil factor limiting plant growth. Selection of genotypes for Al tolerance under controlled conditions has contributed to the breeding of cultivars adapted to acid soils in cereals (Camargo et al., 1980). However, in dicots, especially legumes, correlations between genotypic Al tolerance and genotypic adaptation to acid soils very often were poor (Horst and Klotz, 1990; Sapra et al., 1982). This lack of correlation might be explained by the fact that legumes adapted to acid soils not only require tolerance to excess Al but also to Ca deficiency (Bruce et al., 1988). On the basis of the comparison of 6 cowpea cultivars, Horst (1987) came to the conclusion that Ca efficiency appeared to be a prerequisite for Al tolerance, but that Ca efficiency did not guarantee Al tolerance.

Ca deficiency is not only a problem on acid soils. In many vegetables and fruits physiological Ca disorders seriously reduce quality and, therefore, the economic value (Bangerth, 1979). The control of these disorders by cultural methods is difficult and not reliable. However, considerable differences between cultivars in sensitivity to Ca deficiency exist (English and Maynard, 1981; Hochmuth, 1984). The physiological reasons for these differences are still not clear. A better understanding could contribute to a more efficient identification and breeding of cultivars with improved tolerance to Ca deficiency. In the present paper we study the physiological reasons for the contrasting response to low Ca supply of cowpea cultivars, a crop which is widely grown in the lowlands of the humid and subhumid tropics.

Material and methods

Two cultivars (Solojo, TVu 354) of cowpea (*Vigna unguiculata* (L.) Walp.) were grown in a growth chamber at 30°/25° C day/night temperatures, 16 h daylength, 150 Wm^{-2} light intensity and 70% relative humidity. Seeds were germinated in a peat/sand mixture for 2 to 3 days and then transferred to 22-L plastic vessels containing constantly aerated nutrient solution with the following basic composition (μM): KNO_3 750, $Mg(NO_3)_2$ 325, KH_2PO_4 64, Fe-EDDHA 20, H_3BO_3 8, $CuSO_4$ 0.2, $ZnSO_4$ 0.2, $MnSO_4$ 0.2, $(NH_4)_6Mo_7O_{24}$ 0.2. Ca was added as $CaSO_4$. Therefore, Ca and S supply was varied at the same time. However, analysis of the leaves for SO_4^{2-} after 12 days of treatment clearly showed that the lowest Ca concentration of 10 μM $CaSO_4$ was sufficient to maintain the S nutrition of the plants well in the S-sufficiency range. The Ca concentration was checked periodically and did not vary by more than 10% even at the lowest Ca level of 10 μM. Plants were uniformly pretreated for 1 d at 1 mM Ca supply, before different Ca treatments began. The nutrient solution was completely changed first after 4, then every 3 days. The initial pH of the solution of 5.4 was lowered stepwise to reach 4.0 2 h before the beginning of the Ca treatment and kept constant within 0.1 pH unit by automatic titration with 0.01 M HCl or KOH. For the assessment of the Ca-binding state in the plant tissue, HCl was replaced by HNO_3 in order to keep Cl low in the plants. At harvest plants were fractionated into different organs and dried to constant weight at 80°C. Ca, K, Mg and Mn were measured (in 1/30 HNO_3) by atomic absorption spectrometry (AAS) and P by the molybdate/vanadate method after dry-ashing at 500°C for 6 h.

Ca was extracted from pulverized freeze-dried leaves according to Schilling (1960). The plant tissue was sequentially extracted by 1. distilled H_2O heated to 90°C, 2. 0.5 M NaCl, 3. 1 M CH_3COOH, and 2 M HCl. Cations (Ca, Mg, K) in extracts and residues were determined by AAS, anions in the extracts by ion chromatography (Dionex 2000i/SP).

Uptake and translocation of Ca (^{45}Ca) was studied with plants pretreated for 10 days at 20 or 50 μM Ca supply. Plants were transferred to 500 mL constantly aerated nutrient solution with the same Ca concentrations. ^{45}Ca was applied during the 12-h day and night period either at 74 kBq/500 mL during the light period or at 370 kBq/500 mL during the night period. After the 12-h uptake periods roots were washed for 10 min in 1000 μM non-labelled $CaSO_4$ solution. Plants were fractionated into different plant organs, dried, dry-ashed and ^{45}Ca determined by liquid scintillation. All data presented are means of 3 replicates. The data were subjected to analysis of variance and comparisons of means were made according to Tukey (1949).

Results

When grown at a range of Ca concentrations in the nutrient solution, 50 μM Ca proved to be sufficient to sustain optimum dry matter production during the 14-day experimental period for both cultivars (Fig. 1). However, at 10 μM Ca clear differences between the cultivars appeared: whereas TVu 354 was not affected, Solojo was severely depressed in growth. Typical Ca deficiency symptoms appeared starting on the emerging leaves (marginal necrosis and distortion), but also on expanding and expanded leaves (collapsed and necrotic patches on the leaf blade, weakening of the petioles and finally shedding of the leaves).

After 4 days of treatment, before any Ca deficiency symptoms appeared, Ca concentrations in roots, stem and leaves were similar in both cultivars (Table 1). With the exception of stems, there was a tendency towards higher Ca concentrations in the cultivar Solojo. Also Ca uptake rates (Table 2) were similar for both cultivars. After 14 days of treatment, when contrasting responses to 10 μM Ca supply had occurred, again at lower Ca supply no consistent differences in Ca concentrations in any of the

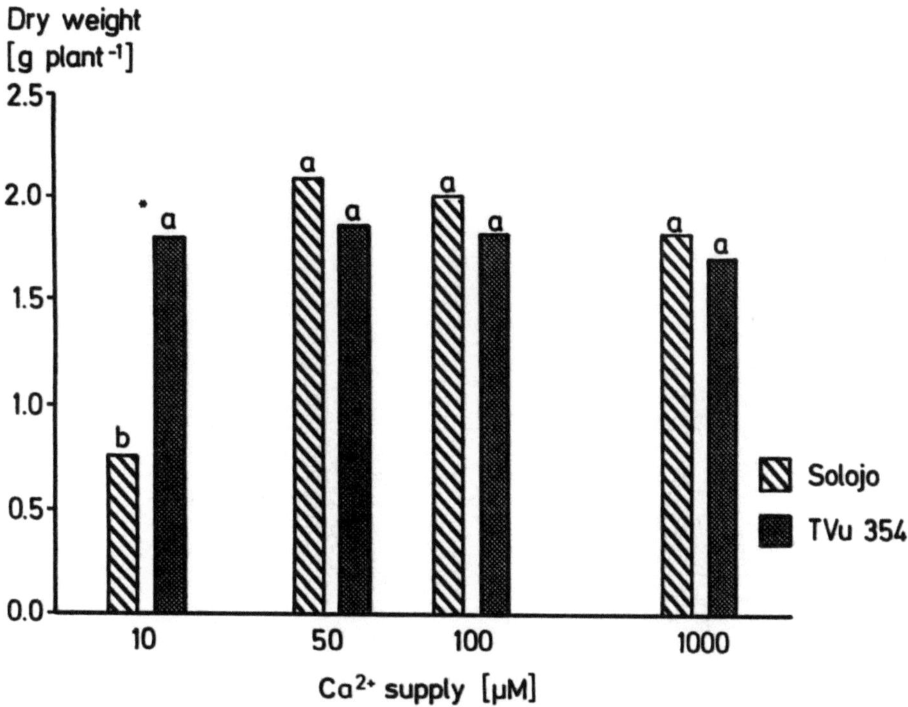

1. Dry mater production of 2 cowpea cultivars grown at a range of Ca concentrations for 14 days. Different letters and * cate significant differences between Ca supplies and cultivars, respectively, at $p<0.05$.

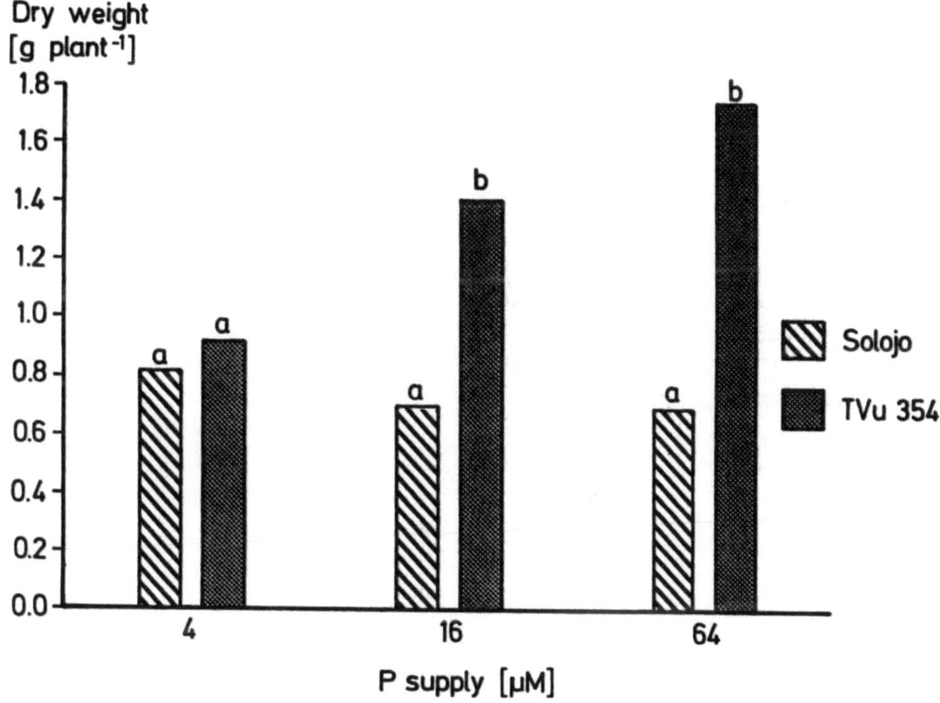

Fig. 2. Dry matter production of 2 cowpea cultivars grown at a range of P concentrations at low Ca supply (10 μM) for 14 days. Different letters indicate significant differences between cultivars at $p<0.05$.

Table 1. Calcium concentrations in different plant parts of 2 cowpea cultivars grown at a range of Ca concentrations for 4 days

Plant part	Ca supply (μM)	Ca concentration (μmol g^{-1} dry weight)		
		Solojo		TVu 354
Roots	10	34 a	*	25 a
	50	37 ab		32 ab
	100	47 b	*	37 b
	1000	91 c	*	75 c
Stem	10	40 a		46 a
	50	58 a		70 ab
	100	74 a		88 b
	1000	234 b	*	322 c
Leaves	10	35 a		34 a
	50	62 b		57 a
	100	94 c		84 b
	1000	368 d		348 c

Different letters and * indicate significant differences between Ca supplies and cultivars, respectively, at $p < 0.05$

Table 2. Mean Ca uptake rate of 2 cowpea cultivars grown at a range of Ca concentrations for 4 days

Ca supply (μM)	Ca uptake rate (μmol cm^{-1} root length d^{-1})		
	Solojo		TVu 354
10	4.6 a		2.8 a
50	5.4 b		4.5 b
100	6.7 c		6.1 c
1000	22.9 d	*	21.2 d

Different letters and * indicate significant differences between Ca supplies and cultivars, respectively, at $p < 0.05$

Table 3. Calcium concentrations in different plant parts of 2 cowpea cultivars grown at a range of Ca concentrations for 14 days

Plant part	Ca supply (μM)	Ca concentration (μmol g^{-1} dry weight)		
		Solojo		TVu 354
Roots	10	23.5 a	*	17.0 a
	50	27.8 a		27.5 b
	100	34.0 b		36.0 c
	1000	88.2 c		84.8 d
Stem	10	27.8 a		25.5 a
	50	39.2 a		39.0 a
	100	63.3 a		71.8 a
	1000	293 b	*	344 b
Petioles	10	45.0 a	*	28.8 a
	50	55.5 ab		56.0 ab
	100	80.0 b		88.0 b
	1000	423 c	*	467 c
Expanded leaves	10	43.8 a		38.8 a
	50	153 a		130 a
	100	223 a		253 a
	1000	1235 b	*	1170 b
Expanding leaves	10	17.8 a		22.0 a
	50	87.8 a		71.3 a
	100	128 a		120 a
	1000	659 b	*	493 b
Emerging leaves	10	68.3 a		21.5 a
	50	38.0 a		56.2 ab
	100	92.0 b		84.8 b
	1000	294 c	*	231 c

Different letters and * indicate significant differences between Ca supplies and cultivars, respectively, at $p < 0.05$

plant parts appeared (Table 3). The high Ca concentration (although not significantly different) in the emerging leaves of cultivar Solojo at 10 μM Ca supply could be due to a concentration effect in the severely injured plant tissue.

The concentrations of other cations in the plant tissues decreased with increasing Ca supply. Only the K, Mg and Mn concentrations in the expanding leaves are presented (Table 4). K concentration of Solojo was decreased at the

Table 4. Potassium (K), Mg and Mn concentrations in expanding leaves of 2 cowpea cultivars grown at a range of Ca concentrations for 14 days

Ca supply (μM)	Concentration in expanding leaves (μmol g^{-1} dry weight)								
	K			Mg			Mn		
	Solojo		TVu 354	Solojo		TVu 354	Solojo		TVu 354
10	1079 a	*	1487 a	315 a		324 a	0.72 a	*	1.22 a
50	1667 b	*	1392 a	330 a	*	233 b	1.03 b	*	0.91 b
100	1567 ab		1444 a	266 b	*	206 b	0.83 ab		0.84 b
1000	1131 a		1121 a	174 c	*	104 c	0.76 a		0.56 c

Different letters and * indicate significant differences between Ca supplies and cultivars, respectively, at $p < 0.05$

lowest Ca supply which might be an indication of tissue injury by Ca deficiency. In TVu 354, Mg concentrations were generally lower whereas Mn concentrations were lower only at high Ca supply, but higher at low Ca supply. The sum of cations increased with increasing Ca supply in both cultivars (not shown). It was higher in Solojo except for the 10 μM Ca supply because of the reduced K concentration.

P concentrations were extremely high especially in the young leaves of cultivar Solojo (Table 5). Since P toxicity as a contributing factor to the sensitivity of Solojo at low Ca supply could not be excluded, P supply was decreased at constant low Ca supply (Fig. 2). Cultivar TVu 354 responded strongly to decreased P supply by reduction of dry matter production owing to P deficiency. Solojo did not respond to variation in P supply, showing that Ca deficiency restricted growth independent of P supply and that P toxicity was not a major reason for poor growth at low Ca supply. Consequently, differences between the cultivars in sensitivity to low Ca supply were only expressed at high P supply. Dry matter production, however, did not sufficiently reflect growth of Solojo at low P supply. Figure 3 clearly shows that Ca deficiency in Solojo was less severe at lower P supply and that the differ-

Table 5. Effect of variation in P supply on P and Ca concentrations in expanding leaves of 2 cowpea cultivars cultivated for 14 days at 10 μM Ca supply

P supply (μM)	P concentration (μmol g^{-1} dry wt)			Ca concentration (μmol g^{-1} dry wt)	
	Solojo		TVu 354	Solojo	TVu 354
4	99.4 a	*	48.7 a	14.2 a	13.2 a
16	307.4 b	*	127.7 b	13.7 a	13.5 a
64	361.6 b	*	205.8 c	15.5 a	14.3 a

Different letters and * indicate significant differences between Ca supplies and cultivars, respectively, at $p < 0.05$

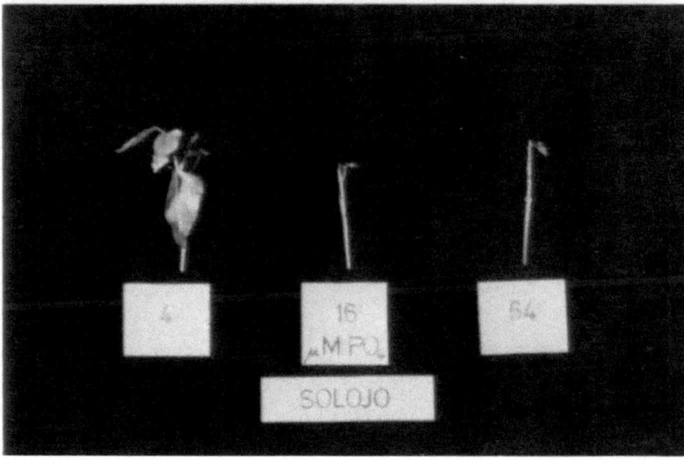

Fig. 3. Response of 2 cowpea cultivars to a range of P concentrations at low Ca supply (10 μM) for 14 days.

ences in Ca efficiency were most dramatic at higher P supply. Raising the P supply increased P concentrations of the emerging leaves to very high levels especially in Solojo (Table 5). Ca concentrations were unaffected by variation in P supply.

Sequential extraction should give an indication of the binding state of Ca in leaf tissue (Table 6). Generally, with increasing Ca supply and increasing Ca concentrations in the tissue the proportion of H_2O-extractable Ca increased, NaCl-exchangeable Ca remained stable, whereas CH_3COOH-soluble Ca decreased. The HCl-and residue fractions were of minor importance. In expanding compared to expanded leaves the percentages of both H_2O- and CH_3COOH-soluble Ca were smaller and higher, respectively. Between the genotypes no consistent differences existed that could indicate genotypic differences in binding state.

In the H_2O extract, in addition to Ca, also other nutrient cations (K, Mg) and anions (NO_3, PO_4, SO_4) were determined. Water-extractable nitrate, phosphate and sulfate concentrations ranged between (μmol g^{-1} dry weight) 220 – 450, 100 – 230, 180 – 260 and 190 – 330, 150 – 380, 50 – 110 in expanded and

Table 7. Cation – anion balance (C – A) in H_2O extracts of leaves of 2 cowpea cultivars grown at a range of Ca concentration for 12 days

Ca supply (μM)	(C – A) (μeq g^{-1} dry weight)				
	Expanded leaves		Expanding leaves		
	Solojo	TVu 354	Solojo		TVu 354
10	810 a	750 a	350 a	*	543 a
20	550 a	590 a	433 a		371 a
50	622 a	705 a	444 a		520 a

Different letters and * indicate significant differences between Ca supplies and cultivars, respectively, at $p < 0.05$

Table 6. Sequential extractability of Ca from leaves of 2 cowpea cultivars grown for 12 days at a range of Ca concentrations

Fraction	Ca supply (μM)	Ca extractability				
		Expanded leaves			Expanding leaves	
		TVu 354		Solojo	TVu 354	Solojo
		(μmol g^{-1} dry weight)				
Total	10	34 a	*	49 a	20 a	17 a
	20	55 a		54 a	31 b	31 b
	50	146 b		148 b	85 c	86 c
		(% of total)				
H_2O	10	14		15	3	0
	20	30		23	13	14
	50	33		43	35	35
NaCl	10	36		33	36	34
	20	31		38	43	34
	50	35		37	38	39
CH_3COOH	10	43		42	52	53
	20	32		33	37	42
	50	28		18	23	22
HCl	10	6		9	7	10
	20	6		5	5	9
	50	4		3	3	3
Residue	10	1		1	2	3
	20	1		1	2	1
	50	0		0	1	1

Different letters and * indicate significant differences between Ca supplies and cultivars, respectively, at $p < 0.05$

expanding leaves, respectively. Whereas nitrate and sulfate concentrations were lower, phosphate concentrations were higher at low Ca supply. No consistent differences existed between cultivars. The difference between cations and anions (C – A) gives an estimate of the concentration of organic anions. C – A was independent of Ca supply and lower in expanding com-

Table 8. Ca (^{45}Ca) concentrations in different plant organs of 2 cowpea cultivars after 12 h ^{45}Ca uptake either during the light or dark period. Ca supply during preculture and ^{45}Ca uptake was 20 or 50 μM Ca

Plant organ	Ca supply (μM)	Ca (^{45}Ca) concentration (μmol g^{-1} dry wt)						
		Light period				Dark period		
		Solojo		TVu 354		Solojo		TVu 354
Roots	20	7.41 a		5.93 a	s	5.22 a		4.12 a
	50	14.26 b	*	9.80 b	s	8.81 b	*	7.12 a
Stem	20	6.67 a	*	7.37 a	s	3.14 a	*	3.68 a
	50	13.67 b	*	16.64 b	s	8.88 b	*	11.70 b
Primary leaf	20	0.29 a	*	0.50 a	s	0.05 a		0.03 a
	50	1.27 a	*	1.63 a	s	0.12 a		0.15 a
1. Trifoliate	20	0.94 a		1.00 a		0.04 a		0.09 a
	50	2.59 a		3.15 a	s	0.15 a		0.53 a
2. Trifoliate	20	1.27 a		1.56 a		0.25 a		0.54 a
	50	4.25 b		5.29 b	s	2.33 b	*	3.10 b
Apex	20	0.77 a		1.33 a		0.60 a		0.62 a
	50	3.14 b	*	3.81 b		4.17 b		4.07 b

Different letters, * and s indicate significant differences for each plant organ between Ca supplies, cultivars and light regimes, respectively, at $p < 0.05$

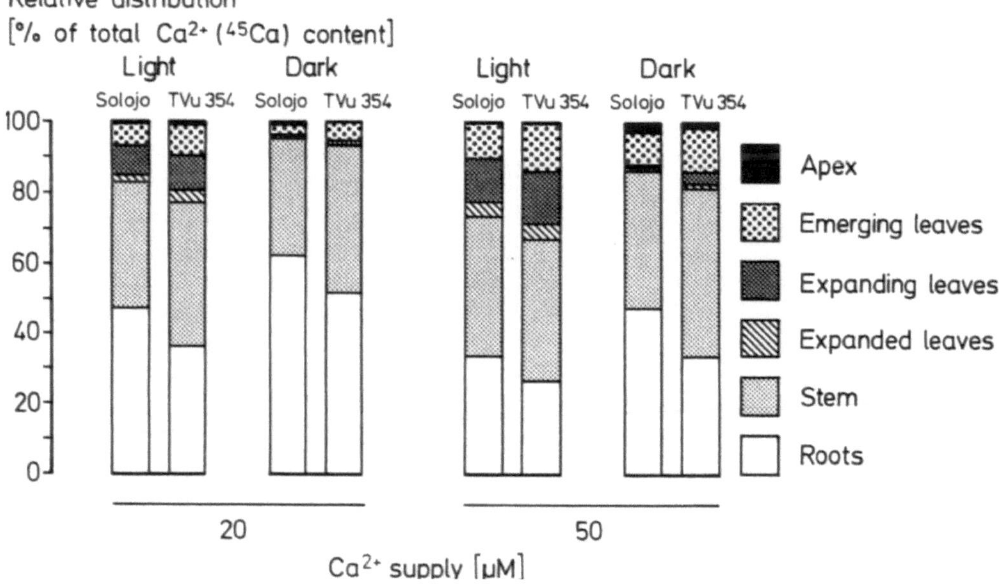

Fig. 4. Relative distribution of Ca (^{45}Ca) between plant organs of 2 cowpea cultivars after 12 h ^{45}Ca uptake either during the light or the dark period. Ca supply during preculture and ^{45}Ca uptake was 20 or 50 μM Ca.

pared to expanded leaves (Table 7). Only at low Ca supply in the expanding leaves C − A was higher in TVu 354 than in Solojo.

Ca uptake and translocation of 10-day-old plants during a 12-h light or dark period was studied using ^{45}Ca. (Table 8). After the uptake period Ca (^{45}Ca) concentrations in the roots but especially in the leaves were much higher during the light period. However, the differences in Ca (^{45}Ca) concentrations between the light regimes decreased in the youngest leaves. In the apex there was even a tendency of higher Ca (^{45}Ca) concentrations when ^{45}Ca was applied during the dark period at 50 μM Ca supply. In the cultivar Solojo Ca (^{45}Ca) concentrations in the roots were higher but were lower in the stem and leaves, especially in the younger leaves, than in cultivar TVu 354. These differences between cultivars were less marked when ^{45}Ca was applied during the night period. The relative distribution of Ca (^{45}Ca) within the plant clearly shows that Ca taken up during the night period is retained more strongly in the roots and less Ca was transported to the leaves (Fig. 4). In cultivar TVu 354 Ca was more readily transported from the roots to the shoot and within the shoot to the younger leaves.

Discussion

The two cowpea cultivars studied differed in tolerance to low Ca supply (Figs. 1, 4). Similar differences in Ca efficiency between cultivars have been described for e.g. tomato (Behling et al., 1989) and cauliflower (Hochmuth, 1984). These differences might be due to characteristics of Ca uptake, translocation or utilization within the plant tissue.

Ca uptake

The higher Ca uptake by dicots compared to monocots (Islam et al., 1987) could be attributed to the higher root cation exchange capacity (CEC) (Haynes, 1980). Although a strict relationship between binding of Ca to exchange sites in the root apoplast and Ca uptake can not be expected, there is evidence that accumulation of Ca and Mg in the apoplast enhances and blocking of exchange sites by Al inhibits uptake of Ca and Mg (Godbold et al., 1988; Klotz and Horst, 1988). A low CEC of the roots has been reported to be characteristic of cultivars with high Al tolerance (Kennedy et al., 1986). However, since Ca is taken up mainly via the apoplastic pathway in non-differentiated root-tip tissue and during side-root emergence (Häussling et al., 1988), differences between cultivars in morphological characteristics of roots might be more important for differences in Ca uptake (Cox, 1980). A relationship between number of root tips and Ca uptake can be expected.

The difference in Ca efficiency between the 2 cowpea cultivars studied can not be explained by differences in Ca uptake. Neither in the long-term studies (4 to 14 days) nor in the short-term study (12 h) using ^{45}Ca, the cultivars differed in Ca uptake (Tables 1–3, Table 8).

Ca translocation

Ca deficiency symptoms first appear in shoots in emerging and expanding leaves and in storage organs. Therefore, transport of Ca from the roots to the shoots and distribution within the shoot is of major importance for the expression of Ca deficiency. Ca transport and distribution in the shoot mainly depends on transpiration. However, for the transport of Ca to low transpiring plant organs (apex, storage tissue) root pressure (Guttridge et al., 1981) and a cellular auxin/Ca antiport (Banuelos et al., 1987) are most important. Banuelos et al. (1988) related differences in sensitivity to Ca deficiency between lettuce cultivars to lower basipetal auxin transport. The short-term Ca (^{45}Ca) uptake study revealed that Ca was more readily transported to the shoot and within the shoot to the younger leaves in the Ca-efficient cultivar TVu 354 (Fig. 4). It is not known whether this was due to differences in transpiration rate. A better Ca transport by root pressure is unlikely, because the differences between the cultivars were less clear during the dark period. The higher Ca (^{45}Ca) retention in the roots by the Ca-inefficient cultivar Solojo suggests that Ca (^{45}Ca) taken up was more readily sequestered either in the cell walls or the vacuoles in the roots. However, Ca concentrations in the leaf tissues well before

(Table 1) and after (Table 3) the occurrence of visible Ca deficiency symptoms do not give any indication that differences in overall Ca transport were decisive for differences among the cultivars in sensitivity to low Ca supply.

Ca utilization

It appears that the main reason for the difference between the cowpea cultivars in sensitivity to low Ca supply are differences in the Ca requirement of the plant tissue to maintain tissue organization and function. High Ca utilization efficiency does not only play an important role in the lower Ca requirement of grasses compared to legumes (Loneragan and Snowball, 1969) and calcifuge compared to calcicole plants (Kinzel, 1983), but also for Ca-efficient tomato (Giordano et al., 1982) and cauliflower (Hochmuth, 1984) cultivars. Differences in Ca utilization efficiency on a tissue basis are not yet well understood. Inactivation of Ca through binding/precipitation to oxalate has been suggested as indicative for low Ca efficiency in tobacco (Brumagen and Hiatt, 1966) and tomato (English and Maynard, 1981). However, water-extractable Ca from dried leaves did not reflect differences between tomato cultivars (English and Barker, 1982). Also in cowpea, the extraction procedure was unsuited to demonstrate major differences in Ca binding state between the cultivars, although effects of leaf age and Ca supply were clearly reflected (Table 6). Also C-A did not indicate a higher sequestration of Ca by organic acids in the inefficient cultivar. Therefore, it can be concluded that the extraction procedure is not sensitive enough to detect in-vivo genotypic differences in Ca binding stage, a conclusion which was already drawn by Ferguson et al., 1980. Inactivation of Ca by precipitation as Ca phosphate could be the consequence rather then the reason for the sensitivity of cultivar Solojo to low Ca supply (Fig. 3). It appears that under Ca deficiency as under Zn deficiency (Marschner and Cakmak, 1986), the ability of the plant to regulate its P uptake and translocation is hampered. A low physiological Ca requirement first of all depends on a high affinity to Ca of binding sites in cell membranes and cell walls, thus maintaining compartmentation and structural stability at low Ca supply (Kinzel, 1983). The results clearly show that considerable genetic potential in tolerance to low Ca supply exists in cowpea. However, a better understanding of the physiological/biochemical reasons for low internal Ca requirement is needed. This would allow the use of tissue culture in further efforts to screen cowpea germplasm for Ca efficiency.

References

Bangerth F 1979 Calcium-related physiological disorders of plants. Annu. Rev. Phytopathol. 17, 97–122.

Banuelos G S, Bangerth F and Marschner H 1987 Relationship between polar basipetal auxin transport and acropetal Ca^{2+} transport into tomato fruits. Physiol. Plant. 71, 321–327.

Banuelos G S, Bangerth F and Marschner H 1988 Basipetal auxin transport in lettuce and its possible involvement in acropetal calcium transport and incidence of tipburn. J. Plant Nutr. 11, 525–533.

Behling J P, Gabelman W H and Gerloff G C 1989 The distribution and utilization of calcium by two tomato (*Lycopersicon esculentum* Mill.) lines differing in calcium efficiency when grown under low-Ca stress. Plant and Soil 113, 189–196.

Bruce R C, Warrell L A, Edwards D G and Bell L C 1988 Effects of aluminium and calcium in the soil solution of acid soils on root elongation of *Glycine max* c. Forrest. Aust. J. Agric. Res. 38, 319–338.

Brumagen D M and Hiatt A J 1966 The relationship of oxalic acid to the translocation and utilization of calcium in *Nicotiana tabacum*. Plant and Soil 24, 239–249.

Camargo C E O, Kronstad W E and Metzger R J 1980 Parent-progeny regression estimates and associations of height level with aluminium toxicity and grain yield in wheat. Crop Sci. 20, 355–358.

Cox E F 1980 Growth of lettuce roots and its possible relationship to tipburn development. Hortic. Res. 20, 61–66.

English J E and Maynard D N 1981 Calcium efficiency among tomato strains. J. Am. Soc. Hortic. Sci. 106, 552–557.

English J E and Barker A V 1982 Water-soluble calcium in Ca-efficient and Ca-inefficient tomato strains. Hortic. Sci. 17, 929–931.

Ferguson J B, Turner N A and Bollard E G 1980 Problems in fractionating calcium in plant tissue. J. Sci. Food Agric. 31, 7–14.

Giordano L B, Gabelman W H and Gerloff G C 1982 Inheritance of differences in calcium utilization by tomatoes under low-calcium stress. J. Am. Soc. Hortic. Sci. 107, 664–669.

Godbold D L, Fritz E and Hüttermann A 1988 Aluminium toxicity and forest decline. Proc. Natl. Acad. Sci. USA 85, 3888–3892.

Guttridge C G, Bradfield E G and Holder R 1981 Dependence of calcium transport into strawberry leaves on positive pressure in the xylem. Ann. Bot. 48, 473–480.

Häussling M, Jorns C A, Lehmbecker G, Hecht-Buchholz Ch and Marschner H 1988 Ion and water uptake in relation to root development in Norway spruce (*Picea abies* (L.) Karst). J. Plant Physiol. 133, 486–491.

Haynes R J 1980 Ion exchange properties of roots and ionic interactions within the root apoplasm: Their role in ion accumulation by plants. Bot. Rev. 46, 75–99.

Hochmuth G J 1984 Variation in calcium efficiency among strains of cauliflower. J. Am. Soc. Hortic. Sci. 109, 667–672.

Horst W J 1987 Aluminium tolerance and calcium efficiency of cowpea genotypes. J. Plant Nutr. 10, 1121–1129.

Horst W J and Klotz F 1990 Screening soybean for aluminium tolerance and adaptation to acid soils. *In* Genetic Aspects of Plant Mineral Nutrition. Eds. N El Bassam, M Dambroth and B C Loughman. pp 355–360. Kluwer Academic Publishers, Dordrecht, The Netherlands.

Islam A K M S, Asher C J and Edwards D G 1987 Response of plants to calcium concentration in flowing solution culture with chloride or sulphate as the counter-ion. Plant and Soil 98, 377–395.

Kennedy C W, Smith W C and Ba M T 1986 Root cation exchange capacity of cotton cultivars in relation to aluminium toxicity. J. Plant Nutr. 9, 1123–1133.

Kinzel H 1983 Influence of limestone, silicates and soil pH on vegetation. *In* Physiological Plant Ecology III Encyclopedia of Plant Physiology; New Series Vol 12C. Eds. O L Lange, P S Nobel, C B Osmond and H Ziegler. pp 201–244. Springer-Verlag, Berlin.

Klotz F and Horst W J 1988 Effect of ammonium- and nitrate-nitrogen nutrition on aluminium tolerance of soybean (*Glycine max* L.). Plant and Soil 111, 59–65.

Loneragan J F and Snowball K 1969 Calcium requirements of plants. Aust. J. Agric. Res. 20, 465–478.

Marschner H and Cakmak I 1986 Mechanism of phosphorus-induced zinc deficiency in cotton. II. Evidence for impaired shoot control of phosphorus uptake and translocation under zinc deficiency. Physiol. Plant. 68, 491–496.

Sapra V T, Mebrahtu T and Mugwira L M 1982 Soybean germplasm and cultivar aluminium tolerance in nutrient solution and Bladen clay loam soil. Agron. J. 74, 687–690.

Schilling G 1960 Strontium in der höheren Pflanze. II. Verteilung und Bindungszustand in der Pflanze. Z. Pflanzenernaehr. Bodenkd. 91, 212–224.

Tukey J W 1949 Comparing individual means in the analysis of variance. Biometrics 5, 99–114.

P. J. Randall et al. (Eds.), Genetic aspects of plant mineral nutrition, 69–73.

PLSO SV18

Wheat growth responses of cultivars to H^+ concentration

J.W. JOHNSON and R.E. WILKINSON
Department of Agronomy, University of Georgia Agricultural Experiment Stations, Georgia Station, Griffin, GA 30223-1797, USA

Key words: acid soils, low pH, *Triticum aestivum*, wheat

Abstract

Shoot length (cm), shoot fresh weight (g/pot), root length (cm), and root fresh weight (g/pot) were measured on six cultivars of wheat (*Triticum aestivum* L. cv Saluda, C9733, Gore, Stacy, FL301, and FL302) grown at pH 6.0, 5.5, 5.0, 4.5, or 4.0 for 14 days in 'white quartz flintshot' sand. Plants were watered on alternate days with pH-adjusted buffer solutions. All measured plant parameters decreased as H^+ concentration increased from pH 6.0 to 4.0. Decreased lengths of shoots and roots were similar among the cultivars as the pH decreased. This indicated a uniform response of wheat cultivars to excess H^+ concentration in the soil solution; however, the decrease in shoot and root length was only about 50% as large as was previously reported for sorghum [*Sorghum bicolor* (L.) Moench.].

Introduction

Acid soils (<pH 4.8) are deleterious to plant growth (Clark, 1982; Foy, 1983; 1984); these soils are characterized as containing excess H^+, Mn^{2+}, and Al^{3+} as well as deficiencies of Ca^{2+}, Mg^{2+}, and PO_4^{3-} (Foy, 1984). Sorghum [*Sorghum bicolor* (L.) Moench] responded to excess H^+ by decreased shoot and root growth (Wilkinson and Duncan, 1989a, b). Shoot growth was highly correlated with gibberellic acid (GA_3) partitioning between water and ethylacetate (Wilkinson and Duncan, 1989a), while root growth was correlated with 3-indoleacetic acid (IAA) diffusion (Wilkinson and Duncan, 1989b). In sorghum seed, there is a two-cell-thick layer of dead waxy cells between the scutellum and the aleurone layer (Doggett, 1988) so that the movement of GA_3 from the scutellum to the aleurone layer could be influenced by the pH of the external solution. In roots, apoplastic IAA translocation from the zone of synthesis (i.e. root tip) (Davies et al., 1976; Feldman, 1980; Pernot and Pilet, 1976; Raven, 1975) to the zone of elongation would be highly susceptible to external pH. If there are very few plasmodesmata between the two regions, symplastic movement of IAA from the zone of synthesis to the zone of elongation would be very slow. Partitioning of IAA into water was very closely correlated with sorghum root growth as influenced by excess H^+ (Wilkinson and Duncan, 1989b).

If the responses of other plant species to excess H^+ are the same as reported for sorghum, then these are more general phenomena. But if the responses of other plant species vary significantly from sorghum responses, then modifying factors must be examined. Wheat (*Triticum aestivum* L.) was chosen as an assay of these parameters.

Methods and materials

'White quartz flintshot' sand (350 g) was used to fill 8 cm × 8 cm × 8 cm plastic pots. Twenty-five seeds of each wheat cultivar in five replications were planted 1-cm deep and the pots were watered (100 mL) every other day with 0.01 *M* sodium acetate adjusted to pH 6.0, 5.5., 5.0, 4.5, or 4.0 with conc. HCl. After fourteen days, shoots were measured from the soil line to the longest leaf tip and weighed. Roots were washed from the sand, blotted, weighed, and measured for root length. Data were subjected to analysis of variance on a randomized complete block design and means were separated by the LSD multiple range test.

Results

Wheat shoot length (cm) decreased as H^+ concentration increased (Fig. 1). Significant differences among cultivars occurred at pH 6.0, 5.5, 5.0, and 4.5 but not at pH 4.0 (Fig. 1). When calculated on a percentage basis with length at pH 6.0 as the base standard, the differences between cultivars were not significant (Fig. 2).

Wheat shoot fresh weight (g/pot) decreased as H^+ concentration increased from pH 6.0 to 4.0 (Fig. 3), and significant differences were found among cultivars (Fig. 3). When wheat shoot fresh weight (g/plant) was calculated as a percentage basis, no significant differences were detected among cultivars (Fig. 4).

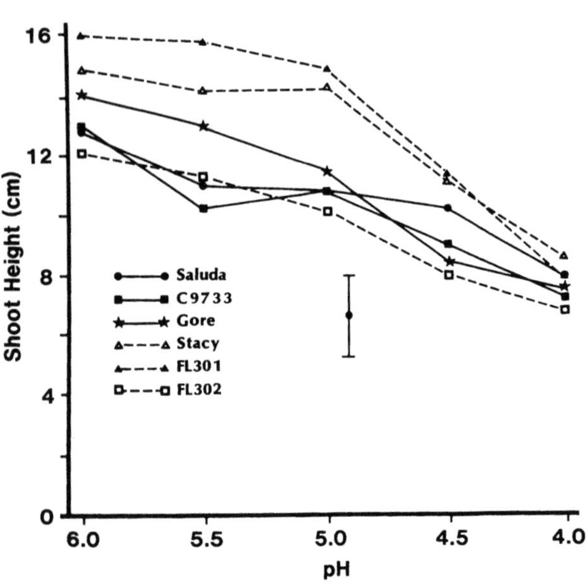

Fig. 1. Seedling root length (cm) of six wheat (*Triticum aestivum* L.) cultivars grown for 14 days at different H^+ concentrations produced by 0.01 *M* sodium acetate and adjusted to pH 6.0, 5.5, 5.0, 4.5, or 4.0 with conc. HCl. Standard errors (2X) are shown.

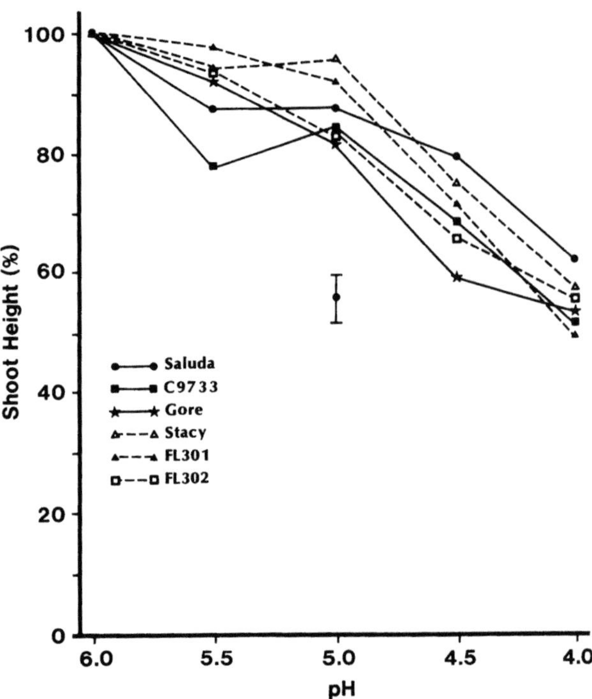

Fig. 2. Wheat (*Triticum aestivum* L.) cultivar length (%) using growth at pH 6.0 = 100%. Standard errors (2X) are shown.

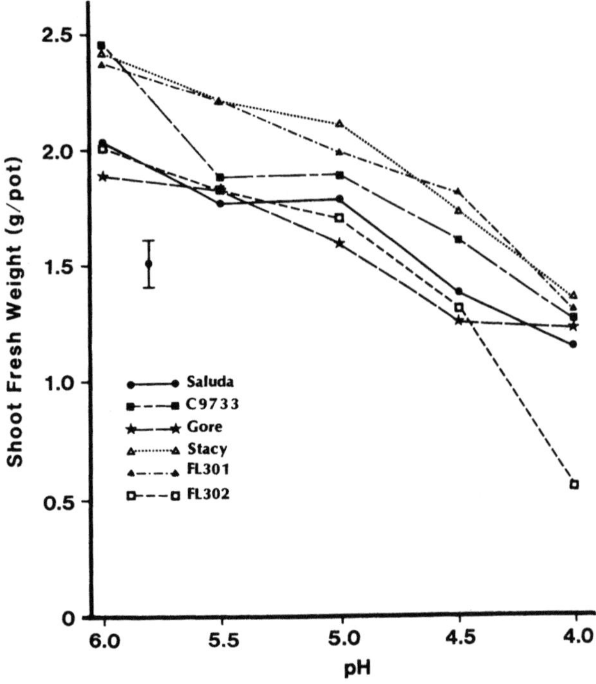

Fig. 3. Seedling shoot fresh weight (g/pot) of six wheat (*Triticum aestivum* L.) cultivars grown for 14 days at pH 6.0, 5.5, 5.0, 4.5, or 4.0. Standard errors (2X) are shown.

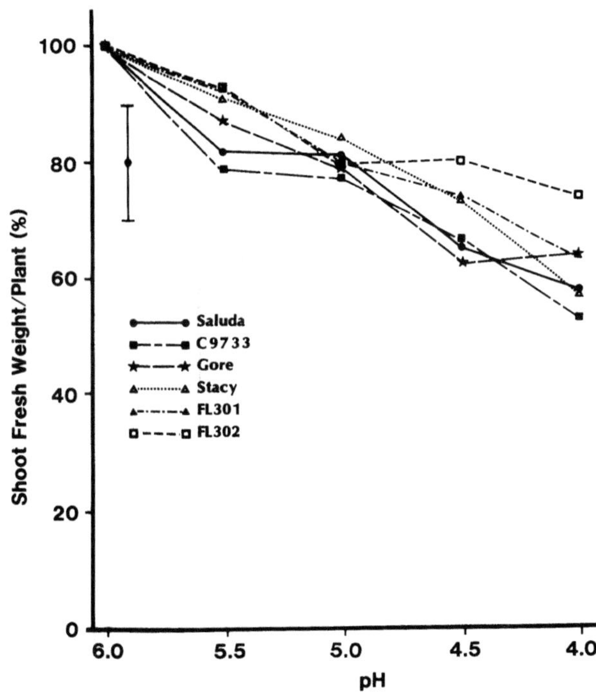

Fig. 4. Wheat (*Triticum aestivum* L.) leaf fresh weight/plant (%) using weight at pH 6.0 = 100%. Standard errors (2X) are shown.

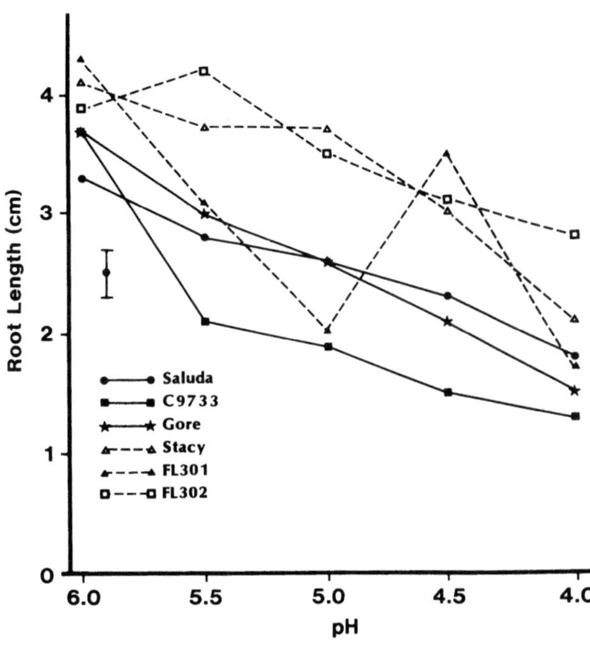

Fig. 5. Wheat (*Triticum aestivum* L.) root length (cm) of six cultivars grown for 14 days at different H^+ concentrations established by 0.01 *M* sodium acetate adjusted to pH 6.0, 5.5, 5.0, 4.5, or 4.0 with conc. HCl. Standard errors (2X) are shown.

Root length (cm) decreased as H^+ concentration increased from pH 6.0 to 4.0 with a large variability detected among cultivars (Fig. 5). Plotted on a percentage basis, the cultivar root lengths were significantly different in response to H^+ concentration (Fig. 6). Starting at pH 6.0 the lines for Stacy, Saluda, and Gore show a relatively uniform decrease in root length as pH changes from 6.0 to 4.0 (Fig. 6). The curve for FL302 would fit a linear line except for the point at pH 5.5; and C9733 was a linear decrease in root length between pH 5.5 and 4.0. FL301 is obviously different (Fig. 6). The decrease in root length appears to decrease to a minimum at approximately pH 5.0; but how to explain the increased length in FL301 at pH 4.5? And significant differences among cultivars in root length were evident at all pH levels (Fig. 6). Similarly, root fresh weight (g/pot) decreased as H^+ concentration increased and differences were found among cultivars (Fig. 7). And again, the root fresh weight (%) data show large significant

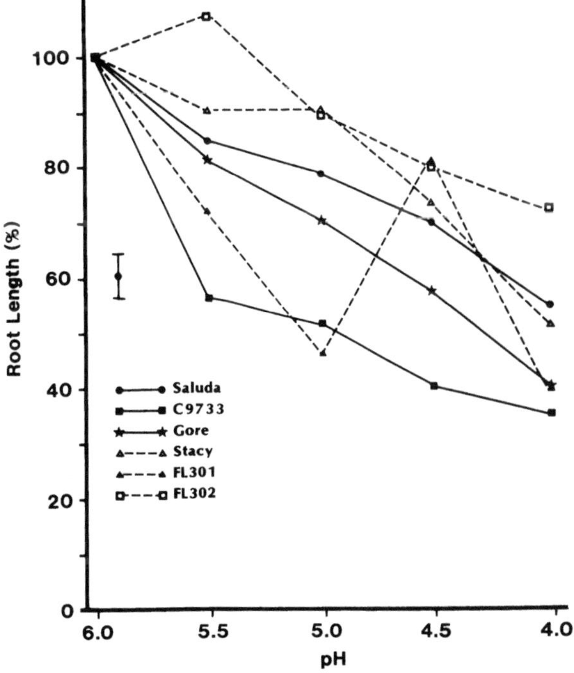

Fig. 6. Wheat (*Triticum aestivum* L.) root length (%) using growth at pH 6.0 = 100%. Standard errors (2X) are shown.

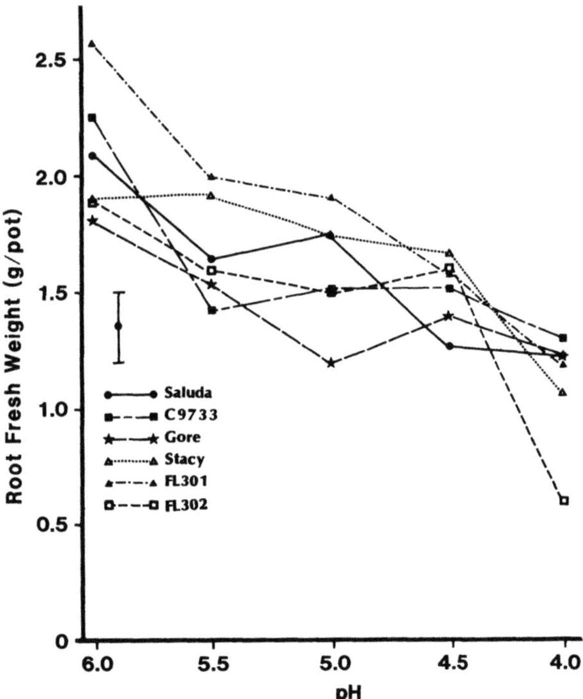

Fig. 7. Root fresh weight (g/pot) of six wheat (*Triticum aestivum* L.) cultivars grown for 14 days at pH 6.0, 5.5, 5.0, 4.5, or 4.0 established with 0.01 *M* sodium acetate and adjusted with conc. HCl. Standard errors (2X) are shown.

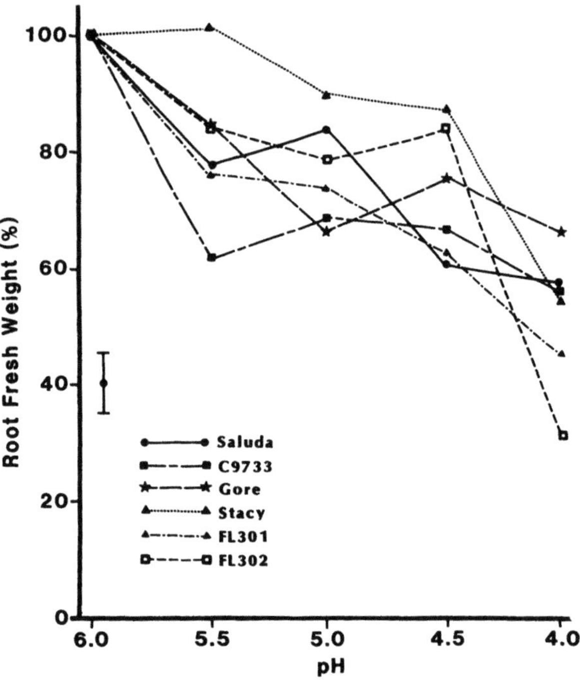

Fig. 8. Wheat (*Ttiticum aestivum* L.) root fresh weight (%) using weight at pH 6.0 = 100%. Standard errors (2X) are shown.

differences among cultivars at all pH levels ≤5.5. (Fig. 8).

Discussion

Differences in basic growth rates of wheat cultivars were present in the shoots (Fig. 1) at pH 6.0, 5.5, 5.0, and 4.5, but these differences were not observed at pH 4.0 (Fig. 1). When plotted as shoot growth percentage, the values showed some differences among cultivars at individual pH's, but the general curves were similar for all cultivars.

In sorghum, relative shoot length decreased from ~90% to ~30% as H^+ concentration increases from pH 5.8 to 4.0 (Fig. 3) (Wilkinson and Duncan, 1989a). The decrease in relative wheat shoot length was from ~100% to ~65% in the same pH range (Fig. 2). Since the sorghum data (Wilkinson and Duncan, 1989a) were explicable on the basis of a complete lack of plasmodesmata across the dead, waxy, two-cell layer between the scutellum and the aleurone layer, the responses seen by wheat in this study might be similarly attributable to reductions in the numbers of plasmodesmata between the wheat scutellum and aleurone layer. Thus, although much of the GA_3 could be moved apoplastically by diffusion at pH 6.0, only that GA_3 that could move symplastically via restricted numbers of plasmodesmata was available to influence wheat shoot growth at pH 4.0.

The results obtained in this study might indicate the presence of more plasmodesmata in wheat roots than in sorghum roots (Figs. 5–8). Sorghum root length decreased by ~90% as the H^+ concentration increased from pH 6.0 to 4.0 (Wilkinson and Duncan, 1989b). Wheat root length decreased by ~55% over the same pH range (Figs. 5, 6) and root fresh weight (Figs. 7, 8) approximated the same decrease. The shoot growth data (Figs. 1–4) might be influenced by a decrease in the width of the water-attainable dead, waxy, cell layer between scutellum and aleurone area. Or a change in the quantity or quality of the wax in this layer might also influence GA_3 diffusion by decreasing the size of the GA_3 absorptive sink at this point. But, apoplastic IAA translocation can only be influenced by: a) the H^+ concentration of the water which greatly influences the quantity of associated IAA at the membrane surface where IAAH is strongly lipophyllic, and b) the quantity of plasmodesmata which would permit an avoidance of IAA apoplastic IAA translocation at pH < 4.5.

A final explanation of these data, then, will depend upon measurement of the number of plasmodesmata in the appropriate zones of elongation. However, wheat seedling growth was similar to sorghum seedling growth, but there were intragenetically-based differences. Influence of H^+ concentration on elimination of IAA apoplastic translocation and entrapment of IAA in membranes and cytosol (pH > 6.0) could explain many responses. Concomitantly, symplastic IAA translocation at low pH would require considerable numbers of plasmodesmata. An extreme paucity of information exists in this area. However, the data available on the presence of plasmodesmata in root tips indicate a severe rarity of plasmodesmata in this root region. Therefore, a strong pH influence should be, and was indeed present.

Acknowledgements

This research was supported by State Hatch funds allocated to Projects H-1386 and H-1432, Georgia Agricultural Experiment Stations.

References

Clark R B 1982 Plant response to mineral element toxicity and deficiency. *In* Breeding Plants for Less Favorable Environments Eds. M N Christiansen and C F Lewis. pp 71–142, Wiley, New York.

Davies P J, Doro J A and Tarbox A W 1976 The movement and physiological effect of indoleacetic acid following point applications to root tips of *Zea mays*. Physiol. Plant. 36, 333–337.

Doggett H 1988 Sorghum. Longman Scientific and Technical. Wiley, New York. pp 70–122.

Feldman L J 1980 Auxin biosynthesis and metabolism in isolated roots of *Zea mays*. Planta 49, 145–150.

Foy C D 1983 The physiology of plant adaptation to mineral stress. Iowa State J. Res. 57, 355–391.

Foy C D 1984 Physiological effects of hydrogen, aluminum, and manganese toxicities in acid soil. *In* Soil Acidity and Liming, Ed. F Adams. pp 57–97. American Society of Agronomy Mongraph No. 12, 2nd Ed., American Society of Agronomy, Madison, WI.

Hare R C 1964 Indoleacetic acid oxidase. Bot. Rev. 30, 129–165.

Pernet J J and Pilet P E 1976 Indoleacetic acid movement in the root cap. Planta 128, 183–194.

Raven J A 1975 Transport of indoleacetic acid in plant cells in relation to pH and electrical potential gradients, and its significance for polar IAA transport. New Phytol. 74, 163–175.

Wilkinson R E and Duncan R R 1989a H^+, Ca^{++}, and Mn^{++} influence on sorghum seedling shoot growth. J. Plant Nutr. 12, 1395–1407.

Wilkinson R E and Duncan R R 1989b Sorghum seedling root growth as influenced by H^+, Ca^{++}, and Mn^{++} concentration. J. Plant Nutr. 12, 1379–1394.

P. J. Randall et al. (Eds.), Genetic aspects of plant mineral nutrition, 75–80.

PLSO SV23

Effect of aluminium on the growth of 34 plant species: A summary of results obtained in low ionic strength solution culture

D. M. WHEELER, D. C. EDMEADES, R. A. CHRISTIE and R. GARDNER[1]
Ruakura Agricultural Centre, Ministry of Agriculture and Fisheries, Private Bag, Hamilton, NZ.
[1]*Department of Molecular Biology, Auckland University, Private Bag, Auckland, NZ*

Key words: aluminium, asparagus, Arabidopsis, bean, cereals, grass, legume, Nicotiana, petunia, relative tolerance, tomato

Abstract

The results from many experiments conducted over 5 years to determine the tolerance of 34 plant species (87 cultivars) to aluminium (Al) are summarised. All experiments were conducted in a temperature-controlled glasshouse using a low-ionic-strength solution culture technique. The activity of Al^{3+} (μM) at which top yields were reduced by 50% (Al_{RY50}) was determined for each cultivar.

The species *Bromus wildenowii*, *Cynosurus cristatus*, *Hordeum vulgare*, *Triticum aestivum* (cvs Warigal, Scout, Sonora-63), *Avena byzantina*, *Arabidopsis thaliana*, *Lycopersicon esculentum* and *Nicotiana plumbaginifolia* were all very sensitive to Al ($Al_{RY50} < 1$). The species *Poa pratense*, *Lolium perenne* (NZ-derived cultivars), *Lotus corniculatus*, *Avena sativa* (cvs West, Carbeen, Camellia and Coolabah), *Triticum aestivum* (cvs Cardinal and Waalt), *Allium cepa* and *Asparagus officinalis* were sensitive to Al (Al_{RY50} 1–2).

The pasture grass species *Lolium perenne* (Australian and European and derived cultivars), *Lolium hybridum* and *Lolium multiflorum*, *Dactylis glomerata* (Apanui and Kara), *Phalaris aquatica*, *Festuca arundinacea* and the pasture legumes species *Trifolium pratense*, *Trifolium repens* and *Trifolium subterraneum* were all moderately sensitive to Al (Al_{RY50} 2–5). Other species that were also moderately sensitive included *Triticum aestivum* (cvs Atlas-66, BH146, and Carazinho), *Avena sativa* (cvs Swan and Blackbutt), *Avena Strigosa*, *Petunia x* and *Phaseolus vulgaris* (cvs Red Kidney, Black Turtle and Haricot).

The most tolerant species ($Al_{RY50} > 5$) werc (in order of increasing tolerance) *Phaseolus vulgaris* (cvs Tendergreen, The Prince and Yatescrop), *Cucurbita maxima*, *Dactylis glomerata* (cv Wana), *Paspalum dilatatum*, *Lotus pedunculatus*, *Ehrharta calycina*, *Medicago sativa*, *Holcus lanatus*, *Festuca rubra*, *Phaseolus lunatus* and *Agrostis tenuis*. *Agrostis tenuis* was at least twice as tolerant as the next most tolerant species ($Al_{RY50} > 30$ compared to 15.6).

Introduction

Soil acidity and as a consequence Al toxicity is a major factor limiting plant growth on many soils. For pastoral soils there is evidence, particularly in Australia where liming is not widely practiced, that soils are becoming increasingly acid (Helyar and Porter, 1989). In New Zealand there are large areas of hill country and upland soils which are acid (pH < 5.5) and where liming is no longer economic (Edmeades et al., 1985).

One alternative strategy to overcome Al toxicity problems associated with acid soils is to breed Al-tolerant cultivars. Differential Al tolerance between cultivars has been reported for a number of species (reviewed by Foy, 1988). Thus, it might be possible to breed Al-tolerant pasture cultivars by conventional plant breeding techniques coupled with the techniques of molecular biology.

A screening technique using a low-ionic-strength nutrient solution (2.7×10^{-3} M) has been developed (Edmeades et al., 1991a) initially to screen grasses and legumes for Al tolerance. The importance of using a low-ionic-strength nutrient solution in terms of quantifying the 'critical' levels of Al tolerance (Edmeades et al., 1991a) and in terms of ranking plants for tolerance to Al (Blamey et al., 1991a) has been discussed. This technique has now been used to quantify the effects of Al on a large number of plant species in order to identify species, cultivars and individual plants that are tolerant to Al.

This paper summarises the results from all experiments to date. Its unique value is that all cultivars and species were examined using solutions of similar composition.

Methods

All results reported in this paper have been derived from 'flowing' culture or 'still' solution culture experiments in temperature-controlled glasshouses at the University of Queensland, Brisbane, Australia, or the Ruakura Agricultural Centre, Hamilton, New Zealand. Details of the experimental methods and techniques are given elsewhere (Edmeades et al., 1991a; b; Wheeler and Follett, 1991; Wheeler et al., 1992a; b). In general, nutrient concentrations were maintained at (μM) 500 Ca, 100 Mg, 300 K, 600 N (150 NH_4, 450 NO_3), 600 SO_4, 2.5 P, 5 Fe, 6 Mn, 1 Zn, 1 BO_3 and 0.1 Cu and the solution pH maintained at 4.7. For a number of species, an additional pH treatment (pH 6.0) was also included. For most species, more than one experiment was conducted at different times.

Nominal Al rates varied between experiments (Edmeades et al., 1991a; b; Wheeler and Follett, 1991; Wheeler et al., 1992a; b) but were generally between 0 and 50 μM Al. Aluminium was added as $Al_2(SO_4)_3$. The activity of solution Al^{3+}, which was calculated using Geochem (Sposito and Matigod, 1980) as modified by Parker et al., 1987, generally varied between 0 and 10.8 μM.

The relationship between Al^{3+} activity (μM) and relative top and root yields were determined for each cultivar using polynomial exponential splines employing Bayesian smoothing techniques (Upsdell, 1985). Relative yields were used to allow the combining of data from a number of experiments for each species and to enable comparison between species. From these splines, the Al^{3+} activity at which top yield was decreased by 50% (Al_{RY50}) and the 95% confidence interval was determined for each cultivar. The data from Wheeler and Follett (1991) and Edmeades et al. (1991a) were reanalysed using polynomial exponential splines to give results in the same format.

Results and discussion

In order to rank plant species for tolerance to Al toxicity we propose various categories based on the Al^{3+} activity required to reduce the yield of plant tops by 50% (Al_{RY50}) (Table 1). These criteria have been applied to the various pasture grasses and legumes tested (Table 2), field crops including cereals (Table 3) and miscellaneous species (Table 4). Most species examined were either very sensitive (12 cultivars), sensitive (18 cultivars) or moderately sensitive (45 cultivars) to Al. Only 5 cultivars were tolerant or very tolerant. The most tolerant species tested was browntop (*Agrostis tenuis* $Al_{RY50} > 30$ (Edmeades et al., 1991b)) which had a Al_{RY50} approximately twice that of the next most tolerant species (*Phaseolus lunatus* Al_{RY50} 15.6 (Wheeler et al., 1992b)). The ranking of each species based on the Al_{Ry50} for roots was similar to that

Table 1. Proposed categories for tolerance to solution aluminium based on the aluminium activity (μM Al^{3+}) required to reduce yields by 50% (Al_{RY50}). The number of cultivars in each category is given

Category	Range in Al_{RY50}	Number of cultivars
Very sensitive	<1	12
Sensitive	1–2	18
Moderately sensitive	2–5	45
Moderately tolerant	5–10	7
Tolerant	10–20	4
Very tolerant	>20	1

Table 2. Relative tolerance to aluminium of 20 species of pastoral grasses and legumes

Category	Species	Cultivar
Very sensitive	*Bromus wildenowii*	Grasslands Matua
	Cynosurus cristatus	Unknown
	Poa pratensis	Baron
Sensitive	*Lolium perenne*	NZ derived cultivars[a]
	Lotus corniculatus	Maitland
Moderately sensitive	*Dactylis glomerata*	Grasslands Apanui
		Grasslands Kara
	Lolium hybridum	G4708
		Grasslands Manawa
		Grasslands Ariki
		North Canterbury
		Corvette
		Concord
		Augusta
	Lolium multiflorum	Grasslands Tama
		Grasslands Paroa
		Grasslands Moata
	Lolium perenne	European and Australian derived cultivars[b]
	Festuca arundinacea	Grasslands Roa
		Triumph
	Trifolium pratense	Grasslands Pawera
		Grasslands Turoa
	Trifolium repens	Grasslands G18
		Grasslands Pitau
		Grasslands Tahora
		Grasslands Huia
	Trifolium subterraneum	Tallarook
		Woogenellup
Moderately tolerant	*Dactylis glomerata*	Grasslands Wana
	Ehrharta calycina	Unknown
	Lotus pedunculatus	Grasslands Maku
	Paspalum dilatatum	Grasslands Raki
	Phalaris aquatica	Grasslands Maru
Tolerant	*Festuca rubra*	Unknown
	Holcus lanatus	Massey Basyn
	Medicago sativa	Hunter River
Very tolerant	*Agrostis tenuis*	Unknown

[a] Nui, Ruanui, Ellet, Droughtmaster, Takapau broad, Takapau persistor, Levin broad, Levin fine.
[b] Meltra, Mantilla, Vigour, Kangaroo Valley, Victoria.

for the tops (Edmeades et al., 1991a; b; Wheeler and Follett, 1991; Wheeler et al., 1992a; b).

The temperate grasses ranged from very sensitive (*Poa pratense*, *Bromus wildenowii*, *Cynosurus cristatus*) to very tolerant (*Agrostis tenuis*) (Table 2). Species normally regarded as 'weed' grasses in productive pastures (*Paspalum dilatatum*, *Holcus lanatus*, *Festuca rubra*, *Agrostis tenuis*) were more tolerant to Al than the normally preferred pasture species (*Lolium* species, *Dactylis glomerata*, *Bromus wildenowii*).

Within the *Lolium* species, the New Zealand-derived cultivars of perennial ryegrass (*Lolium perenne*) were more sensitive to Al than the cultivars derived from Australia and Europe. Furthermore, these New Zealand-derived cul-

Table 3. Relative tolerance to aluminium of 10 species of field crops

Category	Species	Cultivar
Very sensitive	*Avena byzantina*	Acacia
	Hordeum vulgare	Kearney
		Dayton
	Triticum aestivum	Warigal
		Sonora-63
		Scout
Sensitive	*Allium cepa*	Pukekohe Longkeeper
	Asparagus officinalis	Lucullus
	Avena sativa	Coolabah
		Carbeen
		Camellia
		West
	Triticum aestivum	Cardinal
		Waalt
		Atlas-66
Moderately sensitive	*Avena sativa*	Blackbutt
		Swan
	Avena strigosa	Saia
	Phaseolus vulgaris	Red Kidney
		Black Turtle
		Haricot
	Triticum aestivum	BH1146
		Carazinho
Moderately	*Cucurbita maxima*	Delica
	Phaseolus vulgaris	The Prince
		Tendergreen
		Yatescrop 37067
Tolerant	*Phaseolus lunatus*	Unknown

tivars are more sensitive than the *Lolium hybridum* or *L. multiflorum* cultivars. It is possible that these differences could reflect the fact most New Zealand ryegrass material has been selected from and bred on fertile and therefore high-pH soils (Edmeades et al., 1991b).

The three cocksfoot (*Dactylis glomerata*) cultivars examined exhibit a range in tolerance to Al with the cultivar Grasslands Wana being more tolerant than Grasslands Apanui. Grasslands Kara was intermediate in its tolerance. Once again the European origin of Grasslands Wana may explain its greater tolerance (Edmeades et al., 1991b).

The greater tolerance to Al of *Lotus pedunculatus* compared to the other pasture legumes in general (Table 2) is consistent with field trial evidence (Lowther, 1980). The other Lotus species, *Lotus corniculatus* cv. Maitland was quite different, being relatively sensitive to Al. The variation in Al tolerance among and within Lotus lines is discussed elsewhere (Blamey et al., 1991b).

Comparison of the pasture legumes and grasses (Table 2) indicates that grasses were generally no more tolerant to Al than the legumes. While some grasses such as poa (*Poa pratense*), crested dogstail (*Cynosurus cristatus*) and prairie grass (*Bromus wildenowii*) are more sensitive to Al relative to the legumes, there are other grasses such as browntop (*Agrostis tenuis*), Yorkshire fog (*Holcus lanatus*) and chewings fescue (*Festuca rubra*) which are more tolerant. Generally, the ryegrasses (Lolium cultivars) appear to be at least as sensitive as the legumes (*Trifolium pratense* and *T. repens*) normally grown in association with them in New Zealand grassland pastures.

Surprisingly, *Phalaris aquatica* and lucerne (*Medicago sativa*) showed some degree of toler-

Table 4. Relative tolerance to aluminium of 4 species

Category	Species	Cultivar
Very sensitive	*Arabidopsis thaliana*	Unknown
	Lycopersicon esculentum	Royal Ace V.F.
		Moneymaker
	Nicotiana plumbaginifolia	Unknown
Moderately sensitive	*Petunia x*	Blue picotee
		Red picotee
		Grandiflora appleblossum
		Grandiflora supermagic
		Multiflorum maddness
		Multiflorum double

ance to Al (Table 2). However, these species are also very sensitive to low pH (4.7 compared to 6.0) (Edmeades et al., 1991a) which explains their unsuitability for growth in acid soils. None of the other pasture species in which a pH treatment was included were sensitive to low pH (Edmeades et al., 1991a; b).

Within the cereals (Table 3), both barley (*Hordeum vulgare*) cultivars were sensitive to Al, although Kearney was slightly more tolerant than Dayton (Al_{RY50} 0.8 and 0.3, respectively (Wheeler et al., 1992a)). The wheat (*Triticum aestivum*) cultivars Warigal, Scout and Sonora, and the cultivar of *Avena byzantina* were also sensitive to Al. The other cultivars of wheat and oats (*Avena* species) were more tolerant to Al. The range of Al_{RY50} between the more tolerant wheat and oat cultivars were similar (Wheeler et al., 1992a). The cultivars of Brazilian origin (Carazinho in wheat and Saia in oats) were the most tolerant cereal species examined.

The relative ranking of the cereal cultivars produced in these experiments is generally consistent qualitatively with previous assessments of Al tolerance (Foy, 1987; Foy and Brown, 1964; Foy et al., 1965; Larkin, 1987; Lafever, 1988; Taylor and Foy, 1985; Zhang and Taylor 1988). The higher Al tolerance of Brazilian-derived cereal cultivars has been previously reported by Mugwira et al. (1981). However it is important to note that the critical Al concentrations measured in low-ionic-strength solutions are much lower, by an order of magnitude, than those previously reported. The reason for this are discussed in detail elsewhere (Blamey et al., 1991a; Edmeades et al., 1992b).

Of the other field crops examined, onion (*Allium cepa*) and asparagus (*Asparagus officinalis*) were sensitive and squash (*Cucurbita maxima*) moderately tolerant to Al (Table 3). Within the bean (*Phaseolus vulgaris*) cultivars, The Prince, Tendergreen and Yatescrop were more tolerant than Red Kidney, Black Turtle and Haricot (Table 3). The most tolerant field crop species was Lima beans (*Phaseolus lunatus*) with an Al_{RY50} of 15.6 (Wheeler et al., 1992b).

Our interest in the non-agricultural species listed in Table 4 stems from a desire to identify Al-tolerant and -sensitive germplasm from genetically simple plants – that is, plants with a small genome which are readily manipulated using the techniques of molecular biology. Of the species examined to date *Arabidopsis thaliana*, *Lycopersicon esculentum* (tomato) and *Nicotiana plumbaginifolia* were all very sensitive to Al ($Al_{RY50} < 1$) (Table 4). The tomato cultivar Royal Ace was slightly more tolerant than Moneymaker (Wheeler et al., 1992b). Petunia (*Petunia x*) were moderately sensitive to Al. There were no consistent differences between Petunia cultivars, possibly due to variability in the yields as a result of difficulties in germinating and transplanting sufficient seedlings of uniform size (Wheeler et al., 1992b).

The species in which differential Al tolerance between cultivars was observed in these experiments have been previously reported as having differential Al tolerance between cultivars (reviewed by Foy, 1988). The fact that many species show differential Al tolerance between their cultivars indicates that it is probably possible to breed a Al-tolerant cultivar for most species if desired. However, the difference in Al tolerance between cultivars in some species (e.g. barley, tomato) was small. Thus breeding Al-tolerant cultivars in some species may lead to only small increases in growth potential in the field.

Many of the agriculturally desirable species are only, at best, moderately sensitive to Al (Al_{RY50} 2–5). Whether the level of Al tolerance shown by the more tolerant species such as the 'weed' grasses (*Agrostis tenuis*, *Holcus lanatus*, *Festuca rubra*) can be attained in agriculturally desirable species, either by traditional breeding techniques or with the use of molecular biology, could depend on whether the trait of Al tolerance is multigenic and complex, or is linked with other undesirable traits.

References

Blamey F P C, Edmeades D C, Asher C J, Edwards D G and Wheeler D M 1991a Evaluation of solution culture techniques for studying aluminium toxicity in plants. *In* Plant-Soil Interactions at Low pH. Eds. R J Wright, V C Baligar and R P Murrmann. pp 905–912. Kluwer Academic Publishers, Dordrecht, The Netherlands.

Blamey F P C, Wheeler D M, Christie R A and Edmeades D C 1991b Variation in aluminum tolerance among and within Lotus lines. J. Plant Nutr. 13, 745–756.

Edmeades D C, Pringle R M, Mansell G P, Shannon P W Ritchie J and Stewart K W 1985 Effects of lime on pasture production on soil in the North Island of New Zealand. 5. Description of a lime recommendation scheme. N. Z. J. Exp. Agric. 13, 47–58.

Edmeades D C, Blamey F P C, Asher C J and Edwards D G 1991a Effects of pH and aluminium on the growth of temperate pasture species. 1. Temperate grasses and legumes supplied with inorganic nitrogen. Aust. J. Agric. Res. 42, 559–569.

Edmeades D C, Wheeler D M and Christie R A 1991b The effect of aluminium and pH on the growth of a range of temperate grass-species and cultivars. *In* Plant-Soil Interactions at Low pH. Eds. R J Wight, V C Baligar and R P Murrmann. pp 913–924. Kluwer Academic Publishers, Dordrecht, The Netherlands.

Helyar K R and Porter W M 1989 Soil acidification, its measurement and the processes involved. *In* Soil Acidity and Plant Growth. Ed. A D Robson. pp 61–101. Academic Press, Australia.

Foy C D and Brown J C 1964 Toxic factors in acid soils. II. Differential aluminium tolerance of plant species. Soil Sci. Soc. Am. Proc. 28, 27–32.

Foy C D, Armiger W H, Briggle L W and Reid D A 1965 Differential aluminium tolerance of wheat and barley varieties in acid soils. Agron. J. 5, 413–417.

Foy C D 1987 Acid soil tolerances of two wheat cultivars related to soil pH, KCl-extractable aluminum and degree of aluminum saturation. J. Plant Nutr. 10, 609–623.

Foy C D 1988 Plant adaption to acid, aluminum-toxic soils. Commun. Soil Sci. Plant Anal. 19, 959–987.

Lafever H N 1988 Registration of 'Cardinal' wheat. Crop Sci. 28, 377.

Larkin P J 1987 Calmodulin levels are not responsible for aluminium tolerance in wheat. Aust. J. Plant Phys. 14, 377–385.

Lowther W L 1980 Establishment and growth of clovers and lotus on acid soils. N. Z. J. Exp. Agric. 8, 131–138.

Mugwira L M, Sapra V T, Patel S U and Choudry M A 1981 Aluminium tolerance of triticale and wheat cultivars developed in different regions. Agron. J. 73, 470–475.

Parker D R, Zelazny L W, and Kinraide T B 1987 Improvements to the program GEOCHEM. Soil Sci. Soc. Am. J. 51, 488–491.

Sposito G and Mattigod S V 1980 GEOCHEM: A Computer Program for the Calculation of Chemical Equilibria in Soil Solution and other Natural Water Systems. Kearney Foundation of Soil Science, University of California, Riverside.

Taylor G J and Foy C D 1985 Mechanisms of aluminum tolerance in *Triticum aestivum* L. (Wheat). I. Differential pH induced by winter cultivars in nutrient solutions. Am. J. Bot. 72, 695–701.

Wheeler D M, Edmeades D C and Christie R A 1992a Effect of aluminium on relative yield and plant chemical concentrations for cereals grown in solution culture at low ionic strength. J. Plant Nutr. 15, 403–418.

Wheeler D M, Edmeades D C, Christie R A and Gardner R 1992b Effect of Al and pH on relative yield and plant chemical concentrations of 7 dicotyledonus species. J. Plant Nutr. 15, 419–433.

Wheeler D M and Follett J M 1991 Effect of aluminium on onions, asparagus and squash. J. Plant Nutr. 14, 897–912.

Upsdell M P 1985 Bayesian Inference for Functions. Ph. D. Thesis, Nottingham, UK.

Zhang G and Taylor G J 1988 Effect of aluminum on growth and distribution of aluminum in tolerant and sensitive cultivars of *Triticum aestivum* L. Commun. Soil Sci. Plant Anal. 19, 1196–1205.

P. J. Randall et al. (Eds.), Genetic aspects of plant mineral nutrition, 81–89.

PLSO SV10

Callose formation as parameter for assessing genotypical plant tolerance of aluminium and manganese

A.H. WISSEMEIER, A. DIENING, A. HERGENRÖDER, W.J. HORST and G. MIX-WAGNER[1]
Institute of Plant Nutrition, University of Hannover, DW-3000 Hannover 21, Germany and [1]Institute of Crop Science and Plant Breeding, FAL, Bundesallee 50, DW-3200 Braunschweig, Germany

Key words: aluminium, callose, cell culture, cowpea, linseed, manganese, screening, soybean, tolerance, toxicity

Abstract

Callose ((1,3)-β-glucan) formation in plant tissues is induced by excess Al and Mn. In the present study callose was spectrophotometrically quantified in order to evaluate whether it could be used as a parameter to identify genotypical differences in Al and Mn tolerance. Mn leaf-tissue tolerance of cowpea and linseed genotypes was assessed using the technique of isolated leaf tissue floating on Mn solution. Genotypical differences in the density of brown speckles on the leaf tissue (Mn toxicity symptoms) correlated closely with the concentrations of callose for both plant species. In cell suspension cultures Mn excess also induced callose formation. However, differences in tolerance of cowpea genotypes using callose formation as a parameter could only be found in cultured cowpea cells if controls cultured at optimum Mn supply showed low background callose. As soon as after 1 h, Al supply (50 μM) induced callose formation predominantly in the 5-mm root tip of soybean seedlings. Callose concentration in the 0–30 mm root tips was inversely related to the root elongation rate when roots were subjected to an increasing Al supply above 10 μM. Three soybean genotypes differed in inhibition of root-elongation rate and induction of callose formation when treated with 50 μM Al for 8 h. Relative callose concentrations and relative root-elongation rates for these genotypes were significantly negatively correlated.

Introduction

In acid mineral soils aluminium (Al) and manganese (Mn) toxicities are major factors limiting plant growth. Since soil amelioration to overcome Al and Mn toxicities often has ecological and economical limits, it is widely agreed that the use of plant genotypes tolerant of high Al and Mn supplies in the soil might make a substantial contribution to improved crop production. Differences in Al and Mn tolerance not only between, but also within plant species, have been demonstrated, as well as their genetic control (Foy et al., 1978, 1988). However, for efficient progress in plant selection and breeding, screening procedures are needed which allow quick and reliable identification of the desired trait. For cowpea genotypes simple destructive (Wissemeier and Horst, 1991) and nondestructive (Horst, 1982; Wissemeier and Horst, 1991) screening procedures have been described on the basis of symptom development on leaf tissue. Inhibition of root elongation by Al is widely used for assessing genotypical Al tolerance (Hanson and Kamprath, 1979; Horst and Klotz, 1990).

In recent years tissue culture as a screening

tool has obtained increasing interest either to identify existing variability or to use spontaneous and induced genetic variation (Meins, 1983). In particular, tissue culture techniques are required to produce transgenic plants with improved tolerance.

However, it is to be expected that selection on the cellular level will only be successful if the desired trait is not a whole-plant characteristic but is expressed on the cellular level as apparently is the case of B tolerance (Huang and Graham, 1990). There is ample evidence that genotypical Mn tolerance for vegetative growth in species like cowpea (Horst, 1988; Wissemeier and Horst, 1991) and linseed (DeMarco and Randall, 1988) is controlled on the leaf tissue level and, therefore, probably on the cellular level. Also, Al tolerance is believed to have a fundamental cellular basis (Haug, 1984).

Up to date identification of nutritional characteristics of genotypes in tissue culture is based mainly on growth characteristics. Reliable physiological/biochemical parameters could improve screening methods at this level. Therefore, in the present study the induction of callose formation in plant tissues by Al and Mn excess (Wissemeier et al., 1987; Wissemeier and Horst, 1987) was evaluated to diagnose genotypical differences in Al and Mn tolerance.

Materials and methods

Plant material

Seeds of cowpea (*Vigna unguiculata* (L.) Walp.) genotypes TVu 91, TVu 1977, TVu 1987 and TVu 3231 were initially supplied by the International Institute of Tropical Agriculture, Ibadan, Nigeria, and genotype Pipo by Dr. C. Haigis, Argentina. Linseed (*Linum usitatissimum* L.) genotypes W 11, W 53, W 86, W 169, W 364, W 371, Areco, Croxton and Glenelg were kindly provided by Dr P J Randall, CSIRO, Canberra, Australia. The Kleinwanzlebener Saatzucht AG, Einbeck, Germany, and the Institute of Crop Science, University of Hohenheim, Stuttgart, Germany, kindly provided the seeds of the soybean (*Glycine max* L.) genotypes Effi, Gieso, and Ronda.

Growing conditions of intact plants

All plants were grown in growth chambers under controlled environmental conditions with a 16/8 h light/dark regime at 30°C/25°C, respectively, and 60–70% relative humidity. Seeds were germinated between filter papers soaked with 1 m*M* $CaSO_4$ solution for 2 or 3 d and then transferred to constantly aerated nutrient solutions.

For the Mn experiments with cowpea and linseed the composition of the nutrient solution which was renewed every 2 to 3 d, was as follows [μM]: $Ca(NO_3)_2$, 1000; K_2SO_4, 375; $MgSO_4$, 325; KH_2PO_4, 100; Fe-EDDHA, 20; H_3BO_3, 8; $MnSO_4$, 0.25; $ZnSO_4$, 0.2; $CuSO_4$, 0.2; $(NH_4)_6Mo_7O_{24}$, 0.2. Two cowpea or 9 linseed plants were grown in 5-L plastic pots. In order to induce Mn toxicity in intact cowpea plants after growth for 17 d in nutrient solution, 0.25 (control), 20, 30, and 40 μM $MnSO_4$ (Mn excess) was supplied for up to 3 d until harvest. The light intensity was 300 $\mu E\ m^{-2}s^{-1}$.

The nutrient solution for the Al experiments with soybean had the following composition [μM]: KNO_3, 750; $Mg(NO_3)_2$, 325; $CaSO_4$, 500; Fe-EDDHA, 20; KH_2PO_4, 10; H_3BO_3, 8; $CuSO_4$, 0.2; $ZnSO_4$, 0.2; $MnSO_4$, 0.2; $(NH_4)_6Mo_7O_{24}$, 0.2. Up to 20 plants were grown in 22-L pots. After growth for 16 h at pH 5.5, the pH was lowered to 4.2 and kept constant by automatic titration with HCl or KOH. Al was added to half of the pots as $AlCl_3$.

Screening procedure for Mn tolerance of isolated leaf tissue

For screening cowpea and linseed genotypes the floating leaf-disk method described by Wissemeier and Horst (1991) was used. For cowpea, leaf disks (1.13 cm^2) from the oldest trifoliate leaf of 14-day-old plants grown at 0.25 μM Mn supply were excised using a cork borer and placed onto 20 mL of solutions containing both 500 μM $CaSO_4$ and 0 (control) or 500 (for linseed 1000) (Mn excess) μM $MnSO_4$. For linseed, fully expanded whole leaves were placed with the abaxial side on the solutions. The screening was performed in a growth chamber under the conditions described above.

Assessment of Mn and Al toxicity

The severity of Mn toxicity was assessed as density of brown speckles on the leaf tissue. With intact plants the number of brown speckles on leaves was determined by the naked eye at 4 places each 1 cm^2 in size of the abaxial leaf surface. Thereafter, samples of leaf tissue for callose quantification were excised with a cork borer and immediately chemically fixed (see below). On floating tissue, speckles were counted in situ with the aid of a macroscope (linseed) or by the naked eye on a luminous table (cowpea). For each replicate at least three leaf samples were counted.

For the assessment of the response of plants to Al, elongation rate of the primary roots which had been marked with indian ink about 3 cm behind the tip was measured on graph paper.

Cell suspension culture

Soybean cell suspension cultures (kindly provided by Dr Harms, FAL, Braunschweig, Germany) were derived from root meristems of genotype Mandarin which had been cultured since 1970 in liquid B5-medium (Gamborg and Wetter, 1975). Suspension cultures of cowpea genotypes were produced from calli from hypocotyl segments which were later on transferred to liquid modified MS medium (Murashige and Skoog, 1962) with the following composition [μM]: NH_4NO_3, 20600; KNO_3, 18800; $CaCl_2$, 3000; $MgSO_4$, 1500; KH_2PO_4, 1250; H_3BO_3, 100; $MnSO_4$, 100; $ZnSO_4$, 30; $NaMoO_4$, 1; $CuSO_4$, 0.1; Na_2EDTA, 100; $FeSO_4$, 100. Organic components were 3% (w/v) sucrose, and [mg L^{-1}] meso-inositol 100, nicotinic acid 1, pyridoxine 1, thiamine 0.5, glycin 1, IAA 0.5 and NES 0.5. The pH of the medium prior to autoclaving at 1.1 bar for 10 min ranged from 5.5 5.8. The suspension cultures were continuously agitated on a shaker at 150 r.p.m. in 250-mL Erlenmeyer flasks at constant 25°C and 16 h photoperiod (2000 lux). Depending on growth, the suspension cultures were transferred to fresh liquid medium one or two times a week.

Before the beginning of the differential treatments of the suspension cultures with Mn the original Mn supply of 100 μM (MS medium) and 60 μM (B5 medium) was reduced to 6 μM and 1 μM $MnSO_4$, respectively. Mn was supplied as $MnSO_4$ in concentrations up to 20 mM Mn in soybean and 15 mM Mn in cowpea (Mn excess).

Callose quantification

At harvest, roots, leaf tissue or cell samples were fixed in a solution with the final concentration of (v/v) 70% ethanol containing 5% formaldehyde and 5% propionic acid. Prior to determination, all samples were extensively rinsed in distilled water to remove fixative. The weight determined after a standardized drying procedure on filter papers was assigned as the fresh weight of the samples.

The microscopic studies of Galway and McCully (1987) showed that in roots wound callose is formed during chemical fixation at room temperature: this did not occur in deep-frozen material. Therefore, the preservation methods were compared using soybean roots of different genotypes and different Al supply. Measured callose concentrations in chemically fixed roots were only slightly higher than in frozen roots without genotypic interaction to the preservation method.

Callose was extracted in hot 1 N NaOH and measured spectrofluorometrically using aniline blue (Riedel-de Haën 32703) against laminarin (Sigma L 9634) according to the method of Köhle et al. (1985) and Kauss (1989). Blanks were run without aniline blue for each sample. For root tissue, in most cases a total root length of 9 cm was extracted in 2.25 mL NaOH, whilst for leaf tissues, a ratio of about 1.3 to 2.3 cm^2 per 2.25 mL NaOH was found to be adequate. The margins of the fixed leaf disks of cowpea were discarded prior to extraction in order to avoid an influence of wound callose. The whole leaves of linseed were used. About 300 mg fresh weight of cells were extracted in 10 mL NaOH.

Results

Manganese toxicity

Dark brown speckles occurring first on oldest leaves are typical Mn-toxicity symptoms in cowpea and in other plant species. A wide range of densities of brown speckles on leaves of Mn-sensitive cowpea genotype TVu 91 were produced by supplying Mn via the roots from 0.25 μM (optimum Mn supply) up to 40 μM (Mn excess) during the last 2 or 3 d prior to harvest (Fig. 1). The extractable callose concentrations of leaves were closely related to this density of the brown speckles.

In the Mn-sensitive cowpea genotype TVu 91 brown speckles could be observed on leaf disks as early as 1 d after transfer to high Mn supply while in the Mn-tolerant genotypes TVu 3231 and Pipo only a few brown speckles appeared even after 4 d on Mn solution (Fig. 2). The correlation of the density of brown speckles with the callose concentration of the leaf disks was highly significant (Fig. 3), indicating therefore, that callose formation through high Mn supply is indicative of genotypical differences in Mn tolerance in cowpea. Linseed cultivars also differed in formation of brown speckles when screened using the floating-leaf method (Fig. 4). Again, callose concentrations were correlated closely to the visible Mn toxicity symptoms in isolated leaves.

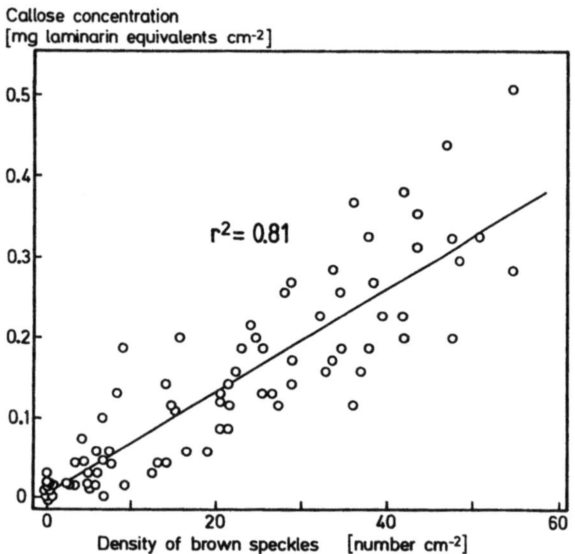

Fig. 1. Relationship between callose concentrations in leaves of cowpea genotype TVu 91 and the density of brown speckles (Mn toxicity symptoms) on leaves. Variation in severity of Mn toxicity was due to differences in Mn supply (0.25, 20, 30 or 40 μM $MnSO_4$), treatment duration (2 or 3 d), leaf age, and shading of plants.

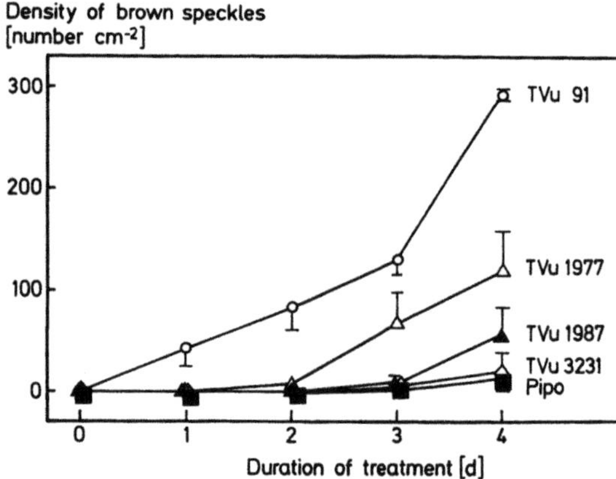

Fig. 2. Development of brown speckles (Mn toxicity symptoms) on leaf disks of five cowpea genotypes floating on solution containing 500 μM $MnSO_4$ and 500 μM $CaSO_4$; n = 3-6, *bars* indicate SD.

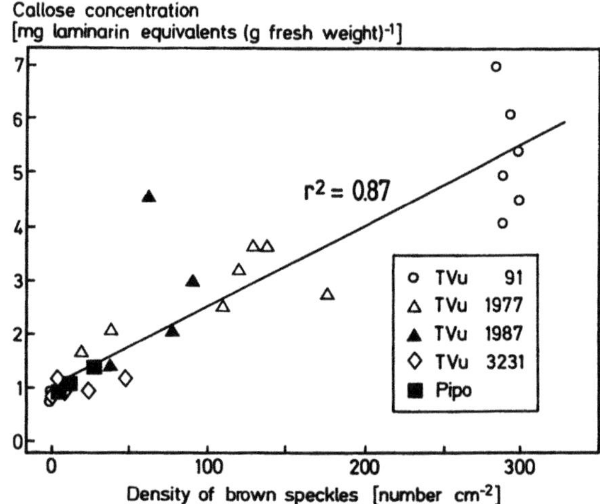

Fig. 3. Relationship between callose concentrations and the density of brown speckles on leaf disks of five cowpea genotypes floating for 4 d on solutions containing 500 μM $CaSO_4$ and 0 or 500 μM $MnSO_4$.

In cell suspension culture, Mn toxicity is more difficult to assess. It was, therefore, investigated whether callose formation could be used as quantitative parameter for cell injury by Mn excess. An established cell suspension culture of soybean was used to study the influence of Mn supply on callose formation with time. Five mM and 20 mM Mn supplies were compared with control cultures supplied with 1 μM Mn which was sufficient to sustain optimal cell growth according to preliminary experiments. The two

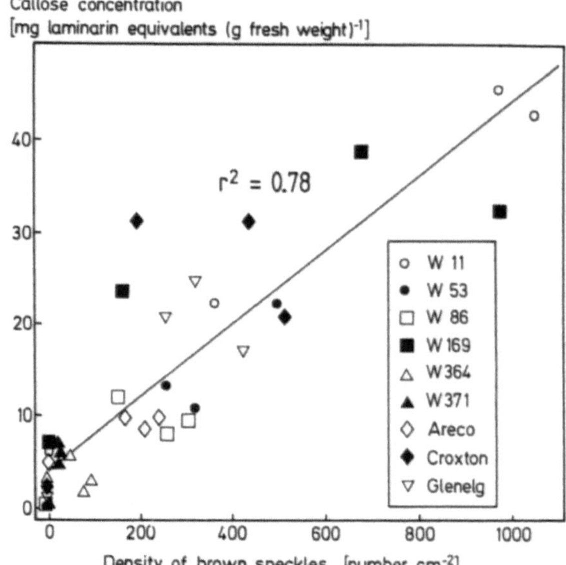

Fig. 4. Relationship between callose concentrations and the density of brown speckles on isolated leaves of 9 linseed genotypes floating for 1 d on solutions containing 500 μM $CaSO_4$ and 0 or 1000 μM $MnSO_4$.

higher Mn concentrations in the growth medium more than doubled the callose concentrations after 48 h treatment when compared to the 1 μM controls or with shorter exposure times (Fig. 5). Callose concentration further increased up to 72 h Mn treatment, especially with 20 mM Mn supply.

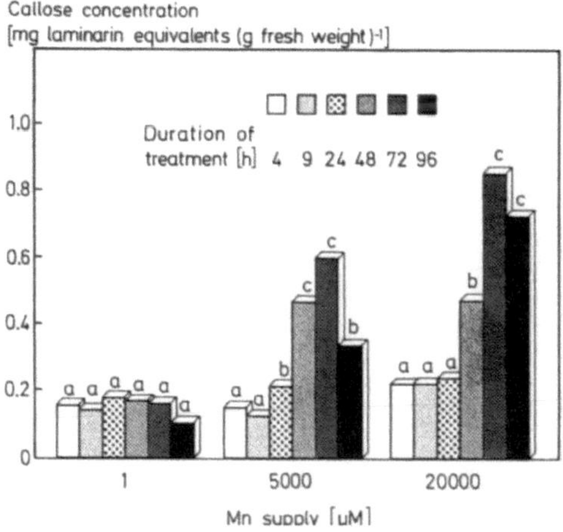

Fig. 5. Callose concentrations of soybean cell-suspension cultures at different Mn supply and treatment duration. Different letters indicate significant differences between treatment duration within Mn supplies at $p = 0.05$.

The same technique was applied to suspension cultures from cowpea genotypes screened for Mn tolerance (see above). However, suspension culture could only be established from three genotypes. Genotypes TVu 1977 and TVu 3231 induced callose formation at high Mn supply, but in genotype Pipo callose concentration was even lower with application of 15 mM Mn (Fig. 6). In Pipo callose concentration in the control treatment was higher than in any other treatment. Controls of genotype TVu 1977 and TVu 3231 had also substantial amounts of callose, a fact that was also evident from examination with fluorescence microscopy.

Comparing TVu 1977 with TVu 3231, both in absolute and relative terms, TVu 1977 produced more callose than TVu 3231 owing to high Mn supply. This is in agreement with the differences in Mn tolerance demonstrated above by the floating-leaf-disk technique (Figs. 2 and 3).

Aluminium toxicity

Callose formation could also be induced by Al treatment (Fig. 7). In soybean roots callose formation was mainly confined to the 0–5 mm root tip. One hour of Al supply was sufficient to induce callose formation. Callose formation also increased with treatment duration up to 8 h.

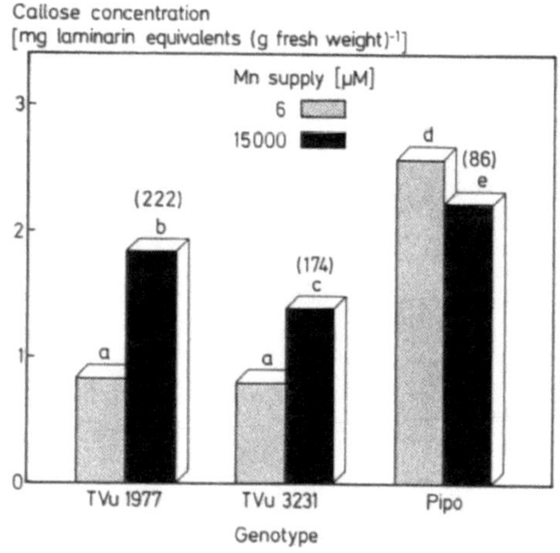

Fig. 6. Callose concentrations of cell suspension cultures of three cowpea genotypes at different Mn supply for 96 h. Relative callose concentrations at high Mn supply (6 μM = 100) in brackets. Different letters indicate significant differences at $p = 0.05$.

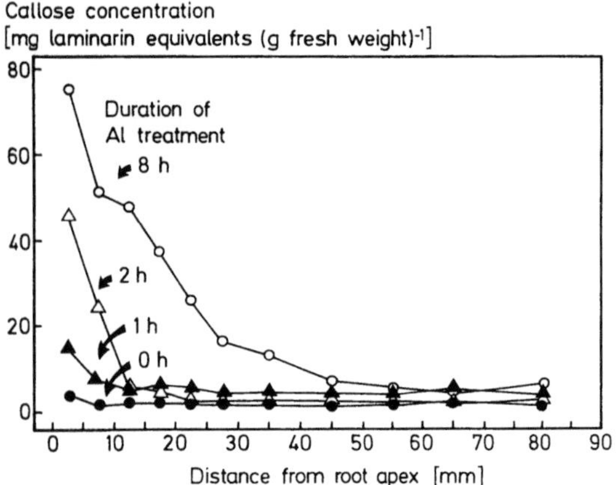

Fig. 7. Effect of 50 μM Al supply on callose concentrations in root segments of soybean genotype Effi as affected by duration of Al treatment; n = 3.

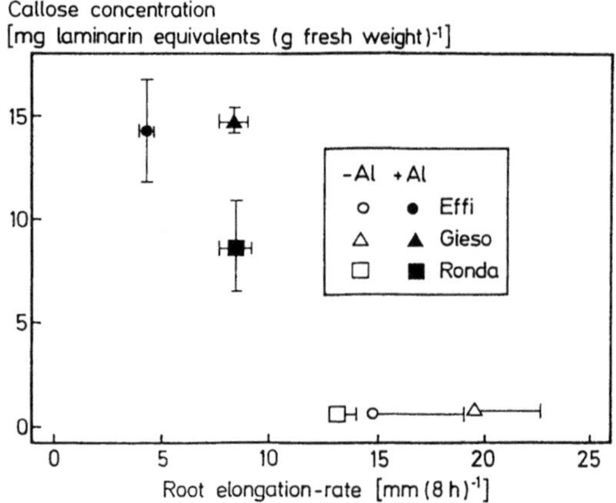

Fig. 9. Relationship between callose concentrations in the 30-mm root tips of three soybean genotypes and root elongation rates. Al supply 0 (−Al) or 50 (+Al) μM Al for 8 h; n = 3, *bars* indicate SD, if larger than symbols for root elongation rate (horizontal) and callose concentration (vertical).

Callose concentrations in the 0–30 mm root tips was inversely related to the root elongation rates when roots were subjected to an increasing Al supply above 10 μM (Fig. 8).

When treated with 50 μM Al three soybean genotypes responded differently in root elongation rate and callose formation (Fig. 9). The genotypes Ronda and Gieso had a less depressed root growth than Effi, which proved to be very Al-sensitive. In agreement with its higher Al tolerance according to root growth, callose formation in the root tips of Ronda was much lower, but was comparably high in Effi and Gieso. However, relative callose concentrations and relative root-elongation rates were negatively correlated (Fig. 10).

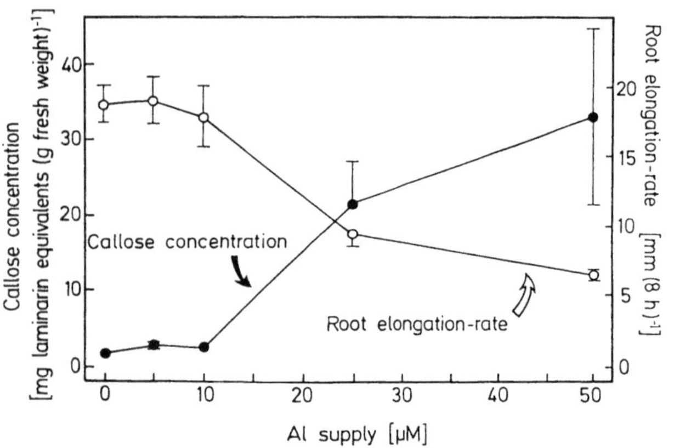

Fig. 8. Effect of different Al supply for 8 h on root elongation rates and callose concentrations in the 30-mm root tips of soybean genotype Effi; n = 3, *bars* indicate SD.

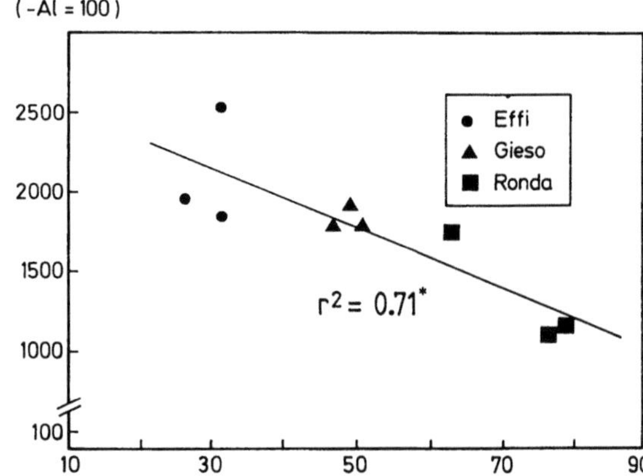

Fig. 10. Relationship between relative callose concentrations in the 30-mm root tips of three soybean genotypes and relative root elongation rates. Al supply 50 μM for 8 h (without Al = 100).

Discussion

The results using the spectrofluorometric method for callose quantification introduced by Köhle et al. (1985) confirmed observations first found by fluorescence microscopy: callose formation is a symptom of Mn and Al toxicity in plant tissues (Wissemeier and Horst, 1987; Wissemeier et al.,

1987). Schaeffer and Walton (1990) showed that Al can induce callose formation in protoplasts.

In Mn-toxicity studies, the spectrofluorometric measurement of callose clearly improved accuracy of determination compared with microscopic methods. This was also true when an image analyser was used for microscopic quantification (Wissemeier and Horst, 1992). For Al-induced callose deposits in roots the method was sensitive enough to document effects of 50 μM Al at least as early as after 1-h exposure time (Fig. 7). However, it must be stressed that the spectrofluorometric method only represents a relative measurement since the fluorescence efficiency depends on the degree of callose polymerisation. The source of (1,3)-β-glucan used for calibration can change fluorescence efficiency by a factor larger than 100 (Kauss, 1989).

Mn toxicity

Quantification of brown speckles in screenings for Mn tolerance in cowpea was well correlated with the reduction in dry matter production of intact plants at high Mn supply (Horst, 1983; Wissemeier and Horst, 1991). Therefore, the results presented strongly suggest that Mn-induced callose formation correlated well with Mn tolerance of whole plants in cowpea. Callose concentration and density of brown speckles were equally useful as indices of tissue injury by Mn toxicity. However, callose formation could be of substantial advantage if typical Mn-toxicity symptoms are absent as occurs in tissue-culture systems. Then, at least in theory, it should be possible to use callose formation even for single-cell selection for Mn-tolerant germplasm under a fluorescent microscope. Problems following such an approach, however, might be (i) that the specific fluorochrome of callose, Siroflour, is at least an inhibitor of the (1,3)-β-glucan synthase (Morrow and Lucas, 1986) and (ii) that the excitation wavelength of 400 nm possibly injures cells during the selection process under a fluorescent microscope.

Practical problems with 'single cell selection' are illustrated by the fact that even control cells cultured at 'optimum' conditions show a high level of callose. High levels of background callose were in particular pronounced in newly established cowpea cultures compared to well established soybean cells (Figs. 5 and 6). Improved culture conditions may overcome this problem. On the other hand, not all cells treated with a high Mn supply were callosed at harvest, not even when the Mn stress was severe enough to greatly restrict growth. In this respect it should be noted that cultured soybean and cowpea cells produced less callose than the floating leaf pieces of cowpea and linseed genotypes, although the cultured cells were exposed to a much higher Mn supply and, hence, had much higher Mn concentrations (data not shown). This can well be explained with the physiological age of undifferentiated, immature cells. In intact plants tissue maturation is an important factor enhancing Mn sensitivity (Horst, 1988).

The time study in cultured cells of soybean indicates that Mn excess induces callose formation, not within hours but within days (Fig. 5). The slow response could be explained by direct stimulation of the plasma membrane (1,3)-β-glucan synthase by Mn (Morrow and Lucas, 1986) from the cytoplasmic side. Normally, cytosolic Mn concentrations are considered to be low (Clarkson, 1988). Since Mn tolerance of tissue is not primarily related to Mn concentrations in the tissue (Horst, 1988) it is possible to conclude that genotypical differences in callose formation at high Mn supply reflect genotypic differences in Mn compartmentation in the cell walls or vacuoles.

Al toxicity

In contrast to Mn excess, Al stimulates callose formation as early as after 1 h supply (Fig. 7). Together with changes in K efflux (Horst et al., 1991), callose formation is one of the earliest physiological reactions to Al, as fast as induction of callose synthesis by fungal elicitors (Kauss, 1985). Work from Kauss and his group indicate that interference of elicitors with the plasma membrane leading to higher levels of free Ca in the cytoplasm triggers callose synthesis (Kauss, 1987). Recent results suggest that elevated Ca concentrations in the cytoplasm are essential, but are not the only signal for callose synthesis (Kauss et al., 1991). Whether Al-induced callose formation is mediated by increased concentrations of cytoplasmic free Ca is not known. There are suggestions that alteration of membrane architecture (Jacob and Northcote, 1985) or changes in the membrane-bound enzyme complex (Delmer, 1987) can also lead to callose formation.

Callose formation induced by Al is confined preferentially to the very root tip (Fig. 7). It is generally agreed that the root tip is the site where Al initially affected plant growth by inhibition of cell division (Clarkson, 1965), cell elongation (Horst and Klotz, 1990) or possibly signal perception and transduction (Bennet and Breen, 1991). Al-induced callose formation is restricted to root-cap cells and the outermost cortical root cells (Wissemeier et al., 1987). It is not known, whether inhibition of root growth and callose formation are causally related. However, Eklund and Eliason (1990) suggested that callose itself might be a factor in preventing 'wall-loosening processes' and, hence, cell wall extension.

The presented results suggest that callose formation in primary roots of soybean seedlings owing to Al supply reflects genotypical differences in Al tolerance. Work by Schaeffer and Walton (1990) on callose formation of protoplast from different Al-sensitive genotypes of oats, barley and wheat is largely in line with the results presented here. Oat and wheat protoplasts from Al-sensitive genotypes had higher rates of callose formation than protoplasts from Al-tolerant genotypes. Only in barley, no clear differences in rate of callose formation existed between protoplasts of Al-tolerant and Al-sensitive genotypes. The results encourage further studies on callose formation as a sensitive biochemical marker of Al sensitivity of plant genotypes.

References

Bennet R J and Breen C M 1991 The aluminium signal: New dimensions to mechanisms of aluminium tolerance. Plant and Soil 134, 153–166.

Clarkson D T 1965 The effect of aluminium and some other trivalent metal cations on cell division in the root apcies of *Allium cepa*. Ann. Bot. 29, 309–315.

Clarkson D T 1988 The uptake and translocation of manganese by plant roots. *In* Manganese in Soils and Plants. Eds. R D Graham, R J Hannam and N C Uren. pp 101–111. Kluwer Academic Publishers, Dordrecht, The Netherlands.

Delmer D P 1987 Cellulose biosynthesis. Annu. Rev. Plant Physiol. 38, 259–290.

DeMarco D G and Randall P J 1988 Tolerance of linseed cultivars to manganese toxicity. *In* International Symposium on Manganese in Soils and Plants: Contributed papers. Eds. M J Webb, R O Nable, R D Graham and R J Hannam. pp 117–118. Manganese Symposium 1988 Inc., Adelaide.

Eklund L and Eliasson L 1990 Effects of calcium ion concentration on cell wall synthesis. J. Exp. Bot. 41, 863–867.

Foy C D, Chaney R L and White M C 1978 The physiology of metal toxicity in plants. Annu. Rev. Plant Physiol. 29, 511–566.

Foy C D, Scott B J and Fisher J A 1988 Genetics and breeding of plants tolerant to manganese toxicity. *In* Manganese in Soils and Plants. Eds. R D Graham, R J Hannam and N C Uren. pp 293–307. Kluwer Academic Publishers, Dordrecht, The Netherlands.

Galway M E and McCully M E 1987 The time course of the induction of callose in wounded pea roots. Protoplasma 139, 77–91.

Gamborg O L and Wetter L R 1975 Plant tissue culture methods. National Research Council of Canada, Prairie Regional Laboratory, Saskatoon.

Haug A 1984 Molecular aspects of aluminium toxicity. CRC Critical Rev. Plant Sci. 1, 345–373.

Hanson W D and Kamprath E J 1979 Selection for aluminium tolerance in soybeans based on seedling-root growth. Agron. J. 71, 581–586.

Horst W J 1982 Quick screening of cowpea genotypes for manganese tolerance during vegetative and reproductive growth. Z. Pflanzenernaehr. Bodenkd. 145, 423–435.

Horst W J 1983 Factors responsible for genotypic manganese tolerance in cowpea (*Vigna unguiculata*). Plant and Soil 72, 213–218.

Horst W J 1988 The physiology of manganese toxicity. *In* Manganese in Soils and Plants. Eds. R D Graham, R J Hannam and N C Uren. pp 175–188. Kluwer Academic Publishers, Dordrecht, The Netherlands.

Horst W J and Klotz F 1990 Screening soybean for aluminium tolerance and adaptation to acid soils. *In* Genetic Aspects of Plant Mineral Nutrition. Eds. N El Bassam, M Dambroth and B C Loughman. pp 355–360. Kluwer Academic Publishers, Dordrecht, The Netherlands.

Horst W J, Asher C J, Cakmak I, Szulkiewicz P and Wissemeier A H 1991 Short-term responses of soybean roots to aluminium. *In* Utilization of Acid Soils for Crop Production. Eds. R J Wright, V C Baligar and R P Murrman. pp 733–739. Kluwer Academic Publishers, Dordrecht, The Netherlands.

Huang C and Graham R D 1990 Resistance of wheat genotypes to boron toxicity is expressed at the cellular level. Plant and Soil 126, 295–300.

Jacob S R and Northcote D H 1985 *In vitro* glucan synthesis by membranes of celery petioles: The role of the membrane in determining the type of linkage formed. J. Cell Sci. Suppl. 2, 1–11.

Kauss H 1985 Callose biosynthesis as a Ca^{2+}-regulated process and possible relations to the induction of other metabolic changes. J. Cell Sci. Suppl. 2, 89–103.

Kauss H 1987 Some aspects of calcium-dependent regulation in plant metabolism. Annu. Rev. Plant Physiol. 38, 47–72.

Kauss H 1989 Fluorometric measurement of callose and other 1,3-β-glucans. *In* Modern Methods of Plant Analysis, New Series Vol. 10, Plant Fibers. Eds. H F Linskens and J F Jackson. pp 127–137. Springer-Verlag, Berlin.

Kauss H, Waldmann T, Jeblick W and Takemoto J Y 1991 The phytotoxin syringomycin elicits Ca^{2+}-dependent callose synthesis in suspension-cultured cells of *Catharanthus roseus*. Physiol. Plant. 81, 134–138.

Köhle H, Jeblick W, Poten F, Blaschek W and Kauss H 1985 Chitosan-elicited callose synthesis in soybean cells as a Ca^{2+}-dependent process. Plant Physiol. 77, 544–551.

Meins Jr F 1983 Heritable variation in plant cell culture. Annu. Rev. Plant Physiol. 34, 327–346.

Morrow D L and Lucas W J 1986 (1→3)-β-D-glucan synthase from sugar beet. I. Isolation and solubilization. Plant Physiol. 81, 171–176.

Murashige T and Skoog F J 1962 A revised medium for rapid growth and bio assays with tobacco tissue cultures. Physiol. Plant. 15, 473–497.

Schaeffer H J and Walton J B 1990 Aluminum ions induce oat protoplasts to produce an extracellular (1→3)-β-D-glucan. Plant Physiol. 94, 13–19.

Wissemeier A H, Klotz F and Horst W J 1987 Aluminium-induced callose synthesis in roots of soybean (*Glycine max*. L.). J. Plant Physiol. 129, 487–492.

Wissemeier A H and Horst W J 1987 Callose deposition in leaves of cowpea (*Vigna unguiculata* (L.) Walp.) as a sensitive response to high Mn supply. Plant and Soil 102, 283–286.

Wissemeier A H and Horst W J 1991 Simplified methods for screening cowpea cultivars for manganese leaf-tissue tolerance. Crop Sci. 31, 435–439.

Wissemeier A H and Horst W J 1992 Effect of light intensity on manganese toxicity symptoms and callose formation in cowpea (*Vigna unguiculata* (L.) Walp.). Plant and Soil 143, 299–309.

P. J. Randall et al. (Eds.), *Genetic aspects of plant mineral nutrition*, 91–96.

PLSO SV20

Interspecific differences in aluminium tolerance in relation to root cation-exchange capacity

F.P.C. BLAMEY, N.J. ROBINSON and C.J. ASHER
Department of Agriculture, The University of Queensland, Brisbane Qld 4072, Australia

Key words: aluminium genotype, root cation-exchange capacity

Abstract

Genotypic differences in aluminium (Al) tolerance hold considerable promise in overcoming an important limitation to plant growth in acid soils. Little is known, however, about the biochemical basis of such differences. Extracellular properties, particularly low root cation-exchange capacity (CEC), have been associated with Al tolerance, since roots of low CEC adsorb less Al than do those of high CEC. A solution culture study was conducted in which 12 plant species (monocots and dicots) were grown in solution culture of low ionic strength (ca 2 mM) for 8 d at four Al concentrations (0, 16, 28 and 55 μM). The species differed significantly in Al tolerance as shown by differences in root length. Root length relative to that of the same species grown in the absence of Al varied from 6 to 117% at 16 μM Al, and from 6 to 75% at 28 μM Al. Species tolerance of Al was not closely associated with differences in root CEC. Although in some species Al sensitivity was associated with high adsorption of Al during a 10 - or 40-min exposure to Al (expressed on a fresh mass or root length basis), this was not a good predictor of Al tolerance across all species studied.

Introduction

Soil acidity has long been recognised as an important limitation to crop and pasture production, affecting tropical, sub-tropical and temperate agriculture (Adams, 1984). Acid soil infertility is a complex problem, and the reasons for poor plant growth are not always clear. However, toxic concentrations of aluminium (Al) in the soil are considered the primary constraint to plant growth in many situations. Two approaches have been proposed to overcome Al toxicity, viz. (i) liming, and (ii) breeding or selecting crop and pasture species tolerant of Al in solution. Liming is an expensive operation, and many countries have limited lime deposits. Genotypic differences in sensitivity to acid soil problems have been reported in many crop and pasture species (Foy, 1984). The magnitude of these differences holds particular promise in overcoming acid soil problems.

To overcome the difficulties related to the complexity of acid soil problems and the complex chemistry of soil Al, many studies on the toxic effects of Al have been conducted in nutrient solution culture. However, many solution culture studies have used nutrient concentrations far in excess of those found in the soil solution. The Al concentration (e.g. 222 to 1000 μM) has also often exceeded that commonly found in soil solutions of acid, Al-toxic soils (Edmeades et al., 1985).

A number of hypotheses have been put forward to explain the differences in Al tolerance among species and among cultivars within species. These may involve detoxification of Al (i) in the rhizosphere, (ii) in the cell wall, and (iii) in the cytoplasm. In support of the hypothesis that the site of Al tolerance is in the cell wall, Foy et al. (1967) found that Al-tolerant species had roots with low cation-exchange capacity (CEC). Recently, Blamey et al. (1990)

found that root CEC was the only factor measured which differed between two *Lotus* species that differed markedly in Al tolerance. Root CEC is related to the properties of pectic substances in the cell wall (Haynes, 1980; Knight et al., 1961). The pectic substances are complex macromolecules, consisting predominantly of α 1,4-linked-D-galacturonic acid residues (McNeil et al., 1984) that may be extensively branched (Fry, 1986). Also, the poly-galacturonic acid may be methylated to varying degrees, and it is suggested that differences in methylation account for differences in CEC. Horst et al. (1982) presented evidence of the protective capacity of the mucilage, the major components of which are pectic substances, that accumulates at the root tip.

On the basis of the above evidence, it was hypothesised that differential Al tolerance results from differences in cell wall properties. In particular, it was proposed that the differences arise from differences in root CEC and the properties of the associated pectic substances.

Materials and methods

Twelve plant species were chosen on the basis of differences in Al tolerance as follows: soybean (*Glycine max* cv. Bethel); mungbean (*Vigna radiata* line 30067); lupin (*Lupinus angustifolius* cv. Unicrop); cotton (*Gossypium hirsutum* cv. Sicot 3); sorghum (*Sorghum bicolor* line var 'H'); peanut (*Arachis hypogaea* cv. McCubbin); triticale (*Triticum* × *Secale* cv. Currency); maize (*Zea mays* cv. XL94); French bean (*Phaseolus vulgaris* cv. Covey); squash (*Cucurbita* spp. cv. Early White Bush); wheat cvv. Egret, Carazinho). Seeds of each species were germinated on paper towel moistened with tap water. When the roots were ca. 30 mm in length, eight seedlings were transferred to 20-L aerated solution cultures. The young seedlings were supported by extending their roots through shade cloth which replaced the bottoms of paper cups, which were in turn supported in the lids of the 20-L pots. Black polyethylene beads were placed in the paper cups for further support, and to exclude light from the solutions. Concentrations of nutrients in the solution were as follows (μM): 600 N (550 as NO_3 and 50 as NH_4); 515 Cl; 500 Ca; 302 K; 252 S; 100 Mg; 5 B; 5 Fe; 2 P; 1 Zn; 1 Mn; 0.2 Cu). The ionic strength of the final solution was ca. 2 mM.

Prior to transplanting the seedlings, aliquots of $Al_2(SO_4)_3$ were added to result in four Al concentrations, viz. 0 (Al_0), 20 (Al_1), 40 (Al_2) and 80 (Al_3) μM Al. Actual concentrations of monomeric Al in solution were determined at the start of the experiment and 8 d after transplanting the seedlings (i.e. at harvest) using a modification of the pyrocatechol-violet method of Dougan and Wilson (1974). Means (±SD) at the end of the experiment were as follows: 0.0 ± 0.0; 15.9 ± 0.4; 27.8 ± 1.1; and 55.4 ± 2.6 μM Al. There were no differences among species in the Al concentration in solution at this time. Following addition of Al, solution pH was adjusted to ca. 4.8, the initial values being 4.8, 4.7, 4.6 and 4.6 at Al_0, Al_1, Al_2, and Al_3, respectively. At the end of the experiment, solution pH ranged from 4.40 to 4.67. Solution pH declined during the experimental period, averaging 4.5 at the end of the experiment. At this time, solution pH was highest in the Al_0 treatment (pH 4.50) compared with solutions to which Al had been added (pH 4.45). Also, there were small (<pH 0.1) but inconsistent differences among species grown at the four Al levels. Treatments were arranged in a 12×4 factorial, randomised complete block design with three replications.

At the end of the experiment (8 d after seedling transfer), roots from each pot were harvested, and placed in 1:3 ethanol:acetic acid solution and stored at 4°C. Prior to measurement of root length, the solution was replaced with deionised water. Root length was determined using an image analyser, and root length per plant calculated.

Additional plants were grown in nutrient solution of the same composition as above (without added Al) for the determination of root CEC using two methods. The first method used a modification of the method of Crooke (1964) (Blamey et al., 1990), root CEC being expressed on a dry-mass (DM) basis. The second method used was that of Rengel and Robinson (1989), root CEC being expressed on a fresh-mass (FM) basis. Additional plants were grown for the production of roots to determine Al sorption of roots as described by Blamey et al. (1990). Approximately 5 g root FM was blotted dry and immersed in 150 mL nutrient solution of the same composition as above and containing 20 μM Al. The concentration of Al in solution was determined after 10 and 40 min, and the quantity of Al sorbed by the roots determined on a FM- and root-length basis. Unfortunately, in these studies the germination of peanut and lupin seed was unsatisfactory, and there was insufficient root material for determination or CEC or Al sorption.

Results and discussion

Within a few days of transferring the seedlings to nutrient solution, it was evident that the appearance of those in solutions containing added Al differed from that of those at Al_0. The roots of plants in solutions containing Al were shorter, thicker and darker than those grown in the absence of Al. Lateral roots were either shorter or absent. Additionally, top growth was visibly poorer in the presence of Al, many plants developing symptoms of chlorosis, necrosis or curling of the leaves.

Species differed in the Al level at which these symptoms became evident, and it was possible to rank species on this basis. Species noticeably affected at Al_1 included cotton, maize, wheat cv. Egret and sorghum. In contrast, wheat cv. Carazinho and triticale only differed from those at Al_0 at the highest level of Al (i.e. 55 μM Al). All other species were affected at Al_2. Lupin, however, appeared intermediate in tolerance, initially showing no symptoms at Al_2, but symptoms had developed in this treatment by the end of the experiment.

In the absence of Al, there were significant differences among species in root length which ranged from 110 to 909 cm plant^{-1} (Table 1). There was a significant reduction in root length when plants were grown in solution containing Al, root length decreasing with increase in Al in solution. On average, root length at Al_1, Al_2 and Al_3 was only 49, 33 and 21% of that in the absence of added Al.

The effect of Al on root length differed among species (Fig. 1). Peanut and wheat cv. Carazinho were the only species whose root length was not reduced significantly with 16 μM Al in solution. Indeed, the root length of peanut was increased slightly, the effect not being significant. Soybean, triticale, lupin and mungbean appeared somewhat tolerant at Al_1, with French bean and maize slightly less tolerant. In the remaining species, sorghum, cotton, wheat cv. Egret and

Table 1. Root length of 12 plant species grown for 8 d in the absence of Al, root cation-exchange capacity (CEC) (determined using two methods) and Al remaining in solution after 10 and 40 minutes of immersion of excised roots of 10 of these species.

Species	Root length (cm/plant)	Root CEC		Al remaining in soln (μM) after	
		cmol$^+$ kg^{-1} DM[a]	mmol$^+$ kg^{-1} FM[b]	10 min	40 min
Soybean	391	14.6	7.79	18.9	14.6
Mungbean	110	17.0	6.99	17.5	15.3
Lupin	112	–	–	–	–
Cotton	202	9.04	3.86	14.7	10.4
Sorghum	246	6.45	5.78	10.4	9.4
Peanut	269	–	–	–	–
Triticale	254	1.10	5.67	14.9	11.5
Maize	909	1.03	4.25	16.4	14.1
French bean	747	8.99	5.59	12.0	11.6
Squash	537	9.46	5.95	14.3	11.9
Wheat cv. Egret	291	2.17	6.41	11.7	8.5
Wheat cv. Carazinho	273	2.02	4.27	7.1	3.3
LSD ($p = 0.05$)	74	13.4	2.10	3.3	4.3

[a]Crooke (1964).
[b]Rengel and Robinson (1989).

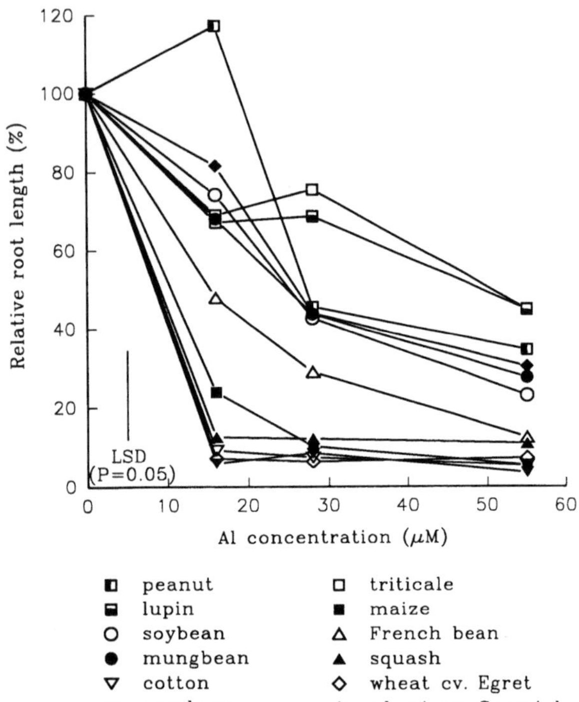

Fig. 1. Effects of monomeric aluminium in solution on relative root length of 12 plant species grown in solution culture.

squash, there was a marked reduction in relative root length at Al_1. Root length of these species was only 6 to 13% of that at Al_0. The differences among species in tolerance of Al was still evident at Al_2, although there were some differences in ranking compared with that at Al_1. For example, triticale and lupin were the most tolerant at Al_2, peanut and wheat cv. Carazinho being similar to soybean and mungbean in Al tolerance. At Al_3, the effects of Al became more severe, and the differences among species less evident. However, even at this concentration (55 μM Al), root length varied from 4 to 45% of that in the absence of Al.

The responses of these species in many instances are similar to those reported in other studies. Adams and Pearson (1970), for example, found that peanut is considerably more tolerant of Al than is cotton. The tolerance of wheat cv. Carazinho has been known for some time, while wheat cv. Egret is known to be sensitive to Al. Also, sorghum has been found to be more sensitive than maize to Al (Brenes and Pearson, 1973), and it has been suggested that

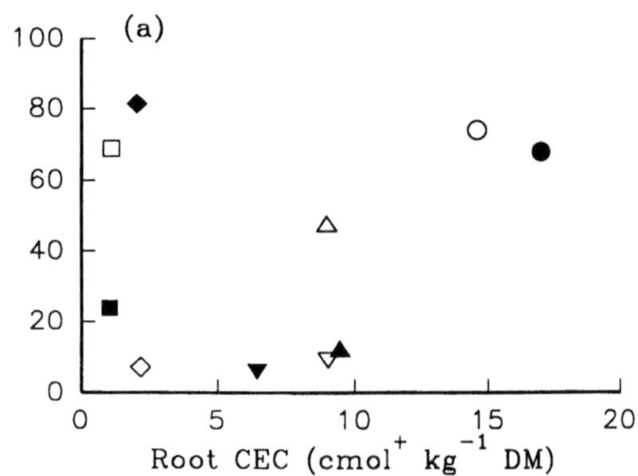

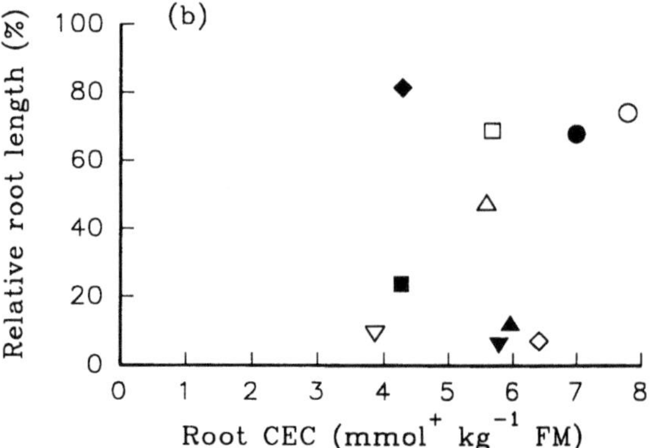

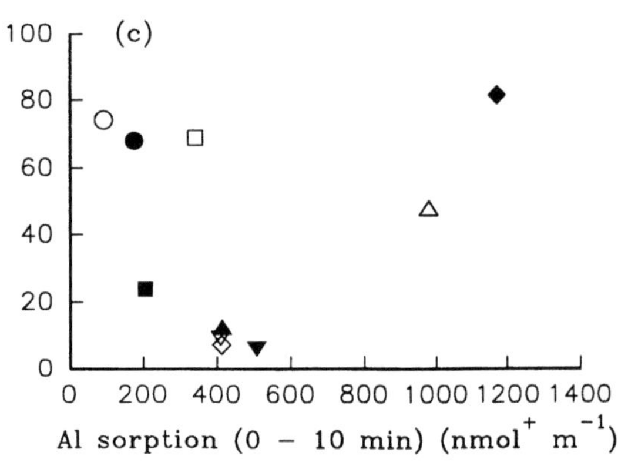

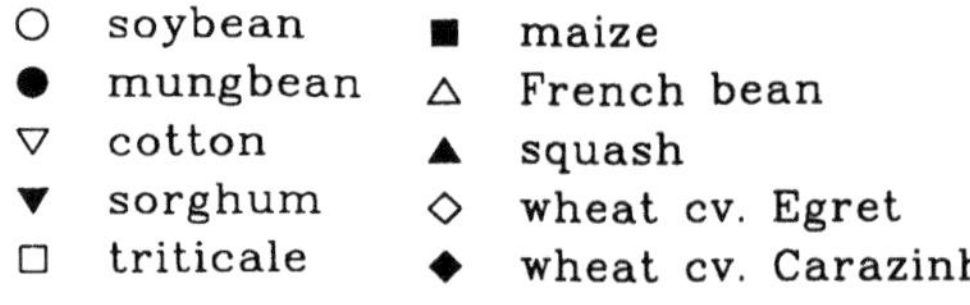

Fig. 2. Associations between relative root length of 10 plant species and root cation-exchange capacity (CEC) using the method of Crooke (1964) (**a**) and Rengel and Robinson (1989) (**b**) and aluminium sorption from solution after 10 minutes' immersion of excised roots (**c**).

maize is more sensitive than is soybean (MPW Farina, personal communication). However, reasons for differences in tolerance have not yet been elucidated.

In a number of studies (Blamey et al., 1990; Foy et al., 1967; Horst et al., 1982; Vose and Randall, 1962), species with low root CEC have been found to be tolerant of Al in the rhizosphere. In the present study, both methods of determination of root CEC showed marked species differences, but there was not a good agreement between the two methods. Also, with neither method was there a close relationship between Al tolerance at 16 μM Al and root CEC (Fig. 2a, b). A similar result was evident at higher Al concentrations (data not shown).

In the present study, there was a rapid decline in Al concentration on immersion of roots in nutrient solutions containing Al (Table 1). Indeed, between 4 and 90% of the initial 20 μM Al disappeared within 10 min. This supports the suggestion that there is an initial, rapid phase of Al adsorption important in Al toxicity (Zhang and Taylor, 1990). In the study of Blamey et al. (1990), roots of an Al-tolerant species of *Lotus* (*L. pedunculatus* cv. Grasslands Maku) adsorbed less Al than did those of an Al-sensitive species (*L. corniculatus* cv. Maitland). While Al tolerance of some species was associated with differences in Al adsorption during a 10- or 40-min exposure to Al (expressed on a root length basis) short-term Al sorption was not a good predictor of Al tolerance across all species studied (Fig. 2c). (Similar results were evident when Al adsorption was calculated on a fresh mass basis.) In particular, French bean (moderate Al tolerance) and wheat cv Carazinho (high Al tolerance) adsorbed large quantities of Al per unit root length. Difficulties were encountered in this study with the measurement of root length, in that no account was taken of very fine roots or of root hairs which would have increased considerably the surface area for adsorption. In this regard, wheat cv. Carazinho had very fine roots with a profusion of root hairs.

Conclusions

With as little as 16 μM Al in solution, root length of sorghum, cotton, wheat cv. Egret and squash was <15% of that in the absence of Al. In contrast, root growth of peanut and wheat cv. Carazinho was little affected by this concentration of Al. Species differences in root length were evident with 28 and 55 μM Al in solution, but differences among species were not as marked, particularly at 55 μM Al (Fig. 1).

Results of this study did not support the hypothesis that plants with roots of low CEC are Al-tolerant, in spite of considerable evidence to support this contention (Blamey et al., 1990; Foy et al., 1967; Horst et al., 1982; Vose and Randall, 1962). Also, the removal of Al from solution by excised roots did not prove a good index of Al tolerance across all species studied. Differences in Al sorption by the root tip and regions remote from the tip may have played a role in masking species differences. However, the marked and rapid reduction in Al in solution on immersion of the roots suggests that there is an important first stage of Al sorption associated with Al toxicity, probably followed by more complex reactions.

Acknowledgements

The authors would like to thank Mr D G Appleton, Ms J I Mercer and Mr G L Walters for technical assistance.

References

Adams F 1984 Soil Acidity and Liming. 2nd Ed. Agronomy No. 12, Am. Soc. Agron. Madison, WI.

Adams F and Pearson R W 1970 Differential response of cotton and peanuts to subsoil acidity. Agron. J. 62, 9–12.

Blamey F P C, Edmeades D C and Wheeler D M 1990 Role of root cation-exchange capacity in differential aluminium tolerance of *Lotus* species. J. Plant Nutr. 13, 729–744.

Brenes E and Pearson R W 1973 Root responses of three gramineae species to soil acidity in an oxisol and an ultisol. Soil Sci. 116, 295–302.

Crooke W M 1964 The measurement of the cation-exchange capacity of plant roots. Plant and Soil 21, 43–49.

Dougan W K and Wilson A L 1974 The absorptiometric determination of aluminium in water: A comparison of some chromogenic reagents and the development of an improved method. Analyst 99, 413–440.

Edmeades D C, Wheeler D M and Clinton O E 1985 The chemical composition and ionic strength of soil solutions from New Zealand topsoils. Aust. J. Soil Res. 23, 151–165.

Foy C D 1984 Physiological effects of hydrogen, aluminum, and manganese toxicities in acid soil. *In* Soil Acidity and Liming. 2nd Ed. Agronomy No. 12. Ed. F Adams. pp 57–97. Am. Soc. Agron., Madison, WI.

Foy C D, Fleming A L, Burns G E and Armiger W H 1967 Characterization of differential aluminum tolerance among varieites of wheat and barley. Soil Sci. Soc. Am. Proc. 31, 513–521.

Fry S C 1986 Cross-linking of matrix polymers in the growing cell walls of angiosperms. Annu. Rev. Plant Physiol. 37, 165–186.

Haynes R J 1980 Ion exchange properties of roots and ionic interactions within the root apoplasm: Their role in ion accumulation by plants. Bot. Rev. 46, 75–99.

Horst W J, Wagner A and Marschner H 1982 Mucilage protects root meristems from aluminium injury. Z. Pflanzenphysiol. 105, 435–444.

Knight A H, Crooke W M and Inkson R H E 1961 Cation-exchange capacities of tissues of higher and lower plants and their related uronic acid contents. Nature, Lond. 192, 142–143.

McNeil M, Darvill A G, Fry S C and Albersheim P 1984 Structure and function of the primary cell walls of plants. Annu. Rev. Biochem. 53, 625–663.

Rengel Z and Robinson D L 1989 Determination of cation exchange capacity of ryegrass roots by summing exchangeable cations. Plant and Soil 116, 217–222.

Taylor G J and Foy C D 1985 Mechanisms of aluminum tolerance in *Triticum aestivum* L. (wheat). I. Differential pH induced by winter cultivars in nutrient solutions. Am. J. Bot. 72, 695–701.

Vose P B and Randall P J 1962 Resistance to aluminum and manganese toxicities in plants related to variety and cation exchange capacity. Nature, Lond. 196, 85–86.

Zhang G and Taylor G J 1989 Kinetics of aluminum uptake by excised roots of aluminum-tolerant and aluminum-sensitive cultivars of *Triticum aestivum* L. Plant Physiol. 91, 1094–1099.

P. J. Randall et al. (Eds.), Genetic aspects of plant mineral nutrition, 97–101.

PLSO SV22

Preliminary results from a microscopic examination on the effects of aluminium on the root tips of wheat

D.M. WHEELER, D.J.C. WILD[1] and D.C. EDMEADES
Ministry of Agricultural and Fisheries, Ruakura Agricultural Centre, Private Bag, Hamilton, New Zealand and [1]Waikato Microscope Unit, MIRENZ, PO Box 617, Hamilton, New Zealand

Key words: aluminium, electron microscope, light microscope, *Triticum aestivum*, wheat

Abstract

Root tips from aluminium (Al) tolerant (Waalt) and Al sensitive (Warigal) wheat (*Triticum aestivum* (L). Thell.) cultivars exposed to low concentrations of Al (10 μM) for 10, 24 and 72 hours were examined under the light and electron microscope. After fixing and embedding, longitudinal and transverse thin and ultrathin sections were cut. There was no evidence of Al damage to the root tips of the Al tolerant cultivar under both the light and electron microscope. For the Al sensitive cultivar, Al had no observable effect on the root tips 10 hours after Al addition when examined under the light microscope. When examined under an electron microscope, electron dense globular deposits were observed between the cell wall and cell membrane of the epidermal cells. There was no obvious damage to the cell cytoplasm. Two or 3 days after Al addition, light microscopy showed that the cells in the root tips had become swollen and extensively vacuolated. The tissues appeared disorganised and degenerate, particularly in the epidermis and outer cortical cells. The electron microscope also revealed a thickening of the cell wall. The cell wall was broken down, particularly in the epidermis in the region 4–6 mm from the root tip. The tissue in the meristematic area was largely intact.

Introduction

It may be possible in the near future to modify the tolerance of plants to soil acidity, and in particular aluminium (Al) toxicity, using the techniques of molecular biology. A program is in progress aimed at identifying gene(s) conferring Al tolerance in wheat (*Triticum aestivum* (L.) Thell.). An understanding of the mechanism(s) of Al tolerance or toxicity could assist in identifying these gene(s). The microscopic examination of the effects of Al on the structure of cells in wheat roots, and in particular the root tips, could help to elucidate the mechanism(s).

The examination of the effects of Al on the morphology of root tips using electron microscopy has been reported for maize (Bennet et al., 1984; Wagatsuma et al., 1987), Norway spruce (Hecht-Buchholz et al., 1987; Hodson and Wilkins, 1991), rice, oats and pea (Wagatsuma et al., 1987) and wheat (Puthota et al., 1991). In all species examined, Al had a detrimental effect on the structure of the root tip, although the effect depended on the species and cultivar examined.

This paper reports the preliminary results of a microscopic examination of wheat root tips 10, 24 and 72 hours after exposure to Al in a low ionic strength nutrient solution.

Methods

In three separate experiments (72, 24 or 10 hours exposure to Al), seeds of 2 wheat cultivars (Waalt (Al-tolerant) and Warigal (Al-sensitive)) were germinated on moistened filter paper. Three seedlings of each cultivar were transferred into separate gauze bottomed pots and the pots suspended over 500 mL of a low ionic strength basal nutrient solution containing 10 μM Al. The composition of the nutrient solution is described by Edmeades et al. (1991). The nominal Al concentration corresponds to an Al activity in solution of 2.2 μM as calculated by Geochem (Sposito and Mattigod, 1980) using the modifications of Parker et al. (1987).

Root tips (6–10 mm sections) were removed and placed in fixative (2.5% glutaraldehyde in 0.01 M sodium cacodylate buffer at pH 7.2) after 10, 24 or 72 hours exposure to solution Al. The root tips were further cut into smaller sections and the sections washed twice with sodium cacodylate buffer, then post-fixed with 1% osmium tetraoxide for 30 minutes, washed with distilled water and enbloc stained in saturated uranyl acetate. The samples were dehydrated with a graded series of acetone and embedded in Spurr's resin.

Longitudinal and transverse sections for light microscopy were cut with glass knifes and stained with 1% Toluidine Blue. Sections for electron microscopy were cut with a diamond knife from root tips exposed to Al for 10 and 72 hours only and stained with uranyl acetate and lead citrate.

Additional sections from the root tips of the sensitive cultivar exposed to solution Al for 10 hours were prepared as above, mounted on gold grids and then immersed in 1% periodic acid for 30 minutes, washed 3 times in distilled water and left in 0.2% thiocarbohydrazide (TCH) in 20% glacial acetic acid for 20 hours. The grids were then transferred through a series of acetic acid solutions (15%, 10%, 5% and 2%) with 20 minutes in each solution, then immersed in 1% silver proteinate in the dark for 30 minutes, washed briefly in distilled water and dried on filter paper. Control grids were put through the same procedure except for immersion in periodic acid.

Results and discussion

For the Al tolerant cultivar (Waalt), there was no evidence of damage to the cells from the root tip after exposure to solution Al for 10, 24 or 72 hours when examined under the light or electron microscope.

For the Al sensitive cultivar (Warigal), 10 hours exposure to solution Al had no observable effects on the structure of cells in the root tip when examined under the light microscope (Fig. 1a). When examined under the electron microscope, solution Al generally had no observable effects on the cell cytoplasm although there was increased vacuolation in some epidermal cells of the root tip (Fig. 1b). At higher magnification, the presence of electron dense globular deposits between the cell wall and cell membrane of the epidermal cells, and in the endoplasmic reticulum of the epidermal cells could be clearly seen (Fig. 1c). These electron dense globular deposits were not observed between the cell wall and cell membrane, nor in the endoplasmic reticulum of the cortical or meristematic cells. Also, they were not observed in the Al sensitive cultivar (Warigal) in the absence of Al, nor in the Al tolerant cultivar (Waalt) in the presence or absence of Al.

After 24 hours exposure to Al, visual observations of the cells in the root tips of the Al sensitive cultivar under the light microscope showed that they were swollen and extensively vacuolated. The cell walls of some of the epidermal cells were disintegrating.

By 72 hours, all the cells in the root tip of the sensitive cultivar exposed to Al were swollen and extensively vacuolated (Fig. 2, left). The epidermal and outer cortical cells 4–6 mm from the root cap were largely disintegrated. The cells in the meristematic region and in the root cap were intact although swollen and vacuolated. When the meristematic region was examined under the electron microscope, the cell walls were thickened and the cytoplasm disorganised (Fig. 2, right) although the nucleus appeared to be intact. The electron dense globular deposits observed between the cell wall and cell membrane in the epidermal cells from roots exposed to Al for 10 hours were generally not found in the meristimatic region of roots exposed to Al for 72 hours. However, there were some electron dense globular deposits in the vacuoles of roots exposed to Al for 72 hours. No observations were made of the epidermal cells of roots exposed to

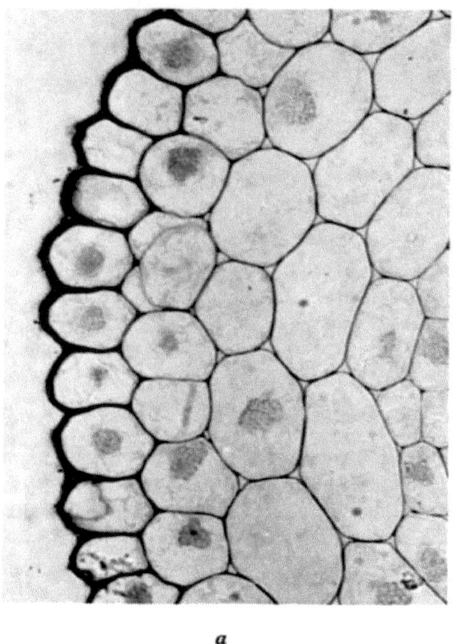

a

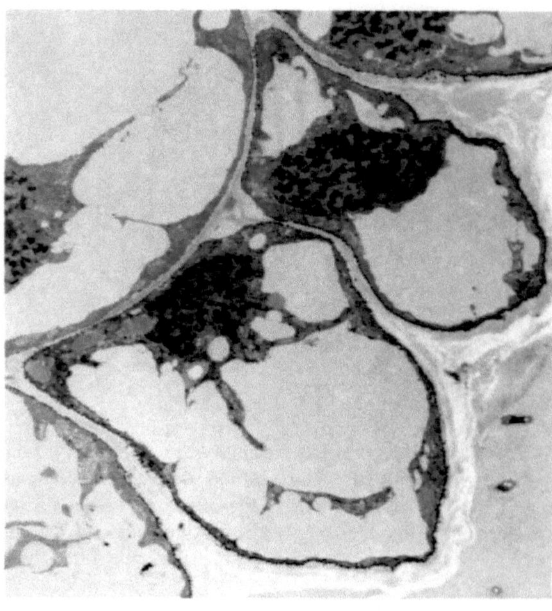

b

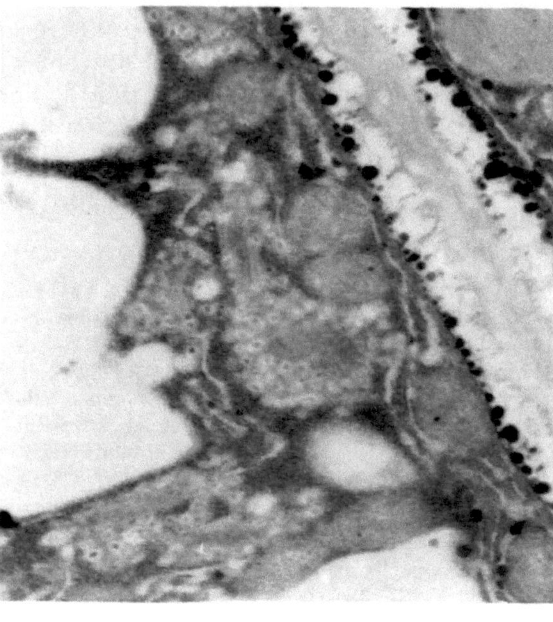

c

Fig. 1. Micrographs of a transverse ultra-thin section taken through the meristematic region from the root tip of an aluminium sensitive wheat cultivar (Warigal) exposed to 10 μM solution aluminium for 10 hours. **a** Light micrograph of the epidermal cells (magnification 420×). **b** Electron micrograph of epidermal cells (magnification 2, 150×). Note the electron dense globular deposits between the cell wall and plasma membrane. **c** Electron micrograph of the cell wall of the epidermal cells (magnification 16, 500×). Note the electron dense globular deposits between the cell wall and cell membrane and in the endoplasmic reticulum.

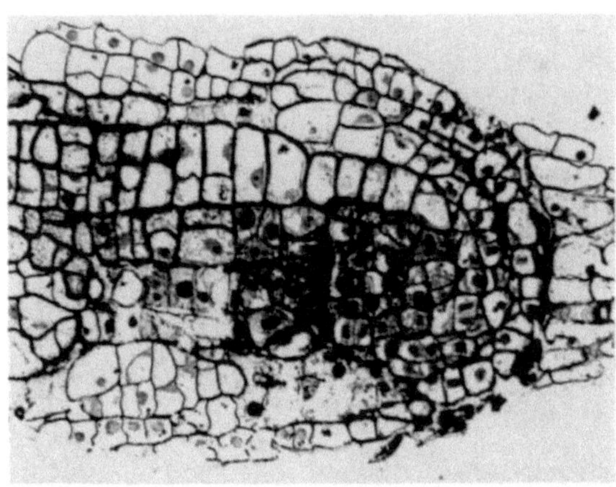

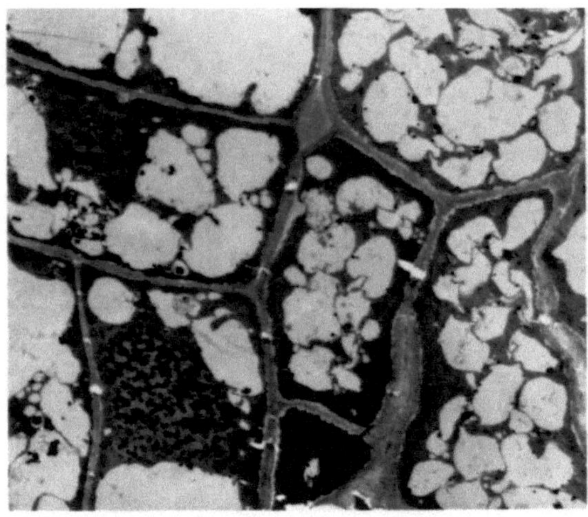

Fig. 2. Micrographs of a longitudinal ultra-thin section taken through the meristematic region from the root tip of an aluminium sensitive wheat cultivar (Warigal) exposed to 10 μM solution aluminium for 72 hours. **Left**: Light micrograph of the root tip (magnification 170×). Note that the cells are swollen and extensively vacuolated. **Right**: Electron micrograph of the meristematic cells (magnification 1650×). Note the thickening of the cell walls.

Al or 72 hours due to the extensive disintegration of these cells.

The electron dense globular deposits seen in the sensitive cultivar in the presence of Al were not stained by silver proteinate.

The occurrence of the electron dense globular deposits between the cell wall and cell membrane found in the Al sensitive cultivar in the presence of Al have not been previous reported. The composition of these globular deposits is not known although the lack of staining by silver proteinate indicates that they were not carbohydrate deposits. However, the fact that they only occurred in the Al sensitive cultivar (Warigal) in the presence of Al suggests that these deposits may contain Al. It is noted that these deposits occurred in the same region of the cell in which Hodson and Wilkens (1991) and Hult et al. (1992) reported the accumulation of Al in roots of Norway spruce (*Picea abies*) and beech (*Fagus sylvatica*) respectively.

If these electron dense globular deposits do in fact contain Al, then the presence of these deposits in the endoplasmic reticulum could also indicate that the sensitive cultivar is capable of secreting some Al from within the cell. Also, the lack of these deposits in the Al tolerant cultivar in the presence of Al could suggest that the Al tolerant cultivar can exclude Al from the cells.

These results at present do not allow the elucidation of the mechanism of Al toxicity in wheat. Further work is continuing to identify the distribution of Al within the root tips of both the Al tolerant and Al sensitive cultivars, to determine the change in root structure in more detail between 10 and 72 hours and to determine the earliest time that the effects of Al are observed.

References

Bennet R J, Breen C M and Fey M V 1984 The primary site of aluminium injury in the root of *Zea mays* L. S. Afr. J. Plant and Soil 2, 8–17.

Edmeades D C, Wheeler D M and Christie R A 1991 The effect of aluminum and pH on the growth of a range of temperate grass species and cultivars. *In* Plant-Soil Interactions at Low pH. Eds. R J Wright, V C Baligar and R P Murrmann. pp 913–924. Kluwer Academic Publishers, Dordrecht, The Netherlands.

Hecht-Buchholz C, Jorns C A and Keil P 1987 Effect of excess aluminium and manganese on Norway spruce seedlings as related to magnesium nutrition. J. Plant Nutr. 10, 9–16.

Hodson M J and Wilkins D A 1991 Localization of aluminium in the roots of Norway Spruce (*Picea abies* (L.) Karst) inoculated with *Paxillus involutus* Fr. New Phytol. 118, 273–278.

Hult M, Bengtsson B, Larsson N P O and Yuang C 1992 PIXE-microanalysis of root samples from Beech (*Fagus Sylvatica*). Scanning Microsc. *submitted.*

Parker D R, Zelazny L W and Kinraide T B 1987 Improvements to the program GEOCHEM. Soil Sci. Soc. Am. J. 51, 488–491.

Puthota V, Cruz-Ortega R, Johnson J and Ownby J 1991 An ultrastructural study of the inhibition on mucilage secretion in the wheat root cap in aluminium. *In* Plant-Soil Interactions at Low pH. Eds. R J Wright, V C Baligar and R P Murrmann. pp 779–787. Kluwer Academic Publishers, Dordrecht, The Netherlands.

Sposito G and Mattigod S V 1980 GEOCHEM: A computer program for the calculation of chemical equilibria in soil solution and other natural water systems. Kearney Foundation of Soil Science, University of California, Riverside.

Wagatsuma T, Kaneko M and Hayasaka Y 1987 Destruction process of plant root cells by aluminium. Soil Sci. Plant Nutr. 33, 161–175.

P. J. Randall et al. (Eds.), *Genetic aspects of plant mineral nutrition*, 103–116.

PLSO SV12

Aluminium toxicity: Towards an understanding of how plant roots react to the physical environment

R. J. BENNET and C. M. BREEN
Grasslands Research Centre, Private Bag X9059, Pietermaritzburg, 3200, Republic of South Africa and Institute of Natural Resources, P.O. Box 375, Pietermaritzburg, 3200, Republic of South Africa

Key words: aluminium, cap secretions, roots, stimulus-response systems, tolerance

Abstract

The root cap has an established but poorly understood role in determining the roots' response to the physical environment (including Al). Data from studies of Al toxicity are combined with more general aspects of root physiology to develop understanding of the basis for Al toxicity as well as the regulatory systems involved in mediating a transmitted response between Al-damaged cap cells and the responding cells of the cap and root.

Pivotal roles are identified for the physiology of the cap mucilages in the stimulus transduction pathway and links are indicated between the secretory activity of peripheral cap cells and primary events involved in signal perception.

This information is used to identify possible mechanisms of Al tolerance notably involving connections between root cation-exchange capacity and the chemistry of cap mucilage and root recovery from Al reflecting signal inactivation and biochemical adjustments directed at alleviating the stress condition.

New research prospects for extending our perception of Al tolerance are discussed.

Introduction

The presence of aluminium (Al) in many soils is an agronomic factor of considerable relevance (Sanchez, 1990). The phytotoxic effects of Al, first noted by Hartwell and Pember (1918) have been extensively researched and the results of these investigations are the subject of numerous modern reviews (Bennet and Breen, 1991a; Foy, 1988; Haug, 1984; Roy et al., 1988; Schaedle et al., 1989; Taylor, 1988). It is therefore, important that the mechanism(s) controlling Al tolerance, observed in some plants, remain largely unresolved. An understanding of how tolerance mechanisms operate and identification of the genes encoding them is required to improve the selectivity of programmes directed at increasing the Al tolerance of crop plants.

Contemporary views of Al tolerance have been concerned principally with the notion that the mitotically active cells of the root (and cap) are primary targets for Al (Clarkson, 1965; Matsumato et al., 1976; Sampson et al., 1965) and that tolerance must in consequence reflect mechanisms whereby Al is either excluded from these cells or is detoxified prior to reaching them (see Foy et al., 1978). Positive growth responses arising from some Al treatments (Aniol, 1984; Bennet et al., 1987) and reports dealing with the recuperative capacity of roots following Al treatment (Bennet and Breen, 1991b) contradict this idea and suggest instead that the functions of meristematic cells are not permanently impaired by Al. Implicit in these proposals is the thesis that the growth reactions (Clarkson, 1965; Horst et al., 1983; Morimura et al., 1978) of plant roots to Al arise from the transduction of a signal between the cap and root (Bennet and Breen,

1991a). The considerably expanded perception of the complexity of the plants' reaction to Al intrinsic to these proposals is discussed in the context of identifying adaptive strategies directed at enabling some plants to cope with the stress condition.

This paper does not include a detailed discussion of evidence leading to the application of stimulus-response systems to Al toxicity (see Bennet and Breen, 1991a). Stimulus-response reactions have, however, been widely used to explain the plants' reaction to the physical environment (see Bentrup, 1979; Shropshire, 1979). Very little is known of the mechanisms which underlie these reactions to external stimuli. Our objectives in dealing with some of the more general aspects of stimulus-response systems in this review are therefore to discover 1) whether the considerable body of literature can be applied to Al studies and 2) whether Al can be used to probe the broader aspects of how plants react to a much wider range of external stimuli.

We also briefly consider that changes initiated in the root affect not only the root, but also indirectly the shoot. This implies that stimuli perceived by the root can lead to a reorientation of shoot growth, suggesting that "root signals" are integrated into much higher levels of functional organisation within the plant.

For the discussion which follows it is important to distinguish between the cell populations comprising the root apex. Figure 1 defines the terminology used.

The signal pathway

Since it is suggested (Bennet and Breen, 1991a) that the site where the Al signal is perceived is spatially removed from the responding cells of the cap and root it follows that events connected to signal perception and transduction may form the basis for mechanisms of Al tolerance. It is therefore necessary to discover something of what constitutes a stimulus-response system and how the development of interacting cell populations is affected by the primary event or signal.

Characterisation of stimulus-response systems

Characterisation of stimulus-modulated reactions by plants is less than satisfactory. The problem is exacerbated by an absence (in plants) of a clearly defined sensory network and growth measurements remain the only criterion for determining that signal perception has occurred. Ideas relating to early events in the stimulus-response chain are therefore derived from indirect correlations and observations and the unambiguous identification of cytoplasmic structures (or pathways) involved in signal perception and transduction remains to be achieved. It is therefore, important that events which define stimulus-response systems can be analysed in three indispensable steps viz.

signal perception→stimulus transduction→ physiological reaction (usually a growth response) .

Criteria associated with the reaction to external stimuli will also predictably include 1) the initial event or signal arises from an altered state in the environment (e.g. gravity, light, toxic chemicals, anoxia, mechanical impedance etc.), 2) signal perception depends on the existence of a receptor which interacts with the energy of the signal, 3) it follows from (2) that the site of signal perception will be located at the plant/ signal interface, 4) Ca^{2+} is ubiquitous in linking signal perception to signal transduction, 5) signal transduction involves the gating or redirection of existing physiological pathway(s), so as to bring about the required reaction (growth) and 6) most signals are intended to be transient events and mechanisms must exist for inactivating the signal once its message has been delivered.

The root cap

For many years it was believed that the root cap functioned solely to protect the sub-terminal apical meristem of the root (Haberlandt, 1914). More recently the functions of the cap have been found to be considerably more diverse (see Feldman, 1984) and it is crucial to the present discussion that all external stimuli affecting root development (including Al), studied thus far, are perceived by the root cap (Audus, 1979; Bennet et al., 1985a; Juniper, 1976; Kays et al., 1974, Pilet and Ney, 1978).

In considering the caps' involvement in coordinating and organising the roots' response to external stimuli it is necessary to recall that caps are present on the roots of all plants (Barlow, 1975) and analogous tissue regions occur nowhere else in the plant (Esau, 1965). It is therefore, a reasonable expectation that events involved in signal perception and transduction will reside in the more conspicuous activities of the root cap. Features of the cap which relate to the present discussion are 1) in contrast to shoot apices the root produces cells not only towards the axial structure, but also away from it (Figure 1), for in so doing it initiates the cap. A complex spatial distribution of the rates of cell growth and cell division is necessary because the cells are generated in two opposing directions (Clowes, 1972), 2) in actively growing roots the number of cap cells remains nearly constant and mature cells are continually lost from the cap periphery (Barlow, 1975) to compensate for the new cells generated by the cap meristem, and 3) in most plants (see Mollenhauer, 1967) the outermost or peripheral cap cells are characterised by quantities of mucilaginous, polysaccharide materials which eventually pass to the cell exterior (apoplasm) as cap mucilages (Mollenhauer and Morré, 1980; Northcote and Picket-Heaps, 1966). These conspicuous secretory activities immediately precede the abscission (Barlow, 1975) of mature cells from the cap.

It is these activities which largely determine the organisation of the root apex. A critical point emerges from the finding that the equilibria between these events in the cap can be upset to produce visible changes in cell cycles within a few minutes, by forcing cells of the cap and root out of their normal cycle of DNA synthesis and mitosis (Clowes, 1972).

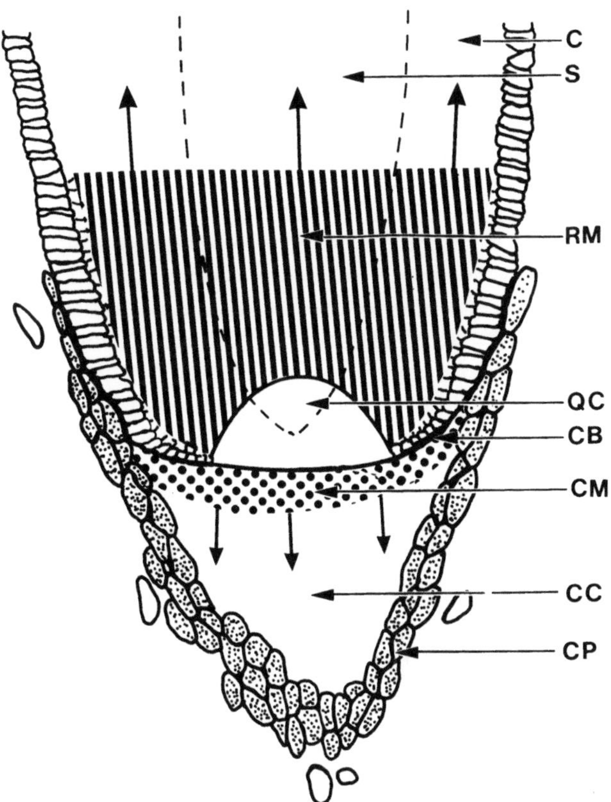

Fig. 1. Diagram of the principal tissue regions of the root of *Zea mays*, indicating the location of the root meristem (RM), cap meristem (CM) and quiescent centre (QC). The root meristem functions as the source of cells for the cortex (C) and stele (S) comprising the axial structure of the root, by adding new cells in the direction of the unlabelled arrows. The cap meristem contributes cells to the root cap in the direction indicated (unlabelled arrows). The central cap cells (CC) are rich in amyloplasts which are believed to be gravity sensors. Peripheral cap (PC) cells secrete mucilagenous, polysaccharide materials to the cell exterior. Mucilages have an indicated role in the stimulus transduction pathway. Mature cap cells are constantly lost from the cap periphery so that the number of cap cells remains nearly constant in actively growing roots. The roots of the grasses are characterised by a distinctive boundary between the cap and distal regions of the root (CB).

Aluminium and cap mucilages

Key observations concern the effects of Al on cap size and construction (Bennet et al., 1987; Bennet and Breen 1991b), where Al-induced reductions in secretory activity of peripheral cap cells, which are not themselves mitotically active (Barlow, 1975), coincided with alterations in root growth rates and interference in mechanisms controlling the number of cap cells. Moreover, experiments which follow the recovery of plant roots from different Al treatments (Bennet and Breen, 1991b) demonstrate that cap repair (size and construction) preceded by some days a return of root elongation rates. The plants (*Zea mays*) used in these experiments are characterised by a "closed" meristem so that the cap is structurally and developmentally distinct from the root (Clowes, 1981) and it was deduced from the distribution of cells in the root apex, that the new cells required for cap repair were derived from reactivation of the cap meristem. While this observation is consistent with the notion that the functions of mitotically active cells are not permanently affected by Al it also serves to indicate that the root meristem (and axial growth) is subject to regulatory controls that do not directly affect the cap. The early importance attached to cap repair during recovery additionally alerts us to the possibility that these regulatory controls may be derived from processes normally resident in the cap.

Aluminium treatment is widely associated with diminished secretory activity (Bennet et al., 1985b; 1987; Hecht-Buchholz and Foy, 1981; Johnson and Bennet, 1991; Puthota et al., 1991) and a point of fundamental importance emerges from the finding (Bennet and Breen, 1991b) that a resumption of root elongation rates during recovery from Al was concomitant with the reappearance in peripheral cap cells of a morphologically distinctive, granular secretory product. Moreover differences in root elongation rates derived from the severity of the different Al treatments coincided with the amount of secretory activity present.

Collectively these observations suggest that the secretory activity of peripheral cap cells is somehow allied to the regulatory activities that determine root growth rates. The information exchange between interacting cell populations is thought to depend on the transport of chemical signals (Wilkins, 1979).

Cap inhibitor complex

In a number of experiments (see Feldman, 1984) caps have been surgically removed from the root to demonstrate a role for the cap in root development. Excision of the cap results in a temporary increase in root elongation rates (Gibbons and Wilkins, 1970; Pilet, 1972). It is argued that removal of the cap removes a source of growth inhibition allowing the root to elongate more rapidly. Low doses of Al produce a similar effect (Bennet et al., 1987). Results of this sort support the idea that the cap functions as the source of an endogenous inhibitor of root growth (Bennet et al., 1987; Feldman, 1981; Gibbons and Wilkins, 1970; Pilet, 1972) and lead to the proposal (Bennet et al., 1987) that the expression of Al toxicity may be directed through the synthesis of this material.

Why then are only some Al treatments stimulatory (Bennet et al., 1987)? A crucial point emerges from the finding that while suboptimal levels of cap activity increase root growth rates, minimum levels of cap activity are required to initiate (or maintain) growth responses (Bennet and Breen, 1991b). This suggests that the cap inhibitor may determine growth rates not only through inhibitory interactions in the root, but that it may also in some way be necessary to maintain the activity of the root meristem. It is pertinent to this view that the increase in root growth rates following cap excision are of short duration (Pilet, 1972).

A temporary loss of gravisensitivity is coincident with the loss of the cap (Barlow, 1974a). Removal of the cap also results in a change in the activities of tissues directly basal to the cap (Barlow, 1974b; Barlow and Grundwag, 1974; Barlow and Sargent, 1978; Bennet et al., 1985c; Feldman, 1976), ultimately leading to the formation of a new cap. Many cytoplasmic features normally characteristic of distinct locations in the undisturbed cap now occur within the cells of the regenerating root apex (Barlow and Grundwag, 1974; Barlow and Sargent, 1978; Bennet et al., 1985c) and may account for the return of gravisensitivity prior to the formation of a new cap (Barlow, 1974b).

These observations lead to the suggestion (Barlow, 1975) that the positional arrangement of cells within the cap (or regenerating apex) determine their function. The initiation of secretory activity in the exposed, outermost cells of the regenerating apex is an early and conspicuous response to cap removal (Barlow and Hines, 1982; Barlow and Sargent, 1978; Bennet et al., 1985c) and may precede the development of amyloplasts (Bennet et al., 1985c) in cells which previously constituted the quiescent center (QC). Surgically bisecting the cap also initiates secretory activity along the cut edge of the cap irrespective of the age of the cells (Ellmore, 1982) so that the outermost cells of the cap (or regenerating apex) located at the root/signal interface appear to "sense" their position and begin secreting mucilage.

Evidence of this nature strengthens the belief that the activities of the peripheral cap cells determine root growth rates and is consistent with requirements outlined in earlier discussion for the transduction pathway. The biphasic response of the root to the cap inhibitor may also help explain the contrasting interpretations of the action and identity of this material (see Jackson and Barlow, 1981; Moore and Evans, 1986). We should also note that the effects of Al on root growth are nearly instantaneous (Clarkson, 1965; Clymo, 1962) indicating that the regulatory messages associated with cap secretory activity can pass between the cap and root with considerable rapidity.

A second cap message

An anomaly exists with respect to the absence of cap regeneration sequences following Al treatment (Bennet and Breen, 1991b) since the action of Al on cap function is in some respects analogous to cap removal. In this connection Barlow (1987) has intimated that caps may be able to "sense" harmful conditions and initiate their own regeneration and it has been established that low temperatures (Barlow and Rathfelder, 1985) lead to cap replacement. Moreover the QC of Al-treated roots is generally resistant to the effects of the treatment (Bennet and Breen, 1991b) so that cap replacement remains possible in Al-treated roots.

The capacity of roots to resume primary growth after Al treatment (Bennet and Breen, 1991b) may be important in identifying an arrested growth phase which may function as an alternative to the cap replacement phenomenon. Under field conditions the intensity of many harmful factors that affect the root (including Al) fluctuate with time. The roots' ability to recommence primary growth, may therefore, be an adaptive strategy enabling the root to survive unfavourable circumstances until improved conditions permit a resumption of active growth. The arrested growth phase also provides the opportunity for root cells to make biochemical and/or structural adjustments directed at alleviating the stress condition.

The differences in the roots' response to cap excision (which deprives the root of all cap activities) and Al treatment are important since the absence of the expected cap replacement sequences implies that these activities are suppressed in Al-treated roots. This observation presupposes the existence of a second cap message which may be coincident with a change in the appearance of the secretory product external to the plasmalemma of peripheral cap cells (Bennet and Breen, 1991b).

The calcium connection

Considerable relevance attaches to experiments involving the Ca^{2+} regulatory materials EGTA and La^{3+} (Bennet et al., 1990; Bennet and Breen, 1991c) which lead to the proposal that changes in Ca^{2+} homeostasis may determine the secretory activity of peripheral cap cells. Pertinent to this proposal is the finding that a lateral redistribution of Ca^{2+} precedes the bending reaction of gravireactive roots (Lee et al., 1983) and it is suggested that Ca^{2+} stimulated differences in cap secretory activity may account for the anisotropic growth occurring in gravireactive roots (Bennet et al., 1990).

It is widely accepted that Ca^{2+} functions ubiquitously in plant and animal sensory systems as a "second messenger" linking primary events involved with signal perception to the transduction phase (see Hepler and Wayne, 1985). The mechanism is believed to depend on exploiting electrochemical gradients of Ca^{2+} across cell membranes (Hepler and Wayne, 1985; Huggins and England, 1985), so that voltage dependent Ca^{2+} ion channels (Hedrich and Schroeder, 1989) function to control cytosolic Ca^{2+} concentrations through changes in membrane polarity.

Signal perception

The receptor for the Al signal has yet to be identified. Our current model for Al toxicity is summarised in Figure 2. The combined data of studies involving 1) Al uptake (Bennet et al., 1985a), 2) morphology of cap cells (Bennet et al., 1985b; 1987), 3) similarities in the response

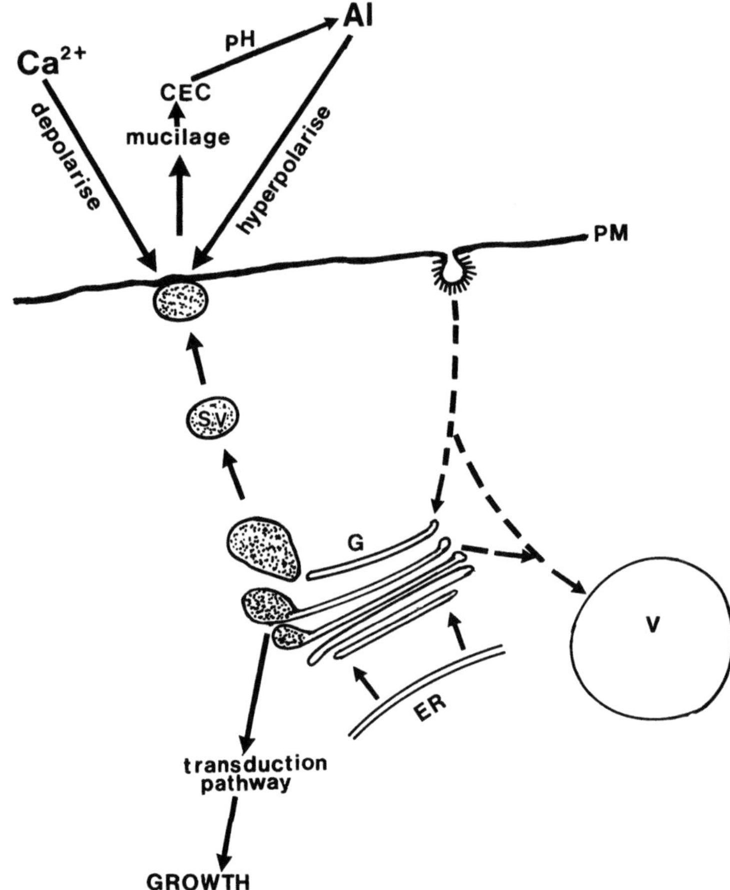

Fig. 2. Diagrammatic interpretation of the basis for Al toxicity. Secretory activity in peripheral cap cells involves the movement of secretory vesicles (SV) from dictyosomes of the Golgi apparatus (G) where they fuse with the plasmalemma (PM) before passing mucilages to extracellular locations in the apoplasm. This exocytotic activity is also linked via the signal transduction pathway to root growth. Calcium homeostasis determines the level of secretory activity and an influx of Ca^{2+} depends on membrane depolarisation. Properties of mucilage (pH), which may be reflected in root CEC measurements determine Al uptake. Al competes with Ca^{2+} and changes in membrane polarity (hyperpolarisation) initiate alterations in secretory activity and consequent growth responses. A possible endocytotic pathway is indicated by the dotted arrows. Endocytosis is an important source of membrane material recycled through the Golgi apparatus-complex. Links must therefore exist to co-ordinate endocytosis with Ca^{2+} stimulated secretions (exocytosis). Membrane material is also derived from the endoplasmic reticulum (ER). Vacuole (V).

of cap cells to Al (Bennet et al., 1985b; 1987) and Ca^{2+} deprivation (Bennet et al., 1990) have led us to propose that the primary reaction to Al arises from competition between Al and Ca^{2+} for sites external to the plasmalemma of peripheral cap cells (Bennet and Breen, 1991a).

Calcium homeostasis determines the secretory activity of peripheral cap cells and an influx of extracellular Ca^{2+} into the cytosol depends on membrane depolarization (Rincon and Hanson, 1986). An Al-induced increase in membrane polarity (Kinraide, 1988) may effectively prevent this and changes in secretory activity are reflected through altered root growth rates. There are many unresolved questions concerning the central, co-ordinating activities of the Golgi apparatus in determining the reaction to Al (Bennet et al., 1985b). The movement of secretory vesicles from the Golgi apparatus to extracellular destinations shows an early response and vesicles in Al-treated cells appear to loose their ability to recognise their targets on the plasmalemma

(Bennet et al., 1985b). Golgi apparatus secretions are normally directional (Morré, 1977) and polarised so that the secretory product accumulates preferentially along the outer cell walls (Barlow and Sargent, 1978).

These observations indicate that a change in membrane polarity may be an early link in the chain of reactions to external stimuli.

A re-evaluation of Al tolerance

The most striking feature of the preceding discussion is the complexity of the mechanisms which underlie the plants' reaction to Al. We are only beginning to understand the significance of these systems which must account for the coordinated response of cells (and their development), many of which are not themselves directly affected by Al (see Bennet and Breen, 1991a).

The natural environment of roots is never uniform and roots must accommodate wide fluctuations in conditions if they are to survive. Homeostasis and stability are nevertheless remarkable features of roots and root systems. In these apparently conflicting requirements of stability and adaptability reside the mechanisms which compensate for the disruptive changes associated with stress and which prevent the disintegration of organisation that characterises the root. Stress has been ubiquitous in the development of plants. Plants are seldom exposed to individual stresses but are more commonly subjected simultaneously to multiple environmental factors. The effects of these different stresses on the plant system are interactive. Notable similarities also exist in the structural and functional reactions by the root to widely different stresses and it has been suggested (Bennet and Breen, 1989; 1991a) that these reactions are directed through a common regulatory mechanism. Relevant questions arise regarding the specificity of tolerance mechanisms so that we must be aware of which mechanisms are directly concerned with Al tolerance and which protect the plant against Al as a consequence of a more general adaptation to stress. We need to be constantly alert to the possibility that adaptation to one stress will carry with it an adaptation to another (Levitt, 1990).

Root CEC and Al tolerance

A major change in perceptions of Al exclusion mechanisms is indicated by the suggestion (Bennet and Breen, 1991a) that the primary lesion to Al involves extracellular targets. This means that Al need not cross a living membrane to initiate a phytotoxic response and that exclusion mechanisms of tolerance will in consequence function at the level of the apoplasm. Cap mucilages are negatively charged (Ray et al., 1988) and contribute to the exchange capacity (Jenny, 1966) of the root. Numerous reports have consistently suggested a relationship between cation-exchange capacity (CEC) and Al tolerance (Blamey et al., 1990; Büscher et al., 1990; Foy et al., 1967; Kennedy et al., 1986; Vose and Randall, 1962), although the actual mechanism(s) remain elusive.

Aluminium uptake

It is suggested (Wacquant, 1977) that the roots' selectivity for polyvalent cations increases with increasing CEC, so that plants with high root CEC's will be expected to preferentially accumulate Al. Intraspecific (Lotus) correlations between root Al and Al tolerance accord with this view (Blamey et al., 1990) and it is necessary to consider the extent to which the exchange laws can account for Al exclusion. It is instructive that dicotyledenous plants, with correspondingly high root CEC's (Haynes, 1980) appear to dominate the list of plants rated as highly Al tolerant (Sanchez and Salinas, 1981) since this observation may indicate that factors additional to physical equilibria determine Al uptake.

Plants differ widely in root CEC (Haynes, 1980) and interest surrounds the finding (Knight et al., 1961) that these differences in CEC are largely accounted for by the uronic acid component of the mucilage. The physiological significance of uronic acid is presently unclear. Aluminium ion speciation is, however, pH dependent (Wright, 1989) and it is the products of hydrolysis reactions which are most frequently implicated in Al toxicity (Foy, 1988; Taylor, 1988). Uronic acid is the major acidic component of cap mucilage (Ray et al., 1988) and the increasingly acidic pH's associated with higher levels of uronic acid could conceivably explain the proposals that 1) properties of mucilage control Al entry (Bennet and Breen, 1991a), 2) provide conditions consistent with maintaining Al solubility in the apoplast of the peripheral cap zone (Bennet and Breen, 1991a) and 3) that mucilage may determine Al ion speciation in the apoplasm and hence rhizotoxicity against conditions existing in the external milieu (Kinraide, 1990).

It is an exciting possibility that the biochemistry of a fundamental physiological pathway involved in stimulus transduction may determine Al uptake and rhizotoxicity. Root CEC is a reproducible measurement (Haynes, 1980) and with a better understanding of the links indicated, root CEC may emerge as a practical means of characterising properties of cap mucilage associated with Al tolerance.

The proposal that Al ion speciation is determined by the properties of cap mucilage argue strongly against the suggestion (Foy et al., 1978; Foy, 1988) that plant-induced pH changes in the rhizosphere are a primary defence against Al. In these cases it is believed that an increase in rhizosphere pH reflects preferential anion uptake (OH^- efflux) in the presence of Al (Haynes, 1990). Conclusions derived from Kinraide's (1990) study clearly require that these pH changes extend beyond the rhizosphere into the apoplasm of the peripheral cap region. There is presently no evidence to suggest that this happens. Literature derived from animal cell studies (Scharschmidt and van Dyke, 1987) imply that vesicle acidification (decrease in mucilage pH) is a more likely consequence of membrane hyperpolarization associated with Al treatments.

Different views on the links between root CEC and Al tolerance are taken by Büscher and colleagues (Büscher et al., 1990) who suggest that root CEC may be a natural ecological factor selecting indirectly for Al tolerance through a primary requirement for Ca^{2+} and Blamey et al., (1990) who believe that Al tolerance reflects properties of cap mucilage involved in the sequestration and detoxification of Al.

The production of cap mucilages

Quantitative effects involving differences in mucilage production between Al-tolerant and Al sensitive plants are reported from several sources (Johnson and Bennet, 1991; Puthota et al., 1991). Mucilage is a major component of root CEC (Oades, 1978); direct correlations between increased mucilage production (increased CEC) and Al tolerance may, however, be elusive for reasons discussed in the preceding section.

Aluminium acts against mucilage production (Bennet et al., 1985b; 1987) and treatment of roots with Al is therefore expected to diminish root CEC. Interest surrounds the data of Vose and Randall (1962) which demonstrates that the reduction in root CEC following Al treatment was proportionately greatest in those lines (Lolium) categorised as being susceptible to Al. The corollary to this finding requires that Al tolerant plants may be better able to maintain normal cap activity in the presence of Al. This suggestion accords with the data of Ownby and co-workers (Puthota et al., 1991) on Al tolerant wheat lines. The fact that Al tolerant plants do show a response to Al (Johnson and Bennet, 1991; Puthota et al., 1991; Vose and Randall, 1962) suggests that Al is not totally excluded from the apoplasm. This may indicate that Al tolerance is derived from differential sensitivities to Al-induced changes in the direction of the transduction pathway. The mechanism(s) involved are presently unknown.

Earlier discussion focused on the suggestion that the effects of Al are directed through the action of inhibitory materials. Does this mean that plants deriving Al tolerance from factors associated with the transduction pathway are naturally slow growing? Consistent with this suggestion are the findings that 1) Al-induced growth stimulation (i.e. release from growth inhibition) was greatest in rice (*Oryza sativa*) cultivars assessed as Al tolerant (Howeler and Cadavid, 1976), and 2) positive correlations which are found to exist between increasing CEC and yield (Crooke and Knight, 1971). Predictably the growth reactions involved with redirection of the transduction pathway (stimulation or inhibition) will depend on the severity of the treatment (Bennet et al., 1987), will be detected soon after the treatment is applied and since the level of stress will remain effectively constant, growth rates will be maintained with time.

Living with aluminium

Most stimuli are designed to be transient events capable of being "switched off" after the signal has been transduced. Mechanisms must therefore exist for inactivating the stimulus after its message has been delivered. Failure to do this will mean that the plant's regulatory systems (and metabolic energies) will be permanently diverted. Aluminium-induced root growth responses directed along these pathways will be characterised by an initial period of growth inhibition followed by a distinct recovery phase (see Hecht-Buchholz, 1983; Randall and Marco, 1990), coinciding with alleviation of the stress or signal.

The endocytotic pathway

It is widely believed (Robinson and Hillmer, 1990) that signalling molecules/receptor complexes are removed from extracellular sites in the apoplast and internalized into the cytosol by endocytosis, involving invagination of (pits) and vesiculation of the plasmalemma, coincident with the recycling of membrane material through the Golgi apparatus complex. Although endocytosis in plants has been controversial, evidence in favour of the endocytotic pathway is now overwhelming (Horn et al., 1989; Hübner et al., 1985; Robinson and Depta, 1988; Robinson and Hillmer, 1990).

Differences in response of root cap cells (Zea) to La^{3+} and Al are important here (Bennet and Breen, 1991b; 1991c). At equivalent levels of ionic charge roots subjected to continuous La^{3+} treatments showed significant levels of recovery of primary growth subsequent to an initial period of growth inhibition. Aluminium treated roots did not recover. This implies that La^{3+} treated roots were able to adjust to the stress condition whereas roots subject to continuous Al treatments were not able to do so.

Hübner et al. (1985) have used the electron opaque properties of La^{3+} to describe the endocytotic pathway in root cap cells (Zea). Differences in the response of these cells to La^{3+} and Al suggest that in the latter case the Al signal remained intact implying either that endocytosis did not occur or that Al and La^{3+} follow different endocytotic routes. Other heavy metals (Pb) are internalized into cell vacuoles (Hübner et al., 1985) as is the case with La^{3+}. Differences in the response of cap cells to Al and La^{3+} and Pb^{2+} suggest that a specific recognition event is involved in initiating endocytosis. The most obvious differences between these metal ions involves atomic mass. Could differences in the sensitivity of the plasmalemma to the recognition event in endocytosis account for Al tolerance in some plants?

The fate of toxic materials internalized by endocytosis may be crucial. Internalized heavy metals are not loosely distributed in the cytosol but are confined to endomembrane components and vacuoles (Hübner et al., 1985; Robinson and Hillmer, 1990). Cap cells are constantly being lost from the cap (Barlow, 1975) so that toxic materials internalized in this way may have only a transient effect on root development before they are returned "encapsulated" to the soil continuum.

Adjustment to the stress condition

Metal binding proteins

Many environmental factors (heat, cold, anoxia: See Barlow, 1987 for references), including Al (Bennet et al., 1985d; Bennet and Breen, 1991b) affect the nucleolar structure of root cells where nucleolar segregation is characteristic of alterations in r RNA synthesis (Barlow, 1987). The effects on nucleolar ultrastructure are rapidly reversible (Barlow, 1987; Bennet and Breen, 1991b,c) and in Al studies (Bennet and Breen, 1991c) appear to be linked directly to the Al stress since reversal of segregation is independent of cap secretory activity. Collectively this evidence argues for the existence of Al-inducible stress proteins and there are reports of a shift in protein synthesis following Al treatment (Delhaize et al., 1991; Foy, 1988; Picton et al., 1991; Rincon and Gonzales, 1991).

Changes in protein synthesis do not always correlate with observable levels of Al tolerance (Delhaize et al., 1991; Rincon and Gonzales, 1991). It is therefore important that two most basic questions on stress-inducible proteins remain unanswered – when and where do they act? Contemporary views (Sachs and Ho, 1986; Steffens, 1990) on the reaction to heavy metals depend on detoxification of excess metal ions by sequestration. This interpretation hardly seems tenable for Al, where the localised nature of Al targets requires that metal binding proteins are transported from meristematic cells into the extracellular apoplasm of the cap periphery. Alternatively the action of stress proteins may be directed at protecting root cells from the effects of the stress. The latter thesis seems more compatible with the diversity of the stresses and the concept of interactive regulatory mechanisms (see Bennet and Breen, 1991a).

We also need to consider that drastic metabolic changes involving a shift in protein synthesis are observed in response to a considerable diversity (heat, cold, anoxia, salt, toxic metals, light, pathogens) of factors (see Sachs and Ho, 1986; Steffens, 1990; van den Bulcke et al., 1989). An understanding of the significance of these changes may depend on establishing whether the altered synthetic activity is a primary response to stress (defence) or whether it reflects disordered metabolism as a consequence of the treatment.

The identification of Al-inducible proteins and the genes encoding them does nevertheless offer exciting possibilities for selectively increasing Al tolerance.

Lateral root initiation

Aluminium-induced changes in root development are limited to alterations in root growth rates and the initiation of lateral roots along the main root axis (Clarkson, 1969). Morphological differences between the caps of primary and lateral roots are associated with a reduction in the gravisensitivity of lateral roots (Moore and Pasieniuk, 1984). Secondary root development, initiated in response to Al may therefore confer a perceptible increase in Al tolerance.

Root signals and shoot development

Aluminium is preferentially retained in the roots of most Al-sensitive plants with little or no movement into the shoot (Foy et al., 1978). This presupposes that Al-induced changes in shoot growth (Pan et al., 1989) depend on a signal passing between the root and shoot. There are also suggestions that roots can 'sense' changes in the moisture status of soils and send signals to the leaves that control their behaviour, overriding any effects of turgor on the leaves (Gowing et al., 1990; Jackson and Kowalewska, 1983; Passioura, 1988; Saab and Sharp, 1989).

In the case of soil moisture it is proposed that changes in root tip turgor affect the synthesis of materials in the root tip, therby influencing the development of the shoot (Davies et al., 1986). The root environment has a marked effect on mucilage production and when water is not available mucilage does not move through the cell wall, remaining instead between the wall and plasmalemma (Oades, 1978). Calcium deprivation during recovery of roots previously treated with Al (Bennet and Breen, 1991c) indicates that 1) recovery is delayed by a lack of Ca^{2+}, 2) reduced root cap activity coincides with a decline in the activity of the shoot apical meristem, 3) a partial resumption of root growth rates (and cap activity), concomitant with the presence of Ca^{2+} in the nutrient solution (probably from leaky membranes) preceded similar changes in the shoot. Differences in the recovery times for the root and shoot meristems indicate that these tissue regions do not compete independently for available Ca^{2+} resources. This accords with the data of Saab and Sharp (1989) who were unable to connect root signals (water stress) to low nutrient supply. Moreover, root excision experiments (Gowing et al., 1990) which are not expected to increase water (or Ca) supply appeared to be associated with "positive" root signals that inhibited shoot growth in drying plants.

It is an interesting possibility that root signals derived from cap secretory activity may be integrated into a much higher level of functional organisation, affecting the development of the whole plant.

Concluding remarks

We have used stimulus-response systems to analyze the plants' reaction to Al. The mechanisms involved have proved to be exceedingly complex. Advances in understanding these regulatory systems is, however, likely to provide worthwhile and unexpected answers to the genetic manipulation of Al tolerance in crop plans. Our present concern is that current procedures screening for Al tolerance on the basis of early seedling root growth rates will not be satisfactory for detecting Al tolerance mechanisms associated with signal inactivation and recovery.

The most widely researched aspect of stimulus-response systems involves the effects of gravity on plant organs. Contemporary views with respect to gravity signal perception (see Audus, 1979) differ from proposals relating to Al. We need to consider that although the complexity of the regulatory systems is beyond dispute, roots can only respond to external signals (irrespective of the source) by altering growth rates and growth direction (apical dominance). These factors together with the notable similarity (see Bennet and Breen, 1989; 1991a) in many of the cellular reactions to external stimuli argue for a common regulatory pathway. If this is so then the challenge lies in determining how plant regulatory systems integrate the effects of widely different signals of fluctuating intensity into the limited range of possible responses by the root.

We have omitted in this review consideration of the identity of plant growth substances involved in the regulatory activities which underlie the roots' reaction to stress. It is presently difficult to relate the information we have, which is sometimes controversial and contradictory. to a system as complex as the root apex. The response to stress of mutant plants deficient in one or more regulatory materials seems to offer an exciting and novel approach to this area of investigation.

Acknowledgements

We would like to thank the Department of Agricultural Development, Natal Region, (N5241/03/1/1 and N5413/03/1/1) and the University of Natal for supporting our investigations.

References

Aniol A 1984 Induction of aluminium tolerance in wheat seedlings by low doses of aluminium in the nutrient solution. Plant Physiol. 75, 551–555.

Audus L J 1979 Plant geosensors. J. Exp. Bot. 30, 1051–1073.

Barlow P W 1974a Regeneration of the cap of primary roots of *Zea mays*. New Phytol. 73, 937–954.

Barlow P W 1974b Recovery of geotropism after removal of the root cap. J. Exp. Bot. 25, 1137–1146.

Barlow P W 1975 The root cap. *In*: The Development and Function of Roots. Eds. J G Torrey and D T Clarkson. Third Cabot Symposium. Ch. 2, pp 21–54. Academic Press, London.

Barlow P W 1987 The cellular organisation of roots and its response to the physical environment. *In* Root Development and Function. Eds. P J Gregory, J V Lake and D A Rose. SEB Seminar Series 30. Cambridge University Press, Great Britain.

Barlow P W and Grundwag M 1974 The development of amyloplasts in cells of quiescent centre of Zea roots in response to removal of the root cap. Z. Pflanzenphysiol. 73, 56–64.

Barlow P W and Hines E R 1982 Regeneration of the root cap of *Zea mays* L. and *Pisum sativum* L.: A study with the scanning electron microscope. Ann. Bot. 49, 521–539.

Barlow P W and Rathfelder E L 1985 Cell division and regeneration in primary root meristems of *Zea mays* recovering from cold treatment. Environ. Exp. Bot. 25, 303–314.

Barlow P W and Sargent J A 1978 The ultrastructure of the regenerating root cap of *Zea mays* L. Ann. Bot. 42, 791–799.

Bennet R J, Breen C M and Fey M V 1985a Aluminium uptake sites in the primary roots of *Zea mays* L. S. Afr. J. Plant Soil 2, 1–7.

Bennet R J, Breen C M and Fey M V 1985b The primary site of aluminium injury in the root of *Zea mays*. S. Afr. J. Plant Soil 2, 8–17.

Bennet R J, Breen C M and Fey M V 1985c Aluminium toxicity and regeneration of the root cap: Preliminary evidence for a Golgi apparatus-derived morphogen in the primary root of *Zea mays*. S. Afr. J. Bot. 51, 363–370.

Bennet R J, Breen C M and Fey M V 1985d Aluminium induced changes in the morphology of the quiescent centre, proximal meristem and growth region of the root of *Zea mays*. S. Afr. J. Bot. 51, 355–362.

Bennet R J, Breen C M and Fey M V 1987 The effects of aluminium on cap function and root development in *Zea mays* L. Environ. Exp. Bot. 27, 91–104.

Bennet R J and Breen C M 1989 Towards understanding root growth responses to environmental signals: The effect of aluminium on maize. S. Afr. J. Sci. 85, 9–12.

Bennet R J, Breen C M and Bandu V H 1990 A role for Ca^{2+} in the cellular differentiation of root cap cells: A re-examination of root growth control mechanisms. Environ. Exp. Bot. 30, 515–523.

Bennet R J and Breen C M 1991a The aluminium signal: New dimensions to mechanisms of aluminium tolerance. Plant and Soil 134, 153–166.

Bennet R J and Breen C M 1991b The recovery of the roots of *Zea mays* L. from various aluminium treatments: Towards elucidating the regulatory processes that underlie root growth control. Environ. Exp. Bot. 31, 153–163.

Bennet R J and Breen C M 1991c Environ. Exp. Bot.
Bentrup F W 1979 Reception and transduction of electrical and mechanical stimuli. *In* Physiology of Movement. Eds. W Haupt and M E Feinleib. Encyclopedia of Plant Physiology. Vol. 7, pp 42–70. Springer-Verlag, Berlin.
Blamey F P C, Edmeades D C and Wheeler D M 1990 Role of root cation exchange capacity in differential aluminium tolerance of Lotus species. J. Plant. Nutr. 13, 729–744.
Büscher P, Koedam N and Van Speybroeck D 1990 Cation-exchange properties and adaptation to soil acidity in Bryophytes. New Phytol. 115, 177–186.
Clarkson D T 1965 The effect of aluminium and some other trivalent metal ions on the cell division in the root of *Allium cepa*. Ann. Bot. 20, 309–315.
Clarkson D T 1969 Metabolic aspects of aluminium toxicity and some possible mechanisms of resistance. *In* Ecological Aspects of the Mineral Nutrition of Plants. Ed. I H Rorison. pp 381–397. British Ecological Symposium No. 9. Blackwell Scientific Publications.
Clowes F A L 1972 The control of cell proliferation within root meristems. *In* The Dynamics of Meristem Cell Populations: Advances in Experimental Medicine and Biology. Eds. M W Miller and C C Kuehnert. Vol. 18, pp 133–145. Plenum Press, New York.
Clowes F A L 1981 The difference between open and closed meristems. Ann. Bot. 48, 761–767.
Clymo R S 1962 An experimental approach to part of the calcicole problem. J. Ecol. 50, 707–731.
Crooke W M and Knight A H 1971 Root cation-exchange capacity and organic acid content of tops as indices of varietal yield. J. Sci. Food Agric. 22, 389–392.
Davies W J, Metcalfe J, Lodge T A and Da Costa A R 1986 Plant growth substances and the regulation of growth under drought. Aust. J. Plant Physiol. 13, 105–125.
Delhaize E, Higgins T H and Randall P J 1991 Aluminium tolerance in wheat: Analyses of polypeptides in the root apices of tolerant and sensitive genotypes. *In* Plant-Soil Interactions at Low pH. Eds. R J Wright, V C Baligar and R P Murrman. pp 1071–1079. Kluwer Academic Publishers, Dordrecht, The Netherlands.
Ellmore G S 1982 The organization and plasticity of plant roots. Scanning Electron Microsc. 111, 1083–1102.
Esau C 1965 Plant Anatomy. Second Edition. Wiley, New York.
Feldman L J 1976 The de novo origin of the quiescent centre in regenerating root apices of *Zea mays*. Planta 128, 207–212.
Feldman L J 1981 Root cap inhibitor formation in isolated root caps of *Zea mays*. J. Exp. Bot. 32, 779–788.
Feldman L J 1984 Regulation of root development. Annu. Rev. Plant Physiol. 35, 223–242.
Foy C D 1988 Plant adaptation to acid, aluminium-toxic soils. Commun. Soil Sci. Plant Anal. 19, 959–987.
Foy C D, Chaney R L and White M C 1978 The physiology of metal toxicity in plants. Annu. Rev. Plant Physiol. 29, 511–566.
Foy C D, Fleming A L, Burns G R and Armiger W H 1967 Characterization of differential aluminium tolerance among varieties of wheat and barley. Agron. J. 31, 513–521.
Gibbons G S B and Wilkins M B 1970 Growth inhibitor production by root caps in relation to geotropic responses. Nature 226, 558–559.
Gowing D J G, Davies W J and Jones H G 1990 A positive root-sourced signal as an indicator of soil drying in apple, *Malus × domestica* Borkh. J. Exp. Bot. 41, 1535–1540.
Haberlandt G 1914 Physiological Plant Anatomy. Translation of 4th German Edition. Macmillan, London.
Hartwell B L and Pember F R 1918 The presence of aluminium as a reason for the difference in the effect of so-called acid soil on barley and rye. Soil Sci. 6, 259–280.
Haug A 1984 Molecular aspects of aluminium toxicity. Crit. Rev. Plant Sci. 1, 345–373.
Haynes R J 1980 Ion exchange properties of roots and ionic interactions within the root apoplasm: Their role in ion accumulation in plants. Bot. Rev. 46, 75–99.
Haynes R J 1990 Active ion uptake and maintenance of cation-anion balance: A critical examination of their role in regulating rhizosphere pH. Plant and Soil 126, 247–264.
Hecht-Buchholz C H 1983 Light and electron microscope investigations of the reactions of various genotypes to nutritional disorders. Plant and Soil 72, 151–165.
Hecht-Buchholz C H and Foy C D 1981 Effect of aluminium toxicity on root morphology of barley. Plant and Soil 63, 93–95.
Hedrich R and Schroeder J I 1989 The physiology of ion channels and electrogenic pumps in higher plants. Annu. Rev. Plant Physiol. 40, 539–569.
Hepler P K and Wayne R O 1985 Calcium and plant development. Annu. Rev. Plant Physiol. 36, 397–439.
Horn M A, Heinstein P F and Low P S 1989. Receptor-mediated endocytosis in plants cells. The Plant Cell 1, 1003–1009.
Horst W J, Wagner A and Marschner H 1983 Effect of aluminium on root growth, cell division rate and mineral element contents of roots of *Vigna unguiculate* genotypes. Z. Pflanzenphysiol. 109, 95–103.
Howeler R H and Cadavid L F 1976 Screening of rice cultivars for tolerance to Al-toxicity in nutrient solutions as compared with a field screening method. Agron. J. 68, 551–555.
Hübner R, Depta H and Robinson D G 1985 Endocytosis in maize root cap cells: Evidence obtained using heavy metal salt solutions. Protoplasma. 129, 214–222.
Huggins J P and England P J 1985 The control of calcium pumps and channels and their role in regulating intracellular calcium. *In* Molecular Mechanisms of Transmembrane Signalling. Eds. P. Cohen and M. Houslay. pp 57–87. Elsevier Science Publishers B.V., Amsterdam.
Jackson M B and Barlow P W 1981 Root geotropism and the role of growth regulators from the cap: a reexamination. Plant Cell and Environment. 4, 107–123.
Jackson M B and Kowalewska A K B 1983 Positive and negative messages from roots induce foliar dessication and stomatal closure in flooded pea plants. J. Exp. Bot. 34, 493–506.
Jenny H 1966 Pathways of ions from soil into root according to diffusion models. Plant and Soil 25, 265–289.
Johnson P A and Bennet R J 1991 Aluminium tolerance of root cap cells. J. Plant Physiol. 137, 760–762.

Juniper B E 1976 Geotropism. Annu. Rev. Plant Physiol. 27, 385–406.

Kays S J, Nicklow C W and Simons D H 1974 Ethylene in relation to the response of roots to the physical impedance. Plant and Soil 40, 565–571.

Kennedy C W, Smith W C and Ba M T 1986 Root cation exchange capacity of cotton cultivars in relation to aluminium toxicity. J Plant Nutr. 9, 1123–1133.

Knight A H, Crooke W M and Inkson R H E 1961. Cation-exchange capacities of tissues of higher and lower plants and their related uronic acid contents. Nature 192, 142–143.

Kinraide T B 1988 Proton extrusion by wheat roots exhibiting severe aluminium toxicity symptoms. Plant Physiol. 88, 418–423.

Kinraide T B 1990 Assessing the rhizotoxicity of the aluminate ion, Al $(OH)_4^-$. Plant Physiol. 94, 1620–1625.

Lee J S, Mulkey T J and Evans M L 1983 Gravity-induced polar transport of calcium across root tips of maize. Plant Physiol. 73, 874–876.

Levitt J 1990 Stress interactions – back to the future. Hort. Sci. 25, 1363–1365.

Matsumoto H, Hirasawa E, Torikai H and Tokohaski E 1976 Localisation of absorbed aluminium in pea root and its binding to nucleic acids. Plant Cell Physiol. 17, 127–137.

Mollenhauer H H 1967 A comparison of root cap cells of epiphytic, terrestrial and aquatic plants. Am. J. Bot. 54, 1249–1259.

Mollenhauer H H and Morré D J 1980 The Golgi apparatus. *In* The Biochemistry of Plants. Ed. N E Tolbert. Vol. 1, pp 437–488. Academic Press, New York.

Moore R and Evans M L 1986 How roots perceive and respond to gravity. Am. J. Bot. 73, 574–587.

Moore R and Pasieniuk J 1984 Graviresponsiveness and the development of columella tissue in primary and lateral roots of *Ricinus communis*. Plant Physiol. 74, 529–533.

Morimura S, Takohasi E and Matsumoto H 1978 Association of aluminium with nuclei and inhibition of cell division in onion (*Allium cepa*) roots. Z. Pflanzenphysiol, 88, 395–401.

Morré D J 1977 Membrane differentiation and the control of secretions: A comparison of plant and animal Golgi apparatus. *In* International Cell Biology, Eds. B R Brinkley and K R Porter. pp 293–303. Rockefeller University Press, New York.

Northcote D H and Pickett-Heaps J D 1966 A function for the Golgi apparatus in polysaccharide synthesis and transport in the root cap cells of wheat. Biochem. J. 98, 159–167.

Oades J M 1978 Mucilages at the root surface. J. Soil Sci. 29, 1–16.

Pan W L, Hopkins A G and Jackson W A 1989 Aluminium inhibition of shoot lateral branches of *Glycine max* and reversal by exogenous cytokinin. Plant and Soil 120, 1–9.

Passioura J B 1988 Root signals control leaf expansion in wheat seedlings growing in drying soil. Aust. J. Plant Physiol. 15, 687–693.

Picton S J, Richards K and Gardner R C 1991 Protein profiles in root-tips of two wheat (*Triticum aestivam* L.) cultivars with differential tolerance to aluminium. *In* Plant-Soil Interactions at Low pH. Eds. R J Wright, V C Baligar and R P Murrman. pp 1063–1070. Kluwer Academic Publishers, Dordrecht, The Netherlands.

Pilet P-E 1972 Root cap and root growth. Planta 106, 169–171.

Pilet P-E and Ney D 1978 Rapid localised light effect on root growth in maize. Planta 144, 109–110.

Puthota V, Cruz-Ortega R, Johnson J and Ownby J 1991 An ultrastructural study of the inhibition of mucilage secretion by aluminium in the wheat root cap. *In* Plant-Soil Interactions at Low pH. Eds. R J Wright, V C Baligar and R P Murrman. pp 779–788. Kluwer Academic Publishers, Dordrecht, The Netherlands.

Randall P J and Marco D G 1990 Inhibition and recovery of root growth of linseed in the presence of aluminium. Second International Symposium Plant-Soil Interactions at Low pH. Beckley, WV.

Ray T C, Callow J A and Kennedy J F 1988 Composition of root cap mucilage polysaccharide from *Lepidium sativum*. J. Exp. Bot. 39, 1249–1261.

Rincon M and Hanson J B 1986 Controls of calcium ion fluxes in injured or shocked corn root cells: Importance of proton pumping and cell membrane potential. Physiol. Plant. 67, 576–583.

Rincon M and Gonzales R 1991 Induction of protein synthesis by aluminium in wheat (*Triticum aestivum* L.). root tips. *In* Plant-Soil Interactions at Low pH. Eds. R J Wright, V C Baligar and R P Murrman. pp 851–858. Kluwer Academic Publishers, Dordrecht, The Netherlands.

Robinson D G and Depta H 1988 Coated vesicles. Annu. Rev. Plant Physiol. Plant Mol. Biol. 39, 53–99.

Robinson D G and Hillmer S 1990 Endocytosis in plants. Physiol. Plant. 79, 96–104.

Roy A K, Sharma A and Talukder A 1988 Some aspects of aluminium toxicity in plants. Bot Rev. 54, 145–178.

Saab I N and Sharpe R E 1989 Non-hydraulic signals from maize roots in drying soil: Inhibition of leaf elongation but not stomatal conductance. Planta 179, 466–474.

Sachs M M and Ho T-H D 1986 Alteration of gene expression during environmental stress in plants. Annu. Rev. Plant Physiol. 37, 363–376.

Sampson M, Clarkson D and Davies D D 1965 DNA synthesis in aluminium treated roots of barley. Science 148, 1476–1477.

Sanchez P A 1990 Advances in the management of acid soils in the tropics. Second International Symposium Plant-Soil Interactions at Low pH. Beckley, WV.

Sanchez P A and Salinas J G 1981 Low-input technology for managing oxisols and ultisols in tropical America. *In*. Advances in Agronomy. Ed. N C Brady Vol 34 pp. 279–406. Academic Press, New York.

Schaedle M, Thornton F C, Rynal D J and Tepper H B 1989 Response of tree seedlings to aluminium. Tree Physiol. 5, 337–356.

Scharschmidt B F and Van Dyke R W 1987 Proton transport by hepatocyte organelles and isolated membrane vesicles. Annu. Rev. Physiol. 49, 69–85.

Shropshire W 1979 Stimulus perception. *In* Physiology of Movement. Eds. W Haupt and M E Feinleib. Ency-

clopedia of Plant Physiology. Vol. 7, pp 10–41. Springer-Verlag, Berlin.
Steffens J C 1990 The heavy metal-binding peptides of plants. Annu. Rev. Plant Physiol. Plant Mol. Biol. 41, 553–575.
Taylor G J 1988 The physiology of aluminium tolerance in higher plants. Commun. Soil. Sci. Plant Anal. 19, 1179–1194.
Van den Bulcke M, Bauw G, Castresana C, Van Montagu M and Vandekerckhove J 1989 Characterisation of vacuolar and extracellular B (1,3)-gluconases of tobacco: Evidence for a strictly compartmentalized plant defense system. Proc. Natl. Acad. Sci. USA 86, 2673–2677.
Vose P B and Randall P J 1962 Resistance to aluminium and manganese toxicities in plants related to variety and cation-exchange capacity. Nature 196, 85–86.
Wacquant J-P 1977 Physiological selectivity for cations and CEC of grass roots. Plant and Soil 47, 257–262.
Wilkins M B 1979 Growth control mechanisms in gravitropism. *In* The Physiology of Movement. Eds W Haupt and M E Feinlieb. Encyclopedia of Plant Physiology. Vol. 7, pp 601–626. Springer-Verlag, Berlin.
Wright R J 1989 Soil aluminium toxicity and plant growth. Commun. Soil. Sci. Plant Anal. 20, 1479–1497.

P. J. Randall et al. (Eds.), Genetic aspects of plant mineral nutrition, 117–126.

PLSO SV34

Increasing salinity tolerance of grain crops: Is it worthwhile?

R.A. RICHARDS
Division of Plant Industry, CSIRO, P.O. Box 1600, Canberra, Australia 2601

Key words: barley, drought, halophytes, productivity, salinity, salt tolerance, sunflower, water-use efficiency, wheat

Abstract

The productivity of wheat and barley was compared in soils of different salt concentrations with a limited water supply. Productivity was assessed as total dry weight or dry weight per unit of water used (water use efficiency, WUE). Barley achieved the highest productivity because it used more of the available water and it had a greater WUE for above-ground dry weight. However, when WUE for total organic weight of roots and shoots was determined, or WUE was corrected for grain production, wheat and barley had the same productivity. In two experiments in drying soils with different salt concentrations but the same amount of soil water, wheat and barley had a higher dry weight than salt-tolerant grasses and they were more productive than C_4 halophytes and non-halophytes when adjusted for water use. In one experiment, sown at a low plant density, barley and wheat used less water than some halophytes and they completed their life cycle leaving some water behind in the soil. Their higher WUE did not compensate for their lower water use. However, when all species were sown at a high density, wheat and barley were either as productive or more productive than the most salt-tolerant species, including a C_4 halophyte, as they used all the available water and had the highest WUE. A sunflower cultivar was similarly more productive than a salt-tolerant relative. The contribution that salt-tolerant relatives of wheat, barley and sunflower can make to genetically improving the productivity of these species in dry saline soils is questioned.

Introduction

Understanding salinity tolerance in plants so as to eventually use this knowledge to genetically increase the tolerance of crop or pasture species is an active research pursuit. As saline soils are extremely variable, salt tolerance is usually assessed by growing plants in salinized nutrient solutions so that their root zone is at a constant salt concentration. These conditions may be satisfactory for studies on marsh plants, or plants growing in tidal zones, or for experiments aimed at understanding the effects of salt on plant growth, but they are not representative of the conditions in which plants grow in their natural or agricultural habitats. Hence they may lead to misleading conclusions and the formulation of inappropriate selection criteria to genetically improve productivity in saline soils.

Economically important species growing in soil experience variable soil water contents depending on irrigation, rainfall, leaf area and evaporative demand. Salinity may reduce the availability of this water because of its effect on soil water potential and it may also reduce total water use because leaf area, transpiration and growth are all reduced by salinity. Productivity of agricultural species on salt-affected soils will depend firstly on whether they are able to use all of the available water, as the more water used the more productive they will be, and secondly on how efficiently the water is used i.e. how much growth per unit of evapotranspiration. Manipulating water use and water-use efficiency genetically or through management are likely to be more important than criteria presently suggested to genetically improve salt tolerance and productivity in saline soils. Criteria suggested to improve salinity tolerance arise from studies in salinised nutrient solution. For example, salt exclusion mechanisms, Na/K discrimination and compartmentation of solutes within cells (Yeo and Flowers, 1986). It is suggested that these are of limited importance when it comes to improving productivity in salt-affected soils.

This study contrasts the productivity of wheat and barley, two species where research efforts to improve their salinity tolerance have been greatest, with numerous halophytic species as well as herbaceous crop and pasture species, when grown in drying saline soils. Productivity was determined on a dry weight basis or the dry weight per unit of water used. The species chosen to contrast with the C_3 wheat and barley were several C_4 species, including two C_4 halophytes that exclude salt and were presumed to have a markedly higher water use efficiency (McCree and Richardson, 1987; Rawson et al., 1977) and grasses with an ability to tolerate extreme salt concentrations, as well as other crop pasture species.

Materials and methods

Three experiments were conducted in a glasshouse maintained at about 25°C during the day and 14°C at night. In all experiments plants were grown in tubes 0.5 m long and 0.11 m diameter. Tubes were filled with about 6 kg of river loam and carefully packed so they had a similar bulk density. They contained a rubber bung in their base to prevent drainage that could be removed if required. Salinised nutrient solutions were prepared by adding a 5 : 1 (g/g) NaCl to $CaCl_2$ mixture to half-strength Hoaglands solution. About 500 mL of the required saline solutions were added to each tube containing the rubber bung depending on the treatment. After several hours bungs were removed and the tubes were allowed to drain. This flushing procedure was repeated several times until the conductivity of the drainage liquid matched that of the salinized nutrient solution added. Different treatments were imposed in each experiment (details below) and there were two replications of all treatments. After seedlings emerged no further water or saline-nutrient solution was added to tubes in any experiment. A 4 cm layer of perlite was placed on the soil surface to prevent evaporation of water. Tubes were weighed each week and plants were harvested when dead. At harvest above-ground plant parts were separated into stems, leaves and reproductive structures (if any), which were then oven dried and weighed. Soil was removed from each tube, weighed and then oven dried at 70°C for 7 days and then reweighed to calculate the percentage of total soil water used by the plant. Water use efficiency (WUE) was calculated as the ratio of total above ground dry weight to water used between emergence and plant death unless otherwise stated.

Experiment 1

Clipper barley and Condor hexaploid wheat (cultivars grown commercially in Australia) were sown in tubes in late November. There were seven salt concentrations with an electrical conductivity (EC) of the drainage water of 0 (no salt), 3, 6, 9, 12, 15, and 18 dS m^{-1}. Three seeds were sown 2 cm deep in each tube and the top 2 cm of soil of all tubes was kept moist with a little tap water so that germination and emergence was uniform. Barley emerged about 1 day earlier than wheat and the highest salt concentrations delayed emergence by about 4 days. Tubes were thinned to 1 healthy plant. Leaf length and width of all main stem leaves and tillers were measured weekly as well as the time when leaves died.

Experiment 2

Thirteen salt tolerant and sensitive species were grown including two genotypes each of barley and hexaploid wheat. These are listed in Table 1. Plants were sown over an extended period beginning in mid-May for the slowest growing species through to July for the fastest growing species to ensure that the period of fastest growth coincided in all species. The sowing order was *Puccinellia*, *Hordeum maritima*, *Atriplex* and *Amaranthus*, *Thinopyrum* and *Trifolium* and

Table 1. Species grown in Experiment 2 with reference to any known characteristics of salt tolerance

Species
Hordeum vulgare cv CM67
Salt tolerant 6-row barley. (Richards et al., 1987; Rawson et al., 1988)
Hordeum vulgare cv Clipper.
Possibly salt sensitive 2-row malting barley. (Rawson et al., 1988)
Hordeum maritima (Sea-barley grass)
Selection from a salt scald in Western Australia. Seed supplied by C. Malcolm, WA Dept of Agriculture.
Triticum aestivum cv Kharchia
Salt tolerant Indian wheat (Kingsbury and Epstein, 1984; Rawson et al., 1988)
Triticum aestivum cv Yecora.
High yield spring wheat.
Thinopyrum elongatum cv Tyrell (tall wheatgrass).
Salt tolerant grass (McGuire and Dvorak, 1981)
Puccinellia ciliata (Saltmarsh grass).
Salt tolerant grass. Seed supplied by C. Malcolm
Trifolium subterranean cv Woogenellup.
Presumed salt sensitive pasture legume.
Trifolium alexandrinum (Berseem clover).
Salt tolerant pasture legume (Wingers and Läuchli, 1982)
Medicago sativum cv Hunter River (lucerne).
Moderately salt tolerant, deep rooted pasture legume (R.W. Downes pers. com.)
Helianthus annuus cv Hysun 31.
High yielding hybrid sunflower.
Helianthus argophyllum
Salt-tolerant relative of sunflower (R.W. Downes pers comm.)
Atriplex nummularia (Old man saltbush).
C4 halophyte with salt glands.
Atriplex lentiformis (Quail bush)
C4 halophyte
Amaranthus edulis (pigweed).
C4 weed

Medicago, *Helianthus argophyllum*, wheat, barley, and sunflower was sown last. Plants in the control treatment were sown up to 21 days later than the salt treated plants.

There were three treatments common to all species, a control flushed with half-strength Hoaglands solution and two salt treatments flushed with salt (NaCl and $CaCl_2$) in half-strength Hoaglands with an EC of either 10 or 20 dS m^{-1}. An extra salt treatment of EC 15 dS m^{-1} was included for the wheat and barley cultivars. In this experiment roots were washed from the soil of all tubes and any extraneous organic matter was removed from the roots. Root samples were dried at 70°C, weighed and then placed in a crucible and ashed at 600°C and then weighed to determine the organic weight of the root sample. Above-ground plant parts were also weighed and ashed to determine mineral content and organic weight.

Experiment 3

Tubes were prepared as before and flushed with one of four salt concentrations. These were half-strength Hoaglands with added NaCl and $CaCl_2$ to give conductivities of 0, 5, 10 and 15 dS m^{-1}. Species grown were *Atriplex nummularia*, *Amaranthus edulis*, *Helianthus argophyllus*, *Helianthus annuus* cv Hysun 31, *Thinopyrum elongatum* cv Tyrell, as in Experiment 2. Also, hexaploid wheat cv. Isis, a winter wheat not expected to reach floral initiation in the glasshouse, cv Songlen, a wheat chosen for its osmotic adjustment (Morgan, 1983) and two 6-row barleys, CM67 as in Experiment 2, and Betzes chosen for its later maturity time. Sowing time was again staggered so that the most rapid growth period of each species coincided. Sowing of *Atriplex* and *Amaranthus* in the 10 and 15 dS m^{-1} salt treatments commenced in August and was followed by *Thinopyrum* and *H. argophyllum* 4 days later, wheat 8 days later and barley 7 days later again. The 0 and 5 dS m^{-1} treatments were sown 7 days after thc 10 and 15 dS m^{-1} treatments. About 5 plants were established in each tube.

Results and discussion

Experiment 1

Despite plants having access to the same amount of water, substantial differences in above-ground dry weight were found between treatments and between Clipper barley and Condor wheat (Fig. 1a). Surprisingly, dry weight in both species increased as salt concentration increased and was only less than the controls at the highest salt concentrations. Barley had a greater weight in all treatments. Variation in dry weight between different salt treatments and between barley and wheat came about primarily because of differences in WUE, viz. the ratio of above-ground

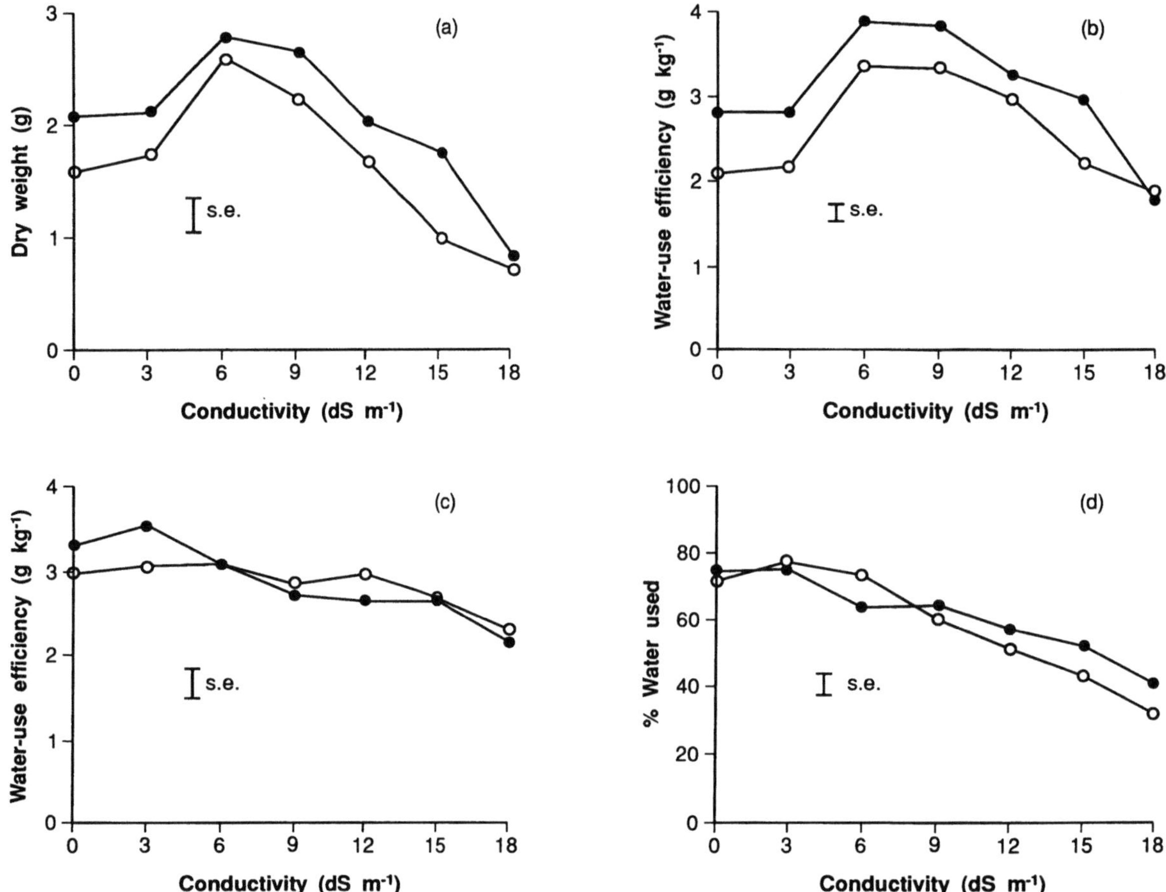

Fig. 1. Comparison of barley (●) and wheat (○) grown in different salt treatments for **a)** above-ground plant weight, **b)** water use efficiency, **c)** water use efficiency after covariate adjustment for grain weight, and **d)** percentage of total soil water used at plant death. The standard error for a difference between wheat and barley at any salt content is shown in each figure. Conductivities are of the drainage water at the time of sowing.

dry weight to total water used, rather than to differences in total water use (Fig. 1b).

Variation in WUE was very similar to variation in dry weight and WUE was highest at the intermediate salt concentrations and only fell below the control at the highest salt level. This increase casts doubt on an increased respiration rate as being important in saline soils except perhaps at the highest salt concentration (Yeo, 1983). Factors that may increase WUE as a result of salinity are firstly, that stomatal conductance may decline without a corresponding fall in assimilation capacity, and secondly, that salt may be sequestered in old leaves, resulting in an apparent increase in WUE. A third factor may be that root growth is less in the salt treatments than in the control and above-ground growth may be correspondingly higher. The latter two factors are examined further in the next experiment.

The increase in WUE attributed to an altered gas exchange could be large. Discrimination against the stable isotope ^{13}C, a measure of WUE (Farquhar and Richards, 1984), can decline by as much as 5×10^{-3}‰ when C_3 species are grown in salinised solution or soils (Brugnoli and Lauteri, 1991; Guy et al., 1988). As a reduction of 1×10^{-3}‰ corresponds to an increase in WUE of about 15% (Farquhar and Richards, 1984), then an altered gas exchange could account for a large part of the increase in WUE, which was about 50% in the intermediate salinity treatments in both wheat and barley (Fig. 1b). Nevertheless, this contrasts with measurements of the gas exchange of wheat and barley grown with and without salinity where no differences in instantaneous WUE were found (Rawson, 1986).

Grain was produced by plans in most salt treatments and hence in the treatments where WUE was high. Grain was presumably produced because plants in the higher salt concentrations had a reduced leaf area and thus lower rate of water use. This in turn extended the duration of growth in the salt treatments and allowed the plants to complete their lifecycle and produce grain. No grain was produced in the control or low salt treatment as plants exhausted the soil water supply before anthesis and died. A reanalysis of WUE using grain weight as a covariate was highly significant (Table 2) and when WUE values are adjusted for the covariate then WUE in fact declined slightly with increasing salt concentration (Fig. 1c). This indicates that the apparent increase in WUE of plants in saline soils arose from a redistribution of carbon from the roots and/or from an increased sink strength that affects gas exchange (Blum et al., 1988). If the latter is important presumably stomatal conductance is reduced more than assimilation capacity in plants setting grain or both. In the covariance analysis differences between wheat and barley were not significant and neither was the interaction between species and treatment.

The other determinant of productivity in drying saline soils is the total amount of water used by plants. In this experiment more water was left behind in the soil as salinity increased (Fig. 1d). This could be due to a reduced leaf area and a reduced leaf area duration as a result of salinity and hence incomplete water use before maturity, or to plants being unable to lower their water potential sufficiently to match the lower potential of the drying saline soil thereby leaving water behind. The former suggestion is favoured for several reasons. In the control and EC3 treatment both wheat and barley averaged 5 tillers (including the main stem) per plant whereas in the EC15 and EC18 treatment wheat and barley plants had a single main stem only and no tillers. Maximum kernel weight achieved was 48 mg for barley and 30 mg for wheat and there was no evidence for a decline in kernel weight at the highest salt levels which was expected if plants died prematurely. Furthermore, although there were no differences between barley and wheat at the low salinity levels, at higher levels there was a consistent trend for barley to use more water than wheat and this was associated with a larger leaf area in barley (data not presented but see Rawson et al., 1988).

Table 2. Analysis of variance for water use efficiency in Experiment 1 with covariance adjustment for grain weight

Source	df	MS	VR
Barley vs wheat	1	0.0027	0.09
Salt treatments	6	0.494	16.9***
Species × Salt	6	0.077	2.6^{NS}
Covariate	1	1.332	45.7***
Residual	13	0.029	

^{NS} Not significant, *** $p < 0.001$.

Experiment 2

This experiment was designed to firstly, contrast wheat and barley with a wider range of species, and secondly, to investigate whether a reduced root mass and a higher salt content in plant tissues may contribute to an apparently higher WUE in wheat and barley in drying saline soils. Because plants used different amounts of water and accumulated different amounts of salt in their tissues, comparisons between species are mainly based on WUE values for the total organic weight of plants rather than total dry weight.

The duration of growth increased in all species as salinity increased (Table 3). Lucerne was the most extreme as it used all the available water in the control treatment and died after 62 days whereas in the highest salt treatment it survived for 261 days despite having no additional water after planting. The longer duration of growth in wheat and barley in the salt treatments enabled plants of both species to produce grain in the EC10 and 15 treatments but in the EC20 treatment only wheat produced grain.

WUE of genotypes in each treatment, determined on the organic weight of roots and shoots rather than total dry weight, are given in Table 4. Wheat again had a greater WUE in the

Table 3. Duration of growth (days) in control and salt treated plants in Experiment 2. Species with a similar growth duration in the three treatments have been grouped together and the number of species in each group are given in parenthesis after each generic name. Standard error for difference between any two values = 8 days

Species	Conductivity (dS m^{-1})		
	0	10	20
Lucerne (1)	62	105	261
Puccinellia (1)	137	158	140
Atriplex (2)	99	99	134
Thinopyrum, *Hordeum* (4)	55	106	135
Trifolium (2)	62	100	96
Triticum (2)	58	70	102
Helianthus, *Amaranthus* (3)	50	70, 110[a]	77, 113[a]

[a] *H. argophyllum*, *Amaranthus*.

Table 4. Water use efficiency (g kg^{-1}) of genotypes grown in saline soils. Values are for whole plants including roots and were calculated from ashed plant parts. Standard error for difference between any two values = 0.3 g kg^{-1}

Species	Conductivity (dS m^{-1})		
	0	10	20
Puccinellia	2.2	1.8	0.7
H. maritima	2.5	2.5	1.8
Thinopyrum	2.4	2.4	1.8
Medicago	2.2	1.9	0.3
T. subterranean	2.4	–	0.2
T. alexandrinum	1.8	1.6	0.7
A. lentiformis	2.7	3.4	3.0
A. nummularia	2.7	3.7	3.0
A. edulis	4.1	3.3	–
H. argophyllum	2.4	1.3	1.4
H. annuus	2.5	2.4	1.1
Kharchia	2.5	3.2	1.5
Yecora	2.6	3.3	1.9
CM67	2.7	2.3	2.4
Clipper	2.7	2.6	1.9

intermediate salt concentrations (data for EC15 not shown) than in the controls but WUE was lower in the highest salt concentration. Values for the barley cultivars were not significantly different to the control values at all salt concentrations. In the control treatment the C_4 species, *Amaranthus edulis* had the highest WUE; wheat, barley and two *Atriplex* species were next highest whereas the legumes and the marsh grass, *Puccinellia*, had the lowest WUE. At the intermediate salt level, WUE of *Atriplex* increased to the same extent as wheat, whereas in the other species WUE either decreased or remained the same. At the highest salt level WUE fell dramatically in all three legumes and in *Puccinellia*, less in both *Helianthus* species, and less again in wheat, barley, *Thinopyrum*, *Hordeum maritima* and the *Atriplex* species. At the highest salt concentration, WUE in both *Atriplex* species was between the control and the intermediate salt concentration whereas in all other species it was less than the control values. The WUE of wheat and barley at EC15 was the same as at EC10 and significantly higher than at EC20.

Table 5 shows how much of the total soil water was used by plants from sowing until their death. Wheat and barley left water behind in the soil at the higher salinity levels as they did in Experiment 1. So did all other short-season determinate species. The only species that used all the available water were lucerne, both *Atriplex* species, *Helianthus argophyllum*, *Thinopyrum* and *Hordeum maritima*; all are long-season indeterminate species. This is consistent with Experiment 1 and supports the suggestion that the leaf area developed in the short-season determinate species in the salt treatments was insufficient to use all the available water. It is also worth noting that the root to shoot ratio (R/S on an organic carbon basis) was highest in the long season species. The R/S for lucerne in the highest salt level was a remarkably high 1.72 com-

Table 5. Percentage of total soil water used by plants grown at different salinities. Standard error for difference between any two values = 4%

Species	Salt concentration (dS m^{-1})		
	0	10	20
Puccinellia	79	72	55
H. maritima	76	71	67
Thinopyrum	74	72	72
Medicago	78	78	77
T. subterranean	77	73	34
T. alexandrinum	77	73	55
A. lentiformis	82	80	78
A. nummularia	82	80	80
A. edulis	74	73	66
H. argophyllum	76	74	71
H. annuus	77	73	36
Kharchia	75	51	35
Yecora	78	44	43
CM67	75	54	55
Clipper	75	57	51

Table 6. Root to shoot ratio at different salinities calculated on a total organic weight basis from ashed plant parts. Standard error for difference between any two values = 0.04

Species	Conductivity (dS m^{-1})		
	0	10	20
Puccinellia	0.04	0.05	0.06
H. maritima	0.34	0.06	0.10
Thinopyrum	0.27	0.26	0.13
Medicago	0.48	0.78	1.72
T. subterranean	0.17	0.20	–
T. alexandrinum	0.21	0.09	0.04
A. lentiformis	0.23	0.27	0.25
A. nummularia	0.24	0.21	0.19
A. edulis	0.30	0.14	0.20
H. argophyllum	0.32	0.22	0.10
H. annuus	0.19	0.11	0.06
Kharchia	0.25	0.07	0.08
Yecora	0.19	0.07	0.05
CM67	0.22	0.13	0.06
Clipper	0.22	0.08	0.08

pared to 0.22 for the *Atriplex* species and 0.06 for wheat and barley (Table 6). With the exception of lucerne and perhaps *Puccinellia* and *A. lentiformis*, the R/S ratio declined with salinity. The higher R/S values in the control could therefore account for part, but not all, of the lower WUE in the control treatment in Experiment 1.

The mineral or ash content of different species provides data on the most effective salt excluders (Table 7). The dicotyledons were far less effective than the grasses. As expected, mineral content of the *Atriplex* species was highest, followed by both *Helianthus* species and the legumes; it was apparent that both *Trifolium* species were unable to exclude salt at the highest salt level as the mineral content of leaves and stems increased. Mineral content in wheat, barley and *Puccinellia* leaves increased in the highest salt concentration whereas in *Thinopyrum* and *H. maritima* mineral content in leaves declined in both salt treatments relative to the control.

Experiment 3

In the previous experiments there was evidence that insufficient leaf area limited the water use of the short season species such as wheat, barley and sunflower. In this experiment plants were grown at a higher density so as to overcome the reduced leaf area. There were also fewer species and more salt treatments than in the previous experiment. The higher plant density had a substantial effect on growth duration and water use in all genotypes. Compared to plants in experiment 2, average growth duration was 9 and 34 days shorter in the control and EC10 treatment. However, salinity still extended growth duration compared to the control. The mean difference in

Table 7. Mineral content (as a % of oven dry weight) of leaves and stems grown at different salinities. Standard error for difference between any two values for both leaves and stems = 5%

Species	Conductivity (dS m^{-1})					
	0		10		20	
	Leaves	Stems	Leaves	Stems	Leaves	Stems
Puccinellia	7	5	11	8	13	12
H. maritima	14	9	10	7	11	9
Thinopyrum	14	10	11	10	11	5
Medicago	14	9	12	7	17	5
T. subterranean	12	14	–	–	28	33
T. alexandrinum	14	15	14	16	31	24
A. lentiformis	23	12	36	16	35	15
A. nummularia	23	15	30	18	31	17
A. edulis	19	19	20	27	–	–
H. argophyllum	21	16	24	19	29	22
H. annuus	18	19	23	25	32	37
Kharchia	13	9	15	14	19	14
Yecora	12	9	16	16	19	15
CM67	14	11	13	13	15	12
Clipper	14	12	15	15	17	16

duration between the control and EC15 treatment was 14 days.

Higher plant density increased the total water use by all plants in the higher salt concentrations compared to the earlier experiments. In contrast to the other experiments there were few differences in water use between treatments; as a percentage of soil water content the mean water use in the control was 72% and 69% in the highest salt treatment. The only genotypes in the high salt treatment unable to use all the available water were Songlen wheat and *A. edulis* that only used 50% of the total soil water. This data therefore confirms the suspicion that in the short-season species, soil water extraction was previously limited by leaf area rather than the low soil water potential. It indicates that the combined osmotic effect of both salinity and soil dryness should not prevent non-halophytes such as wheat, barley and *Helianthus annuus* from using all the available soil water. With the exception of Songlen wheat, they dried the soil to the same extent as the salt tolerant species of *Atriplex* and *Thinopyrum*. The difference arose before because the former group were determinate annuals and salinity decreased leaf growth and hence water use. The increased plant density in the annual species compensated for the reduced leaf growth per plant and enabled plants to use all the available water.

The WUE and above-ground dry weight (AGDW) increased as salinity increased (Table 8) and there was little evidence of a reduction in WUE at the highest salt concentration. As total water use was similar among the species, differences in WUE reflected differences in total above-ground dry weight (Table 8). Surprisingly, differences in WUE or AGDW between C_3 and C_4 species were not substantial. *Thinopyrum* and *H. argophyllum* had the lowest WUE and AGDW at most salinity levels whereas wheat and barley had the highest, being on average as high or higher than *Atriplex*. Root weights were not determined in this experiment; if they were of the same order as in Experiment 2 then the WUE of *Atriplex* would have been higher than in wheat and barley but this would have been offset by the higher mineral content of *Atriplex*.

General discussion

Wheat, barley and sunflower were more productive per unit of water used or on a dry weight basis than C_3 halophytes at all salt concentrations and were as productive as C_4 halophytes at all except the highest salt concentrations. Furthermore, plants grown in saline soils were often heavier than control plants grown in non-saline soils and they generally had a lower R/S ratio than the controls. These results are at variance with the usual findings for plants grown in salinised nutrient solutions and they lead one to question the value of results from nutrient solutions when determining salinity tolerance of different genotypes and when identifying the factors responsible for increasing salinity tolerance of crop species.

In drying saline soils the determinants of dry matter production are firstly, how much soil

Table 8. Above-ground dry weight (AGDW, g) and water-use efficiency (WUE, g kg^{-1}) of genotypes at different salinities in Experiment 3. Standard error for difference between any two values for AGDW = 0.4 g and for WUE = 0.4 g kg^{-1}

	Conductivity (dS m^{-1})							
	0		5		10		15	
	AGDW	WUE	AGDW	WUE	AGDW	WUE	AGDW	WUE
Amaranthus	3.0	4.4	3.3	4.3	4.2	4.6	3.3	4.8
Thinopyrum	2.9	3.7	3.1	4.0	3.4	4.3	3.8	5.0
Atriplex	2.6	3.5	3.9	4.8	5.3	6.2	4.5	5.5
H. argophyllum	2.3	3.0	2.9	3.9	3.9	3.8	3.6	4.9
H. annuus	3.3	4.1	3.6	4.0	4.2	5.0	4.4	6.0
Songlen	2.8	3.8	4.3	4.9	5.3	5.6	4.9	6.4
Isis	3.1	4.1	3.0	3.5	3.7	4.3	3.9	4.7
Betzes	2.9	3.8	4.2	4.8	4.8	5.5	5.4	5.9
CM67	3.4	4.1	4.7	5.0	5.4	7.1	5.5	6.7

water is used, and secondly, how efficiently is it used i.e. the WUE. In experiments reported here both were found to be important. When plants were grown at low density in the intermediate and high salt treatments, short-season determinate species such as wheat, barley and sunflower, completed their lifecycle before using all the available soil water. This was because salinity reduced leaf area, leaf area duration and hence water use. In contrast, the long-season, indeterminate species, whose leaf area was also reduced, continued producing new leaves and ultimately used all of the available water. Thus the long season species ultimately produced more dry matter despite the finding in most species that their WUE was lower than in the determinate species. However, when plants were grown at a high density such that leaf area and transpiration in all species was higher, there was little variation among genotypes in total water use and even in the highest salt treatment of 15 dS m^{-1} the short-season non-halophytes used the same amount of water as the halophytes and other long-season species. The WUE then became the most important factor contributing to variation in dry matter production. Surprisingly, wheat and barley had about the same or a higher WUE and hence were more productive than all other species including the salt tolerant grasses and even C_4 species such as *Atriplex nummalaria*, although it is possible that the WUE of C_4 species was lower than expected due to light and temperature being suboptimal for them during these experiments (Pearcy and Ehleringer, 1984). Sunflower also was more productive than its salt-tolerant relative.

The high productivity achieved by wheat, barley and sunflower in drying saline soils compared to their more salt-tolerant relatives and in relation to the C_4 halophyte *Atriplex*, raises the question of whether attempts to genetically increase their salinity tolerance further, with the use of salt tolerant relatives, are likely to be successful. This has been raised before from a different perspective where it was argued that in saline fields further selection for salinity tolerance may not result in higher yields (Richards, 1983). It was argued that since salt-affected soils are highly variable in their salinity and that most of the yield comes from the least salt-affected areas, then increasing yield potential in favourable areas should result in higher field yields than increasing yield in the salt-affected areas. A different question is raised here. That is, whether the salt-tolerant relatives of wheat, barley and sunflower are likely to contribute to crop improvement in saline soils? The relatives have often been suggested as a source of increased salt tolerance for wheat and barley and there has been extensive research conducted on them (Forster et al., 1990; Gorham, 1990; McGuire and Dvorak, 1981). This study raises the possibility that, for productivity in saline soils that are not frequently irrigated, commercial varieties of wheat, barley and sunflower already exist that have superior productivity than their supposedly salt-tolerant relatives. Although the latter survive for longer in saline soils, they are less productive and it is unlikely that they will contribute to enhanced growth and yield of commercial varieties in saline soils.

It is likely that the main limitation to the yield of wheat, barley and sunflower in drying saline soils is an inadequate leaf area that prevents them from using all of the available water. By increasing the rate of leaf canopy development or duration either genetically (by increasing vigour or extending the duration of leaf development) or by management (by increasing sowing density) should overcome some of this limitation and result in higher yields in salt-affected regions.

Acknowledgement

I thank Bernie Mickelson for expert technical assistance during this study.

References

Blum A, Mayer J and Golan G 1988 The effect of grain number per ear (sink size) on source activity and its water-relations in wheat. J. Exp. Bot. 39, 106–114.

Brugnoli E and Lauteri M 1991 Effects of salinity on stomatal conductance, photosynthetic capacity, and carbon isotope discrimination of salt-tolerant (*Gossypium hirsutum* L.) and salt-sensitive (*Phaseolus vulgaris* L.) C_3 non-halophytes. Plant Physiol. 95, 628–635.

Farquhar G D and Richards R A 1984 Isotopic composition

of plant carbon correlates with water use efficiency of wheat genotypes. Aust. J. Plant Physiol. 11, 539–552.

Forster B P, Phillips M S, Miller T E, Baird E and Powell W 1990 Chromosome locations of genes controlling tolerance of salt (NaCl) and vigour in *Hordeum vulgare* and *H. chilense*. Heredity 65, 99–107.

Gorham J 1990 Salt tolerance in Triticeae : K/Na discrimination in *Aegilops* species. J. Exp. Bot. 41, 615–621.

Guy R D, Warne P G and Reid D M 1988 Stable carbon isotope ratio as an index of water-use efficiency in C_3 halophytes – possible relationship to strategies for osmotic adjustment. *In* Stable Isotopes in Ecological Research. Eds. P W Rundel, J R Ehleringer and K A Nagy. pp 55–75. Springer-Verlag, New York.

Kingsbury R W and Epstein E 1984 Selection for salt resistant spring wheat. Crop Sci. 24, 310–315.

McCree K J 1986 Whole-plant carbon balance during osmotic adjustment to drought and salinity stress. Aust. J. Plant Physiol. 13, 33–43.

McCree K J and Richardson S G 1987 Salt increases the water use efficiency in water stressed plants. Crop Sci. 27, 543–547.

McGuire P E and Dvorak J 1981 High salt-tolerance potential in wheatgrasses. Crop Sci. 21, 702–705.

Morgan J M 1983 Osmoregulation as a selection criterion for drought tolerance in wheat. Aust. J. Agric. Res. 34, 607–614.

Pearcy R W and Ehleringer J R 1984 Comparative ecophysiology of C_3 and C_4 plants. Plant Cell Environ. 7, 1–13.

Rawson H M 1986 Gas exchange and growth in wheat and barley grown in salt. Aust. J. Plant Physiol. 13, 475–489.

Rawson H M, Begg J E and Woodward R G 1977 The effect of atmospheric humidity on photosynthesis, transpiration and water use efficiency of leaves of several plant species. Planta 134, 5–10.

Rawson H M, Richards R A and Munns R 1988 An examination of selection criteria for salt tolerance in wheat, barley and triticale. Aust. J. Agric. Res. 39, 759–772.

Richards R A 1983 Should selection for yield in saline regions be made on saline or non-saline soils? Euphytica 32, 431–438.

Richards R A, Dennet C W, Qualset C O, Epstein E, Norlyn J D and Winslow M D 1987. Variation in yield of grain and biomass in wheat, barley and triticale in a salt-affected field. Field Crops Res. 15, 227–287.

Winter E and Läuchli A 1982 Salt tolerance of *Trifolium alexandrinum* L. I. Comparison of salt response of *T. alexandrinum* and *T pratense*. Aust. J. Plant Physiol. 9, 221–226.

Yeo A R 1983 Salinity resistance: Physiologies and prices. Physiol. Plant. 58, 214–222.

Yeo A R and Flowers T J 1986 Salinity resistance in rice (*Oryza sativa* L.) and a pyramiding approach to breeding varieties for saline soils. Aust. J. Plant Physiol. 13, 161–173.

P. J. Randall et al. (Eds.), *Genetic aspects of plant mineral nutrition*, 127–135.

PLSO SV33

Arguments for the use of physiological criteria for improving the salt tolerance in crops

C.L. NOBLE and M.E. ROGERS
Institute for Sustainable Agriculture, Victorian Department of Agriculture, Tatura 3616, Victoria, Australia

Key words: breeding, glycophytes, ion exclusion, legumes, salt tolerance, selection

Abstract

Efforts to develop new crop varieties with improved salt tolerance have been intensified over the past 15–20 years. Despite the existence of genetic variation for salt tolerance within species, and many methods available for expanding the source of genetic variation, there is only a limited number of varieties that have been developed with improved tolerance. These new varieties have all been based upon selection for agronomic characters such as yield or survival in saline conditions. That is, based upon characters that integrate the various physiological mechanisms responsible for tolerance. Yet over the same time period, knowledge of physiological salt responses has increased substantially.

Selection and breeding to increase salt tolerance might be more successful if selection is based directly on the physiological mechanisms or characters conferring tolerance. Basic questions associated with using physiological selection criteria are discussed in the paper. These are centred around the need for genetic variation, the importance of the targeted mechanism, the ease of detection of the physiological mechanism (including the analytical requirements) and the breeding strategy. Many mechanisms, including ion exclusion, ion accumulation, compatible solute production and osmotic adjustment have been associated with genetic variation in salt tolerance. Yet their successful use in improving salt tolerance, via physiological selection criteria, is largely non-existent. Consideration is given to the role of physiological criteria in the short and long term in improving salt tolerance. In several glycophytic species, particularly legumes, physiological selection based on ion exclusion from the shoots shows promise. Recent results for white clover indicate the potential for using a broad physiological selection criterion of restricted Cl accumulation in the shoots, with scope for future refinement based upon the specific physiological characters that combined result in ion exclusion.

Introduction

Maximising the salt tolerance of crop species is an important component of any long term, integrated system for farming in areas affected by soil or water salinity. Crop salt tolerance has been the subject of extensive research, particularly over the past 30 years. Studies of the plant kingdom have shown substantial variation at the family, genus and species levels for tolerance to salt. Comparisons at these taxonomic levels occupied most salt tolerance studies up to probably the early 1970's and permitted initial identification of suitable crop types for various salinity conditions. Further refinement in identifying more salt tolerant crops requires the presence of variation in tolerance at lower taxonomic levels. Such investigations gathered impetus in the 1960's and have increased ever since, although many evaluations remain to be done.

Whilst the search for genotypes with useful levels of salt tolerance has concentrated primarily on screening the available range of current (mainly commercial) genotypes, efforts have expanded in the last 20 years to develop new genotypes possessing improved salt tolerance, via selection and breeding. These efforts have concentrated on using agronomic selection criteria, such as yield or survival, but only limited success has been reported. One suggestion to accelerate the development of tolerant genotypes has been to use criteria based upon physiological mechanisms responsible for salt tolerance. Many authors have embraced this idea and discussed its theoretical basis (Epstein et al., 1980; Greenway, 1973; Shannon, 1985; Tal, 1985), yet no selection method using physiological criteria has been successfully employed in the release of a new variety.

Identified variation in salt tolerance

Comparative evaluations of the salt tolerance of species have been numerous. Most of our agricultural species have been 'classified' for salt tolerance. Comprehensive lists are available (such as Maas and Hoffman, 1977) specifying threshold soil salinities for initial yield decline and subsequent rates of yield loss for each species.

Intraspecific variation in salt tolerance has been reported widely (for example Brown and Hayward, 1956; Marschner et al., 1981; Shannon, 1978; Taylor et al., 1975; Yeo and Flowers, 1983). In some species, however, little or no variation between varieties has been reported. In such cases the lack of variation may be due to a narrowing of the genetic base of the species via domestication and selection for favourable environments, by which genes for adaptation to adverse conditions are lost or are at very low frequencies. In other cases, the lack of variation can be due to screening only a small portion of the available germplasm. Ayers et al. (1951) found little variation between six varieties of lettuce, while Shannon et al. (1983) screened 85 lettuce cultivars and breeding lines and identified significant variability in salt tolerance.

For agricultural crops, after screening the available commercial varieties, further enhancement of tolerance requires expansion of the germplasm base – identifying 'alternative variation'. A range of methods are available for achieving this expansion. Some of these are noted here. Further discussion can be found in papers by Epstein (1985), Noble (1983), Pasternak (1987) and Shannon (1985).

Attention can be turned to screening ecotypes (Hannon and Bradshaw, 1968; Venables and Wilkins, 1978). Additionally, several crops have substantial plant-to-plant variation for tolerance within varieties, primarily in cross-pollinated species such as alfalfa (Noble et al., 1984) and sugarbeet (Ulrich, 1961). Variation can also be generated artificially by mutagenic agents (Gottschalk, 1981) or the use of tissue culture (somoclonal variation, Larkin and Scowcroft, 1981). Further, closely related species or wild progenitors of our target species can be used. Limited variation for tolerance in the tomato (*Lycopersicum esculentum*) can be overcome by interspecific crosses with the wild related species *L. cheesmani* and *L. peruvianum* (Rush and Epstein, 1981a; Tal and Shannon, 1983) that possess superior salt tolerance although inferior agronomic characteristics. Dvorak et al. (1985) and Dvorak and Ross (1986) showed that the salt tolerance of wheat could be increased by the addition of *Elytrigia* chromosomes into the wheat genetic background. Protoplast fusion might also be useful as it provides a means of overcoming natural interspecific barriers to hybridisation.

It is the use of 'alternative variation' towards increasing the tolerance of crops (requiring selection and breeding for tolerance rather than the more applied screening of germplasm) that has been recognised for many years as offering the next stage in developing salt tolerant crop varieties. Yet despite increased activity in this area, there has been limited success. For the vast majority of crops, the most tolerant commercial varieties have not been the result of deliberate selection and breeding for salt tolerance. Despite evidence of genetic control for variation in salt tolerance (Abel, 1969; Akbar and Yabuno, 1977; Norlyn, 1980; Venables and Wilkins, 1978) there are very limited examples of crop types that have been released with improved tolerance. Few varieties have been selected for improved tolerance (Table 1) and an even smaller number of (released) varieties exist that have been specifically

Table 1. Documented accounts of commercial varieties selected or bred for improved salt tolerance

Species	Variety	Strategy	Reference
Avocado	Arsola 1–18	Varietal cross	Cooper, 1951
Wheatgrass	Nebraska 10	Strain selection	Dewey, 1962
Rice	Giza 159	Varietal cross	Gad El Hak, 1966
Alfalfa	Az Germ Salt 1	Intra-varietal selection	Dobrenz et al., 1983
Red fescue	Saltol	Ecotype selection	Cordukes, 1981
Maize	Arizona 8601	Natural selection program	Day, 1987

bred for improved tolerance (Shannon and Noble, 1990).

The lack of success in developing new, more tolerant genotypes can be attributed to many factors. These include spatial and temporal variability in salinity, variation in tolerance with ontogeny, tolerance being a polygenic trait, inadequate selection criteria and uncoordinated efforts to understand the genetic, physiological, morphological and environmental effects on the tolerance of a species. However, the many natural, and few induced, examples where improved tolerance has occurred indicate that these limitations can be overcome and useful salt tolerant genotypes obtained. Each of the above (and other) limitations could be discussed in separate papers. As the selection criteria employed is fundamental to any program to increase a plant's tolerance, it is appropriate to consider some of our current options in this area.

Agronomic or physiological selection criteria

Agronomic characters, such as yield, survival, leaf damage and plant height, have been the most commonly used criteria for identifying tolerance. This is largely due to their ease of measurement and because, in the end, yield (both absolute and relative) under saline conditions is usually the ultimate requirement. These characters represent the combined genetic and environmental effects on plant growth and include the integration of the physiological mechanisms that confer tolerance.

There is no doubt that such agronomic characters are readily affected by factors other than salinity. Environmental conditions, which can vary markedly between sites or experimental conditions, can have a marked effect on the expression of these characters. Some authors have suggested that these effects can be so large as to limit the usefulness of selection and breeding for tolerance based on such characters. Yet in the absence of demonstrated alternatives agronomic characters cannot and should not be dismissed. Confoundment by environmental variation might to a large extent be overcome by targeting breeding programs to specific conditions or locations. At the very least, agronomic characters represent viable starting criteria for salt tolerance breeding programs. This approach has been successful in developing varieties tolerant to other environmental stresses such as heat, cold and drought.

Selection and breeding approaches to increase tolerance might be more successful, with respect to achieving maximum attainable tolerance, if selection is based directly on the relevant physiological mechanism(s). However, selection based on physiological criteria is fraught with uncertainties and unknowns. Several authors (Flowers and Yeo, 1986; Greenway and Munns, 1980; Läuchli and Epstein, 1990) have reviewed our knowledge of the physiology of salt tolerance. In all cases many mechanisms are implicated. Further, the relative importance of different mechanisms can vary between closely related species (Rush and Epstein, 1981b) and even varieties (Yeo and Flowers, 1983). There is also a lack of studies on the genetic control of these mechanisms.

To use the efficiency of a physiological mechanism as a selection criterion for improving salt tolerance, several basic questions must be considered:

1) Is there genetic variation in the efficiency of the mechanism and is it sufficiently heritable to permit advances via selection and breeding?

This is probably the most critical question and

requires a positive answer. In particular, some stability of genotype performance will be required.

2) Is the targeted mechanism important in affecting the overall, whole plant tolerance?

As salt tolerance is usually the result of a combination of physiological mechanisms, the targeted mechanism(s) must be of major importance to tolerance. Additionally, for a particular physiological mechanism (process), such as ion exclusion, many physiological characters (such as uptake at the root plasmalemma, the tonoplast) can be involved in a hierarchical arrangement involving causal relationships among characters. A choice is required between using a broadly defined mechanism (e.g. ion exclusion) or a specific component character (e.g. plasmalemma uptake) as our criterion. As our understanding of the component characters improves it may be possible to use them as selection criteria. However, the more specific the mechanism, the less related it may be to the observed whole plant tolerance. The frequently poor relationship between selecting for cell tolerance to salt via tissue culture and the subsequent whole plant response (Dracup, 1991) is perhaps an example of such an outcome. The mechanisms for salt tolerance often depend on the anatomical and physiological complexity of the organised plant rather than on tolerance at the cellular level (Smith and McComb, 1981). Hence despite the interest in determining the specific physiological characters of tolerance, their usefulness in selection and breeding programs is uncertain and yet to be demonstrated.

3) Is the physiological criterion capable of rapid detection permitting the screening of large numbers of plants without requiring large and sophisticated resources?

Agronomic selection, such as for leaf damage or survival, can handle large numbers of plants. This is frequently necessary to maximise selection pressures, maintain variability in subsequent generations and permit the use of large germplasm bases. Physiological measures, virtually without exception, require destructive plant harvesting and subsequent tissue analysis. These analyses are expensive in their use of time, people and physical resources. If any of these resources are limiting, as is often the case, compromises might be made in factors such as plant number and accuracy. These compromises may well nullify any selection efficiency sought by using physiological as compared with agronomic criteria.

4) Can several separate selection cycles be made, each using a different physiological criterion, and subsequently the optimised, selected lines recombined?

Ramage (1980) and Yeo and Flowers (1986) proposed a pyramiding approach to breeding a more salt tolerant genotype. In this scheme, individual physiological characters that would each aid increased salt tolerance would separately undergo selection (ie efficiency of the mechanism optimised). Subsequently the mechanisms (lines) are recombined to give optimal tolerance improvements. Assuming no linkage or segregation problems, this approach may well offer greater improvements, but the obvious requirements for knowledge, time and resources may prevent it being fruitful in the near future.

Ion exclusion as a selection criterion

In the most simplified description, salinity has osmotic and specific ion effects on plant growth. The primary responses of glycophytes (most agricultural plants) to salinity are the restricted transport of salt to the shoots and the maintenance of a favourable water balance by the synthesis of organic solutes (Läuchli and Epstein, 1990). Within a particular glycophytic species, variation in salt tolerance might be expected to be associated with variation in one or both these basic responses. The reader is referred to papers by Wainwright (1980) and Wyn Jones and Gorham (1983) for consideration of osmoregulation, the discussion here is restricted to ion accumulation.

For many glycophytes, but not all, differences in salt tolerance between varieties or lines have been closely associated with reduced uptake and accumulation of Na and/or Cl ions at the whole plant, shoot or leaf level (Bernstein et al., 1969; Läuchli and Wieneke, 1979; Salim, 1989). In such cases, tolerance improvement might be facilitated by using ion exclusion from certain tissues as a broad physiological selection criterion. However, Läuchli (1984) noted examples where higher salt tolerance was not related to a higher degree of Na or Cl exclusion. Läuchli and Epstein (1990) believed salt exclusion to be of limited usefulness as a mechanism of salt tolerance in the whole plant.

Legumes are either sensitive or moderately sensitive to salinity (Maas and Hoffman 1977) and are among the most important crop species. Läuchli (1984) noted that most legumes respond to salinity by salt exclusion from the leaves. Tolerance has been associated with Na and/or Cl exclusion from the shoots in many species, including soybean (Abel and Mackenzie, 1964; Wieneke and Läuchli, 1979), alfalfa (Brown and Hayward, 1950; Noble et al., 1984), *Trifolium alexandrinum* (Winter and Läuchli, 1982). An evaluation of white clover (*Trifolium repens*) cultivars by the authors has shown similar results. Five cultivars were grown in a half-strength Hoagland solution (Hoagland and Arnon, 1950), modified after Karmoker and Van Steveninck (1978), with added NaCl of 0, 20, 40 and 60 mol m^{-3}. After exposure to these conditions for 150 days in a greenhouse, significant variation in salt response was observed (Fig. 1). The more tolerant cultivars were found to have lower shoot Cl concentrations (Fig. 2). Across all cultivars, the dry weight and chloride concentrations of the shoot were related by the equation: Dry weight = 12.71 (+/−0.83 se.) − 1.92(+/−0.22 se.) Cl concn., $r^2 = 0.60$,

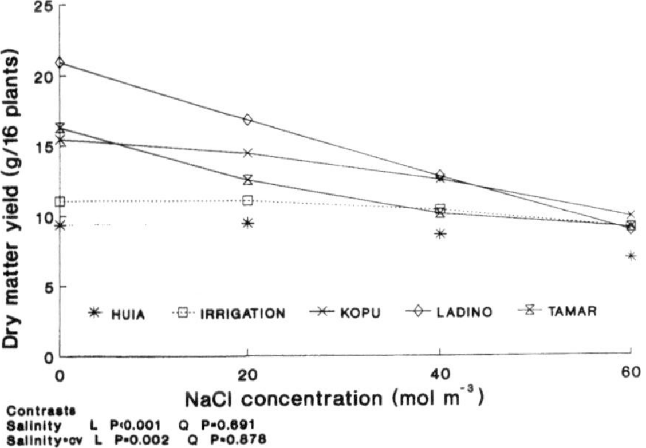

Fig. 1. Dry matter of shoots of five cultivars of *Trifolium repens* when grown in solution culture (greenhouse) with a range of NaCl concentrations. (Contrasts: L = Linear, Q = Quadratic).

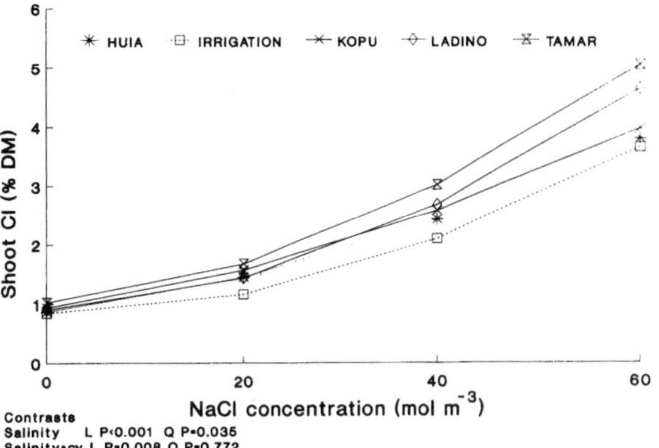

Fig. 2. Concentration of Cl in the shoots of five cultivars of *Trifolium repens* when grown in solution culture (greenhouse) with a range of NaCl concentrations.

where Cl concentration is in percent dry weight and dry weight is in mg per 16 (number of plants per replicate) plants (Rogers and Noble, unpublished).

Similarly, intra-varietal differences linked with Na or Cl exclusion have been reported for alfalfa (Noble et al., 1984) and white clover (Noble and Shannon, 1988). When a population of more than 500 plants of the white clover cultivar Haifa were grown in salinised nutrient solution at 40 mol m^{-3} NaCl in a greenhouse, substantial plant-to-plant variation in yield (shoot dry weight) and shoot Cl concentration was observed. Co-efficients of variation were 48% and 31% respectively, with yield and shoot Cl concentration closely associated and described by the relationship: Dry weight = 1.52 (+/−0.30 se.) − 0.49(+/−0.02 se.) Cl concn., $r^2 = 0.52$, where Cl concentration is in percent dry weight and dry weight is in mg per plant (Rogers and Noble, unpublished).

For the forementioned leguminous species, ion exclusion from the shoot might be an effective selection criterion. This approach has been used in the woody species citrus (Furr and Ream, 1969; Sykes, 1985a) and grapes (Downton, 1977; Sykes, 1985b). However, although Abel (1969) showed Cl exclusion from soybean shoots was controlled by a single gene pair and Flowers and Yeo (1990) selected for high and low sodium transporting lines of rice, there is a lack of studies for non-woody crops using Na or Cl exclusion as a selection criterion for breeding increased salt tolerance.

A capacity to restrict shoot Cl accumulation is no doubt only a broad physiological mechanism, the integrated result of many physiological characters. These include control of uptake at the root plasmalemma and tonoplasts of the cortex (Cram, 1973; Douglas and Walker, 1984), the accumulation and release of ions from the stele to the xylem, reabsorption of ions by xylem parenchyma cells (Läuchli and Wieneke, 1979), phloem translocation and compartmentation in older leaves (Greenway and Munns, 1980; Lessani and Marschner, 1978). Nevertheless, shoot Cl concentration is a tangible parameter that can be used for selection and breeding. We do not yet understand the operation of each component character, and may not be aware of all components, or how they interrelate for a particular species. Even further distant is an understanding of their genetic control. When the specific physiological characters are identified and understood, and their genetic control determined, this might provide opportunities for molecular biologists to utilise techniques such as recombinant DNA to effect rapid advances in tolerance breeding.

It is advocated that given the evidence linking maintenance of low shoot Cl and/or Na concentrations with tolerance for many species, particularly legumes, that this be more widely evaluated as a (broad) physiological selection criterion. Following the identification of both intraspecific and intravarietal variation for salt tolerance in white clover, and the apparent linkage of reduced accumulation of Cl in the shoot, the authors have investigated the approach of using ion exclusion as a selection criterion in this species. This follows previous selection using yield as a criterion that proved, at least initially, to be ineffective (Noble and Shannon, 1988). These authors also found tissue Cl concentration had a more significant, negative relationship with yield than did tissue Na concentration. Using a solution culture at 40 mol m^{-3} NaCl as noted previously, selections have been made within the cultivar Haifa for plants that consistently maintain low shoot Cl levels over 3 cut-regrowth cycles (approximately 21 day intervals). Two cycles of recurrent selection have been made and the resulting line (called 'Low-Cl') has shown a significant reduction in shoot Cl levels of the population (Table 2). Yields have been increased, although the increase just failed to be significant. The selected line and cultivar Haifa have also been evaluated over a range of salinities in solution culture (three weeks exposure to NaCl) and show increasing differences in shoot Cl levels (Fig. 3). These results show that the capacity to control shoot Cl concentrations in

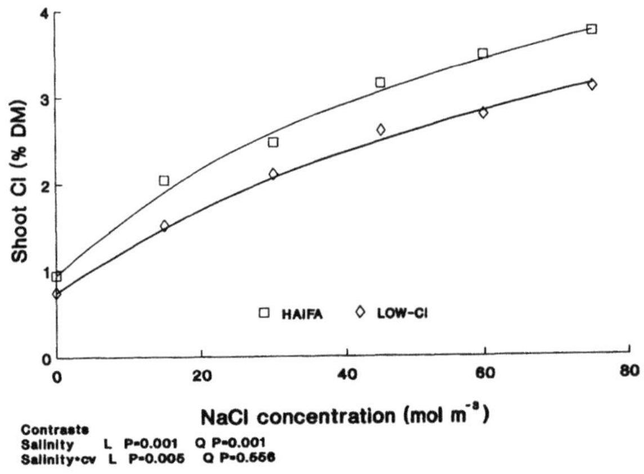

Fig. 3. Concentration of Cl in the shoots of the *Trifolium repens* cultivar Haifa and selection derived line Low-Cl when grown in solution culture (growthroom) with a range of NaCl concentrations.

Table 2. Mean plant dry weight and Cl concentration of the shoots for the *Trifolium repens* cultivar Haifa and selection derived line Low-Cl when grown at 0 and 40 mol m^{-3} added NaCl

Salinity (mol m^{-3} NaCl)	Dry weight (mg)		Shoot Cl (% dry weight)	
	Haifa	Low-Cl	Haifa	Low-Cl
0	577	592	0.567	0.530
40	406	517	2.77	1.90
LSD ($p = 0.05$)	Salinity * line = 135		Salinity * line = 0.70	

white clover is a heritable character and improvements in tolerance may be obtained if it is used as a selection criterion. Further, whilst restricted accumulation of Cl might be associated with yield at low or moderate salinities, it might also be expected to provide plants with better survival potential at high salinities and an improved capacity to tolerate at least short term variations in soil salinities. However, in the regression equations noted earlier, shoot Cl concentration didn't explain all the variability observed. Also, the reduction in shoot Cl levels via selection was not reflected in a similarly large increase in yield. Thus control of Cl levels is only one component of tolerance in white clover. Maximising tolerance will require selection for other components, such as ion distribution between leaf tissue of different ages and/or the type and levels of organic solutes synthesised to aid osmoregulation. Results for other legumes hold promise in this area. Wyn Jones and Storey (1981) found proline-betaine accumulated more in a salt tolerant cultivar of alfalfa than in a salt sensitive one. In essence, a pyramiding approach (after Yeo and Flowers, 1986), where shoot Cl control is one component, would appear applicable in white clover.

Analytical requirements for physiological criteria

At present, the more specific the physiological mechanism under investigation the more difficult and refined the measurement technique and the more specialised the equipment that is required. Time requirements also increase.

New equipment development to speed analysis and minimise sample preparation would substantially benefit the breeder. An important part of efforts to develop more tolerant genotypes that is largely ignored is equipment to facilitate rapid analysis of our parameters. Analytical chemists and equipment developers need to be involved in our programs and kept abreast of our requirements. We should optimise measurement of parameters we do understand or can use now rather than wait until we have clarified specific mechanisms.

Conclusion

Agronomic criteria for salt tolerance currently offer the quickest and most integrated approach to developing selection and breeding programs to increase salt tolerance in most species. Physiological criteria probably offer more scope for maximising tolerance improvements and offer various levels of integration with respect to the operative physiological characters. However, they can have constraints of time and resource inputs and knowledge deficiencies that can preclude their effectiveness. There is a need for concurrent breeding programs that evaluate the effectiveness and input requirements of using selection criteria that are based upon agronomic, broad physiological mechanisms (processes) or specific physiological characters. It has been suggested here that for many glycophytes, particularly legumes, a broad physiological criterion of shoot Na and/or Cl exclusion might provide useful improvements in the short term. A pyramiding approach using a range of (broad) physiological criteria is likely to provide further advances. Concurrently, physiological research should investigate the component characters of ion exclusion, including the hierarchy and causal relationships involved, with a view to refining our selection criteria in the longer term. Such refinement, if it is to be practical, must include the development of techniques and analytical hardware that permit rapid detection. This will require active involvement of expertise from a range of disciplines throughout the breeding program.

References

Abel G H 1969 Inheritance of the capacity for chloride inclusion and exclusion in soybeans. Crop Sci. 9, 697–699.

Abel G H and MacKenzie A J 1964 Salt tolerance of soybean varieties during germination and later growth. Crop Sci. 4, 157–160.

Akbar M and Yabuno T 1977 Breeding for salt-resistant varieties of rice. IV. Inheritance of delayed-type panicle sterility induced by salinity. Jpn. J. Breed. 27, 237–240.

Ayers A D, Wadleigh G H and Bernstein L 1951 Salt tolerance of six varieties of lettuce. Proc. Am. Soc. Hortic. Sci. 57, 237–242.

Bernstein L, Ehlig C F and Clark R A 1969 Effect of grape rootstocks on chloride accumulation in leaves. J. Am. Soc. Hortic. Sci. 94, 584–590.

Brown J W and Hayward H E 1956 Salt tolerance of alfalfa varieties. Agron J. 48, 18–20.
Cram W J 1973 Chloride fluxes in cells of the isolated root cortex of *Zea mays*. Aust. J. Biol. Sci. 26, 757–779.
Cooper W C 1951 Salt Tolerance of Avocados on Various Rootstocks. Texas Avocado Soc. Yearbook 24–80, Weslaco, TX.
Cordukes W E 1981 Saltol, a salt tolerant red fescue. Can. J. Plant Sci. 61, 761–764.
Day A D 1987 Registration of Arizona 8601 maize germplasm for saline environments. Crop Sci. 27, 1096.
Dewey D R 1962 Breeding crested wheatgrass for salt tolerance. Crop Sci. 2, 403–407.
Dobrenz A K, Schonhorst M H, Stone J E, Thompson R K, Allen S and Smith D 1983 AZ-Germ Salt-1 nondormant alfalfa germplasm. Crop Sci. 23, 807.
Douglas T J and Walker R R 1984 Phospholipids, free sterols and adenosine triphosphatase of plasma membrane-enriched preparations from roots of citrus genotypes differing in chloride exclusion ability. Physiol. Plant. 62, 51–58.
Downton W S 1977 Photosynthesis in salt-stressed grapevines. Aust. J. Plant Physiol. 4, 183–192.
Dracup M 1991 Increasing salt tolerance of plants through cell culture requires greater understanding of tolerance mechanisms. Aust. J. Plant Physiol. 18, 1–15.
Dvorak J, Ross K and Mendlinger S 1985 Transfer of salt tolerance from *Elytrigia pontica* (Podp.) Holub to wheat by the addition of an incomplete *Elytrigia* genome. Crop Sci. 25, 306–309.
Dvorak J and Ross K 1986 Expression of tolerance of Na^+, K^+, Mg^{2+}, Cl^-, and SO_4^{2-} ions and sea water in the amphiploid of *Triticum aestivum* × *Elytrigia elongata*. Crop Sci. 26, 658–660.
Epstein E 1985 Salt tolerant crops: Origins, development and prospects of the concept. Plant and Soil 89, 187–198.
Epstein E, Norlyn J D, Rush D W, Kingsbury R W, Kelley D B, Cunningham G A and Wrona A F 1980 Saline culture of crops: A genetic approach. Science 210, 399–404.
Flowers T J and Yeo A R 1986 Ion relations of plants under drought and salinity. Aust. J. Plant Physiol. 13, 75–91.
Flowers T J and Yeo A R 1990 Salinity and rice: A physiological approach to breeding for resistance. *In* Proc. Int. Plant Physiol Congress. Eds. S K Sharma, P V Sabe, S C Bhargava and P K Agrawal. pp 953–959. Soc. Pl. Physiol Biochem., New Delhi, India.
Furr J R and Ream C L 1969 Breeding citrus rootstocks for salt tolerance. *In* Proceedings of 1st Int. Citrus Symposium, 1968. Ed. H D Chapman. Vol. 1, pp 373–380. University of California, Riverside, CA.
Gad El-hak M 1966 Breeding a new variety of rice adapted to recently reclaimed saline soils. Agric. Res. Rev. (Cairo) 44, 1–13.
Gottschalk W 1981 Mutation: Higher plants. Prog. Bot. 43, 139–152.
Greenway H 1973 Salinity, plant growth and metabolism. J. Aust. Inst. Agric. Sci. 39, 24–34.
Greenway H and Munns R 1980 Mechanisms of salt tolerance in nonhalophytes. Annu. Rev. Plant Physiol. 31, 149–190.
Hannon N J and Bradshaw A D 1968 Evolution of salt tolerance in two coexisting species of grasses. Search 3, 259–260.
Hoagland D R and Arnon D I 1950 The water culture method for growing plants without soil. Calif. Agric. Exp. Stn. Circ. No. 347.
Karmoker J L and Van Steveninck R F M 1978 Stimulation of volume flow and ion flux by abscisic acid in excised root systems of *Phaseolus vulgaris* L. cv. Redland Pioneer. Planta 141, 37–43.
Larkin P J and Scowcroft W R 1981 Somaclonal variation – a novel source of variability from cell cultures for plant improvement. Theor. Appl. Genet. 60, 197–214.
Läuchli A 1984 Salt exclusion: An adaptation of legumes for crops and pastures under saline conditions. *In* Salinity Tolerance in Plants: Strageiges for Crop Improvement. Eds R C Staples and G H Toenniessen. pp 171–187. Wiley, New York.
Läuchli A and Epstein E 1990 Plant responses to saline and sodic conditions. *In* Agricultural Salinity Assessment and Management. Ed. K K Tanjii. pp 113–137. ASCE, New York.
Läuchli A and Wieneke J 1979 Studies on growth and distribution of Na, K and Cl in soybean varieties differing in salt tolerance. Z. Pflanzenernaehr. Bodenkd. 142, 3–13.
Lessani H and Marschner H 1978 Relation between salt tolerance and long-distance transport of sodium and chloride in various crop species. Aust. J. Plant Physiol. 5, 27–37.
Maas E V and Hoffman G J 1977 Crop salt tolerance – current assessment. J. Irrig. Drainage Div. ASCE 103, 115–134.
Marschner H, Kylin A and Kuiper P J C 1981 Differences in salt tolerance of three sugar beet genotypes. Physiol. Plant. 51, 234–238.
Noble C L 1983 The potential for breeding salt tolerant plants. Proc. Royal Soc. Vict. 95, 133–138.
Noble C L, Halloran G M and West D W 1984 Identification and selection for salt tolerance in lucerne (*Medicago sativa* L.) Aust. J. Agric. Res. 35, 239–252.
Noble C L and Shannon M C 1988 Salt tolerance selection of forage legumes using physiological criteria. *In* Proc. Int. Plant Physiol Congress. Eds. S K Sharma, P V Sabe, S C Bhargava and P K Agrawal. pp 989–994. Soc. Pl. Physiol Biochem., New Delhi, India.
Norlyn J D 1980 Breeding salt tolerant crop plants. *In* Genetic Engineering of Osmoregulation. Eds. D W Rains, R C Valentine and A Hollaender. Basic Life Sci. 14, 293–309. Plenum Press, New York.
Pasternak D 1987 Salt tolerance and crop production – a comprehensive approach. Annu. Rev. Phytopathol. 25, 271–291.
Ramage R T 1980 Genetic methods to breed salt tolerance in plants. *In* Genetic Engineering of Osmoregulation. Eds. D W Rains, R.C. Valentine and A. Hollaender. Basic Life Sci. 14, 311–318.
Rush D W and Epstein E 1981a Breeding and selection for salt tolerance by the incorporation of wild germplasm into a domesticated tomato. J. Am. Soc. Hortic. Sci. 106, 669–704.
Rush D W and Epstein E 1981b Comparative studies on the

sodium, potassium, and chloride relations of a wild halophytic and a domestic salt-senstive tomato species. Plant Physiol, 68, 1308–1313.

Salim M 1989 Salinity effects on growth and ionic relations of two *Triticale* varieties differing in salt tolerance. J. Agron. Crop Sci. 162, 35–42.

Shannon M C 1978 Testing salt tolerance variability among tall wheatgrass populations. Agron. J. 70, 719–722.

Shannon M C 1985 Principles and strategies in breeding for higher salt tolerance. Plant and Soil 89, 227–241.

Shannon M C, McCreight J D and Draper J H 1983 Screening tests for salt tolerance in lettuce. J. Am. Soc. Hortic. Sci. 108, 225–230.

Shannon M C and Noble C L 1990 Genetic approaches for developing economic salt-tolerant crops. *In* Agricultural Salinity Assessment and Management. Ed. K K Tanjii. pp 161–185. ASCE, New York.

Smith M K and McComb J A 1981 Effect of NaCl on the growth of whole plants and their corresponding callus cultures. Aust. J. Plant Physiol. 8, 267–275.

Sykes S R 1985a A glasshouse screening procedure for identifying citrus hybrids which restrict chloride accumulation in shoot tissues. Aust. J. Agric. Res. 36, 779–789.

Sykes S R 1985b Variation in chloride accumulation by hybrid vines from crosses involving the cultivars Ramsay, Villard blanc and Sulatana. Am. J. Enol. Vitic. 36, 30–37.

Tal M 1985 Genetics of salt tolerance in higher plants, theoretical and practical considerations. Plant and Soil 89, 199–226.

Tal M and Shannon M C 1983 Salt tolerance in the wild relatives of the cultivated tomato, responses of *Lycopersicon esculentum*, *L. cheesmanii*, *L. peruvianum*, *Solanum pennellii* and F_1 hybrids to high salinity. Aust. J. Plant Physiol. 10, 109–117.

Taylor R M, Young E F Jr and Rivera R L 1975 Salt tolerance in cultivars of grain sorghum. Crop Sci. 15, 734–735.

Ulrich A 1961 Variability of sugar beet plants grown in pots without competition for light, water and nutrients. J. Am. Soc. Sugar beet Technology. 11, 595–604.

Venables A V and Wilkins D A 1978 Salt tolerance in pasture grasses. New Phytol. 80, 613–622.

Wainwright S J 1980 Plants in relation to salinity. Adv Bot. Res. 8, 221–261.

Winter E and Lauchli A 1982 Salt Tolerance of *Trifolium alexandrinum* L. I. Comparison of the salt response of *T. alexandrinum* and *T. pratense*. Aust. J. Plant Physiol. 9, 221–226.

Wyn Jones R G and Gorham J 1980 Osmoregulation. *In* Encyclopedia of Plant Physiology, New Series, Vol 12C. Physiological Plant Ecology III. Eds. O L Lange, P S Nobel and C B Osmond. pp 35–58. Springer-Verlag, Berlin.

Wyn Jones R G and Storey R G 1981 Betaines. *In* The Physiology and Biochemistry of Drought Resistance in Plants. Eds. L G Paleg and D Aspinall. pp 171–204. Academic Press, New York.

Yeo A R and Flowers T J 1983 Varietal differences in the toxicity of sodium ions in rice leaves. Physiol. Plant. 59, 189–195.

Yeo A R and Flowers T J 1986 Salinity resistance in rice (*Oryza sativa* L.) and a pyramiding approach to breeding varieties for saline soils. Aust. J. Plant Physiol. 13, 161–173.

P. J. Randall et al. (Eds.), Genetic aspects of plant mineral nutrition, 137–142.

PLSO SV38

Why does in vitro cell selection not improve the salt tolerance of plants?

MILES DRACUP
WA Department of Agriculture, South Perth, Western Australia 6151

Key words: cell culture, Cl^-, growth, Na^+, plant breeding, salinity, salt, salt tolerance, salt toxicity, selection, suspension-culture, water deficit

Abstract

For over a decade cell culture has been advocated in selection programmes for salt tolerance. Selecting cultured cells for survival at high NaCl potentially offers a fast means for generating, evaluating and selecting genotypes with superior salt tolerance. Furthermore, little space is required and the environment can be controlled. However, plants regenerated from selected cells have not shown unequivocal increases in salt tolerance. Clearly the role of cell culture needs to be reconsidered. This logically begins with an examination of their mechanisms of salt tolerance and how these mechanisms relate to salt tolerance in whole plants. Unfortunately little is known about the mechanisms of salt tolerance in cultured cells, largely due to poor methodology and reluctance to search for mechanisms and test hypotheses. Nonetheless there are ample grounds to propose that tolerance mechanisms in cultured cells are inappropriate or even deleterious to salt tolerance in whole plants. This is partly because salt tolerance is multi-genic and depends on the structural and physiological integrity of the whole plant, and partly because of differences in mechanisms of salt tolerance between cells in culture and cells in whole plants. In vitro selection should therefore be confined to traits where whole plant tolerance is cell based and then only with caution.

Introduction

In vitro selection of cells has frequently been suggested as a means of improving the salt tolerance of plants (e.g. Downton, 1984; Larkin and Scowcroft, 1981; Raghava Ram and Nabors, 1985). This involves subjecting a population of cells, either as callus or cell suspensions to high concentrations of NaCl, recovering variant cells which tolerate the salt and regenerating plants from these selected cells (Fig. 1). Potential advantages of the procedure over those traditionally used for plant breeding include:

1. Variation may be generated during culture;
2. Relatively little space is required during selection in the laboratory;
3. Time lag between generations is considerably reduced; and
4. The environment can be controlled.

However, obviously tolerance to the stress must operate both in vitro and in the regenerated plants.

After selection in culture, cells have often tolerated far higher NaCl concentrations than the original cells were capable of, but despite numerous attempts there are no known instances of increased salt tolerance being expressed unequivocally in regenerated plants (e.g. Rowland et al., 1989; Watad et al., 1991). This can not be dismissed as a lack of stability of the in vitro salt tolerance characteristics: Although plants regenerated from cells of *Nicotiana tabacum/gossii* selected for salt tolerance showed no increase in salt tolerance, a high level of salt tolerance was retained in cell cultures subsequently derived from these plants (Watad et al., 1991). Nabors et al. (1980) claimed increased salt tolerance in plants regenerated from selected cells but did not

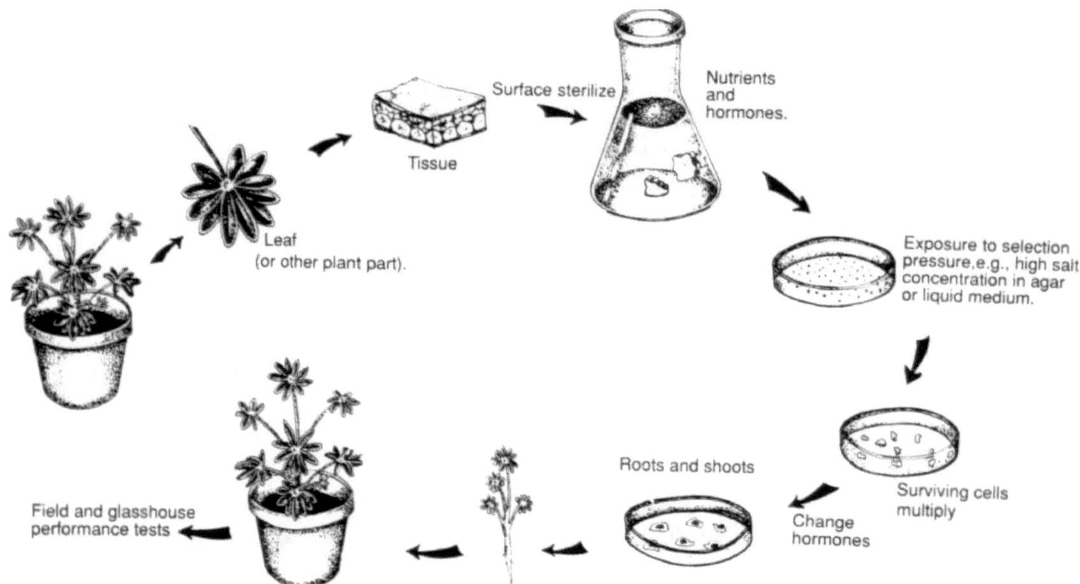

Fig. 1. In vitro selection for salt tolerance.

establish the salt tolerance of plants regenerated from unselected cells or the range of salt tolerance in the original population of plants.

Here I will examine the basis for the lack of success of in vitro selection for salt tolerance in non-halophytes. This necessarily involves comparing mechanisms of salt tolerance in cultured cells with limiting processes in whole plants at high NaCl. The growth of non-halophytic plants is most likely limited by water deficit in the roots in the short term (Munns and Termaat, 1986) and ion excess or water deficit in leaves in the long term, particularly the mature leaves (Flowers and Yeo, 1986; Munns and Termaat, 1986). Data will mainly be drawn from studies on suspension-cultured cells, primarily since they live in a more homogeneous environment than do calluses; for instance concentration gradients of NaCl develop between the medium and the plasmalemma of callus cells (Gibbs et al., 1989).

Given the limited data available and the past reluctance to search for mechanisms and test hypotheses on salt tolerance of cultured cells this examination is necessarily rather speculative. However, apart from its value in the present context, it provides a framework for less pragmatic future use of the technique and for focusing future work on salt tolerance of cultured cells. Improving salt tolerance of plants using other in vitro techniques but where selection does not occur during the culture phase is not considered; such techniques include embryo culture, clonal propagation, generation of somaclonal variation and gene transfer.

Differences between salt tolerance in plants and cultured cells

Ignoring the free space is a fundamental technical problem with most studies on salt tolerance in cultured cells. Cell pellets usually have a free space of over 50% (Dracup et al., 1986; Lorences and Fry, 1991; Reuveni et al., 1985) so if ignored, solutes and solution in the free space can introduce substantial errors into estimates of growth, concentrations of cellular solutes and cellular osmotic pressure. Furthermore, the errors could be confounded by salinity treatments which may affect cell size (Flowers et al., 1985) and thus the free space volume. The magnitude of the errors can be illustrated by considering a hypothetical example of a culture medium containing 150 mol m^{-3} NaCl (osmotic pressure = 1.2 MPa and dissolved solutes = 43 Kg m^{-3}) and cells with a fresh weight of 10 g, dry weight of 0.7 g, internal Na^+ concentration of 50 mol m^{-3} and turgor pressure of 0.5 MPa. Ignoring a 50% free space volume would lead to protoplast mass being overestimated by more than 100% (fresh weight) or 60% (dry weight), cellular Na^+ being overestimated by 100% and turgor pressure being underestimated by 50% if determined by sap expression. Ignoring the free space and failure to wash cell pellets therefore makes much of the data on salt tolerance in cultured cells uninterpretable, although Dracup (1991) was able to recalculate some of the data where the composition of the free space was known and by assuming a 50% free space volume.

Further limitations of approaches to assessing salt tolerance in cultured cells and which are largely conceptual in nature will be examined below.

Growth

Batch cell cultures cease to grow when the most limiting nutrient becomes depleted so final cell mass relates to the initial concentration of the most limiting nutrient and the cellular requirements for it. However, this final mass has frequently been the basis of (equivocal) assessments of salt tolerance of cell lines. Assessing salt tolerance from measurements of cell mass at a single point in time are also equivocal since it ignores the stage of the culture cycle. Since the duration of the various phases of culture growth can be modified by high concentrations of NaCl (Dracup, 1991) spurious comparisons are easily made. Growth rate is also an inappropriate measurement since it is sensitive to the size of the original inoculum. Depending on the aims of the selection programme these methods of assessing salt tolerance may be relevant to whole plants but need to be reconsidered for cultured cells which are heterotrophic and grow under markedly different constraints to whole plants. Tolerance of cultured cells to high NaCl is best assessed from relative growth rates during the exponential phase of growth.

Relative growth rates of cells of halophytes appear less affected by high NaCl than do those of non-halophytes (Dracup, 1991). The distinction between cells from the two groups of plants may reflect cell-based differences in uptake and compartmentation of Na^+ and Cl^-, sensitivity to sudden increases in NaCl (i.e. osmotic pressure) and/or high $Na^+:Ca^{2+}$ ratios (which halophytes appear more tolerant of than non-halophytes (Greenway and Munns, 1980)).

Unfortunately few comparisons are possible between the salt tolerance of cultured cells and whole plants. These few show that for all species tested, except *Suaeda maritima*, growth rates of cultured cells have similar levels of tolerance as do growth rates of the whole plants (Dracup, 1991). This is surprising since there are a number of instances of cells *surviving* relatively high NaCl concentrations in vitro, particularly after selection (e.g. Raghava Ram and Nabors, 1985), far higher than the plants would survive. For instance cultured cells of rice survived 600 mol m^{-3} NaCl yet seedlings die at only 50 mol m^{-3} NaCl (Flowers et al., 1985). Furthermore, at 200 mol m^{-3} NaCl respiration rates decreased by only 38% in roots of rice (Yeo and Flowers, 1986) and were not affected in cultured cells of rice (Flowers et al., 1985). This suggests that cells are much more tolerant of high NaCl than is the whole plant.

Ion excess

Shortly after exposing whole plants and cultured cells to the same concentration of NaCl, leaf cells are bathed in far lower concentrations of Na^+ than are cultured cells; Na^+ concentrations in the transpiration stream of barley are less than 6 mol m^{-3}, even when the soil solution contains 200 mol m^{-3} (Munns and Termaat, 1986). However, with time, transpirational delivery of Na^+ leads to a dilemma. For instance, Yeo and Flowers (1986) calculated that ion concentrations in expanded rice leaves would increase at about 75 mol m^{-3} day^{-1} with just 50 mol m^{-3} NaCl in the soil solution. The apoplast is only about 3% of the mesophyll volume (Flowers and Yeo, 1986) so virtually any salt exclusion would quickly lead to excessive build up of salt in the apoplast, leading to water deficit (Greenway and Munns, 1980). Alternatively, the cytoplasm is only 5–10% of the protoplast volume (Flowers and Yeo, 1986) so Na^+ uptake, coupled with inefficient compartmentation of Na^+ in the vacuole which may be a property of non-halophytes (Greenway and Munns, 1980) (although there is no evidence for poor compartmentation in rice (Yeo and Flowers, 1986)) may lead to salt build up in the cytoplasm, i.e. specific ion toxicity. There is little evidence to suggest that the phloem is important in removing ions from leaves (Flowers and Yeo, 1986). In the long term this excessive build up of ions in transpiring tissue probably restricts productivity of non-halophytes at high salinity (Munns and Termaat, 1986); progressive death of leaves caused by the salt build-up leads to declining photosynthetic area until carbohydrate production is insufficient to support growth.

Cultured cells do not face such a dilemma: external salt concentrations remain relatively constant so there is no penalty for excluding salt. Further, unlike cells in mature leaves of non-halophytes, cultured cells are growing, thus diluting salt which gets taken up. Cultured cells are therefore likely to be able to tolerate far higher external concentrations of salt than transpiring tissue.

Water deficit

After transfer to high NaCl, turgor tends to be maintained in both cultured cells Dracup (1991) and whole plants (Munns, 1988) but this should not be confused with turgor regulation (Dracup, 1991; Munns, 1988). However, the solutes used to maintain turgor could differ between cultured cells and whole plants with important consequences to the relevance of in vitro selection.

In the previous section the build up of ions in transpiring tissue of non-halophytic plants grown at high NaCl was discussed. Excluding ions from the protoplast of these tissues will inevitably lead to water deficit, although this would be delayed by the build up of organic osmotica inside the protoplast (Munns and Termaat, 1986). Organic solutes are also needed to balance ions compartmented in the vacuole (Greenway and Munns, 1980). However, organic solutes can only be supplied at the expense of growth (Munns, 1988); using monosaccharides for just 0.1–0.2 MPa of osmotic adjustment would reduce relative growth rate by 30% (Munns, 1988). Therefore, in whole plants osmotic adjustment with ions is the only way to combine productivity with salt tolerance (Greenway and Munns, 1980). This necessitates ion compartmentation in the vacuole, balanced by organic osmotica in the relatively small cytoplasm (Greenway and Munns, 1980).

In contrast, organic solutes can be used for osmotic adjustment in cultured cells with no penalty to growth since sugars are supplied in the culture medium (Flowers et al., 1985). Interestingly, Dracup and Greenway (1988) found that glucose is involved in turgor regulation of cultured tobacco cells. Cultured cells are therefore under no pressure to take up ions for osmotic adjustment, while reducing Na^+ uptake reduces the likelihood of specific ion toxicity.

Growing cells are more susceptible to water deficit than non-growing cells since osmotica also need to be taken up or synthesized at a rate sufficient to maintain expansion. Similarly, cultured cells grow considerably slower than cells in the root meristem (Bayliss, 1985) so, assuming similar uptake mechanisms and treatments, cultured cells should be less sensitive than roots to water deficit. However, water deficit in expanding tissue does not appear to be the primary cause of reduced long term productivity of non-halophytes (Munns and Termaat, 1986).

The high osmotic pressure of the medium for culturing cells could also increase the sensitivity to water deficit. Cell culture media commonly have osmotic pressures of 0.3–0.5 MPa (Dracup et al., 1987), whereas Hoagland's solution used for growing plants has an osmotic pressure of only 0.07 MPa (Epstein, 1972). Thus, at equivalent salinity treatments the osmotic pressure of the treatment solution would be higher in cultured cells but in leaves this would be at least partly offset by the water potential gradient between the roots and leaves generated by transpiration.

Assuming either specific ion toxicity or the rate of sugar uptake limits the salt tolerance of cultured cells it is possible that in vitro selection for salt tolerance is actually selecting for ion exclusion coupled with sugar uptake (Dracup, 1991; Flowers et al., 1985). However, increased reliance on organic osmotica in regenerated whole plants would reduce growth while decreased uptake of Na^+ and Cl^- in transpiring tissue may exacerbate development of water deficit in the leaves and accentuate demands for organic osmotica, although it may also prolong the life of the leaf by reducing the rate of NaCl accumulation. Clearly in vitro selection for salt tolerance could actually reduce the tolerance of regenerated plants! This may explain findings by Flowers et al. (1985) where, of two cultivars of rice, the most tolerant in vitro was the least tolerant in vivo.

Conclusions

Despite a large research effort, relatively little is known about the mechanisms of salt tolerance in cultured cells or how to select for salt tolerance in vitro that will be expressed in regenerated whole plants. The lack of knowledge largely stems from naive adoption of techniques used for studying and selecting for salt tolerance in whole plants. Work with whole plants has concentrated on the tops, yet cultured cells are more akin to cells in the root apex; they are growing, heterotrophic and bathed in the treatment solution. On a technical level, the very large free space of pellets of cultured cells has been largely ignored and makes a large percentage of the literature uninterpretable. Discussion in this review has therefore been largely a theoretical appraisal of in vitro selection for salt tolerance.

Salt tolerance in cultured cells potential differs in a number of ways from salt tolerance in whole plants and probably accounts for the lack of success of in vitro selection and the frequent survival of cultures at relatively high NaCl concentrations. Notable differences are:

1. Many responses of plants to high NaCl are associated with the functional integrity of the whole plant, particularly transport over long distances such as from roots to leaves, rather than merely cellular responses to high NaCl.
2. The nature and availability of solutes is very different for cultured cells than for cells in whole plants. Furthermore, the composition of the apoplast of suspension-cultured cells (although not of calluses) is similar to the culture medium but is very different in plant tops, particularly the Na^+ concentrations after prolonged growth at high NaCl.
3. Slow growth of cultured cells would make them less sensitive to water deficit than growing cells and less sensitive to ion toxicity than non-growing cells in whole plants.
4. Cell culture media have lower water potential than do culture solutions used for whole plants, although the transpiration-induced difference in water potential between leaves and roots is often of a similar magnitude.

We need more understanding of the mechanisms of salt tolerance in cultured cells. Without knowledge of the limiting process at high NaCl it is difficult to postulate what is being selected for in vitro; salt exclusion? salt compartmentation into vacuoles? sugar uptake? Furthermore, we do not know how and where traits exhibited by cultured cells are expressed in whole plants where the extracellular environment of cells is very different.

If in vitro selection is to be continued, then fresh approaches should be taken which involve selection for specific physiological traits which operate at the cellular level of organisation. For instance, some genotypic differences in salt tolerance of non-halophytes may be related to differences in ion compartmentation (Flowers and Yeo, 1986) so cultured cells could be selected to accumulate and compartment Na and Cl ions in the vacuole. The vacuole is the only compartment with substantial storage capacity and the contents are separated from the important metabolic functioning in the cytoplasm. Complications arising from high sugar supply could be avoided by developing photoautotropic culture systems. Selection pressure could be generated using a non-permeating osmoticum such as melibiose to lower the water potential whilst supplying NaCl at physiologically relevant concentrations of 5–10 mol m^{-3} (the concentration delivered to transpiring tissue (Munns and Termaat, 1986)). Selection for tolerance to high $Na^+:Ca^{2+}$ ratios is another avenue where in vitro selection might have a role.

Acknowledgements

The organising committee of the Fourth International Symposium on Genetic Aspects of Plant Mineral Nutrition provided support to present this paper.

References

Bayliss M W 1985 Regulation of the cell division cycle in cultured plant cells. *In* The Cell Division Cycle in Plants.

Eds. J A Bryant and D Francis. pp 157–177. Cambridge University Press, Cambridge.

Downton W J 1984 Salt tolerance of food crops: Perspectives for improvements. Crit. Rev. Plant Sci. 1, 183–201.

Dracup M 1991 Increasing salt tolerance of plants through cell culture requires greater understanding of tolerance mechanisms. Aust. J. Plant Physiol. 18, 1–15.

Dracup M and Greenway H 1988 Regulation of turgor pressure by suspension-cultured tobacco cells. J. Exp. Bot. 39 1591–1603.

Dracup M Gibbs J and McComb J A 1987 Media with low osmotic pressures but adequate concentrations of nutrients for batch culture of tobacco cells. J. Plant Physiol. 127, 281–291.

Dracup M, Gibbs J, Stuiver C E E, Greenway H and Flowers T J 1986 Determination of free space, growth, solute concentrations and parameters of water relations of suspension-cultured tobacco cells. Plant Cell Environ. 9, 693–701.

Epstein E 1972 Mineral Nutrition of Plants: Principles and Perspectives. Wiley, New York. p 39.

Flowers T J, Lachno D R, Flowers S A and Yeo A R 1985 Some effects of sodium chloride on cells of rice cultured in vitro. Plant Sci. Lett. 39, 205–221.

Flowers T J and Yeo A R 1986 Ion relations of plants under drought and salinity. Aust. J. Plant Physiol. 13, 75–91.

Gibbs J, Dracup M, Greenway H and McComb J A 1989 Effects of high NaCl on growth, turgor and internal solutes of tobacco callus. J. Plant Physiol. 134 61–69.

Greenway H and Munns R 1980 Mechanisms of salt tolerance in nonhalophytes. Annu. Rev. Plant Physiol. 31, 139–190.

Larkin P J and Scowcroft W R 1981 Somaclonal variation: A novel source of variability from cell cultures for plant improvement. Theor. Appl. Genet. 60, 197–214.

Lorences E P and Fry S C 1991 Absolute measurement of cell expansion in plant cell suspension cultures. Plant Cell Tiss. Org. Cult. 24, 211–215.

Munns R 1988 Why measure osmotic adjustment? Aust. J. Plant Physiol. 15, 717–726.

Munns R and Termaat A 1986 Whole-plant responses to salinity. Aust. J. Plant Physiol. 13, 143–160.

Nabors M W, Gibbs S E, Bernstein C S and Meis M E 1980 NaCl-tolerant tobacco plants from cultured cells. Z. Pflanzenphysiol. 97, 13–17.

Raghava Ram N V and Nabors M 1985 Salinity tolerance. *In* Biotechnology: Applications and Research. Eds. P N Cheremisinoff and R P Quellette. pp 623–642. Technomic, Lancaster.

Reuveni M, Lerner H R and Poljakoff-Mayber A 1985 Changes in membrane potential as a demonstration of selective pore formation in the plasmalemma by poly-1-lysine treatment. Plant Physiol. (Bethesda) 79, 406–410.

Rowland G G, McHughen A and McOnie C 1989 Field performance at saline-affected sites of a somaclonal variant of McGregor flax selected for salt tolerance in vitro. Can. J. Plant Sci. 69, 49–60.

Watad, A A, Swartzberg D, Bressan, R A, Izhar S and Hasegawa P M 1991 Stability of salt tolerance at the cell level after regeneration of plants from a salt tolerant tobacco cell line. Physiol. Plant. 83, 307–313.

Yeo A R and Flowers T J 1986 Salinity resistance in rice (*Oryza sativa* L.) and a pyramiding approach to breeding varieties for saline soils. Aust. J. Plant Physiol. 13, 161–173.

P. J. Randall et al. (Eds.), Genetic aspects of plant mineral nutrition, 143–150.
 PLSO SV30

Variation and inheritance of sodium transport in rice

A. R. YEO
School of Biological Sciences, The University of Sussex, Brighton BN1 9QG, UK

Key words: *Oryza sativa* L, plant breeding, rice, salinity, selection, sodium transport

Abstract

Sodium transport in rice is characterised by large variability between individual plants, and large environmental interaction. As a result of these two factors, plant sodium content is a continuous variable which is not distributed normally. This applies both to the quantity of sodium in the plant and to the concentration of sodium on a unit fresh or dry weight basis. This variability is in part because the transpirational by-pass flow, dependent upon root anatomy and development, contributes to sodium uptake. Variability in sodium content within designated cultivars is heritable and line selections diverge during recurrent selection, suggesting that selection is working on residual heterozygosity rather than on a family of homozygous lines. Varieties differ in average sodium uptake into the plant but the direct correlation of this with survival is weak. This is because other independent characters are important (and these have not been combined by natural selection nor by chance) and because overall performance is confounded by the spurious advantage of the tall (non-dwarf) plant type. This advantage is spurious because much of it is due to plant size rather than to any genetic information for salt tolerance. The benefit deriving from plant size will not be heritable in crosses with genotypes of the improved (dwarf), high-yielding plant type because the dwarfing genes are dominant. Sodium transport is heritable in crosses, and the results presented show that both low sodium transport and low sodium to potassium ratio can be selected independently of plant type. This allows the selection of dwarf plants (which are agronomically desirable) with low sodium transport (which will improve salt tolerance).

Introduction

Many traits contribute to the resistance of salinity in rice. They all act by reducing or mitigating the principal problem; which is that the influx of sodium chloride with the transpiration stream is excessive, and leads to internal concentrations of salt which are toxic (Yeo and Flowers, 1986). Concentrations of salt in the shoot can be reduced by lowering sodium transport to the shoot, by plant vigour (which provides for dilution of salt by growth) and by high water use efficiency (which reduces influx with the transpiration stream per unit growth). Compartmentation of salt within the plant (from leaf to leaf) and within the tissue (between protoplast and apoplast and between cytoplasm and vacuole) can greatly alter the impact of a given salt load in the shoot (Flowers et al., 1991). Salt damage occurs at low *external* salinities which do not, per se, reduce growth; damage results from the *internal* accumulation of salt (Yeo and Flowers, 1986; Yeo et al., 1991). The immediate aim in breeding for increased salt resistance in rice is for a reduction in the salt concentration in the shoot.

Vigour and water-use efficiency are associated with the tall, unimproved plant type whilst modern varieties invariably include dwarfing genes (Flowers et al., 1988; Yeo et al., 1990). The dwarf plant type has a higher harvest index, better light interception because of its erect leaves, and much reduced lodging. Vigour is essential in plants that are going to be productive in saline environments (Richards, 1983), so some compromise may be necessary over the degree of dwarfing in varieties intended for saline soils. A clear goal is to develop a plant of, at most, moderate stature with low sodium transport.

Sodium transport in rice is characterised by large individual variability which is not distributed normally (Flowers and Yeo, 1981; Yeo et al., 1988); the range in sodium concentration in the leaves of a population of even a designated cultivar grown with 50 m*M* NaCl is generally two orders of magnitude or more. The distribution is usually skewed and is often multimodal (Flowers and Yeo, 1981). Despite this, variability in sodium content within designated cultivars (which are expected to be almost entirely homozygous) is heritable and line selections diverge during recurrent selection (Yeo et al., 1988). Jones and Wilkins (1984) also reported an increase in the frequency of salt-resistant individuals during recurrent selection within rice cultivars. The limited data available implies that sodium transport is heritable, even though environmental effects are large.

The studies reported here compare the inheritance of sodium transport in intravarietal selections with that in doubled haploid lines derived from anther culture to assess the relative contribution of genetic and environmental effects. The distribution of sodium transport in the progeny of a four-way cross between parents differing in sodium transport and other physiological characters is then described. The data presented show that both sodium transport and sodium:potassium accumulation ratio (which are here indistinguishable) can be selected for independently of plant type (tall or dwarf).

Materials and methods

General

Seeds of cultivars, breeding lines and land races of *Oryza sativa* L. were obtained initially from the International Rice Research Institute, Los Baños, Philippines. Plants which were grown for seed were raised in a glasshouse which was heated to a minimum of 20°C by night and 25°C by day. Maximum daytime temperatures were limited to 28 ± 3°C by automatic venting and supplementary lighting provided a minimum of 400 μmol m^{-2} s^{-1} (photosynthetically active radiation) for 12 h per day. Sodium transport experiments were conducted in a controlled environment chamber with a 12 h photoperiod of 400–500 μmol m^{-2} s^{-1} (photosynthetically active radiation) at 27°C and 1.5 kPa saturation vapour pressure deficit. The dark period was of 25°C and 0.6 kPa saturation vapour pressure deficit. In both glasshouse and controlled environment conditions the mean air velocity was approximately 0.5 m s^{-1}.

Selection based on sodium transport

Seeds were germinated on nylon mesh floating on nutrient solution (Yoshida et al., 1986) and transplanted at seven days into boxes of the same nutrient solution. Plants were salinised with a low concentration of sodium chloride (25–50 m*M*) at 12–14 days. About 6 days later, the third leaf was excised for analysis and the plants returned to non-salinised nutrient solution and subsequently grown to maturity in non-salinised conditions (see Yeo et al., 1988). The leaves were analysed for sodium (and potassium as well in some cases). In some experiments the data are given as a sodium concentration on a dry weight basis, in others the data is in the form of a sodium content (μmol per leaf). Individuals were ranked according to the sodium content of (or concentration in) the sampled leaf, and selected on this basis. The range in overall sodium transport between individuals is very much greater than differences in the distribution of sodium between leaves within the same plant, and the correlation between leaf-3 sodium and whole shoot sodium is very high.

Intravarietal selection

From an original population of *Oryza sativa* cv. IR36, groups of plants with the lowest and highest sodium transport were selected and grown for seed. At subsequent generations, selection was made for low sodium transport within the low-sodium population and for high sodium transport within the high sodium population. The frequency distribution of sodium transport in the population diverged over five generations of recurrent selection (Yeo et al., 1988) after which there was no further change. The material used here had passed through six generations of recurrent selection, then random samples of the resulting populations were bulked once without selection. Two hundred plants of each population were grown in nutrient solution, salinised, and the shoots harvested for analysis.

Doubled haploid lines derived from anther culture

Seed of the parental population of *Oryza sativa* cv. IR43 and anther culture derived lines were kindly supplied by Dr F Zapata (International Rice Research Institute, Los Baños, Philippines). One hundred plants of the parental population and of each of a number of anther culture derived lines were grown in nutrient solution, salinised, and the shoots harvested for analysis.

Breeding population

A four-way cross (IR59462) was constructed at IRRI by Dr D Senadhira from land races and breeding lines which differed in their performance in physiological traits, including a range of average sodium transport (for detailed information see Table 4 in Yeo et al., 1990). Part of the F_2 was handled as shown in Figure 1. Each seed was planted and grown to maturity. Growth was in restricted conditions to conserve space, as is normal practice. Only those plants which would be totally unacceptable in agronomic terms were rejected; these were all both tall and daylength-sensitive and had failed to set seed within six months. Seed was collected from 371 plants and this is referred to as the 'unselected' population, it was comprised of tall and dwarf types in approximately equal numbers. A single seed from each plant was grown to maturity. One seed from each plant was screened for the sodium content of the third leaf (as described above for intravarietal selection) and ranked; the lowest one hundred plants were taken as the 'selected' population and grown to maturity.

Both populations are being advanced in non-saline conditions by single seed descent without selection (other than rejection of plants that fail to set seed). The comparisons reported here were made after one further generation. At this time the breeding population will retain appreciable levels of heterozygosity; lines are considered 'near-homozygous' at about F_6. The longer-term aim is to compare these methods with other conventional plant breeding strategies, including early selection for agronomic traits and initiation of line selections from early generations.

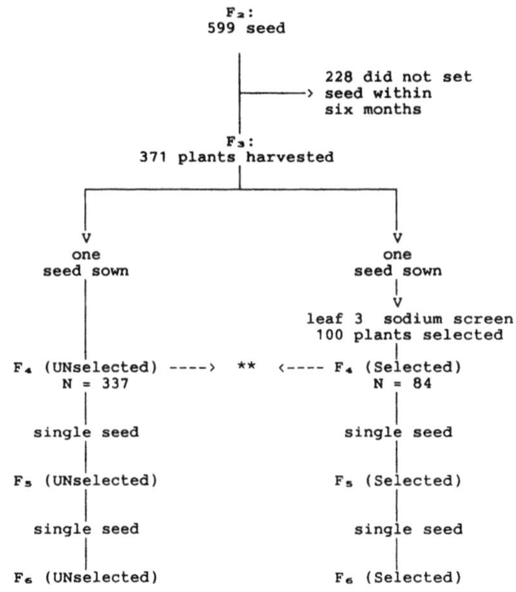

Fig. 1. Flow diagram showing selection within the four way cross IR59462.

Ion analysis

Excised leaves or whole shoots were dried and extracted in acetic acid (100 m*M*) for 2 h at 90°C and sodium (and potassium where this was also measured) were determined in the extract by atomic absorption spectrophotometry (Pye Unicam SP9 800).

Statistical analysis

In most cases, the ion concentrations are not distributed normally. Comparisons were made using the Kruskal-Wallis generalisation of the Mann-Witney-Wilcoxon test using the statistical analysis software MINITAB (Minitab Inc).

Results

Variation in intravarietal selections and in anther culture derived lines

The frequency distributions of sodium transport in populations selected within *Oryza sativa* cv. IR36 through six recurrent selections for low and high sodium transport, followed by one cycle of multiplication without selection, differ at $p = 0.001$ (Fig. 2). However, both selected populations still exhibit large variability. Figure 3 shows frequency distributions of sodium transport for a parental population (in this case of *Oryza sativa* cv IR43) and two doubled haploid lines derived from it via anther culture. The anther culture

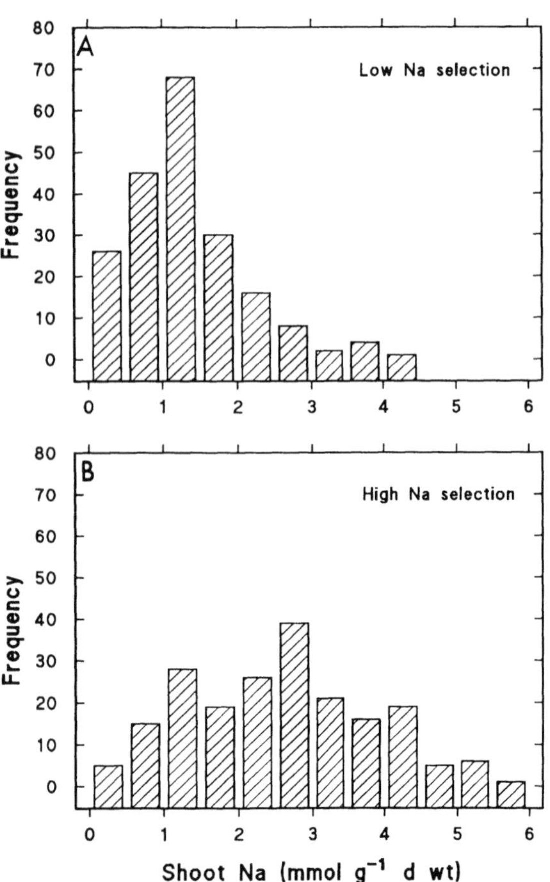

Fig. 2. Frequency distribution of sodium concentration in the shoot of intravarietal selections within cv. IR36. Lines were selected recurrently for low (**A**) and high (**B**) sodium transport for six generations and then bulked once without selection. The populations differ at $p < 0.001$.

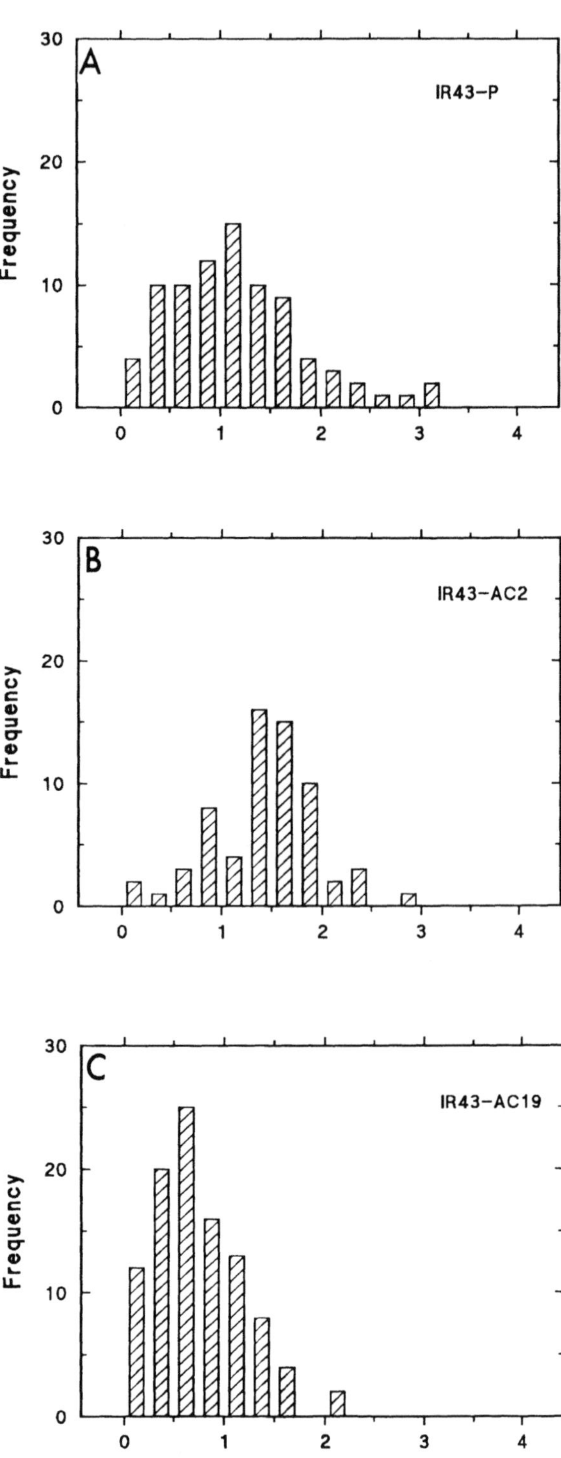

Fig. 3. Frequency distribution of shoot sodium concentration in cv IR43 (IR43-P, parental line (**A**)) and in two anther culture lines derived from it (IR43-AC2 (**B**) and 19 (**C**)). The lines and parent all differ at $p < 0.001$.

lines differ from the parental line and from each other at $p = 0.001$, but both still show very large variability.

Selection in breeding populations

Sodium transport
Since there is very little retranslocation of sodium in rice, the sodium content directly reflects the quantity of sodium transported. The sodium transport by the selected and unselected populations (see Fig. 1) differed significantly at F_4. The range of sodium transport in the unselected population was very extended with a suggestion of being bimodal. By contrast, the selected population was heavily skewed towards low transport and the median approximately half of that in the unselected population (Fig. 4).

Sodium/potassium selectivity
The sodium:potassium ratio in the leaves followed a similar pattern, except that the bimodality of the distribution in the unselected population was more clearly pronounced (with peaks below 1.0 and about 2.0; Fig. 5). The second peak was practically absent in the selected population (Fig. 5).

Plant type
There was a wide range in plant size because dwarfism was present in two of the four parents, while the other two were non-dwarfed land races. To assess the impact of selection for low sodium transport the individuals in the populations were visually classified as tall or dwarf. The unselected population had a small majority of tall types, and tall and dwarf types were almost

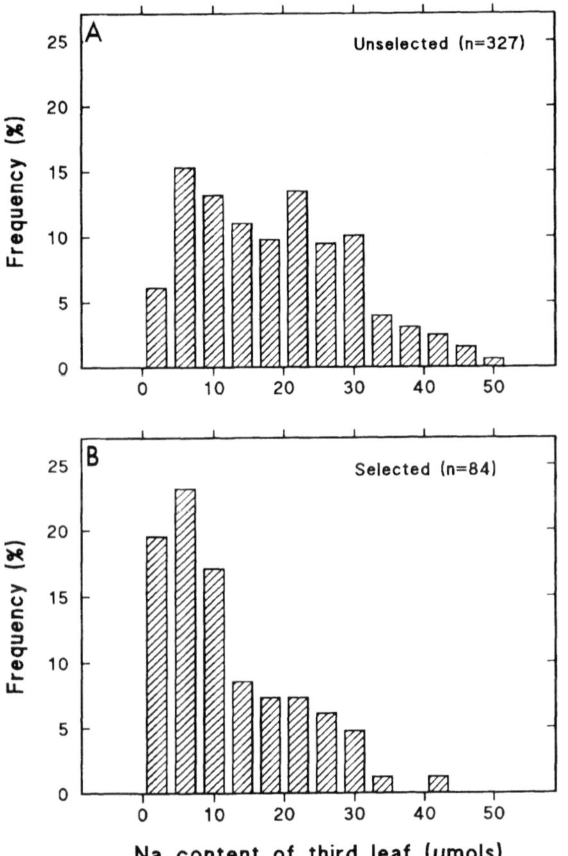

Fig. 4. Frequency distribution of the sodium content of the third leaf in unselected (**A**) and selected (**B**) populations from the four way cross IR59462. The populations differ at $p < 0.001$.

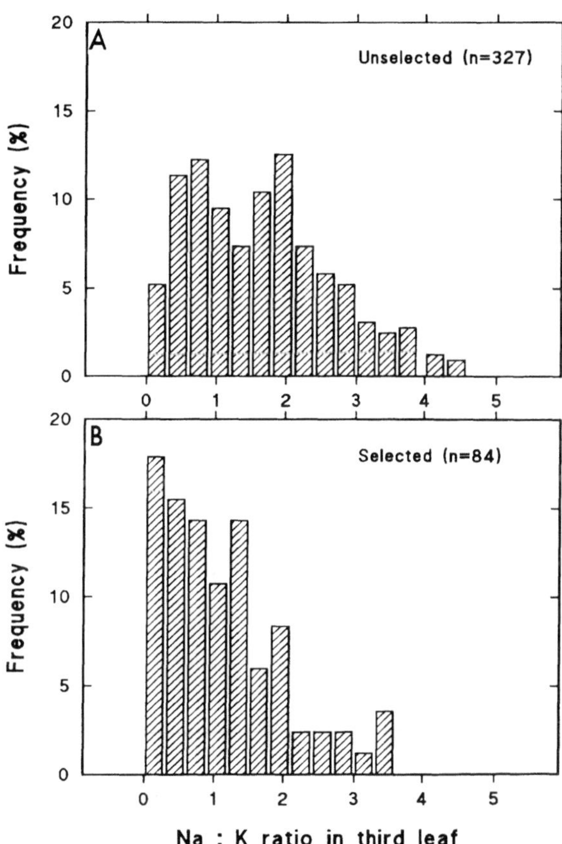

Fig. 5. Frequency distribution of sodium:potassium ratio in the third leaf of the unselected (**A**) and selected (**B**) populations from the four way cross IR59462. The populations differ at $p < 0.001$.

Table 1. The frequency of 'tall' and 'dwarf' plant types in the unselected and selected populations from IR59462. Plants with abnormal (stunted) growth were excluded

Population	Tall		Dwarf	
	n	%	n	%
Unselected	175	53.5	134	41.0
Selected	40	47.6	42	50.0

equal in the selected population (Table 1). Sodium transport and sodium:potassium ratio were not dependent upon plant type; the frequency distribution of Na:K ratio in the different types in the unselected population is shown in Figure 6. The distributions do not differ significantly, although the bimodality is more pronounced in the tall than in the dwarf sub-population. The frequency distributions of sodium content alone are not shown, but the general appearance was similar.

Discussion

Source of variation

The distributions in Figure 2 are closely comparable with those determined at an earlier generation of recurrent selection (Yeo et al., 1988). This suggests that a dividing line between the heritable and non-heritable components of sodium transport had been reached; the limit of divergence during recurrent selection being taken to mark the limit of the heritable component. The differences in the populations were preserved during multiplication without selection. Change during recurrent selection in highly-developed cultivars implies that the variability is due to residual heterozygosity rather than that the cultivar is a family of homozygous lines; in which case all change should be achieved in a single generation (Yeo et al., 1988). Cultivars of inbreeding species are not so homozygous as has often been supposed (Simmonds, 1979); and heterozygosity may have accumulated for characteristics which have not been subjected to selection pressure whilst it may have been practically eliminated by repeated selection in characters of agronomic interest.

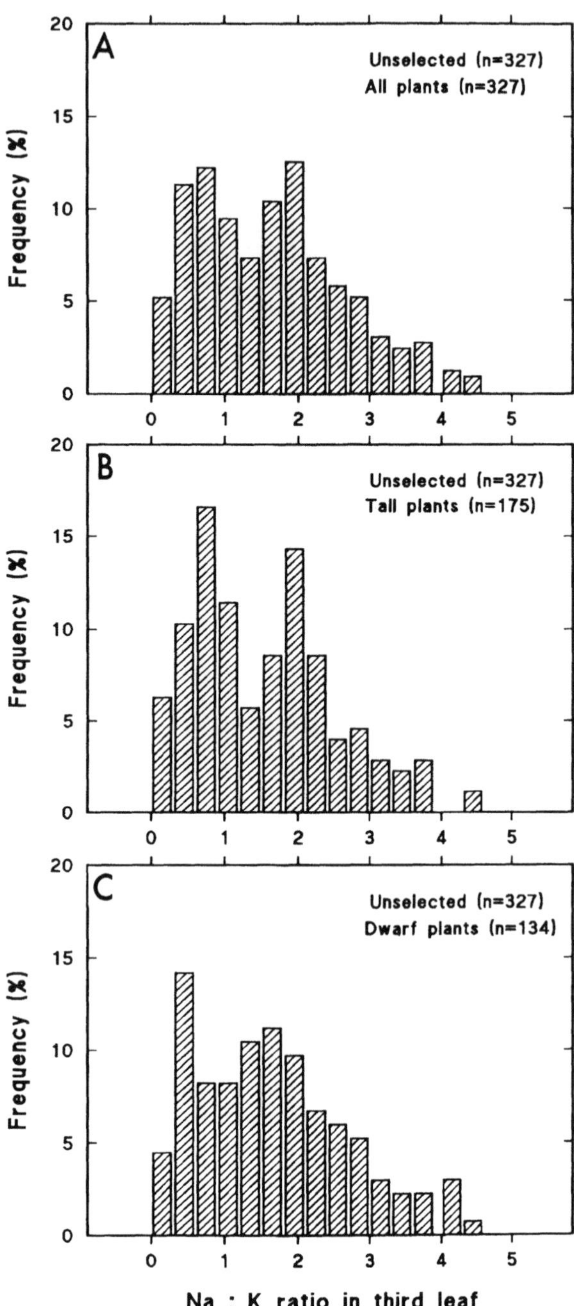

Fig. 6. Frequency distribution of the sodium:potassium ratio in the third leaf of the unselected population of IR59462 (**A**), classified according to plant type (tall (**B**) and dwarf (**C**)).

In the production of anther culture derived lines, anthers are cultured in vitro to generate haploid plantlets. The chromosome number of the plantlet is then doubled to produce a diploid plant, completely homozygous at all loci, called a doubled haploid (Poehlman, 1987). The plant is then increased and evaluated in the same way as a breeding line. These homozygous lines also exhibited large variability (Fig. 3). This supports the conclusion that recurrent selection had utilised the heritable component of sodium transport and that the variability remaining was a large environmental effect.

Two factors contribute to the extent of the variability. Sodium uptake is mediated, in part at least, by the transpirational by-pass flow (Yeo et al., 1987). The bypass-flow is leakage along a direct apoplastic contact from the external medium to the xylem in regions of the root where the endodermis has not yet differentiated or has been disrupted, such as by the development of lateral roots (Dumbroff and Pierson, 1971; Peterson et al., 1981). This is a small and very variable percentage of the transpiration stream but it becomes important as a feature in ion transport at high transpiration rates and high external concentrations (Pitman, 1982; Sanderson, 1983). Transpiration rates in rice are substantial, as befits a tropical C_3 plant which, together with most of its genus, is a native of freshwater marshes (Oka, 1988). The bypass-flow is dependent upon root morphology and developmental anatomy and so will be influenced by the way in which a particular root system has grown. In addition, any factor which affects the transpiration stream will affect sodium uptake. Most environmental variables (such as light, temperature, air movement) affect plant transpiration. For these two reasons, a large, non-heritable component of sodium transport is to be expected.

Sodium transport and sodium : potassium selectivity

The ion content can also be expressed as a sodium:potassium ratio (Figs. 5, 6). Much of this apparent sodium:potassium selectivity is due to a variable sodium uptake superimposed upon a very much more constant, and normally distributed potassium uptake (data not shown, but see Flowers and Yeo, 1981). If the sodium and potassium are entering via different pathways then this apparent selectivity may bear little relation to the well-documented case for the Triticeae (Gorham et al., 1987). In the Triticeae, leakage pathways (such as the by-pass flow) are presumeably much less significant than in rice. There was a suggestion that selectivity in rice does have an additional component; the frequency distributions for Na:K selectivity were more strongly bimodal that was sodium transport alone (Fig. 5). But it will be difficult to recognise a selectivity character in rice against the variability in sodium transport per se.

Whatever the physiological basis of the individual plant differences in sodium and potassium transport, selection for low sodium:potassium ratios (i.e. high potassium:sodium selectivity) is achieved empirically (Fig. 5).

Relationship to plant type

When a large number of accessions of rice were studied, plant vigour had the strongest correlation with overall performance as assessed by survival (Yeo et al., 1990). The most vigorous accessions are of course those which have not been dwarfed, and the accessions identified as the most salt resistant in mass screening trials (of which more than 90,000 have been made; Akbar, 1986) are non-dwarf land races. Low sodium transport is not, however, a characteristic of the tall plant type; their advantage lies with their growth rate which dilutes the salt (Yeo et al., 1990). They also have a higher water-use efficiency, which is conferred by their morphology rather than differences within the leaf (Flowers et al., 1988). The tall plant type is not favoured in modern plant breeding because of the greater yield potential of the dwarfed types. This would leave an insoluble problem if plant type and salinity resistance were not separable. Studies of growth showed that both tall and dwarf types responded similarly to salt, and it was concluded that there was no inherent disadvantage in dwarf types provided that sodium transport could be reduced to compensate for their reduced vigour (Yeo et al., 1991).

Screening to date has selected as donor parents vigorous accessions of the tall plant type, but the need is for parents of dwarf or moderate stature with low sodium transport. The selected population showed a dramatic reduction in sodium transport and in sodium:potassium ratio

(Figs. 4, 5), but was composed almost equally of dwarf and tall plant types (Table 1). These results demonstrate that variability in sodium transport can be produced in a cross, and selected for in a breeding population. independently of plant type. Although screening of existing genotypes has mostly found tolerance which is associated with the tall plant type, these results show that selection within new breeding populations can produce the desired combination of low sodium transport with dwarf plant type.

Intensity of selection

An important question in the handling of breeding populations is how strongly to select and how early. The earlier and the stronger the selection pressure the simpler the logistics. But the variation seen in recurrently selected intravarietal lines and anther culture derived lines shows that the predictive value of individual performance is heavily modulated by environmental and developmental factors. Selection of single individuals would not, in these circumstances, be expected to be an ideal procedure. A selection of one hundred plants at F_3 (just over a quarter of the population; see Fig. 1) appears at this stage to have been effective. A comparison of the populations produced by a range of selection methods will first be made at F_5 and will be reported in due course.

Acknowledgements

This work is supported by the Overseas Development Administration of the United Kingdom and forms a cooperative research programme with the International Rice Research Institute. I am grateful to the Royal Society for a travel grant and to the organisers of the Fourth International Symposium on Genetic Aspects of Plant Mineral Nutrition for financial support in Australia.

References

Akbar M 1986 Breeding for salinity tolerance in rice. *In* Salt-affected Soils of Pakistan, India and Thailand. Eds. IRRI. pp 39–63. International Rice Research Institute, Los Baños, The Philippines.

Dumbroff E B and Pierson D R 1971 Probable sites for passive movement of ions across the endodermis. Can. J. Bot. 49, 35–38.

Flowers T J and Yeo A R 1981 Variability in the resistance of sodium chloride salinity within rice varieties. New Phytol. 88, 363–373.

Flowers T J, Hajibagheri M A and Yeo A R 1991 Ion accumulation in the cell walls of rice plants growing under saline conditions: Evidence for the Oertli hypothesis. Plant Cell Environ. 14, 319–325.

Flowers T J, Salama F M and Yeo A R 1988 Water use efficiency in rice (*Oryza sativa* L.) in relation to resistance to salinity. Plant Cell Environ. 11, 453–459.

Gorham J, Hardy C A, Wyn Jones R G, Joppa L R and Law C N 1987 Chromosomal location of a K/Na discrimination character in the D genome of wheat. Theor. Appl. Gen. 74, 584–588.

Jones M P and Wilkins D A 1984 Screening for salinity tolerance by rapid generation advance. I.R.R.I. Newsl. 9, 9–10.

Oka H I 1988 Origin of Cultivated Rice. Japan Scientific Societies Press, Tokyo. 254 p.

Peterson C A, Emanuel M E and Humphreys G B 1981 Pathways of movement of apoplastic dye tracers through the endodermis at the site of secondary root formation in corn (*Zea mays*) and broad bean (*Vicia faba*). Can. J. Bot. 59, 618–625.

Pitman M G 1982 Transport across plant roots. Q. Rev. Biophys. 15, 481–554.

Poehlman J M 1987 Breeding Field Crops (Third Edition). Van Nostrand Reinhold, New York.

Richards R A 1983 Should selection for yield in saline regions be made on saline or non-saline soils? Euphytica 32, 431–438.

Sanderson J 1983 Water uptake by different regions of the barley root: Pathways of radial flow in relation to development of the endodermis. Plant Cell Environ 8, 309–315.

Simmonds N W 1979 Principles of Crop Improvement. Longman, London and New York.

Yeo A R and Flowers T J 1986 Salinity resistance in rice (*Oryza sativa* L.) and a pyramiding approach to breeding varieties for saline soils. Aust. J. Plant Physiol. 13, 161–173.

Yeo A R, Yeo M E and Flowers T J 1987 The contribution of an apoplastic pathway to sodium uptake by rice roots in saline conditions. J. Exp. Bot. 38, 1141–1153.

Yeo A R, Yeo M E and Flowers T J 1988 Selection of lines with high and low sodium transport within varieties of an inbreeding species; rice (*Oryza sativa* L.). New Phytol. 110, 13–19.

Yeo A R, Yeo M E, Flowers S A and Flowers T J 1990 Screening of rice (*Oryza sativa* L.) genotypes for physiological characters contributing to salinity resistance, and their relationship to overall performance. Theor. Appl. Gen. 79, 377–384.

Yeo A R, Lee K-S, Izard P, Boursier P J and Flowers T J 1991 Short- and long-term effects of salinity on leaf growth in rice (*Oryza sativa* L.). J. Exp. Bot. 42, 881–889.

Yoshida S, Forno D A, Cock J H and Gomez K A 1976 Laboratory Manual for Physiological Studies of Rice (Third Edition). International Rice Research Institute, Los Baños, The Philippines.

P. J. Randall et al. (Eds.), Genetic aspects of plant mineral nutrition, 151–158.

PLSO SV27

Genetics and physiology of enhanced K/Na discrimination

JOHN GORHAM
School of Biological Sciences, Memorial Building, University of Wales, Bangor, Gwynedd, Wales, LL57 2UW, UK

Key words: barley, potassium, salt, sodium, wheat

Abstract

Salt tolerance in crops such as wheat and barley, and in their wild relatives, is largely determined by their ability to exclude sodium and chloride from their shoots, and their ability to maintain high shoot potassium concentrations. This implies the operation of transport processes which discriminate against sodium and for potassium. While such discrimination is common to all plants, there are differences in the effectiveness of discrimination at various stages of growth and of transport to the shoots. This paper is concerned with the differences in K/Na discrimination which are found within the Triticeae, and specifically with the distribution and genetics of one particular character, the *enhanced K/Na discrimination trait*. The significance of the presence or absence of this trait to salt and sodicity tolerance is discussed for wheat and barley.

Introduction

Salt stress is a complex phenomenon, and one which does not occur in isolation from other stresses. The responses of plants to salt stress are also complicated, and some responses (e.g. intracellular solute compartmentation) may not be immediately obvious or easily measured. The cryptic nature of these responses, and the complex, integrated nature of the response of the whole plant, has led to the suggestion that breeding for salt tolerance might best be achieved by 'pyramiding' desirable characters (Yeo and Flowers, 1986) in such a way that there is a positive interaction between different characters. In order to create such a salt tolerant ideotype we need detailed and quantitative models of how individual mechanisms of salt tolerance operate, and how they are integrated into the response of the whole plant. In most cases we have a very poor understanding of these mechanisms.

This paper is concerned with only one aspect of salt tolerance – K/Na discrimination – and how it is controlled in the Triticeae. In higher plants there is a large range of ability to discriminate between Na and K, ranging from the halophytic Chenopodiaceae, which accumulate large amounts of Na, to some tropical grasses in which Na is excluded to the extent that it contributes very little to osmotic adjustment. The Triticeae are mostly good excluders of Na, but the extent of this ability varies with species and external conditions.

Enhanced K/Na discrimination

The *enhanced K/Na discrimination character* was first observed in a study of ion accumulation in the ancestors of modern bread wheat. Hexaploid wheat and the D-genome ancestor, *Aegilops squarrosa*, accumulated less Na and retained more K than tetraploid wheat (Wyn Jones et al., 1984) when grown at 100 mol m^{-3} NaCl. Synthetic hexaploid wheats, in which the D genome of *Ae. squarrosa* was combined with B and A genomes of tetraploid wheat, also showed enhanced K/Na discrimination (Shah et al., 1987). Analysis of aneuploid and substitution lines of tetraploid and hexaploid wheats revealed that the character was located on the long arm of chromosome 4D (Gorham et al., 1987). Physiological investigations showed that the character affected transport of K and Na from the root to the shoot, but did not affect root cation or anion concentrations or leaf anion concentrations (Gorham et al., 1990). Evidence that xylem loading was the probable site of action of this character came from X-ray microanalyses of frozen and fractured roots, ^{22}Na flux studies with intact and excised roots, and analyses of roots stripped of their cortical layers (Gorham and Rains, unpublished observations). A particular feature of this trait is that it is most obvious at low salt concentrations (<50 mol m^{-3} NaCl). At higher salinities other mechanisms may obscure the picture. Thus a strict definition of the *enhanced K/Na discrimination character* must include references to its chromosomal location, its major effects on K and Na transport to the shoots (but not on anion transport or root cation concentrations) and the fact that it operates (and is most clearly demonstrable) at low salinities. The main features of the *enhanced K/Na discrimination trait* are summarized in Table 5.

A re-examination of the occurrence of the enhanced K/Na discrimination trait in the A genome of wheat revealed that, although it is not expressed in durum wheat, it did occur in A-genome diploid wheats, and probably on chromosome 4A (Gorham, 1990b; Gorham et al., 1991). Since it does not occur in S-genome *Aegilops* species (the ancestors of the B and G genomes of tetraploid wheats) and appears to be dominant in intergeneric hybrids, the enhanced K/Na discrimination character (or at least its expression) seems to have been lost during the evolution of durum wheats. Loss of expression may be associated with the considerable chromosomal rearrangements involving chromosome 4A which have occurred during the evolution of most polyploid wheats (Dvorak and Appels, 1982; Gill and Chen, 1987; Naranjo et al., 1987). This suggestion is supported by the occurrence of the trait in GGAA-genome tetraploid wheats *T. araraticum* and *T. timopheevi* (Table 1). In Table 1, all of the wheats except the BBAA-genome tetraploid (*T. durum*) have low leaf Na concentrations and high K concentrations. The differences between species with and without the trait are also seen in the comparison between *Ae. squarrosa* (D-genome) and *Ae. bicornis* (S-genome). The data from amphiploids involving tetraploid wheats is, however, not entirely consistent. The crosses *T. boeoticum* × *T. araraticum* and *T. timopheevi* × *T. monococcum* have the trait, as indicated by the low Na and high K concentrations in the leaves. This is expected since both the diploid and tetraploid wheats appear to have the enhanced K/Na discrimination character. The cross *T. timopheevi* × *Ae. bicornis* also has the trait, although the Na concentrations are somewhat higher than in the two previous crosses. In this hybrid the trait in *T. timopheevi* appears to be dominant. In the two crosses involving both *T. timopheevi* and *T. durum*, there is little evidence of the dominance of the trait from *T. timopheevi*. In both hybrids the Na concentrations are high, and in *T. timopheevi* × *T. durum* the K concentrations are low. A possible explanation of this result could be the chromosomal instability of *T. durum* × *T. timopheevi* hybrids (Badaeva et al., 1986), perhaps resulting in loss of expression of the character. Further anomalous results involving *Aegilops polyploids* are described below.

Table 1. Monovalent cation concentrations (mol m^{-3} plant sap) in the youngest, fully-emerged leaves of various amphiploids grown for 14 days in 50 mol m^{-3} NaCl plus 2.5 mol m^{-3} $CaCl_2$. Values are the means of 4 replicates

Species or amphiploid	Genome[a]	Na	K
Triticum boeoticum	A^dA^d	11	170
Triticum aestivum	BBA^tA^tDD	10	205
Triticum durum	BBA^tA^t	145	130
Triticum araraticum	GGA^aA^a	16	193
Triticum timopheevi	GGA^aA^a	22	194
Aegilops squarrosa	DD	8	230
Aegilops bicornis	SS	107	110
T. boeoticum × *T. araraticum*	$A^dA^dGGA^aA^a$	4	163
T. timopheevi × *Ae. bicornis*	GGA^aA^aSS	37	184
T. timopheevi × *T. monococcum*	$GGA^aA^aA^dA^d$	8	165
T. timopheevi × *T. durum*	$GGA^aA^aBBA^tA^t$	102	120
T. durum × *T. timopheevi*	$BBA^tA^tGGA^aA^a$	72	151

[a] where A^d, A^t and A^a represent the A genomes of diploid wheat, tetraploid BBAA genome wheats and tetraploid GGAA genome wheats respectively.

The enhanced K/Na discrimination trait has

been measured in terms of shoot ion concentrations in plants growing at low salinities (50 mol m^{-3} NaCl). In an attempt to find a quicker (and confirmatory) test of the presence of this character, the accumulation of ^{22}Na in the shoots of 'low-salt' seedlings was examined. It is not practicable to measure the uptake of ^{22}Na into normal shoots growing at high salt concentrations, but uptake into low-salt seedlings exposed to low concentrations of salt is feasible. Table 2 shows that there is, at the species level, a good correlation between enhanced K/Na discrimination (as originally defined) and ^{22}Na accumulation in low salt seedlings. Table 2 also confirms the presence of the trait in the GGAA-genome tetraploid wheats. A good correlation can also be seen in the data for the *T. durum* cv. Cappelli recombinant lines in Figure 2 (see below). Within species, however, there appears to be more variation for ^{22}Na accumulation than for leaf Na concentrations. The extent of this variation has been examined in some detail for *Aegilops squarrosa* and for *T. dicoccoides* (Gorham and Wyn Jones, 1990; Nevo et al., 1992). The relationship between ^{22}Na accumulation and leaf Na concentrations within species does not appear to be as useful for predicting the presence of the *enhanced K/Na discrimination trait* as the same relationship between species.

Table 2. Correlation at the species level between ^{22}Na accumulation in leaves of 'low-salt' seedlings and the presence of the enhanced K/Na discrimination character. ^{22}Na uptake data are the means of 9 replicates

Species	Presence of enhanced K/Na discrimination	^{22}Na accumulation (Bq/mg)
Triticum monococcum	+	11
Triticum boeoticum	+	6
Triticum urartu	+	5
Triticum durum	–	230
Triticum araraticum	+	7
Triticum timopheevi	+	28
Triticum aestivum	+	35
Triticum aestivum (synthetic)	+	12
Hordeum vulgare	–	322
Hordeum chilense	+	53
Hordeum secalinum	?	292
Aegilops squarrosa	+	13
Secale cereale	+	17
T. durum × *Thinopyrum distichum*	–	470
T. aestivum × *Thinopyrum bessarabicum*	+	61

A good correlation between these parameters can be seen for most of the wild barley species in Figure 1. There are, however, two species for which the relationship does not hold. Both *Hordeum stenostachys* and *H. secalinum* have high leaf Na concentrations (but not the low K concentrations usually found in species lacking the enhanced K/Na discrimination trait). *H. secalinum* also has high ^{22}Na accumulation. Most of the other wild *Hordeum* species seem to have the enhanced K/Na discrimination character, but to prove this it is necessary to demonstrate that the character is expressed in interspecific or intergeneric hybrids with species lacking the character. Unfortunately wide hybridization in *Hordeum* is very difficult. One hybrid which is available for such a study is that between *H. chilense* and *T. durum*. The results from such a 'Tritordeum' produced by Antonio Martin are shown in Table 3. In this case the leaf Na and K concentrations in the hexaploid 'Tritordeum' were intermediate between those of the *T. aestivum* (which has the enhanced K/Na discrimination character) and *T. durum* and *H. vulgare* (which lack the character). The *H. chilense* accession which was used to produce this hybrid was not available for comparison, but examination of a large number of *H. chilense* accessions revealed a wide range of variation for K/Na discrimination within this species. Results of experiments on ^{22}Na accumulation in shoots of low-salt seedlings of 'Tritordeum' also indicated a level of K/Na discrimination which was intermediate between species which had, or which lacked, the enhanced K/Na discrimination character. Thus the presence or absence of the enhanced K/Na discrimination character in wild barley species was not resolved in these experiments.

Another series of hybrids which should be useful in resolving the question of the occurrence of enhanced K/Na discrimination in the wild barleys are the *H. bulbosum* × *H. vulgare* crosses. Some of these are available as vegetative material and are currently being tested for the presence of the enhanced K/Na discrimination character. Unfortunately the results from the parental *H.*

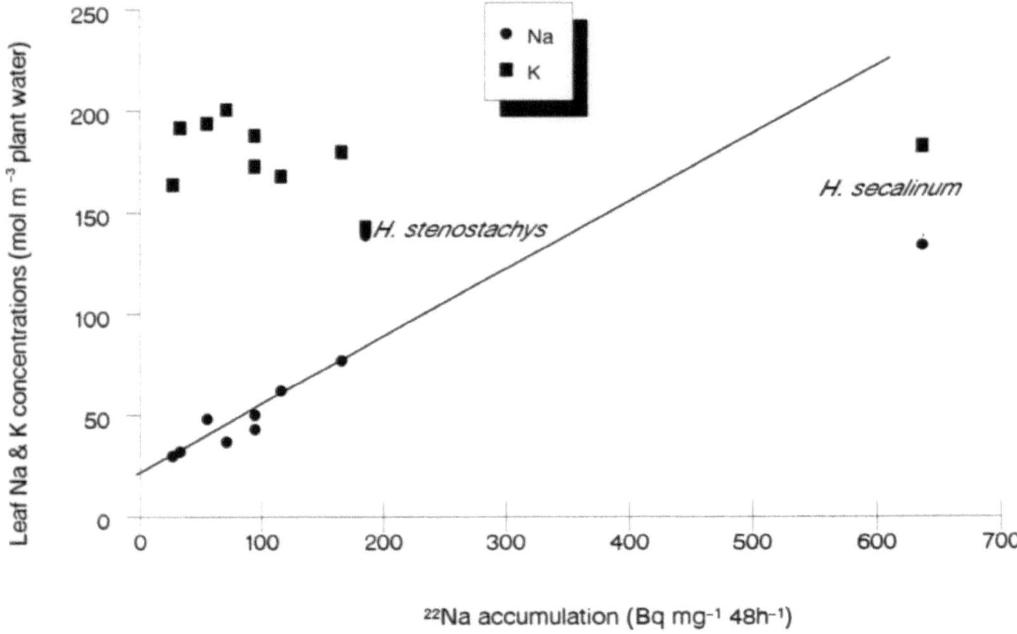

Fig. 1. Relationship between leaf Na and K concentrations in plants grown at 50 mol m^{-3} NaCl and ^{22}Na accumulation in 'low-salt' seedlings of *Hordeum* species.

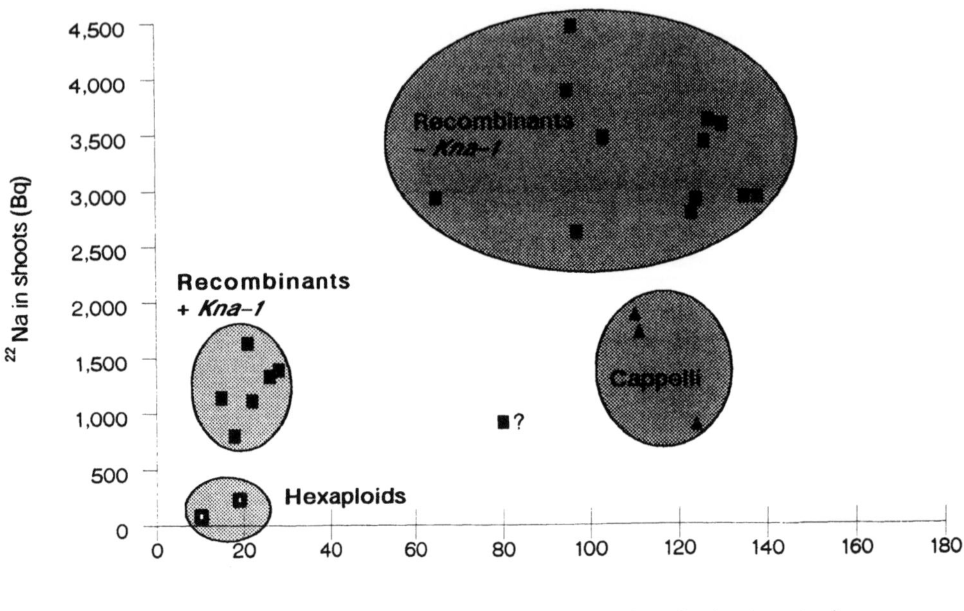

Fig. 2. Grouping of recombinant lines derived from the tetraploid wheat cultivar 'Cappelli' based on leaf Na concentrations of plants grown at 50 mol m^{-3} NaCl and ^{22}Na accumulation in shoots of 'low-salt' seedlings. The two groups include those with and those without the enhanced K/Na discrimination trait (*Kna-1*).

Table 3. Leaf Na and K concentrations (mol m^{-3} expressed sap) in hexaploid *Tritordeum* and related species grown at 50 mol m^{-3} NaCl (+2.5 mol m^{-3} $CaCl_2$). Values are the means of 4 replicates

Species	Na	K
Triticum aestivum	61	208
Triticum durum	191	118
Hordeum vulgare	175	116
Tritordeum (hexaploid)	103	142

bulbosum accessions and the hybrids do not agree with results obtained from the accessions of *H. bulbosum* used in the earlier experiments (Gorham, 1991). This inconsistency is being investigated in further experiments. Although the evidence from leaf Na concentrations and ^{22}Na accumulation quite strongly supports the presence of the enhanced K/Na discrimination character in most wild *Hordeum* species, clear, confirmatory evidence from interspecific and intergeneric hybrids is still lacking.

Some anomalous results were also encountered in an investigation into the occurrence of the enhanced K/Na discrimination character in the genus *Aegilops* (Gorham, 1990a). Most species in which the D or U genomes are present exhibit enhanced K/Na discrimination, since the character is generally dominant in hybrids and allopolyploids. The exceptions to this rule were *Ae. ventricosa* (genome DDNN), *Ae. biuncialis* (UUMM), *Ae. kotschyi* (UUSS) and *Ae. variabilis* (UUSS). Thus there are issues about the genetic control of this character which still need to be resolved. A summary of the known distribution of the enhanced K/Na discrimination character is shown in Table 4.

What is the importance of the enhanced K/Na discrimination character to salt tolerance? There is some evidence that, within cultivated wheats, there is a correlation between yield and leaf K/Na ratios (Joshi et al., 1979; Rashid, 1986). Conversely, cultivated barley is the most salt-tolerant of the cultivated cereals but lacks the enhanced K/Na discrimination character (Gorham, 1992). Attempts to answer this question by measuring the salt tolerance of D-genome substitution lines in tetraploid wheat and ditelosomic lines of hexaploid wheat were frustrated by the poor performance of the aneuploid lines compared to the euploid controls (Wyn Jones and Gorham, 1991). This problem has been addressed by producing a number of recombinant lines, based on the the tetraploid wheat variety 'Cappelli', which contain small fragments of the long arm of chromosome 4D from the hexaploid wheat variety 'Chinese Spring' (Dvorak and Gorham, 1992). When these lines were grown in 50 mol m^{-3} NaCl and their leaf Na concentrations examined, or when ^{22}Na accumulation in low-salt seedlings was measured, two groups of recombinants were detected (Fig. 2). The first group behaved in a similar way to hexaploid wheats and appeared to have the enhanced K/Na discrimination trait. The second group did not have the trait and were similar to the parent tetraploid wheat cv. 'Cappelli'. These lines are currently being tested for their tolerance to

Table 4. Distribution of the enhanced K/Na discrimination trait in the Triticeae. Genome A^a, A^d and A^t denote the A genomes of diploid and tetraploid BBAA-genome and GGAA-genome wheats respectively

Genomes	Species	Trait
A^a	*Triticum monococcum*	+
	T. boeoticum, *T. urartu*	+
BA^d	*T. dicoccoides*, *T. durum*	−
GA^a	*T. araraticum*, *T. timopheevi*	+
BA^tD	*T. aestivum*	+
BA^tA^d	synthetic hexaploids	+
D	*Aegilops squarrosa*	+
U	*Ae. umbellulata*	+
M	*Ae. comosa*	+
T	*Ae. mutica*	+
S	*Ae. speltoides*, *Ae. searsii*	−
C	*Ae. caudata*	−
N	*Ae. uniaristata*	−
I	*Hordeum vulgare*, *H. spontaneum*	−
H	*H. jubatum*, *H. marinum* etc.	+ (?)
R	*Secale cereale*	+
V	*Dasypyrum villosum*	−
BA^tR	*Triticosecale*	+
BA^tDR	*Triticosecale*	+
E	*Lophopyrum elongatum*, *L. ponticum*	−
J	*Thinopyrum bessarabicum*, *T. junceum*	−
	T. junceiforme, *T. distichum*, *T. scirpeum*	
$BA^tE(J)$	*Tritipyrum*	−
$BA^tDE(J)$	*Tritipyrum*	+

Table 5. Characteristics of the enhanced K/Na discrimination character

Genetics

1. The trait is usually expressed in interspecific hybrids.
2. In hexaploid wheat the trait is located on the long arm of chromosome 4D.
3. The trait is present on chromosome 4A of diploid wheat, but not on chromosome 4A of tetraploid wheat.
4. It is only one of many traits affecting K and Na concentrations.
5. The effect of the trait on tolerance to salinity and sodicity is currently being investigated in *T. durum* cv. 'Cappelli' recombinants containing sections of the long arm of chromosome 4D.

Physiology

6. The trait controls K and Na transport from roots to shoots, possibly at the point of xylem loading.
7. The trait has no major effect on root cation concentrations.
8. The trait has no major effect on anion concentrations in the root or the shoot.
9. The trait operates at all salinities, i.e. it is constitutive.
10. The trait is most obvious at low salinities where other mechanisms are less important.
11. *ALL* plants discriminate between K and Na – the trait ENHANCES this discrimination.

salinity and sodicity, but it is clear that enhanced K/Na discrimination is not the only factor affecting tolerance.

A summary of what is known about the enhanced K/Na discrimination trait is given in Table 5. Although we know more about the genetics and physiology of this particular trait than any other single identified response to salt stress, it is only one of many mechanisms which ultimately contribute to leaf K and Na concentrations and to salt tolerance.

Other sources of variation for K/Na discrimination

An investigation into the variability for, and the inheritance of, K/Na discrimination ability in hexaploid wheat has been undertaken at Bangor (Salam et al., in preparation). Data for leaf Na concentrations for four wheat varieties are given in Table 6. Clear differences between varieties are seen more clearly at 150 mol m^{-3} NaCl than at 50 mol m^{-3} NaCl. Within varieties there are significant differences between families derived from single parent plants. Reciprocal F1 hybrids between selected lines of two varieties had leaf Na concentrations intermediate between those of the parent lines (data not shown). This variation for leaf Na concentration, both between and within varieties, is most pronounced at higher salt concentrations. This, together with the lack of major differences in K concentrations (data not shown), suggests that factors other than the enhanced K/Na discrimination character are involved.

Table 6. Mean sodium concentrations (mol m^{-3} plant sap) the youngest fully-expanded leaves of *T. aestivum* lines grown at two external NaCl concentrations

Cultivar	Original parental mean	Inbred lines (single plant progenies)					
		1	2	3	4	5	6
50 mol m^{-3} external NaCl							
Blue silver	18.4	18.3	20.0	20.4	20.8	23.3	14.3
Tobari	15.3	15.5	16.2	15.6	16.7	16.2	15.3
Pato	16.8	14.4	15.4	15.5	13.7	15.6	17.0
Lyp-73	15.0	11.1	13.1	12.0	12.0	12.2	11.0
150 mol m^{-3} external NaCl							
Blue silver	129.0	139.0	117.5	152.5	238.0	214.4	213.3
Tobari	30.6	47.9	43.6	50.7	42.4	45.8	49.7
Pato	98.3	123.4	174.1	146.9	108.3	87.5	96.0
Lyp-73	74.7	83.3	62.7	105.0	61.1	76.2	93.4

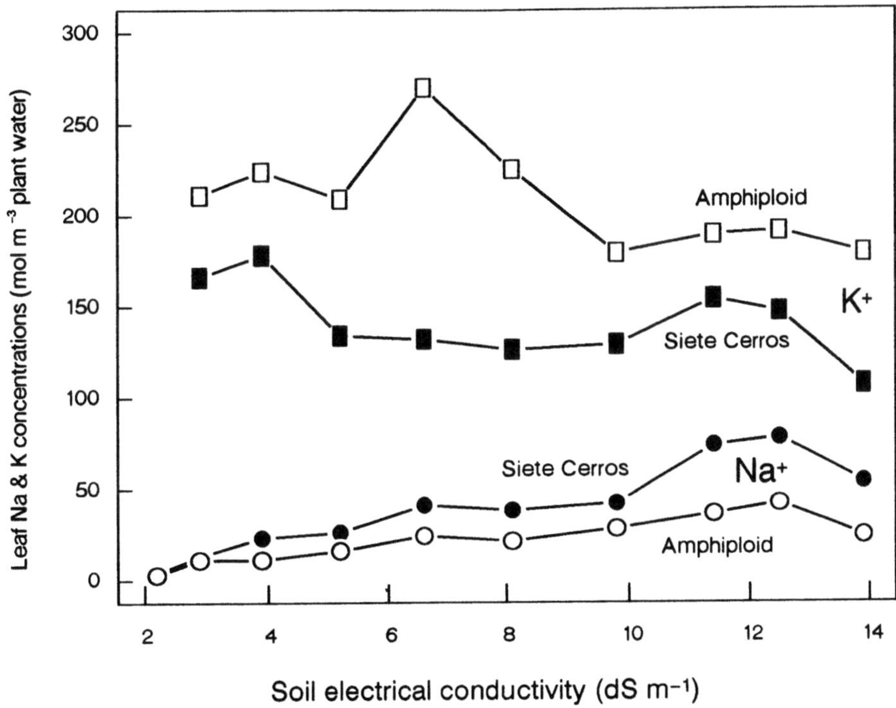

Fig. 3. Leaf Na and K concentrations in the second youngest leaves of plants of the *T. aestivum* cultivar 'Siete Cerros' and the *T. aestivum* × *Thinopyrum bessarabicum* amphiploid (produced at the Plant Breeding Institute, Cambridge). The data is for one side of a triple line source sprinkler system at Zaragoza, Spain in 1991.

Although results from intergeneric hybrids suggests that the wild perennial wheatgrass *Thinopyrum bessarabicum* does not have the enhanced K/Na discrimination trait (as strictly defined), both the wild wheatgrass and its hybrid with hexaploid wheat show good Na exclusion (Gorham et al., 1985, 1986). Experiments in the field using a Triple Line Source Sprinkler System (TLSSS) at Zaragoza in Spain confirmed that Na exclusion of the *Th. bessarabicum* × *T. aestivum* amphidiploid (produced by Terry Miller and Brian Forster at Cambridge) was better than for the hexaploid wheat variety 'Siete Cerros', and that K concentrations in the leaves were maintained at a higher value in the amphiploid (Fig. 3). More detailed experiments on derivatives of an amphiploid between *Lophopyrum elongatum* and hexaploid wheat have also shown that salt tolerance is under complex genetic control (Omielan et al., 1991). In *Th. bessarabicum* and *L. elongatum* there is a mechanism or mechanisms other than the *enhanced K/Na discrimination trait* (as strictly defined) for maintaining high K/Na ratios, and moreover mechanism(s) which are particularly important at high salinities.

Conclusions

From the above it is clear that several mechanisms contribute to K/Na selectivity, and that K/Na ratios need to be treated with caution as a guide to salt tolerance. We need to understand how other mechanisms interact with those controlling leaf monovalent cation concentrations and compartmentation, especially in cultivated barley. The results presented above show that it is possible to dissect out and study individual mechanisms of salt tolerance using the techniques of comparative physiology, but there is much work still to do.

Acknowledgements

I acknowledge the financial support of the Overseas Development Administration of the United Kingdom, the supply of genetic stocks from numerous sources, and permission from Abdus Salam and Philip Hollington to present unpublished data.

References

Badaeva E D, Shkutina F M, Bogdevich I N and Badaev N S 1986 Comparative study of *Triticum aestivum* and *T. timopheevi* genomes using C-banding technique. Plant System. Evol. 154, 183–194.

Dvorak J and Appels R 1982 Chromosome and nucleotide sequence differentiation in genomes of polyploid *Triticum* species. Theor. Appl. Genet. 63, 349–360.

Dvorak J and Gorham J 1992 Methodology of gene transfer by homoeologous recombination into *Triticum turgidum*: Transfer of K^+/Na^+ discrimination from *Triticum aestivum*. Genome 35, (*In press*).

Gill and B S and Chen P D 1987 Role of cytoplasm-specific introgression in the evolution of the polyploid wheats. Proc. Natl. Acad. Sci. USA 84, 6800–6804.

Gorham J 1990a Salt tolerance in the Triticeae: K/Na discrimination in *Aegilops* species. J. Exp. Bot. 41, 615–621.

Gorham J 1990b Salt tolerance in the Triticeae: K/Na discrimination in synthetic hexaploid wheats. J. Expt. Bot. 41, 623–627.

Gorham J 1992 Stress tolerance and mechanisms behind tolerance in barley. *In* Proc. 6th Int. Barley Genet. Symposium. Ed. L Munck. pp 1035–1049. Munksgaard, Copenhagen, Denmark.

Gorham J and Wyn Jones R G 1990 Utilization of Triticeae for improving salt tolerance in wheat. *In* Wheat Genetic Resources: Meeting Diverse Needs. Eds. J P Srivastava and A B Damania. pp 269–278. Wiley, Chichester, UK.

Gorham J, Forster B P, Budrewicz E, Wyn Jones R G, Miller T E and Law C N 1986 Salt tolerance in the Triticeae: Solute accumulation and distribution in an amphidiploid derived from *Triticum aestivum* cv. Chinese Spring and *Thinopyrum bessarabicum*. J. Expt. Bot. 37, 1435–1449.

Gorham J, Bristol A, Young E M and Wyn Jones R G 1991 Presence of the enhanced K/Na discrimination trait in diploid *Triticum* species. Theor. Appl. Genet. 82, 729–736.

Gorham J, McDonnell E, and Wyn Jones R G 1985 Salt tolerance in the Triticeae; Growth and solute accumulation in leaves of *Thinopyrum bessarabicum*. J. Expt. Bot. 36, 1021–1031.

Gorham J, Hardy C A, Wyn Jones R G, Joppa L R and Law C N, 1987 Chromosomal location of a K/Na discrimination character in the D genome of wheat. Theor. Appl. Genet. 74, 584–588.

Gorham J, Wyn Jones R G and Bristol A 1990 Partial characterization of the trait for enhanced K/Na discrimination in the D genome of wheat. Planta 180, 590–597.

Joshi Y C, Qadar A and Rana R S 1979 Differential sodium and potassium accumulation related to sodicity tolerance in wheat. Ind. J. Plant Physiol. 22, 226–230.

Naranjo T, Roca A, Goicoechea P G and Giraldez R 1987 Arm homoeology of wheat and rye chromosomes. Genome 29, 873–882.

Nevo E, Gorham J and Beiles A 1992 Variation for ^{22}Na uptake in wild emmer wheat, *Triticum dicoccoides* in Israel: Salt tolerance resources for wheat improvement. J. Expt. Bot. 43, 511–518.

Omielan J A, Epstein E and Dvorak J 1991 Salt tolerance and ionic relations of wheat as affected by individual chromosomes of salt-tolerant *Lophopyrum elongatum*, Genome 34, 961–974.

Rashid A 1986 Mechanisms of salt tolerance in wheat (*Triticum aestivum* L.). Ph.D. Thesis, University of Agriculture, Faisalabad, Pakistan.

Shah S H, Gorham J, Forster B P and Wyn Jones R G 1987 Salt tolerance in the Triticeae: The contribution of the D genome to cation selectivity in hexaploid wheat. J. Expt. Bot. 38, 254–269.

Wyn Jones R G and Gorham J 1991 Physiological effects of salinity Scope for genetic improvement. *In* Improvement and Management of Winter Cereals under Temperature, Drought and Salinity Stresses. Eds. E Acevedo, E Fereres, C Gimenez and J P Srivastava. pp 177–201. National Institute of Agricultural Research, Madrid, Spain.

Wyn Jones R G, Gorham J and McDonnell E 1984 Organic and inorganic solute contents as selection criteria for salt tolerance in the Triticeae. *In* Salinity Tolerance in Plants: Strategies for Crop Improvement. Eds. R C Staples and G H Toenniessen pp 189–203. Wiley, New York.

Yeo A R and Flowers T J, 1986 The physiology of salinity resistance in rice (*Oryza sativa* L.) and a pyramiding approach to breeding varieties for saline soils. Aust. J. Plant Physiol. 13, 161–173.

P. J. Randall et al. (Eds.), Genetic aspects of plant mineral nutrition, 159–163.

PLSO SV29

Association between genes controlling flowering time and shoot sodium accumulation in the Triticeae

M. TAEB[1], R.M.D. KOEBNER[1], B.P. FORSTER[2] and C.N. LAW[1]
[1]*Cambridge Laboratory, Norwich NR4 7UJ, UK and* [2]*Scottish Crop Research Institute, Invergowrie, Dundee DD2 5DA, UK*

Key words: flowering time, photoperiodic genes, *Ppd*, salt stress, Triticeae, vernalisation genes, *Vrn*

Abstract

Saline hydroponic studies of cytogenetic stocks of wheat have shown that near isogenic lines carrying contrasting alleles *Vrn* (vernalisation requirement) or *Ppd* (photoperiod requirement) genes accumulate less sodium when the dominant allele is present. These dominant alleles also confer early flowering. The genetic control of response to salt stress is discussed with respect to *Vrn* and *Ppd* genes. The data suggest that both these genes have pleiotropic effects on sodium accumulation. Salt treatment did not appear to switch on any genes which control sodium accumulation and it is concluded that the intrinsic genetic make-up of the plant determines fitness under salt stress conditions.

Introduction

Salt is one of the most important environmental stresses in irrigated agriculture, and the development of cultivars capable of growth and production on agricultural land affected by salinity is of increasing importance. Tolerance to environmental stresses such as salinity, drought, waterlogging and chilling is expected to become one of the most important priorities of breeding programs in the near future. It is therefore of interest to understand the genetics of stress tolerance in order to be effective in crop improvement for such traits.

There are generally two approaches to the study of salt tolerance in crop species. The physiological approach aims to identify physiological processes associated with salt tolerance. Once such characteristics are established, it is possible to search for useful genetic variation from different sources, and accumulate the various relevant genes within the target species. However to date there is no single physiological trait which is sufficiently strongly correlated to salt tolerance to attempt this.

In contrast, the genetical approach aims to analyse the inheritance of tolerance in segregating material, with a view to define chromosome segments important in conferring tolerance but without attempting to understand the physiological mechanisms involved. Once chromosomes and/or chromosome segments are identified these can be manipulated and introduced into the crop. Recently, the two disciplines of physiology and genetics have been brought close together by the finding that stress tolerance is often associated with flowering time. In our work on salt tolerance of wheat a link was established between flowering time and plant death (Forster et al., 1987). In these two studies flowering time (whether determined by genotype or vernalisation treatment) was positively correlated with plant death so that all plants appeared to die at a similar stage in the life cycle, and this was thought to be close to the transition from vegetative to generative growth. Subsequent genetic work therefore focused on the effect on genes controlling flowering time and plant response to salt stress. In this paper we describe the effect on salt tolerance, as measured by sodium (Na^+) accumulation of the two major gene families in wheat, controlling the requirement for vernalisation (*Vrn*) and those for photoperiod (*Ppd*). These genes are located on wheat homologous chromosomes 5 and 2, respectively, and are the two most important genetic determinants of days to flowering in wheat. This is the first report implicating *Vrn* and *Ppd* with response to salt stress. The results are discussed in a genetics context and comparisons are made with a salt-tolerant wheat/Thinopyrum amphiploid.

Materials and methods

Experiment 1

The following plant materials were analysed: euploid wheat, *Triticum aestivum* cv. Chinese Spring (CS) (carrying the alleles *Vrn1*, *Vrn3* and *Ppd2, ppd1*) and the amphiploid between CS and *Thinopyrum elongatum* (Rommel and Jenkin, 1959). Paired wheat near-isogenic lines for *Vrn1/vrn1* and *Ppd2/ppd2* were also used. The *Vrn1/vrn1* pair represents two single chromosome recombinant lines for chromosome 5A from *T. spelta* in a Hobbit 'sib' background (Snape et al., 1976) and *Ppd2/ppd2* two single chromosome recombinant lines for chromosome 2B from cv. Marquis in a CS background (Scarth and Law, 1983).

Seeds were germinated and seedlings transferred to a hydroponic system as described by Forster et al. (1987). Two treatments were given, a control with no added salt and 175 mol m^{-3} NaCl added to the half-strength Hewitt solution used. No vernalisation treatment was given. Plants were grown in a completely randomised design under an appropriate day-length regime aimed at manipulating the number of days to flowering. *Vrn1/vrn1* lines and amphiploid were grown under long days, provided by natural daylight, whereas the *Ppd2/ppd2* lines were grown under short days of eight-hour duration, in a specially constructed glasshouse. Chinese Spring was grown under both short-and long-day regimes.

Twenty replicates of each genotype were grown in both control and salt treatments. Measurements commenced 52 days after germination. Thereafter, at irregular intervals, depending upon how fast the particular genotypes grew, two plants were removed from each treatment for analysis and Na^+ was measured in the main shoot, the aim being to sample the plants at various stages of the life cycle up to anthesis to see if there were differences in sodium accumulation and development. All aerial parts of the main stem (leaves, sheaths and stems) were analysed individually for sodium ([Na]) concentration by flame photometry. This was done by first ashing the samples at 550°C in a muffle furnace and then extracting the cations by dissolving the ash in concentrated HNO_3. [Na] was calculated on a dry-weight basis. Data for the various tissue [Na] were added together on a per-plant basis, for statistical comparisons and the data were square-root transformed. Each isogenic pair was analysed separately, and CS was compared with the amphiploid. The data presented in Figure 1 represent mean values of shoot [Na], average over the life cycles up to anthesis of the genotypes.

Experiment 2

In this experiment the effect of neutralising the recessive alleles of *vrn1* and *vrn3* was studied. The plant material analysed was the same as that in Experiment 1, except that the *Ppd2/ppd2* lines were not included. Instead, two single chromosome recombinant lines for chromosome 5D of cv. Hope in CS were studied which varied for contrasting alleles of *Vrn3/vrn3*. Vernalisation was given by placing four-day-old seedlings at 4°C with 18 hours daylight for five weeks. Unvernalised plants were sown three weeks later, to synchronise the growth of vernalised and unvernalised plants such that when the vernalisation treatment was over all the seedlings, vernalised and unvernalised, were at about the three-leaf stage. At this stage the seedlings were transferred to a hydroponic system (as previous described) under long days. There was one salt treatment of 175 mol m^{-3} added NaCl and six replications per genotype per treatment. Leaf five was harvested for ion measurement when leaf six was fully expanded. The hydroponic system and ion measurement were carried out as described in Experiment 1.

Results

Experiment 1

There are genotypic differences in [Na] between plants grown under control and salt-stressed conditions. In the salt treatment, the genotypes had significantly higher [Na] in their shoots than the respective controls, averaged over growth stages up to anthesis. Plants with the dominant alleles, *Vrn1* and *Ppd2*, were significantly lower in [Na] than the respective near-isogenic lines carrying the recessive alleles (Fig. 1). Chinese Spring (CS) grown under short days had significantly higher [Na] than under long days. The CS/*Th. elongatum* amphiploid had the lowest [Na] of all genotypes studies. The ranking of the genotypes for leaf [Na] in the salt treatment was maintained in the control treatment, such that genotypes carrying the dominant alleles were always associated with a lower [Na] than those lines carrying recessive alleles regardless of the treatment. A linear relationship was found between days to flowering and [Na] of the lines both under salt-stress and control conditions (Fig. 2). However, since the regression was not statistically significant the implication is that Na^+ accumulation is not a function of length of the life cycle per se. Hence, the observed differences are not though to have been caused by differences in growth rate of the genotypes.

Experiment 2

The vernalisation treatment had the effect of reducing leaf [Na] content. Means of [Na] of the lines studied are presented in Figure 3. Isogenic lines with the recessive alleles *vrn3* and *vrn1* had a higher [Na] than their respective isogenic with the dominant allele in both the vernalised and unvernalised treatments. CS did not show any significant response to vernalisation.

Discussion

Plant response to photoperiod or vernalisation resulting in flowering is a complicated physiological process involving changes in hormonal activity and metabolism. In this study we have concen-

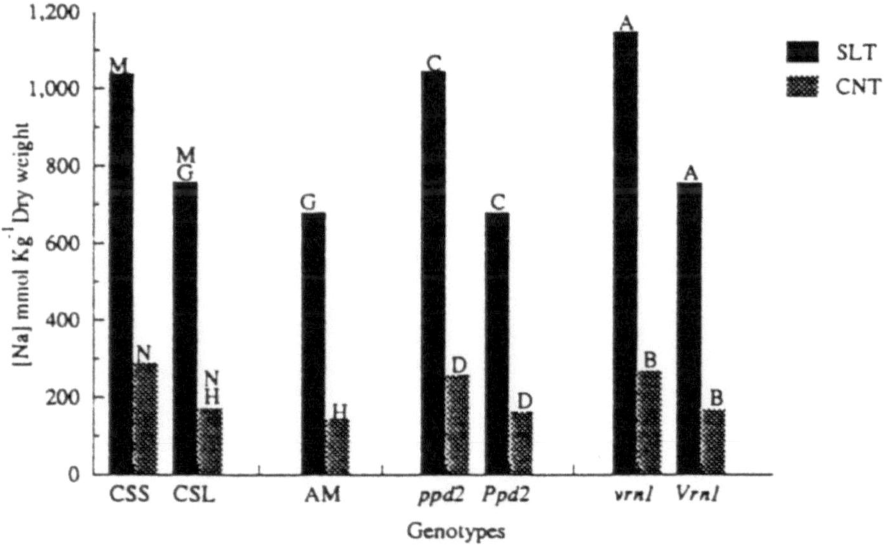

Fig. 1. Average shoot [Na] of near isogenic lines and control plants grown in 175 mol m^{-3} and 0 mol m^{-3} of added NaCl (Experiment 1). Chinese Spring under short days (CSS), Chinese Spring under long days (CSL), amphidiploid (AM), near isogenic lines of *ppd2*, *Ppd2*, *vrn1* and *Vrn1* statistical comparisons based on transformed data; A vs A, B vs B, C vs C, H vs H, differ significantly ($p < 0.05$), D vs D, G vs G, M vs M, N vs N, do not differ significantly.

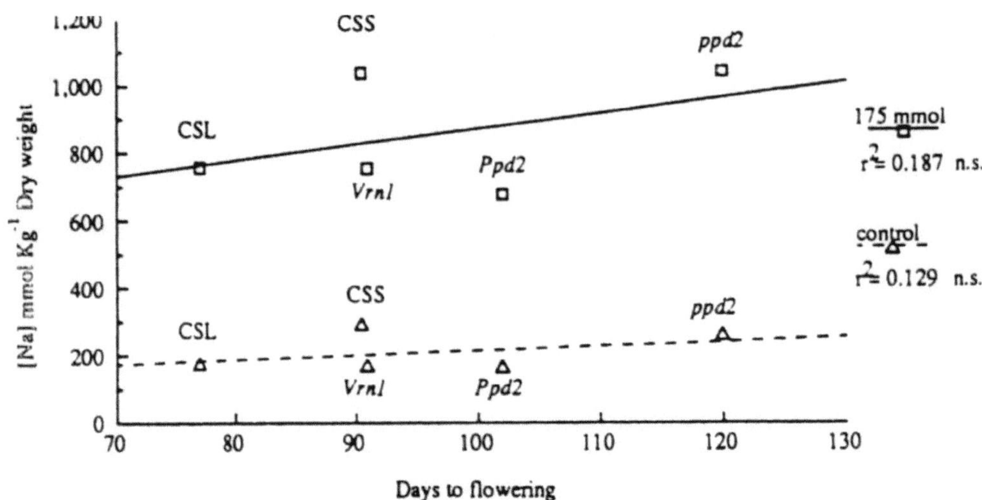

Fig. 2. Linear regression of [Na] against days to flowering of near-isogenic lines and control plants (Experiment 2).

trated on the effect of genes controlling flowering time and sodium uptake under salt-stress conditions.

The genetic control of Na^+ accumulation is likely to be complex, but the observation that CS accumulates more Na^+ when grown under short-day conditions as compared to long days, suggests that whatever controls days to flowering also has an effect on Na^+ accumulation. Similarly the influence of the *Vrn* genes and *Ppd* on flowering time has also an effect on Na^+ accumulation, those alleles reducing flowering time giving lower Na^+ levels. This control of course could be due to close linkage of these genes with specific gene(s) controlling Na^+ accumulation, but this seems very unlikely since the chance of there being similar linkages separately involving *Vrn1*, *Vrn3* and *Ppd2* must be very low. The use of near-isogenic pairs of lines is therefore strongly supportive of the pleiotropic effect of *Vrn1*, *Vrn3* and *Ppd2* on Na^+ accumulation acting through the control of flowering time.

It is of course possible that the activities of related alleles of *Vrn* and *Ppd* in the wheat × *Th. elongatum* amphiploid influence Na^+ uptake (Fig. 3). However, this can not be the sole reason for its low Na^+ accumulation because the amphiploid was very late-flowering. Other genes excluding salt must therefore be active in this line. The correlation between flowering time and

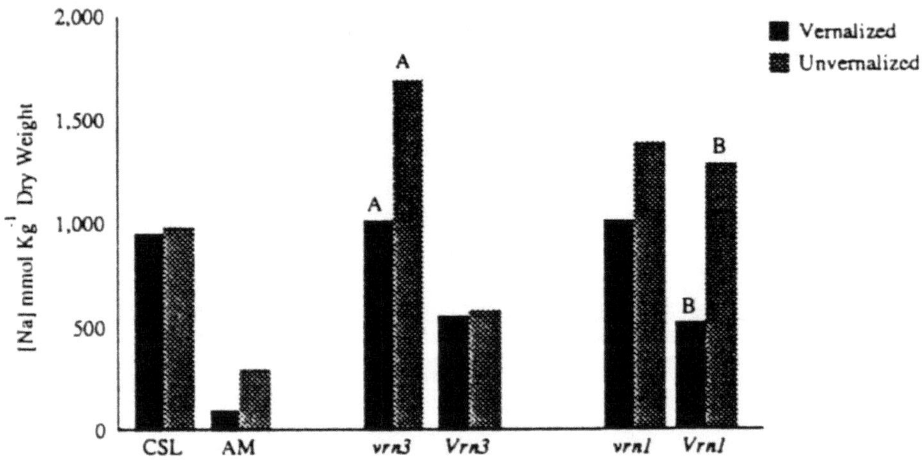

Fig. 3. Effect of vernalisation on Na^+ accumulation of near-isogenic lines and control plants grown in 175 mol m^{-3} added NaCl (Experiment 2). Statistical comparisons based on transformed data show A v A and B v B to be significant ($p < 0.05$).

Na^+ accumulation does however put into question the cause of the improved salt tolerance of the two alien addition lines for chromosome $5E^b$ and $2E^b$ of *Th. bessarabicum* previously reported (Forster, 1988b). Both these chromosomes could carry alleles of *Vrn* (group 5) and *Ppd* (group 2). So both the effects on Na^+ accumulation could be due to the influence these genes have on flowering rather than to genes directly influencing uptake. Similarly, recent experiments using wheat/*Hordeum vulgare* and wheat/*H. chilense* disomic addition lines have shown that chromosome 4 of both barely species has positive effects on plant response to salt (Forster et al., 1990); chromosome 4 of barley carries the *Vrn1* gene, so that this could also account for the observed tolerance.

These studies therefore need to be re-examined in this light. However, just as is the case with *Th. elongatum*, in this study *Th. bessarabicum* and the amphiploid with wheat must possess genes controlling Na^+ accumulation over and above any genes concerned with maturity, since both genotypes are late-flowering. Further studies are therefore needed to confirm whether genes directly controlling salt tolerance are present on $5E^b$ and $2E^b$ and if not to establish their true location.

In the present study an interesting feature is the absence of interaction between treatment and genotype, a positive correlation ($r = 0.67$) being found between [Na] in controls and [Na] in salt-stressed plants. This indicates that genes controlling sodium uptake are not switched on by exposure to salt stress since apparently genes are active in both stressed and non-stressed environments. This agrees with Gluick and Dvořák (1987) who found no novel mRNA in salt-stressed plants of wheat compared with wheat × *Th. elongatum* amphiploid and of Hurkman et al. (1989) in barley comparing a salt-tolerant with a salt-susceptible cultivar. This, therefore, suggests that the salt tolerance of wheat in comparison to its wild relatives might be due to allelic variation at the loci conferring salt tolerance rather that to the presence of one or more genes in the alien species, absent in wheat. This could pose a problem if attempts were to be made at gene isolation following the approach of differential mRNA production.

Acknowledgements

The authors are grateful to A J Worland, J W Snape, T E Miller, and S M Reader for making available various genetic stocks used in this study. This work was supported in part by the Scottish Office Agriculture and Fisheries Department, and the Agricultural and Food Research Council.

References

Forster B P 1992 Genetic engineering for stress tolerance in the Triticeae. *In* Problems and Opportunities in Plant Biotechnology. Proc. of the Royal Society of Edinburgh. Eds. W Powell, J R Hillman and J A Raven. (*In press*)

Forster B P, Gorham J and Miller T E 1987 Salt tolerance of an amphiploid between *Triticum aestivum* and *Agropyron junceum*. Plant Breeding 98, 1–8.

Forster B P, Gorham J and Taeb M 1988a The use of genetic stocks in understanding and improving the salt tolerance of wheat. *In* Cereal Breeding Related to Integrated Cereal Production. Eds. M L Jorna and L A J Sloatmaker. pp 87–91. Proc. of the Conference of the Cereal Section of EUCARPIA. Pudoc, Wageningen.

Forster B P, Miller T E and Law C N 1988b Salt tolerance of two wheat/*Agropyron junceum* disomic addition lines. Genome 30, 559–564.

Forster B P, Philllips M S, Miller T E, Baird E and Powell W 1990 Chromosome location of genes controlling tolerance to salt (NaCl) and vigour in *Hordeum vulgare* and *H. chilense*. Heredity 65, 99–107.

Gulick P and Dvořák J 1987 Gene induction and repression by salt treatment in roots of salinity sensitive Chinese Spring wheat and the salinity tolerant Chinese Spring × *Elytrigia elongata* amphiploid. Proc. Natl. Acad. Sci. USA 74, 99–103.

Hurkman W J, Fornari C S and Tanaka C K 1989 A comparison of the effect of salt on polypeptides and translatable mRNAs in roots of a salt tolerant and a salt sensitive cultivar of barley. Plant Physiol. 90, 1444–1456.

Rommel R and Jenkin B C 1959 Amphiploids in Triticeae produced at the University of Manitoba from March 1958 to December 1959. Wheat Inf. Serv. 9–10, 23.

Scarth R and Law C N 1983 The location of the photoperiod gene, *Ppd2* and an additional genetic factor for ear emergence time on chromosome 2B of wheat. Heredity 51, 607–619.

Shannon M G 1978 Testing salt tolerance variability among tall wheatgrass lines. Agron. Journal 70, 719–722.

Snape J W, Law C N and Worland A J 1976 Chromosomal variation for loci controlling ear emergence time on chromosome 5A. Heredity 37, 335–340.

P. J. Randall et al. (Eds.), Genetic aspects of plant mineral nutrition, 165–171.

PLSO SV32

The inheritance of salt exclusion in woody perennial fruit species

S.R. SYKES
CSIRO, Division of Horticulture, Merbein, Victoria, 3505, Australia

Key words: citrus, Cl-ion exclusion, grapevines, hybrids, inheritance, Na-ion exclusion, rootstocks

Abstract

Citrus and grapevines are salt-sensitive perennial crops. Damage caused by salinity is due mostly to accumulation of excessive concentrations of salt (Na- and Cl ions) in shoot tissues. In both crops, however, some rootstock varieties can restrict the accumulation of salt in scion leaves and stems. Salt-excluding rootstocks, however, are often deficient with regard to other desirable characteristics and as such their use is limited. As a consequence, we have conducted a range of crosses within both crops to select new salt-excluding hybrids which may have potential as new rootstocks and also to investigate the inheritance of salt exclusion in these woody perennials.

In citrus, both Cl-ion and Na-ion exclusion has been observed as a continuous character and progenies segregate widely for their ability to restrict the accumulation of these ions in shoot tissues. The ability to exclude Cl ions appears to be independent of the ability to exclude Na ions. Thus a good Cl-ion excluder is not necessarily a good Na-ion excluder and vice versa. It has been possible, however, to select new salt-excluding citrus hybrids which perform as well as and sometimes better than parent varieties when grafted with a common scion and grown in artificially salinised field plots.

In grapevines, the research has concentrated on the ability for Cl-ion exclusion and depending on the Cl-ion-excluding parent chosen this is inherited as either a polygenic or monogenic trait. In crosses between *Vitis champini* (Cl-ion excluder) and *Vitis vinifera* (Cl-ion accumulator), the ability to restrict Cl-ion accumulation in shoot tissues segregates widely with some offspring transgressing the performance of either parent. In crosses and backcrosses involving *V. berlandieri* and *V. vinifera*, however, hybrids segregate as either Cl-ion excluders or accumulators suggesting the effect of a major dominant gene for Cl-ion exclusion from *V. berlandieri*. This was evident from both field and glasshouse experiments although possible modifying genes from *V. vinifera* may affect the expression of this gene under glasshouse conditions in some crosses.

Introduction

While citrus trees are considered to be salt-sensitive glycophytes (Greenway and Munns, 1980), they exhibit a wide variation for Cl-ion exclusion within species and cultivars (Grieve and Walker, 1983; Sykes, 1985a; Walker and Douglas, 1983). This variation has stimulated breeding programs aimed at producing new salt-tolerant rootstocks (Furr and Ream, 1969; Sykes, 1985b).

High leaf Cl-ion concentrations in citrus scions can lead to physiological disturbances and eventually visible leaf damage (Cooper and Shull, 1953). High leaf Na-ion concentrations also have detrimental effects on photosynthesis and transpiration (Behboudian et al., 1986). When grown under conditions of low to moderate root zone salinities (25–50 mM NaCl), nucellar seedlings of *Poncirus trifoliata* Raf. are able to maintain lower leaf Na-ion concentrations than other citrus rootstocks (Grieve and Walker, 1983; Walker, 1986). *Poncirus trifoliata*, sometimes referred to as *Citrus trifoliata* (Bailey and Bailey, 1976), is an important citrus rootstock which is sexually compatible with *Citrus* species. *Pon-*

cirus × *Citrus* crosses produce viable progeny, e.g. Carrizo citrange (*C. sinensis* × *P. trifoliata*). The variation between nucellar seedling populations of *Poncirus* and *Citrus* suggests a heritable basis for Na-ion exclusion, and so screening for this has been a component of CSIRO's salt-tolerant citrus rootstock breeding program.

Grapevines are moderately salt-tolerant if Cl-ion injury can be avoided (Ehlig, 1960). Leaf Cl-ion concentrations have been shown to be associated with symptoms of salt injury and the health of irrigated vines in the Murray Valley (Thomas, 1934; Woodham, 1956). High leaf Cl-ion concentrations in vines have been accompanied by decreased rates of photosynthesis (Downton, 1977a) and reduced yields (Grieve, 1984).

Rootstocks will reduce leaf Cl-ion concentrations in grapevines (Bernstein et al., 1969; Sauer, 1968) and there is considerable variation for Cl-ion exclusion both between (Antcliff et al., 1983; Downton, 1977b) and within (Ehlig, 1960; Groot Obbink and Alexander, 1973) *Vitis* species. This variation has led to a search for new Cl-ion-excluding grapevine rootstocks through breeding and selection (Newman and Antcliff, 1984; Sykes, 1985c, 1987).

During the course of breeding and selecting new salt-tolerant citrus and grapevine rootstocks, observations have been made concerning the mode of inheritance of salt exclusion. Most of the results discussed in this paper have been published (Newman and Antcliff, 1984; Sykes, 1985b; c; 1987; Sykes and Newman, 1987). New data, however, are presented concerning the variation for Na-ion exclusion between and within *Citrus* × *Poncirus* hybrids.

Materials and methods

Citrus

Hybrids between citrus rootstock varieties were produced using techniques described by Soost and Cameron (1975) during 1980, '81 and '82. The combinations produced in 1980 and 1981 have been published (Sykes, 1985b). The hybrids from these crosses were screened initially for leaf Cl-ion concentrations following a short-term salt test. Since then, leaf materials of eight progenies retained from these tests have been analysed for Na-ion concentrations. Sykes (1985b) described an irrigated pot trial using two hybrids from progeny 80–02, Rangpur lime (*Citrus* × *limonia* Osbeck) × Trifoliate orange (*Poncirus trifolia* Raf.). This trial confirmed the Cl-ion-excluding abilities for these hybrids, viz. 80-02-08 and 80-02-38. Another glasshouse trial has since been conducted to assess the Na-ion excluding abilities of these hybrids. An additional hybrid, 80-02-02, which was morphologically similar to 80-02-08 and 80-02-38 but accumulated high leaf Cl-ion concentrations in the original screening experiment, was also included.

Twelve-month-old nucellar seedlings (see Frost and Soost, 1968) of Rangpur lime and Trifoliate orange and twelve-month-old plants of hybrids 80-02-02, -08 and -38 grown from stem cuttings, were transferred from a standard potting mix to half-strength No. 1 nutrient solutions (Hoagland and Amon, 1950) held in 576-litre fibreglass tanks (1.2×0.6 m surface area; 0.8 m deep). Solutions were made with rainwater and were maintained at pH 6.5. Iron ($2.5 - 10^{-5}$ *M* Fe as ethylenediametetracetic acid ferric monosodium salt), trace elements (4.6×10^{-5} *M* B, 9.1×10^{-6} *M* Mn, 7.6×10^{-6} *M* Zn, 3.1×10^{-7} *M* Cu and 1.0×10^{-7} *M* Mo) and benomyl (4.5×10^{-6} *M* methyl 1-(butylcarbamoyl)-2-benzimidazole carbamate applied as 'Benalate', a 50:50 wettable powder; Dupont Chemical Company) were added to solutions weekly throughout the experiment. Three plants of each genotype, supported by PVC covers, were randomised within each of four tanks and shoots pruned to 20 cm. One shoot was allowed to develop from the youngest bud. After 30 days solutions were renewed and made up to full strength and the youngest leaf on each plant was labelled with a loosely tied piece of cotton thread. Salt (Nacl) was added to two tanks at a rate of 25 m*M* per day until the concentration reached 50 m*M*. The other tanks were salt-free as controls. After 50 days the plants were pruned back to the original stem length of 20 cm. Prunings were divided into stems and leaves above (distal) and below (proximal) labelled leaves. These tissues were dried at 60°C for 120 h, crushed in a hammermill to pass a 1-mm mesh and retained for ion analyses.

Vitis

Under both vineyard and glasshouse conditions vines of *Vitis champini* and *V. berlandieri* have low Cl-ion concentrations in leaves. Consequently, hybrids involving both species (Newman and Antcliff, 1984; Sykes, 1985c, 1987) crossed with varieties of *V. vinifera* were screened for Cl-ion accumulation as unreplicated vineyard plantings and in replicated glasshouse trials.

Cl-ion exclusion by unreplicated hybrid vines was assessed by measuring Cl-ion concentrations in petioles collected from leaves opposite fruit bunches during January, which is around the time of veraison in the Murray Valley. Petioles were collected from two progenies involving *Vitis champini* (Ramsey), plus an additional progeny from a cross between the other two parents, during 1981 and 1982 (Sykes, 1985c). Petioles were collected from seven progenies from crosses between *V. berlandieri* and *V. vinifera* and four backcross progenies, i.e. (*V. berlandieri* × *V. vinifera*) × *V. vinifera*, during 1983 (Newman and Antcliff, 1984). All petioles were dried at 60°C, crushed in a hammermill to pass a 1-mm mesh and analysed for Cl-ion concentrations.

Replicated glasshouse trials employed nutrient solution cultures containing NaCl (25 m*M*) similar to those used for citrus screening experiments. The full details have been published and the trials involved all hybrids from the cross between *V. champini* (Ramsey) × *V. vinifera* (Sultana) (Sykes, 1985c) and selected hybrids between *V. berlandieri* × *V. vinifera* and backcrosses (Sykes, 1987).

Ion analysis

All tissues were prepared for ion analysis following the methods described by Walker (1986); Na ions were analysed by atomic absorption spectrophotometry and Cl ions by silver ion titration using a Buchler-Cotlove chloridometer.

Results

Citrus

Leaf Na-ion concentrations varied widely between hybrids within each progeny tested. This variation was continuous over the range (e.g. Fig. 1). In all progenies tested, leaf Cl-ion con-

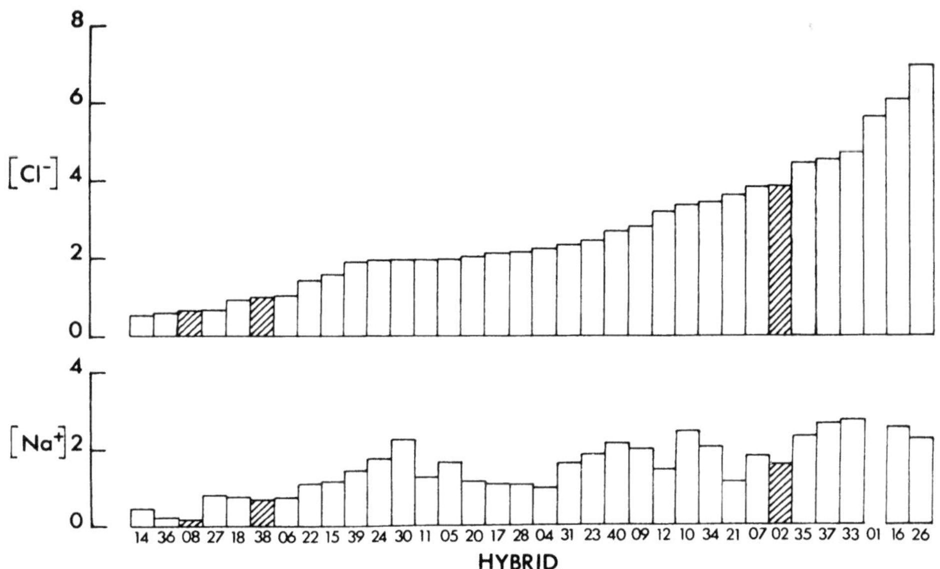

Fig. 1. Leaf Cl-ion and Na-ion concentrations (% of dry weight) for 14-month-old hybrids between Rangpur lime and Trifoliate orange (80-02) when grown in nutrient solution culture containing NaCl (50 m*M*) for 56 days. There was insufficient leaf material for Na-ion analysis for hybrid 80-02-01. Hatched *bars* identify hybrids propagated for a replicated trial (see Table 1).

Table 1. Relationship, expressed as a linear regression, between leaf Na-ion (y axis) and Cl-ion (x axis) concentrations (% dry wt) for hybrids from eight citrus progenies. Leaves were sampled from middle sections of stems of fourteen-month-old seedlings which had grown in nutrient solution cultures containing NaCl (50 m*M*) for 56 days

Progeny	Parents	n[a]	y = a + bx[b]	b + s.d.[c]	t[d]	r^2
80-02	Rangpur lime × Trifoliate orange	33	y = 0.575 + 0.359x	0.359 + 0.049	7.395***	0.64
80-04	Clementine mandarin × Trifoliate orange	39	y = 1.695 + 0.181x	0.181 + 0.080	2.269*	0.12
81-01	Clementine mandarin × Trifoliate orange	613	y = 0.539 + 0.400x	0.400 + 0.009	45.65***	0.77
81-02	Clementine mandarin × Rangpur lime	444	y = 0.652 + 0.560x	0.560 + 0.014	38.15***	0.77
81-03	Clementine mandarin × Carrizo citrange	77	y = 0.565 + 0.484x	0.484 + 0.059	8.212***	0.47
81-04	Rangpur lime × Trifoliate orange	18	y = 0.439 + 0.442x	0.442 + 0.185	2.389*	0.26
82-01	Rangpur lime × Trifoliate orange	33	y = 0.035 + 0.452x	0.452 + 0.048	9.518***	0.74
82-07	Ellendale tangor × Trifoliate orange	158	y = 0.166 + 0.509x	0.509 + 0.031	16.218***	0.63

[a]n = number of hybrids in progeny.
[b]Equation of linear regression where y = leaf Na^+ concentration, x = leaf Cl^- concentration, a = intercept on y axis and b = slope.
[c]s.d. = sample standard deviation of regression coefficient, b.
[d]t test for significance of b (* p 0.05, *** p 0.001).

centrations were positively correlated with leaf Na-ion concentrations and linear regressions could be fitted to these data (Table 1). This association, however, was stronger for some progenies than for others. For example progenies 80-04 and 81-01, which were from the same cross, viz. Clementine mandarin × Trifoliate orange, had r^2 values of 0.12 and 0.77, respectively. A low leaf Cl-ion concentration did not always correspond to a low Na-ion concentration and vice versa, e.g. hybrids 80-02-04, −21 and −30 (Fig. 1).

With the exception of Na-ion concentrations for leaves and distal stems of hybrid 80-02-08, salt treatment (NaCl 50 m*M*) increased Cl-ion and Na-ion concentrations of tissues harvested from nucellar seedlings of Rangpur lime and Trifoliate orange and the vegetatively propagated plants of the three Rangpur lime × Trifoliate orange hybrids. Analyses of variance of the data gave no significant ($p = 0.05$) effects due to tanks and so the data were pooled for each genotype and analysed as single classifications (Table 2). For all tissues, there were signifi-

Table 2. Mean (n = 6) leaf Cl-ion and Na-ion concentrations (% of dry weight) for nucellar seedlings of Rangpur lime and Trifoliate orange and vegetatively propagated plants of three hybrids between these rootstock varieties when grown in nutrient solution cultures containing NaCl (50 m*M*) for 50 days

Ion	Tissue	Genotype					LSD ($p = 0.05$)	F[a]	sig[b]	r_I[c]
		Rangpur lime	Trifoliate orange	80-02-02	80-02-08	80-02-38				
Cl	Proximal leaves	0.32	1.61	2.07	0.34	0.46	0.44	25.7	***	0.80
	Distal leaves	0.27	2.51	2.09	0.30	0.48	0.69	25.0	***	0.80
	Proximal stem	0.59	0.82	1.18	0.39	0.50	0.19	27.4	***	0.81
	Distal stem	0.52	1.08	1.44	0.31	0.47	0.32	21.3	***	0.77
Na	Proximal leaves	0.46	1.01	0.75	0.07	0.49	0.31	10.8	***	0.62
	Distal leaves	0.52	0.68	0.83	0.01	0.61	0.36	6.3	**	0.47
	Proximal stem	0.58	0.75	0.84	0.46	0.52	0.12	14.7	***	0.70
	Distal stem	0.84	0.48	1.34	0.07	0.90	0.24	34.8	***	0.85

a)[a]The variance ratio (F) from analysis of variance comparing the 5 genotypes.
b)[b]The significance of F (** $p < 0.01$, *** $p < 0.001$).
c)[c]The intra-class correlation coefficient (Sokal and Rohlf, 1969) was obtained from the Analysis of Variance as follows:

Source	df.	e.m.s
Between genotypes	4	$\sigma_b^2 + 6\sigma_w^2$
Between individuals within genotypes	5(6 − 1)	σ_w^2

$$r_i = \frac{\sigma_b^2}{\sigma_b^2 + \sigma_w^2}$$

cant genotypic differences for both Cl-ion and Na-ion concentrations (Table 2). These differences were larger for Cl ions with Rangpur lime, 80-02-08 and 80-02-38 all having low concentrations. Hybrid 80-02-02 behaved as predicted by the screening test and had Cl-ion concentrations similar to those of Trifoliate orange. Differences between genotypes for Na-ion concentration were not so large. Nevertheless, all three hybrids accumulated Na ions in the order of ranking as predicted by the screening test. Hybrid 80-02-08 was exceptional in that it only accumulated Na ions in proximal stems. With the exception of Na-ion concentrations for distal leaves, the intra-class correlation coefficient (Table 2) was higher than 0.5 for all ions in all tissues. This indicated that genotypic variation was larger than that between individuals within genotypes.

Vitis

The results from the unreplicated vineyard surveys of hybrid and backcross vines and the replicated glasshouse trials which are pertinent to the following discussion have all been published (Newman and Antcliff, 1984; Sykes, 1985c; 1987).

Discussion

The experience of plant breeders involved with asexually propagated crops is that these species are highly heterozygous and segregate widely upon sexual reproduction (Allard, 1960). While inheritance studies in citrus are scarce, wide segregation has been noted for some characteristics, e.g. root rot resistance (Hutchinson, 1985). An ability for Cl-ion exclusion by citrus hybrids from crosses involving Cl-ion-excluding parents varies widely suggesting it is a polygenic trait (Furr and Ream, 1969; Sykes, 1985b). The results presented here demonstrate this (Fig. 1) and also that the ability to restrict Na ions accumulating in shoots varies widely suggesting that it too is a polygenic trait. The superior nature of hybrid 80-02-08 in excluding Na ions from leaves compared to its parents indicates transgressive segregation.

The data presented in Table 2 support the data provided for Cl-ion exclusion by replicated hybrids by Sykes (1985b) and also support the conclusion that a test period of 56 days using saline nutrient solution cultures is adequate to select new salt-excluding citrus rootstock hybrids. The value of a rapid test to screen young woody perennials in a costly breeding program has been discussed (Sykes, 1985a; b).

In vegetatively propagated crops, the intra-class correlation coefficient (Table 2) provides an estimate of the ratio between the genetic and the phenotypic variance of a particular trait (see Hansche, 1983). This ratio provides a quantitative statement of the relative importance of genetic versus environmental factors affecting observed differences among phenotypic measurements. The intra-class correlation coefficients in Table 2 suggest a strong heritable basis for salt exclusion in Rangpur lime, Trifoliate orange and their hybrids. These statistics provide an estimate of the maximum value of heritability under the test conditions described. The variance between plants (individuals) within a genotype estimates non-heritable effects while the variance between genotypes estimates heritable and non-heritable effects plus any effects due to a common environment amongst cloned individuals. Two examples of possible sources of variation due to common environment being differences in nutritional reserves of cuttings used to produce individuals within a clone or competition between several nucellar seedings emerging from one polyembryonic seed. As the data in Table 2 were obtained for nucellar seedlings of parent varieties and for plants grown from cuttings of the hybrids, the effect of a common environment may have inflated the intra-class correlation coefficient. Nevertheless, the relatively high values for this statistic lend further support to the feasibility of breeding new salt-excluding citrus rootstocks.

were positively correlated within citrus progenies, which may suggest the ability to exclude one or both ions could be considered the same characteristic, the strength of this correlation varied. It was possible to identify individuals which were either good Cl-ion excluders but poor Na-ion excluders and vice versa. This supports the data of Walker (1986) who reported that Cleopatra mandarin is a good Cl-ion excluder but a poor Na-ion excluder. This suggests that the ability to exclude these two ions is due to two different mechanisms although further experiments are needed to show this definitively.

Continuous variation for petiole Cl-ion concentrations in progenies of Ramsey (*V. champini*) × Sultana (*V. vinifera*), Ramsey × Villard blanc (12375-SV, complex hybrid) and Villard blanc × Sultana under vineyard conditions (Sykes, 1985c) suggested that the greater ability for Cl-ion exclusion by Ramsey is a polygenic trait. The variation between replicated hybrids of Ramsey × Sultana under glasshouse conditions supported this. Both vineyard and glasshouse data suggested transgressive segregation. It was therefore possible to identify new genotypes which were better Cl-ion excluders than Ramsey.

Data from the vineyard survey of *V. berlandieri* hybrids and backcrosses (Newman and Antcliff, 1984) suggested that a single dominant gene was the major factor governing the inheritance of Cl-ion exclusion in *V. berlandieri* vines. All hybrids between *V. berlandieri* and *V. vinifera* had low petiole Cl-ion concentrations and backcrosses segregated in a 1:1 ratio for high and low petiole Cl-ion concentrations. Data from glasshouse trials (Sykes, 1987) supported this for two of the four backcrosses. Vines from MF77-13 (*V. berlandieri* × Sultana) × Biancone or Koshu Sanjaku could be identified as either Cl-ion excluders or accumulators when grown in solution cultures containing NaCl (25 m*M*). The variation in two other backcrosses, however, did not support the vineyard data. This may have been due to the action of modifying genes derived from *vinifera* parents (Sykes, 1987). A similar situation can arise when breeding for disease resistance where the expression of a major gene in interspecific crosses is affected by minor genes from *vinifera* parents (Becker and Zimmerman, 1977).

It has been suggested that there has been a trend towards higher salinities throughout the River Murray in SE Australia and there have been several solutions suggested to overcome the effects of salinity on salt-sensitive irrigated horticultural crops (Blesing and Tuffley, 1977). One proposal has been to breed new salt-tolerant varieties. The results of the research described herein clearly support this philosophy.

Selection in any breeding program can only act effectively on inherited characteristics. The results presented here demonstrate that it is possible to select new salt-excluding citrus and grapevine hybrids and thus salt exclusion must, by inference, be inherited. In citrus, Cl-ion and Na-ion exclusion appears to be inherited polygenically, whereas in grapevines Cl-ion exclusion may be expressed either as a polygenic or monogenic trait depending on parents.

The next stage in breeding new salt-excluding rootstocks is to evaluate hybrids as rootstocks under field conditions. This has almost been completed for some citrus hybrids and the results have been very encouraging; hybrid 80-02-08 is a good salt excluder. However, before new salt-tolerant citrus rootstocks can be released, hybrids have to be assessed for many other characteristics. These include disease and pest resistance, polyembryony, scion compatibility, lime tolerance and horticultural performance. As far as the grapevine hybrids are concerned, the progenies described herein have been screened and selected for ease of rooting and grafting. As a result, selections have been entered as rootstocks grafted to Sultana in trials using salinised field plots to assess their effects on yield and fruit quality. The outcome of this research will hopefully lead to the release of a new salt-tolerant grapevine rootstock.

References

Allard R W 1960 Principles of Plant Breeding. Wiley, Toppan, NY; London, Tokyo. 485 p.

Antcliff A J, Newman H P and Barrett H C 1983 Variation in chloride accumulation in some American species of grapevine. Vitis 22, 357–362.

Bailey L H and Bailey E Z 1976 Hortus Third: A Concise Dictionary of Plants Cultivated in the United States and Canada. Macmillan, New York. 1290 p.

Becker N J and Zimmerman H 1977 Breeding of yield varieties resistant to downy mildew. *In* Génétique et amélioration de la vigne. Proc. 2nd Intern. Symp. on Amelioration of the Vine, pp 209–214. INRA, Paris.

Behboudian M H, Törökfalvy E and Walker R R 1986 Effects of salinity on ionic content, water relations and gas exchange parameters in some citrus scion-rootstock combinations. Sci. Hortic. 28, 105–116.

Bernstein L, Ehlig C F and Clark R A 1969 Effect of grape rootstock on chloride accumulation in leaves. J. Am. Soc. Hortic. Sci. 94, 561–571.

Blesing N V and Tuffley J R 1977 Study of potential

economic benefits from reduction in salinity of River Murray water. Amdel Report No. 1179, Adelaide, South Australia.

Cooper WC and Gorton B S 1952 Toxicity and accumulation of chloride salts in cirtus on various rootstocks. Proc. Am. Soc. Hortic. Sci. 59, 143–146.

Cooper W C and Shull A V 1953 Salt tolerance and accumulation of sodium and chloride ions in grapefruit on various rootstocks grown in a naturally saline soil. Proc. 7th Rio Grande Val. Hortic. Inst. 7, 107–117.

Downton W J S 1977a Photosynthesis in salt-stressed grapevines. Aust. J. Plant Physiol. 4, 183–192.

Downton W J S 1977b Chloride accumulation in different species of grapevine. Sci. Hortic. 8, 249–253.

Ehlig E F 1960 Effects of salinity on four varieties of table grapes grown in sand culture. Proc. Am. Soc. Hort. Sic. 76, 323–331.

Frost H B and Soost R K 1968 Seed reproduction: Development of gametes and embryos. *In* The Citrus Industry. Vol. II. Eds. W Reuther, L D Batchelor and H J Webber. University of California, Berkeley.

Furr J R and Ream C L 1969 Breeding citrus rootstocks for salt tolerance. Proc. 1st Int. Citrus symposium Vol. 1, 373–380.

Grieve A M 1984 The effects of salinity on grapevine productivity. *In* Irrigation, Salinity and Grape quality. Eds. D C Lester and T H Lee. pp 5–13. Proc. of a seminar, Loxton, S.A., Nov. 1984.

Grieve A M and Walker R R 1983 Uptake and distribution of chloride, sodium and potassium ions in salt-treated citrus plants. Aust. J. Agric. Res. 34, 133–143.

Greenway H and Munns R 1980 Mechanisms of salt tolerance in nonhalophytes. Annu. Rev. Plant Physiol. 31, 149–190.

Groot Obbink J and Alexander D McE 1973 Response of six grapevine cultivars to a range of chloride concentrations. Am. J. Enol. Vitic. 24, 65–68.

Hansche P E 1983 Response to selection. *In* Methods in Fruit Breeding. Eds. J N Moore and J Janick. pp 154–171. Purdue University Press, West Lafayette, IN.

Hoagland D R and Arnon D I 1950 The Water Culture Method for Growing Plants Without Soil. Circ. Calif. Agric. Exp. Stn. Berkeley, FL, No. 347.

Hutchinson D J 1985 Rootstock development screening and selection for disease tolerance and horticultural characteristics. Fruit Varieties J. 39(3), 21–25.

Newman H P and Antcliff A J 1984 Chloride accumulation in some hybrids and backcrosses of *Vitis berlandieri* and *Vitis vinifera*. Vitis 23, 106–112.

Sauer M R 1968 Effects of vine rootstocks on chloride concentration in sultana scions. Vitis 7, 223–226.

Sokal R R and Rohlf F J 1969 Biometry: The Principles and Practice of Statistics in Biological Research. Freeman, San Francisco. 776 p.

Soost R K and Cameron J W 1975 Citrus. *In* Advances in Fruit Breeding. Eds. J Janick and J N Moore. pp 507–540. Purdue University Press, West Lafayette, IN.

Sykes S R 1985a Effects of seedling age and size on chloride accumulation by juvenile citrus seedlings treated with sodium chloride under glasshouse conditions. Aust. J. Exp. Agric. 25, 943–953.

Sykes S R 1985b A glasshouse screening procedure for identifying citrus hybrids which restrict chloride accumulation in shoot tissues. Aust. J. Agric. Res. 36, 779–789.

Sykes S R 1985c Variation in chloride accumulation by hybrid vines from crosses involving the cultivars Ramsey, Villard blanc and Sultana. Am. J. Enol. Vitic. 36, 30–37.

Sykes S R 1987 Variation in chloride accumulation in hybrids and backcrosses of *Vitis berlandieri* and *Vitis vinifera* under glasshouse conditions. Am. J. Enol. Vitic. 38, 313–320.

Sykes S R and Newman H P 1987 The genetic basis for salt exclusion in grapevines. Aust. Grapegrower Winemaker. April 1987, 75–78.

Thomas J E 1934 The diagnostic value of the chloride content of the vine leaf. J. Counc. Sci. Indust. Res. Aust. 7, 29–38.

Walker R R 1986 Sodium exclusion and potassium-sodium selectivity in salt-treated Trifoliate orange (*Poncirus trifoliata*) and Cleopatra mandarin (*Citrus reticultata*) plants. Aust. J. Plant Physiol. 13, 293–303.

Walker R R and Douglas T J 1983 Effects of salinity level on uptake and distribution of chloride, sodium and potassium ions in citrus plants. Aust. J. Agric. Res. 34, 145–153.

Woodham R C 1956 The chloride status of the irrigated Sultana vine and its relation to vine health. Aust. J. Agric. Res. 7, 414–427.

P. J. Randall et al. (Eds.), Genetic aspects of plant mineral nutrition, 173–178.

PLSO SV44

Variation in growth and ion accumulation between two selected populations of *Trifolium repens* L. differing in salt tolerance

M.E. ROGERS and C.L. NOBLE
Institute for Sustainable Agriculture, Tatura, Vic., 3616, Australia

Key words: salt tolerance, *Trifolium repens*, white clover

Abstract

Two divergent populations of *T. repens* cv. Haifa developed from two generations of recurrent selection for shoot chloride concentration, were grown in the greenhouse at 0 and 40 mol m^{-3} NaCl. Over two harvest cycles at 40 mol m^{-3} NaCl, the population selected for a low concentration of chloride in the shoot maintained a significantly lower chloride and sodium concentration compared with those plants selected for a high shoot chloride concentration. The distribution of chloride in the shoots was further examined in a subsample of plants from both populations. In all plants, concentrations of chloride were lower in the expanding and fully expanded leaves than in the older leaf tissue or petioles.

While there were no significant differences in the photosynthetic rates between lines, shoot yields and relative leaf expansion rates were higher in the low chloride population. Plant death was greater in plants selected for high shoot chloride. These results suggest that selections based on measurements of low shoot chloride concentrations may be successful in developing a cultivar of *T. repens* with improved salt tolerance.

Introduction

In northern Victoria, Australia, white clover (*Trifolium repens* L.) is a major component of perennial pastures that are affected by rising water tables and increasing levels of soil salinity. As a species, *T. repens* is sensitive to NaCl (Gauch and Magistad, 1943), but it is polyploid and cross-fertilizing with considerable plant-to-plant variation for salt tolerance (Noble and Shannon, 1987), indicating that there is potential to improve its salt tolerance by selection and breeding.

Exclusion of sodium and chloride from actively growing tissue is a salt resistance mechanism common in salt-sensitive species such as white clover (Läuchli, 1984). Individual plants may differ in their capacity to regulate and control ion transport and accumulation and, hence, in their salt tolerance (Noble et al., 1983; Winter and Läuchli, 1982). With a view to improving the salt tolerance of *T. repens*, efforts have concentrated on selecting individual plants for low rates of transport of chloride. Two divergent populations of *T. repens* cv. Haifa have been developed based on the concentration of chloride in the shoots. These populations have undergone two generations of recurrent selection following evaluation and selection at 40 mol m^{-3} NaCl and are providing the material for investigations of the physiological and genetic basis for salt tolerance in *T. repens*.

Materials and methods

The effect of NaCl on the growth and tissue ion concentration in two divergent populations of *T. repens* (viz. 'high Cl' plants and 'low Cl' plants) and the parent cultivar Haifa was assessed in a naturally lit greenhouse at Tatura (at night and day temperatures of 10°C ± 3°C and 25°C ± 3°C, respectively). Seeds with a uniform seed weight were first germinated under non-saline conditions in trays of vermiculite and seedlings were transplanted at the second trifoliate leaf stage into cells of polystyrene 'speedling' trays filled with vermiculite. These trays were floated on modified half-strength Hoagland solution (Karmoker and Van Steveninck, 1978) in stainless steel tanks with a volume of 160 L. Salinity treatments of 0 and 40 mol m^{-3} NaCl were imposed after two weeks, the latter being reached in increments of 20 mol m^{-3} NaCl over two days. Plants were grown hydroponically in continuously aerated solutions and the pH and electrical conductivity of the solutions were monitored and adjusted as necessary every three days. The solutions were replenished every two weeks. The experiment was a randomised block-split plot design. The NaCl treatment was applied to the main plots or tanks and there were four replicates. The split plots in each main plot were rows of 10 plants of each of the 'high Cl', 'low Cl' and cv. Haifa populations.

There were two harvests of the shoots at three-week intervals commencing three weeks after the salinity treatment had been imposed. At harvest, the shoots of all individual plants were cut and plant material was dried at 70°C for 48 hours and weighed before ashing at 460°C overnight. Shoot chloride concentration was measured on individual plants using a Buchler chloridometer based on titration with silver ions, and sodium, potassium, calcium and magnesium were measured using an Inductively Coupled Plasma Optical Emission Spectrophotometer (Labtam Plasma Scan).

The effect of tissue Na or Cl concentration on specific growth mechanisms such as leaf photosynthesis and leaf expansion were measured on plants from the three populations. Leaf expansion rates were measured on three plants of each line in each replicate between weeks 4 and 6 after the salinity treatments had been imposed. Leaves were identified at development stage 0.4 (Carlson, 1966) and their areas were measured non-destructively using a series of templates based on the rating procedure developed by Williams et al., (1964) for *T. subterraneum* (i.e. rating = 10 $\log_e$ 10A where A is the leaf area in square centimetres). Relative leaf expansion rates were then calculated as ($\text{Log}_e A_2 - \text{Loge} A_1/t_2 - t_1$) where A_1 and A_2 are leaf areas at time t_1 and t_2. Leaf areas were measured over five days until leaves had fully expanded.

Leaf photosynthetic rates, at full light and CO_2 levels equal to approximately 340 ppm, were measured using a portable photosynthesis system (Licor 6200) with a chamber attached (volume = 0.6 L) at week 6 after salt had been imposed. One leaf was selected from four plants in each treatment and replicate. The leaf areas were measured on individual leaves before being enclosed into the chamber.

At the completion of the experiment, a subsample of ten plants that were known, from a previous harvest, to cover a range of mean shoot chloride concentrations, were destructively harvested and divided into leaves (old, fully expanded, expanding), and petioles (senesced, old, and young). The yield of these plant parts and their tissue chloride concentrations were measured using the techniques described earlier.

Yield, tissue ion data, relative leaf expansion rates and leaf photosynthetic rates were analysed by anova using Genstat 2.1 (Lawes Agricultural Trust). Plant death was analysed using a generalised linear model with binomial error distribution.

Results

After six weeks exposure to 40 mol m^{-3} NaCl, concentrations of Cl in the shoots of the three populations increased significantly ($p < 0.05$, Table 1), however the 'low Cl' plants maintained a significantly lower concentration of Cl in their shoots ($p < 0.05$, Table 1) than either the 'high Cl' plants or the Haifa population. There was no difference between populations at 0 mol m^{-3} NaCl. Results for shoot concentrations of Na were similar to those of Cl (Table 2), with concentrations being significantly lower in the 'low Cl' population ($p < 0.05$) than in the 'high Cl' and Haifa populations at 40 mol m^{-3} NaCl. Concentrations of potassium in the shoots of 'high Cl' plants were significantly lower than those in the 'low Cl' or Haifa populations ($p <$ 0.1, Table 2). K:Na ratios tended to be more favourable in the 'low Cl' plants than the 'high

Table 1. Shoot yield and tissue chloride concentration following a second cut-regrowth cycle in individual, second generation plants of *T. repens* selected for high and low shoot chloride concentration and plants of the parent cultivar Haifa when grown at 0 and 40 mol m^{-3} NaCl in the greenhouse

Dry weight (g/plant)				Shoot Cl concentration (mol m^{-3} kg^{-1} dry weight)		
NaCl mol m^{-3}	Haifa	High Cl	Low Cl	Haifa	High Cl	Low Cl
0	0.58	0.59	0.59	160	178	150
40	0.41	0.41	0.52	782	893	537

Salinity * line, n = 214; $LSD_{(p=0.05)} = 0.13$; $LSD_{(p=0.1)} = 0.11$.
Salinity * line, n = 214; $LSD_{(p=0.05)} = 188$; $LSD_{(p=0.1)} = 157$.

Table 2. Shoot sodium and potassium concentrations following a second cut-regrowth cycle in individual, second generation plants of *T. repens* selected for high and low shoot chloride concentration and plants of the parent cultivar Haifa when grown at 0 and 40 mol m^{-3} NaCl in the greenhouse

Shoot Na concentration (mol m^{-3} kg^{-1} dry weight)				Shoot K concentration (mol m^{-3} kg^{-1} dry weight)		
	Haifa	High Cl	Low Cl	Haifa	High Cl	Low Cl
0	99	80	87	1074	1168	1157
40	796	934	635	952	907	1043

Salinity * line, n = 214; $LSD_{(p=0.05)} = 127$; $LSD_{(p=0.1)} = 106$.
Salinity * line, n = 214; $LSD_{(p=0.05)} = 156$; $LSD_{(p=0.1)} = 131$.

Cl' plants at 40 mol m^{-3} NaCl although these values were not significantly different (viz. 1.95 compared with 1.20) and both ratios were above the value (1.0) required for optimal efficiency (Greenway and Munns, 1980). Concentrations of calcium in the shoots decreased at 40 mol m^{-3} NaCl ($p < 0.05$) but there was no difference between populations ($p = 0.208$) (viz. mean concentrations were 399 mol m^{-3} Ca at 0 mol m^{-3} NaCl compared with mean concentrations of 267 mol m^{-3} Cl at 40 mol m^{-3} NaCl for the three populations combined). There were no significant differences in the shoot concentrations of magnesium between plant populations ($p = 0.324$) or at either NaCl concentration ($p = 0.693$) (viz. mean concentrations for the three populations combined were 105 mol m^{-3} Mg at 0 mol m^{-3} NaCl compared with 102 mol m^{-3} Mg at 40 mol m^{-3} NaCl).

Shoot yields at harvest 2, both in the 'high Cl' and Haifa populations, decreased significantly at 40 mol m^{-3} NaCl ($p < 0.05$, Table 1) and were lower than the yield of 'low Cl' plants at this concentration ($p < 0.1$, Table 1). The yield results from harvest one also revealed that the 'low Cl' plants were higher yielding than the Haifa plants ($p < 0.1$, data not presented).

Information on Cl distribution in three plants with different mean shoot Cl concentrations is shown in Table 3. In plants with a lower overall mean shoot Cl concentration, concentrations were lower in every plant part compared with those plants with a higher mean shoot Cl concentration. Concentrations also tended to be higher in the older leaves and petioles than in the expanding leaves.

Leaf expansion rates were also sensitive to increased levels of NaCl and were significantly lower at 40 compared with 0 mol m^{-3} NaCl ($p < 0.05$, Table 4). At 40 mol m^{-3} NaCl, relative leaf expansion rates for the 'low Cl' and Haifa plants were significantly higher ($p < 0.05$, Table 4) than those for the 'high Cl' plants which would account for some of the observed differ-

Table 3. The distribution of chloride in the shoots of three plants of *T. repens* differing in mean shoot chloride concentration when grown at 40 mol m^{-3} NaCl for six weeks

	Mean plant	Chloride concentration (mol m^{-3} kg^{-1} dry weight) Leaves			Petioles		
		Old	Fully expanded	Expanding	Senesced	Old	Young
Low Cl	708	1354	564	211	846	1721	592
Haifa	1027	1336	846	367	1185	2623	715
High Cl	1577	2228	1100	370	1628	2376	821

Table 4. Relative leaf expansion rates and leaf photosynthesis rates in second-generation plants of *T. repens* selected for high and low shoot chloride concentration and plants of the parent cultivar Haifa when grown at 0 and 40 mol m^{-3} NaCl in the greenhouse

NaCl mol m^{-3}	Relative leaf expansion rates			Leaf photosynthesis rate μ mol^{-2} s^{-1}		
	Haifa	High Cl	Low Cl	Haifa	High Cl	Low Cl
0	0.271	0.302	0.290	7.60	6.68	7.91
40	0.142	0.065	0.166	7.59	7.35	8.43

Salinity * line, n = 25
$LSD_{(p=0.05)} = 0.040$;
$LSD_{(p=0.1)} = 0.033$.

Salinity * line, n = 25;
$LSD_{(p=0.05)} = 3.033$;
$LSD_{(p=0.1)} = 2.475$.

Table 5. Plant death in second-generation plants selected for high and low shoot chloride concentration and in plants of the parent cultivar Haifa when grown at 0 and 40 mol m^{-3} NaCl in the greenhouse for six weeks

NaCl (mol m^{-3})	Predicted plant mortalities (proportion of plants that died					
	Haifa	s.e.	High Cl	s.e.	Low Cl	s.e.
0	0.0242	(0.0155)	0.0271	(0.0154)	0.0288	(0.0177)
40	0.1258	(0.0516)	0.3029	(0.0715)	0.1462	(0.0549)

ences in yield between populations. However, despite differences in leaf expansion rates, there were no significant differences in individual leaf photosynthesis rates between populations or between NaCl concentrations (Table 4). Individual leaf photosynthesis values tended to be higher at 40 mol m^{-3} NaCl, but rates varied greatly between leaves even of the same treatment despite similar leaf areas.

Throughout the experiment, the number of plants that died was significantly larger in the 'high Cl' plants at 40 mol m^{-3} NaCl than in the 'low Cl' or Haifa plant populations ($p < 0.001$, Table 5).

Discussion

For all plant characteristics measured in this experiment the 'low Cl' plants were superior to those of the 'high Cl' or Haifa plants in the presence of 40 mol m^{-3} NaCl. The concentrations of Cl and Na in the shoots were lower, the concentrations of K were higher rendering a more favourable K:Na ratio, plant death was lower and shoot yield was higher. This suggests that firstly salt tolerance in *T. repens*, in common with other salt-sensitive species such as *Festuca* (Hannon and Barber, 1972), grapevine (Downton, 1977), soybean (Abel, 1969; Läuchli and Wieneke, 1979) and rice (Yeo and Flowers, 1982), is correlated with restricted and regulated Cl and Na translocation in the shoot. Secondly, that there is significant variation in Cl uptake and distribution within *T. repens* and that this variation can be selected for and incorporated into a breeding program.

To date, there has been little research on the salt tolerance of *T. repens*. This species is generally classified as salt-sensitive (Gauch and Magistad, 1943; Smith and McComb, 1983) although it is recognised as being genetically and phenotypically variable (Burdon, 1980). Ab-Shukor et al., (1987) demonstrated that several natural populations of *T. repens* exhibited high to very high salt tolerance (comparable to that of *T. alexandrinum*) in terms of root growth at 150 to 200 mol m^{-3} NaCl, but made no measurements of tissue ion concentrations. Our limited research examining Cl distribution throughout the plant revealed that the Cl concentration in the leaves tended to be about one third to one half that of the petioles. These results are similar to those of research on ion distribution in *T. alexandrinum* by Winter and Läuchli (1982), where chloride concentrations per gram dry weight in the petioles was about three times that of the leaves (allowing for differences in water content of the two organs). These authors concluded that *T. alexandrinum* uses several mechanisms to cope with moderate salinity levels including retranslocation of Na and Cl out of young leaves. In *T. repens*, further research is now possible using these divergent plant selections to identify how Cl and Na exclusion is regulated and to study ion compartmentation. Early indications using a scanning electron microscope with X-ray microprobe suggest that chloride is concentrated in the vacuoles of the palisade and spongy mesophyll cells within the leaf tissue (Rogers and Noble, unpublished).

The effects of NaCl on other physiological mechanisms in *T. repens* were varied. Measurements of leaf expansion rates were sensitive to shoot Na and Cl concentrations, whereas rates of photosynthesis in expanding leaves were insensitive. This suggests that concentrations were not sufficiently high to affect photosynthesis or chlorophyll activity. Other authors have drawn similar conclusions in cereals (Rawson et al., 1988) and spinach (Robinson et al., 1983). Although there were significant differences in leaf expansion rates between populations, the difference in overall yield was less significant, implying that the allocation of assimilates between roots, stems and leaves may differ between selected populations.

In this experiment we were concerned with the effect of NaCl on shoot growth and ion relations in *T. repens*. Subsequent research will examine ion distribution and growth in both roots and shoots, and the heritability of salt tolerance. To date, the results are encouraging indicating that selections based on measurements of shoot Cl may be successful in developing a cultivar of *T. repens* with improved salt tolerance.

Acknowledgements

We thank Mrs M Filip and Mr I Treacy for providing excellent technical assistance. This research was funded by the Victorian State Salinity Program.

References

Abel G H 1969 Inheritance of the capacity for chloride inclusion and chloride exclusion by soybeans. Crop Sci. 9, 697–698.

Ab-Shukor N A, Kay Q O N, Stevens D P and Skibinski D O F 1988 Salt tolerance in natural populations of *Trifolium repens* L. New Phytol. 109, 483–490.

Burdon J J 1980 Intra-specific diversity in a natural population of *Trifolium repens*. J. Ecol. 68, 717–735.

Carlson G E 1966 Growth of clover leaves: Developmental morphology and parameters at ten stages. Crop Sci. 6, 293–294.

Downton W J S 1977 Photosynthesis in salt-stressed grape leaves. Aust. J. Plant Physiol. 4, 183–192.

Gauch H G and Magistad O C 1943 Growth of strawberry clover varieties and of alfalfa and ladino clovers as affected by salt. J. Am. Soc. Agron. 35, 871–880.

Greenway H and Munns R 1980 Mechanisms of salt tolerance in nonhalophytes. Annu. Rev. Plant Physiol. 31, 149–190.

Hannon N J and Barber H N 1972 The mechanism of salt tolerance in naturally selected populations of grasses. Search 3, 259–260.

Karmoker J L and Van Steveninck R F M 1978 Stimulation of volume flow and ion flux by abscisic acid in excised root systems of *Phaseolus vulgaris* L. cv. Redland Pioneer. Planta 141, 37–43.

Läuchli A 1984 Salt exclusion; an adaptation of legumes for crops and pastures under saline conditions. *In* Salinity Tolerance in Plants: Strategies for Crop Improvement. Ed. R C Staples. pp 171–187. Wiley, New York.

Läuchli A and Wieneke J 1979 Studies on the growth and distribution of Na^+, K^+ and Cl^- in soybean varieties differing in salt tolerance. Z. Pflanzenernaehr. Bodenkd. 124, 3–13.

Noble C L and Shannon M C 1987 Strategies for irrigation of white clover with saline water on the heavy clay soils of northern Victoria. Proc. Fourth Aust. Agron. Conf. Melbourne, 305 p.

Noble C L, Halloran G M and West D W 1984 Identification and selection for salt tolerance in lucerne (*Medicago sativa* L.). Aust. J. Agric. Res. 35, 239–252.

Rawson H M, Richards R A and Munns R 1988 An examination of selection criteria for salt tolerance in wheat, barley and triticale genotypes. Aust. J. Agric. Res. 39, 759–772.

Robinson S P, Downton W J S and Millhouse J A 1983 Photosynthesis and ion content of leaves and isolated chloroplasts of salt-stressed spinach. Plant Physiol. 73, 238–242.

Smith M K and McComb J A 1981 Use of callus cultures to detect NaCl tolerance in cultivars of three species of pasture legumes. Aust. J. Plant Physiol. 8, 437–442.

Williams R F, Evans L T and Ludwig L J 1964 Estimation of leaf area for clover and lucerne. Aust. J. Agric. Res. 15, 231–233.

Winter E 1982 Salt tolerance of *Trifolium alexandrinum* L. II. Ion balance in relation to its salt tolerance. Aust. J. Plant Physiol. 9, 227–237.

Yeo A R and Flowers T J 1982 Accumulation and localisation of sodium ions within the shoots of rice (*Oryza sativa*) varieties differing in salt resistance. Physiol. Plant. 56, 343–348.

P. J. Randall et al. (Eds.), Genetic aspects of plant mineral nutrition, 179–186.

PLSO SV41

The role of ion channels in plant nutrition and prospects for their genetic manipulation

S.D. TYERMAN and D.P. SCHACHTMAN
School of Biological Sciences, Flinders University, Bedford Park, SA 5042 Australia and Biology Department, UC-San Diego, La Jolla, CA 92093-0116, USA

Key words: genetics, ion channel, membranes, physiology, plants, review, transport

Abstract

Ion channels can function in three physiological modes through their ability to: 1) accommodate osmotically significant fluxes over short periods; 2) propagate signals along or across membranes; 3) control the membrane potential. With respect to mineral nutrition it is via the control of the membrane potential that ion channels are probably most significant. In this paper the physiology and prospects for molecular biology of plant ion channels are discussed. It is concluded that identifying and altering the primary structures that determine functional characteristics of plant ion channel genes could result in changes in the transport characteristics of higher plants.

Physiology of ion channels

Modes of ion channel function

Ion channels facilitate passive fluxes of ions across a membrane. The electrochemical gradients driving the passive fluxes are ultimately established by pumps. Ion channels are capable of dissipating electrochemical gradients both in terms of the electrical potential difference (V_m) and the ion concentrations in cellular compartments. However, ion channel currents are tightly controlled and are important in the following types of plant processes:

(1) Large net fluxes which may be osmotically significant, for example, the reduction of turgor in a guard cell (Hedrich and Schroeder, 1989; MacRobbie, 1988);
(2) Electrical signals via action potentials (Davies, 1987) or chemical signals via ligand gated channels, e.g. ABA dependent Ca^{2+} permeable channels in plasma membrane of guard cells (Schroeder and Hagiwara, 1990);
(3) Maintenance of a certain membrane potential through low background ion channel activity, perhaps to maintain a constant electrochemical gradient. Background anion channels in animal cells can maintain a negative membrane potential (Franciolini and Petris, 1990), while in *Chara* an anion channel seems to be closely matched to a constant electrochemical potential for protons (Tyerman et al., 1986).

On the basis of the types of ion transport processes listed above, three modes of ion channel operation in plants can be postulated: an osmotic flux mode, signalling mode, and gradient control mode. It is the ability of ion channels to control electrochemical gradients via the membrane potential that is probably most significant in plant mineral nutrition. This is particularly so in roots where ion gradients rapidly change as roots explore the soil and as convective ion flow occurs during transpiration.

Much of our recent knowledge of the physiology of ion channels in higher plants has been gained by the use of the patch-clamp technique on isolated protoplasts or vacuoles (see Hedrich and Schroeder, 1989 for methods). The technique allows the measurement of single channel currents in the native membrane (e.g. Fig. 1) which can then be compared with the electrophysiology of the entire membrane either in protoplasts (Schachtman et al., 1991) or preferentially in intact cells (Coleman, 1986).

Functional properties of ion channels relevant to mineral nutrition

Selectivity

Generally channels can be classified as cation or anion selective, with variations observed between species in the extent of the selectivity (e.g. Hedrich et al., 1988). The variation in selectivity may not be surprising given that selectivity may be controlled by single amino acids within the protein pore (Yool and Schwarz, 1991). Some channels show high selectivity between similar ions (e.g. permeability ratio $K^+/Na^+ = 30$ in wheat, Schachtman et al., 1991), while others can be nonselective amongst certain cations or anions, e.g. a non-selective cation channel in endosperm cells (Stoeckel and Takeda 1989). Anion channels in Amaranthus and Chara have a similar permeability to NO_3^- and Cl^- (Terry et al., 1991; Tyerman and Findlay, 1989).

Regulation

Initially it is important to realise that the whole membrane current (I_m) (at steady state) resulting from ion channels at one particular voltage is determined by the channel density in the membrane (D) times the proportion of time that any channel will be open (P_0, or open probability) times the current through a single open channel (I_{ch}). It turns out that most channels show more than one open conductance so that I_{ch} does not have a discrete value at a particular V_m. An example of this is shown in Figure 1 for an anion channel in the plasma membrane of an *Amaranthus* cotyledon/hypocotyl protoplast. Sudden steps to different current levels at constant V_m indicates a number of open states of the channel which may be related to different conformations of the channel protein. For a channel which displays more than one value of I_{ch} (i.e. substates), the membrane current would be given by D times the sum of $P_0 \times I_{ch}$ over all substates:

$$I_m = D \cdot \sum_{i=1}^{n} P_0^i \cdot I_{ch}^i$$

The three terms in Eq. 1 could be varied to alter membrane current through an ensemble of ion channels. Normally it is the open probability (P_0) which is affected by various factors but D and I_{ch} may also vary. P_0 of a particular channel may be affected by one or more of the following factors: membrane voltage (V_m) (e.g. Keller et al., 1989), a hormone (Blatt, 1990; Schroeder and Hagiwara 1990), cytoplasmic chemical sig-

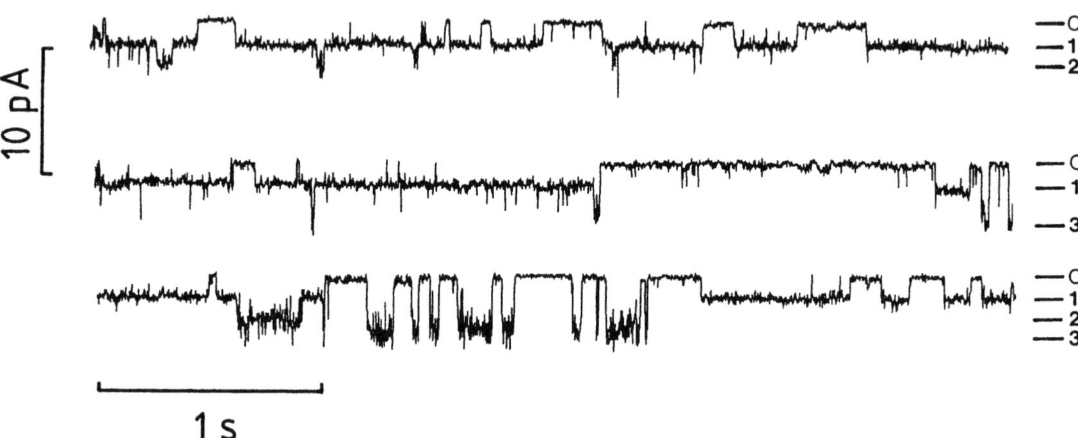

Fig. 1. Current flow as a function of time through an anion channel in the plasma membrane of *Amaranthus tricolor* recorded using patch-clamping. The membrane was in the outside-out configuration and was held at −95 mV (V_m corrected for liquid junction potential). Steps in current downwards represent opening of the channel and the efflux of Cl^- from the pipette equivalent to efflux from the cell. This channel has multiple conductance levels, three of which can be clearly seen in the figure (1, 2, 3; C = closed). This channel, activated at negative V_m, has a very high conductance in the fully open state (200 pS) and shows complex transitions between conductance levels. See Terry et al. (1991) for further details.

nals including Ca^{2+} (Alexandre et al., 1990; Hedrich et al., 1990), the presence of nucleotide phosphates (Hedrich et al., 1990; Terry et al., 1991), pH (Ilan et al., 1991; Tyerman et al., 1986b) and permeant ion concentrations (Lew, 1991).

Ion fluxes mediated by channels could show saturation

The concentration at which saturation occurs is usually high (e.g. for a K^+ channel in Chara tonoplast $K_m = 87$ mol m^{-3} at 100 mV (Laver et al., 1989)) so that in the normal physiological range a channel mediated flux as a function of concentration may appear as a linear component (but c.f. anion channel in Lew, 1991). A cation channel could account for the linear component of K^+ uptake into roots at high K^+ concentrations. This linear component is blocked by the K^+ channel blocker tetraethylammonium (Kochian et al., 1985). It should be noted that in these experiments V_m is most likely altered by K^+ concentration which would in turn alter P_0 of the channel.

The possibility also exists for saturation to occur at much lower concentrations. This is because P_0 of an ion channel may be modified either directly or indirectly by permeant ions (e.g. Cl^-, Lew, 1991, or NO_3^- Tyerman and Findlay, 1989). The proportion of the time that a channel remains open may be decreased as the concentration of permeant ion is increased either by direct allosteric interaction with the channel or indirectly via a change in V_m. It is also possible that different conductance levels are preferred at different concentrations particularly for channels which show more than one open channel conductance (e.g. Figure 1 and Tyerman et al., 1989). The different conductance levels may also have different selectivities (e.g. Terry et al. 1991). These properties of ion channels serve to highlight their complexity; they are not simple pores!

Ion channels in the plasma membrane of parenchyma cells

It is well established that ion channels are present in plant membranes (Tester, 1990). The physiological role of some of these channels is becoming clearer particularly in relation to the "net flux mode" in cells which undergo large changes in turgor pressure during their normal physiological function (Hedrich and Schroeder, 1989; Hedrich et al., 1990). Parenchyma cells and algal cells that may not be expected to normally undergo large changes in osmotic pressure reveal an interesting array of ion channels. Listed below are commonly observed ion channels found in the plasma membrane of parenchyma cells of higher plants. Some reference to channels in fresh water algal cells is also made.

Outward rectifier (OR) cation channels

These are found in most plant cells and are characterised by activation (higher P_0) at depolarised V_m. Their activation by depolarisation results in the outward flow of cations (usually K^+) from the cell (Fairley et al., 1991; Ketchum et al., 1989; Terry et al., 1991) but this does not preclude ion influx under certain conditions (Schachtman et al., 1991). Higher plant cells and algae are often observed to have their V_m near the equilibrium potential for K^+ (E_K) (about -100 mV in 1 mol m^{-3} K^+). This situation can occur transiently or for long periods under a variety of conditions (Bisson, 1984; Blatt, 1988). The cation outward rectifier channel and perhaps the cation inward rectifier channel (see below) are most likely responsible for this state of the plasma membrane.

There are diverse types of cation OR, based on the cation selectivities that have been observed. The single channels in wheat roots have a conductance of 32 pS (100 mol m^{-3} KCl) and are highly selective for K^+ (permeability: $K^+ \gg Na^+ > Cl^-$ as 1:0.033:0.02) (Schachtman et al., 1991). The function of these channels in root cortex cells is not really understood but at high ion concentrations the channels may serve to hold V_m near E_K. It has been proposed that these channels may be the molecular pathway for Na^+ entry into root cells under saline conditions (Schachtman et al., 1991).

Inward rectifier (IR) cation channels

Inward rectifying cation channels (cation IR) have been reported in several plant cells (Ketchum et al., 1989; Schroeder 1988; Schroeder et al., 1987). They conduct cations into the cell, and are activated by V_m negative of the channel reversal potential provided the cytoplasmic calcium concentration is low (i.e. <1 mol m^{-3} [Ca^{2+}]cyt) (Schroeder and Hagiwara, 1989). The cation IR may also be sensitive to cation and anion concentrations in the cytoplasm. In *Amaranthus* cotyledon/hypocotyl cells the K^+/Na^+ selectivity of this channel is less than the cation OR (permeability: $K^+ > Na^+ \gg Cl^-$ as 1 ~0.3:0.03) and the channel conductance is about 10 pS (10/100 mol m^{-3} K^+) (B R Terry personal communication).

The cation IR is proposed to function in K^+ uptake when the membrane potential is hyperpolarised by the operation of the proton pump (Hedrich and Schroeder 1989). It is not clear how this may be related to active K^+ uptake by K^+-H^+ symport described in *Neurospora* (Rodriguez-Navarro et al. 1986) or K^+-Na^+ symport in *Chara* (Walker et al., 1989, see discussion in Hedrich and Schroeder, 1989). It is possible for the channel/pump combination to accumulate K^+ provided V_m remains more negative than the E_K. For barley roots this occurs for a K^+ concentration range from 0.1 mol m^{-3} to 100 mol m^{-3} external K^+ (Reid et al., 1985).

Hyperpolarisation activated anion channels

In the plasma membranes of *Amaranthus*, *Chara* and *Asclepias* an anion channel opens when V_m becomes hyperpolarised (more negative) (Coleman, 1986; Terry et al., 1991; Tyerman et al., 1986; Schauf and Wilson, 1987). Under normal circumstances the opening of anion channels in the plasma membrane will result in the efflux of anions (often Cl^-) since the electrochemical equilibrium for anions usually occurs at depolarised V_m. For nitrate starved roots it may be possible for uptake of nitrate to occur via this channel (R A Leigh, personal communication) particularly since it is more permeable to nitrate than chloride.

In wheat root protoplasts the anion channel was often observed in outside-out patches when the cytoplasmic face of the membrane was exposed to less than 10^{-8} *M* Ca^{2+} and ATP (Tyerman et al., unpublished). However, the activity of the channel gradually declined with time after forming the patch. In protoplasts of *Amaranthus tricolor* the anion channels have very short open times in cell attached and whole-cell configurations, but in detached patches the channels show extremely long open times and multiple conductance levels. Clearly some other type of cellular control must be exerted on this channel.

The function of the anion channel in *Chara* is proposed to be in the maintenance of a constant electrochemical gradient for protons ($\Delta \mu H^+$) across the plasma membrane (Tyerman, 1992; Tyerman et al., 1986). With variations in external pH the maintenance of a constant $\Delta \mu H^+$ by variations in V_m would mean that H^+-gradient-coupled transport would be operating with a less variable driving force. This gradient-control function is suggested by the sensitivity of the channel to external pH. In *Chara*, the whole cell I–V curve shifts positive along the V_m axis as pH_0 declines. All other currents being equal, this would depolarise V_m as pH_0 is decreased to about the extent needed to keep the $\Delta \mu H^+$ constant. The pump current may also decline as pH_0 is decreased contributing to the depolarisation (Blatt et al., 1990). It is not yet known whether the same function can be assigned to the anion channel in higher plants since a detailed study of the effects of pH_0 have not been reported. Preliminary work on *Amaranthus* has shown that the anion channel activates with decreased pH_0 (B R Terry and S D Tyerman unpublished) indicating qualitative similarity with the *Chara* channel.

Genetics of ion channels

Molecular characterization of ion channels

Channel proteins are excellent targets for the genetic manipulation of ion transport systems in plants using classical and/or genetic engineering techniques. Although the genes encoding ion channels in plants are unknown at present there is a large amount of information on the genes that encode channel proteins in animal systems. These studies have demonstrated that the in vitro functional characteristics of ion channels can be altered.

Genes encoding voltage-sensitive cation (Jan and Jan 1989) and anion channels (Blachly-Dyson et al., 1990; Jentsch et al., 1990) have been isolated. A Na^+ channel protein was purified from *Electrophorus electricus*, and a cDNA was isolated using antibodies and degenerate oligonucleotide probes (Noda et al., 1984). A potassium channel gene was isolated using *Shaker* mutants in *Drosophila* and chromosome walking (Papazian et al., 1987). A delayed rectifier type K^+ channel was isolated from rat brain by the functional expression of total mRNA in oocytes from *Xenopus laevis* (Frech et al., 1989). Functional expression in oocytes was also used to clone a chloride channel from *Torpedo marmorata*.

Oocytes that are excised from the frog *Xenopus laevis* are particularly useful for cloning and characterizing ion channel proteins (Sigel, 1990). In this system mRNA is microinjected and then translated by the oocyte's abundant protein synthesis machinery. Plasma membrane ion channel proteins are subsequently incorporated into the oocyte plasma membrane. Expression cloning has been used to isolate several membrane bound transport proteins including ion channels (Frech et al., 1989; Hediger et al., 1987; Jentsch et al., 1990; Takumi et al., 1988). Messenger RNA may then be fractionated and these fractions assayed in oocytes to isolate a single mRNA transcript coding for a specific gene. The gene for a specific channel is identified by its functional properties and its "electrical signature". This method of gene cloning does not depend on prior knowledge of the DNA or protein sequence which are so far unknown for plants. The oocyte system has also been used extensively as discussed below to study the functional characteristics of wild type and mutagenized ion channel genes. Ion channels in plants should be amenable for study in oocytes because other plant proteins have been translated and targeted correctly in *Xenopus oocytes* (Matthews et al., 1981, Wallace et al., 1988).

Changes in ion channel structure alter function

Several regions in the primary structure in at least one class of ion channel protein have been identified that determine functional properties such as selectivity, kinetic properties and sensitivity to inhibitors and voltage. Mutagenesis of ion channel genes is one approach that has been used to identify the structures that determine functional properties. Site-directed mutagenesis of the *Shaker* K^+ channel from *Drosophila* showed that changes in a single amino acid in the H5 region alter the ion selectivity (Yool and Schwarz, 1991). One prominent kinetic characteristic of the *Shaker* K^+ channels is the inactivation over time of the current. Mutagenesis revealed that the first 22 amino acids in the amino terminus of this protein were responsible for inactivation of the current through the channel (Hoshi et al., 1990). These results are consistent with a model of a cytoplasmic region termed the "ball on a chain" that inactivate the channel (Armstrong and Bezanilla, 1977). Addition of a synthetic peptide ("the ball on a chain region") to non-inactivating mutant channels expressed in *Xenopus oocytes* restored inactivation of the K^+ channel (Zagotta et al., 1990). Tetraethylammonium (TEA), a cation channel blocker, was used to putatively identify the amino acid residues that line the narrow region of a K^+ channel pore (MacKinnon and Yellen, 1990). One amino acid substitution in the H5 region of the *Shaker* K^+ channel protein was sufficient to increase the channel's sensitivity to blockade by TEA by fifty times. This work confirms observations made on a family of genes isolated from rat brain (Stuhmer et al., 1989). The members of this K^+ channel gene family in rat brain differ in their amino acid composition in at least one position in the S5-S6 linker and also differ in their sensitivity to TEA.

A second approach used to determine the primary structures that are responsible for ion channel function is to compare DNA or amino acid sequences of ion channels that differ in a particular functional characteristic. This approach was used by Lichtinghagen et al. (1990) who compared sequences from *Drosophila* ion channel mutants and wild type sequences to identify the region of a K^+ channel gene that was responsible for inactivation.

Ion channel genes in plants

Little is currently known about the genes coding for ion channels in plant membranes or the molecular structure of ion channel proteins in plants (Sussman and Harper, 1989). Genes for a

H^+-ATPase, another important membrane bound transport protein, have recently been isolated from *Arabidopsis* and tobacco using DNA sequences from highly conserved regions of the protein (Harper et al., 1989; Pardo and Serrano, 1989).

Evolutionary models of ion channels suggest that there are similarities between ion channel proteins in different organisms (Franciolini and Petris, 1989; Hille, 1984). For example K^+ channels are found in a number of different organisms from different sections of the proposed phylogenetic tree. Plants have at least two different types of voltage activated K^+ channels, i.e. inward and outward rectifiers. Time course analyses of K^+ channel currents in higher plant cells (Schroeder, 1989) show similar properties to K^+ channels in heart muscle and algae.

Similarities in nucleic acid sequence between a K^+ channel gene in *Drosophila* and wheat have also been demonstrated (Schachtman, 1991). In an attempt to correlate ion channel characteristics with plant tolerance of high NaCl in soils, a cDNA encoding a portion of the K^+ channel gene from the *Drosophila Shaker* mutant "ShB" was used as an RFLP marker in a gene mapping study. This cDNA was mapped using an F_2 population (Lagudah et al., 1991) to the long arm of chromosome 4 of *Triticum tauschii* (a progenitor of wheat). Ion channel proteins and genes appear to be conserved between different organisms. Therefore the extensive amount of information gathered in other systems should serve as a starting point for the molecular and genetic characterization of plant ion channels.

Conclusion

At low external ion concentrations (e.g. $<1\ mol\ m^{-3}$) ion channels are probably not the molecular route for ion uptake with the possible exception of the cation IR channel. The major roles of ion channels are more likely to be in signalling and the control of electrochemical gradients. The OR cation channel and the hyperpolarization activated anion channel will keep the membrane potential within a limited range of negative values (e.g. between -100 and $-250\ mV$). Furthermore, it is likely that ion channels have a greater degree of fidelity in the control of V_m than previously thought, particularly since they can be sensitive to gradients of other ions across the membrane. More work is required on the anion channel in higher plants to determine its possible role in the control of $\Delta\ \mu H^+$.

Identifying and altering the primary structures that determine functional characteristics in plant ion channel genes could result in changes in the transport characteristics of higher plants. Molecular approaches such as site directed mutagenesis could be used to alter a functional characteristic of an ion channel. Mutant ion channel genes could then potentially be transferred into the plant genome using transformation techniques. Another approach to altering ion channel function with molecular techniques will be to use antisense RNA (Bird et al., 1991; Inouye, 1988) to decrease the expression of the gene and possibly alter channel density. At this time it is not known if the introduction of mutagenized ion channel genes or a reduction in ion channel density will significantly alter whole plant ion transport characteristics (see Eq. 1). Classical genetics may also be used to alter the functional characteristics of ion channels in plants. This approach depends on the ability to easily recognize the naturally occurring variation in ion channel function. The classical genetic approach will certainly need to be augmented by either electrophysiological characterization or molecular markers that can be used to identify the differences in ion channel function that have been introgressed.

Acknowledgements

Funding was provided by the Australian Research Council and a CSIRO/Flinders cooperative grant. Thanks to J I Schroeder for comments on the second section, The genetics of ion channels.

References

Alexandre J, Lassalles J P and Kado R T 1990 Opening of Ca^{2+} channels in isolated red beet root vacuole membrane by inositol 1,4,5-trisphosphate. Nature 343, 567–570.

Armstrong C M and Bezanilla F 1977 Inactivation of the sodium channel. II. Gating current experiments. J. Gen. Physiol. 70, 567–590.

Bird C R, Ray J A, Fletcher J D, Boniwell J M, Bird A S, Teulieres C, Blain I, Bramley P M and Schuch W 1991 Using antisense RNA to study gene function: Inhibition of carotenoid biosynthesis in transgenic tomatoes. Bio/technology 9, 635–639.

Bisson M A 1984 Calcium effects on electrogenic pump and passive permeability of the plasma membrane of *Chara corallina*. J. Membrane Biol. 81, 59–67.

Blachly-Dyson E, Peng S, Colombi M and Forte M 1990 Selectivity changes in site-directed mutants of the VDAC ion channel: Structural implications. Science 247, 1233–1236.

Blatt M R 1988 Potassium-dependent bipolar gating of K^+ channels in guard cells. J. Membrane Biol. 102, 235–246.

Blatt M R 1990 Potassium channel currents in intact stomatal guard cells: Rapid enhancement by abscisic acid. Planta 180, 445–455.

Blatt M R, Beilby M J and Tester, M 1990a Voltage dependence of the Chara proton pump revealed by current-voltage measurement during rapid metabolic blockade with cyanide. J. Membrane Biol. 114, 205–223.

Coleman H A 1986 Chloride currents in Chara – a patch clamp study. J. Membrane Biol. 83, 109–18.

Davies E 1987 Action potentials as multifunctional signals in plants: A unifying hypothesis to explain apparently disparate wound response. Plant Cell and Environment 10, 623–631.

Fairley K, Laver D and Walker N A 1991 Whole cell and single-channel currents across the plasmalemma of corn shoot suspension cells. J. Membrane Biol. 121, 11–22.

Franciolini F and Petris A 1989 Evolution of ionic channels of biological membranes. Mol. Biol. Evol. 6, 503–513.

Franciolini F and Petris A 1990 Chloride channels of biological membranes. Biochim. Biophys. Acta 1031, 247–59.

Frech G C, Van Dongen A M J, Schuster G, Brown A M and Joho R H 1989 A novel potassium channel with delayed rectifier properties isolated from rat brain by expression cloning. Nature 340, 642–645.

Harper J F, Surowy T K and Sussman M R 1989 Molecular cloning and sequence of cDNA encoding the plasma membrane proton pump (H^+-ATPase) of *Arabidopsis thaliana*. Proc. Natl. Acad. Sci. USA 86, 1234–1238.

Hediger M A, Coady M J, Ikeda T S and Wright E M 1987 Expression cloning and cDNA sequencing of the Na^+/glucose co-transporter. Nature 330, 379–381.

Hedrich R, Barbier-Brygoo H, Felle H, Flågge U I and Låttge U 1988 General mechanisms for solute transport across the tonoplast of plant vacuoles: A patch-clamp survey of ion channels and proton pumps. Botanica Acta 101, 7–13.

Hedrich R, Busch H and Raschke K 1990 Ca^{2+} and nucleotide dependent regulation of voltage dependent anion channels in the plasma membrane of guard cells. EMBO Journal 9, 3889–3892.

Hedrich R and Schroeder J I 1989 The physiology of ion channels and electrogenic pumps in higher plants. Annu. Rev. Plant Physiol. 40, 539–569.

Hille B 1984 Ionic channels in excitable membranes. Sinauer Assoc. Inc., Sunderland, MA.

Hoshi T, Zagotta W N and Aldrich R W 1990 Biophysical and molecular mechanisms of *Shaker* potassium channel inactivation. Science 250, 533–538.

Ilan N, Scheartz A and Moran N 1991 pH effect on K^+ channels in plasmalema of guard cell protoplasts. Plant Physiol. 96, 912.

Inouye M 1988 Antisense RNA: Its functions and applications in gene regulation – a review. Gene 72, 25–34.

Jan L Y and Jan Y N 1989 Voltage-sensitive ion channels. Cell 56, 13–25.

Jentsch T J, Steinmeyer K and Schwarz G 1990 Primary structure of isolated by expression cloning in *Xenopus oocytes*. Nature 348, 510–514.

Keller B U, Hedrich R and Raschke K 1989 Voltage-dependent anion channels in the plasma membrane of guard cells. Nature 341, 450–453.

Ketchum K A, Shrier A and Poole R J (1989) Characterisation of potassium-dependent currents in protoplasts of corn suspension cells. Plant Physiol. 89, 1184–1192.

Kochian L V, Xin-Zhi J and Lucas W J 1985 Potassium transport in corn roots. IV. Characterization of the linear component. Plant Physiol. 79, 771–776.

Lagudah E S, Appels R, Brown A H D and McNeil D 1991 The molecular-genetic analysis of *Triticum tauschii* – the D genome donor to hexaploid wheat. Genome 34, 375–386.

Laver D R, Fairley K A and Walker N A 1989 Ion permeation in a K^+ channel in *Chara australis*: Direct evidence for diffusion limitation of ion flow in a maxi-K channel. J. Membrane Biol. 108, 153–164.

Lew R R 1991 Substrate regulation of single potassium and chloride ion channels in *Arabidopsis* plasma membrane. Plant Physiol. 95, 642–647.

Lichtinghagen R, Stocker M, Wittka R, Boheim G, Stuhmer W, Ferrus A and Pongs O 1990 Molecular basis of altered excitability in *Shaker* mutants of *Drosophila melanogaster*. EMBO J. 9, 4399–4407.

MacKinnon R and Yellen G 1990 Mutations affecting TEA blockade and ion permeation in voltage-activated K^+ channels. Science 250, 276–279.

MacRobbie E A C 1988 Control of ion fluxes in stomatal guard cells. Botanica Acta 101, 140–48.

Matthews J A, Brown J W S and Hall T C 1981 Phaseolin mRNA is translated to yield glycosylated polypeptides in Xenopus oocytes. Nature 294, 175–176.

Noda M, Shimizu S, Tanabe T and Many O et al. 1984 Primary structure of *Electrophorus electricus* sodium channel deduced from cDNA sequence. Nature 312, 121–127.

Papazian D M, Schwarz T L, Tempel B L, Jan Y N and Jan L Y 1987 Cloning of genomic and complementary DNA from *Shaker*, a putative potassium channel gene from *Drosophila*. Science 237, 749–753.

Pardo J M and Serrano R 1989 Structure of a plasma membrane H^+-ATPase gene from the plant *Arabidopsis thaliana*. J. Biol. Chem. 264, 8557–8562.

Reid R J, Dejaegere R and Pitman M G 1985 Regulation of electrogenic pumping in barley by pH and ATP. J. Exp. Bot. 36, 535–549.

Rodriguez-Navarro A, Blatt M R and Slayman C L 1986 A

potassium proton symport in *Neurospora crassa*. J. Gen. Physiol. 87, 649–674.

Schachtman D P 1991 Physiology and Genetics of Sodium Accumulation and Salt Tolerance in *Triticum* species. PhD Thesis. Australian National University.

Schachtman D P, Tyerman S D and Terry B R 1991 The K^+/Na^+ selectivity of a cation channel in the plasma membrane of root cells does not differ in salt-tolerant and salt-sensitive wheat species. Plant Physiol. (*In press*.)

Schauf C L and Wilson K J 1987 Properties of single K^+ and Cl^- channels in Asclepias tuberosa protoplasts. Plant Physiol. 85, 413–18.

Schroeder J I 1988 K^+ transport properties of K^+ channels in the plasma membrane of *Vicia faba* guard cells. J. Gen. Physiol. 92, 667–683.

Schroeder J I 1989 Quantitative analysis of outward rectifying K^+ channel currents in guard cell protoplasts from *Vicia faba*. J. Membrane Biol. 107, 229–235.

Schroeder J I and Hagiwara S 1989 Cytosolic calcium regulates ion channels in the plasma membrane of *Vicia faba* guard cells. Nature 338, 427–430.

Schroeder J I and Hagiwara S 1990 Repetitive increases in cytosolic Ca^{2+} of guard cells by abscisic acid activation of nonselective Ca^{2+} permeable channels. Proc. Nat. Acad. Sci. USA 87, 9305–9309.

Schroeder J I, Raschke K and Neher E 1987 Voltage dependence of K^+ channels in guard cell protoplasts. Proc. Nat. Acad. Sci. USA 84, 4108–4112.

Sigel E 1990 Use of Xenopus oocytes for the functional expression of plasma membrane proteins. J Membrane Biol. 117, 201–221.

Stoeckel H and Takeda K 1989 Calcium-activated, voltage-dependent, non-selective cation currents in endosperm plasma membrane from higher plants. Proc. R. Soc. Lond. B 237, 213–231.

Stuhmer W, Ruppersberg J P, Schroter K H, Sakmann B, Stocker M, Giese K P, Perschke A, Baumann A and Pongs O 1989 Molecular basis of functional diversity of voltage-gated potassium channels in mammalian brain. EMBO 8, 3235–3244.

Sussman M R and Harper J F 1989 Molecular biology of the plasma membrane of higher plants. The Plant Cell 1, 953–960.

Takumi T, Hiroaki O and Nakanishi S 1988 Cloning of a membrane protein that induces a slow voltage-gated potassium current. Science 242, 1042–1045.

Terry B R, Tyerman S D and Findlay G P 1991 Ion channels in the plasma membrane of Amaranthus protoplasts: One cation and one anion channel dominate the conductance. J. Membrane Biol. 121, 223–236.

Tester M 1990 Plant ion channels: Whole-cell and single-channel studies. New Phytol. 114, 305–340.

Tyerman S D 1992 Anion channels in plants. Annu Rev. Plant Physiol. Plant Mol. Biol. 43, in press.

Tyerman S D and Findlay G P 1989 Current-voltage curves of single Cl^- channels which coexist with two types of K^+ channel in the tonoplast of *Chara corallina*. J. Exp. Bot. 40, 105–118.

Tyerman S D, Findlay G P and Patterson G J 1986 Inward membrane current in *Chara* inflata. II. Effects of pH, Cl^--channel blockers and NH_4^+, and significance for the hyperpolarised state. J. Membrane Biol. 89, 153–161.

Tyerman S D, Findlay G P and Terry B R 1989 Behaviour of K^+ and Cl^- channels in the cytoplasmic drop membrane of *Chara* corallina using a transient detection method of analysing single-channel recordings. *In* Plant Membrane Transport: The Current Position. Eds. J Dainty, M I De Michelis, E Marre and F Rasi-Caldogno. pp 173–178. Elsevier, Amsterdam.

Walker N A, Smith F A and McCulloch S R 1989 Plasmalemma transport in the life of the cell: *Chara* uses both sodium and protons as chemiosmotic drivers. *In* Plant Membrane Transport: The Current Position. Eds. J Dainty, M I De Michelis, E Marre and F Rasi-Caldogno. pp 19–24. Elsevier, Amsterdam.

Wallace J C, Galilli G, Kawata G G, Cuellar R G, Shotwell M A and Larkins B A Aggregation of lysine-containing zeins into protein bodies in Xenopus oocytes. 1988 Science 240, 662–664.

Yool A J and Schwarz T L 1991 Alteration of ionic selectivity of a K^+ – channel by mutation of the H5 region. Nature 349, 700–704.

Zagotta W N, Hoshi T and Aldrich R W 1990 Restoration of inactivation in mutants of *Shaker* potassium channels by a peptide derived from ShB. Science 250, 568–571.

P. J. Randall et al. (Eds.), Genetic aspects of plant mineral nutrition, 187–193.

PLSO SV35

Effect of salt stress on plant gene expression: A review

WILLIAM J. HURKMAN
United States Department of Agriculture, Agricultural Research Service, Western Regional Research Center, Plant Development-Productivity Research Unit, Albany, CA 94710, USA

Key words: barley, gene expression, ice plant, rice, salt stress, tobacco, tomato

Abstract

Soil salinity is an important agricultural problem, particularly since the majority of crop plants have low salt tolerance. The identification of genes whose expression enables plants to adapt to or tolerate salt stress is essential for breeding programs, but little is known about the genetic mechanisms for salt tolerance. Recent research demonstrates that salt stress modulates the levels of a number of gene products. Although the detection of gene products that respond specifically to salt stress is a significant finding, they must be identified, functions assigned, and their relation to salt tolerance determined. This article focuses on a few of the salt-responsive proteins and mRNAs that have been discovered and the methods employed to identify and characterize them.

Introduction

Soil salinity is an important agricultural problem, especially in farm lands dependent on irrigation. The problem is compounded by the relatively low salt tolerance of most crop plants. The development of more salt-tolerant crop plants is a reasonable goal for plant breeding programs, because salt tolerance exists in some cultivated species and their halophytic relatives. Defining salt tolerance is difficult because of the complex nature of salt stress and the wide range of plant responses. A broad definition is that salt tolerance is a multigenic trait that allows plants to grow and maintain economic yield in the presence of nonphysiologically high and relatively constant levels of salt, in particular NaCl. The identification of genes whose expression enables plants to adapt to or tolerate salt stress is essential for breeding programs, but little is known about the genetic mechanisms for salt tolerance. Recent research demonstrates that salt stress modulates the levels of a number of polypeptides and mRNAs in cultured cells adapted to high levels of salt (citrus: Ben-Hayyim et al., 1989; *Distichlis spicata*: Zhao et al., 1989; rice: Shirata and Takagishi, 1990; sugarcane: Ramagopal and Carr, 1991; tobacco: Singh et al., 1985; tomato: King et al., 1986), in callus cultures adapted to high levels of salt (alfalfa: Winicov et al., 1989; 1990; barley: Ramagopal, 1988b; maize: Ramagopal, 1986), in relatively salt-tolerant glycophytes (barley: Hurkman et al., 1989; Hurkman and Tanaka, 1987; Ramogopal 1987a; b; 1988a; Robinson et al., 1990; wheat amphiploid: Gulick and Dvořák, 1987), and in halophytes (*Mesembryanthemum crystallinum*: Ostrem et al., 1987; Michalowski et al., 1989). Although the detection of gene products that respond specifically to salt stress is a significant finding, they must be identified, functions assigned, and their relation to salt tolerance determined. This article focuses on a few of the salt-responsive proteins and mRNAs that have been discovered and the methods employed to identify and characterize them. Additional information on salt stress and gene expression can be found in a recent review by Cushman et al. (1990).

Methods used in studies on the effect of salt stress on gene expression

Analysis of protein patterns in polyacrylamide gels is the method of choice for detecting polypeptides whose levels are altered by salt stress. By comparing patterns from control and salt-treated plants, polypeptides can be identified that increase, decrease, appear, or disappear with salt treatment. Visual analysis of one- (Ericson and Alfinito, 1984; Singh et al., 1985) and two-dimensional (Hurkman and Tanaka, 1987; 1991; Ramagopal, 1988a; Singht et al., 1985; Winicov et al., 1989) gels and computer-assisted analysis of two-dimensional gels (Ben-Hayyim et al., 1989; Hurkman et al., 1989) demonstrate that salt stress alters the synthesis and accumulation of a number of polypeptides. An advantage of two-dimensional gels is that individual polypeptides can be isolated for characterization studies involving antibody preparation and N-terminal amino acid sequence analysis.

Polyacrylamide gels can also be utilized to measure indirectly changes in mRNA populations by comparing polypeptide levels in polyacrylamide gels of translation products synthesized in vitro by total RNA or poly(A^+)RNA isolated from control and salt-treated plants. A number of genes are differentially expressed, some in a genotype specific manner (Gulick and Dvořák, 1987; Hurkman et al., 1989; Ramagopal, 1987a; Ramagopal and Carr, 1991; Robinson et al., 1990). Although confirming data are required, these results suggest that gene expression is regulated during salt stress by both transcriptional and posttranscriptional mechanisms. One disadvantage of this approach is that the in-vivo mRNA population, and its synthetic capacity, are not accurately represented due to translation in heterologous systems. Another is that it is not practical to isolate the small quantities of polypeptides present in gels of translation products for characterization studies. However, the approach is valuable for selecting an experimental system and defining parameters that elicit a response to salt stress.

The effects of salt stress on levels of specific mRNAs (transcripts) can be measured with greater sensitivity using cDNA probes. Information obtained from analysis of proteins can be used to identify and isolate appropriate cDNAs. cDNA libraries can be probed with synthetic oligonucleotides, based on N-terminal or internal amino acid sequences, to identify recombinants (King et al., 1988). Antibodies can also be used to screen for recombinants in expression libraries, cDNA libraries that produce fusion proteins containing the amino acid sequence specified by the DNA insert fragment (Singh et al., 1989). Foregoing protein analysis altogether, differential screening, where replica filters of cDNA libraries are probed with mRNAs from control and stressed plants, can be used to identify recombinants containing cDNAs that represent mRNAs regulated by salt stress (Goday et al., 1990; Mundy and Chua, 1988).

Thaumatin-related polypeptides

Considerable information has been obtained about 26-kD polypeptides associated with cultured tobacco and tomato cells adapted to high levels of NaCl. In cultured tobacco cells adapted to 171 m*M*NaCl, one-dimensional polyacrylamide gels revealed that a 26-kD protein increased (Ericson and Alfinito, 1984) and constituted 10% of the total cellular protein (Singh et al., 1985). The 26-kD polypeptide also increased significantly in PEG-adapted (25–30%) lines (Singh et al., 1985). Two-dimensional polyacrylamide gels revealed that the cells synthesize two 26-kD polypeptides with pls of 7.8 and 8.2 (Singh et al., 1985). Unadapted cells accumulate low levels of both forms and adapted cells accumulated almost entirely the pl 8.2 form (LaRosa et al., 1989). The 26-kD polypeptide was named osmotin because it is synthesized and accumulates in cells undergoing gradual osmotic adjustment, but not in cells subjected to salt shock (Singh et al., 1987a). Sequence comparisons revealed that osmotin has extensive homology with thaumatin, a tobacco pathogenesis-related (PR) protein, and maize trypsin/α-amylase inhibitor (Singh et al., 1987a; 1989). Osmotin is synthesized as a precursor protein that contains a signal polypeptide and is concentrated in dense inclusion bodies within vacuoles in adapted cells (Singh et al., 1987a). Osmotin was found in cultured cells of different species, i.e. tomato, soybean, and bindweed, following adaptation to salt (Singh et al., 1987b). The rate of adaptation of tobacco cells is accelerated by ABA (LaRosa, 1985). Although the synthesis of osmotin is induced by ABA, accumulation is dependent on the presence of NaCl (Singh et al., 1987b). In tobacco plants, much more osmotin was found in the outer stem than in root or leaf; the distribution of osmotin in salt-treated plants was not reported (Singh et al., 1987b). Exogenous ABA stimulated the synthesis of osmotin more in the root than in the outer stem or leaf (Singh et al., 1987b).

A clone for osmotin was isolated from a λgt11 expression library using osmotin antibodies (Singh et al., 1989). Identity of the clone was confirmed by in-vitro translation of hybrid selected poly(A^+) RNA and immunoprecipitation of translation products corresponding in size to the 29-kD precursor of osmotin. Osmotin transcripts were present in unadapted cells, but at lower levels than in adapted cells. ABA increased the levels of osmotin transcripts in both unadapted and adapted cells. Salt shock (171 m*M* NaCl) of unadapted cells did not increase the level of message or protein.

A 26-kD protein that accumulates in tomato cells adapted to grow on high levels of NaCl (342 m*M*) is immunologically related to osmotin (King et al., 1986). The protein decreased when salt-adapted cells were transferred to medium without salt. It increased following other stresses that lower water potential, e.g. PEG and KCl, but not heat shock or Cd, a metal that influences the accumulation of heat shock-specific mRNAs. In tobacco and alfalfa plants, the concentration of the 26-kD protein was highest in roots. The protein accumulated in roots of *Lycopersicon esculentum* plants grown in the presence of 171 m*M*NaCl. The protein was also present in other Solanaceae (potato, petunia, datura) and Leguminoseae (green bean). When protein profiles in control roots of *L. esculentum* were compared to those of salt-tolerant lines of tomato, *Solanum pennellii* and *L. cheesmanii*, levels of 26-kD polypeptide were the same in all three lines. This observation prompted King et al. (1986) to suggest that the 26-kD polypeptide was not correlated with salt tolerance in tomato, although protein profiles of salt-treated roots for the three lines were not compared.

The tomato 26-kD polypeptide was purified and the N-terminal amino acid sequence determined (King et al., 1988). A synthetic oligonucleotide was used to obtain a cDNA, NP24, from a λgt10 library. Sequence comparison of the deduced amino acid sequence of NP24 with the N-terminal sequence of the purified protein confirmed the authenticity of the clone. Further sequence comparisons revealed an extensive homology to thaumatin. NP24 transcripts were absent from unadapted tomato cells. The presence of NP24 transcripts was dependent on stage of cell cycle; they were presented at low levels in adapted cells in early log phase and present at low levels in unadapted cells and high levels in adapted cells in late log phase. In roots of tomato plants, NP24 transcripts were not detectable until several weeks after germination, but were induced earlier and to higher levels in roots of plants stressed with 171 m*M* NaCl.

Germin-related polypeptides

In roots of barley, the most salt-tolerant grain of agricultural importance, two-dimensional polyacrylamide gels revealed that the synthesis (^{35}S[met] in-vivo labeling) of 26-kD polypeptides increased significantly when plants were grown in the presence of 200 m*M* NaCl. Compared to the 26-kD polypeptides of tobacco, the barley polypeptides had more acidic pls of 6.3 and 6.5. The synthesis of these polypeptides increased when plants were germinated and grown in 200 m*M* NaCl for 6 d and also when 5-d-old plants were treated with NaCl for 24 h (Hurkman and Tanaka, 1987). When plants were transferred to medium without salt, the synthesis of the pl 6.3 and 6.5 polypeptides decreased to near control levels. The synthesis of the pl 6.3 and 6.5 polypeptides increased as early as 4 h following treatment with NaCl (Robinson et al., 1990). Preliminary experiments indicated that synthesis of the pl 6.3 and 6.5 polypeptides increased at NaCl concentrations as low as 50 m*M* (Hurkman and Tanaka, 1987). In addition, synthesis was not affected significantly by desiccation, by PEG or mannitol (Hurkman and Tanaka, 1988), or by heat shock (Hurkman, 1990). Immunoblots probed with antibodies to the tobacco 26-kD protein indicated that the barley 26-kD polypeptides were not related to those of tobacco. This was not surprising because the barley 26-kD polypeptides are more acidic and actually slightly larger than the tobacco polypeptides (Hurkman and Tanaka, 1987). The pI 6.3 and 6.5 polypeptides also accumulate (increase quantitatively in silver-strained 2-D gels) during salt stress (Hurkman et al., 1991).

The pI 6.3 and 6.5, 26-kD polypeptides from barley roots were isolated from two-dimensional gels and the N-terminal sequences determined using a pulsed liquid-phase amino acid sequencer (Hurkman et al., 1991). Sequence comparisons revealed that the two polypeptides were isoforms and that they shared strong residue identity with germin, a protein that increases significantly in germinating wheat seeds (Grzelczak and Lane, 1984; Grzelczak et al., 1982; Thompson and Lane, 1980). The pI 6.3 polypeptide was named Gs1 (salt-responsive, germin-like polypeptide 1) and the pI 6.5 polypeptide was named Gs2 (salt-responsive, germin-like polypeptide 2). Like germin (Grzelczak and Lane, 1984; Lane et al., 1987), Gs1 and Gs2 were resistant to proteases and were glycosylated (Hurkman et al., 1991). Immunoblots probed with antibodies to Gs1 and Gs2 revealed that Gs1 and Gs2 were present in roots and coleoptiles, but not in leaves. Gs1 and Gs2 increased in roots, but decreased in coleoptiles in response to salt stress. Gs1 and Gs2 were distributed among soluble, microsomal, and cell wall fractions, but the majority was in the soluble fraction. When the microsomal fraction was separated in discontinuous sucrose gradients, Gs1 and Gs2 were found in membrane fractions enriched in endoplasmic reticulum, tonoplast, and plasma membrane (Hurkman et al., 1988). Because these polypeptides were resistant to hydrolysis by proteases, it was not possible to determine if they were specifically associated with particular membrane fractions (Hurkman et al., 1988). Although Gs1 and Gs2 were heat-stable, their synthesis was not affected by ABA. Gs2 accumulated during ABA treatment, whereas Gs1 did not. However, a 25.5-kD, pI 6.1 polypeptide that was immunologically related to Gs1 did accumulate during ABA treatment (Hurkman et al., 1991).

Evidence indicates that Gs1 and Gs2 differ from germin. Gs1 and Gs2 are not immunologically related; antibodies to germin did not react with Gs1 or Gs2 and antibodies to Gs1 and Gs2 did not react with germin (Hurkman et al., 1991). The majority of Gs1 and Gs2 are located in the soluble fraction (Hurkman et al., 1991), whereas germin is principally in the cell wall fraction (Hurkman and Tanaka, unpubl. observ.). Gs1 and Gs2 accumulate during salt stress, whereas germin decreases (Hurkman et al., 1991). Like germin (Grzelczak et al., 1985), Gs1 and Gs2 are present in roots and shoots (Hurkman et al., 1991). The distribution of Gs1 and Gs2 varies within roots; they are not detected in the tip, but are present in the mature region. In contrast, germin was most abundant in distal (to the seed) root sections of wheat seedlings.

Phosphoenolpyruvate carboxylase

Mesembryanthemum crystallinum (common ice plant), a facultative halophyte, responds to salt stress by switching the primary path of CO_2 fixation from C_3 photosynthesis to Crassulacean acid metabolism (CAM) (Ostrem et al., 1987). The change to the CAM photosynthetic pathway provides an opportunity to examine molecular mechanisms involved in the perception and response of plants to salt stress. The activity of PEPCase (phosphoenolpyruvate carboxylase), a key enzyme in CAM, increases, dependent on plant age, during CAM induction by salt stress. In leaves and axillary shoots, PEPCase levels, revealed by immunoblots, and PEPCase mRNA, assayed by immunoprecipitation of PEPCase from the in-vitro translation products, increased during salt stress (Ostrem et al., 1987). When plants were taken off of stress, PEPCase mRNA levels decreased rapidly followed by a slower decline in PEPCase activity and polypeptide levels (Vernon et al., 1988). Transcripts for ppc (PEPCase) and ppdk (pyruvate orthophosphate dikinase) increased and transcripts for rbcS (ribulose-1,5-bisphosphate carboxylase, small subunit) and cab (chlorophyll a/b-binding protein) decreased during salt stress (Michalowski et al., 1989). Treatments with PEG (15%) or ABA (10 μM) also induced large increases in the activity and amount of PEPCase (Chu et al., 1990).

Rab 21

As a first step in isolating genes whose expression is affected by ABA, poly(A^+) RNA was isolated from developing rice seeds and translation products analyzed in polyacrylamide gels (Mundy and Chua, 1988). Polypeptides were detected that accumulated late in embryogenesis, disappeared during germination, and returned with ABA treatment. A cDNA library was constructed using poly(A^+) RNA isolated from ABA-treated seeds as the template. ABA-responsive cDNA clones were obtained by differential screening and one clone, rab 21, was chosen for further characterization. Rab 21 cDNA hybridized to mRNAs that encoded polypeptides of 20-, 21-, and 23-kD. Immunoblots probed with antibodies raised against β-galactosidase/rab 21 fusion protein revealed that the rab 21 protein was present only in the soluble fraction and not in crude organellar or nuclear fractions. Rab 21 transcripts and proteins accumulated in rice embryos, roots, shoots, and callus-derived suspension cells upon treatment with 200 m*M* NaCl and/or 10 μM ABA. Desiccation of rice plants also increased rab 21 levels. The cDNA was used to isolate a genomic clone, and it was found that the gene encodes a basic, glycine-rich protein.

A cDNA library was constructed using, as the template, poly$(A)^+$ RNA from tomato seedlings treated with 10 μM ABA and 170 m*M* NaCl for 24 h. Responsive cDNA clones were isolated by differential screening and one clone, TAS14, was chosen for further characterization (Godoy et al., 1990). The nucleotide sequence predicted an open reading frame coding for a highly hydrophilic and glycine-rich protein of 130 amino acids. TAS14 selected mRNA encoded a single 16-kD, pl 6.5 polypeptide. The sequence of the predicted protein showed four structural domains similar to rice rab 21, as well as cotton LEA D11 and barley and maize dehydrin genes. TAS14 transcripts accumulated in tomato seedlings during NaCl, ABA, or mannitol treatments, but not with cold treatment or wounding. Transcripts were induced in roots, stems, and leaves by NaCl. Transcripts were induced slightly in stems and substantially in roots by ABA.

Conclusions

The genetic mechanisms of salt tolerance must be defined in order to efficiently develop crop plants that are more salt-tolerant. There is no question that salt stress alters gene expression in a tissue-specific and time-dependent manner. However, with the exception of PEPCase, a key enzyme in CAM, the functions of the gene products that increase during salt stress are not known. Osmotin is induced not only by salt stress, but also by water deficit and ABA treatment. Because osmotin levels are not affected by salt shock and it accumulates in the vacuoles of cells undergoing gradual osmotic adjustment, osmotin may have a role in this process (Singh et al., 1987a). Osmotin and NP-24 are related to thaumatin and maize trypsin/α-amylase inhibitor, as is a class of PR proteins (Linthorst, 1991). The relationship to trypsin/α-amylase inhibitor, a protein that inhibits protein hydrolysis, has prompted suggestions that the PR proteins have a role in the resistance of plants to pathogens that secrete hydrolytic enzymes (Cushman et al., 1990; Linthorst, 1991). Rab 21 is a late embryogenesis-abundant (LEA) protein, a class of proteins that accumulates in maturing embryos during seed desiccation. LEA proteins can be induced prematurely in seeds with ABA or induced in seedlings by water deficit and/or ABA treatment, or by salt stress. Wheat Em protein, the most abundant protein of dry wheat embryos (Morris et al., 1990), and the dehydrins, dehydration-induced proteins (Close et al., 1989), are related to the LEA proteins (Dure et al., 1989). These proteins are very hydrophilic and remain soluble or heat stable upon boiling. Biophysical properties indicate that they confer desiccation protection through a stabilizing interaction with cellular proteins (Dure et al., 1989). Like the Em protein and the dehydrins, Gs1 and Gs2 are heat-stable, but are not induced by water deficit or ABA treatment. Gs1 and Gs2 are related to germin, a protein in wheat embryos that increases significantly during seed germination. Based on the increase in carbohydrate synthesis and the presence of adventitious arabinoxylans in germin, it was suggested that germin has a role in cell wall expansion (Jaikaran et al., 1990).

Use of molecular techniques, in conjunction with physiological and biochemical analyses to identify potential genetic mechanisms for salt tolerance is a valid approach to development of salt-tolerant plants. Identification of transcripts whose expression is correlated with salt stress is a first step in identifying genes that contribute to the salt tolerance of plants. Structural analysis of isolated genes will provide clues to elements that control their expression. Identification of cis-acting elements, and trans-acting factors and their genes will provide information on the regulatory properties of these genes and the signals that control them. Transgenic plants that over-express these genes or modified genes, or express antisense genes will provide information about the role of these genes in salt tolerance. The use of RFLP (restriction fragment length polymorphism) maps to determine gene loci will aid in the identification of biochemical markers or genes linked with salt tolerance.

Understanding how plants survive salt stress, however, is complicated by a number of factors. Salt stress consists of several components – reduced water potential, specific ion stress or toxicity, ion imbalance or nutrient deficiency (Yeo, 1983). Information on physiological, biochemical, and metabolic processes indicates that the response to salt stress varies with salt concentration, type of salt, length of exposure, presence of other ions, and environmental conditions, including temperature, humidity, and available soil moisture. In addition, the stress response is dependent upon stage of plant development, hormonal balances, and diurnal rhythms. Cultured cells adapted to high levels of salt are often used to limit the analysis to cellular mechanisms. However, some traits correlated with salt tolerance may be unstable in cell culture, while others may not be retained or be useful in regenerated plants. The effects of salt stress on gene expression in relatively salt-tolerant glycophytes can identify responses that allow these plants to cope with salt stress. This analysis could be improved by comparison of the salt stress response between isogenic lines or substitution lines that differ in salt tolerance. Studies of salt-tolerant wild relatives and halophytes provide insights into programmed mechanisms that allow them to grow in the presence of high salt concentrations. Whether mechanisms operative in such plants are useful to glycophytes remains to be determined. Whatever the approach, it is readily apparent that collaboration between physiologists, biochemists, molecular biologists, geneticists, and breeders is essential to tailor crops that are more salt-tolerant.

References

Ben-Hayyim G, Vaadia Y and Williams B G 1989 Proteins associated with salt adaptation in citrus and tomato cells: Involvement of 26 kDa polypeptides. Physiol. Plant. 77, 332–340.

Chu C, Dai Z and Edwards G E 1990 Induction of crassulacean acid metabolism in the facultative halophyte *Mesembryanthemum crystallinum* by abscisic acid. Plant Physiol. 93, 1253–1260.

Close T J, Kortt A A and Chandler P M 1989 A cDNA-based comparison of the dehydration-induced proteins (dehydrins) in barley and corn. Plant Mol. Biol. 13, 95–108.

Cushman J C, DeRocher E J and Bohnert H J 1990 Gene expression during adaptation to salt stress. *In* Environmental Injury to Plants. Ed. F Katterman. pp 173–203. Academic Press, Inc., San Diego, CA.

Dure III L, Crouch M, Harada J, Ho T-H D, Mundy J, Quatrano R, Thomas T and Sung Z R 1989 Common amino acid sequence domains among the LEA proteins of higher plants. Plant Molec. Biol. 12, 475–486.

Ericson M C and Alfinito S H 1984 Proteins produced during salt stress in tobacco cell culture. Plant Physiol. 74, 506–509.

Godoy J A, Pardo J M and Pintor-Toro J A 1990 A tomato cDNA inducible by salt stress and abscisic acid: Nucleotide sequence and expression pattern. Plant Molec. Biol. 15, 695–705.

Grzelczak Z F and Lane B G 1984 Signal resistance of a soluble protein to enzymic proteolysis: An unorthodox approach to the isolation and purification of germin, a rare growth-related protein. Can. J. Biochem. Cell Biol. 62, 1351–1353.

Grzelczak Z F, Rahman S, Kennedy T D and Lane B G 1985 Germin. Compartmentation of the protein, its translatable mRNA, and its biosynthesis among roots, stems, and leaves of wheat seedlings. Can. J. Biochem. Cell Biol. 63, 1003–1013.

Grzelczak Z F, Sattolo M H, Hanley-Bowdoin L K, Kennedy T D and Lane B G 1982 Synthesis and turnover of proteins and mRNA in germinating wheat embryos. Can. J. Biochem. 60, 389–397.

Gulick P and Dvořák J 1987 Gene induction and repression by salt treatment in roots of the salinity-sensitive Chinese Spring wheat and the salinity-tolerant Chinese Spring × *Elytrigia elongata* amphiploid. Proc. Natl. Acad. Sci. USA 84, 99–103.

Hurkman W J 1990 Use of two-dimensional gel electrophoresis to characterize changes in gene expression assocated with salt stress in barley. *In* Environmental Injury to Plants. Ed. F Katterman. pp 205–229. Academic Press, San Diego, CA.

Hurkman W J, Fornari C S and Tanaka C K 1989 A comparison of the effect of salt on polypeptides and translatable mRNAs in roots of a salt-tolerant and a salt-sensitive cultivar of barley. Plant Physiol. 90, 1444–1456.

Hurkman W J and Tanaka C K 1987 The effects of salt on the pattern of protein synthesis in barley roots. Plant Physiol. 83, 517–524.

Hurkman W J and Tanaka C K 1988 Polypeptide changes induced by salt stress, water deficit, and osmotic stress in barley roots: A comparison using two-dimensional gel electrophoresis. Electrophoresis 9, 781–787.

Hurkman W J, Tao H P and Tanaka C K 1991 Germin-like polypeptides increase in barley roots during salt stress. Plant Physiol. 97, 366–374.

Hurkman W J, Tanaka C K and DuPont F M 1988 The effect of salt stress on polypeptides in membrane fractions from barley roots. Plant Physiol. 88, 1263–1273.

Jaikaran A S, Kennedy T D, Dratewka-Kos E and Lane B G 1990 Covalently bonded and adventitious glycans in germin. J. Biol. Chem. 265, 12503–12512.

King G J, Hussey Jr. C E and Turner V A 1986 A protein induced by NaCl in suspension cultures of *Nicotiana tabacum* accumulates in whole plant roots. Plant Molec. Biol. 7, 441–449.

King G J, Turner V A, Hussey Jr. C E, Wurtele E S and Lee M S 1988 Isolation and characterization of a tomato cDNA clone which codes for a salt-induced protein. Plant Molec. Biol. 10, 401–412.

Lane B G, Gzelczak Z, Kennedy T, Hew C and Joshi S 1987 Preparation and analysis of mass amounts of germin: Demonstration that the protein which signals the onset of growth in germinating wheat is a glycoprotein. Biochem. Cell. Biol. 65, 354–362.

LaRosa P C, Handa A K, Hasegawa P M and Bressan R A 1985 Abscisic acid accelerates adaptation of cultured tobacco cells to salt. Plant Physiol. 79, 138–142.

LaRosa P C, Singh N K, Hasegawa P M and Bressan R A 1989 Stable NaCl tolerance of tobacco cells is associated with enhanced accumulation of osmotin. Plant Physiol. 91, 855–861.

Linthorst H M 1991 Pathogenesis-related proteins. Crit. Rev. Plant Sci. 10, 123–150.

Michalowski C B, Olson S W, Piepenbrock M, Schmitt J M and Bohnert H J 1989 Time course of mRNA induction elicited by salt stress in the common ice plant *Mesembryanthemum crystallinum*. Plant Physiol. 89, 811–816.

Morris P C, Kumar A, Bowles D J and Cuming A C 1990 Osmotic stress and abscisic acid induce expression of the wheat Em genes. Eur. J. Biochem. 190, 625–630.

Mundy J and Chua N-H 1988 Abscisic acid and water-stress induce the expression of a novel rice gene. EMBO J. 7, 2279–2286.

Ostrem J A, Olson S W, Schmitt J M and Bohnert H J 1987 Salt stress increases the level of translatable mRNA for phosphoenolpyruvate carboxylase in *Mesembryanthemum crystallinum*. Plant Physiol. 84, 1270–1275.

Ramagopal S 1986 Protein synthesis in a maize callus exposed to NaCl and mannitol. Plant Cell Reports 5, 430–434.

Ramagopal S 1987a Differential mRNA transcription during salinity stress in barley. Proc. Natl. Acad. Sci. USA 84, 94–98.

Ramagopal S 1987b Salinity stress induced tissue-specific proteins in barley seedlings. Plant Physiol. 84, 324–331.

Ramagopal S 1988a Regulation of protein synthesis in root, shoot, and embryonic tissues of germinating barley during salinity stress. Plant Cell Environ. 11, 501–515.

Ramagopal S 1988b Sodium chloride effects on dedifferentiation and protein synthesis in root meristem cultures of two contrasting barley genotypes. J. Plant Physiol. 132, 245–249.

Ramagopal S and Carr J B 1991 Sugarcane proteins and messenger RNAs regulated by salt in suspension cells. Plant Cell Environ. 14, 47–56.

Robinson N L, Tanaka C K and Hurkman W J 1990 Time-dependent changes in polypeptide and translatable mRNA levels caused by NaCl in barley roots. Physiol. Plant. 78, 128–134.

Shirata K and Takagishi H 1990 Salt-induced accumulation of 26 and 27 kD proteins in cultured cells of rice plant. Soil Sci. Plant Nutr. 36, 153–157.

Singh N K, Bracker C A, Hasegawa P M, Handa A K, Buckel S, Hermodson A, Pfankoch E, Regnier F E and Bressan R A 1987a Characterization of osmotin, a thaumatin-like protein associated with osmotic adaptation in plant cells. Plant Physiol. 85, 529–536.

Singh N K, Handa A K, Hasegawa P M and Bressan R A 1985 Proteins associated with adaptation of cultured tobacco cells to NaCl. Plant Physiol. 79, 126–137.

Singh N K, LaRosa C, Handa A K, Hasegawa P M and Bressan R A 1987b Hormonal regulation of protein synthesis associated with salt tolerance in plant cells. Proc. Natl. Acad. Sci. USA 84, 739–743.

Singh N K, Nelson D E, Kuhn D, Hasegawa P M and Bressan R A 1989 Molecular cloning of osmotin and regulation of its expression by ABA and adaptation to low water potential. Plant Physiol. 90, 1096–1101.

Thompson E W and Lane B G 1980 Relation of protein synthesis in imbibing wheat embryos to the cell-free translational capacities of bulk mRNA from dry and imbibing embryos. J. Biol. Chem. 255, 5965–5970.

Vernon D M, Ostrem J A, Schmitt J M and Bohnert H J 1988 PEPCase transcript levels in *Mesembryanthemum crystallinum* decline rapidly upon relief from stress. Plant Physiol. 86, 1002–1004.

Winikov I, Waterborg J H, Harrington R E and McCoy T J 1989 Messenger RNA induction in cellular salt tolerance of alfalfa (*Medicago sativa*). Plant Cell Rep. 8, 6–11.

Yeo A R 1983 Salinity resistance: Physiologies and prices. Physiol. Plant. 58, 214–222.

Zhao Z, Heyser J W and Bohnert H J 1989 Gene expression in suspension culture cells of the halophyte *Distichlis spicata* during adaptation to high salt. Plant Cell Physiol. 30, 861–867.

P. J. Randall et al. (Eds.), Genetic aspects of plant mineral nutrition, 195–203.

PLSO SV37

Protein synthesis in halophytes: The influence of potassium, sodium and magnesium in vitro

T.J. FLOWERS and D. DALMOND
School of Biological Sciences, University of Sussex, Falmer, Brighton, Sussex, BN1 9QG, UK

Key words: glycophytes, halophytes, protein synthesis, salinity, *Suaeda maritima*, wheat germ

Abstract

The amino acid (^{35}S-methionine) incorporating activity of an in vitro wheat germ translation system was found to be maximal in 80 to 125 mol m^{-3} K with 2 to 4 mol m^{-3} Mg both as the acetate. Substitution of Na for K, or chloride for acetate at concentrations above 80 mol m^{-3} inhibited incorporation. When the K acetate concentration was raised to 200 mol m^{-3}, no incorporation of radioactive methionine occurred.

Translation by polysomes extracted from leaf tissue of *S. maritima*, supplemented with post-ribosomal supernatant from wheat germ, showed activity which was optimal in the presence of 225 mol m^{-3} K acetate and 8 mol m^{-3} Mg acetate. However, the translation system was not directly comparable with the wheat germ system, as studies with an initiation inhibitor, aurintricarboxylic acid, suggested that the *S. maritima* system was essentially elongation-dependent, while initiation occurred in the wheat germ system.

Elongation-dependent polysomal preparations were extracted from leaves of the glycophytes *Pisum sativum*, *Triticum aestivum*, *Oryza sativa* and *Hordeum vulgare*, and from the halophytes *Atriplex isatidea* and *Inula crithmoides*. Translation by polysomes from the salt-tolerant plants was optimal at higher K and Mg concentrations, than by polysomes from the glycophytes. Furthermore, NaCl was better able partially to substitute for the role of K in polysomal preparations from halophytes than glycophytes.

Introduction

Halophytes, plants adapted to growing in saline conditions, must adjust to the low external water potentials that are a characteristic of their environment – salt concentrations as high as 1000 mol m^{-3} have been measured on salt marshes (see Flowers, 1985). For dicotyledonous halophytes this generally means the accumulation of Na and Cl ions, the ions commonly at highest concentration in their immediate surroundings. However, at concentrations high enough to adjust the plant's water relations, Na and chloride are toxic: it was an early observation that a number of enzymes isolated from halophytes were sensitive, when assayed in vitro, to lower concentrations of NaCl than were present in their tissues as a whole (see Flowers et al., 1977). It is not only individual enzymes that are sensitive to the salt concentrations to be found in the vacuoles of halophytes. In vitro translation systems from plants are completely inhibited in 200 mol m^{-3} monovalent cation (Carlier and Peumans, 1976; Marcu and Duddock, 1974; Peumans et al., 1980).

Translation systems obtained from animal and plant sources show, in vitro, a narrow concentration range of ions essential for efficient translation of mRNA. Although there are exceptions, the ionic requirement for translation by ribosomes from genetically diverse sources varies from 1–4 mol m^{-3} Mg and from 80–120 mol m^{-3}

K (see Dalmond, 1988). Ionic shifts of as little as 10 mol m^{-3} K or 1 mol m^{-3} Mg above or below the optimum decrease the absolute rate of translation in vitro (e.g. Weber et al., 1977; Wyn Jones et al., 1979) and can cause significant changes in the efficiencies with which mRNAs are translated (Igarashi et al., 1982; Konecki et al., 1975).

The inorganic ion concentrations of plant cell cytoplasm can, however, only be guessed at in spite of sophisticated analytical procedures for measuring the elemental composition, because of uncertainties over the water content. Estimates for the cytoplasmic concentrations of K, Na, chloride and Mg in plants growing in non-saline environments are: K, 100–120 mol m^{-3}; Na less than 20 mol m^{-3}; chloride less than 20–30 mol m^{-3} and Mg, 1–4 mol m^{-3} (Wyn Jones et al., 1979). Estimates from published analytical data of the cytoplasm from cells of halophytes growing in saline conditions suggests cytoplasmic concentrations of K of 20–200 mol m^{-3}, Na of 100–200 mol m^{-3} even when it reaches 500 mol m^{-3} in the vacuole, and a chloride: (K + Na) ratio of less than one (Flowers et al., 1986). These analyses indicate the cytoplasmic Na plus K concentration in the cells of halophytes could be as high as 300 to 450 mol m^{-3} (Flowers et al., 1986) and certainly may be 250 mol m^{-3} (Hajibagheri and Flowers, 1989). However, initiation-dependent in vitro translation on wheat germ ribosomes is completely inhibited in 140 mol m^{-3} KCl or 200 mol m^{-3} K acetate, irrespective of the salt-tolerance of the source of the mRNA (Gibson et al., 1984), while above 125 mol m^{-3} KCl, polysomal stability was inversely proportional to salt concentration in a range of species (Brady et al., 1984). These facts are difficult to reconcile, unless protein synthesis on halophyte ribosomes is more tolerant of salts than is protein synthesis on wheat germ ribosomes.

The main aim of the investigation whose results are reported here was to attempt to develop and characterize an in vitro translation system from the halophyte *Suaeda maritima*, a species whose overall response to salt has been widely described. Since the the vitro system for translation by wheat germ has been fairly well characterized, it was investigated first, and then used as a reference in order to make comparisons with *S. maritima*. Ribosomes were also extracted from the leaves of some glycophytic plants, wheat, peas, rice and barley and from the halophytes *Atriplex isatidea* and *Inula crithmoides*.

Materials and methods

Wheat germ

Wheat germ, which was stored over silica gel at 4°C, was either purchased from Sigma or was a generous gift from General Mills, Vallejo, California.

The preparation of wheat germ lysates was essentially based on the method of Marcu and Duddock (1974). Wheat germ (2 g) was ground to a fine powder in extraction buffer (4 mL; Hepes/KOH, 20 mol m^{-3}, pH 7.6; K acetate, 100 mol m^{-3}; Mg acetate, 1 mol m^{-3}; $CaCl_2$, 2 mol m^{-3}; DTT, 1 mol m^{-3}) and the thick paste centrifuged in a Beckman HS 21 centrifuge at 30,000 *g* for 12 min. After passing through a Sephadex G-25 column (medium; 21 × 1 cm) equilibrated with elution buffer (Hepes/KOH, pH 7.6, 20 mol m^{-3}; K acetate, 90 mol m^{-3}; Mg acetate, 2 mol m^{-3}; DTT, 1 mol m^{-3}), the most turbid fractions containing ribosomal material were pooled and centrifuged at 30,000 G for 20 min. The resultant supernatant was dispensed in small aliquots (200 μL) into polypropylene tubes, frozen in liquid nitrogen and stored at −80°C: this preparation is subsequently described as WG S30.

In an attempt to reduce the endogenous pool of mRNA, an extract obtained from the first 30,000 *g* centrifugation was made up to 3.5 mol m^{-3} Mg acetate. This homogenate was preincubated with ATP, 1 mol m^{-3}; GTP, 30 mmol m^{-3}; DTT, 2 mol m^{-3}; creatine phosphate, 8 mol m^{-3} and creatine phosphokinase, 40 μg mL^{-1} for 15 min at 30°C (Roberts and Paterson, 1973) and applied to a Sephadex column; the lysate was collected and stored as described for the non-incubated preparation. In order to inactivate endogenous mRNA, WG S30 was, on some occasions, incubated for 10 minutes at 0°C with micrococcal Ca-dependent nuclease.

To obtain post-ribosomal supernatant (WG S100), WG S30 prepared in extraction buffer containing KCl, 500 mol m^{-3}, was applied to a Sephadex column equilibrated with elution buffer so as to reduce the K level and then centrifuged at 100,000 *g* for 60 min (14 × 8 mL rotor, MSE 65) and stored as described for WG S30.

Protein synthesis was carried out based on the procedures of Schleif and Wensink (1981) and BRL (1981). The final concentrations in a 30 μL assay volume were: ATP, 1 mol m^{-3}; GTP, 20 mmol m^{-3}; creatine phosphate, 8 mol m^{-3}; DTT, 2 mol m^{-3}; Hepes, 20 mol m^{-3}; spermidine (when present or not being varied), 250 mol m^{-3} and an amino acid mixture, 50 mmol m^{-3}. Incorporation of amino acids into protein was stopped by transferring the tubes into an ice bath and the amount of radioactivity incorporated determined. An aliquot (2 to 5 μL) of each reaction mixture was applied to Whatman filter discs, which were dropped into ice-cold 10% trichloroacetic acid (TCA), boiled for 10 min in a large volume (approximately 50 mL per disc) of 5% TCA at 90°C to hydrolyze charged tRNA (unlabelled methionine, 0.1%, was included in the hot TCA wash to reduce non-specific binding of soluble radioactive amino acid to the precipitates) and after twice washing with TCA (5%) at room temperature, twice with ethanol and a final rinse in ether, the discs were dried with hot air. Radioactivity incorporated was determined in toluene based scintillant containing 0.35% (w/v) PPO and 0.02% (w/v) POPOP using a liquid scintillation spectrophotometer (LKB Rack-beta 1217) at a counting efficiency of 80%.

Other systems

Suaeda maritima plants were grown as described previously (Yeo and Flowers, 1980). *Pisum sativum* L. cv Meteor, *Triticum aestivum* L. cv. Maris Huntsman, *Oryza sativa* L. cv Amber and *Hordeum vulgare* seeds were hydrated overnight and grown under artificial light at 25°C with a 12 h day/night cycle, being watered daily with half-strength culture solution. *Atriplex isatidea* (Accension WA 867/86, a gift from Clive Malcolm) and *Inula crithmoides* (from Cuckmere Haven, Sussex) seeds were hydrated and raised in growth cabinets under the same conditions as used for *S. maritima*. The nutrient solution used at harvest time for these salt-grown plants included NaCl (340 mol m^{-3}).

For *S. maritima*, polysome preparations were generally made from 15 or 46 day-old plants: leaves (10–20 g) were immersed in saturated Ca hypochlorite for 5–10 minutes, rinsed, frozen in liquid nitrogen and ground to a powder. Thawed powder was mixed in a buffer (4 mL per g tissue: Tris-HCl/pH 8.5, 200 mol m^{-3}; KCl, 200 mol m^{-3}; Mg acetate, 25 mol m^{-3}; sucrose, 200 mol m^{-3}; β mercaptoethanol, 5 mol m^{-3} and heparin, 1 mol mL^{-1}) and after filtering through two layers of cheesecloth centrifuged at 1,000 *g* for 5 minutes. The detergent Nonidet p-40 (0.4%) was added to the supernatant and following centrifugation at 30,000 *g* for 20 min, the supernatant layered over buffered sucrose and centrifuged for a further 4 h at 200,000 *g*. The pellet was resuspended in buffer (Hepes KOH, pH 7.6, 20 mol m^{-3}; K acetate, 20 mol m^{-3}; Mg acetate, 1 mol m^{-3}) and stored as described previously for the WG S30. The procedure used for ^{35}S-methionine incorporation into protein was that used for the WG S30 system. Polysomes were also isolated from the young leaves (5 g) of 10-day-old peas, wheat, rice and barley and from 40-day-old leaves of Atriplex and Inula using the methods described above; extracts were stored at −80°C after freezing in liquid nitrogen. When used, these polysomes replaced *S. maritima* in in vitro translation assays.

Initially assays were performed in duplicate or triplicate, but as the counts differed by less than 1%, subsequent experiments were carried out in single assays. All experiment were at least repeated.

Results and discussion

Wheat germ

Wheat germ ribosomes were active in the translation of exogenous and endogenous mRNA (Table 1). Studies with inhibitors showed the

Table 1. The optimal Mg and K concentrations for the incorporation of ^{35}S-methionine into protein by various preparations (see below) of WG S30 and the effects of three inhibitors (ATA, aurintricarboxylic acid, 50 mmol m^{-3}; CH, cycloheximide, 355 nmol m^{-3} and CP, chloramphenicol, 620 nmol m^{-3})

WG S30	Optimal		Inhibition (%) by		
	Mg (mol m^{-3})	K	ATA	CH	CP
1[a]	2–4	100–120	80	93	2
2	2	100	–	–	–
3	3	125	90	97	4
4	2	100	–	–	–

[a] 1 – pre-incubated, but using endogenous mRNA;
2 – pre-incubated, and using haemoglobin 10 S mRNA;
3 – nuclease treated plus *S. maritima* RNA;
4 – nuclease treated plus haemoglobin 10 S RNA.

characteristics of incorporation on 80S rather than any contaminating 70S ribosomes: there was was a 93–97% reduction in incorporation by cycloheximide (CH, 355 nmol m^{-3}), an inhibitor of protein synthesis on 80S but not 70S ribosomes; chloramphenicol (CP, 620 nmol m^{-3}), a specific inhibitor of protein synthesis on 70S ribosomes inhibited by 2–4%; aurintricarboxylic acid (ATA, a triphenylmethane dye; 50 mmol m^{-3}), an inhibitor of initiation complex formation (Markus et al., 1970), inhibited by 80–90% (Table 1), confirming that initiation was occurring.

The optimal concentrations of Mg and K for the synthesis of protein were determined. Incorporation of labelled methionine had a broad optimum for Mg of 2 to 4 mol m^{-3} as the acetate (and similar to that previously reported for WG S30; e.g. Weber et al., 1977). There was a progressive inhibition at higher concentrations of Mg, with 70% inhibition at about 6 mol m^{-3} (Fig. 1A). In the presence of the polyamine, spermidine (at 250 mmol m^{-3}), incorporation was increased by about 13% over the rate in 3 mol m^{-3} Mg alone when Mg was at 0.8 mol m^{-3} (Fig. 1A). Spermidine reduces the requirement for Mg and in many cases increases the fidelity of the translation system (Konecki et al., 1975; Igarashi et al., 1982). The translation of endogenous mRNAs (with no mRNA added) by WG S30 showed a well defined optimum of ^{35}S-methionine incorporation for K at a concentration of 100–120 mol m^{-3} with acetate as the anion and 80 mol m^{-3} with chloride (Fig. 1B). There was a decrease in incorporation at optimal KCl of 20% relative to optimal K acetate, whereas at the optimal KNO_3 concentration (80 mol m^{-}3) incorporation was still lower – by 28% in comparison with K acetate. Optimal Mg and K concentrations remained similar whether the WG S30 was pre-incubated or nuclease treated, or translating foreign mRNA (Table 1). It was noteworthy that the optimal ion concentrations remained the same when the WG S30 was translating mRNA from *S. maritima*, grown in the presence or absence of NaCl: the halophyte mRNA did not require higher ionic

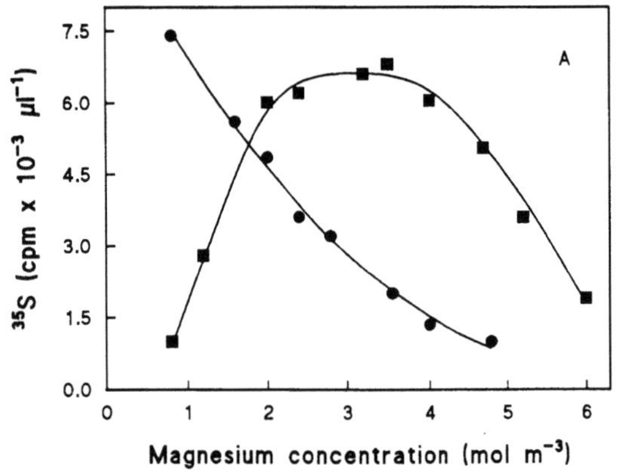

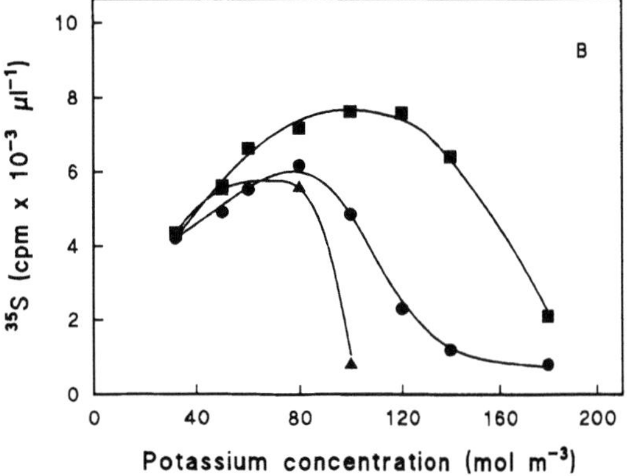

Fig. 1. The effect of change in the concentration of magnesium acetate (**A**) and potassium (**B**) on the incorporation of ^{35}S-methionine into protein by WG S30. In (A) the concentration of K acetate was 100 mol m^{-3} and the assays were performed in the presence (circles) or absence (squares) of spermidine (250 mmol m^{-3}). In (B) Mg acetate was 0.8 mol m^{-3} and spermidine 250 mmol m^{-3}; potassium was provided either as the acetate (squares), chloride (circles) or nitrate (triangles).

concentrations of Mg and K than the wheat germ itself confirming the earlier conclusions of Gibson et al., (1984).

Since plants growing in the presence of NaCl commonly have reduced K contents and elevated Na:K ratios, the consequence of replacing K with Na on in vitro protein synthesis was investigated. Where the total cation concentration was maintained at 80 mol m^{-3} (solid symbols, Fig. 2), increasing concentrations of Na and chloride reduced incorporation of ^{35}S-methionine (Fig. 2). In the absence of chloride, Na up to 30 mol m^{-3} with a Na:K ratio of 0.6 had no effect on the activity (solid circles, Fig. 2). The combination of Na, 84 mol m^{-3}, plus K, 36 mol m^{-3}, with total cations of 120 mol m^{-3} and a Na:K ratio of 2.3:1 inhibited the system almost completely with either acetate or a combination of acetate and chloride (Fig. 2); K acetate at 120 mol m^{-3} supported optimal incorporation. Complete inhibition has been previously reported with this ratio of Na:K (Wyn Jones et al., 1979).

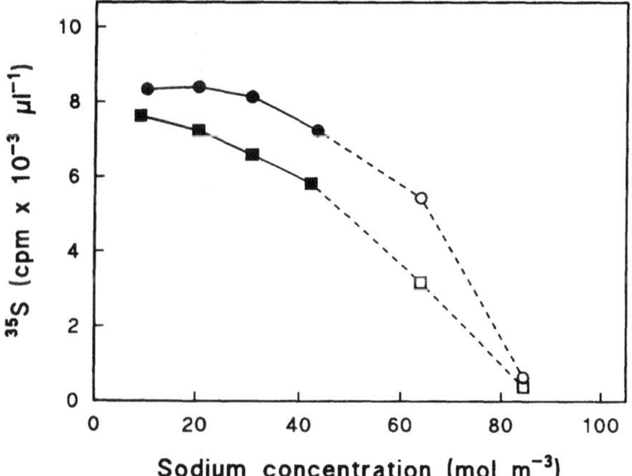

Fig. 2. The effect of various sodium/potassium ratios on the incorporation of ^{35}S-methionine into protein by WG S30. The measurements represented by the circles were made in the absence of chloride (i.e. in Na and K acetates): the closed circles had a total Na + K concentration of 80 mol m^{-3} and the open circles 100 (64 mol m^{-3} Na) and 120 mol m^{-3} (84 mol m^{-3} Na). The squares represent data where Na, K, chloride and acetate were varied: the solid squares all represent measurements at 80 mol m^{-3} Na + K with a chloride at 44 mol m^{-3} and acetate at 36 mol m^{-3}. The open squares are for Na:K:chloride:acetate of 64:36:64:36 and 84:36:84:36.

Suaeda maritima

All attempts to produce from seeds and leaves of *Suaeda maritima* a system analagous to that from wheat germ failed. Subsequently, attempts were made to produce polysomal systems from the leaves: these were partially (see below) successful. Incorporation of ^{35}S-methionine into protein by polysomes isolated from *S. maritima* leaves had an absolute requirement for an exogenous supply of energy, and the WG S100. A similar supernatant protein fraction prepared from *S. maritima* leaves was only 9.5% as effective as the WG S100 fraction (data not presented).

Using standard assays, the optimal concentrations for acetates of K and Mg were found to vary a little with different polysomal preparations. Typically, the Mg optimum was 8 mol m^{-3} (range 7–9 mol m^{-3}) with K at 125 or 225 mol m^{-3} (Fig. 3A): in the WG S30 endogenous system, activity was maximal at 2–4 mol m^{-3} Mg acetate. In the presence of 250 mmol m^{-3} spermidine, the optimal Mg concentration was 2–3 mol m^{-3} (Fig. 3A). The Mg optimum remained at 8 mol m^{-3} (in the absence of spermidine) when the K concentration was varied over the range 25–125 mol m^{-3} (data not presented). In the presence of 8 mol m^{-3} Mg, the optimal potassium (as acetate) concentration was 225 mol m^{-3} (Fig. 3B), nearly twice the optimal potassium acetate concentration for the wheat germ system.

The anion associated with K had a significant effect on the incorporation in standard assays with Mg at a concentration of 8 mol m^{-3}. With acetate as the anion, the optimum K concentration was between 225 and 275 mol m^{-3} (Fig. 3B); KCl, With its optimum at 175 mol m^{-3}, reduced maximal incorporation by 26%; in KNO_3, with an optimum of 150 mol m^{-3}, the maximum rate of incorporation was reduced by 60% relative to K acetate (Fig. 3B). The sensitivity of the wheat germ and of the Suaeda polysomes appeared to be similar to chloride, but the Suaeda system was more sensitive to nitrate than the wheat germ. Why the latter should be so is unknown: nitrate concentrations have not, to our knowledge, been estimated in the cells of halophytes. Chloride appears more inhibitory than sodium and is at a lower concen-

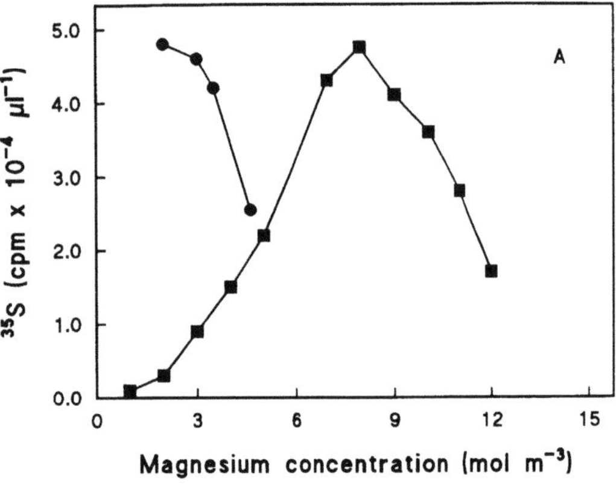

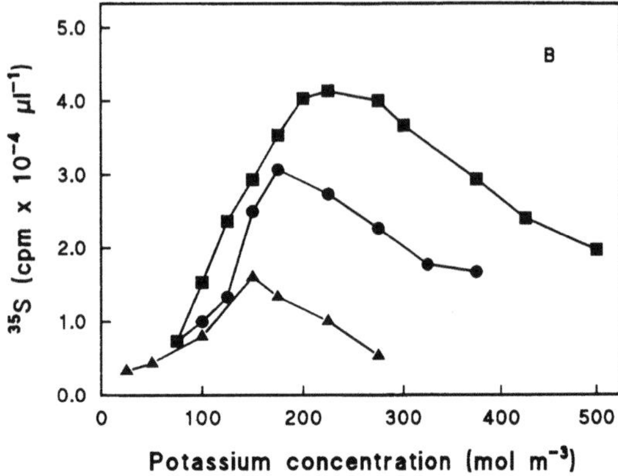

Fig. 3. The effect of change in the concentration of magnesium (**A**) and potassium (**B**) on the incorporation of ^{35}S-methionine into protein by a polysomal preparation from *Suaeda maritima*. In (A) the concentration of K acetate was 125 mol m^{-3} and the assays were performed in the presence (circles) or absence (squares) of spermidine (250 mmol m^{-3}). In (B) Mg acetate was 8 mol m^{-3}; potassium was provided either as the acetate (squares), chloride (circles) or nitrate (triangles).

tration than sodium in the cytoplasm of *S. maritima* (Hajibaheri and Flowers, 1979; Harvey et al., 1981). Chloride also appears more toxic than sodium or potassium for enzymes with anionic substrates (Gimmler et al., 1984): the precise nature of the inhibition of protein synthesis by chloride is unknown.

The concentrations of Na, K and Cl have all been estimated in the mature leaves of *S. maritima*. Flowers et al. (1986) calculated (from the data in Harvey et al. 1981) the cytoplasm to include Na at 166 mol m^{-3}, K at 27 mol m^{-3} and Cl at 77 mol m^{-3}, while the vacuolar concentrations of the same ions were 494, 20 and 352 mol m^{-3}, respectively. The cytoplasmic and vacuolar osmotic potentials are balanced by the presence of glycinebetaine in the cytoplasmic compartment (Flowers and Hall, 1978). Glycinebetaine did not affect the rate of ^{35}S-methionine incorporation into protein at concentrations as high as 660 mol m^{-3} in the presence of either 24 or 225 mol m^{-3} K acetate (the WG S30 is similarly tolerant of glycine betaine in 125 mol m^{-3} K acetate). The reported cytoplasmic concentrations of potassium, at about 30 mol m^{-3}, are very much lower than the optimal (225 mol m^{-3}) concentrations for in vitro protein synthesis (Fig. 3B). In order to determine whether protein synthesis might be expected to occur with such a low K concentration, the cytoplasmic ion concentrations were mimicked in the test tube. Sodium in the form of acetate or chloride, was added to the assays containing polysomes from young (20-d-grown in the absence of NaCl) and mature leaves (46-d-grown in a final concentration of 340 mol m^{-3} NaCl). The concentration of K in the preparations, as determined by atomic absorption spectrophotometry prior to any additions, was 37 mol m^{-3}. The ionic concentrations in the control treatment were 8 mol m^{-3} Mg acetate and 230 mol m^{-3} K acetate. The results (Table 2) indicate that a suboptimal K concentration of 37 mol m^{-3} can support protein synthesis in the

Table 2. Incorporation of ^{35}S-methionine by polysomes isolated from 'mature' leaves of 48-day old plants of *Suaeda maritima* (grown in culture solution containing 340 mol m^{-3} NaCl) and 'young' leaves (20-day-old plants grown in culture solution alone). The various concentrations of solutes used constituted an attempt to mimic cytoplasmic ion concentrations

Ions					
Salt (mol m^{-3})					
K acetate	37	37	37	37	237
Na acetate	–	60	130	–	–
Na chloride	–	70	–	130	–
Activity					
Mature leaves					
cpm $\mu L^{-1} \times 10^{-3}$	5	32	37	25	37
relative (%)	15	86	99	69	100
Young leaves					
cpm $\mu L^{-1} \times 10^{-3}$	4	19	22	10	25
relative (%)	15	75	86	40	100

presence of Na at a concentration of 130 mol m^{-3}. With acetate as the counter anion for Na the system incorporated between 86% (young leaves) to 99% (mature leaves) of the optimum with K acetate: Na acetate at 60 mol m^{-3} and Na chloride at 70 mol m^{-3} was 75 (young) to 86% (old leaves) as effective as 237 mol m^{-3} K acetate. When Na chloride was at 130 mol m^{-3}, incorporation was between 40 to 69% of the rate in optimal K acetate. With only 37 mol m^{-3} K acetate present in the system, incorporation was down by 85%.

The response of the system to varying Na:K ratio was further investigated by measuring the effect of the concentration of NaCl on incorporation when K was at a sub-optimal concentration. At 25 mol m^{-3} K, incorporation was enhanced by the addition of NaCl up to an optimum of 100 mol m^{-3} after which there was a progressive inhibition (Fig. 4). There was significant activity with a Na:K ratio of 4 (cf WG S30 and Wyn Jones et al., 1979), although K at 25 mol m^{-3} plus NaCl at 100 mol m^{-3} was only 68% as effective as 225 mol m^{-3} K acetate. When 100 mol m^{-3} K acetate was present in the assay medium, addition of 100 mol m^{-3} NaCl restored incorporation to the level seen with 225 mol m^{-3} K acetate. Thus the total amount of monovalent cation, which in this case was K plus Na, that made up the optimum of 225 mol m^{-3} was important. Addition of NaCl when the system contained 225 mol m^{-3} K acetate was inhibitory.

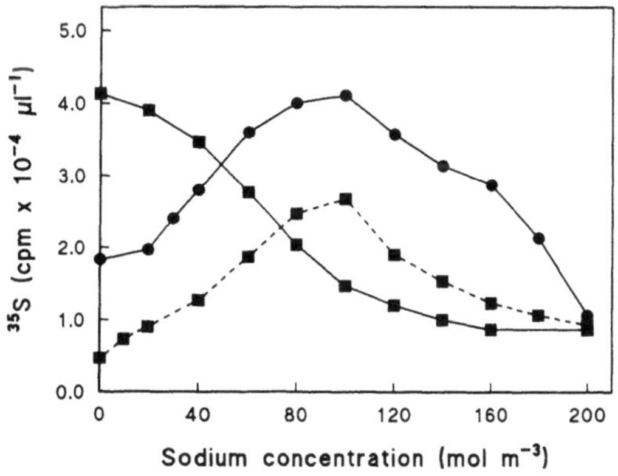

Fig. 4. The effect of Na chloride on the incorporation of ^{35}S-methionine into protein by a polysomal preparation from *Suaeda maritima*. The incubations contained either 25 mol m^{-3} (broken line), 100 mol m^{-3} (circles) or 225 mol m^{-3} (squares joined by a solid line) K acetate. Mg acetate was present at 8 mol m^{-3}.

The measurements of the rates of incorporation of ^{35}S-methionine into protein in vitro in various Na and K concentrations suggest that the concentrations estimated to occur in the cytoplasm of *S. maritima* are not inimical for protein synthesis in vivo. However, it must be recognised that the conditions within the cytoplasm of the highly vacuolated mature cells may be different from those in meristematic tissues. In the latter, Na:K ratios are lower than in the mature cells (see Hajibagheri et al., 1985), conditions that should promote the rate and fidelity of protein synthesis since translational efficiency declines with an increasing ratio of Na to K (Wyn Jones et al., 1979).

The two systems, from wheat germ and *S. maritima* leaves appear very different in their optimal requirements for protein synthesis in vitro. There were, however, also differences between the two systems which suggest that it may not be valid to make direct comparisons. While the polysomal system from *S. maritima* was inhibited by just 2% by chloramphenicol (620 nmol m^{-3}) and by 80% by cycloheximide (355 nmol m^{-3}), a slightly lesser inhibition than for the WG S30 (Table 1), ATA (50 mmol m^{-3}), an inhibitor of initiation, only reduced the incorporation of ^{35}S-methionine into protein by about 24%: in the WG S30 system ATA inhibited by 90%. These results suggest that the *S. maritima* polysomal system is simply elongating message, not initiating the translation of new message. This view is supported by the finding that addition of polysomal RNA prepared from *S. maritima* did not stimulate ^{35}S incorporation. If the *S. maritima* polysomal system is simply completing the translation of existing message and not initiating the translation of new message, then comparison with the wheat germ system (an initiating system) may not be truly valid. Consequently, comparisons were made between elongating systems like that from the *S. maritima* prepared from a number of glycophytes and halophytes.

Polysomes were isolated from the halophytes Atriplex and Inula and the glycophytes Pisum, Triticum, Oryza and Hordeum. These polysomes were subjected to in vitro translation conditions

Table 3. The ionic requirements for optimal incorporation of ^{35}S-methionine into protein, the consequences of alterations in the Na:K ratio and the effect of various inhibitors determined in vitro with polysomal systems from a range of halophytes and glycophytes and in WG S30

Species	Optimal		Incorporation relative to optimum			Incorporation in 25 K + 100 Na rel to 25 K (%)
	Mg[a]	K[a]	ATA[b]	CH[c]	CP[d]	
	(mol m^{-3})		(%)			
Pisum	2.5–3.5	125	94(78)	29	89	118
Oryza	2.5–3.5	100–150	nd	nd	nd	nd
Hordeum	4	150	nd	nd	nd	nd
Triticum leaf	3	125	nd(84)	33	97	82
Triticum WG S30	2–4	100–120	20	7	98	nd
Atriplex	6–8	125–200	nd(74)	23	89	569
Inula	6	125–175	nd(79)	30	90	720
Suaeda	8	200–275	76(0)	20	98	615

[a]as the acetate.
[b]aurintricarboxylic acid, 50 mmol m^{-3} (100 mmol m^{-3}, in parentheses).
[c]cycloheximide, 355 nmol m^{-3}.
[d]chloramphenicol, 620 nmol m^{-3}.

similar to those used for *Suaeda maritima*, and the incorporation of ^{35}S-methionine into protein determined. The results, which are summarized in Table 3, show some of the characteristics of the systems isolated. The Table includes data obtained from earlier experiments on the translation of *S. maritima* polysomes and wheat germ S30. Inhibition by ATA in the polysomal systems was between 6 and 24% (Table 3). These values again indicate that the ribosomes were completing the translation of nascent polypeptide chains (elongation and termination), although there might have been some reinitiation on the discharged mRNA by the ribosomes. In support of the fact that the ribosomes involved were of cytoplasmic origin, the presence of 355 nmol m^{-3} cycloheximide caused an inhibition of generally over 70%, whereas incorporation of label was largely insensitive to chloramphenicol at 620 nmol m^{-3} (Table 3). The optimal ionic concentration for in vitro translation was somewhat elastic. In the glycophytes, the Mg optimum ranged between 2 to 4 mol m^{-3} and the K optimum between 125 to 150 mol m^{-3}. In contrast, the optimal Mg concentration in the halophytic in vitro systems was 6 to 8 mol m^{-3}, and the optimal K concentration was 125 to 275 mol m^{-3}. It appears that there was a higher requirement for Mg and tolerance of K in the halophytes than for the glycophytes across a range of families.

As was found for *Suaeda maritima*, the addition of NaCl to the Atriplex and Inula polysomal systems brought about an increase in incorporation when K was kept at a minimum concentration of 25 mol m^{-3}. At optimal K concentration, additional NaCl was inhibitory. In contrast, wheat and pea leaf polysomes at suboptimal K concentration of 25 mol m^{-3} were not stimulated by the additions of NaCl. It appears that significant differences in the requirements for the tolerance of ions have evolved during the adaptation of halophytes to the saline environments in which they live.

Acknowledgement

TJF would like to acknowledge the receipt of a grant from the Royal Society for assistance with the cost of travel to Canberra.

References

Brady C J, Gibson T S, Barlow E W R, Speirs J and Wyn Jones R G 1984 Salt-tolerance in plants. I. Ions, compatible organic solutes and the stability of plant ribosomes. Plant, Cell Environ. 7, 571–578.

BRL 1981 BRL in vitro translation systems: Nuclease-treated and untreated wheat germ extract systems. Bethesda Research Laboratories, Gaithersburg, MD.

Carlier A R and Peumans W J 1976 The rye embryo system as an alternative to the wheat system for protein synthesis in vitro. Biochim. Biophys. Acta 447, 436–444.

Dalmond D 1988 The Effect of Ions on in vitro Protein Synthesis in the Halophyte *Suaeda maritima*. D. Ph. Thesis, University of Sussex, Brighton, UK.

Flowers T J 1985 Physiology of halophytes. Plant and Soil 89, 41–56.

Flowers T J and Hall J L 1978 Salt tolerance in the halophyte, *Suaeda maritima* (L.) Dum.: The influence of salinity of the culture solution on the content of various organic compounds. Ann. Bot. 42, 1057–1063.

Flowers T J, Troke P F and Yeo A R 1977 The mechanism of salt tolerance in halophytes. Annu. Rev. Plant Physiol. 28, 89–121.

Flowers T J, Hajibagheri M A and Clipson N J 1986 Halophytes. Quart. Rev. Biol. 61, 313–337.

Gibson T S, Speirs J and Brady C J 1984 Salt tolerance in plants. II. In vitro translation of m-RNAs from salt-tolerant and salt-sensitive plants on wheat germ ribosomes: Responses to ions and compatible solutes. Plant, Cell Environ. 7, 579–587.

Gimmler H, Kaaden R, Kirchner U and Weyand A 1984 The chloride sensitivity of *Dunaliella parva* enzymes. Z. Pflanzenphysiol. 114, 131–150.

Hajibagheri M A, Yeo A R and Flowers T J 1985 Salt tolerance in *Suaeda maritima* (L.) Dum.: Fine structure and ion concentrations in the apical region of roots. New Phytol. 99, 331–343.

Hajibaheri M A and Flowers T J 1989 X-ray microanalysis of ion distribution within root cortical cells of the halophyte *Suaeda maritima*. Planta 177, 131–134.

Harvey D M R, Hall J L, Flowers T J and Kent B 1981 Quantitative ion localization within *Suaeda maritima* mesophyll cells. Planta 151, 555–560.

Igarashi K, Hashimoto S, Miyake A, Kashiwagi K and Hirose S 1982 Increase of fidelity of polypeptide synthesis by spermidine in eukaryotic cell-free systems. Eur. J. Biochem. 128, 597–604.

Konecki D, Kramer G, Pinphanichakarn P and Hardesty B 1975 Polyamines are necessary for maximum in vitro synthesis of globin peptides and play a role in chain initiation. Arch. Biochem. Biophys. 169, 192–198.

Marcu K and Duddock B 1974 Characterization of a highly efficient protein synthesizing system derived from commercial wheat germ. Nucl. Acids Res. 1, 1385–1397.

Markus A, Bewley J D and Weeks D P 1970 Aurintricarboxylic acid and initiation factors of wheat embryos. Science 167, 1735–1736.

Peumans W J, Carlier A R and Delaey B M 1980 Preparation and characterization of a highly active cell-free protein-synthesizing system from dry pea primary axes. Plant Physiol. 66, 584–587.

Roberts B E and Paterson B M 1973 Efficient translation of tobacco mosaic virus RNA and rabbit globin 9S RNA in a cell-free system from commercial wheat germ. Proc. Natl. Acad. Sci. USA 70, 2330–2334.

Schleif R F and Wensink P C 1981 Practical Methods in Molecular Biology. pp 161–3. Springer-Verlag, Berlin.

Weber L A, Hickey E D, Maroney P A and Baglioni C 1977 Inhibition of protein synthesis by Cl^-. J. Biol. Chem. 252, 4007–4010.

Wyn Jones R G, Brady C J and Speirs, J 1979 Ionic and osmotic relations in plant cells. *In* Recent Advances in the Biochemistry of the Cereals, Annu. Proc. Phytochem. Soc. Eur. No. 16. Eds. D L Laidman and R G Wyn Jones. pp 63–104. Academic Press, London.

Yeo A R and Flowers T J 1980 Salt tolerance in the halophyte *Suaeda maritima* (L.) Dum.: Evaluation of the effect of salinity upon growth. J. Exp. Bot. 31, 1171–1183.

P. J. Randall et al. (Eds.), Genetic aspects of plant mineral nutrition, 205–213.

PLSO SV50

Nutrient efficiency – what do we really mean?

GRAEME BLAIR
Department of Agronomy and Soil Science, University of New England, NSW, 2351, Australia

Key words: efficiency, nutrients, phosphorus, wheat, white clover

Abstract

There are many ways in which plants can adapt or respond to nutrient stress. These can be via alterations to root branching and root extension rates, rate of uptake per unit root length or root dry weight, partitioning between roots and shoots and between shoots and grain and the amount or concentration of the nutrient required for plants to function. Each of these can be altered to some extent by selection, breeding or biotechnology.

In undertaking such programs the question arises as to how to measure nutrient efficiency. Examples of the effect of alterations in the definition of P efficiency in white clover and wheat are presented. The efficiency ranking of seven accessions of white clover grown under P stress was found to alter according to the definition used. In addition to the definition of efficiency affecting ranking, the level of P under which the efficiency is defined, ontogeny, and plant competition effects can also affect rankings. In wheat, selection for P harvest index was not found to be related to efficiency.

Clearly, selection or manipulation programs need to consider the end product and for this reason in food and forage production systems the definition of grain yield (crop) or edible dry matter yield (forages) per unit nutrient applied is considered the most appropriate definition. Other definitions are more appropriate in other systems.

Introduction

Plants growing in apparently infertile environments are often classified as "efficient" rather than being viewed in the ecological context as the plant best capable of utilising the available resources.

Most Australian soils are inherently deficient in N and P. Native shrubs such as Banksia, and grasses such as *Themeda australis* (Kangaroo grass), which have evolved under these conditions, are capable of surviving on these soils by a combination of slow growth rate (Chapin, 1980) and efficient recycling of nutrients, both within the plant and in leaf and root debris. Whilst such adaptation is appropriate under conditions of low grazing pressure by native fauna, species such as *Themeda* are unable to persist and produce under higher grazing pressures imposed by domesticated livestock.

Such deficiencies in the native grass flora of Australia as agricultural plants led to the search for forage species which are capable of production and persistence under the variable soil nutrient status conditions and grazing pressures created by the combination of climatic and economic circumstances which dominate the Australian agricultural industries.

Plant P

P uptake by plants occurs from soil solution with a concentration of usually much less than $10\ \mu mol\ m^{-3}$ and often as low as $0.01\ \mu mol$ $P\ m^{-3}$ and is incorporated into plant tissue with concentrations of about $10\ mmol\ m^{-3}$ P (Bieleski, 1976). Not only does the uptake process occur against this concentration gradient of 10^{-3}–$10^{-6}:1$ but also against an electrical potential of $-120\ mV$ which effectively means the P concentration gradient to be overcome is probably of the order of 10^{-5}–10^{-7} : 1 (Bieleski, 1976).

From measured potential differences across the cell membranes, Bowling and Dunlop (1978) were able to calculate the approximate driving force on P transported to the cell. The calculations showed that white clover roots expend 0.024 kJ hr^{-1} g root $tissue^{-1}$ on P uptake.

Studies on barley (Clarkson et al., 1968) show that distance from the root tip or root age has no effect on P uptake and translocation. Due to the slow fluxes of P to the root, older zones of root will be in contact with soil solutions of low P content.

Large differences in P influx rate between species are apparent and studies by Schenk and Barber (1980) have shown large varietal differences in uptake characteristics in maize. Factors in addition to P supply rate, genotype and age that can influence uptake are: rate of transpiration (Russell and Shorrocks, 1959), temperature and light (Barber, 1980), competition from other ions (Nye and Tinker, 1977), and root aeration (Nye and Tinker, 1977).

Phosphorus in plants is present as a range of organic and inorganic compounds and the amount of the various compounds varies markedly. Depending on the P nutrition of the plant shoot, total P concentration is in the range of 0.05 to 1.0% (dry wt. basis). The conversion of orthophosphate to organic forms occurs rapidly within the plant. Jackson and Hagen (1960) found that 80% of the phosphate absorbed was incorporated into organic forms within 10 minutes. The level of inorganic P in the tissue is more sensitive to P nutrition than the levels of the various organic fractions (DNA, RNA, lipid-P and ester-P) (Bieleski, 1973, Chisholm et al., 1981, Chisholm and Blair, 1988 a,b). In P conditions ranging from acute deficiency to excess the concentrations of ester-P, RNA and P-lipid in shoots and roots may change 5 fold while inorganic P can change 50-fold.

An understanding of the P cycle in grazed pastures indicates that plant P utilization is a small part of the overall P cycle and any gains in plant P utilization must be translated through the soil/plant/animal/system if the overall P utilization rate is to be increased.

In grazed pastures, the maximum P utilization rate within the system will likely be in systems which accumulate the maximum amounts of P in plants. The rationale for this is that in such systems less P remains in the soil where it is subjected to chemical changes which alter the rate of availability. This contrasts with the generally rapid recycling of P via plant litter (Till and Blair 1978) and dung. This rate of recycling declines where considerable amounts of plant litter and organic debris accumulate on the soil surface where, because of moisture stress, the P is often positionally unavailable to the plant.

P efficiency – How to define it!

At the outset, a distinction needs to be made between efficiency and response. Efficiency is defined as the ability of a genotype to acquire plant nutrients from the rhizosphere solution and/or to incorporate or utilize them in the production of total above and/or below ground level plant biomass or utilizable plant material (seed, grain, fruit, forage). Response, on the other hand, is the capacity of the genotype to increase uptake or yield as the supply of the nutrient to the root is increased. These may be quite different from each other in a particular genotype.

A major difficulty in the selection or breeding of nutrient efficient plants is to define what is efficiency. To a plant physiologist, an increase in the efficiency of a biochemical process within the plant constitutes a gain, whereas to an agronomist, an increase in efficiency means more plant and/or animal production for the same input or the same output for a lower input.

Gerloff (1977) classified plants into 4 response groups as follows:

1. Efficient responders – plants which produce high yields at low levels of nutrition and which respond to nutrient additions.
2. Inefficient responders – plants with low yields at low levels of nutrition which have a high response to added nutrients.
3. Efficient non-responders – plants with high yields at low levels of nutrition but which do not respond to nutrient addition.
4. Inefficient non-responders – plants with low

yields at low levels of nutrition and which do not respond to nutrient additions.

A diagrammatic representation of these 4 groups is presented in Figure 1.

P efficiency criteria

Many definitions of P efficiency have been proposed and these have been divided by Wilson (1976) into three categories namely: a) uptake efficiency, b) incorporation efficiency and c) utilisation efficiency.

Uptake efficiency criteria are based on root parameters, incorporation efficiency relates to shoot yields whereas utilization efficiency incorporates the whole plant (shoot and roots). The ideal plant would be efficient in all three categories. Scott (1977) has identified white clover (*Trifolium repens*) ideotypes presently in use, which could form the basis for a breeding program that would ideally produce a genotype which had higher yields at low P levels relative to the maximum (Type 3 in Fig. 2).

Various estimates of efficiency have been suggested as follows:

a) Uptake efficiency

P uptake per unit root length (μg P cm root^{-1})
P uptake per unit root dry weight (μg P mg root d wt^{-1}) (Barrow 1975)
P uptake rate per unit root length (μg P cm root^{-1} day^{-1}) (Williams 1948)
P uptake rate per unit root weight (μg P mg root d wt^{-1} day^{-1}) (Williams 1948)
Low concentration at which maximum uptake occurs (K_m) and low concentration to which the plant can deplete the solution (C_{min}) (Barrow 1975a, Edwards & Barber 1976)
High solution concentration where the plant produces 50% of its maximum yield (V_{max}) (Cacco et al., 1976)

b) Incorporation efficiency

Shoot DM per unit P in shoot (g DM mg P^{-1}) (Jones 1974, Blair and Cordero 1978)
Utilisation quotient (shoot P %$^{-1}$) (Steenbjerg and Jakobsen 1963)
Shoot P % under P deficient conditions (Clark and Brown 1974)

c) Utilisation efficiency

Shoot DM per unit P applied (g DM kg P^{-1}) (Blair and Cordero 1978)
Shoot DM per unit total P uptake (g DM mg P^{-1}) (Blair and Cordero 1978)
Shoot DM at the same % P (Blair and Cordero 1978)
Plant DM (shoot + roots) per unit P uptake (g DM mg P^{-1}) (McLachlan 1976)
Critical P concentration (% P) (Jones 1974, Blair and Cordero 1978)
Relative yield over P rates (Mitscherlich c coefficient)

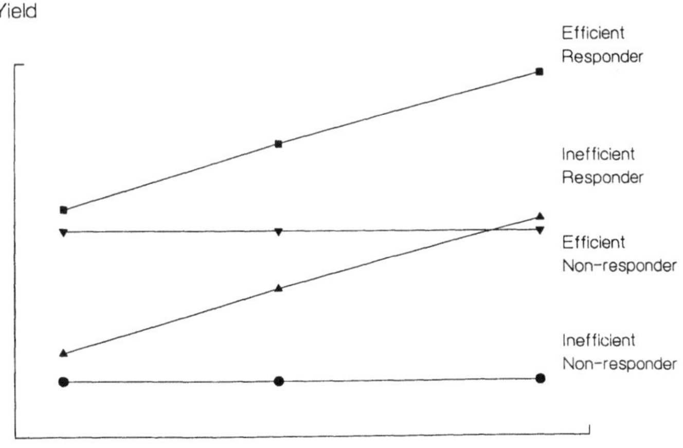

Fig. 1. Diagrammatic representation of response classes as defined by Gerloff (1977).

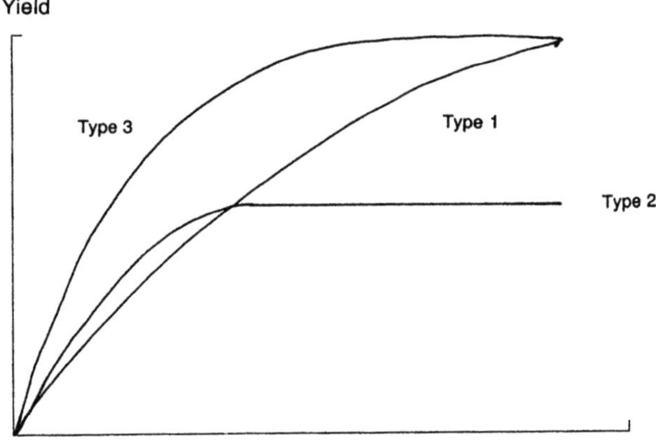

Fig. 2. Types of yield response to P in present, required breeding lies and the ideal outcome as proposed by Scott (1977). Type 1: Present lines, responsive to high phosphorus. Type 2: Genotype required for breeding, responsive to low phosphorus. Type 3: Ideal combination.

Yield under P deficiency/yield at P adequate (Caradus and Dunlop 1978)
Application rate of P required to reach maximum yield (Ozanne et al. 1976, Jones 1974)

It is not the intention here to discuss the biochemical, physiological or agronomic arguments for or against any of the above definitions of efficiency as this has been done elsewhere (Godwin, 1981; Wilson, 1976). The purpose of this paper is to raise the issue of the need to clearly define what is meant by efficiency and to be satisfied that any selection, conventional breeding or gene transfer program has in mind the end product of such manipulations.

A case study under P deficiency

In a study with white clover (Blair and Godwin, 1991; Godwin, 1981; Godwin and Blair, 1991), seven accessions were grown in a soil deficient in P in pots. The accessions were the commercial cultivars Ladino, Haifa and Grasslands Huia, an introduced non-commercial accession Algerian, and three ecotypes collected from the Northern Tablelands of NSW, namely Newholme, Point Lookout and Chiswick.

In the experiment, P was applied to the pots in amounts equivalent to 0, 5, 10, 20, 40 and 80 kg P ha^{-1} and the plants grown for 40 days. At harvest, dry weights of shoot and roots and their P concentration were determined, and root length estimated by the intercept method so that the range of P efficiency parameters listed above could be calculated. The definitions based on K_m, C_{min} and V_{max} require knowledge of the P concentrations in the solution in contact with the root and could not be calculated for this soil experiment. All the calculations in Table 1, which are made at a single level of applied P, are for severely deficient plants grown where 5 kg P ha^{-1} was applied. Yields of shoot dry matter for the experiment are presented in Figure 3.

Table 1 shows, the ranking of the genotypes and how this changes depending on the criteria used. For the utilization efficiency criteria Ladino is ranked 1 or 2 in all except critical P % and the P required for maximum yield. This contrasts with Algerian which changes from most (1) to least efficient (7) across this range of definitions. When comparisons are made between utilization, uptake and incorporation efficiencies marked variations are apparent. Whilst Ladino produces the highest shoot DM/unit P applied (utilization efficiency) it is the least efficient in terms of P uptake/unit root length. Point Lookout, which ranks medium to low in utilization efficiency is a high ranking accession in terms of uptake efficiency.

In a subsequent experiment Blair and Godwin (1991) examined the relations between yield and P uptake and root parameters in Ladino and Chiswick grown with P deficient and P adequate supplies. There were marked changes in response to P in the length, diameter and number of roots in the two accessions with time. Chiswick tended to produce many short roots whilst Ladino fewer long roots. Only small differences in P uptake per unit root length were measured, which suggest that total root length or root extension rate is the primary determinant of total P uptake in these accessions of white clover.

Other factors which affect P efficiency rankings

The rankings presented in Table 1 are those for P deficient plants and the data in Figure 3 indicate crossovers in yield as application rates of P change. This affects some rankings of P efficiency and examples of this are presented in Table 2.

Using the criterion of shoot DM per unit P

Table 1. Ranking of 7 accessions of white clover using a range of efficiency criteria

	Ladino	Haifa	Grassl. Huia	Algerian	Newholme	Point Lookout	Chiswick
Uptake efficiency							
P upt/unit root length	7[a]	2	5	4	3	1	6
P upt/unit root wt	6	2	5	7	2	1	3
Inst. P upt/unit root length	6	2	4	7	3	1	5
Inst. P upt/unit root d.wt.	1	3	5	7	4	3	6
Incorporation efficiency							
Shoot DM/unit P in shoot	1	5	3	2	4	6	7
Utilisation quotient (% P)$^{-1}$	1	5	3	2	4	6	7
Shoot P % under P stress	1	5	3	2	4	6	7
Utilisation efficiency							
Shoot DM/unit P applied	1	3	6	7	2	4	5
Shoot DM/unit P uptake	2	5	7	1	3	6	4
Shoot DM at const. % P	1	3	5	6	2	4	7
Plant DM/unit P uptake	2	5	7	1	3	6	4
Critical P %	4	7	5	3	2	1	6
Rel DM over P rates	2	6	4	3	1	5	7
Yield at low P/high P	1	3	6	4	2	5	7
P req for max Y	4	7	6	3	1	2	5

[a] 1 = most efficient 7 = least efficient.

applied Ladino is the most efficient and Grasslands Huia the second least efficient accession at both 5 and 40 kg P ha^{-1}. However, the ranking of accessions such as Haifa and Chiswick change markedly. A more marked re-ordering of the rankings is apparent when the Shoot DM at constant % P criterion is used.

When the accessions are compared on the basis of the Gerloff (1977) grouping (Fig. 1) Ladino is an efficient responder, Point Lookout and Chiswick are inefficient responders, Haifa is an efficient non-responder and Newholme, Grasslands Huia and Algerian are inefficient non-responders. Clearly on this basis Ladino is the most favoured accession with the capacity to produce higher amounts of edible dry matter than the other accessions over a wide range of P availabilities.

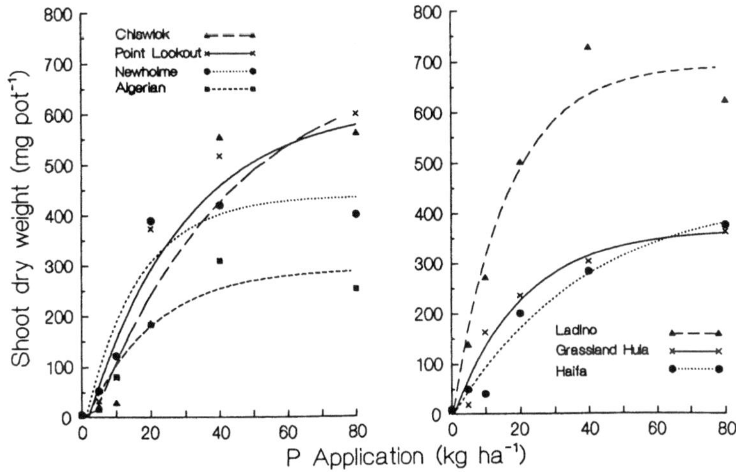

Fig. 3. Shoot dry weight (g pot^{-1}) response to added P in several accessions of white clover (Godwin and Blair 1991). Lines are fitted curves.

Table 2. The effect of P application rate (5 or 40 k P ha^{-1}) on the ranking of P efficiency in white clover

Parameter	P	Ladino	Haifa	Grassl. Huia	Algerian	Newholme	Point Lookout	Chiswick
Shoot DM per unit P applied	P5	1	3	6	7	2	4	5
	P40	1	7	6	5	4	3	2
Shoot DM at constant % P	P5	1	3	5	6	2	4	7
	P40	3	7	6	5	1	2	4

All the above comparisons have been made on whole plants harvested at the same time. A number of authors have used an identifiable plant part of known physiological age, such as the youngest fully expanded leaf, for determining P concentration. Such studies by Bates (1971) and Jones et al. (1972) in subterranean clover (*T. subterraneum*) and Hart and Jessop (1982) in white clover and lotus (*Lotus corniculatus*) have shown that this procedure can substantially remove the effects of changes in P concentration that occur with changes in physiological age.

A limitation of this procedure is that although the selected tissue is of the same physiological age the plant can be at a different ontogenetic stage and therefore in a different state of P requirement (e.g. flowering, fruiting). Also, repeated defoliation can alter the P concentration of plant parts of the same physiological age (Jones et al. 1972). A further limitation of this procedure is that P concentration is but one of the measures of efficiency as outlined above.

Kemp (1985) approached this problem in a pot study that examined P efficiency in Tama ryegrass and *Sirosa phalaris* with plants grown under P deficient (5 kg P ha^{-1}) or P adequate (20 kg P ha^{-1}) conditions. These studies showed that at any harvest between 15 and 115 days after sowing P uptake was higher in ryegrass than Phalaris (Fig. 4a). Using the P efficiency definition of DM produced per unit P applied, ryegrass is a more efficient plant than Phalaris at any harvest time.

Tama ryegrass is an annual or bi-annual whereas Phalaris is a perennial species with different flowering times. Kemp (1985) adapted Zadok's growth analysis used in cereals to these pasture species to enable comparisons to be made at the same physiological age. When this was done no differences were recorded between the two species grown at the same P rate except at maturity where 20 kg P ha^{-1} was applied (Fig. 4b). When comparisons were made between species at the same P level with the same number of leaves no differences in P uptake between the species was recorded (Fig. 4c).

Plant competition and P efficiency

In mixed grass-legume swards the problem of defining P efficiency becomes more difficult. The efficiency with which the applied P is used by both the legume and the grass will have a bearing on the efficiency of recycling in the system. In addition, the efficiency with which N is fixed by the clover will have an impact on the capacity of the grass to utilize the applied P.

This means that an analysis of the P efficiency of the legume component alone has little value in determining the overall efficiency of the grass/legume system. If a very P efficient legume is unable to tolerate grazing and is outcompeted by the grass then its contribution to the overall system will be limited.

P efficiency in cereals

The situation regarding definitions of P efficiency in grain crops is even more complex than in forage plants. It can be argued that the criteria for forage plants listed in Table 1 relating to DM yield per unit P applied or P incorporated are appropriate for grain crops during their vegetative phase where a capital of leaf area and photosynthetic capacity is established.

The additional complication in grain crops is the amount of P translocated to grain. The most efficient grain crops would be those which produce high grain yields with a low P concentration

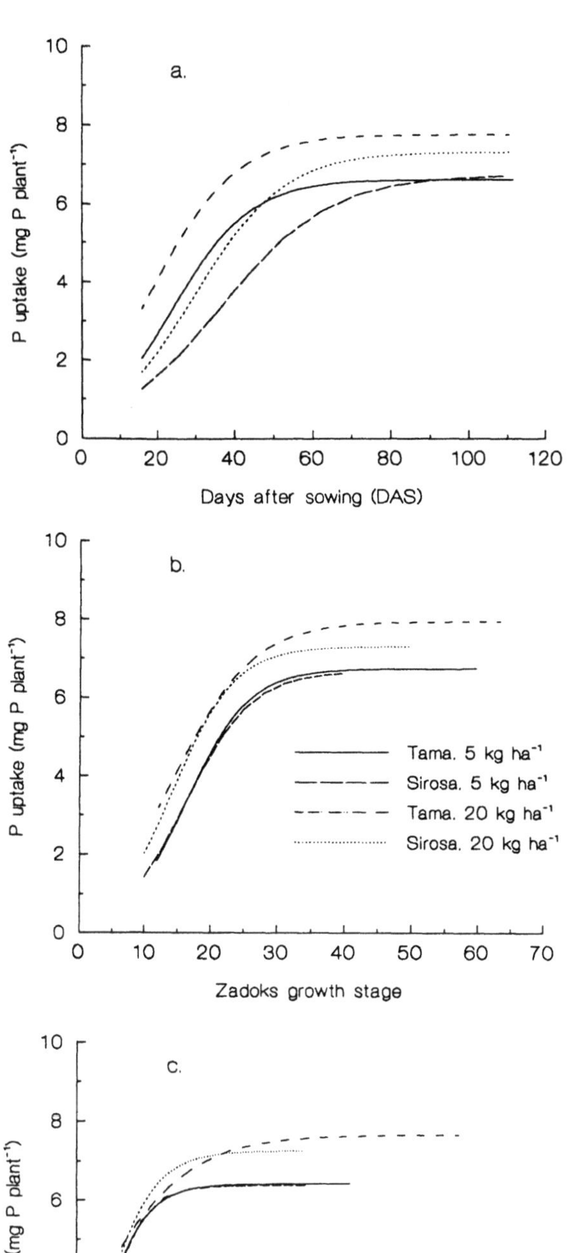

Fig. 4. Fitted curves of P uptake of Sirosa phalaris and tama ryegrass grown at two rates of P application compared on the basis of *a*) days after sowing *b*) Zadok's growth stage and *c*) number of leaves per plant (Kemp, 1985).

and hence remove as little P as possible from the cropping system, with the P in the residues potentially available for recycling.

Batten (1984) suggested that inadvertent selection for P efficiency may already exist in wheat. In his study of diploid, tetraploid and hexaploid wheats he found a high negative correlation between dry matter harvest index and grain P concentration (i.e. selection for increased harvest indices has lead to inadvertent selection for low grain P concentrations which potentially lessens the removal of P from the cropping system). The results of Batten (1984) may be spurious in that they were obtained from plants in which tillers had been removed during growth. This would have severely restricted the translocation of P from infertile tillers to the grain. In a subsequent study Jones (1986) found no such relation in 23 wheat cultivars which included semi-dwarf and standard varieties (Fig. 5).

Jones (1986) investigated the potential for selection of cultivars having a low P harvest index with a high grain yield per unit of P uptake as a means of increasing P efficiency. In his study of 23 wheat varieties a strong linear correlation ($r^2 = 0.85$) was found between P harvest index (%) (Y) and grain yield unit P uptake^{-1} (g mg P^{-1}) (x) as follows

$$Y = 6.52 + 227x$$

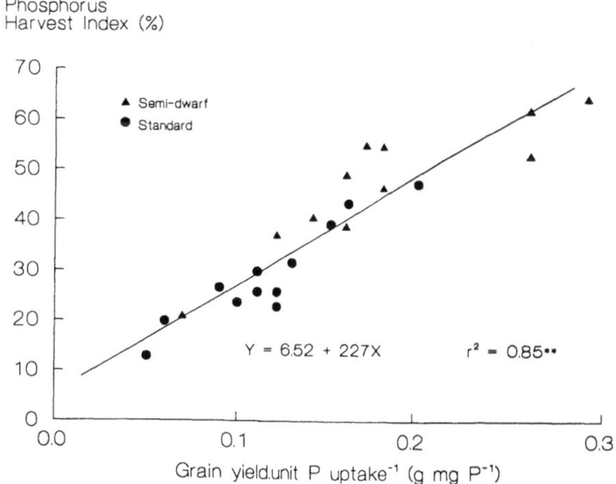

Fig. 5. Relationship between grain yield unit P uptake^{-1} and P harvest index (%) for 23 wheat cultivars grown with 40 kg P ha^{-1} (Jones, 1986).

The lack of genotypes with a high grain yield per unit P uptake with a low P harvest index among the wide range of cultivars studied suggests that the scope for selection of such P efficient wheat cultivars is poor.

Palmer and Jessop (1977) suggested that root fineness or branching was an important determinant of P efficiency in wheat. In the study of Jones (1986) the semi-dwarf variety Israel M68 had a lower root weight than the standard variety Olympic but higher P uptake per unit root weight and total P uptake which could be accounted for by the finer and more branched root system in Israel M68. A similar difference was observed between semi-dwarf and standard varieties by Matheson (1971).

Jones (1986) concluded that some of the breeding objectives for a P efficient wheat plant would be:

1. High agronomic performance via high grain yields and high harvest indices and the ability to produce near maximal yields at low P levels.
2. An extensive root system to access nutrients and water.
3. High P uptake per unit root length or weight.
4. High grain yields per unit P uptake.
5. Low P harvest index.

Conclusions

A clear distinction needs to be made between the efficiency and responsiveness of a particular genotype. The definition of efficiency used needs to be carefully selected to be one which best reflects the end use of the study.

Rankings in P efficiency between genotypes made in plants of the same age may be altered by factors such as plant P status, physiological age, ontogeny and the competiveness of the genotype in mixed communities.

In wheat, the scope for selection for improved P efficiency based on P harvest index appears limited. Available data suggests that root exploration capacity may be the most satisfactory selection criteria.

Clearly, selection, breeding or gene manipulation programs need to consider the end product of the production system. In food or forage production systems the efficiency definition of grain yield (crop) or edible dry matter yield (forages) per unit nutrient applied is considered the most appropriate.

Where the risk of environmental damage from P runoff or leaching is high the P efficiency criteria of maximum plant incorporation of applied P and its subsequent recycling via litter and/or animals would be more approriate. In such situations manipulation of fertiliser P release rate may lead to greater overall P utilization efficiency than manipulation of the plant genotype.

References

Barber S A 1980 Soil interactions in the phosphorus nutrition of plants. *In* The Role of Phosphorus in Agriculture. Eds F E Khasawneh, E C Sample and E J Kamprath. pp 591–615. ASA, CSSA, SSSA, Madison, WI.

Barrow N J 1975 The response to phosphate of two annual pasture species. II. The specific rate of uptake of phosphate, its distribution and use for growth. Aust. J. Agric. Res. 26, 145–156.

Bates T E 1971 Factors affecting critical nutrient concentrations in plants and their evaluation: A review. Soil Sci. 112, 116–130.

Batten G D 1984 Grain development in wheat in relation to phosphorus. Ph.D. Thesis. Australian National University, Canberra, Australia.

Bieleski R L 1973 Phosphate pools, phosphate transport, and phosphate availability. Annu. Rev. Plant Physiol. 24, 225–252.

Bieleski R L 1976 Passage of phosphate from soil to plant. *In* Prospects for Improving Efficiency of Phosphorus Utilization. Reviews in Rural Science. III. Ed G J Blair. pp. 9–19. The University of New England, Armidale, Australia.

Blair G J and Cordero S 1978 The phosphorus efficiency of three annual legumes. Plant and Soil 50, 387–398.

Blair G J and Godwin D C 1991 Phosphorus efficiency in pasture species. VII. Relationships between yield and P uptake and root parameters in two accessions of white clover. Aust. J. Agric. Res. 42, 1271–1283.

Bowling D J F and Dunlop J 1978 Uptake of phosphate by white clover. J. Exp. Bot. 29, 1139–1146.

Cacco C, Ferrari G and Lucci G C 1976 Uptake efficiency of roots in plants at different ploidy levels. J. Agric. Sci. Camb. 87, 585–589.

Caradus J R and Dunlop J 1978 Screening white clover plants for efficient phosphorus use. *In* Plant Nutrition 1978. Proceedings of the 8th International Colloquium on Plant Analysis and Fertilizer Problems, Auckland, New Zealand, 28 August–1 September 1978. Eds. A R Ferguson, R L Bieleski and I B Ferguson. pp 75–82. DSIR, Wellington.

Chapin, F S 1980 The mineral nutrition of wild plants. Ann. Rev. Ecol. Syst. 11, 233–260.

Chisholm R H E, Blair G J, Bowden J W and Bofinger V J 1981 Improved estimates of "critical" phosphorus concentration from considerations of plant phosphorus chemistry. Comm. Soil Sc. Pl. Anal. 12, 1059–1065.

Chisholm R H and Blair G J 1988a Phosphorus efficiency in pastures species. II. Differences in the utilization of P between major chemical fractions. Aust. J. Agric. Res. 39, 817–826.

Chisholm R H and Blair G J 1988b Phosphorus efficiency in pastures species. III. Correlations of dry matter accumulation with phosphorus pool sizes and their net transfer rate. Aust. J. Agric. Res. 39, 827–836.

Clark R B and Brown J C 1974 Differential mineral uptake by maize inbreds. Comm. Soil Sci. Plant Anal. 5, 213–227.

Clarkson D T, Sanderson J and Scott–Russell R 1968 Ion uptake and root age. Nature 220, 805–806.

Edwards J H and Barber S A 1976 Phosphorus uptake rate of soybean roots as influenced by plant age, root trimming, and solution P concentration. Agron. J. 68, 973–975.

Gerloff G C 1977 Plant efficiencies in the use of N, P and K. *In* Plant Adaptation to Mineral Stress in Problem Soils. Ed. M J Wright. pp 161–174. Cornell University Press, New York.

Godwin D C 1981 Uptake efficiency of utilization of phosphorus by white clover. M. Rur. Sci. Thesis. University of New England, Armidale, Australia.

Godwin D C and Blair G J 1991 Phosphorus efficiency in pasture species. V. A comparison of white clover accessions. Aust. J. Agric. Res. 42, 531–540.

Hart A L and Jessop D 1982 Concentration of total, inorganic, and lipid phosphorus in leaves of white clover and stylosanthes. N.Z.J. Agric. Res. 25, 69–76.

Jackson W A and Hagen C E 1960 Products of orthophosphate absorption by barley roots. Plant Physiol. 35, 326–332.

Jones G P D 1986 Genotypc variation in phosphorus efficiency and utilization by wheat. M. Rur. Sci. Thesis. University of New England, Armidale, Australia.

Jones M B, Ruckman J E and Lawler P W 1972 Critical levels of P in subclover (*Trifolium subterraneum* L.). Agron. J. 64, 695–698.

Jones R R 1974 A study of the phosphorus responses of a wide range of accessions from the genus *Stylosanthes*. Aust. J. Agric. Res. 25, 847–862.

Kemp P D 1985 The efficiency of acquisition and utilization of phosphate by four temperate pasture species. Ph.D. Thesis. University of New England, Armidale, Australia.

Matheson E M 1971 Some aspects of the whole plant physiology of selected short stature wheat cultivars. Ph.D. Thesis, University of New England, Armidale, Australia.

McLachlan K O 1976 Comparative phosphorus responses in plants to a range of available phosphorus situations. Aust. J. Agric. Res. 27, 323–341.

Nye P H and Tinker P B 1977 Solute Movement in the Soil-root System. Blackwell Scientific Publications, Oxford. 342 p.

Ozanne P G, Howes K M W and Petch A 1976 The comparative phosphate requirements of four annual pastures and two crops. Aust. J. Agric. Res. 27, 479–488.

Palmer B and Jessop R S 1977 Some aspects of wheat cultivar response to applied phosphate. Plant and Soil 47, 63–73.

Russell R S and Shorrocks V M 1959 The relationship between transpiration and absorption of inorganic ions by intact plants. J. Exp. Bot. 10, 301–316.

Schenk M R and Barber S A 1980 Potassium and phosphorus uptake by corn genotypes grown in the field as influenced by root characteristics. Plant and Soil 54, 65–76.

Scott R S 1977 The phosphate nutrition of white clover. Proc. N. Z. Grassl. Assn. 38, 151–159.

Steenbjerg G and Jakobsen S T 1963 Plant nutrition and yield curves. Soil Sci. 95, 69–88.

Till A R and Blair G J 1978 The utilization by grass of sulfur and phosphorus from clover litter. Aust. J. Agric. Res. 29, 235–242.

Williams R F 1948 The effects of phosphorus supply on the rates of intake of phosphorus and nitrogen upon certain aspects of phosphorus metabolism in gramineous plants. Aust. J. Sci. Res. Ser. B. 1, 336–361.

Wilson E J 1975 The phosphorus utilisation efficiency of two white clovers in solution culture. M. Sci. Agric. Thesis, University of New England, Armidale, Australia.

P. J. Randall et al. (Eds.), Genetic aspects of plant mineral nutrition, 215–220.

PLSO SV55

A review of phosphorus efficiency in wheat

GRAEME D. BATTEN
Yanco Agricultural Institute, Yanco, NSW 2703, Australia

Key words: genetic variability, NIR, ploidy, phosphorus, wheat

Abstract

More efficient utilization of phosphorus by wheat plants is needed to extend the useful life of the phosphate reserves in the world, to reduce the cost of producing crops, and to improve the value of the grain and the straw produced. In this paper definitions of efficient use of phosphorus by wheat are reviewed, genotypic variation in phosphorus efficiency is reported, some consequences of breeding for greater efficiency are discussed, and ways to select more efficient genotypes are suggested.

Introduction

The production of cereals throughout the world relies on the use of fertilizers to correct natural deficiencies of plant-essential elements in the soil and to replace elements which are removed in the products harvested. After nitrogen, phosphorus is the most widely used fertilizer throughout the world. In 1988–89 the world consumption of nitrogen was 79.6 million tons N compared to the consumption of 16.6 million tons of elemental phosphorus (Constant and Sheldrick, 1991).

Phosphorus makes a significant contribution to the cost of producing cereals and the price of phosphate fertilizer can be expected to increase in the future as some new users of phosphate fertilizer, principally in Asia, enter the market; the high grade phosphate rock reserves are being depleted and lower grade ores are being mined. The cost of transporting ores and fertilizer products may also increase along with depletion of oil reserves. At the same time there will be increasing demands on world food production by the natural increase in the population of the world (Constant and Sheldrick, 1991; Strangel and Von Uexkull, 1990).

The amount of food produced using the world reserves of phosphate can be increased by producing more food from each kg of applied phosphorus but the food products must remain of high quality.

This paper reviews the utilization of phosphorus by wheat plants using some of the possible definitions of efficiency; examines genotypic variability in phosphorus efficiency; and proposes the use of two techniques which could be used to select cereal genotypes which use phosphorus more efficiently.

Definitions of phosphorus efficiency

Grain yield per kg of fertilizer phosphorus applied to the crop

Plant breeders unconsciously test wheat genotypes for phosphorus efficiency if they apply less phosphate than the crop requires to produce the optimum yield at the experimental site. The practice of wheat breeders in New South Wales Australia has been to apply 10 kg P ha^{-1} to dryland crops in the 450–550 mm average annual rainfall zone (M A Khan and J A Fisher, pers. commun.). These rates of phosphorus are less than required for optimum crop yields in most seasons, especially where high average yields are achieved or soils are very acidic (Batten and Khan, 1987a; Smith and Batten, 1976). Field studies by Jessop and Palmer (1976) and Batten

Table 1. Average uptake and utilization of phosphorus by wheats grown in three seasons under natural rainfall

Season	Growing season rainfall (mm)	Phosphorus ($kg\ ha^{-1}$)		Yield ($t\ ha^{-1}$)	PE ($kg\ kg^{-1}$)	HIP (%)	Grain P (%)
		Applied	In shoots				
1977	202	4	2.5	1.37	543	84	0.156
		8	3.2	1.63	525	83	0.160
		16	4.0	1.99	504	82	0.167
		se	0.1	0.04	7	ns	0.002
1978	584	2	11.0	3.58	394	89	0.229
		6	12.1	3.53	378	89	0.239
		18	14.9	3.76	330	86	0.265
		se	0.1	0.04	14	ns	0.006
1979	340	2	7.5	2.56	457	85	0.189
		6	10.7	3.10	410	85	0.212
		18	13.5	3.34	334	81	0.249
		se	0.6	0.02	15	1	0.005

From Batten et al. (1984) and Batten and Khan (1987a).

et al. (1984) have shown significant differences in the response to fertilizer phosphate by wheat genotypes but in each study there were large differences in response between seasons. Semi-dwarf genotypes produce more grain per kg of applied phosphorus than older, taller genotypes especially at higher rates of applied phosphorus.

Care needs to be taken when comparing the efficiency of wheat using the definition of yield per unit of fertilizer applied. In some seasons more phosphorus is taken up by the plant than was applied (Table 1). It is not clear in yield comparisons if the more efficient genotypes take up more phosphorus from the fertilizer applied and from the soil, or if the phosphorus taken up by the plants is utilized more efficiently. Figure 1 shows two genotypes with similar dry matter production but different phosphorus uptake rates, especially during the pre-heading period of development.

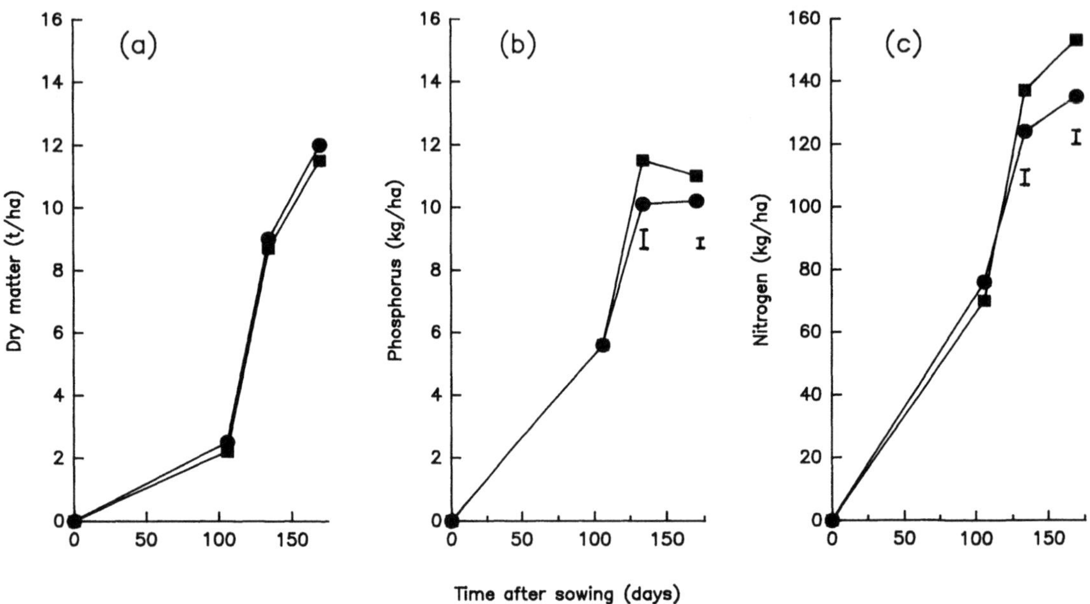

Fig. 1. Accumulation of dry matter (**a**), phosphorus (**b**), and nitrogen (**c**) in the shoots of two wheat genotypes (■ Kite and ● Egret) grown under dryland conditions. Days after sowing 106, 134 and 170 correspond to the growth stages of early stem elongation, anthesis and maturity respectively. Vertical *bars* indicate standard errors of the difference between means where differences were significant (Batten and Khan, 1987a).

Selection pressure for phosphorus efficiency has been imposed consciously in the American tropics. There, under low input technology, wheats have been selected which can produce economic yields on acid, infertile Oxisols and Ultisols (Salinas and Sanchez, 1976; Sanchez and Salinas, 1981).

Grain yield per unit of phosphorus contained in the shoots of the plant at harvest

This is referred to as the phosphorus efficiency ratio (PE) and is expressed in terms of kg grain per kg phosphorus in shoots for plants grown in the field or in terms of g grain per g phosphorus in shoots for pot experiments.

The phosphorus efficiency ratio attempts to describe the utilization of phosphorus, which a plant extracts from soil and fertilizer sources, to produce grain. The PE ratios reported for plants grown in the field under dryland conditions range from about 250–565 kg grain kg P^{-1} in shoots (Batten and Khan, 1987a; Piper and de Vries, 1964; Smith, 1965). Plants grown under controlled environmental conditions have PE ratios ranging from as low as 40 to as high as 715 g grain g P^{-1} in shoots (Batten, 1986a; Jessop et al., 1984; Lipsett, 1964).

Higher PE ratios are usually obtained at lower rates of applied phosphorus (Batten, 1986a; Batten and Khan, 1987a, Table 1; Jones et al., 1989), but within a phosphorus rate treatment the PE ratio is correlated with the ratio of grain to total shoot dry weight – the harvest index ratio (Batten, 1986a; Batten and Khan, 1987a, Table 2). These studies showed that at a constant harvest index there were significant differences in PE between some genotypes which indicates that it would be possible to select for this type of phosphorus efficiency.

Grain yield per unit of phosphorus translocated to the grain

This term for the concentration of phosphorus in the grain is expressed as mg phosphorus per gram dry matter or as a percentage.

The concentration of phosphorus in the harvested grain is most important for several reasons. It indicates the amount of phosphorus used to produce a kg of grain. There are economic reasons for seeking genotypes which achieve high yields of grain with a low concentration of phosphorus. These would remove less phosphorus from the soil and so reduce the cost to produce each tonne of grain. Whilst protein has an overriding influence on the volume of a loaf of bread there is a negative association with grain phosphorus (Bequette et al., 1963). In addition, if grain phosphorus is reduced the grain will contain a lower concentration of the anti-nutritional factor phytic acid (Batten, 1986b).

Table 2. Correlations (r) between dry matter harvest index (HI) and phosphorus efficiency ratio (PE), harvest index for phosphorus (HIP), and grain phosphorus concentration (%P) in wheats grown under dryland conditions at three levels of applied phosphorus

		Applied P (kg ha^{-1})		
		2	6	18
PE	1978	0.33	0.38	0.21
	1979	0.40	0.49	0.58
HIP	1978	0.24	0.50	0.42
	1979	0.62	0.40	0.50
%P	1978	−0.28	−0.18	0.04
	1979	−0.09	−0.35	−0.40

r values >0.205 are significant at $p < 0.05$. From Batten and Khan (1987a).

But there may also be advantages if grain phosphorus is increased. Although it has been suggested that a higher phosphorus content in grain may provide safer or healthier products for humans, due to the ability of phytate to inhibit aflotoxin production and impart other benefits (Graf 1983), this advantage of high phosphorus is unlikely to be of great importance in human nutrition. An increase in grain phosphorus may also be associated with more vigorous seedlings and a higher grain yield if the seed is used to grow the following crop (Bolland, 1988; DeMarco, 1990). Each of these affects of high grain phosphorus requires additional study.

Glasshouse studies by Batten (1986a) indicated that the concentration of phosphorus in the grain of a genetically-diverse range of wheats, which included diploids, tetrapolids and hexaploids, was negatively correlated to the distribution of dry matter between grain and straw in the plant (Fig. 2). Harvest index has also been shown to be negatively correlated with grain

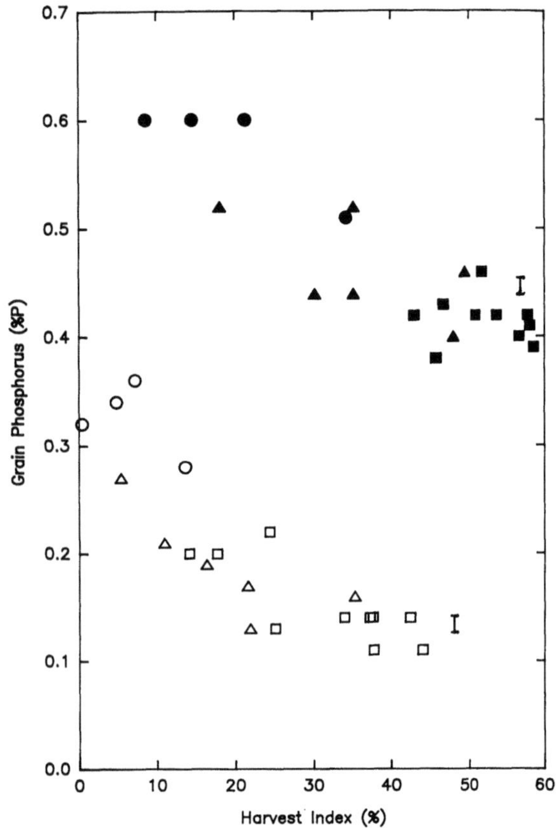

Fig. 2. Relation between harvest index (100 · dry weight grain/grain + straw) and grain phosphorus concentration for diploid (●, ○), tetraploid (▲, △), and hexaploid (■, □) wheats grown with a control (solid symbols) and a low (open symbols) phosphorus regime. Vertical bars indicate the standard error of the difference between means within a phosphorus regime (Batten, 1986a).

phosphorus concentration in plants grown in the field (Batten and Khan, 1987a) although the strength of the relations varied between seasons and the level of applied phosphate fertilizer (Table 2).

These limited data from glasshouse and fields studies indicate that higher or lower grain phosphorus concentrations can be achieved in wheat grains if adequate selection pressure is applied.

The proportion of shoot phosphorus in the grain at maturity or phosphorus harvest index (PHI)

Wheat plants accumulate as much of the 90 per cent of the phosphorus in shoots into the grain. This measure of phosphorus efficiency is, by nature of the values used in its calculation, positively correlated with the yield of grain and highly correlated with the harvest index. Such correlations have been found for plants grown in pots (Batten, 1986a) and in the field (Batten and Khan, 1987a). The proportion of phosphorus in the grain of mature wheat plants ranges from less than 30%, for plants which produce low yields, to 90% for plants grown under favourable conditions (Batten, 1986a; Batten and Khan, 1987a; Jones et al., 1989).

Jessop et al. (1984) suggested that plants with a low PHI ratio would yield straw which is more valuable to livestock and more easily decomposed. A low PHI could also be desirable in phosphorus-deficient plants if the retention of phosphorus in photosynthetic tissues prolonged leaf green area duration and so increased the supply of assimilate available for grain development. Evidence reviewed by Batten and Wardlaw (1987) suggests that prolonging the leaf area duration alone may not increase grain yield and recent studies by Batten and Slack (1990) indicate that a reduction in the supply of phosphorus to developing grains results in smaller grains. It therefore appears that attempts to retain phosphorus in vegetative tissues may be counterproductive.

Phosphorus uptake per unit of root dry weight or root efficiency ratio (RE)

Few workers have reported RE ratios for cereals. Jones et al. (1989) found that RE ratios varied from 5.0 to 10.6 mg P g^{-1} among plants grown with a 2 kg P ha^{-1}. Taller standard genotypes were found to take up more phosphorus on average than semi-dwarf genotypes. The RE ratio ranged from 9.7 to 21.2 mg P g^{-1} for plants grown with 40 kg P ha^{-1} but no significant differences were detected between tall and semi-dwarf genotypes.

Other measures of root phosphorus efficiency are discussed by other authors in these proceedings.

Choosing a selection technique

The definitions of phosphorus efficiency described above may be used to select genotypes which are more phosphorus efficient for particular situations. In each of the studies above only a limited number of genotypes were compared. This indicates the large amount of work which is required to compare genotypes at a range of phosphorus levels.

Unconscious selection of wheats which use phosphorus more efficiently is likely to occur in breeding programs which achieve higher yields. This is more likely if the yield advantage is due to an increase in the harvest index ratio as this is positively correlated with yield per unit of shoot phosphorus and negatively correlated with grain phosphorus concentration.

Where genotypes are selected for yield at sub-optimal level of applied phosphorus it may not be clear if the plant has achieved a high yield because of an efficient phosphorus uptake mechanism or by efficient internal utilization of phosphorus. These can only be ascertained by chemical analysis. In most breeding programs chemical analysis is regarded as being prohibitively expensive.

Progress in the identification and conscious selection of efficient genotypes will require a dedicated effort over many seasons and involve the analysis of grain and straw samples. The analysis of these samples cannot be achieved effectively using classical methods. However samples could now be analysed using near infrared reflectance spectroscopy (Williams and Norris, 1989; McClure, 1984).

Grain samples are already being tested for nitrogen in most breeding programs using near infrared reflectance spectroscopy. These samples could be tested for other nutrients without additional expense to screen genotypes for phosphorus uptake and partitioning. Near infrared reflectance is being used to measure nitrogen in wheat crops in eastern Australia at the tillering to early stem elongation stages of growth. Calibrations are being developed to determine phosphorus simultaneously on the same samples (Batten and Blakeney, unpubl. data) and these could be used to assess the amount of phosphorus taken up by different genotypes in the early stages of crop development.

It is impractical to compare a large number of genotypes at more than one level of applied phosphate in each of several seasons in order to assess the phosphorus efficiency of potential new cultivars. Not only is the cost of field work prohibitive, but there are a number of confounding environmental factors. The uptake and utilization of phosphorus in the field are influenced by genotype and the factors which affect water use efficiency (French and Schultz, 1984) including soil temperature (Barrow, 1974), and planting date (Batten and Khan, 1987b; Smith and Batten, 1976).

Table 3. Effect of phosphorus supplied to cultured ears during grain development on the growth of central spikelet grains of 11 wheat genotypes

Genotype	Grains (ear^{-1})	Kernel weight (mg)		
		nil-P	Control[a]	nil%control
R14137	19.4	15.8	19.4	82
AUS1490	17.9	20.5	30.8	68
AUS1408	18.2	22.9	28.3	81
Kenya59	20.3	14.3	16.3	88
Kenya321	24.5	20.8	30.1	69
Suneca	22.4	15.6	19.1	82
Halberd	24.8	15.0	18.3	82
Cook	21.3	14.4	19.9	73
Sonora64A	18.7	29.1	33.9	86
SUN56A	15.0	38.1	42.7	89
Dollarbird	25.3	16.0	26.6	60
se	2.5	2.0[b]	2.0[b]	5.4

[a]A solution containing phosphorus at 2.27 m*M*.
[b]Standard error for the interaction = 1.95.

To avoid the confounding effects of the environment and management on the selection of genotypes studies may be conducted in pots or solution culture. To study the role of phosphorus in the developing grains and to limit the supply of phosphorus to the grain from vegetative tissues it is necessary to culture the ears or grains of the plant *in vitro*. Present studies by Batten (unpubl. data) using the head culture technique indicate that there are genotypic differences in the requirement for external phosphorus by developing wheat grains (Table 3).

Conclusions

More phosphorus-efficient wheat genotypes can and will be found either by conscious or unconscious selection pressure. There have been relatively few studies of the phosphorus utilization of wheat genotypes possibly because of the amount of work required to fully describe phosphorus uptake and utilization. However new analytical techniques such as near infrared reflectance spectroscopy will facilitate studies in the field and culturing individual heads in growth media will facilitate studies under controlled conditions.

References

Barrow N J 1974 Factors affecting the long-term effectiveness of phosphate and molybdenum fertilizer. Comm. Soil Sci. Plant Anal. 5, 355–387.

Batten G D 1986a The uptake and utilization of phosphorus and nitrogen by diploid, tetraploid and hexaploid wheats (*Triticum* spp). Ann. Bot. 58, 49–59.

Batten G D 1986b Phosphorus fractions in the grain of diploid, tetraploid, and hexaploid wheat grown with contrasting phosphorus supplies. Cereal Chem. 63, 384–387.

Batten G D, Khan M A and Cullis B R 1984 Yield responses by modern wheat genotypes to phosphate fertilizer and their implications for breeding. Euphytica 33, 81–89.

Batten G D and Khan M A 1987a Uptake and utilization of phosphorus and nitrogen by bread wheats grown under natural rainfall. Aust. J. Exp. Agric. 27, 405–410.

Batten G D and Khan M A 1987b Effect of time of sowing on grain yield, and nutrient uptake of wheats with contrasting phenology. Aust. J. Agric. Res 27, 881–887.

Batten G D and Slack K 1990 Grain development in wheat (*Triticum aestivum*) ears cultured in media with different concentrations of phosphorus and sucrose. *In* Plant Nutrition – Physiology and Applications. Ed, M L van Beusichem. pp 185–187. Kluwer Academic Publishers, Dordrecht, The Netherlands.

Batten G D and Wardlaw I F 1987 Senescence of the flag leaf and grain yield following late foliar and root applications of phosphate on plants or differing phosphate status. J. Plant Nut. 10, 735–748.

Bequette R K, Watson C A, Miller B S, Johnson J A and Shrenk W G 1963 Mineral composition of gluten, starch and water-soluble fractions of wheat flour and its relationship to flour quality. Agron. J. 55, 537–542.

Bolland M D A and Baker M J 1988 High phosphorus concentrations in seed of wheat and annual medic are related to higher rates of dry matter production of seedlings and plants. Aust. J. Exp. Agric. 28, 765–770.

Constant K M and Sheldrick W F 1991 An outlook for fertilizer demand, supply, and trade, 1988/89–1993/94. World Bank Technical Paper No. 137, Asia Technical Department Series. The World Bank, Washington, DC.

DeMarco D G 1990 Early growth of wheat seedlings as affected by seed weight, seed phosphorus and seed nitrogen. Aust. J. Exp. Agric. 30, 545–547.

French R J and Schultz J E 1984 Water use efficiency of wheat in a Mediterranean-type environment. II. Some limitations to efficiency. Aust. J. Agric. Res. 35, 765–775.

Graf E 1983 Applications of phytic acid. JAOCS 60, 1861–67.

Jessop R S and Palmer B 1976 Seasonal dependence of wheat variety response to superphosphate. J. Agric. Sci., Camb. 87, 307–314.

Jessop R S, Jones G P and Blair G J 1984 Performance of 22 wheat varieties under low and high phosphorus conditions and implications for the selection of P-efficient varieties. *In* Proc. 3rd Int. Cong. Phos. Compounds, Brussels, 4–6 Oct. 1983. pp 445–454.

Jones G P D, Blair G J and Jessop R S 1989 Phosphorus efficiency in wheat – a useful selection criterion?. Field Crops Res. 21, 257–264.

Lipsett J 1964 The phosphorus content and yield of grain of different wheat cultivars in relation to phosphorus deficiency. Aust. J. Agric. Res. 15, 1–8.

McClure W F 1984 Status of NIR in the tobacco industry. *In* NIR-84. Eds. D Miskelly, D P Law and T Clucus. pp 127–133. Cereal Chem. Divn., Royal Aust. Chem. Inst., Parkville, Vic.

Piper C S and de Vries M P C 1964 The residual value of superphosphate on a red-brown earth soil in South Australia. Aust. J. Agric. Res. 15, 234–272.

Salinas J G and Sanchez P A 1976 Soil-plant relationships affecting varietal and species differences in tolerance to low available soil phosphorus. Cien. Cult. 28, 156–168.

Sanchez P A and Salinas J G 1981 Low-input technology for managing Oxisol and Ultisols in tropical America. Adv. Agron. 34, 279–406.

Smith A N 1965 The influence of superphosphate fertilizer on the yield and uptake of phosphorus by wheat. Aust. J. Exp. Agric. Anim. Husb. 15, 152–157.

Smith A N and Batten G D 1976 The efficiency, accuracy and reliability of soil testing for phosphorus in wheat: A study in the south west wheat belt. Tech. Bull. No. 11, Dept. Agric. NSW.

Strangel P J and Von Uexkull 1990 Regional food security: Demographic and geographic implications. *In* Phosphorus Requirements for Sustainable Agriculture in Asia and Oceania. Ed. S J Banta. pp 21–43. International Rice Research Institute, Manila, Philippines.

Williams P and Norris K 1987 Eds. Near-Infrared Technology in the Agricultural and Food Industries. Am. Assoc. Cereal Chemists Inc., St. Paul, MN. 330 p.

P. J. Randall et al. (Eds.), Genetic aspects of plant mineral nutrition, 221–231.

PLSO SV51

The involvement of mycorrhizas in assessment of genetically dependent efficiency of nutrient uptake and use

S. E. SMITH[1], A. D. ROBSON[2] and L. K. ABBOTT[2]
[1]*Department of Soil Science, Waite Institute, University of Adelaide, South Australia and* [2]*Department of Soil Science and Plant Nutrition, School of Agriculture, University of Western Australia, Nedlands, Western Australia*

Key words: efficiency, genetic control, mycorrhiza, nutrition, phosphate, vesicular-arbuscular

Abstract

This article summarises the way in which mycorrhizal infection of roots affects the mineral nutrition of plants and how the symbiosis may interact with the evaluation of efficiency of nutrient uptake and use by plants. A brief account of the processes of infection and the way they are affected by host genotype and environmental conditions is given and the relationships between this and mineral nutrition (especially phosphate nutrition) are outlined.

The interactions between mycorrhizal infection and P efficiency are considered at two levels. Mycorrhizas may act as general modifiers of efficiency regardless of the extent to which the plants are infected and in some mycorrhiza-dependent plants infection may change the ranking of genotypes. The extent of infection is also under genetic control and shows considerable variability between genotypes in some species. This variation could be used in programs to select varieties in which infection is rapid and nutrient uptake from nutrient deficient or low input systems is, in consequence, increased.

Introduction

Plant species differ in the efficiency with which they acquire and utilise nutrients. Efficiency of uptake is frequently defined in terms of total uptake (mg $plant^{-1}$), inflow or specific uptake rate (mol cm^{-1} s^{-1} or mol g^{-1} s^{-1}) and efficiency of use either as dry matter production per unit of nutrient absorbed (mg dw mg^{-1}) or per unit nutrient applied to the soil (mg dw mg^{-1}). Growth analysis techniques have also been advocated as a means of comparing different genotypes or the effects of different environmental conditions, with respect to relative growth rate and root weight ratio (see Baas and van Beusichem 1990; Poorter and Lewis 1986). This may overcome problems of differences in seed weight, size and nutrient content between cultivars. Furthermore, the experiments may need to be conducted over a range of different nutrient availabilities and at a number of harvests in order to unravel complex interactions. The methodological approach adopted will depend upon the particular aims and scale of the investigation and whether it is to be carried out in the field or in pots.

Screening programs designed to select genotypes and identify the genes involved in the processes contributing to efficiency must both target quantitatively important processes and take into account environmental factors which may modify them. One biological factor which has the potential to affect efficiency of a number of nutrients is the formation of mycorrhizas. Mycorrhizas, which develop on around 90% of plant species including most major crops, are well known to affect both uptake and accumulation of nutrients and may therefore act as important biological factors contributing to efficiency of both nutrient uptake and use.

The interactions between fungus and plant are

complex at both the developmental and physiological levels and are under genetic control, being affected by plant species (or cultivar) as well as environment. This review will consider the ways in which vesicular-arbuscular (VA) mycorrhizas may be involved in plant nutrition and why it is important to consider them in genetic studies of mineral nutrition. The importance of particular experimental approaches in mycorrhizal studies will be highlighted.

Mycorrhizal development and function: a brief overview

Development

Typical infection is initiated as appressoria on the epidermis of the roots and from these inter- and intracellular hyphae colonise the root cortex. Branches from the intercellular hyphae penetrate the cortical parenchyma cells and branch profusely, giving rise to intracellular arbuscules, which are thought to be the main sites of nutrient exchange between the symbionts. As the infections age, swollen storage vesicles containing lipid are produced. Concurrently with the intraradical colonisation, extensive external mycelium develops in soil and plays an important role in nutrient acquisition (see Harley and Smith, 1983; Smith and Gianinazzi-Pearson, 1988 for further details).

The development of VA mycorrhizal infection is under control of plant genes. Some plant species (notably members of the Cruciferae, Proteaceae and *Lupinus* spp) never form mycorrhizas, while in others (e.g. Gentianaceae) the way in which the infection develops and the structure (and possibly function) of the mycorrhiza differs from the most commonly described form. However, the majority of plant species (including a large number of major crops) do form typical mycorrhizas and, in contrast to pathogenic associations, these mutualistic symbioses are non-specific. A single plant species can associate with most (if not all) known VA mycorrhizal fungi, while a single fungal isolate can infect a vast array of host species, including pteridophytes, gymnosperms and angiosperms.

The symbiotic interaction must involve genetically controlled recognition events and signals which lead to the development of characteristic structures in different plant tissues and to the physiological integration of the symbionts. However, resistance mechanisms leading to close specificity between particular taxa of plants and fungi (such as the gene-for-gene system in biotrophic plant pathogens) have not been identified and would not be expected in mutualistic associations (see Harley and Smith, 1983). The broad specificity and widespread occurrence of mycorrhizal fungi in soil means that mycorrhizas (not uninfected roots) are the typical nutrient-absorbing organs of most plants.

The extent to which typical VA mycorrhizal fungi colonise root systems (usually measured as the % or fraction of the root system infected) varies with species of plant. Differences in the extent of infection have also been observed between genotypes of the same species (see below), although the differences have not as yet been related to specific genes.

Recently, mutant plants derived from species which normally form mycorrhizas (*Pisum*, *Vicia* and *Medicago*) have been identified which either do not form mycorrhizas or which form atypical mycorrhizas (Bradbury et al., 1991; Duc et al., 1989; see Gianinazzi, 1991). In some cases, only appressoria are formed and the subsequent development of the mycorrhiza does not occur, so that intercellular hyphae and arbuscules are absent. Preliminary work indicates that at least three host genes are involved in the control of these early stages of infection. The mutations are recessive and at present they have only been identified in legumes, where the genes appear to be closely linked to nodulation genes (see Duc et al., 1989). This raises the possibility that symbiotic function may involve some common genes. We can expect rapid progress in the field of genetic control of mycorrhiza formation and increased understanding will lead to the possibility of including mycorrhiza development in breeding programs which will be of particular relevance to mineral nutrition.

The extent of mycorrhizal infection in root systems is also influenced by environmental conditions. In the context of this paper the most important of these are the age of the plants, the level of phosphate (P) in the soil relative to the requirements of the plant and the capacity of the population of mycorrhizal propagules in the soil to form mycorrhizas. The fraction of the root system infected increases with time sigmoidally, and measurement at a number of harvests may be required for proper comparison of infection in different genotypes. Percentage infection is frequently inversely related to the P supply (Fig. 1), although small additions of P to very deficient soil may increase infection (see Bolan et al., 1984). The effects may be mediated via growth of both roots and fungus and the responses vary in different plant species. Other environmental factors, such as photon irradiance, also influence the infection processes.

In some plants high soil P may effectively eliminate infection, whereas in others infection is maintained over a wide range of soil P (e.g. Baon et al., 1992; Bolan et al., 1987). The interaction between P supply and infection is therefore important in the interpretation of data obtained from P response curves with mycorrhizal and non-mycorrhizal plants.

The infectivity of mycorrhizal propagules in the soil can also have a marked effect on the rate at which infections are initiated on the root and hence the rate at which the cortex is colonised and the extent to which mycelium develops in the soil (e.g. Baon et al., 1992; Smith and Walker, 1981; and see Fig. 1).

As both the rate and extent of mycorrhizal development affect P uptake in young plants (see below), soil infectivity is an important consideration in screening programs and in investigations of the interactions between mycorrhizal infection and mineral nutrition. Most studies of genetic variation in colonisation have involved relatively mature plants, so that comparisons have been made of the maximum extent of infection, rather than its rate of development. In mature plants death of the root cortex may result in very low ratings for mycorrhizal infection which do not necessarily reflect the earlier history of the infection (see Young et al., 1983). Early infection is very important, because it results in rapid uptake of nutrients during the period of high demand immediately following germination. Future studies should consider the rates of development of mycorrhizas within the roots as a characteristic which may be influenced by plant genotype.

Many plant breeding programs (other than those specifically related to mineral nutrition) have utilised high levels of soil nutrients. Under these conditions mycorrhizal infection and its effects on host growth are likely to be minimal and the potential for selection of genotypes which both become rapidly infected and which respond positively to infection has not only been ignored, but actually mitigated against (but see Krishna et al., 1985). In breeding and selection programs mycorrhiza development not only has potential as a beneficial characteristic, but may also introduce significant problems in the interpretation of results.

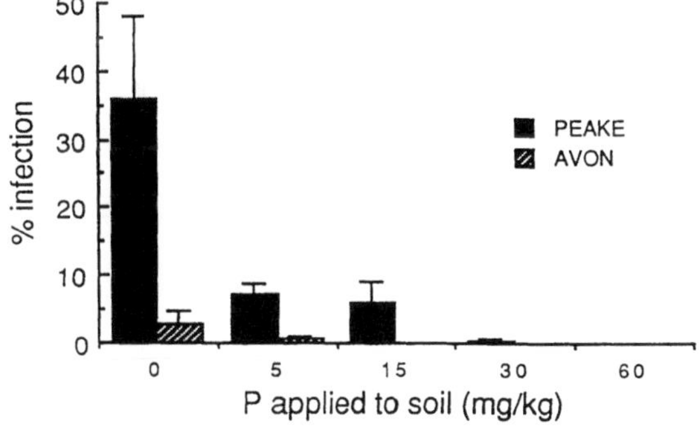

Fig. 1. The effect of applied phosphate on mycorrhizal infection in *Triticum aestivum* cv. Spear grown in two soils with different propagule densities (Most Probable Numbers estimates). Solid bars Peake soil; hatched bars Avon soil. The two soils contained 26 and 0.9 propagules g^{-1}, respectively. Means and standard errors of means of three replicate plants. Results of Baon et al., 1992.

Plant nutrition

Mycorrhizal infection increases both the uptake of nutrients by roots and the concentration of nutrients in the plant tissues. The magnitude of the latter effect will depend on to what extent growth of the plant is limited by nutrient availability. At low soil nutrient levels a marked growth response can be expected, which may not be apparent when nutrients are readily available, although increases in the rates of nutrient uptake may still take place (see Smith and Gianinazzi-Pearson, 1988). These complex and interrelated mycorrhizal effects will influence the determination of the efficiencies of nutrient uptake and utilisation.

It is important to determine the relative contributions of plant and mycorrhizal fungus, so that breeding programs can target the organisms and processes that make the largest contribution to nutrition. It must be stressed that the mycorrhizal contribution will vary from one plant species to another and from one soil to another.

The nutrient most influenced by mycorrhizal infection is P, but mycorrhizas can also increase the uptake of other nutrients both directly (e.g. Zn, Cu and possibly N) and indirectly (e.g. N, cation/anion ratio, etc) (see Abbott and Robson, 1984; Harley and Smith, 1983). This discussion will concentrate on P, for which most information is available.

Uptake

The factors affecting P uptake and utilisation by plants have been reviewed on numerous occasions (e.g. Barber, 1984; Clarkson, 1985; Koide, 1991). It is clear that rate of root growth (or root length) and root radius, which affect the volume of soil exploited, are more important than kinetic characteristics of the uptake processes (Km and Vmax). Many of the non-mycorrhizal species have evolved strategies which enhance nutrient acquisition, including the formation of cluster roots (e.g. Gardner et al., 1982; Lamont, 1982), long or numerous root hairs (Allen et al., 1989; Itoh and Barber, 1983) and production of acids, enzymes or chelating agents (Hedley et al., 1982; McLachlan et al., 1976). The formation of an extensive network of mycorrhizal mycelium in soil also increases the volume of soil accessible to plants and absorption by this mycelium effectively depletes soil P (Kothari et al., 1991) and increases the inflow or specific uptake rate of nutrients by infected roots. Furthermore, species or genotypes which are dependent on mycorrhizas for adequate nutrient acquisition are more likely to have root systems with poor exploitation efficiency (e.g. Azcon and Ocampo, 1981; Bolan et al., 1987; Hall, 1978; Manjunath and Habte, 1991), but this correlation is not always observed (e.g. Crush and Caradus, 1980; Rajapakse and Miller, 1988).

In mycorrhiza-dependent plants, infection increases the efficiency with which roots absorb both P and Zn from soil (see Harley and Smith, 1983). Mycorrhizal involvement in nutrient uptake may well explain some discrepancies between predicted and observed uptake of P from low-P soils (Stribley et al., 1980a). Effective prediction of the potential benefits of P uptake via mycorrhizas in field situations is a prerequisite for the development of management strategies and breeding programs for maximising the efficiency of the symbiosis (see Abbott and Robson, 1991).

The fungal contribution to uptake can be estimated by comparing the inflow (uptake per unit length of root per unit time) in mycorrhizal and non-mycorrhizal plants. It is frequently greatest in young plants growing in severely phosphate-deficient soil. However, even when soil P is adequate for the growth of non-mycorrhizal plants (so that a mycorrhizal growth response does not occur, see Figs. 2a and 3), there may still be a hyphal contribution to P uptake as long as infection persists in the root system (Smith et al., 1986; and see Smith and Gianinazzi-Pearson, 1988). Inflow can be increased as much as nine-fold, although values of three- to four-fold are more usual (Sanders et al., 1977; Smith et al., 1986). This is a significant contribution to uptake, which should not be ignored in genetic studies.

Utilisation

Increased efficiency of P uptake by mycorrhizal plants frequently leads to higher critical concentrations of P in the tissues (Fig. 2b). This may reflect luxury accumulation in young plants and may also be a consequence of carbon (rather than P) limitation (Stribley et al., 1980b; Son and Smith, 1988). Rating of the efficiency of P use based on relationships between shoot weight and P concentration in the shoots (i.e. determination of critical concentration for 90% of maximum shoot weight) may therefore be affected by mycorrhizal colonisation if the plants are harvested before the enhanced phosphate uptake is reflected in increased plant growth or if C drain to the fungus affects dry matter production.

Plant growth

In P-deficient soil the development of mycorrhizal infection (which will be maximal in that situation) and consequent increase in P uptake results in relief of P stress, increased photosynthetic rate and increased plant growth. The magnitude of

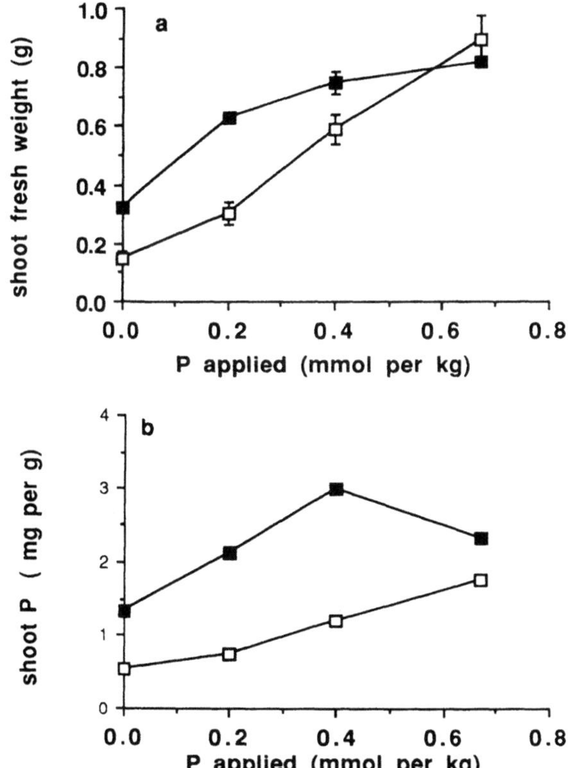

Fig. 2. The response of *Trifolium subterraneum* cv Mt Barker to applied P in the presence and absence of mycorrhizal infection. **a**) shoot fresh weight, **b**) shoot P concentration. Solid symbols, mycorrhizal plants; open symbols non-mycorrhizal plants. The percentage of the root length infected ranged from 74 (no P addition) to 53 (0.67 mmol P kg^{-1}) in the mycorrhizal plants. Means and standard errors of means of three replicate plants. Data from Oliver et al., 1983.

the growth response depends both on the P supply, the rate of P uptake and P requirement of the plant (see Fig. 2). Both these factors contribute to the mycorrhizal dependency of the plant (see Koide, 1991). Variation in mycorrhiza-dependency has been observed both between and within species (Estaun et al., 1987; Hall, 1977; Krikun and Levy, 1980; Lambert et al., 1980; Manjunath and Habte, 1991; Menge et al., 1978; Ollivier et al., 1983) and although the physiological bases for the differences are still not altogether clear, it is an area of importance in genetic studies. Mycorrhiza development as a strategy for nutrient acquisition appears less important when alternative strategies (see above) are well developed. However, many plants have plastic responses to changes in environmental conditions, showing alterations in such characteristics as rate of root growth, root : shoot ratio and contribution of root hairs to uptake when

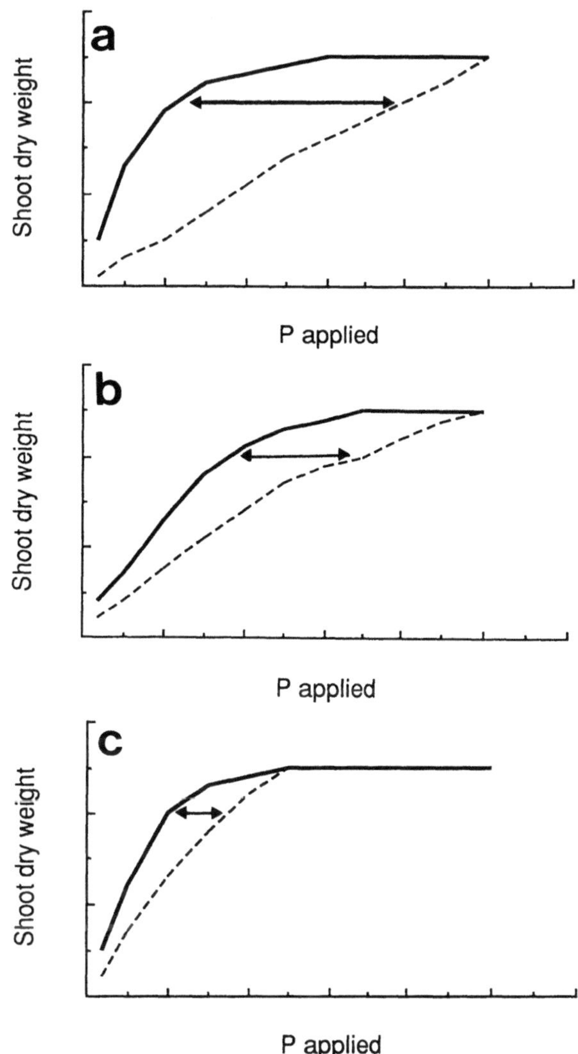

Fig. 3. Generalised phosphate response curves for plants of different types, illustrating the effect of mycorrhizal infection on the external P requirements of the plants to achieve maximum growth. Solid lines, mycorrhizal plants; dashed lines, non-mycorrhizal plants. **a**) Non-mycorrhizal plants have a high P requirement, which is greatly reduced in mycorrhizal plants. **b**) Non-mycorrhizal plants have a high P requirement which is only slightly modified following mycorrhizal infection and **c**) Non-mycorrhizal plants have a low P requirement. In some species mycorrhizal infection has little effect on this requirement, in others it may be more significant. The horizontal lines indicate the differences in P requirement associated with mycorrhizal infection ('horizontal' comparison).

the plants are mycorrhizal. This an area of research which has attracted considerable interest, but in which much care is necessary in the design of experiments involving species or cultivars with different rooting characteristics. The data need to be evaluated in terms of the effects of a particular characteristic on nutrient uptake or response to nutrient supply, rather than on shoot growth. In P response curves this means making "horizontal" comparisons in terms of effectiveness of P applied, rather than "vertical" comparisons of plant growth (see Fig. 3). Alternatively, actual measurements of P uptake (as inflow or specific uptake rate) will give a direct measure of effect of root characteristics or mycorrhiza development. Again, the differential "control" of infection by plant genotype is not well understood, but could be mediated via carbohydrate supply in conjunction with modifications in membrane transport processes (see Azcon and Ocampo, 1981; Schwab et al., 1991; Smith and Smith, 1990; Thomson et al., 1986). The fungal partner in the symbiosis is completely dependent on the plant for supplies of carbohydrate, but under conditions of nutrient deficiency the increased carbon demand is offset by the advantages of increased P uptake. When P supply is non-limiting a growth response to mycorrhizal infection is usually not observed, although other effects on the physiology of the plants such as increased P uptake and P concentration in tissues and decreased root : shoot ratio may be apparent (see Smith and Gianinazzi-Pearson, 1988).

Mycorrhizal interactions with genetically determined phosphorus efficiency

The potential interactions of mycorrhizas in the determination of P efficiency operate both with respect to uptake (positive effects) and utilisation (negative effects).

The involvement of mycorrhizas can be considered at two levels: 1) As general modifiers of P efficiency which will interact with plant determinants of efficiency and may complicate (and under some circumstances confound) the interpretation of experiments and selection programs so that effects mediated by plant genes may be masked and 2) As modifiers of efficiency under the control of specific plant genes, which may be identified in future research and included in specific plant selection or breeding programs with the aim of increasing the capacity of plants to absorb nutrients from deficient soils.

Mycorrhiza development as a general modifier of nutrition

The account of mycorrhizas given above makes it clear that their development can have significant effects on the efficiency with which plant roots absorb nutrients from non-sterile soil and that the contribution of mycorrhizas may not be constant, even within a single experiment. The complex interactions have usually been ignored. For example, Schenck and Barber (1979) compared actual P uptake by a number of maize cultivars with uptake predicted by a model. The model matched P uptake from high-P soil well, but underestimated uptake from low-P soil by a factor of 2. Neither root hairs nor mycorrhizas were included in the model, but mycorrhizal infection is well known to affect maize growth (e.g. Hall, 1978) and could have accounted for the discrepancy.

The extent of infection will also vary with infectivity of the soil, so that in trials conducted at different sites the plants may be differently infected, or not infected at all (Baon et al., 1992). Effects of P on the extent of infection will mean that the plants may be differently infected at different parts of a P response curve and hence the factors determining efficiency of P uptake will also be different, as will the mycorrhizal contribution to yield.

Several investigations have shown that differences between plant genotypes in response to P were modified when the plants were mycorrhizal. Lambert et al. (1980) found that 6 cultivars of *Medicago sativa* did not differ in their response to P when non-mycorrhizal, but were quite variable when they developed mycorrhizas in non-sterile soil, suggesting that mycorrhizal interaction with plant genotype may be important. At low P all yield responses to mycorrhizas were positive, while at high P, negative as well as positive responses were observed (see also Berthau et al., 1980). The application of P did extent to which the roots were infected. Other workers have also found that mycorrhizal infection altered the ranking of maize cultivars for P efficiency (see Hall, 1977; Toth et al., 1984) and that cultivars of *Pisum* displayed a full range of responses to P and/or mycorrhizal infection (positive to both or positive to either P or infection, Estaun et al., 1987).

Plant breeders and others interested in plant P efficiency should consider:

1) Whether the species in question forms mycorrhizas. Crucifers, including *Arabidopsis*, do not and therefore work on these plants will not be complicated by mycorrhizal interactions.

2) For a mycorrhizal species, the magnitude of its mycorrhiza-dependency. Some species are strongly or obligately dependent, even in high-P soils (e.g. *Stylosanthes*, *Cassava*, oil palm; see Fig. 3a) and in these it may be very important to consider the potential of mycorrhizas. Others species (although infected) respond little to mycorrhizas (e.g. *Lolium*, Fig. 3c) and infection will not be a significant factor. The majority of plants are intermediate between these two extremes and experiments will be required to quantify their response to mycorrhizas over a range of soil P levels and to determine whether mycorrhizas make a quantitatively significant contribution, compared with other factors influencing nutrient uptake and utilisation. (see Fig. 3b).

3) The infectivity of the soil and differences between trial sites in this respect. This will affect the rate and extent of infection and hence the magnitude of response in a dependent species and possibly the ranking of cultivars at different sites.

4) The effects of nutrient application (especially P) on the rate and extent of mycorrhizal infection in the species in question.

5) The need to compare effects of mycorrhizas in inefficient as well as efficient genotypes, in order to obtain information on the mechanisms underlying the interactions and their genetic bases.

Mycorrhiza development as a genetically determined character

Several investigations have shown variation between genotypes of a number of species in the % of root length infected by mycorrhizal fungi and, in some investigations, related factors such as spore production and alkaline phosphatase activity (see Table 1). These differences may influence P uptake, utilisation and plant growth, but need to be distinguished from them in gen-

Table 1. Examples of crop species in which variation in the extent of mycorrhizal infection has been demonstrated between genotypes. The number of genotypes screened and criteria used to select them is given in column 2, where appropriate

Species	No. of cultivars	Reference
Medicago	20, half sib	Lackie et al., 1988
Medicago	4, dormancy type	O'Bannon et al., 1980
Vigna	101, diverse	Mercy et al., 1990
Vigna	root morphology	Rajapakse and Miller, 1988
Arachis	15, disease resistance	Heckman and Angle, 1987
Arachis	9, various, non-nod	Kesava Rao et al., 1990
Trifolium	2, P efficiency	Hall, 1977
Citrus	8, root stocks	Levy et al., 1983
Citrus	6, MD root stocks	Menge et al., 1978
Zea	13, disease resistance	Toth et al., 1990
Zea	6, p Use	Toth et al., 1990
Hordeum	7, hulled/hull-less	Tilak and Murthy, 1987
Pennisetum	30, various	Krishna et al., 1985
Triticum	8	Veirheilig and Ocampo, 1991
Triticum	18, agronomic	Berthau et al., 1980
Triticum	22	Sreenivasa and Rajashekhara, 1989
Triticum	13	Azcon and Ocampo, 1981
Triticum	60, commercial/breeder	Young et al., 1983
Triticum	22, high yield	Manske, 1989
Triticum	22, land race	Manske, 1989

etic studies. Experiments are difficult to design to give unequivocal results and few studies have shown any clear relationship between differences in extent of colonisation (related to genotype) and differences in yield or efficiency. For example, Robson and Collins (unpublished) found that differences in P response within *Trifolium subterraneum* were not related to differences in mycorrhizal infection, which was low in both efficient and inefficient genotypes, but they did not compare genotypes in the non-mycorrhizal condition.

It is important to make comparisons on the basis of real differences in fungal colonisation and to distinguish these from differences brought about by environmental factors like rates of root growth. The development of arbuscules (see Alexander et al., 1988; Guillemin, 1989) and other features of infection are also under genetic control and together with such biochemical markers as mycorrhiza-specific alkaline phosphatase, membrane-bound, H^+ ATPase and root acid phosphatase deserve increased attention (e.g. Gianinazzi-Pearson et al., 1991; Kesava Rao et al., 1990; Rubio et al., 1990). The identification of the genes involved in the infection processes and the study of their regulation and expression will be an interesting area of study, leading to a deeper understanding of plant/fungus interactions at the molecular level.

The genetic control of symbiotic function and its relationship to the extent of infection is also important. Data are mainly available for whole plants, but there is a need to target processes at the cellular level, including cell wall and membrane modifications, concentration of carbohydrates in the apoplast, plant "defence" reactions or signals in the symbiotic interface (see Gianinazzi, 1991; Graham et al., 1991; Smith and Smith, 1990) and a measure of "symbiotic function" such as P flux across the symbiotic interface (see Smith and Dickson, 1991).

One aim of understanding the genetic control of mycorrhizal infection is to breed or select plants with maximum symbiotic efficiency for P uptake in low-P soils and possibly in soils where the natural population of mycorrhizal fungi is low. At present it appears that rapid rates of infection and production of numerous arbuscules in young plants are important determinants of efficient uptake (e.g. Abbott and Robson, 1982; Menge, 1984; Smith et al., 1979). If this rapid, early spread of infection is also correlated with a high plateau value for % infection then the latter will be a useful parameter in screening genotypes for variations in mycorrhiza development (see Abbott and Robson, 1991). It remains to be seen whether a "super-mycorrhizal" plant in which infection spreads very rapidly and reaches high values under all environmental conditions, would be efficient with respect to P uptake and plant growth, or whether, as in the supernodulating genotypes of legumes (Carroll et al., 1985), the lack of control of the symbiosis would result in poor yields. Like nodulation, plant shoots appear to be involved in the control of mycorrhizal infection, as shown by grafting experiments (see Gianinazzi, 1991).

Breeding for more efficient mycorrhizal symbioses

The potential benefits of mycorrhiza development in mineral nutrition have been emphasised in this review and it is clear that there is scope for including mycorrhizas in breeding programs aimed at maximising the efficiency of nutrient uptake and use by crops which show genotypic variability in either infection or response to infection. If mycorrhizas are ignored the potential for exploiting this variability will be lost. Elimination of mycorrhizal infection either by inappropriate soil management practices or by breeding crops which become poorly infected will eliminate beneficial effects on soil structure, water relations and disease tolerance, as well as on plant nutrition.

It has been argued that inclusion of mycorrhizas in programs aimed at increasing yields in broad-scale agriculture is premature because inoculation is not feasible, being limited to horticultural production systems at present (see Gianinazzi et al., 1989; 1990). If plant breeding continues to ignore the contribution of mycorrhizas the likelihood is that variability will be lost and the potential for exploiting inoculation or management practices in the future will be reduced.

There is considerable interest in the development of management practices which maintain high soil infectivity and hence promote rapid and extensive infection without inoculation (see Abbott and Robson, 1991). The symbiosis could also be managed via selection of genotypes which are rapidly colonised from low inoculum densities and this could be a further factor in developing substainable cropping systems with efficient use of low fertiliser inputs.

Understanding the genetic control of fungal growth and physiological integration between symbionts may have important long-term outcomes leading to techniques for axenic culture of infective VA mycorrhizal fungi and hence production of inoculum suitable for broad-scale agriculture.

References

Abbott L K and Robson A D 1982 The role of vesicular-arbuscular mycorrhizal fungi in agriculture and the selection of fungi for inoculation. Aust. J. Agric. Res. 33, 389–408.

Abbott L K and Robson A D 1984 The effect of mycorrhizae on plant growth. *In* VA Mycorrhizae. Eds. C L Powell and D J Bagyaraj. pp 113–130. CRC Press, Boca Raton, FL.

Abbott L K and Robson A D 1991 Field management of VA mycorrhizal fungi. *In* The Rhizosphere and Plant Growth. Eds. D L Keister and P B Cregan. pp 355–362. Kluwer Academic Publishers, Dordrecht, The Netherlands.

Alexander T, Meier R, Toth R and Weber H C 1988 Dynamics of arbuscule development and degeneration in mycorrhizas of *Triticum aestivum* L. and *Avena sativa* L. with reference to *Zea mays* L. New Phytol. 110, 363–370.

Allen M F, Allen E B and Friese C F 1989 Responses of the non-mycotrophic plant *Salsola kali* to invasion by vesicular-arbuscular mycorrhizal fungi. New Phytol. 111, 45–49.

Azcon R and Ocampo J A 1981 Factors affecting the vesicular-arbuscular infection and mycorrhizal dependency of thirteen wheat cultivars. New Phytol. 87, 677–685.

Baas R and van Beusichem M L 1990 Genetic differentation in *Plantago major* L. in growth and P uptake under conditions of P limitation. Plant and Soil 123, 185–192.

Baon J B, Smith S E, Alston A M and Wheeler R D 1992 Phosphorus efficiency of three cereals as related to indigenous mycorrhizal infection. Aust. J. Agric. Res. 43, 1–15.

Barber S A 1984 Soil Nutrient Bioavailability. Wiley, New York. 398 p.

Berthau Y, Gianinazzi-Pearson V and Gianinazzi S 1980 Development et expression de l'association endomycorhizienne chez le Ble. I. Mise en evidence d'un effet varietale. Annales d'Amelioration des Plantes 30, 67–78.

Bolan N S, Robson A D and Barrow N J 1984 Increasing phosphorus supply can increase the infection of plant roots by vesicular-arbuscular mycorrhizal fungi. Soil Biol. Biochem. 16, 419–420.

Bolan N S, Robson A D and Barrow N J 1987 Effects of vesicular-arbuscular mycorrhiza on the availability of iron phosphates to plants. Plant and Soil 99, 401–410.

Bradbury S M, Peterson R L and Bowley S R 1991 Interactions between three alfalfa nodulation genotypes and two *Glomus* species. New Phytol. 119, 115–120.

Carroll B J, McNeil D L and Gresshoff P M 1985 A supernodulation and nitrate-tolerant symbiotic (nts) soybean mutant. Plant Physiol. 78, 34–40.

Clarkson D T 1985 Factors affecting mineral nutrient acquisition by plants. Annu. Rev. Plant Physiol. 36, 77–115.

Crush J R and Carradus J R 1980 Effect of mycorrhizas on growth of some white clovers. N. Z. J. Agric. Res. 23, 233–237.

Duc G, Trouvelot A, Gianinazzi-Pearson V and Gianinazzi S 1989 First report of non-mycorrhizal plant mutants (myc-) obtained in pea (*Pisum sativum* L.) and faba bean (*Vicia faba* L.). Plant Science 60, 215–222.

Estaun V, Calvet C and Hayman D S 1987 Influence of plant genotype on mycorrhizal infection: Response of three pea cultivars. Plant and Soil 103, 295–298.

Gardner W K, Parbery D G and Barber D A 1982 The acquisition of phosphorus by *Lupinus albus* L. II. The effect of varying phosphorus supply and soil type on some characteristics of the soil-root interface. Plant and Soil 68, 33–41.

Gianinazzi S 1991 Vesicular arbuscular (endo-) mycorrhizas: Cellular, biochemical and genetic aspects. Agric. Ecosystems Environ. 35, 105–119.

Gianinazzi S, Gianinazzi-Pearson V and Trouvelot A 1989 Potentialities and procedures for the use of endomycorrhizas with special emphasis on high value crops. *In* Biotechnology of Fungi for Improving Plant Growth. Eds. J M Whipps and R D Lumsden. pp 41–54. CUP, Cambridge, UK.

Gianinazzi S, Trouvelot A and Gianinazzi-Pearson V 1990 Conceptual approaches for the rational use of VA endomycorrhizae in agriculture: Possibilities and limitations. Agric. Ecosystems and Environ. 29, 153–161.

Gianinazzi-Pearson V, Smith S E, Gianinazzi S and Smith F A 1991 Enzymatic studies on the metabolism of vesicular-arbuscular mycorrhizas. V. Is H^+-ATPase a component of ATP-hydrolysing enzyme activities in plant-fungus interfaces? New Phytol. 117, 61–74.

Graham J H, Eissenstat D M and Drouillard D L 1991 On the relationship of a plants mycorrhizal dependency and rate of mycorrhizal colonisation. Func. Ecol. 5, 773–779.

Guillemin J P 1989 Mise en evidence chez le pois de nouveaux mutants resistants aux champignons endomycorhizogenes: Analyses genetiques et cellulaires. Diplomes d'Etudes Approfondies, Université de Bourgogne. 25 p.

Hall I R 1978 Effect of vesicular-arbuscular mycorrhizas on two varieties of maize and sweetcorn. N. Z. J. Agric. Res. 21, 517–519.

Hall I R, Scott R S and Johnstone P. D 1977 Effect of vesicular-arbuscular mycorrhizas on response of 'Grass lands Huia' and 'Tamar' white clovers to phosphorus. N. Z. J. Agric. Res. 20, 349–355.

Harley J L and Smith S E 1983 Mycorrhizal Symbiosis. Academic Press, London and New York.

Heckman J R and Angle J S 1987 Variation between soybean cultivars in vesicular-arbuscuar mycorrhiza fungi colonization. Agron. J. 79, 428–430.

Hedley M J, Nye P H and White R E 1982 Plant induced changes in the rhizosphere of rape (*Brassica napus* var. Emerald) seedlings. II. Origin of the pH change. New Phytol. 91, 31–44.

Itoh A and Barber S A 1983 Phosphorus uptake by six plant species as related to root hairs. Agron. J. 75, 457–461.

Kesava-Rao P S, Tilak K V B R and Arunachalam V 1990 Genetic variation for VA mycorrhiza-dependent phosphate mobilisation in groundnut (*Arachis hypogaea* L.). Plant and Soil 122, 137–142.

Koide R T 1991 Nutrient supply, nutrient demand and plant response to mycorrhizal infection. New Phytol. 117, 365–386.

Kothari S K, Marschner H and Römheld V 1991 Contribution of the VA mycorrhizal hyphae in acquisition of phosphorus and zinc by maize grown in a calcareous soil. Plant and Soil, 131, 177–185.

Krishna K R, Shetty K G, Dart P J and Andrews D J 1985 Genotype dependent variation in mycorrhizal colonisation and response to inoculation of pearl millet. Plant and Soil 86, 113–125.

Lackie S M, Bowley S R and Peterson R L 1988 Comparison of colonisation among half-sib families of *Medicago sativa* L. by *Glomus versiforme* (Daniels and Trappe) Berch. New Phytol. 108, 477–482.

Lambert D H, Cole H and Baker D E 1980 Variation in the response of alfalfa clones and cultivars to mycorrhizae and phosphorus. Crop Sci. 20, 615–618.

Lamont B B 1982 Mechanisms for enhancing nutrient uptake in plants, with particular reference to mediterranean, South Africa and Western Australia. Bot. Rev. 48, 597–689.

Levy Y, Dodd J and Krikun J 1983 Effect of irrigation, water salinity and rootstock on the vertical distribution of vesicular-arbuscular mycorrhiza in citrus roots. New Phytol. 95, 397–403.

Manjunath A and Habte M 1991 Root morphological characteristics of host species having distinct mycorrhizal dependency. Can. J. Bot. 69, 671–676.

Manske G G B 1989 Genetical analysis of the efficiency of VA mycorrhiza with spring wheat. Agric. Ecosystems Environ. 29, 273–280.

McLachlan K D 1976 Comparative phosphorus responses in plants to a range of available phosphorus situations. Aust. J. Agric. Res. 27, 323–341.

Menge J 1984 Inoculum production. *In* VA Mycorrhizae. Eds. C L. Powell and D J Bagyaraj, pp 187–201. CRC press, Boca Raton, FL.

Mercy M A, Shivashankar G and Bagyaraj D J 1990 Mycorrhizal colonisation in cowpea is host dependent and heritable. Plant and Soil 121, 292–294.

O'Bannon J H, Evans D W and Peaden R N 1980 Alfalfa varietal response to seven isolates of vesicular-arbuscular mycorrhizal fungi. Can. J. Plant Sci. 60, 859–863.

Oliver A J, Smith S E, Nicholas D J D, Wallace W and Smith F A 1983 Activity of nitrate reductase in *Trifolium subterraneum*: effects of mycorrhizal infection and phosphate nutrition. New Phytol. 94, 63–79.

Poorter H and Lewis C 1986 Testing differences in relative growth rate: A method avoiding curve fitting and pairing. Physiol. Plant. 67, 223–226.

Rajapakse S and Miller J C 1987 Intraspecific variability for V.A. mycorrhizal symbiosis in cowpea (*Vigna unguiculata* [L.] Walp.). *In* Genetic Aspects of Plant Mineral Nutrition. Eds. W H Gabelman and B C Loughman. pp 523–536. Martinus Nijhoff, Dordrecht, The Netherlands.

Rubio R, Moraga E and Borie F 1990 Acid phosphatase activity and vesicular-arbuscular mycorrhizal infection associated with roots of four wheat cultivars. J. Plant Nutr. 13, 585–598.

Sanders F E, Tinker P B, Black R L and Palmerley S M 1977 The development of endomycorrhizal root systems I. Spread of infection and growth promoting effects with four species of vesicular-arbuscular mycorrhizas. New Phytol. 78, 257–268.

Schenk M K and Barber S A 1979 Root charcteristics of corn genotypes as related to phosphorus uptake. Agron. J. 71, 921–924.

Schwab S M, Menge J A and Tinker P B 1991 Regulation of nutrient transfer between host and fungus in vesicular-arbuscular mycorrhizas. New Phytol. 117, 387–398.

Smith S E and Dickson S 1991 Quantification of active vesicular-arbuscular mycorrhizal infection using image analysis and other techniques. Aust. J. Plant Physiol. 18, 637–648.

Smith S E and Gianinazzi-Pearson V 1988 Physiological interactions between symbionts in vesicular-arbuscular mycorrhizal plants. Annu. Rev. Plant Physiol. Mol. Biol. 39, 221–244.

Smith S E and Smith F A 1990 Structure and formation of the interfacaes in biotrophic symbioses as they relate to nutrient transport. New Phytol. 114, 1–38.

Smith S E and Walker N A 1981 A quantitative study of mycorrhizal infection in Trifolium: Separate determination of the rates of infection and of mycelial growth. New Phytol. 89, 225–240.

Smith S E, Nicholas D J D and Smith F A 1979 The effect of early mycorrhizal infection on nodulation and nitrogen fixation in *Trifolium subterraneum*. Aust. J. Plant Physiol. 6, 305–311.

Smith S E, St John B J, Smith F A and Bromley J-L 1986 Effect of mycorrhizal infection on plant growth, nitrogen and phosphorus nutrition of glasshouse-grown *Allium cepa* L. New Phytol. 103, 359–373.

Son C L and Smith S E 1988 Mycorrhizal growth responses: Interactions between photon irradiance and phosphorus nutrition. New Phytol. 108, 305–314.

Sreenivasa M N and Rajashekhara E 1989 Influence of the stage of host development and genotypes on mycorrhizal colonization of field grown wheat. Z. Mikrob. 144, 381–384.

Stribley D P, Tinker P B and Rayner J H 1980b Relation of internal phosphorus concentration and plant weight in plants infected by vesicular-arbuscular mycorrhizas. New Phytol. 86, 261–266.

Stribley D P, Tinker P B and Snellgrove R C 1980a Effect of vesicular-arbuscular mycorrhizal fungi on the relations of plant growth, internal phosphorus concentration and soil phosphate analyses. J. Soil Sci., 31, 655–672.

Thomson B D, Robson A D and Abbott L K 1986 Effects of phosphorus on the formation of mycorrhizas by *Gigaspora calospora* and *Glomus fasciculatum* in relation to root carbohydrates. New Phytol. 103, 751–765.

Tilak K V B R and Murthy B N 1987 Association of vesicualr-arbuscular mycorrhizal fungi with the roots of different cultivars of barley (*Hordeum vulgare*). Curr. Sci. 56, 1114–1115.

Toth R, Page T and Castleberry R 1984 Differences in mycorrhizal colonisation of maize selections for high and low ear leaf phosphorus. Crop Sci. 24, 994–996.

Toth R, Toth D, Starke D and Smith D R 1990 Vesicular-arbuscular mycorrhizal colonization in *Zea mays* affected by breeding for resistance to fungal pathogens. Can. J. Bot. 68, 1039–1044.

Vierheilig H and Ocampo J A 1991 Receptivity of various wheat cultivars to infection by VA-mycorrhizal fungi as influenced by inoculum potential and the relation of VAM-effectivness to succinic dehydrogenase activity of the mycelium in the roots. Plant and Soil 133, 291–296.

Young J L, Davis E A and Rose S L 1985 Endomycorrhizal fungi in breeder wheats and Triticale cultivars field grown on fertile soil. Agron. J. 77, 219–224.

P. J. Randall et al. (Eds.), Genetic aspects of plant mineral nutrition, 233–239.

PLSO SV67

Screening maize inbred lines for tolerance to low-P stress condition

ÁLVARO ELEUTÉRIO DA SILVA and WARREN H. GABELMAN
EMBRAPA/CNPAF, C.P. 179, Goiania, Goiás, 74.001-790, Brazil and Department of Horticulture, University of Wisconsin, Madison, WI 53706, USA

Key words: genetics, maize genotype, mineral nutrition, sand-alumina, stress

Abstract

Genotype screening and selection for tolerance to low-phosphorus stress conditions is an important strategy for the development of cultivars growing on soils low in available P. This study was conducted to adjust an existing screening methodology which provides stable, diffusion-limited low-P concentrations for use with maize, and then use the modified methodology to select maize inbreds tolerant to low-P stress during the vegetative stage of development. Low and high-P concentration levels were established that provided a reproducible and diffusion-controlled availability of P to the plants at concentrations of 8–10 μM and 40–50 μM, respectively, of culture medium solution P at the time of transplanting. The procedure was also effective in providing low concentrations of P during the period of plant growth. The sand-alumina culture medium was used to screen 20 maize inbreds known to be efficient in P uptake or accumulation and/or tolerant to aluminum toxicity. The inbreds B37, Oh40B, NY821, Pa36, and MS1334 were selected as tolerant to low-P and WH, H99, H84, Pa32, and W37A were selected as intolerant to low-P in sand-alumina medium.

Introduction

Recognition of the complexity of soil-root relationships has helped in developing new methodologies to screen plants for efficiency at low levels of mineral nutrients. Due to immobility of P, diffusion is the most important mechanism involved in movement of this ion to absorbing roots in soils (Barber, 1980; Marschner, 1986; Tisdale et al., 1985). Difficulties should be anticipated when trying to relate the ability of plants to absorb P in solution cultures with their ability to acquire P under field conditions. Traditional sand culture and nutrient solution cultures are relatively easy to prepare, and can be used effectively to screen for efficiencies based on physiological or anatomical factors other than nutrient acquisition. However, they are not effective in screening for P acquisition (Gerloff, 1987). The limited number of studies using soil as culture medium to detect phenotypic differences in maize to P efficiency have not been conclusive. No attempt has been made to select maize genotypes in a slow release, diffusion limited system such as the sand-alumina culture medium developed by Coltman et al. (1982).

The primary objectives of the work reported here were to modify the sand-alumina system to meet the specific needs of monocots, and to select maize inbreds tolerant to low-P stress conditions using that system.

Materials and methods

Sand-alumina culture medium

The solid-phase sand-alumina culture system consists of a mixture of pure quartz sand and activated alumina (Al_2O_3). Preparation of the culture medium followed the procedure of Coltman et al. (1982) with modifications. A 28–48 Mesh Alcoa F-1 activated alumina was prepared by sieving and washing with distilled water. The washed alumina was allowed to adsorb phosphate (loading procedure) from a 0.01 *M* NaCl solution containing P, in the form of KH_2PO_4. Loading concentrations (LC) ranged between 100 and 600 m*M* P. Loading was conducted in 10 L PVC tubes containing 2 kg of washed activated alumina and P-loading solution. Tubes were mounted on a tumbling machine holding up to 6 tubes, and rotated at about 20 rpm for 7 minutes every 3 hours for 15 days. After loading, the alumina was rinsed with distilled water and dried for 48 h at 67°C.

A homogeneous mixture of 40 g alumina and 2400 g of autoclaved quartz sand was poured into 2 L plastic pots, prepared previously with a plastic screen and gravel on the bottom to facilitate drainage. Excess phosphate ('desorption procedure') was leached from the pots by pouring 500 mL of distilled water twice a day, for 18 to 25 days, to bring the P concentrations to stable levels ('desorption procedure'). Solution samples were extracted from the pots with a vacuum pump to analyse the pH and the P concentration at each loading level. Pot solution P concentration was determined by the molybdenum blue method (Murphy and Riley, 1962).

Seed preparation – plant growth

Maize seeds were disinfected with a 0.1% solution of $HgCl_2$ for 2 minutes and rinsed several times with distilled water, then germinated in rolled paper towels moistened with water at 29°C under fluorescent and high intensity sodium lights. The seedlings were grown in the paper towels for 7 days with a nutrient solution free of P, then 2 seedlings were transplanted into each pot and 2 days later one of the seedlings was removed, allowing the development of only one seedling per pot. The nutrient solution used was a modification of the Hoagland's solution (Clark, 1982; Hoagland and Arnon, 1938; Magnavaca, 1982). It was applied at $\frac{1}{2}$ concentration, twice a day, 100 mL each time up to 15 days after transplanting, and a full strength, three times a day thereafter. The nutrient solution had no P, so that the P concentrations in the sand-alumina medium were dependent exclusively on P desorbing from the alumina. Iron was a difficult nutrient to control in the sand-alumina medium and additional Fe in the form of FeHEDTA (50 μ*M*) had to be applied in the pots, twice during seedling development (fourth to fifth days and twelfth to fifteenth days after transplanting) to overcome Fe deficiency.

The experiments were carried out in a controlled environment growth room with a mixture of fluorescent, metal halide, and high intensity sodium light, yielding a photosynthetic photon flux density of 400 μ*M* m^{-2} s^{-1} at 80 cm below the lights. Air temperature was maintained at 32°C during the light period and at 20°C during the dark period. The daylength was 16 hours up to 15 days of plant growth and 13 hours during the remaining period of plant growth.

Treatments and experimental design

In experiment 1, four inbreds of maize were grown with alumina loaded at 6 P loading concentrations (100, 200, 300, 400, 500, 600 m*M*). The inbreds used included: (1) Pa36, known to be tolerant to high levels of Al in nutrient solution (Clark and Brown, 1974; Clark, 1983) and tolerant to low-P content in nutrient solution (Baker et al., 1970; Clark and Brown, 1974) and in soil conditions (Gorsline et al., 1964); (2) B57, known to be tolerant to high levels of Al in nutrient solution (Clark, 1983); (3) WH and (4) W153R, known to be intolerant to high levels of Al, but tolerant to low-P in nutrient solution (Clark and Brown, 1974; Clark, 1983; Nielsen and Barber, 1978; Rhue and Grogan, 1976). Inbred line WH was also reported as intolerant to low P in nutrient solution (Baker et al., 1970) and soil conditions (Gorsline et al., 1964). The experimental design was a randomized complete block in a 4 × 6 factorial arrangement with 3 replications.

In experiment 2, inbreds of maize obtained from various public breeding programs were screened in sand-alumina cultures providing a low-P stress condition (LC 200 m*M* P) and a non-limiting concentration of P (LC 400 m*M* P). Twenty inbreds were included for these studies based on previous nutritional studies of tolerance to Al toxicity and/or tolerance to low-P stress in different growth conditions (liquid nutrient solution and soil conditions) (Silva, 1990). A randomized complete block design with 3 replications was used for the 20 maize inbreds grown under each P concentration.

Plant measurements and analyses

In experiment 1, pot solution samples were extracted at 7, 13, 18, and 25 days before transplanting, and 3, 19, and 24 days after transplanting the seedlings. Pot solution P concentration and pot solution pH were measured. Shoots and roots were oven dried for 72h at 67°C, then weighed.

In experiment 2, measurements were taken for shoot and root dry weight and shoot and root P concentration. Tissue P concentration analysis followed the vanadomolybdate yellow method (Bertramson, 1942, Jackson, 1958).

Results and discussion

Pot solution P concentration decreased sharply during the first 15 to 18 days of leaching with distilled water (Fig. 1) with a trend to become more stable after 18 days (Fig. 2). This period is longer than reported by Coltman (1983) by four to six days and closer to the 17 days reported by Pereira (1987). From previous research with the sand-alumina medium there is a general agreement about the higher P loading concentrations requiring a longer leaching time than the lower P loading concentrations (Buso, 1986; Coltman,

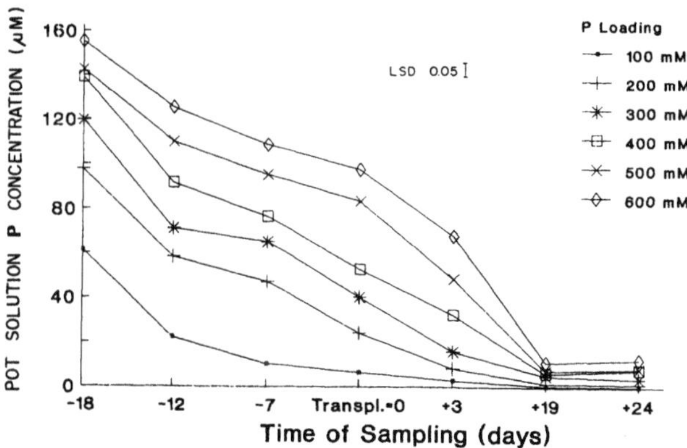

Fig. 1. Culture medium P concentration sampled at periodical intervals (before and after transplanting) for treatments containing 40 g per pot of Alumina Loaded to six P concentrations.

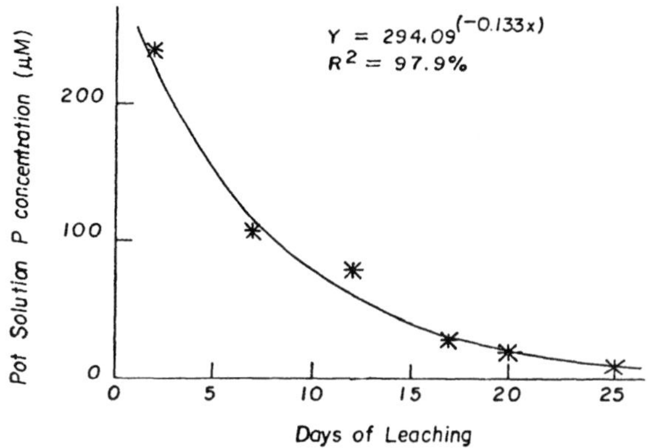

Fig. 2. Sand-alumina culture medium P concentration over time of leaching with distilled water for treatment containing 40 g/pot of alumina loaded at 200 m*M* P concentration.

1983; Pereira, 1987). Although lower levels of P loading concentrations reached a more stable condition earlier, a decrease in the pot solution P concentration from transplanting (zero time) to harvesting (24 days after transplanting) was observed for all P concentrations (Fig. 1). This is not in agreement with the steady-state equilibrium reported by Coltman et al. (1982), but agrees with results obtained by Buso (1986), Elliott et al. (1983), and Pereira (1987). The data from time zero (transplanting) to time +3 (3 days after transplanting) indicated that near steady state conditions were not attained. No more leaching occurred after time zero, and the maize seedlings were too young to be able to absorb that much of P. The decrease in pot solution P concentration was observed for every P loading level. It is likely that when leaching ceased at time zero, the addition of nutrient solution in the medium to feed the seedlings caused a shift in the point zero charge of the alumina particles, causing a disruption in the balance of the system with a consequent increase in the binding ability of the alumina particles (White, 1980). When the leaching process continued with distilled water and no nutrient solution was added at the corresponding time zero (Fig. 2), the decrease in pot solution P concentration was smoother suggesting a near steady state condition in the medium. The data from Figure 2 showed that pot solution P concentration response to time of leaching seems to be exponential ($R^2 = 87.6\%$ using the original data and 97.9% using log e transformation of the original data) with the leveling off and more stable concentrations reached at 18 to 20 days. The pH of the sand-alumina culture medium remained between 6.8 and 7.2 from transplanting to harvest time.

Increasing pot solution P concentration did increase plant dry matter production differentially among the maize inbreds evaluated. The analysis of variance for shoot dry matter accumulation and plant dry matter accumulation showed a highly significant interaction between inbreds and P concentrations. As anticipated, this is a clear indication that different maize inbreds accumulate dry matter differently when grown at different P concentrations. Because the results from shoot dry weight were similar to the ones obtained for plant dry weight, only the shoot dry weight is discussed. The average reduction of dry matter accumulation of the 4 maize inbreds grown at the 100 m*M* P loading concentration was 66.4% that of inbreds grown at 400 m*M* P loading concentration (Table 1). Pa36, an inbred rated in previous studies as Al tolerant when evaluated in nutrient solution (Clark and Brown, 1974; Clark, 1983) as well as tolerant to low-P in nutrient solution and soil conditions (Clark and Brown, 1974; Baker et al., 1970; Gorsline et al., 1964), was the most responsive inbred to increasing levels of P concentration. Inbreds B57 and W153R had similar increases in dry matter production with increasing P concentration up to 500 m*M* of loading P concentration , followed by a decrease in production at the highest level of P concentration (600 m*M*). This drop in dry matter production was also observed for the inbred WH at the 500 m*M* P loading concentration.

At the higher levels of P concentrations Fe deficiency symptoms were observed for all four inbred lines and Zn deficiency was observed mainly with the inbred B57. Phosphorus-zinc interactions in plants are well known. Several experimental results indicate that inhibition of zinc translocation from roots to shoot and physiological inactivation of zinc within the shoots can occur (Boawn and Brown, 1968; Burleson and Page, 1967; Marschner, 1986). The latter suggestion is based on the observation that symptoms of zinc deficiency are related to the P:Zn ratio rather than to the zinc concentration in the shoots (Millikan, 1963; Boawn and Brown, 1968). Therefore, even with equivalent

Table 1. Means for shoot dry weight and plant dry weight (g) of maize inbreds evaluated at 6 P loading concentrations using the sand-alumina medium

Loading P concentration (m*M*)	Maize inbreds			
	WH	B57	W153R	Pa36
	Shoot dry weight			
100	0.73a*	0.54a	0.36a	0.62a
200	0.93a	0.93b	0.55a	1.17b
300	1.34b	1.20bc	0.94b	2.36c
400	2.20d	1.69d	1.35c	2.97d
500	1.67c	2.18e	1.86d	3.73e
600	2.36d	1.23c	1.51c	3.98e
Control**	2.71	2.68	2.58	4.63
	Plant dry weight			
100	1.62a	0.87a	0.62a	1.06a
200	1.70a	1.46b	0.96a	1.96b
300	2.02a	1.71b	1.40b	3.69c
400	3.76d	2.45c	1.87c	4.34d
500	2.39ab	2.98d	3.23d	5.20e
600	3.18c	1.66b	2.03c	4.92e
Control**	3.67	4.28	3.27	6.23

* Means with different letters at the same column and same trait are significantly different at 0.05 probability level by the LSD test.

** Control = Sand only (alumina free) fed with 1/2 Hoagland's solution amounts of zinc at both P concentrations, the maize seedlings could be stressed and suffering from Zn deficiency at the higher levels of P.

Elliot and Lauchli (1985) reported Fe deficiency symptoms for all inbreds of maize supplied with greater than 20 mM m^{-3} P, grown in a sand-alumina medium. Both Fe absorption by roots and partitioning of Fe to shoots decreased with increasing concentration of P in the nutrient solution. Phosphate inhibition of Fe uptake has also been reported previously for other species (Brown, 1972; Chaney and Coulombe, 1982; Marschner, 1978). Although Fe tissue concentration for both levels of loading P were similar (data not shown), there is a possibility that most of the Fe was retained in the roots and may not be translocated to the sites of utilization in the shoots causing the general Fe deficiency symptoms observed during plant growth.

Twenty inbreds of maize obtained from various public breeding programs were evaluated using the sand-aluminum medium. Tolerance to low-P stress based on evaluations of shoot dry weight production can be regarded as the product of the plant mechanisms involving acquisition, translocation, and utilization of P in inbreds grown in the sand-alumina culture medium. The criterium used to select the maize inbreds as tolerant and intolerant to low-P stress was based on the similar performance of the inbreds at high-P concentration (400 mM of P loading concentration) for shoot dry matter accumulation and different dry matter accumulation at low-P (200 mM of P loading concentration), grown in sand-alumina culture medium. Total P content in the plant was regarded as a phenotypic measure of P acquisition in the sand-alumina medium.

Separate statistical analyses for shoot dry matter accumulation were done for each P concentration due to heterogeneity of error variances at the two P levels. Highly significant inbred effects were detected at both concentrations of P. The data reported in Table 2 are ranked by decreasing shoot dry weight at low P concentration. Although there were significant differences among inbreds for dry matter accumulation at high P, inbreds with similar dry weights at high P yet very dissimilar dry weights at low P were apparent (Table 2). Variation in shoot dry weight at low P were almost threefold indicating wide variation for this characteristic among genotypes.

Table 2. Means for shoot dry weight (g) of the 20 maize inbreds grown under low- and high-P concentrations in the sand-alumina culture medium

Maize inbreds	Shoot dry weight	
	Low-P	High-P
B37	1.69 a*	1.99 abc
NY821	1.63 ab	1.63 abcde
MS1334	1.54 ab	2.56 a
Pa36	1.34 abc	1.38 bcdef
Oh40B	1.30 bcd	1.80 abcd
Oh51A	1.20 cde	1.06 efg
W117	1.05 def	1.91 abc
W59M	1.01 efg	1.65 abcde
W64A	0.98 efg	1.48 bcde
B57	0.98 efg	1.48 bcde
A632	0.94 efg	0.72 gh
Pa32	0.93 fg	1.40 bcde
W37A	0.91 fg	1.52 bcde
H84	0.88 fgh	1.80 abcd
Oh43	0.84 fgh	1.08 defg
W153R	0.83 fghi	0.63 h
H99	0.80 ghi	1.18 cdefg
WH	0.69 hij	1.23 bcdef
CMD5	0.65 ij	0.51 h
Wf9	0.58 j	0.79 fgh

*Means with different letters at the same column are significantly different at 0.05 probability level by the LSD calculated from log e of shoot dry weight.

Based on data in Table 2, five inbreds, B37, NY821, MS1334, Pa36, and Oh40B were selected as tolerant, and five inbreds, Pa32, W37A, H84, H99, and WH were selected as intolerant to low-P stress in sand-alumina. The maize inbreds selected as tolerant based on shoot dry weight were also found to be tolerant based on total plant dry weight and total P content in the plant. The inbreds selected as intolerant based on shoot dry weight were also found to be intolerant based on total plant dry weight and total P content in the plant (data not shown).

In conclusion, the sand-alumina culture medium was effective in providing different and stable solution P concentrations at transplanting, highly predictable from loading P concentrations. The procedure was also effective in providing low concentrations of P during the period of plant growth. Care should be taken whenever any of the parameters of the system are changed and standardization would be required for the growth of different species. Variations in plant growth based on the requirements for micronutrients, for example, can become a source of variation upsetting the results to be obtained.

By comparing the results obtained here with results from the literature (Baker et al., 1970; Clark and Brown, 1974; Gorsline et al., 1964; Nielsen and Barber, 1978), the data support that the sand-alumina culture medium more closely approximates the results obtained with the diffusion-limited environment in naturally occurring P-deficient soils as opposed to liquid nutrient culture medium. The genotypes determined to be tolerant to low-P stress using the sand-alumina medium may have the greatest genetic potential to improve those traits most important in acquisition of P in naturally occurring P-deficient soils. The correct choice of the screening methodology will permit the expression of some characteristics better than others. Understanding the inheritance of each of these characteristics is a prerequisite to rational development of plant genotypes that have a greater capability to absorb P from the soil.

Acknowledgements

The authors are grateful to Prof. Sergio Volkweiss (in memorium), Prof. Richard B Corey, Prof. Gerald C Gerloff, and Dr Pedro A A Pereira for technical advices and EMBRAPA for financial support.

References

Baker D E, Jarrel A E, Marshall L E and Thomas W I 1970 Phosphorus uptake from soils by corn hybrids selected for high and low phosphorus accumulation. Agron. J. 62, 103–106.

Barber S A 1980 Soil-plant interactions in the phosphorus nutrition of plants. *In* The Role of Phosphorus in Agriculture. Eds. F E Khasawneh and E C Sample. Soil Sci. Soc. Am. Madison, WI.

Bertramson B R 1942. Phosphorus analysis of plant material. Plant Physiol. 17, 447–455.

Boawn L C and Brown J C 1968 Further evidence for P-Zn imbalance in plants. Soil Sci. Soc. Am. Proc. 32, 94–97.

Brown J C 1972 Competition between phosphate and the plant for Fe from Fe^{2+} ferrozine. Agron. J. 64, 240–243.

Burleson C A and Page N R 1967 Phosphorus and zinc interactions in flax. Soil Sci. Soc. Am. Proc. 31, 510–513.

Buso G S C 1986 Variability among lettuce grown under diffusion limited conditions of phosphorus supply. M.Sc. Thesis, Univ. of Wisconsin, Madison, WI.

Chaney R L and Coulombe B A 1982 Effect of phosphate on regulation of Fe-stress response in soybean and peanut. J. Plant Nutr. 5, 469–487.

Clark R B 1982 Nutrient solution growth of sorghum and corn in mineral nutrition studies. J. Plant Nutr. 5, 1039–1057.

Clark R B 1983 Plant genotype differences in the uptake, translocation, accumulation, and use of mineral elements required for plant growth. Plant and Soil 72, 175–196.

Clark R B and Brown J C 1974 Differential phosphorus uptake by phosphorus-stressed corn inbreds. Crop Sci. 14, 505–508.

Coltman R R 1983 Intraspecific variation in tomato for dry matter accumulation under maintained and diffusion-controlled phosphorus deficiency. Ph.D. Thesis, University of Wisconsin, Madison, WI.

Coltman R R, Gerloff G C and Gabelman W H 1982 A sand culture system for simulating plant responses to phosphorus in soil. J. Amer. Soc. Hort. Sci. 107, 938–942.

Elliott G C and Lauchli A 1985 Phosphorus efficiency and phosphate-iron interaction in maize. Agron. J. 77, 399–403.

Elliott G C, Carlson R M, Lauchli A and Rosen C J 1983. A solid-phase buffer technique to maintain low concentrations of phosphate in nutrient solutions. J. Plant Nutr. 6, 1043–1058.

Gerloff G C 1987 Intact-plant screening for tolerance of nutrient-deficiency stress. *In* Genetic Aspects of Plant Mineral Nutrition. Eds. W H Gabelman and B C Loughman. pp 55–68. Martinus Nijhoff Publishers, Dordrecht, The Netherlands.

Gorsline G W, Thomas W I and Baker D E 1964 Inheritance of P, K, Mg, Cu, B, Zn, Al and Fe concentrations by corn (*Zea mays* L.) leaves and grain. Crop Sci. 4, 207–210.

Hoagland D R and Arnon D I 1938. The Water-culture Method for Growing Plants without Soil. Univ. California, Agric. Exp. Station, Berkeley, CA. Circular 347.

Jackson M L 1958 Soil Chemical Analysis. Prentice Hall Inc., Englewood Cliffs, NJ.

Magnavaca R 1982 Genetic variability and the inheritance of aluminum tolerance in maize (*Zea mays* L.). Ph.D. Thesis, University of Nebraska, Lincoln, NE. 135 p.

Marschner H 1978 Role of the rhizosphere in iron nutrition of plants. Iran. J. Agric. Res. 6, 69–80.

Marschner H 1986 Mineral Nutrition of Higher Plants. Academic Press, London. 674 p.

Millikan C R 1963 Effects of different levels of zinc and phosphorus on the growth of subterranean clover (*Trifolium subterraneum* L.). Austr. J. Aric. Res. 14, 180–205.

Murphy J and Riley J P 1962 A modified single solution method for the determination of phosphate in natural waters. Anal. Chem. Acta 27, 31–36.

Nielsen N E and Barber S A 1978 Differences among genotypes of corn in the kinetics of P uptake. Agron. J. 70, 695–698.

Pereira P A A 1987 Improvement of N_2 fixation in common bean (*Phaseolus vulgaris* L.) at different levels of available phosphorus. Ph.D. Thesis, University of Wisconsin, Madison, WI.

Rhue R D and Grogan C O 1977 Screening corn for Al tolerance using different Ca and Mg concentrations. Agron. J. 69, 755–760.

Silva A E 1990 Inheritance studies of phosphorus acquisition and utilization efficiency in maize (*Zea mays* L.). Ph.D. Thesis, University of Wisconsin, Madison, WI. 149 p.

Tisdale S L, Nelson W L and Beaton J D 1985 Soil Fertility and Fertilizers. Macmillan, New York. 754 p.

White R E 1980 Retention and release of phosphate by soil and soil constituents. *In* Soil and Agriculture: Critical Reports on Applied Chemistry. Vol. 2. Ed. P B Tinker. pp. 71–114. Blackwell Scientific Publications, Oxford.

P. J. Randall et al. (Eds.), Genetic aspects of plant mineral nutrition, 241–249.

PLSO SV57

Inheritance studies of low-phosphorus tolerance in maize (*Zea mays* L.), grown in a sand-alumina culture medium

ÁLVARO ELEUTÉRIO DA SILVA, WARREN H. GABELMAN and JAMES G. COORS
EMBRAPA/CNPAF, C. P. 179, Goiania, Goiás 74001-790, Brazil, Department of Horticulture, University of Wisconsin, Madison, WI 53706, USA and Department of Agronomy, University of Wisconsin, Madison, WI 53706, USA

Key words: diallel, low-P stress, maize, sand-alumina

Abstract

Inbred lines of maize selected as tolerant and intolerant to low-P stress using a sand-alumina culture medium were used to obtain F_1 hybrids and advanced generations to be evaluated in diallel mating schemes and generation means analyses for the inheritance studies. Sand-alumina, a solid culture medium, which simulates a slow release, diffusion-limited P movement in soil solution was used in the inheritance studies. Tolerance to low-P stress conditions in maize seedlings is controlled largely by additive gene effects, but dominance is also important.

Introduction

With the agricultural frontier moving to acid, less fertile soils, an increase in price of fertilizers, and a growing concern about environmental pollution, interest in genotypic differences in mineral nutrition has recently increased (Gerloff, 1976; Gerloff and Gabelman, 1983).

By using a sand-alumina system to provide slow-release of P to the root system, approximating more closely the conditions of P availability in soils, phenotypic variation for P acquisition has been detected in tomato plants (Coltman et al., 1985), lettuce (Buso and Bliss, 1988), dry beans (Pereira and Bliss, 1989), and maize (da Silva, 1990). Identification of genetic differences in P efficiency by plants can be used in a breeding program aiming to develop cultivars more tolerant to low availability of P.

In maize, genetic variation in P absorption and utilization has been reported by several authors (Baker et al., 1964; Gorsline et al., 1964; Harvey, 1939; Lyness, 1936; Thomas, 1930; Smith, 1934). There have been few studies to determine the inheritance of P in maize. Gorsline et al. (1964) concluded that additive gene action was important for ear-leaf and grain concentration of P. Non-additive gene was indicated for ear-leaf concentration of P. Barber et al. (1967) showed that hybrids of maize developed by crossing inbred lines with low ability for accumulating P in plant tissue yielded predominantly low accumulators of P. High-P × high-P crosses yielded predominantly high-P accumulating hybrids.

General combining ability effects exceeded specific combining ability effects in 49 hybrids of forage sorghum for all mineral elements except P, Cl, and Fe in females, and S and Cl in males (Gorz et al., 1987).

Generation means analysis of six bean families derived from crosses between efficient and inefficient strains showed that P utilization efficiency was controlled by few genes with significant dominance and epistatic gene effects (Fawole et al., 1982). The successful introgression of alleles conferring P efficiency from a plant introduction line into an adapted variety of beans has also been demonstrated (Schettini et al., 1987).

Biometric analysis of a single-cross between tolerant and intolerant tomato strains indicated that P-acquisition efficiency had a high broad-sense heritability (genetic variance/phenotypic variance) with dominance effects being important (Coltman et al., 1987). More recently, Reiter (1989) using two inbreds of maize grown in a sand-alumina medium under low P concentration demonstrated by the use of restriction fragment length polymorphism (RFLP) analysis, that additive gene action was predominant for all quantitative trait loci identified.

The objective of this study was to obtain preliminary information on the inheritance of low-P tolerance in maize grown in a sand-alumina culture medium using the diallel mating system.

Materials and methods

The genotypes used in this study were produced in the nursery of the maize breeding program at the University of Wisconsin, Madison. Two sets of diallel matings involving 9 maize inbreds (Table 1) were designed to study the inheritance of tolerance to low-P stress conditions.

Diallel 1

The first diallel (Diallel 1) involved four of the selected inbreds from previous screening experiments (da Silva, 1990). The low-P tolerant parents, B37 and NY821, and low-P intolerant parents, H84 and WH, were chosen based on shoot dry weight as well as total P content in the plant. The resulting F_1's, reciprocal F_1's, and parental inbreds (16 entries) were grown under low-P conditions in the sand-alumina medium. A high-P treatment was included as a control. A randomized complete block design was used with six replications of each entry grown at each P level. Dry weights for shoots were obtained from the six replications. Root dry weight and P concentration were obtained from three replications. Shoot and root dry weights and P concentration were used to calculate total P content in the plant (TPP) and the ratio root dry weight:shoot dry weight (R:S ratio). General combining ability (GCA), specific combining ability (SCA), and reciprocal effects were obtained by using Method 3, Model I analysis of Griffing (1956). An unbiased estimate of combining ability should exclude the parents from the analysis. A general analysis of variance, using the GLM sub-routine from the SAS statistical package (SAS Institute, 1985), provided the variation due to parents, crosses, and a contrast parents vs crosses, the latter being an estimate of the general heterosis.

The small number of inbreds sampled and the systematic selection for P efficiency characterized a fixed model restricted to estimation of GCA, SCA, and reciprocal mean squares and effects. Thus, it was possible to determine whether reciprocal differences as well as general and specific combining ability effects were important in tolerance to low-P stress conditions.

Table 1. Parents used in the inheritance studies, their origin and state source

Maize inbreds	Origin	State source
B37	Iowa Stiff Stalk Synthetic	Iowa
B57	Midland	Iowa
H84	(B37 × GE440)	Indiana
H99	Illinois Synthetic 60C	Indiana
MS1334	[(Golden Glow × Maiz Amargo) × (Golden Glow)]	Michigan
NY821	Pa Disease Resistant Synthetic	New York
Oh40B	Lancaster Sure Crop (2nd Cycle)	Ohio
WH	Wisconsin #25	Wisconsin
W64A	(Wf9 × 187-2)	Wisconsin

Diallel 2

The second diallel mating scheme (Diallel 2) involved seven maize inbreds from previous screening experiments (da Silva, 1990). Inbreds B37, Oh40B, and MS1334 were rated as low-P tolerant, inbreds B57 and W64A as intermediate and inbreds H84 and H99 as low-P intolerants based on shoot dry weight and total P content in the plant. The 21 F_1 families resulting from all possible crosses among inbreds (no parents, no reciprocals) were evaluated under low-P conditions in the sand-alumina medium. A randomized complete block design was used with five replications. Dry weight for shoots was obtained from all five replications. Root dry weight and P concentrations were obtained from only three replications. Shoot and root dry weights and P concentration were used to calculate total P content in the plant (TPP) and the ratio root dry weight:shoot dry weight (R:S ratio). A randomized complete block analysis of variance was used to determine differences among genotypes. The GCA and SCA information on crosses were obtained by using Method 4, Model I analysis of Griffing (1956).

The small number of inbreds sampled and the systematic selection for P efficiency required a fixed model, restricted to estimation of GCA and SCA mean squares and effects. All discussion generated from these diallel mating experiments shall be restricted to the set of maize inbreds involved.

Plant growth procedures

The procedures for the sand-alumina preparation, seed germination, transplanting, and environmental plant growth conditions were the same as described for the screening experiments (da Silva, 1990).

Briefly, 200 mM and 400 mM loading P concentrations were used as the low and high P levels, respectively, and will be referred to as low-P and high-P levels. The seedlings were 7 days old at the time of transplanting. Plants were grown for 25 days after transplanting in a controlled environmental growth room. The light was provided by a mixture of fluorescent, metal-halide, and high intensity sodium light, yielding a photosynthetic photon flux density of 400 μM m^{-2} s^{-1} at 80 cm below the lights. Each pot contained only one plant, and this was considered the experimental unit. At harvest time, maize shoots were excised at substrate level. Roots were gently removed from the sand-alumina medium by turning the pots upside down in a bowl of water and shaking them in a careful manner so that the whole root system was recovered. Shoots and roots were dried in a forced-air oven, weighed, and ground to pass a 40 mesh screen. Root and shoot tissues were ashed in a furnace at 550°C for 6 hours. Tissue P concentration was determined following the vanado-molybdate yellow method according to Bertramson (1942) and Jackson (1958).

Results and discussion

Variation among parents and crosses

For diallel 1 separate statistical analyses were done for each P concentration due to heterogeneity of error variances among low- and high-P treatments for most of the traits. Tables 2 and 3 present the partial combined analyses of variance for low- and high-P, respectively, according to Method 3, Model I of Griffing (1956). Highly significant genotype effects were detected at low-P concentration for shoot dry weight, root dry weight, total P content in the plant (TPP), and

Table 2. Partial analyses of variance for shoot and root dry weight, total P content in the plant (TPP), and the ratio root dry weight:shoot dry weight (R:S) from a complete diallel mating study involving four maize inbreds grown under low-P concentration (Diallel 1)

Source of variation	d.f.	Mean squares			
		Shoot dry wt. (g)	Root dry wt. (g)	TPP (mg)	R:S (g/g)
Genotypes	15	0.754**	0.154**	2.869**	0.053**
Parents (P)	3	1.085**	0.188**	2.721**	0.079**
Crosses (C)	11	0.332**	0.128**	2.127**	0.048**
GCA	3	0.759**	0.095	4.772**	0.044*
SCA	2	0.282*	0.018	0.249	0.022
Reciprocal	6	0.135	0.182**	1.430*	0.059*
P vs C	1	4.401**	0.327**	11.478**	0.032
Error	30	0.064†	0.034	0.523	0.014
C.V. (%)		18.8	28.2	19.8	23.7

*, ** Significant at the 0.05 and 0.01 probability levels, respectively by the F-test.
†Error degrees of freedom for shoot dry weight = 75.

Table 3. Partial analyses of variance for shoot and root dry weight, total P content in the plant (TPP), and the ratio root dry weight:shoot dry weight (R:S) from a complete diallel mating study involving four maize inbreds grown under high-P concentration (Diallel 1)

Source of variation	d.f.	Mean squares			
		Shoot dry wt. (g)	Root dry wt. (g)	TPP (mg)	R:S (g/g)
Genotypes	15	3.361**	0.397**	22.705**	0.012*
Parents (P)	3	0.051	0.032	0.399	0.028**
Crosses (C)	11	1.566**	0.305**	11.583**	0.007
GCA	3	3.074**	0.595**	24.974**	–
SCA	2	0.197	0.028	1.682	–
Reciprocal	6	1.268**	0.253**	8.189**	–
P vs C	1	33.042**	2.499**	211.963**	–
Error	30	0.179†	0.040	1.732	0.006
C.V. (%)		19.5	22.6	14.6	20.5

*, ** Significant at the 0.05 and 0.01 probability levels, respectively by the F-test.
† Error degrees of freedom for shoot dry weight = 75.

R:S ratio. Shoot dry weight, root dry weight and TPP genotype effects were also highly significant at high-P level, and R:S ratio was significant (Table 3). Variation due to parents included in the diallel 1 low-P, was highly significant for shoot dry weight, root dry weight, TPP, and R:S ratio. At high-P, variation due to parents was significant only for R:S ratio (Table 3).

As expected the low-P tolerant inbreds (B37 and NY821) and the intolerant inbreds (H84 and WH) had contrasting shoot dry weights in diallel 1 at low-P concentration (Table 4), but there were no differences at high-P concentration (Table 3). The mean shoot dry weights for crosses (Table 4) ranged from 1.07 g for H84 × B37 to 2.01 g for WH × NY821.

In diallel 2, highly significant effects due to crosses were detected at the low-P concentration for shoot dry weight, total P content in the plant (TPP), and R:S ratio (Table 5).

Table 4. Means for shoot and root dry weight, total P content in the plant (TPP), and root dry weight:shoot dry weight (R:S ratio) from a complete diallel mating study involving four maize inbreds grown at low-P concentration (Diallel 1)

Genotypes	Shoot dry wt. (g)	Root dry wt. (g)	TPP (mg)	R:S (g/g)
B37	1.28	0.86	3.40	0.65
NY821	1.30	0.51	3.67	0.38
H84	0.91	0.39	2.59	0.43
WH	0.39	0.29	1.55	0.71
B37 × NY821	1.40	0.94	4.34	0.70
NY821 × B37	1.46	0.66	4.69	0.44
B37 × H84	1.41	0.55	3.35	0.40
H84 × B37	1.07	0.71	3.02	0.71
B37 × WH	1.42	0.56	3.32	0.42
WH × B37	1.46	0.80	4.17	0.49
NY821 × H84	1.69	0.67	5.18	0.41
H84 × NY821	1.62	0.64	3.65	0.37
NY821 × WH	1.43	0.54	3.90	0.35
WH × NY821	2.01	1.18	5.45	0.55
H84 × WH	1.21	0.79	3.10	0.59
WH × H84	1.27	0.41	3.03	0.36
$LSD_{(0.05)}$	0.39	0.39	0.85	0.20

Table 5. Partial analyses of variance for shoot dry weight, total P content in the plant (TPP), and the ratio root dry weight : shoot dry weight (R:S) from a diallel mating study involving seven maize inbreds (no parentals, no reciprocals) grown under low-P concentration (Diallel 2)

Source of variation	d.f.	Mean squares		
		Shoot dry wt. (g)	TPP (mg)	R:S (g/g)
Crosses	20	0.562**	7.427**	0.019**
GCA	6	1.090**	12.096**	0.054**
SCA	14	0.336**	5.426	0.004
Error	40	0.119†	3.020	0.006
C.V. (%)		12.0	18.7	15.0

*, ** Significant at the 0.05 and 0.01 probability levels, respectively by the F-test.
† Error degrees of freedom for shoot dry weight = 80.

Reciprocal differences

Complications such as reciprocal effects must be thoroughly considered in the planning of future breeding programs. Highly significant reciprocal effects detected in diallel 1 high-P for shoot dry weight and TPP were not expected. Highly significant reciprocal effects were also detected for root dry weight at both P concentrations (Tables 2 and 3).

Maternal effects for overall K efficiency were reported to be insignificant for tomatoes by Makmur (1977). Maternal influences for both K efficiency and Na substitution capacity were not important for tomatoes (Figdore, 1986). Maternal effects on the efficiency of Ca utilization were also reported as unimportant for tomatoes (Giordano, 1980; Li, 1989). The same conclusion was reached in studies of the efficiency of K and P utilization in snapbean and tomato (O'Sullivan et al., 1974; Shea et al., 1967; Whiteaker et al., 1976).

It is important to note that all these studies have in common the evaluation of genotypes using a liquid nutrient solution, which does not provide a continuous, stable, low nutrient stress environment to the roots nor any physical barrier to the development of the root system throughout plant development. The sand-alumina culture medium, therefore, may detect reciprocal effects due to environmental stress not encountered in previous studies. The detection of highly significant reciprocal differences at both levels of P for root dry weight in diallel 1 (Tables 2 and 3) may be responsible for the reciprocal differences obtained for other characteristics as well. Corroborating this idea, maternal effects for total P content in the plant were detected for F_1 and backcross generations in tomatoes when grown in a sand-alumina medium (Coltman et al., 1987).

Combining ability effects

In diallel 1, general combining ability (GCA) was significant for shoot dry weight and total P content in the plant (TPP) at low-P and for shoot dry weight, root dry weight, and TPP at high-P, but only shoot dry weight had significant specific combining ability (SCA) mean square at low-P. Thus, variation among the crosses was due mainly to additive rather than non-additive effects. However, non-additive gene action may also be important as shown by the significant variation obtained for parents vs crosses (a measure of heterosis) detected for shoot dry weight, root dry weight, and TPP for both P concentrations.

Although B37 has favorable alleles involved in tolerance to low-P stress, it had a negative estimate of GCA for shoot dry weight (Table 6). The cross B37 × NY821 had a large negative SCA effect for shoot dry weight, hybrids B37 × H84 and B37 × WH had positive SCA effects of smaller magnitudes. Inbred NY821 had a relatively high positive GCA effect. Hybrids NY821 × H84 and NY821 × WH had positive SCA effects. Inbred WH had a positive GCA effect of relatively small magnitude and the hybrid WH × H84 had a negative SCA effect.

Table 6. Estimates of specific combining ability ($\hat{s}_{ij}$) for shoot dry weight and general combining ability ($\hat{g}_i$) for shoot dry weight, total P content in the plant (TPP), and root dry weight:shoot dry weight (R:S ratio), from a complete diallel mating study involving four maize inbreds grown under low-P concentration (Diallel 1)

Maize inbreds	NY821	H84	WH	General c. ability ($\hat{g}_i$)		
	Shoot (g)	dry wt.	($\hat{s}_{ij}$)	Shoot dry wt. (g)	TPP (mg)	R:S (g/g)
B37	−0.103	0.015	0.088	−0.139	−0.176	0.066
NY821		0.088	0.015	0.211	0.903	−0.019
H84			−0.103	−0.102	−0.567	−0.014
WH				0.031	−0.160	−0.033
S.E.	$\hat{s}_{ij}$ = 0.103			$\hat{g}_i$ = 0.110	0.313	0.051

The results suggest that inbreds B37 and NY821 may have genes in common for low-P tolerance.

Significant variation for GCA was also observed for total P content in the plant and for R:S ratio at low-P concentration (Table 2). The means for TPP in crosses (Table 4) ranged from 3.02 mg for H84 × B37 to 5.45 mg for WH × NY821. The only positive GCA effect for TPP was observed for inbred NY821 (Table 6). A low R:S ratio when associated with high TPP in the same genotype, may be desirable inasmuch as a reduced R:S ratio suggests that more carbohydrates might be allocated to the shoot growth. For this particular case, negative GCA effects are desirable for R:S ratio. Negative GCA effects were observed for inbreds NY821, H84, and WH (Table 6).

The results from diallel 1 showed that response to selection for tolerance to low-P stress should be effective if the screening methodology is representative of conditions encountered in the field and genotypes with strong general and specific combining abilities are identified.

In diallel 2, general combining ability (GCA) effects were highly significant for all traits. Specific combining ability (SCA) effects were highly significant only for shoot dry weight and non significant for the other traits. The significant SCA effects for shoot dry weight would be expected to be followed also by a significant SCA effect for one or more traits related to tolerance to low-P stress. SCA effects for TPP was at the margin of 0.05 significance level.

In diallel 2, shoot dry weight means ranged from 1.91 g for W64A × MS1334 to 3.44 g for B57 × Oh40B (Table 7). The three inbreds with high positive GCA effects (B57, Oh40B, and B37), were parents of top-yielding crosses for shoot dry weight (Table 7). Inbred H99 had a negative GCA for shoot dry weight of similar

Table 7. F_1 means for shoot dry weight in grams (above diagonal) and specific combining ability effects (below diagonal) for the 21 crosses, and general combining ability effects for each inbred from a diallel mating study involving seven inbreds (no parents, no reciprocals) grown under low-P concentration (Diallel 2)

Maize inbreds	Maize inbreds							General combining ability ($\hat{g}_i$)
	B57	Oh40B	H99	W64A	MS1334	B37	H84	
B57		3.44[a]	2.58	2.69	3.26	3.09	3.33	0.24[b]
Oh40B	0.097[c]		2.41	2.85	3.17	3.27	3.24	0.24
H99	−0.177	−0.349		2.37	2.20	3.23	2.68	−0.35
W64A	−0.043	0.121	0.221		1.91	3.02	2.50	−0.37
MS1334	0.203	0.118	−0.269	−0.539		3.26	3.15	−0.05
B37	−0.259	−0.075	0.465	0.285	0.199		2.54	0.24
H84	0.179	0.087	0.07	−0.046	0.287	−0.615		0.05

[a] $LSD_{(0.05)} = 0.57$ for cross means.
[b] S.E. $= \hat{g}_i - \hat{g}_j = 0.048$.
[c] S.E. $= \hat{s}_{ij} - \hat{s}_{ik} = 0.190$; $\hat{s}_{ij} - \hat{s}_{kl} = 0.143$.

magnitude as W64A, and MS1334 had a negative GCA effect (Table 7). Inbreds B57, Oh40B, and B37 had positive GCA effects of the same magnitude for shoot dry weight, and H84 had a positive, but relatively low GCA for shoot dry weight (Table 7). Inbred Oh40B, a low-P tolerant inbred, had a negative SCA with B37, another low-P tolerant inbred, but a positive SCA when crossed with MS1334. Inbred H84, a low-P intolerant inbred, had negative SCA effects for shoot dry weight, only when crossed with W64A and B37. Due to the close relationship between H84 and B37 the negative SCA between both inbreds for shoot dry weight was expected.

In diallel 2, means for TPP ranged from 6.95 mg for W64A × MS1334 to 12.40 mg for H99 × B37 (Table 8). Positive GCA effects were detected for inbreds Oh40B and B37. Inbreds Oh40B, B37, and MS1334 were ranked 3rd, 1st, and 4th in TPP efficiency in previous experiments (da Silva, 1990). An examination of the means (Table 8) shows that MS1334 had the lowest TPP means when crossed with inbreds W64A (6.95 mg) and H99 (7.29 mg), which may have been responsible for a negative estimate of GCA effects for MS1334. Inbreds W64A and H99 were ranked intermediate for TPP in previous experiments (da Silva, 1990), suggesting that inbred MS1334 may have alleles in common with W64A and H99 for TPP.

Means for R:S ratio ranged from 0.39 for W64A × H84 to 0.69 for H99 × B37 (Table 9). Positive GCA effects were obtained for inbreds B57, Oh40B, H99, and B37. As discussed before, a negative GCA effect for R:S ratio might be desirable in association with a high TPP. Inbreds W64A, MS1334, and H84 had negative GCA effects for R:S ratio.

The results obtained for diallel 2 agreed closely with results from diallel 1 inasmuch as variation among the crosses was due mainly to additive rather than non-additive effects for both studies. Total P content in the plant (TPP) and

Table 8. F_1 means[a], and general combining ability[b] ($\hat{g}_i$) for total P content in the plant (TPP) from a partial diallel study involving seven maize inbreds (no parents, no reciprocals), grown under low-P concentration (Diallel 2)

Maize inbreds	Oh40B	H99	W64A	MS1334	B37	H84	GCA ($\hat{g}_i$)
			TPP (mg)				
B57	9.40	7.33	7.60	9.17	10.87	8.18	−0.620
Oh40B		9.62	10.14	10.67	11.78	10.45	1.284
H99			8.01	7.29	12.40	9.05	−0.390
W64A				6.95	9.64	7.39	−1.183
MS1334					11.02	9.55	−0.201
B37						8.25	1.665
H84							−0.556

[a] $LSD_{(0.05)} = 2.87$.
[b] S.E. $(\hat{g}_i - \hat{g}_j) = 0.635$.

Table 9. F_1 means[a] and general combining ability[b] ($\hat{g}_i$) effects for R:S ratio from a partial diallel mating study involving seven maize inbreds (no parents, no reciprocals), grown under low-P concentration (Diallel 2)

Maize inbreds	Oh40B	H99	W64A	MS1334	B37	H84	GCA ($\hat{g}_i$)
		R:S ratio (g/g)					
B57	0.58	0.68	0.53	0.56	0.59	0.56	0.057
Oh40B		0.61	0.51	0.53	0.63	0.48	0.023
H99			0.48	0.56	0.69	0.53	0.068
W64A				0.49	0.51	0.39	−0.061
MS1334					0.53	0.40	−0.030
B37						0.42	0.031
H84							−0.088

[a] $LSD_{(0.05)} = 0.133$.
[b] S.E. $(\hat{g}_i - \hat{g}_j) = 0.002$.

R:S ratio seem to be the traits explaining most of the variation obtained for shoot dry weight at low-P concentration. Differences in shoot dry weight production in segregating populations of tomatoes grown in sand-alumina, low-P stress condition, were reported to be related to differences in abilities to acquire P than differences in internal P utilization (Coltman et al., 1987). Differences in shoot and root dry weight and total P in the plant, but no differences in P utilization efficiency, were also reported for 40 maize inbreds grown in low-P nutrient solution medium (Furlani et al., 1984).

The genetic studies showed that tolerance to low-P stress in maize seedlings grown in a sand-alumina culture medium is controlled by additive gene effects, although dominance also was important. The diallel studies, complemented by generation means analysis studies (data not reported here), suggested that a small number of loci might be controlling the expression of the various traits related with tolerance to low-P stress in maize. Reiter (1989) used two inbreds identified as tolerant and intolerant to low-P in my studies (NY821 and H99) and reported that five RFLP marker loci were associated with performance of maize under low-P stress, based on shoot dry matter production. The author also concluded that additive gene action on shoot dry matter production was predominant for all quantitative trait loci identified.

The heritable differences demonstrated by the maize inbreds in tolerating low-P stress conditions may warrant the use of these inbreds, as well as the screening methodology outlined here, in a maize breeding program aimed at improving maize for tolerance to low-P stress.

Acknowledgements

The authors are grateful to Enrico Colosimo and Dr. Sook-Fwe Yap for the statistical advices, Marcio Ferreira for field assistance and discussions and EMBRAPA for financial support.

References

Baker D E, Thomas W I and Gorsline G W 1964 Differential accumulation of strontium, calcium, and other elements by corn (*Zea mays* L.) under greenhouse and field conditions. Agron. J. 56, 352–355.

Barber W D, Thomas W I and Baker D E 1967 Inheritance of relative phosphorus accumulation in corn (*Zea mays* L.). Crop Sci. 7, 104–107.

Bertramson B R 1942 Phosphorus analysis of plant material. Plant Physiol. 17, 447–455.

Buso G S C and Bliss F A 1988 Variability among lettuce cultivars grown at two levels of available phosphorus. Plant and Soil 11, 67–74.

Coltman R R, Gerloff G C and Gabelman W H 1985 Differential tolerance of tomato strains to maintained and deficient levels of phosphorus. J. Am. Soc. Hort. Sci. 110, 140–144.

Coltman R R, Gabelman W H, Gerloff G C and Barta S 1987 Genetics and physiology of low-phosphorus tolerance in a family derived from two differentially adapted strains of tomato (*Lycopersicon esculentum* Mill.). *In* Genetic Aspects of Plant Mineral Nutrition. Eds. W H Gabelman and B C Loughman. pp. 309–315. Martinus Nijhoff Publishers, Dordrecht, The Netherlands.

Fawole I, Gabelman W H, Gerloff G C and Nordheim E V 1982 Heritability and efficiency in phosphorus utilization in beans (*Phaseolous vulgaris* L.) grown under phosphorus stress. J. Am. Soc. Hort. Sci. 107, 94–97.

Figdore S 1986 Physiology and inheritance of potassium efficiency and sodium substitution capacity in tomatoes (*Lycopersicon esculentum* Mill.) grown under low-K stress. Ph.D. Thesis, University of Wisconsin, Madison, WI.

Furlani A M C, Bataglia O C and Lima M 1984 Eficiência de linhagens de milho na absorção e/ou utilização de fósforo em solução nutritiva. Bragantia 43, 129–147.

Gerloff G C 1976 Plant efficiencies in the use of nitrogen, phosphorus, and potassium. *In* Proc. Workshop Plant Adaptation to Mineral Stress in Problem Soils. Ed. M J Wright. pp 162–173. Cornell Univ. Agric. Exp. Sta., Ithaca, NY.

Gerloff G C and Gabelman W H 1983 Genetic basis of inorganic plant nutrition. *In* Encyclopedia of Plant Physiology, New Series, vol. 15B. Eds. A Lauchli and R L Bieleski. pp 453–480. Springer-Verlag, Berlin and New York.

Giordano L B 1980 Inheritance and physiology of calcium utilization in tomatoes (*Lycopersicon esculentum* Mill.) grown under low-calcium stress. Ph.D. Thesis, University of Wisconsin, Madison, WI.

Gorsline G W, Thomas W I and Baker D E 1964 Inheritance of P, K, Mg, Cu, B, Zn, Mn, Al, and Fe concentrations by corn (*Zea mays* L.) leaves and grain. Crop Sci. 4, 207–210.

Gorz H J, Haskins F A, Pederson J F and Ross W M 1987 Combining ability effects for mineral elements in forage sorghum hybrids. Crop Sci. 27, 216–219.

Griffing B 1956 Concept of general and specific combining ability in relation to diallel crossing systems. Austr. J. Biol. Sci. 9, 463–493.

Harvey P H 1939 Hereditary variation in plant nutrition. Genetics 24, 437–461.

Jackson M L 1958 Soil Chemical Analysis. Prentice-Hall Inc., Englewood Cliffs, NJ.

Li Y 1989 Inheritance of efficiency in calcium utilization in tomatoes (*Lycopersicon esculentum* Mill.) grown under low-calcium stress. Ph.D. Thesis, University of Wisconsin, Madison, WI.

Lyness A S 1936 Varietal differences in the phosphorus feeding capacity of plants. Plant Physiol. 11, 665–688.

Makmur A 1977 Physiology and inheritance of efficiency in potassium utilization in tomatoes (*Lycopersicon esculentum* Mill.) grown under potassium stress. Ph.D. Thesis, University of Wisconsin, Madison, WI.

O'Sullivan J, Gabelman W H and Gerloff G C 1974 Variations in efficiency of nitrogen utilization in tomatoes (*Lycopersicon esculentum* Mill.) grown under nitrogen stress. J. Am. Soc. Hort. Sci. 99, 543–547.

Pereira P A A and Bliss F A 1987 Nitrogen fixation and plant growth of common bean (*Phaseolus vulgaris* L.) at different levels of phosphorus availabilty. Plant and Soil 104, 79–84.

Reiter R S 1989 Genetic analysis of tolerance to low-phosphorus stress in maize using restriction fragment length polymorphisms. Ph.D. Thesis, University of Wisconsin, Madison, WI.

SAS Institute Inc. 1985 SAS User's Guide: Statistics, Version 5 Edition. Cary, NC: SAS Institute Inc. 956 p.

Schettini T, Gabelman W H and Gerloff G C 1987 Incorporation of phosphorus efficiency from exotic germplasm into agriculturally adapted germplasm of common bean (*Phaseolus vulgaris* L.). *In* Genetic Aspects of Plant Mineral Nutrition. Eds. W H Gabelman and B C Loughman. pp 559–568. Martinus Nijhoff Publishers, Dordrecht, The Netherlands.

Shea P F, Gabelman W H and Gerloff G C 1967 The inheritance of efficiency in potassium utilization in snapbeans (*Phaseolus vulgaris* L.). Proc. Am. Soc. Hort. Sci. 91, 286–293.

Silva A E 1990 Inheritance studies of phosphorus acquisition and utilization efficiency in maize (*Zea mays* L.). Ph.D. Thesis, University of Wisconsin, Madison, WI.

Smith S N 1934 Response of inbred lines and crosses in maize to variations of nitrogen and phosphorus supplied as nutrients. J. Am. Soc. Agron. 26, 785–804.

Thomas W 1930 The feeding power of plants. Plant Physiol. 5, 443–489.

Whiteaker G, Gerloff G C, Gabelman W H and Lindgren D 1976 Intraspecific differences in growth of beans at stress levels of phosphorus. J. Am. Soc. Hort. Sci. 101. 472–475.

P. J. Randall et al. (Eds.), Genetic aspects of plant mineral nutrition, 251–260.

PLSO SV56

Inheritance of phosphorus response in white clover (*Trifolium repens* L.)

J.R. CARADUS[1], A.D. MACKAY[1], S. WEWALA[2]*, J. DUNLOP[1], A. HART[1], J. VAN DEN BOSCH[1], M.G. LAMBERT[1] and M.J.M. HAY[1]
[1]*DSIR Grasslands and* [2]*DSIR Physical Sciences, Private Bag, Palmerston North, New Zealand*

Key words: breeding, combining ability, dominance, genetics, heritability, inheritance, phosphorus response, *Trifolium repens*, white clover

Abstract

Genotypes of white clover that exhibited divergent responses to P were identified in a glasshouse pot trial. Six high P-responding genotypes were selected from previously identified high P-responding cultivars and 5 low P-responding genotypes were selected from previously identified low P-responding cultivars. These were crossed in a full diallel design without selfing and reciprocals were kept separate. The P-response of progeny lines was compared with parents. High P-response was dominant over low P-response with progeny from crosses between high and low P-response genotypes being similar to the high P-response parent. Reciprocal effects were not significant. The general combining abilities of high P-response genotypes were generally greater than that of the low P-response genotypes, although there were significant specific combining abilities. Narrow sense heritabilities for P response were moderate, 0.46 based on the linear coefficient and 0.33 based on the quadratic coefficient of the fitted response curves.

The mode of inheritance, feasibility of manipulating differences in P response by breeding and future directions of this work are discussed.

Introduction

In New Zealand phosphatic fertilizers are applied primarily to stimulate white clover (*Trifolium repens* L.) growth, because many soils are phosphorus (P) deficient. For the pastoral farmer application of phosphatic fertiliser is a major expense. A cultivar of white clover which required less P to sustain the same production as that of present cultivars or producing more dry matter with the same amount of phosphate would be of considerable value in maintaining the profitability of pastoral farming in New Zealand. A programme designed to develop such a cultivar was begun (Dunlop et al., 1990) with the initial aim of identifying germplasm differing in response to P. Subsequently, heritability for differences in response to P was determined and an understanding of the mode of inheritance of P response gained before undertaking field studies to verify the significance of selections.

A previous study (Mackay et al., 1990) had shown considerable variation for P response among a world collection of white clover cultivars. Eleven of these cultivars were selected to represent the range of variation. The present study describes the variation for P response among genotypes from these 11 cultivars, the identification and crossing of genotypes selected for high and low P-response and determines the heritability and mode of inheritance of P response in white clover.

P response is defined here as the change in dry matter yield with increasing level of P supply such that high P-responses are associated with a rapid increase in dry weight or plant size with small increases in P supply, the maximum yields being reached at lower P levels than for germ-

* Deceased.

plasm with a low P-response. Quadratic curves were fitted to describe response to P so that genotypes or lines designated as high P-response had a larger positive linear coefficient and a larger, but negative quadratic coefficient than low P-response genotypes or lines. To our knowledge there have been no previously published estimates of heritability or determinations of mode of inheritance for response to added P in any crop or forage species.

Materials and methods

The P response of white clover genotypes and progeny lines was determined in pots in a glasshouse at Palmerston North. The soil used was a Wainui silt loam (Typic Dystrochrept), with a pH of 5.0 and Olsen P of 6 mg kg^{-1} soil. It was collected from the field to a depth of 100 mm and passed moist through a 4-mm sieve. Phosphorus added as $Ca(H_2PO_4)$ was mixed into soil along with sulphur added as $CaSO_4$ (500 mg S kg^{-1} soil) and potassium as KCl (500 mg K kg^{-1} soil) and incubated at a soil moisture content of 50% for 30 days before packing into 120-mm diameter pots at a bulk density of 0.7 Mg m^{-3}. There were 7 P levels, 0, 60, 150, 250, 350, 450 and 550 mg P kg^{-1} soil with 3 replicates. A *Rhizobium trifolii* inoculum cocktail containing strains NZP 540, 541, 547, 548 and 560 was added to each pot one week after planting. Three times a week pots were watered to weight so that soil was at a gravimetric moisture content of 55%. At each watering pots were re-randomised within blocks, and blocks were removed within the glasshouse.

Phosphorus response of white clover genotypes

Ten genotypes (labelled A to J) were randomly selected from 11 cultivars (labelled 1 to 11) shown to vary in their response to added P (Table 1) (Mackay et al., 1990). These were grown in potting mix for 6 months until a minimum of 30 stolon tips were produced. Stolon tips consisting of 1 or 2 nodes were removed and pre-rooted in sand for 10 days before weighing and transplanting singly into pots. The study was conducted over a 6 week period during December and January 1988/89.

Selection and crossing programme

Selections were made of the six highest P-responding genotypes that came from high P-responding cultivars (i.e. Gwenda, Trifo, Ladino Giganteum Lodigiano, Viglasska and Huia) and five lowest P-responding genotypes that came from low P-responding cultivars (i.e. Isolation V, Dusi, Crau, El Lucero, G.23 and Luclair) (Table 1). A diallel cross, without selfing, was carried out with the 11 selected genotypes, keeping the reciprocals separate.

Table 1. Description of cultivars from which genotypes were randomly selected to identify high and low P response genotypes

No.	Cultivar	Type[a]	Origin	P response[b]	
				Fitted curve	Category
1	Gwenda	Intermediate	UK	$y = -3.68 + 0.200x - 0.00147x^2$	High
2	Trifo	Large	Denmark	$y = -3.41 + 0.166x - 0.00083x^2$	High
3	Ladino Giganteum Lodigiano	Ladino	Italy	$y = -3.42 + 0.132x - 0.00051x^2$	High
4	Viglasska	Intermediate	Czechoslovakia	$y = -3.38 + 0.139x - 0.00064x^2$	High
5	Huia	Intermediate	NZ	$y = -3.28 + 0.112x - 0.00037x^2$	Medium-high
6	Isolation V	Small	NZ	$y = -3.17 + 0.108x - 0.00024x^2$	Medium-low
7	Dusi	Ladino	South Africa	$y = -3.00 + 0.088x - 0.00015x^2$	Low
8	Crau	Large	France	$y = -3.14 + 0.067x - 0.00020x^2$	Low
9	El Lucero	Large	Argentina	$y = -3.01 + 0.071x - 0.00002x^2$	Low
10	G.23	Large	NZ	$y = -3.10 + 0.073x - 0.00002x^2$	Low
11	Luclair	Intermediate	France	$y = -3.28 + 0.081x - 0.00041x^2$	Medium-low

[a] from Caradus et al., 1989.
[b] quadratic fitted to principal component 1 from the trial reported by Mackay et al., 1990, where y = principal component 1 and x = P level.

Progeny test

The 110 progeny lines from the diallel cross were compared with the 11 parent genotypes to determine mode of inheritance of P response. From each line 63 seedlings were sown to provide stolon tips for direct comparison with the cloned parent genotypes. One stolon tip was taken from each seedling and 63 from each parent genotype and planted into sand for 10 days to form roots. Rooted stolon tips were weighed before being transplanted in sets of three into pots. The study was conducted over a 5 week period during December and January 1989/90.

Measurements and data analyses

Measurements were made of leaf number, stolon length, total shoot, leaf and stolon dry weight, proportion of leaf to total weight and individual leaf weight. It was argued that measurements other than just shoot dry weight might identify germplasm that would perform better in the field than if selection was based solely on shoot dry weight responses. It is known (Caradus and Snaydon, 1986) that pot studies poorly predict both production and persistence of white clover in the field, especially when shoot yield is the only criteria measured. For white clover to be effective in pasture it must not only be high yielding but also have good persistence. In many low-P soils persistence may be more important than productivity. These two characters, production and persistence, are generally negatively related. Production relates to the amount of leaf produced and persistence to the number of stolon growing points and stolon length per unit area. Hence in the field these two characters must be balanced to obtain a plant that is both reasonably productive and yet persistent. For these reasons measurement of the seven listed characters was made and summarised using principal component analysis.

Data from the genotype comparison were analysed by analysis of covariance with initial stolon weight as the covariate. For the progeny lines it was not necessary to analyse by covariance analysis since initial stolon weights were not correlated with final plant weights. Therefore analysis of variance was used. Where appropriate, data were log-transformed to improve homogeneity of variance.

Principal component analysis was carried out on log-transformed data (except where data were in the form of a proportion) to summarise the data while losing as little information as possible in the process. Quadratic response curves ($y = a + bx + cx^2$) were fitted to the first principal component. The linear and quadratic functions were ranked to compare genotype response. The linear (p) and quadratic (p^2) response functions were taken as measures of P response to compare progeny lines from the diallel cross with each other and their parents. A quadratic response curve was used since many cultivars and genotypes showed a reduction in yield at the highest levels of P applied. The form of the quadratic response curve used allowed indentification of not only genotypes that differ in their P response (based on the quadratic coefficient) but also those that differ in their tolerance of low levels of applied P, i.e. those that respond rapidly to only small additions of P (based on the linear coefficient).

General combining abilities (g.c.a.) and specific combining abilities (s.c.a.) were calculated for parent genotypes for p and p^2 using Experimental method 3, Model 1 of Griffing (1956). G.c.a. designates the average performance of a parental genotype in hybrid combination, and s.c.a. designates those cases in which certain combinations do relatively better or worse than would be expected on the basis of the average performance (i.e. g.c.a) of the genotypes involved (Sprague and Tatum, 1942).

Broad sense heritability for P response was calculated by equating the mean square expectations to the mean squares in the analysis of variance of genotypes. Heritability was determined by $\sigma^2_{gp}/\sigma^2 + \sigma^2_g + \sigma^2_r + \sigma^2_p + \sigma^2_{gp}$ where σ^2 was the error variance, σ^2_g the genotype variance, σ^2_r the replicate variance, σ^2_p the P level variance, and σ^2_{gp} the genotype $\times$ P level variance. Broad sense heritability includes all additive and non-additive genetic effects. Narrow sense heritabilities for the P response functions (p and p^2) were estimated by calculating mid-parent versus progeny regressions (Falconer, 1981). Narrow sense heritability is therefore the proportion of observed variability that is due to heredity and in particular the additive effects of genes, rather than that due to environmental causes.

Principal genetic components were calculated using the method of Hayman (1954) which is further explained by Mather and Jinks (1982). This method assumes a simple additive-dominance model of inheritance which can be tested by (a) regression analysis of the array (row or column) variance (V_r) versus the parent offspring covariance (W_r) which must show a significant linear regression of unit slope, and (b) analysis of variance of $W_r - V_r$ which should not be significant for the line differences.

Table 2. Eigenvectors of variables for the first and second principal components for (**a**) comparison of genotypes and (**b**) comparison of progeny lines

Variable	Principal component	
	1	2
(a)		
Log-shoot dry weight	0.36	−0.02
Log-leaf dry weight	0.35	−0.01
Log-stolon dry weight	0.35	0.00
Log-leaf number	0.32	−0.31
Log-stolon length	0.34	−0.22
Log-individual leaf weight	0.21	0.44
Proportion of leaf	−0.27	0.10
(b)		
Log-shoot dry weight	0.41	0.16
Log-leaf dry weight	0.39	−0.14
Log-stolon dry weight	0.41	−0.05
Log-leaf number	0.40	−0.07
Log-stolon length	0.41	0.12
Log-individual leaf weight	0.32	0.64
Proportion of leaf	−0.29	0.73

Results and discussion

P response of genotypes

There was a large amount of variation ($p < 0.001$) among genotypes for P response for all characters measured. The first and second principal component accounted for 69% and 15% of the variation respectively. The first principal component was a measure of plant size; eigenvectors associated with dry weight, leaf number and stolon length had the largest magnitudes (Table 2a). The second principal component was dominated by individual leaf dry weight. After ranking on both linear and quadratic response functions 65% of the top 20 and 70% of the top 10 genotypes came from high or medium-high-P-response cultivars; 70% of the bottom 20 and 80% of the bottom 10 genotypes came from low or medium-low P-response cultivars. This would suggest that repeatability for P response was reasonably good. Correlation of P response values of parent genotypes in the genotype screening trial and the progeny test were, however, not significant with $r = 0.51$ for the linear coefficient and $r = 0.24$ for the quadratic coefficient.

The fitted quadratic P-response curves of the 11 selected genotypes are given in Figure 1. The extent of the variation among genotypes for P-response was less than that previously observed among cultivars (Mackay et al., 1990), since coefficients of variation for the quadratic coefficient of the fitted P response curves were 84% for cultivars and 46% for genotypes.

Progeny test

Once again there was a large amount of variation ($p < 0.001$) among progeny lines and parent genotypes for P-response of all characters measured.

The first and second principal component accounted for 84% and 11% of the variation, respectively. The first principal component was again a measure of plant size and the second principal component was again a measure of plant type (Table 2b).

Principal component 1 scores of parent genotypes were analysed by response type group, i.e. either high P-response or low P-response. There was a significant ($p < 0.001$) difference between groups and a significant ($p < 0.01$) P-response group × P level interaction. The mean fitted quadratic curves of the two groups were however significantly ($p < 0.05$) different for the linear coefficient only (Table 3a, Fig. 2).

Principal component 1 scores of progeny were analysed according to the response type of their parents, i.e. progeny from crosses between

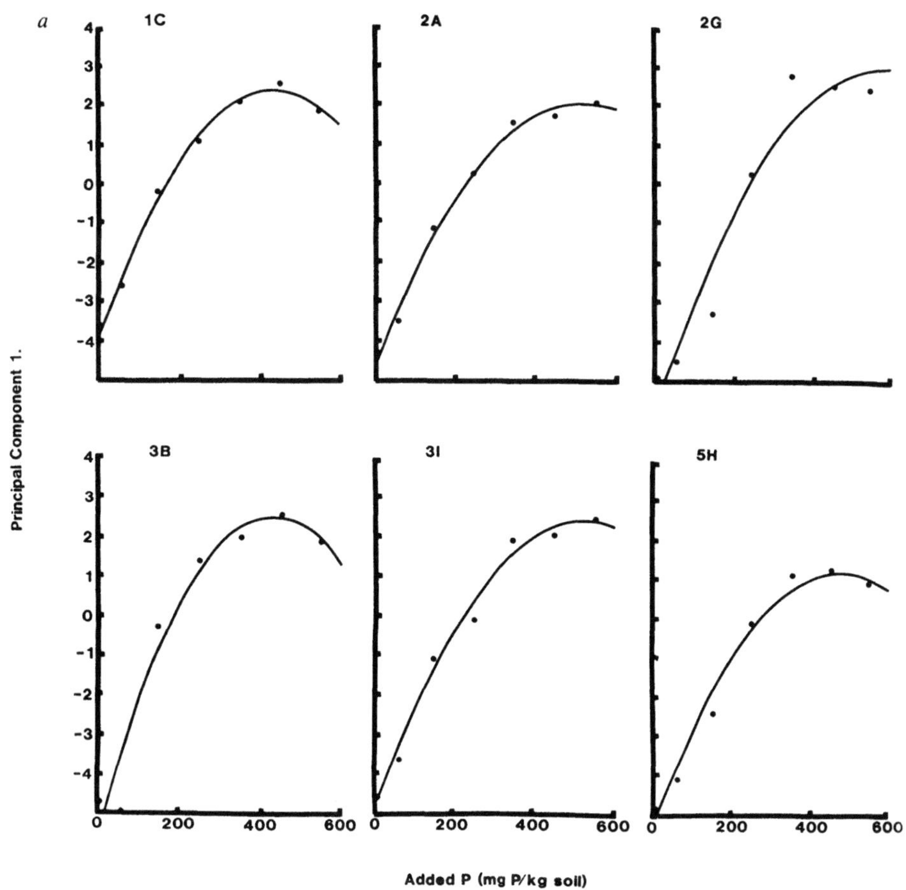

Fig. 1. Phosphorus response of white clover genotypes selected for (**a**) high P-response and (**b**) (*next page*) low P-response. The number in the parent genotype identifier refers to the cultivar from which it came (Table 1) and the letter to the genotype within that cultivar.

either high P-responding genotypes, low P-responding genotypes, or high P- and low P-responding genotypes. There was a significant ($p < 0.001$) difference between groups and a significant ($p < 0.001$) cross-response-type × P level interaction. Comparison of the coefficients of the mean fitted quadratic response curves of the three groups showed a difference between progeny from crosses between high P-responding parent genotypes and those from crosses between low P-responding parent genotypes (Table 3b, Fig. 3). Progeny from crosses between high P-responding and low P-responding parent genotypes were intermediate but closer to the high P-responding parent genotype group.

Combining abilities and analysis of reciprocals

There was no significant effect of reciprocals suggesting that maternal effects were non-existent or unimportant. Reciprocals were combined for analysis of g.c.a. and s.c.a. Both g.c.a. and s.c.a. effects were significant ($p < 0.05$) for the linear and quadratic coefficients of fitted P response curves. The g.c.a. of the high P-response parent genotypes tended to be higher than those of the low P-response parent genotypes; this was particularly the case for the quadratic coefficient (Table 4). S.c.a. effects for the linear P response coefficient were closely correlated with s.c.a. effects for the quadratic coefficient (r = 0.96,

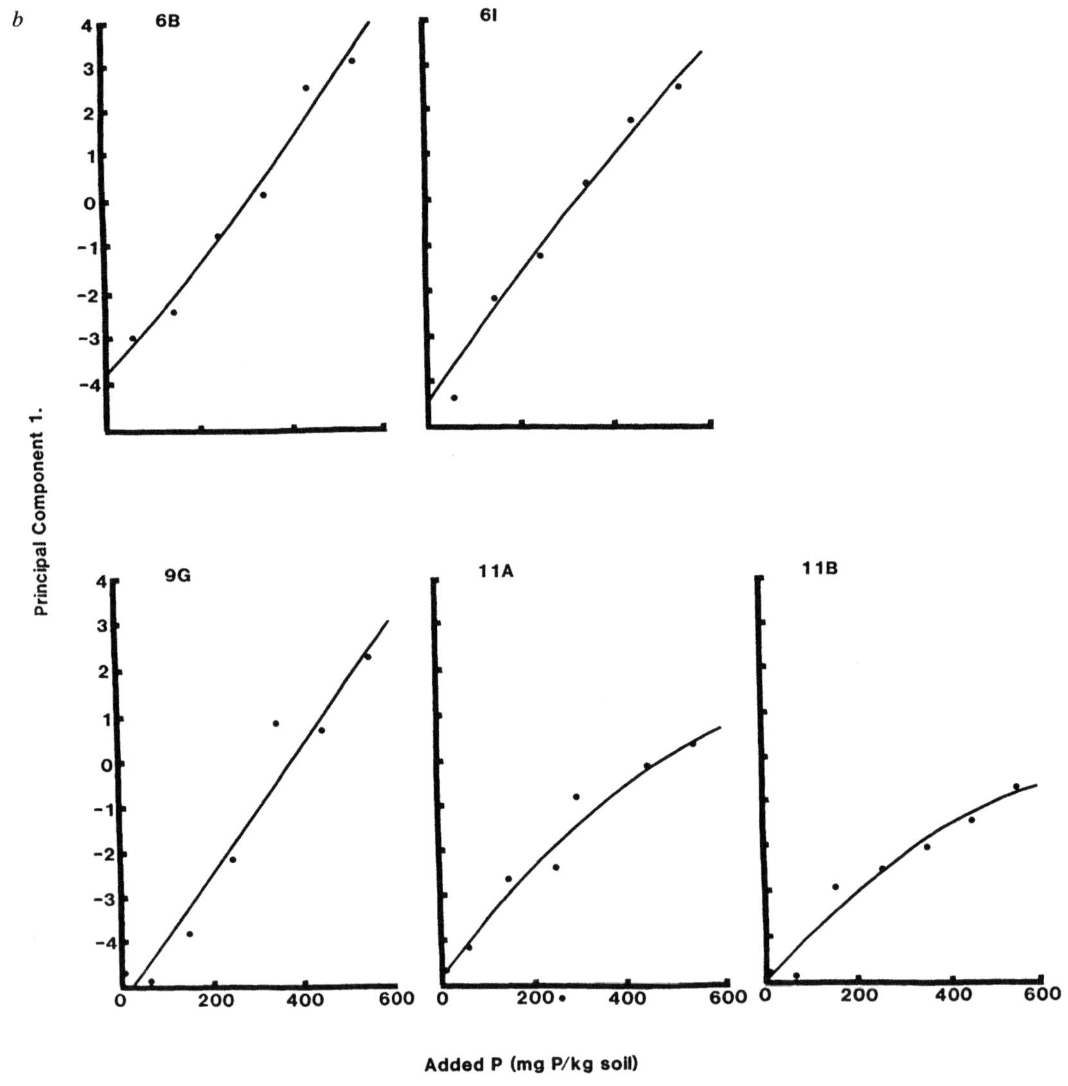

Fig. . (Continued)

$p < 0.001$). The largest negative s.c.a's were for 1C × 2G, 2A × 3B, 9G × 11A, 11A × 11B, 1C × 6I and 1C × 6B and the largest positive s.c.a's were for 1C × 11A, 1C × 11B and 2A × 11B.

Heritability of P response

Estimates of variance components from analysis of variance among genotypes for principal component 1 were used to calculate a broad sense heritability for P response (i.e. the interaction between P level and genotype). The variance components for replicate, genotype, P × genotype and error were 0.039, 0, 0.644 and 0.566 respectively. The proportion of the total variance due to the P × genotype interaction, and hence broad sense heritability for P response, was 0.52.

To obtain narrow sense heritabilities progeny coefficients were regressed on mid-parent coefficients, and calculated with either reciprocal combined or kept separate (Table 5). Narrow sense heritabilities were moderate, ranging from 0.30 to 0.46.

Table 3. Comparison of P response for principal component 1 of (**a**) parent genotypes selected as high P-responding (H) or low P-responding (L) and (**b**) progeny lines derived from crosses between high P-responding genotypes (HH), high and low P-responding genotypes (HL) and low P-responding genotypes (LL)

Genotype or progeny line group	No. of genotypes or lines	Quadratic coefficients	
		p	p^2
(**a**)			
H	6	0.0318	−0.0000411
L	5	0.0290	−0.0000375
Significance level		*	NS
$LSD_{0.05}$		0.0020	0.0000043
(**b**)			
HH	30	0.0348	−0.0000488
HL	60	0.0344	−0.0000474
LL	20	0.0326	−0.0000436
Significance level		*	*
$LSD_{0.05}$		0.016	0.0000041

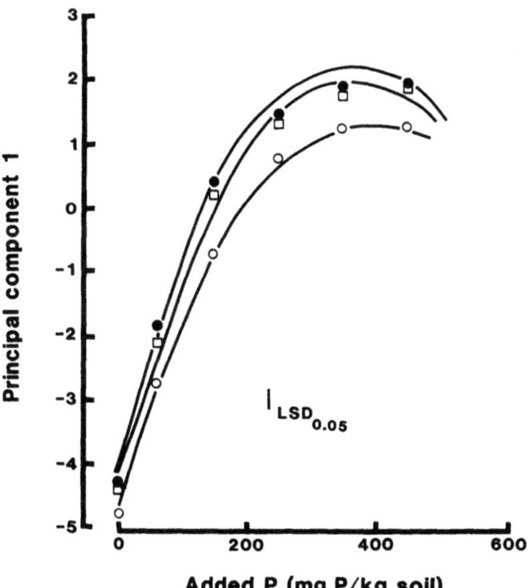

Fig. 3. Mean phosphorus response of progeny from crosses between high P-response genotypes (●) and between low P-response genotypes (○), and between high P-response and low P-response genotypes (□).

Table 4. General combining ability effects for P response based on the linear (p) and quadratic (p^2) coefficient of fitted quadratic P response curves. The number in the parent genotype identifier refers to the cultivar from which it came (Table 1) and the letter to the genotype within that cultivar

Parent genotype	Quadratic coefficient	
	p	p^2
1C	0.091	0.0165
2A	−0.022	0.0196
2G	0.046	0.0062
3B	0.137	0.0276
3I	−0.026	0.0014
5H	0.093	0.0019
6B	0.028	−0.0065
6I	−0.003	−0.0041
9G	−0.117	−0.0279
11A	−0.064	−0.0110
11B	−0.163	−0.0237
Significance level	*	*
$LSD_{0.05}$	0.105	0.0198

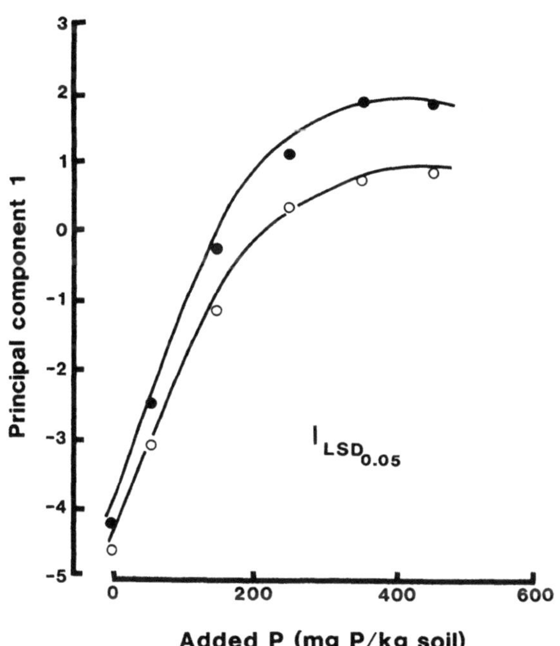

Fig. 2. Mean phosphorus response of white clover genotypes selected for high P-response (●) and low P-response (○) when grown in the progeny testing trial.

Mode of inheritance for P response

P response estimates determined by the linear (p) and quadratic (p^2) coefficients of the fitted quadratic response curves fulfilled the assumption of a simple additive-dominance model of

Table 5. Narrow sense heritabilities for P response functions calculated from progeny versus mid-parent regressions, with and without reciprocals combined. Standard errors and significance levels (in brackets) are given for each heritability estimate

P Response function	Reciprocals combined	Reciprocals separate
Linear coefficient (p)	0.30 ± 0.13 (0.028)	0.46 ± 0.12 (0.0002)
Quadratic coefficient (p^2)	0.36 ± 0.15 (0.016)	0.33 ± 0.14 (0.0177)

inheritance, thus allowing the use of Hayman's (1954) analysis for determining the significance of principal genetic components. For both the linear and quadratic coefficients the W_r/V_r regression coefficients were not significantly different from unity (Fig. 4a and b, respectively), and the analyses of $W_r - V_r$ not significant. A significant difference among arrays (i.e. all the progenies of a common parent) for $W_r + V_r$ (parental order of dominance) ($p = 0.04$ for 'p' and $p = 0.02$ for 'p^2'), but not for $W_r - V_r$ ($p > 0.05$) would indicate non-additive genetic variation due to dominance effects of genes, which are independently distributed among the parental genotypes. This is further supported by the slopes of W_r versus V_r plots (Fig. 4) being significantly different from zero but not significantly ($p < 0.05$) different from unity.

The genetic components of variation (Table 6) estimated from procedures given by Hayman (1954) can be used to calculate derived values. The ratio $\sqrt{H_1/D}$ measures the degree of dominance independently of gene frequency (Hayman, 1960) and being greater than 1 (Table 6) indicates at least complete dominance rather than over-dominance since the magnitude of P response coefficients from fitted quadratic curves were similar for parent genotypes and progeny lines. The correlation between Y_r (mean of progeny from a parent, plus the parent mean) and $W_r + V_r$ (parental order of dominance) was negative for both p and p^2 ($r = -0.56$ ns and -0.77 $p < 0.01$, respectively) indicating that high P-response was dominant over low P-response. This is confirmed by the closer proximity of the P response curve for progeny from crosses between high P-response and low P-response genotypes to the curve for progeny from crosses between high P-response genotypes than that for crosses between low P-response genotypes (Fig. 3).

The array (progenies from a common parent) points are distributed along the regression line (W_r versus V_r) in the order of the number of dominant alleles carried by their common par-

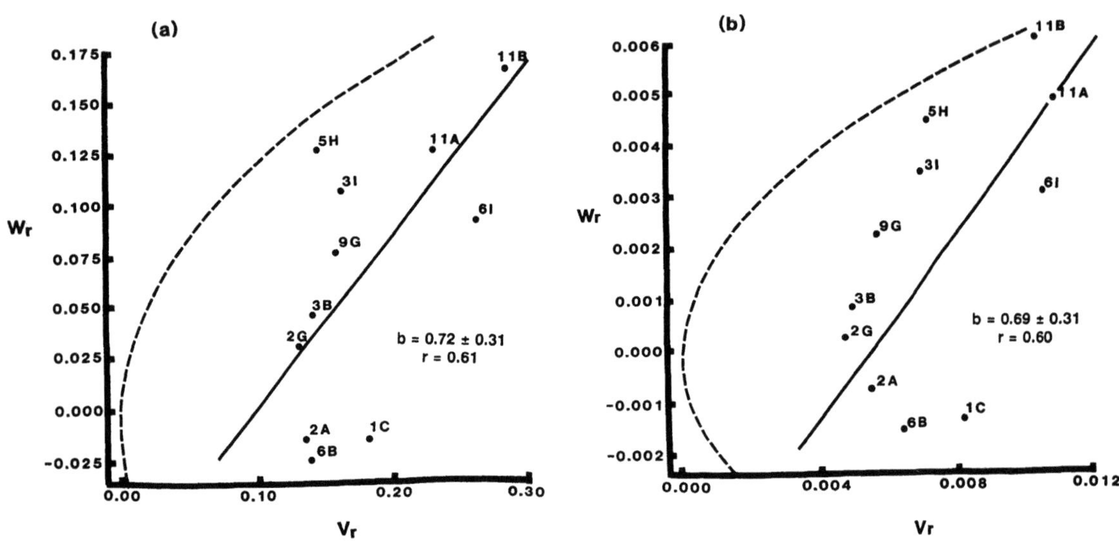

Fig. 4. Covariance (W_r) versus variance (V_r) regressions of P response estimated as (**a**) the linear coefficient and (p) and (**b**) the quadratic coefficient (p^2) of fitted quadratic functions. The number in the parent genotype identifier refers to the cultivar from which it came (Table 1) and the letter to the genotype within that cultivar.

Table 6. Estimated genetic components of variation for P response determined by linear (p) and quadratic (p^2) coefficients of fitted quadratic response curves

	Quadratic coefficients	
	p	p^2
Component of variation (± standard error)		
D	0.1363 ± 0.1044	0.0040 ± 0.0035
H_1	0.1493 ± 0.1730	0.0061 ± 0.0056
H_2	0.1253 ± 0.1069	0.0046 ± 0.0040
F	0.1103 ± 0.1878	0.0044 ± 0.0045
h^2	0.4611 ± 0.1806	0.0186 ± 0.0084
Derived values		
$\sqrt{H_1/D}$ (Dominance ratio)	1.05	1.22
$H_2/4H_1$ (Asymmetry of loci)	0.21	0.19
$F/2\sqrt{D(H_1 - H_2}$ (Variation in dominance level across loci)	0.96	0.92
$(\sqrt{4DH_1} + F)/(\sqrt{4DH_1} - F)$ (Ratio of dominant to recessive alleles)	2.24	2.57
h^2/H_2 (Number of effective factors)	3.68	3.96

ent. This order is relatively similar for the two measures of P response used; thus for the linear response coefficient (p) the order from the highest to the least number of dominant alleles is 6B, 2A, 1C, 2G, 3B, 9G, 2G, 3I, 6I, 5H, 11A and 11B and for the quadratic response coefficient (p^2) the order is 6B, 1C, 2A, 2G, 3B, 9G, 3I, 6I, 5H, 11A and 11B (Fig. 4).

Since $H_2/4H_1$ was less than 0.25 (Table 6) the frequency of dominant and recessive alleles was not equal over all loci that might affect P response (Mather and Jinks, 1982). However since

$$F/2\sqrt{D(H_1 - H_2)}$$

was approximately 1.0 dominance was the result of dominance at all loci rather than dominance at some and not at others. Since F is positive (Table 6) dominant alleles are more frequent than recessive alleles. The ratio of dominant to recessive genes in all parents ($[\sqrt{4DH_1} + F]/[\sqrt{4DH_1} - F]$) was approximately 2 for both linear and quadratic estimates of P responses (Table 6). The ratio of h^2/H_2 gives an estimate of the number of effective factors, which may or may not be synonymous with number of genes, or groups of genes, which not only influence but exhibit some degree of dominance for the character being studied. Since it provides no information about effective factors or genes exhibiting little or no dominance it should be considered an underestimate. In the genotypes of white clover examined here at least 4 individual or groups of effective factors or genes, are involved in P response (Table 6).

Breeding implications

P response in white clover is genetically controlled. Heritability is moderate but at a level where selection for P response will be possible. Additionally, high P-response was dominant over low P-response and genotypes with high P-response had higher g.c.a's than those with low P-response. This study has shown that it is possible to select for high and low P-response in a relatively controlled environment and that it should be feasible to incorporate high P-response characteristics into agronomically suitable germplasm so that the advantages and adaptive significance of this character can be evaluated in the field.

Acknowledgements

To Ian Black, Philip Budding, Renee Carson, Wendy Collier, Des Costall, Sarah Gardiner, Margaret Carline, Allison MacKay, Vanessa Pokaia and Jocelyn van der Sar for technical assistance, Cherry Liddane, DSIR Fruit and Trees, for supplying *Rhizobium* inoculum, and Gary Thomas, DSIR Physical Sciences, for statistical advice after the untimely death of Siri Wewala.

References

Caradus J R and Snaydon R W 1986 Response to phosphorus of populations of white clover. 3. Comparison of experimental techniques. N.Z.J. Agric. Res. 29, 169–178.

Caradus J R, MacKay A C, Woodfield D R, van den Bosch J and Wewala G S 1989 Classification of a world collection of white clover cultivars. Euphytica 42, 183–196.

Dunlop J, Lambert M G, van den Bosch J, Caradus J R, Hart A L, Wewala G S, Mackay A D and Hay M J M 1990 A programme to breed a cultivar of *Trifolium repens* L. for more efficient phosphate. *In* Genetic Aspects of Plant Mineral Nutrition. Eds. N El Bassam, M Dambroth and B C Loughman. pp 547–552. Kluwer Academic Publishers, Dordrecht, The Netherlands.

Falconer D S 1981 Introduction to Quantitative Genetics. Oliver and Boyd, Edinburgh. 365 p.

Griffing B 1956 Concept of general and specific combining ability in relation to diallel crossing systems. Aust. J. Biol. Sci. 9, 463–493.

Hayman B 1954 The theory and analysis of diallel crosses. Genetics 39, 789–809.

Mackay A D, Caradus J R, Dunlop J, Wewala G S, Mouat M C H, Lambert M G, Hart A L and van den Bosch J 1990 Response to phosphorus of a world collection of white clover cultivars. *In* Genetic Aspects of Plant Mineral Nutrition. Eds. N El Bassam, M Dambroth and B C Loughman. pp 553–558. Kluwer Academic Publishers, Dordrecht, The Netherlands.

Mather K and Jinks J L 1982 Biometrical Genetics. Chapman and Hall Ltd., London. 382 p.

Sprague E F and Tatum L A 1942 General versus specific combining ability in single crosses of corn. J. Am. Soc. Agron. 34, 923–932.

P. J. Randall et al. (Eds.), Genetic aspects of plant mineral nutrition, 261–269.

PLSO SV66

Heritability of, and relationships between phosphorus and nitrogen concentration in shoot, stolon and root of white clover (*Trifolium repens* L.)

J.R. CARADUS
DSIR Grasslands, Private Bag, Palmerston North, New Zealand

Key words: heritability, nitrogen concentration, plant parts, phosphorus concentration, *Trifolium repens* L., white clover

Abstract

Ninety eight white clover genotypes were cloned and grown in pots at two levels of phosphorus (P) supply in soil. After harvest the nitrogen (N) and P content of shoot (leaf, petiole and unrooted stolon), stolon and root tissue was determined. Broad sense heritabilities for %N, %P, and proportion of total N or P in each tissue type were calculated. Heritabilities ranged from 0.22 to 0.68. They were generally higher for %P than %N; and higher in shoot and stolon tissue than root tissue for %P, %N, and proportion of N or P. Level of P in which plants were grown had little effect on heritability values. Genotypes from bred cultivars differed from those collected from hill country pastures for plant size, and partitioning of N and P to shoot, stolon and root. Relationships between plant characters were examined to determine the consequences of selection.

Introduction

Plant nutrient concentration is often used to determine the extent of nutrient deficiencies (Mengel and Kirkby, 1982), efficiency of nutrient use (Blair and Cordero, 1978; Clark, 1990; Longeragan and Asher, 1967; Ramirez, 1982) or quality of plant products (Miller et al., 1987). White clover (*Trifolium repens* L.) is the most important forage legume of grazed temperate pastures, contributing significantly to pasture quality. Two important plant nutrients associated with quality are phosphorus (P) and nitrogen (N). Nitrogen is an important component of protein and since white clover is grown as an animal feed, P content needs to be in the range of 0.15 to 0.5% for adequate animal nutrition depending on stock class, animal maturity and reproductive status (Lambert and Toussaint, 1978; Underwood, 1981).

Intraspecific variation for plant nutrient concentration has been frequently observed (Krstic and Sarić, 1990). However, there are relatively few estimates of heritability. The aim of the present study was to estimate the heritability for P and N concentration, ratio and the distribution of P and N within a range of plant tissues of white clover grown under P deficient or sufficient conditions. Additionally, correlation analyses were used to identify characters that vary together since when selecting for a trait it is important to be aware of associated characters that may similarly change in response to selection. This paper is the second in a series which firstly, examined (i) the extent of variation for %P, %N and partitioning of N, P and dry matter to plant components, (ii) the effect of P supply on these characters, and (iii) the effect on these characters of selection for white clover plants able to tolerate grazing in low soil P conditions (Caradus, 1986), and secondly, in this paper, examines (iv) the heritability of %P, %N and partitioning of N and P to plant components, (v) the effect of P supply on heritability estimates and (vi) relationships between plant characters and the consequences of selection for high and low %P.

Materials and method

Plant material

A total of 98 white clover genotypes varying in origin and shoot morphology were studied. Twenty genotypes were collected from a range of soils on roadside cuttings in the North Island; 30 genotypes were from five cultivars (Kent Wild White, Kopu, Huia, S.100, and Regal)(6 genotypes from each cultivar); and 48 genotypes were from an extensive collection of hill country white clovers (Suckling and Forde, 1978). A full description of genotypes is given by Caradus (1986). Genotypes were initially grown in potting mix until sufficient stolon tips were generated to allow vegetative propagation.

Soil

The soil was a B horizon Egmont sandy loam, a free draining, low nitrogen, highly P retentive soil (N.Z. Soil Bureau, 1968). Soil was sieved through a 4 mm mesh before use. This soil had a pH of 6.4, total N of 0.15% and Olsen P of 1 ppm.

Conduct of study

Phosphorus as superphosphate (9%P) was mixed into the soil at 300 mg P kg^{-1} soil (low-P treatment) and 2000 mg P kg^{-1} soil (high-P treatment) three weeks before planting. Adequate amounts of all other nutrients, except nitrogen were also added. Soil was packed into pots (23 × 23 cm by 30 cm deep) to give a bulk density of 0.7 Mg m^{-3}. Soil was not sterilized and no extra Rhizobium inoculum was added since good nodulation of white clover on this soil was known to occur (Caradus et al. 1980). In early spring pre-rooted stolon tips were planted individually into pots. There were three replicates and pots were completely randomised in high-P and low-P pairs in a glasshouse. Pots were watered daily. Growing conditions are given in Caradus (1986). At the end of the growing period the low soil P level was 26 ppm Olsen P and the high soil P level 140 ppm Olsen P.

Measurement and data analysis

Shoots comprising leaf, petiole and unrooted stolon were harvested at 8 and 12 weeks and discarded. This shoot material was considered equivalent to the component of the plant available to the grazing animal. At 16 weeks shoot (leaf, petiole and unrooted stolon), rooted stolon and root were harvested separately, and weighed dry. Dried shoot, stolon and root material from the 16 week harvest was ground, digested using a Kjeldhal mixture and analysed for nitrogen and P content, for each plant individually. Before the final harvest the width of one middle leaflet, of a second open leaf back from the stolon tip was measured, per plant.

Data for each P level were analysed separately using a completely randomised design. The expectations of the mean squares for this analysis are given in Table 1. Estimates of total genetic variance (σ^2g) and error variance (σ^2) were calculated by equating the mean square expectations to the mean squares in the analysis of variance. Phenotypic variance (σ^2p) was calculated as σ^2 and σ^2g. Heritability (h^2_β) determined by $\sigma^2 g/\sigma^2 p$ (Scossiroli et al., 1963) should be considered as broad sense because all genetic effects are included and descriptive since the genotypic variance is the total variance due to the average effect of genes, as distinct from

Table 1. Analysis of variance and mean square expectations where r = number of replicates and g = number of genotypes

Source of variation	df	Components of variation
Total	gr − 1	
Between genotypes	g − 1	$\sigma^2 + r\sigma^2_g$
Between genotypes within genotypes	(gr − 1) − (g − 1)	σ^2

actual inherited genetic variability. Broad sense heritability should be regarded as an estimate of the upper limit of heritability in the narrow sense (Robinson, 1963) and the degree of genetic determination of the character (Falconer, 1961). Heritability has value primarily as a method for quantifying the concept of whether progress from selection for a plant character is relatively easy or difficult to make in a breeding programme (Hanson, 1963).

The genetic coefficient of variation, calculated as the square root of the genetic variance ($\sigma^2 g$) expressed as a percentage of the mean (Burton and DeVane, 1953), provides an index of the potential genetic advance that might be made with populations similar to the one being studied.

Phenotypic correlation coefficients were calculated separately for plants grown at high-P and low-P, for all character combinations using genotype means. Data were also analysed by analysis of variance with plant component (i.e. shoot, leaf and root) as a factor using a split-plot plant component design which is analogous to a split plot in time design.

Genotypes were grouped by a origin and breeding history into roadside (n = 20), hill country (n = 48) and cultivar (n = 30) and compared by one way analysis of variance.

Results

Heritability and genetic variation of characters

Broad sense heritabilities for leaflet width and dry weight of each component were high irrespective of P level at which genotypes were grown (Table 2). Heritabilities for %P, %N, and proportion of P and N tended to be lower for the root component than for shoot and stolon, except for %P at low-P supply (Table 2). Lowest heritabilities were for proportion of P and N in root and root %N, at both P levels. Genetic coefficients of variation were higher for dry weight than leaf size, which in turn were higher than for all nutrient related characters (Table 2).

Genetic coefficients of variation were higher for %P than %N irrespective of plant part.

Table 2. Mean, genetic variance ($\sigma^2 g$), broad sense heritability (h^2_β) and genetic coefficient of variation ($CVg = (\sqrt{\sigma^2} \times 100) / x$) for characters of genotypes grown at 300 mg P kg^{-1} soil (low-P) and 2000 mg P kg^{-1} soil (high-P)

Plant character	Low-P level				High-P level			
	$\bar{x}$	σ^2_g	h^2_β	CVg	$\bar{x}$	σ^2_g	h^2_β	CVg
Leaflet width (mm)	9.4	19.61	0.92	47.1	10.0	20.79	0.90	45.6
Shoot dry weight (g)	2.39	10.69	0.76	136.8	4.26	27.59	0.77	123.3
Stolon dry weight (g)	1.03	1.22	0.82	107.2	1.78	2.42	0.75	87.4
Root dry weight (g)	1.32	2.49	0.78	119.5	1.79	3.66	0.83	106.9
Shoot %P	0.38	0.0018	0.47	11.2	0.58	0.0036	0.65	10.3
Stolon %P	0.41	0.0075	0.56	21.1	0.78	0.0108	0.68	13.3
Root %P	0.42	0.0044	0.49	15.8	0.88	0.0169	0.46	14.8
Shoot %N	4.48	0.089	0.56	6.7	4.67	0.065	0.56	5.5
Stolon %N	2.69	0.119	0.57	12.8	2.71	0.052	0.45	8.4
Root %N	2.75	0.022	0.29	5.4	2.82	0.027	0.33	5.8
Prop^n P shoot	0.44	0.0046	0.42	15.5	0.40	0.0038	0.42	15.4
Prop^n P stolon	0.25	0.0033	0.54	23.0	0.30	0.0029	0.53	18.0
Prop^n P root	0.31	0.0011	0.23	10.7	0.30	0.0014	0.37	12.5
Prop^n N shoot	0.58	0.0027	0.29	9.0	0.61	0.0027	0.36	8.5
Prop^n N stolon	0.19	0.0018	0.45	22.3	0.20	0.0017	0.45	20.6
Prop^n N root	0.23	0.0009	0.23	13.0	0.19	0.0005	0.26	11.8

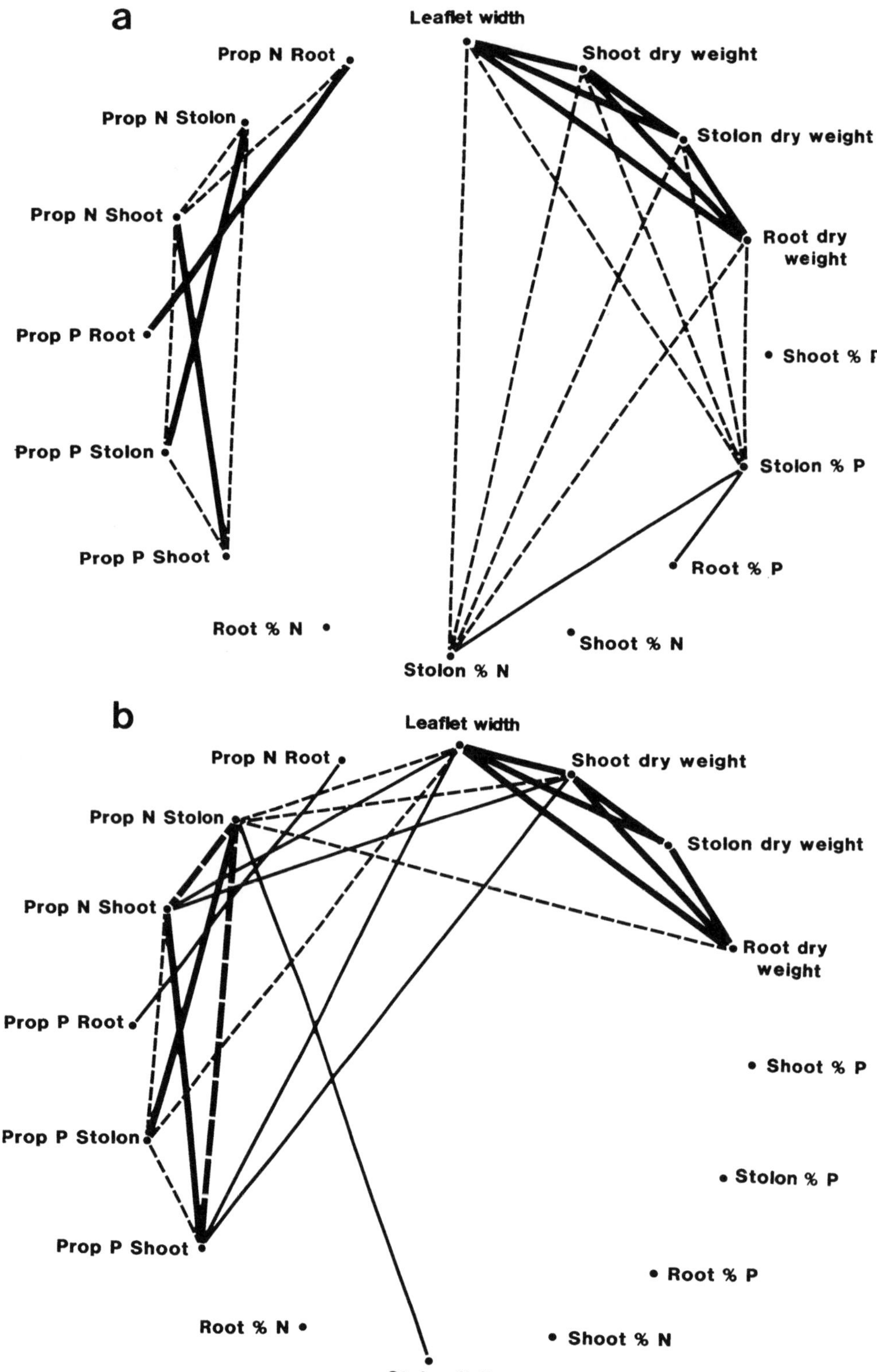

Fig. 1. The degree to which characters varied together among genotypes at (**a**) 300 mg P kg^{-1} soil and (**b**) 2000 mg P kg^{-1} soil. *Broken lines* indicate negative correlations and *solid lines* indicated positive correlations when $r \geq 0.87$ (——) or $r \geq 0.71 < 0.87$ (——).

Relationships between plant characters

Only comparisons where the degree to which characters varied together was greater than 50% (i.e. $r > 0.71$) will be described (Fig. 1).

At both P levels the largest leaved plants had the highest dry weight of shoot, stolon and root (Fig. 1a, b). Variation for shoot %P, shoot %N and root %N appeared to be independent of variation observed for most other characters. Shoot, stolon and root %P and %N as a group varied independently at both P levels, except at low-P where genotypes with high stolon %P also had high root %P and high stolon %N. At low-P only, both stolon %P and %N were negatively associated with leaf size and dry weight, such that genotypes with a high stolon %P and %N were generally small leaved with low dry weights.

Shoot, stolon and root %P and %N varied independently of proportion of N and P in shoot, stolon and root at both P levels, with the exception, that at high-P genotypes with high stolon %N had a high proportion of N in stolon (Fig. 1a, b). At high-P only was there an association between proportion of N and P in shoot and both leaf size and dry weight, such that genotypes with a high proportion of N and P in shoot were large leaved and had high dry weights (Fig. 1b).

A high proportion of N in shoot was associated with a high proportion of P in shoot; likewise for proportion of N and P in stolon and proportion of N and P in root. At both P levels a high proportion of P in shoot was associated with a low proportion of P in stolon; the same applied for proportion of N. A high proportion of P in shoot was associated with a low proportion of N in stolon, and similarly a high proportion of N in shoot was associated with a low proportion of P in stolon, at both P levels.

The interaction between genotype and plant part

There were highly significant ($p < 0.001$) genotype × plant part (i.e. shoot, stolon and root) interactions for %P, %N, proportion of N and P and dry weight, measured at both P levels. Only examples of the types of differences observed due to interactions for %P and %N will be described.

At low-P, analysed over all genotypes %P of stolon and root were significantly ($p < 0.001$) higher than for shoot. However, for 55% of genotypes there was no significant ($p > 0.05$) difference between plant parts for %P (e.g. genotype 36, Fig. 2a). For 13% of genotypes both root % and stolon %P were higher ($p < 0.05$) than shoot %P (e.g. genotype 26, Fig. 2a); for 12%, only stolon %P was higher ($p < 0.05$)

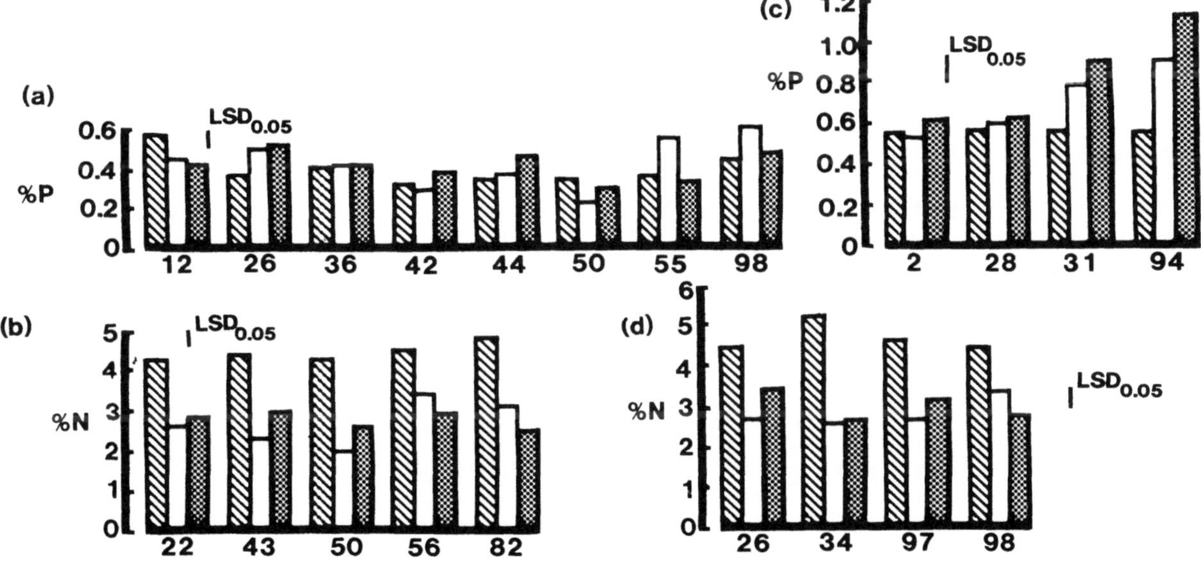

Fig. 2. Comparison of example genotypes for variation between shoot ▧, stolon □, and root ▩ components at low-P (300 mg P kg^{-1} soil) for (a) %P and (b) %N, and at high P (2000 mg P kg^{-1} soil) for (c) %P and (d) %N.

than shoot and/or root %P (e.g. genotypes 55 and 98, figure 2a); for 13%, only root %P was higher ($p < 0.05$) than shoot and/or stolon %P (e.g. genotypes 42 and 44, Fig. 2a); and, for 5%, only shoot %P was greater ($p < 0.05$) than stolon and/or root %P (genotypes 12 and 50, Fig. 2a).

At low-P, shoot %N was higher ($p < 0.01$) than stolon %N and root %N for all genotypes. Also, 74% of genotypes showed no significant ($p > 0.05$) difference between stolon and root for %N (e.g. genotype 22, Fig. 2b). Of the remaining genotypes root %N was significantly ($p < 0.05$) higher than stolon %N for 16 genotypes, of which 12 come from the five cultivars studied, 3 from the hill country collection and 1 from the roadside collection (e.g. genotypes 43 and 50, Fig. 2b). Stolon %N was significantly ($p < 0.05$) higher than root %N for only 10 genotypes, eight came from the hill country collection, and two from the roadside collection (e.g. genotypes 56 and 82, Fig. 2b). Significantly no genotypes from the bred cultivars had a %N in stolons higher than %N in roots.

At high P, shoot %P was significantly ($p < 0.05$) lower than stolon %P for 75 genotypes, and root %P for 90 genotypes (e.g. genotypes 31 and 94, Fig. 2c, respectively). For six genotypes there were no significant differences between the three plant parts for %P (e.g. genotypes 2 and 28, Fig. 2c).

For all genotypes at high-P, shoot %N was significantly ($p < 0.01$) higher than either stolon %N or root %N. For the majority of genotypes (80%) there was no significant difference between stolon %N and root %N (e.g. genotype 34, Fig. 2d). Root %N was significantly ($p < 0.05$) higher than stolon %N for 16 genotypes, of which 13 came from the cultivars and three from the hill country collection (e.g. genotypes 26 and 97, Fig. 2d). For the remaining five genotypes, stolon %N was significantly ($p < 0.05$) higher than root %N; of these four were from the hill country collection and one from the roadside collection (e.g. genotype 98, Fig. 2d). As at low-P no genotypes from the bred cultivars had a %N in stolon higher than %N in roots.

Comparison of genotypes by group

Genotypes of bred cultivars as a group were larger leaved and heavier with a higher proportion of P and N in shoots and lower proportion of P and N in stolon and roots than hill country

Table 3. Means and standard errors for characters analysed by group – either roadside genotypes (n = 20), hill country genotypes (n = 48) or cultivar genotypes (n = 30) at 300 mg P kg^{-1} soil (low-P) and 2000 mg P kg^{-1} soil (high-P)

Plant character	Low-P				High-P			
	Roadside	Hill country	Cultivar	P	Roadside	Hill country	Cultivar	P
Leaflet width (mm)	8.8 ± 0.6	6.8 ± 0.2	13.9 ± 1.0	* * *	9.1 ± 0.6	7.4 ± 0.2	14.7 ± 1.0	* * *
Shoot dry weight (g)	1.23 ± 0.18	0.60 ± 0.04	6.02 ± 0.80	* * *	2.49 ± 0.46	1.30 ± 0.11	10.17 ± 1.23	* * *
Stolon dry weight (g)	0.62 ± 0.08	0.34 ± 0.02	2.42 ± 0.21	* * *	1.17 ± 0.14	0.79 ± 0.05	3.77 ± 0.29	* * *
Root dry weight (g)	0.72 ± 0.09	0.40 ± 0.02	3.20 ± 0.35	* * *	1.02 ± 0.12	0.67 ± 0.04	4.08 ± 0.40	* * *
Shoot %P	0.40 ± 0.01	0.39 ± 0.01	0.34 ± 0.01	* * *	0.56 ± 0.01	0.57 ± 0.01	0.60 ± 0.01	ns
Stolon %P	0.42 ± 0.02	0.46 ± 0.01	0.33 ± 0.02	* * *	0.77 ± 0.03	0.78 ± 0.01	0.79 ± 0.02	ns
Root %P	0.44 ± 0.01	0.45 ± 0.01	0.37 ± 0.02	* * *	0.82 ± 0.03	0.89 ± 0.02	0.90 ± 0.03	ns
Shoot %N	4.43 ± 0.06	4.56 ± 0.07	4.41 ± 0.03	ns	4.59 ± 0.05	4.71 ± 0.04	4.65 ± 0.07	ns
Stolon %N	2.76 ± 0.04	2.90 ± 0.04	2.32 ± 0.02	* * *	2.74 ± 0.04	2.83 ± 0.03	2.50 ± 0.05	* * *
Root %N	2.74 ± 0.07	2.80 ± 0.06	2.68 ± 0.05	ns	2.74 ± 0.04	2.82 ± 0.03	2.87 ± 0.04	ns
Prop^n P shoot	0.46 ± 0.01	0.40 ± 0.01	0.48 ± 0.02	* * *	0.40 ± 0.01	0.36 ± 0.01	0.45 ± 0.01	* * *
Prop^n P stolon	0.24 ± 0.01	0.27 ± 0.01	0.22 ± 0.01	* * *	0.31 ± 0.01	0.32 ± 0.01	0.26 ± 0.01	* * *
Prop^n P root	0.30 ± 0.01	0.33 ± 0.01	0.30 ± 0.01	*	0.29 ± 0.01	0.32 ± 0.01	0.29 ± 0.01	*
Prop^n N shoot	0.59 ± 0.01	0.55 ± 0.01	0.62 ± 0.01	* * *	0.61 ± 0.01	0.57 ± 0.01	0.66 ± 0.01	* * *
Prop^n N stolon	0.19 ± 0.01	0.21 ± 0.01	0.16 ± 0.01	* * *	0.21 ± 0.01	0.23 ± 0.01	0.16 ± 0.01	* * *
Prop^n N root	0.22 ± 0.01	0.24 ± 0.01	0.22 ± 0.01	*	0.18 ± 0.006	0.20 ± 0.005	0.18 ± 0.004	*

genotypes at both P levels (Table 3). Roadside genotypes tended to be intermediate for these characters. At low-P, shoot, stolon and root %P and stolon %N of cultivar genotypes were lower than that for both roadside and hill country genotypes. However, at high-P there was no significant ($p > 0.05$) difference between the three groups for shoot, stolon and root %P. At both P levels, stolon %N of cultivar genotypes were lower than that of hill country and roadside genotypes but there was no significant difference between the three groups for shoot and root %N.

Discussion

Heritability of N and P content

The broadsense heritabilities (h^2_β) of leaflet width, shoot and root dry weight were higher than those measured previously for white clover (Caradus and Woodfield, 1990; Woodfield and Caradus, 1990), a result of the larger genetic variation observed in the present study (Table 2). This large genetic variation was due to the very wide range of plant types investigated, from prostrate very low yielding types typical of some of the hill country genotypes through to very upright high yielding genotypes of the cultivars. There was a 98 and 86 fold range in shoot dry weights among the genotype at low-P and high-P, respectively. Heritabilities for %P of shoot were also slightly higher than those observed for lucerne which ranged from 0.02–0.42 (Miller et al., 1987) and perennial ryegrass ($h^2_\beta = 0.30$) (Butler et al., 1962), but comparable to those for wheat (h^2_β = up to 0.88) (Gorsline et al., 1968; Kolmakova et al., 1983).

Heritabilities for %N of shoot were slightly higher than those observed for tall fescue ($h^2_\beta = 0.34$) (Pietrzak and Rafalski, 1982) and at the upper end of the range observed for wheat ($h^2_\beta = 0.24–0.57$) (Cox et al., 1985), but lower than those for maize ($h^2_\beta = 0.65$) (Mollaretti et al., 1987) and the narrow sense heritabilities for switchgrass (*Panicum virgatum*) ($h^2_N = 0.68–0.84$) (Talbert et al., 1983). Heritabilities have been generally high for %N of barley grain ($h^2_\beta = 0.56$, $h^2_N = 0.41$)(Rivoire and Ecochard, 1989), soybean seed ($h^2_\beta = 0.2–0.87$) (Ronis et al., 1985) and sweet potato (*Ipomoea batatas*) tubers ($h^2_\beta = 0.66$) (Shiga et al., 1985).

While level of P supply had a major effect on the magnitude of some of the characters examined (Caradus, 1986) it did not appear to have a large influence on broad sense heritabilities (Table 2). Selection at either P level could therefore be considered feasible, although it can not be assumed that the same genotypes could be selected at each P level.

In lucerne (Hill and Lanyon, 1983) and maize (Baker et al., 1971), selection for increases and decreases in %P level of shoots has been achieved with realised heritabilities for lucerne of 0.17 to 0.36 (Miller et al., 1987). Similarly in lucerne selections for high and low %N levels have been successful (Teuber et al., 1984).

Manipulation of %P and %N levels in plant tissues is a feasible option for plant breeders in an attempt to improve either plant quality or the efficiency of nutrient use. Improvements in plant quality are legitimate and attempts have been made in lucerne by selecting for high %P (Hill and Lanyon, 1983; Miller et al., 1987). However selections for low %P and/or %N to improve nutrient use efficiency is fraught with conceptual difficulties (Caradus, 1991). Often low shoot %P, or high phosphorus use efficiency, is associated with inefficiencies elsewhere such as poor translocation from roots to shoots or low P uptake per unit root. Apparent high P use efficiency may simply be the result of P deficiency. Selection for low %P or low %N should not be considered as the first choice or principal means to achieving improved nutrient use efficiency.

Relationship between plant characters

If the requirement was for high shoot %P to improve feed quality, such genotypes tended to be low yielding (Table 3). However, since the correlation between %P and shoot dry weight was not high (Fig. 1) it would appear possible to select higher yielding genotypes with a high shoot %P. Selection for high %P in lucerne resulted in concomitant increases in levels of other nutrients (Miller et al., 1984) but lower germination and Ca:P ratio (Hill and Lanyon, 1983; Miller et al., 1984, 1987). Selection for high shoot %N would appear feasible without detrimentally affecting yield (Fig. 1 and Table 3). This has been demonstrated in alfalfa where selection for both high shoot dry weight and %N was possible (Teuber et al., 1984).

Variation among plant parts

Intraspecific variation for %P and %N among plant parts was considerable. For example, genotypes with high shoot %P did not necessarily have high stolon or root %P – compare genotypes 12 and 26 at low-P (Fig. 2c) and genotypes 2, 28, 31 and 94 at high-P (Fig. 2d) which all had similar shoot %P levels but markedly different stolon and root %P levels. Phenotypic correlations between %P and %N of plant parts were generally low particularly at high-P (Fig. 1) suggesting that selections could be made for any combination of high and low %P in shoot, stolon and root. It would therefore be possible to identify genotypes which have a low %P in the grazed portion of the plant, 'conserving' P for further growth in the stolon and root (e.g. genotype 26, Fig. 2a and genotypes 31 and 94, Fig. 2c). However, genotypes with low shoot %P tended to have high shoot yields (Table 3) and a low proportion of dry matter as stolon (Caradus, 1986) resulting in poor persistance (Williams and Caradus, 1979) and a high proportion of P in shoot that is lost to the plant when defoliated. To select for a high proportion of P below grazing height (ie. in stolon and root), direct selection for a low proportion of P in shoot may be more appropriate though it is necessary to consider that proportion of P in a plant tissue component is largely determined by dry matter allocation and therefore in this instance will result in a low dry matter harvest index. This, however, may be the necessary trade-off for 'conserving' P in a grazed forage species (Caradus, 1986; Caradus and Williams, 1981).

Conclusion

There was considerable variability among plant parts for %P and %N. Broad sense heritabilities were moderate to high for these characters, particularly for the shoot and stolon components. High correlations between a number of plant characters suggest that selection for any single character may have other consequences. For example, at low-P supply, selection for high stolon %P may result in unproductive, small-leaved plants. Variation among genotypes for differences in nutrient concentration among plant component parts was considerable.

References

Baker D E, Wooding F J and Johnson M W 1971 Chemical element accumulation by populations of corn (*Zea mays* L.) selected for high and low accumulation of P. Agron. J 63, 404–406.

Bean E W 1972 Clonal evaluation for increased seed production in two species of forage grasses, *Festuca arundinancea* Schreb. and *Phleum pratense* L. Euphytica 21, 377–383.

Blair G J and Cordero S 1978 The phosphorus efficiency of three annual legumes. Plant and Soil 50, 387–398.

Burton G W and DeVane E H 1953 Estimating heritability in tall fescue (*Festuca arundinacea*) from replicated clonal material. Agron. J. 45, 478–481.

Butler G W, Barclay P C and Glenday A C 1962 Genetic and environmental differences in the mineral composition of ryegrass herbage. Plant and Soil 16, 214–228.

Caradus J R 1986 Variation in partitioning and percentage nitrogen and phosphorus content of the leaf, stolon and most of white clover genotypes. N.Z. J. Agric. Res. 29, 367–379.

Caradus J R 1991 The inadequacy of using tissue phosphorus concentration as an indicator of efficiency of phosphorus use. *In* Proceedings of a workshop on Soil and Plant Testing for Nutrient Deficiencies and Toxicities. Eds. R E White and L D Currie. pp 33–41. Palmerston North Fertilizer and Lime Research Centre, Massey University, Occas. Report No. 5.

Caradus J R, Dunlop J and Williams W M 1980 Screening white clover (*Trifolium repens* L.) plants for different responses to phosphate. N.Z. J. Agric. Res. 23, 211–217.

Caradus J R and Williams W M 1981 Breeding for improved white clover production in New Zealand hill country. *In* Plant Physiology and Herbage Production. Ed. C E Wright. pp 163–168. Occas. Symp. British Grasslands Society 13.

Caradus J R and Woodfield D R 1990 Estimates of heritability for, and relationships between, root and shoot characteris of white clover. I. Replicated clonal material. Euphytica 46, 203–209.

Clark R B 1990 Physiology of cereals for mineral nutrient uptake, use, and efficiency. *In* Crops as Enhancers of Nutrient Use. Eds. V C Baligar and R R Duncan. pp 131–210. Academic Press, San Diego, CA.

Cox M C, Qualset C O and Rains D W 1985 Genetic variation for nitrogen assimilation and translocation in wheat. I. Dry matter and nitrogen. Crop Science 25, 430–435.

Falconer D S 1961 Introduction to quantitative genetices. Oliver and Boyd, Edinburgh.

Gorsline G W, Thomas W I and Baker D E 1968 Major gene inheritance of Sr-Ca, Mg, K, P, Zn, Cu, B, Al-Fe, and Mn concentration corn. Pennsylvania State University Agriculture Experiment Station Bulletin 746. 47 p.

Hanson W D 1963 Heritability. *In* Statistical Genetics and Plant Breeding. Eds. W D Hanson and H F Robinson. pp 125-140. National Academy of Science, National Research Council, Washington DC. Publication 982.

Hill R R and Lanyon L E 1983 Phosphorus fertilizer response in experimental alfalfas selected for different phosphorus concentrations. Crop Science 23, 973–976.

Kolmakova I R, Gamzikova O L, Kalashnik N A and Gamzikov G P 1983 Combining ability and heritabiity for phosphorus fertilizer utilisation in varieties of spring wheat (from data of studies using ^{32}P). Genetika, USSR 19, 808–814.

Krstić B and Sarić M R 1990 Concentrations of N, P, and K and dry matter in maize inbred lines. *In* Genetic Aspects of Plant Mineral Nutrition. Eds. N El Bassam, M Dambroth and B C Loughman. pp 25–32. Kluwer Academic Publishers. Dordrecht, The Netherlands.

Lambert J and Touossaint B 1978 An investigation of the factors influencing the phosphorus content of herbage. Phosphorus Agric. 73, 1–12.

Loneragan J F and Asher C J 1967 Response of plants to phosphate concentration in solution culture II. Rate of phosphate absorption and its relation to growth. Soil Science 103, 311–318.

Mengel K and Kirkby E A 1982 Principles of Plant Nutrition. 3rd Edition. International Potash Institute, Bern. 655 p.

Miller D J, Melton B A, McCaslin B D, Waissman N and Olivares E 1984 Breeding alfalfa for phosphorus concentration in southern New Mexico. Report 29th Alfalfa Improvement Conference. p 64.

Miller D, Waissman N, Melton B, Currier C and McCaslin B 1987 Selection for increased phosphorus in alfalfa and effects on other characteristics. Crop Science 27, 22–26.

Mollaretti G, Bosio M, Gentinetta F and Motto M 1987 Genotypic variability for N-related traits in maize; Identification of inbred lines within high or low levels of NO_3-N in the stalks. Maydica 32, 309–323.

New Zealand Soil Bureau 1968. Soils of New Zealand. New Zealand Soil Bureau Bulletin 26, 84–85.

Pietrzak H and Rafalski A 1982 Variability and heritability of the content of N compounds in grasses. III. Heritability of N compounds in tall fescue (*Festuca arundinacea*). Biuletyn Instytutu Hodowli i Aklimatyzacji Róslin 147, 35–42.

Ramirez R 1982 Efficient use of nitrogen, phosphorus and potassium by corn (*Zea mays* L.) inbreds. *In* Plant Nutrition 1982. Ed. A Scaife. pp 515–520. Proceedings 9th International Plant Nutrition Colloquium, Warwick University, UK.

Rivoire P and Ecochard R 1989 Genetic determinism of the protein content of barley: A diallel study. Cereal Research Communications 17, 35–41.

Robinson P 1963 Heritability: A second look. *In* Statistical Genetics and Plant Breeding. Eds. W D Hanson and H F Robinson. pp 609–616. National Academy of Sciences – National Research Council, Washington DC. Publication 982.

Ronis D H, Sammons D J, Kenworthy W J and Meisinger J J 1985 Heritability of total and fixed N content of the seed in two soybean populations. Crop Science 25, 1–4.

Scossirolli R E, Ferrari A and Haussmann G 1963 Genetic variability for quantitative characters in alfalfa. *In* Statistical Genetics and Plant Breeding. Eds. W D Hanson and H F Robinson. pp 597–607. National Academy of Science – National Research Council, Washington, DC. Publication 982.

Shiga T, Kato S and Ishikawa H 1985 Utilisation of K_2O/N ratio as selection in tuber yield of sweet potato Jap. J. Breed. 35, 41–49.

Suckling F E T and Forde M B 9178 Genetic resources in high-rainfall hill pastures of New Zealand. I. Collection of ryegrass, browntop and white clover. N.Z. J. Agric. Res. 21, 499–508.

Talbert L E, Timothy D H, Burns J C, Rawlings J O and Moll R H 1983. Estimates of genetic parameters in switchgrass. Crop Science 23, 725–728.

Teuber L R, Levin R P, Sweeney T C and Phillps D A 1984 Selection for nitrogen concentration and forage yield in alfalfa. Crop Science 24, 553–558.

Underwood E J 1981 The Mineral Nutrition of Livestock. Commonwealth Agricultural Bureau, Farnham Royal, Bucks.

Williams W M and Caradus J R 1979 Performance of white clover lines on New Zealand hill country. Proceedings of the New Zealand Grassland Association 40, 162–169.

Woodfield D R and Caradus J R 1990 Estimates of heritability for, and relations between, root and shoot characters of white clover. II. Regression of progeny on mid-parent. Euphytica 46, 211–215.

P. J. Randall et al. (Eds.), Genetic aspects of plant mineral nutrition, 271–277.

PLSO SV58

Polymorphism and physiology of arsenate tolerance in *Holcus lanatus* L. from an uncontaminated site

ANDREW A. MEHARG and MARK R. MACNAIR
Department of Biological Sciences, Hatherly Laboratories, University of Exeter, Prince of Wales Road, Exeter EX4 4PS, UK

Key words: arsenate, *Holcus lanatus* L., polymorphism, tolerance, uptake mechanisms

Abstract

The polymorphism of arsenate tolerance in a *Holcus lanatus* L. population from an uncontaminated soil was investigated and a high percentage of tolerant individuals (65%) was found in the population studied. Influx of arsenate was highly correlated to arsenate tolerance within the population, with the most tolerant individuals having the lowest rates of arsenate influx. Isotherms for the high affinity arsenate uptake systems were determined in six tolerant and six non-tolerant genotypes. Tolerant plants had the lowest rates of arsenate influx. This was achieved by adaptation of the V_{max} of arsenate influx with the V_{max} of the high affinity uptake system saturating at lower substrate concentrations in the tolerant plants. The polymorphism is discussed with relation to adaptation to the extreme environments to which the plants are subjected on mine-spoil soils.

Introduction

The study of the mechanisms of metal tolerance in angiosperms is problematic in that plants collected from metal-contaminated soils may differ in their physiology in aspects other than their adaptation to the metals concerned. In mine-spoil soils plants may have to adapt to drought due to poor soil structure, soil compaction low pH, and low mineral nutrient status (Meharg and Macnair, 1990). Only by making comparisons between genotypes that differ solely in their tolerance to metal excesses, such as tolerant mutants screened from an otherwise non-tolerant population, or by breeding can conclusions be drawn about the physiological basis of tolerance (Baker and Walker, 1990; Macnair, 1981; Macnair and Watkins, 1983).

Arsenate tolerance has been reported for a number of angiosperms (Meharg and Macnair, 1990). Tolerance to arsenate has been shown to be a heritable trait in *Agrostis capillaris* (Watkins and Macnair, 1991) and in *Holcus lanatus* (Macnair et al., 1992). Macnair et al., (1992) have shown that there is a major gene for arsenate tolerance in *Holcus lanatus*, which appears to be polymorphic in normal populations. Meharg and Macnair (1990) showed that limited influx of arsenate by the excised roots of *H. lanatus* was due to adaptation of the phosphate uptake system (as arsenate behaves as a phosphate analogue and is taken up by the same uptake system) and that this adaptation was a mechanism of arsenate tolerance. The high-affinity phosphate uptake system (the uptake system dominant at low substrate concentrations) was suppressed in tolerant plants leading to greatly reduced short-term influx of arsenate into excised roots. This short-term influx of arsenate into excised roots was also reflected in long-term uptake in intact plants (Meharg and Macnair, 1991a). Reduced influx of arsenate also occurred in the arsenate-tolerant grasses *Deschampsia cespitosa* and *Agrostis capillaris* (Meharg and Macnair, 1991b). Reduced arsenate accumulation has also been observed in arsenate-tolerant *Silene vulgaris* (Paliouris and Hutchinson, 1991). The adaptation of the phosphate uptake system may also be an adaptation to the poor phosphate nutrient status of the mine soils. Reduced influx may be the only mechanism of tolerance, or it may be one of a number of mechanisms such as biochemical detoxification of arsenate within the plant cell. In crosses between tolerant and non-tolerant clones of *H. lanatus* the altered arsenate uptake system co-segregated with phenotypic arsenate tolerance (determined by root growth in arsenate) in F_2 segregating families, with tolerant progeny having reduced rates of arsenate and phosphate influx (Meharg and Macnair, 1992). This provided further evidence that reduced arsenate influx was a mechanism of arsenate tolerance. Another approach to determine if the proposed mechanism is a mechanism of tolerance is to screen tolerant individuals from uncontaminated populations. The gene coding for arsenate tolerance was found to be polymorphic in uncontaminated environments (Macnair et al., 1992). In this study the nature of the polymorphism for arsenate tolerance in a population of *H. lanatus* from an uncontaminated site was investigated to determine if the mechanisms of arsenate tolerance in plants from arsenate contaminated soils were due to the mine environment or to elevated levels of arsenate. The relationship between arsenate tolerance and the rate of short-term arsenate influx was determined in excised roots. Short-term arsenate influx parameters were determined in individual genotypes which had a range of tolerances. Results were interpreted with regards to the physiological basis of tolerance and the evolutionary process leading to the colonization of highly contaminated mine soils.

Materials and methods

Plant material

Ripe seed of *H. lanatus* was collected from the Hoopern valley, University of Exeter campus. The site was uncontaminated with arsenate (soil arsenic 0.27 mmol g^{-1}) and was not in the proximity of any contaminated soils. Plants were tested for tolerance as seedlings and then pricked out into potting compost and grown in a greenhouse until sufficient tillers were available for tolerance testing and the determination of the rate of arsenate uptake.

Rooting tests

In the tolerance tests, both seedlings and unrooted tillers of the genotypes were placed in a nutrient solution containing 0.2 mol m^{-3} $Ca(NO_3)_2$, 0.2 mol m^{-3} KNO_3 and 0.1 mol m^{-3} $MgSO_4.7H_2O$ and 0.133 mol m^{-3} Na_2HAsO_4 for 7d. For tolerance testing of tillers, tillers were rooted under a light bank with a 16 h daylength in a single 12 dm^3 container containing 10 dm^3 of nutrient solution. The container was fitted with a lid with 240 tubes into which the tillers were inserted. The length of the longest root was determined after 7d. For seedling testing, seeds were pregerminated for 3d on moist filter paper in Petri dishes, and only seeds from which the root apex had just emerged were tested. Seedlings were rooted in polystyrene cups containing 100 cm^3 of the test solution. Seedlings were supported by black alkathene beads floating on the surface of the test solution. The cups were kept on the windowsill of the laboratory and grown at ambient temperature. Seedling root length was determined after 7d growth in the arsenate culture solution.

Incubation procedure

To determine arsenate influx into excised roots, tillers were rooted in arsenate-free nutrient solution for 7d. Then roots (excised at the node) were incubated in 100 cm^3 of aerated arsenate test solution for 20 min at ambient temperature. All test solutions contained 10 mol m^{-3} 2-[N-Morpholino] ethanesulfonic acid (MES) and 0.5 mol m^{-3} $Ca(NO_3)_2$, and the test solutions were adjusted to pH 5 using KOH. Arsenate was added to the test solution as Na_2HAsO_4. In all experiments on termination of incubation in test solution the roots were rinsed in an ice-cold solution containing 1 mol m^{-3} K_2HOP_4, 10 mol m^{-3} MES and 0.5 mol m^{-3} $Ca(NO_3)_2$. The roots were then incubated for 10 min in ice-cold solution of the same composition to ensure desorption of arsenate from the roots' "free space". Fresh weights of the roots were determined before analysis.

Analysis

Arsenic was determined by digesting roots in 2 cm^3 concentrated nitric acid (Aristar grade). Samples were digested by heating on a block, digester for 1 h at 180°C and 1 h at 200°C, to evaporate the samples to dryness. The residue was taken up in 10 cm^3 of 5% HCl (Analar grade) containing 20 mol m^{-3} KI (to reduce arsenate to arsenite). Arsenic was then determined by a hydride generation technique using a Philips PU9060 continuous-flow vapour system which was interfaced with a Philips SP9 series atomic absorption spectrophotometer. The technique is based on the reduction of arsenite by a 1% m/V solution of $NaBH_4$, the reductant and sample are pumped (by means of a peristaltic pump) into a chamber where AsH_3 is evolved at a constant rate and swept into a heated quartz cell mounted on top of the burner using N_2 as the carrying gas. Arsenic from the breakdown of the hydride was determined by absorbance at a wavelength of 193.7 nm.

Results

Tolerance testing

The results of tolerance testing of seeds collected from the Hoopern valley are shown in Figure 1. The histogram shows the segregation into tolerant and non-tolerant plants, with the segregation at about 15 mm root growth after 7d growth in 0.133 mol m^{-3} arsenate. In this population 65% of the individuals tested were tolerant to arsenate. The relationship between tolerance as seedlings and tolerance as adults is illustrated in a scatter plot (Fig. 2). Again there is segregation into two groups, with non-tolerants clumped at the bottom left of the plot. A few individuals with poor root growth as seedlings produce long roots as tillers. The reverse is also true. The majority of plants that score as tolerants as seedlings score as tolerant as tillers.

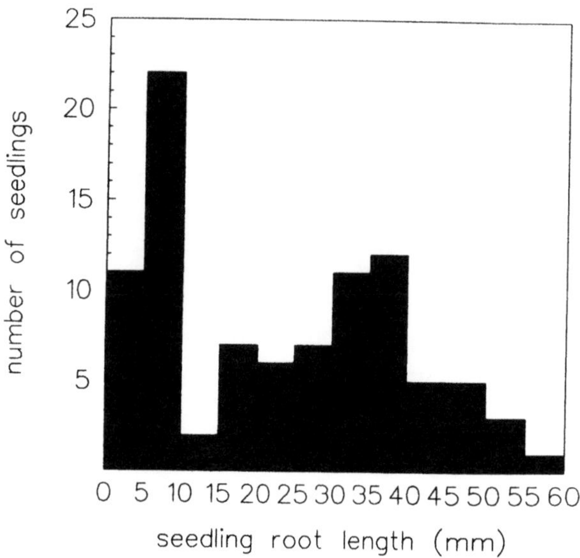

Fig. 1. Seedling root length after 7d growth in 0.133 mol m^{-3} arsenate.

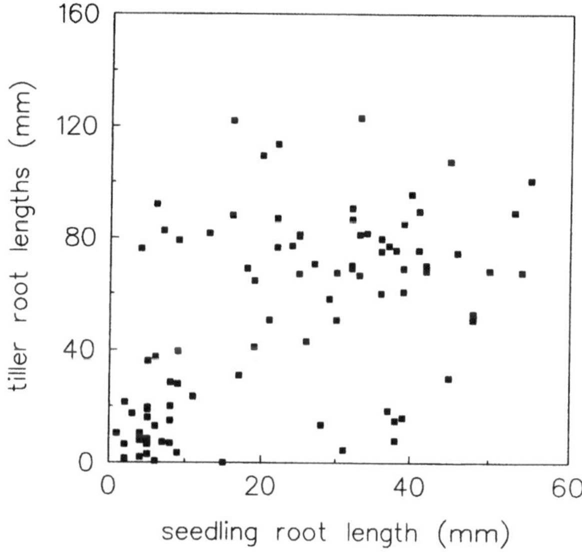

Fig. 2. Relationship between seedling root length and tiller root length after 7d growth in 0.133 mol m^{-3} arsenate. Each data point is the average of 2 replicates of tiller root length.

Relationship between arsenate tolerance and arsenate influx

The short-term influx (20 min) of 0.05 mol m^{-3} arsenate was determined in excised roots of 38 genotypes chosen at random from the 100 plants used in the seedling test. Influx was plotted against seedling growth in 0.133 mol m^{-3} arsenate and the results shown in Figure 3. Arsenate influx and tolerance (as measured by root growth) were highly correlated, with an R^2 of 35.5%, and this correlation was significant (at the 0.1% level). A range of influx rates was found for both tolerant and non-tolerant plants. Tolerant plants had influx rates ranging from 10-110 nmol g^{-1} f.wt h^{-1} and non-tolerants ranging from 50–165 mmol g^{-1} f.wt h^{-1}, but the plants that were most tolerant had the lowest rates of arsenate influx.

Dose response curves were determined for 6 tolerant and 6 non-tolerant plants with a range of influx rates. The average dose response curves for the tolerant and non-tolerant plants are reported in Figure 4. The dose response curves showed that in the non-tolerant plants root growth was severely inhibited at 5 mmol m^{-3} but in tolerant plants there was substantial root growth at 100 mmol m^{-3}. Analysis of covariance showed that differences between tolerant and non-tolerant plants were highly significant, while differences between genotypes within these classes were not significant at the 5% level (analysis not shown).

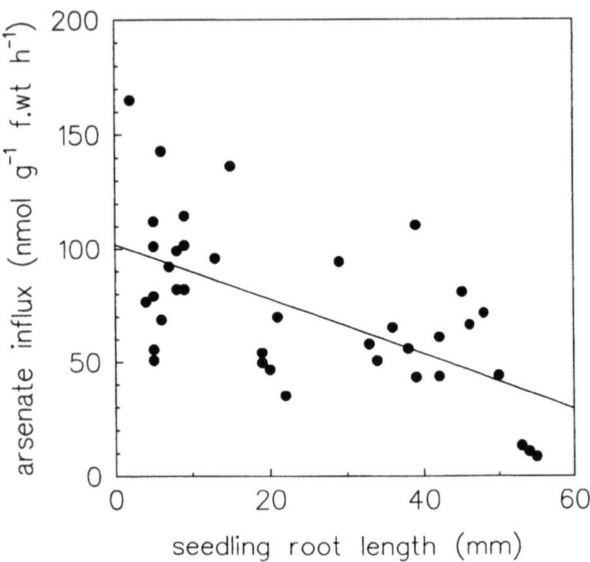

Fig. 3. Correlation between arsenate influx by excised roots and seedling root length after 7d growth in 0.133 mol m^{-3} arsenate. Each data point is the average of 2 replicates of influx determination.

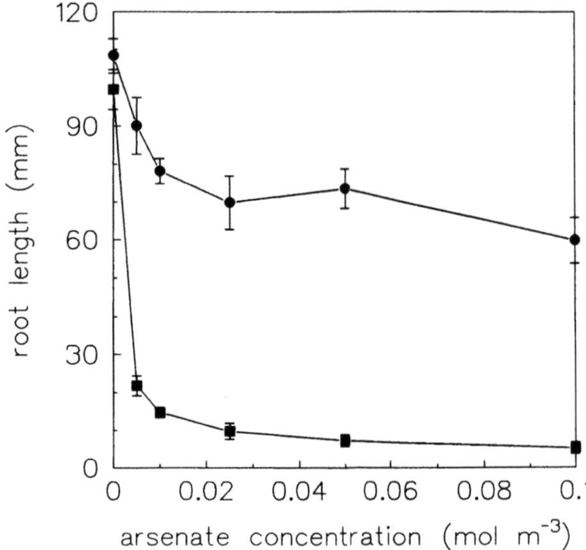

Fig. 4. Dose response curves for tiller root growth in arsenate. Tolerant, circles; non-tolerant, squares. Each point is the average of six genotypes.

Kinetics of arsenate uptake

Influx isotherms were determined for the high-affinity uptake system in the 6 tolerant and the 6 non-tolerant genotypes used to obtain the dose response curves in Figure 4. Short-term (20 min) influx was determined in excised roots over a range of arsenate concentrations. The resulting isotherms for average influx in the tolerants and in the non-tolerants are shown in Figure 5. Influx parameters for the isotherms were determined for each clone using an iterative computer program based on the algorithm of Marquardt (1963). The data were fitted to the Michaelis-Menten function. The kinetic parameters for ion influx for all twelve genotypes are reported in Table 1. There are large differences in uptake parameters between genotypes, with V_{max} ranging from 201.8 to 540 nmol g^{-1} f.wt h^{-1} and K_m ranging from 0.0185 to 0.049 mol m^{-3}.

Differences in V_{max} between tolerants and non-tolerants were significant at the 1% level while there was no significant difference at the 5% level in the K_m of ion uptake. Although there was variation in uptake between tolerant

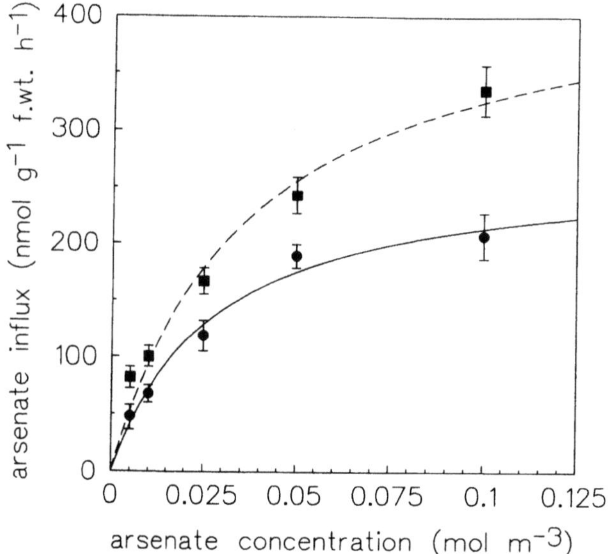

Fig. 5. Arsenate influx isotherms for the high-affinity uptake system. Tolerant, circles: non-tolerant, squares. Each point is the average of six genotypes.

Table 1. Kinetic parameters for arsenate influx

Genotype	V_{max} (nmol g^{-1} f.wt h^{-1})	K_m (mol m^{-3})
Tolerant		
a	201.8 ± 43.9	0.025 ± 0.014
b	351.8 ± 70.6	0.042 ± 0.020
c	316.5 ± 68.3	0.040 ± 0.020
d	306.1 ± 45.3	0.018 ± 0.008
e	246.3 ± 37.4	0.030 ± 0.012
f	286.4 ± 44.2	0.025 ± 0.010
Average	284.8 ± 21.8	0.030 ± 0.004
Non-tolerant		
g	299.0 ± 17.6	0.022 ± 0.007
h	413.3 ± 54.9	0.035 ± 0.011
i	497.8 ± 84.4	0.045 ± 0.017
j	494.2 ± 87.9	0.036 ± 0.015
k	540.0 ± 115.0	0.038 ± 0.019
l	453.0 ± 101.1	0.049 ± 0.023
Average	449.6 ± 34.9**	$0.037 \pm 0.004^{n.s.}$

Figures represent kinetic parameters ± standard error of the mean. Parameters estimated by least-squares procedure Marquardt (1963). Influx isotherms were determined for each genotype; with influx at each substrate concentration replicated 3 times. Each uptake isotherm was determined using 5 substrate concentrations (see Fig. 5). To determine the mean kinetic parameters for the tolerant and non-tolerant genotypes the kinetic parameters of the 6 tolerant and non-tolerant genotypes were averaged. The means of the kinetic parameters between tolerants and non-tolerants were compared using an unpaired t-test; **; $0.010 > p > 0.001$; n.s., non-significant at the 5% level.

and non-tolerant plants the average of the kinetic parameters for tolerant and non-tolerant plants showed that there was a large difference in influx parameters based on tolerance (Fig. 5). Tolerant plants had both a lower K_m and V_{max} of arsenate influx than non-tolerant plants (Table 1). The main difference was in the V_{max}, with tolerants having an average V_{max} 284.8 nmol g^{-1} f.wt h^{-1} and non-tolerants 449.6 nmol g^{-1} f.wt h^{-1}.

Discussion

Arsenate tolerance is polymorphic in uncontaminated soil, and in the population studied 65% of the plants were tolerant. Copper-tolerant individuals are present in a number of grass populations from uncontaminated soil, but the percentage of tolerant individuals was low, ranging from 0.005% in *Lolium perenne* to 0.16% in *Holcus lanatus* (Bradshaw, 1984). The species that possessed tolerant individuals in the populations from uncontaminated sites were generally found on contaminated sites, but those that did not show the required variation in normal populations were not found on mine-spoil soils. Variability within populations for zinc tolerance was present in a number of grass species growing on uncontaminated soil which enabled them to adapt to soil contaminated by electricity pylons (Al-Hiyaly et al., 1990). Selection can only act on heritable variation and if the variation is not present then evolutionary change will not occur (Bradshaw, 1984; Macnair, 1990). Arsenate tolerance has been shown to be heritable in *Holcus lanatus* (Macnair et al. 1992) and in *Agrostis capillaris* (Watkins and Macnair, 1991). The percentage of arsenate-tolerant individuals found in the population studied here was much higher than that reported for copper tolerance.

The high percentage of arsenate-tolerant individuals may be due to generally elevated levels of arsenate in soils in the Dartmoor area of SW England. Elevated levels of arsenate in the soils are associated with mineralised zones around a granite intrusion and its metamorphic aureole around Dartmoor (Colbourn et al., 1975). The population studied was on the edge of this area, with the most highly contaminated sites (associated with arsenic mining) some distance away. Levels of arsenate present in the soils around the Tamar and Dartmoor areas of SW England show considerable enhancement compared to control soils, soil arsenic levels where as high as 2500 μg g^{-1} (Colbourn et al., 1975). The soils have probably been contaminated by air-borne deposits from mining and smelting processes, with the mining area extending up to 140 km^2. The actual concentrations of arsenic within the Dartmoor area is highly variable due to different extents of contamination and variation in parent material. The high percentage of tolerant individuals may be due to gene flow from the contaminated mine sites described by Porter and Peterson (1975, 1977). It is clear that there is the potential for adaptation of the population studied to colonize arsenate-contaminated sites.

There is a gradation in arsenate tolerance within the population and this is highly correlated to arsenate influx (Fig. 3). The most tolerant plants have the lowest rates of arsenate influx but analysis of variance of the dose response curves for both tolerant and non-tolerant plants with high and low rates of arsenate influx show that influx does not affect the plants response as measured by root growth to increasing arsenate concentrations (analysis not shown). This suggests that restricted influx of arsenate is not the only mechanism of arsenate tolerance operating within this species, but that possession of an altered arsenate uptake system confers increased tolerance. There are difficulties in interpreting root growth data as a measure of tolerance (Macnair, 1990). Root length is governed by both genetic factors (other than tolerance) and by environmental factors (such as the nutritional status of the plant).

Arsenate behaves as a phosphate analogue, and is taken up by the phosphate uptake system in *H. lanatus* (Meharg and Macnair, 1990, 1991a). There may be some selective disadvantage to having reduced rates of phosphate uptake, although models of phosphate acquisition by plants from soil suggest that kinetic parameters for phosphate uptake such as K_m and V_{max} are much less important in determining phosphate accumulation compared to rates of root growth, root diameter and factors affecting phosphate diffusion to the roots (Silberbush and Barber, 1983). This is because phosphate diffusion to the root is the rate-limiting step in phosphate uptake and not the transfer of phosphate across the plasmalemma (Bieleski, 1973). If there was a high cost for tolerance a high percentage of tolerant plants from an uncontaminated site would be difficult to explain. Wilson (1988) suggests that it is expected that there would be a cost of tolerance, as if there were no cost of tolerance the tolerance gene(s) would be selectively neutral on normal soil, and since there is appreciable gene flow between mine and non-mine populations, these genes would be expected to reach high overall rates of fixation throughout. This suggests that the cost of possessing an altered phosphate uptake system is low as a high percentage of tolerant individuals are present on uncontaminated soil. Costs of tolerance have been illustrated for plants grown in competition on uncontaminated soil (Cook et al., 1972; Nicholls and McNeilly, 1985), and in relative growth rates under optimal growing conditions (Wilson, 1988).

Kinetic analyses of uptake isotherms showed that there was a large variation in arsenate uptake kinetics within the population studied (Table 1). The arsenate (phosphate) uptake system appears to be highly plastic with regards to both the K_m and V_{max} of ion uptake. The results compare to those of phosphate and arsenate uptake in mine (tolerant) and non-mine (non-tolerant) plants where the rate of uptake of both ions were suppressed at low substrate concentrations in the tolerant plants (Meharg and Macnair, 1990). The suppression of the uptake system is not as severe in the tolerant plants collected from uncontaminated soils as compared to tolerant plants from contaminated sites. It is possible that plants adapted to grow on highly contaminated sites are selected from normal populations to have the lowest rates of arsenate uptake. This is consistent with the results shown in Figure 3 where the most tolerant plants (as measured by the rooting test) had the lowest rates of arsenate uptake.

Chapin (1980) suggests that species from infertile habitats would generally have lower V_{max} and perhaps lower apparent K_m of nutrient absorption than species from more fertile habitats. Atwell et al. (1980), investigated phosphate uptake in 6 *Carex* species collected from differing habitats and showed that no clear relationship between phosphate uptake parameters and environment could be found. That is, species from nutrient-poor habitats did not have lower or higher rates of phosphate uptake than plants from more fertile habitats. When uptake parameters for potassium were determined in 12 *Taraxacum* microspecies from soils of differing nutritional status, again no trend in kinetic parameters with environment could be determined (Hommels et al., 1990). Thus, in *Taraxacum* microspecies and *Carex* species there appears to be no adaptation of the potassium and phosphate uptake systems, respectively. The genotypes investigated in this study where all collected from the same habitat and thus differences in uptake parameters would appear not to be related to soil nutrient status. This further confirms the correlation between influx and tolerance.

Acknowledgements

This research was funded by SERC grant no. GR/F18947. Thanks are due to Dale Sanders and Ian Jennings of the Biology Department, University of York for the curve-fitting program.

References

Al-Hiyaly S A K, McNeilly T and Bradshaw A D 1990 The effect of zinc contamination from electricity pylons: Contrasting patterns of evolution in five grass species. New Phytol. 114, 183–190.

Atwell B J, Veerkamp M T, Stuiver B C E E and Kuiper P J C 1980 The uptake of phosphate by *Carex* species from oligotrophic to eutrophic swamp habitats. Physiol. Plant. 49, 487–494.

Baker A J M and Walker P L 1990 Ecophysiology of metal uptake by tolerant plants. In Heavy Metal Tolerance in Plants: Evolutionary Aspects. Ed. A J Shaw. pp 155–177. CRC Press, Boca Raton, FL.

Bieleski R L 1973 Phosphate pools, phosphate transport, and phosphate availability. Annu. Rev. Plant Physiol. 24, 225–252.

Bradshaw A D 1984 The importance of evolutionary ideas in ecology – and vice versa. Evolutionary Ecology Ed. B Shorrocks. pp 1–25. Blackwell Scientific Publications, Oxford.

Chapin F S 1980 The mineral nutrition of wild plants. Ann. Rev. Ecolog. Syst. 11, 233–260.

Colbourn P, Alloway B J and Thornton I 1975 Arsenic and heavy metals in soils associated with regional geochemical anomalies in South-West England. Sci. Total Environ. 4, 359–363.

Cook S A, Lefebvre C and McNeilly T 1972 Competition between metal tolerant and normal plant populations on normal soil. Evolution 26, 366–372.

Hommels C H, Saat T A W and Kuiper P J C 1990 Characterization of the high affinity K^+ (Rb^+) uptake system in roots of intact *Taraxacum* microspecies: Comparison of 12 microspecies in relation to the mineral ecology. New Phytol. 114, 695–701.

Macnair M R 1981 The uptake of copper by plants of *Mimulus gutattus* differing in genotype primarily at a single major copper tolerance locus. New Phytol. 88, 723–730.

Macnair M R 1990. The genetics of metal tolerance in natural populations. Heavy Metal Tolerances in Plants: Evolutionary Aspects Ed. A J Shaw. pp 235–253. CRC press, Boca Raton, Fl.

Macnair M R, Cumbes Q C and Meharg A A 1992 The genetics of arsenate tolerance in Yorkshire fog, *Holcus lanatus* L. Heredity (*In press*).

Macnair M R and Watkins A D 1983 The fitness of the copper tolerance gene of *Mimulus gutattus* in uncontaminated soil. New Phytol. 95, 133–137.

Marquardt D W 1963 An algorithm for least-squares estimation of nonlinear parameters. J. Soc. Ind. App. Math. 11, 431–441.

Meharg A A and Macnair M R 1990 An altered phosphate uptake system in arsenate tolerant *Holcus lanatus*. New Phytol. 16, 29–35.

Meharg A A and Macnair M R 1991a Uptake, accumulation and translocation of arsenate in arsenate tolerant and non-tolerant *Holcus lanatus* L. New Phytol. 117, 225–231.

Meharg A A and Macnair M R 1991b Mechanisms of arsenate tolerance in *Deschampsia cespitosa* L. (Beauv.) and *Agrostis capillaris* L.: Adaptation to the arsenate uptake system. New Phytol. 119, 291–297.

Meharg A A and Macnair M R 1992 Genetic correlation between arsenate tolerance and the rate of influx of aresenate and phosphate in *Holcus lanatus* L. Heredity (*In press*).

Nicholls M K and McNeilly T 1985 The performance of *Agrostis capillaris* L. genotypes, differing in copper tolerance, in rye grass swards on normal soil. New Phytol. 101, 207–217.

Paliouris G and Hutchinson T C 1991 Arsenic, cobalt and nickel tolerances in two populations of *Silene vulgaris* (Moench) Garcke from Ontario, Canada. New Phytol. 117, 449–459.

Porter E K and Peterson P J 1975 Arsenic accumulation by plants on mine waste (United Kingdom). Sci. Total Environ. 4, 365–371.

Porter E K and Peterson P J 1977 Arsenic tolerance in grasses growing on mine waste. Environ. Poll. 14, 255–265.

Silberbush M and Barber S A 1983 Sensitivity of simulated phosphorus uptake to parameters used by a mechanistic-mathematical model. Plant and Soil 74, 93–100.

Watkins A D and Macnair M R 1991 Genetics of arsenic tolerance in *Agrostis capillaris* L. Heredity 66, 47–54.

Wilson B J 1988 The cost of heavy-metal tolerance: An example. Evolution 42, 408–413.

P. J. Randall et al. (Eds.), Genetic aspects of plant mineral nutrition, 279–288.

PLSO SV52

The role of piscidic acid secreted by pigeonpea roots grown in an Alfisol with low-P fertility

N. AE[1], J. ARIHARA[2], K. OKADA[3], T. YOSHIHARA[4], T. OTANI[1] and C. JOHANSEN[5]
[1]*National Institute of Agro-Environmental Sciences, Tsukuba, 305, Japan,* [2]*Hokkaido Agriculture Experimental Station, Sapporo, 060, Japan,* [3]*Tropical Agriculture Research Center, Tsukuba, 305, Japan,* [4]*Hokkaido University, Sapporo, 060 Japan and* [5]*ICRISAT (International Crops Research Institute for the Semi-Arid Tropics), Patancheru, A.P. 502324, India*

Key words: alfisol, iron phosphate, phosphorus uptake, pigeonpea, piscidic acid, root exudate, sorghum, vertisol

Abstract

In India, pigeonpea (*Cajanus cajan* (L.) Millsp.) has been traditionally grown as an intercrop, mainly with cereals such as sorghum (*Sorghum bicolor* (L.) Moench) and pearl millet (*Pennisetum americanum* (L.) Leeke) under low inputs of fertilizers. The response of pigeonpea to applied phosphorus (P) is generally low even in low-P Alfisols where a major fraction of inorganic P is in the iron-associated form (Fe-P). Pigeonpea has a special ability to take up P from low-P Alfisols on which other crops (sorghum, maize (*Zea mays* L.), soybean (*Glycine max* (L.) Merrill), and pearl millet) cannot survive. This characteristic is attributed to piscidic acid and its derivative, which is secreted from the roots of pigeonpea, but not by those of the other crop species. These substances can release P from Fe-P by chelating Fe^{3+}. From results of both the composition of mineral contents and the growth stimulated by the inoculation of VAM fungi we propose a mechanism of P acquisition by pigeonpea from an Alfisol.

Introduction

In the semi-arid tropics, the main soil types are Alfisols and Vertisols. Pigeonpea is widely intercropped with sorghum, pearl millet and other crops in these soil types in the Indian sub-continent (Aiyer, 1949). Many field experiments have shown that pigeonpea has little response to phosphorus (P) fertilizer even on Alfisols with low-P availability, compared to sorghum and pearl millet (Ae et al., 1991b; Johansen, 1990). To explain this, three possible reasons are usually given: 1) deeper root penetration, 2) association with VAM fungi, 3) special mechanisms of P uptake from insoluble P sources which are unavailable to other crop species.

As to the first reason, traditional pigeonpea genotypes have a large, deep tap root system (Sheldrake and Narayanan, 1979). It may be that pigeonpea can survive by taking up water and nutrients from deeper soil layers in the post-rainy season but cannot absorb P fertilizer nearer the surface in the dry season.

A second possible reason why pigeonpea responds less to P application than sorghum, is its dependence on vesicular-arbuscular mycorrhizal (VAM) associations that would permit better utilization of available soil P (Johansen, 1990). Most higher plant species examined have their nutrient uptake enhanced by mycorrhizal hyphae acting as an extension of the nutrient absorptive area (Mosse, 1981). But this explanation would depend on mycorrhizal fungi being more strongly associated with pigeonpea than with sorghum.

Thirdly, comparative studies by Itoh (1987) on kinetic parameters, such as maximum uptake rate (Imax) and Michaelis constant (Km), showed that they were of the same order among four crop species – soybean, chickpea (*Cicer arietinum* L.), maize and pigeonpea. He thus proposed further investigation on the development of root hairs, mycorrhizal associations and root activity to solubilize soil P. This paper further examines mechanisms of P uptake by pigeonpea from soils with low-P availability. Comparisons are made with other crop species normally grown on Alfisols and Vertisols in peninsular India.

Materials and methods

Deep-root and distribution

From ICRISAT Center (Patancheru, Andhra Pradesh, India), an Alfisol and a Vertisol with low P availability, on the basis of Olsen's bicarbonate extraction method (Olsen et al., 1954), were carefully chosen for field and pot experiments. The P status of these soils is shown in Table 1. Single superphosphate at rates of 0, 9, 18, 35, 70 kg P ha^{-1} was applied to determine the response of pigeonpea (ICPL87) and a sorghum hybrid (CSH5) to P fertilizer in the field, in both an Alfisol and a Vertisol. Nitrogen and potassium were applied as basal dressings at the rate of 100 kg N ha^{-1} as ammonium sulphate and 52 kg K ha^{-1} as potassium chloride, respectively. Only K, but not N, was applied to pigeonpea. A split plot design was used with crop species in main plots and P fertilizer treatment in sub-plots. There were three replications. Plot size was 8 × 5 m. The experiment was conducted during the 1987 rainy season.

The effect of restricted soil volume for root exploitation was examined in several crop species (pigeonpea, sorghum, maize, pearl millet, soybean, chickpea) in a pot experiment using 2.5 kg of each of the Alfisol and the Vertisol without addition of P fertilizer. Nitrogen was added at 100 mg kg^{-1} soil as ammonium sulphate and potassium at 52 mg kg^{-1} soil as potassium chloride. Plants were grown in a greenhouse and the above-ground portions harvested at the grain filling stage (for each crop). For the P analysis, dried tissues were digested with concentrated H_2SO_4 containing 0.5% selenium. Phosphorus concentrations in plant digests were then measured by the method of Murphy and Riley (1962).

Mycorrhizal association

The response of pigeonpea and sorghum to mycorrhizal inoculation was compared in a pot experiment using the same soils with low P fertility and at two levels of P application, 0 and 8.7 mg P kg^{-1} soil as single superphosphate. Nitrogen and potassium were applied as for the abovementioned experiment. Fifty gram of soil containing a mixed culture of VAM fungi (*Glomus constricum*, *G. fasciculatum*, *G. epigaerum*, *G. monosporum*, and *Acaulospora morroweae*) was used as an inoculation treatment into 2.5 kg of the sterilised potted soil. Sorghum (CSH5) and pigeonpea (ICPL87) were then sown and grown in a greenhouse until their harvest at the grain filling stage.

Better utilization of iron-associated P by pigeonpea

To compare the ability of different crop species to absorb P from different sources, a sand-vermiculite nutrient culture experiment was carried out in a greenhouse. Phosphorus was applied in the form of either $CaHPO_4$, $AlPO_4$, or $FePO_4$, to simulate the three forms of inorganic P found in soils. To 2.5 kg of the sand-vermiculite mix in pots, either $CaHPO_4$ was added at the rates of 11.3, 34, and 102 mg P kg^{-1}, $AlPO_4$ at the rates of 10.7, 32.96, mg P kg^{-1} or $FePO_4$ at the rates of 8.7, 26, and 76 mg P kg^{-1}. Nutrients other than P were applied as per Arnon (1938). Pigeonpea, maize, pearl millet, sorghum, and chickpea were grown in these pots under near-optimal environmental conditions in a greenhouse. At the flowering stage, the plants were harvested and P concentrations were measured. The solubility of these chemical forms of P in the absence of plants was measured after they were incorporated in the sand-vermiculite mix at levels of 117 mg P kg^{-1}.

Root exudates

To collect root exudates of several crop species, plants were grown in the same sand-vermiculite culture as described above, at a low P level (5 mg kg^{-1}). The roots of 2-month old plants were washed in water and then soaked in 2 m*M* $CaCl_2$ for collection of root exudates. The collected root exudates were passed through two ion-exchangers (Dowex 50 and 1) and the acid fractions were extracted from Dowex 1 resin with 6 *M* formic acid. After esterification with methyl alcohol, the acid fractions were analyzed by gas-chromatography (GC) and nuclear magnetic resonance (NMR) for identification of substances in the acid fraction of pigeonpea root exudates.

To test the ability of piscidic acid and related compounds to specifically release P from $FePO_4$, piscidic acid from *Narcissus poeticum* bulbs (Smeby et al., 1954) and derivatives of fukiic acid from *Petasites japonicus* (Yoshihara et al., 1974) were prepared. These chemical compounds were solubilized with ethylacetate to a concentration of 2.5 m*M*. One ml of this solution was mixed in a centrifuge tube with 1.0 ml suspension of 5 mg $FePO_4$ at pH 4.5 of acetate buffer. The solubilized P in the water layer was then measured colorimetrically after shaking for 30 min.

Results

Deep root distribution

In the Alfisol, most of the P is associated with iron (Fe-P), whereas in the Vertisol there is a large fraction of calcium-bound P (Ca-P) (Table 1). Responses of pigeonpea to P fertilizer were compared to those of other species in Alfisol and Vertisol fields with low P availability, as evaluated by Olsen's method. Despite the lower value of Olsen's P in the Vertisol (1.5 mg kg^{-1}) than in the Alfisol (3.5 mg kg^{-1}), grain yield of sorghum obtained in the Alfisol (87 kg ha^{-1}) was much lower than in the Vertisol (3043 kg ha^{-1}) without application of P fertilizer (Table 2). With increasing P applied, grain yield of sorghum increased in the Alfisol to a much greater extent than in the Vertisol. On the other hand, pigeonpea is comparatively less responsive to P application in the Alfisol and 929 kg ha^{-1} of grain yield was harvested in the Alfisol even where no P fertilizer was added. With this treatment, pigeonpea could take up 3.18 kg P ha^{-1}

Table 1. Some chemical characteristics of virgin soil taken from an Alfisol and a Vertisol field at ICRISAT Center, and of the type used for field and pot experiments

Soil	$pH(H_2O)$	Electrical conductivity (mS cm^{-1})	P fixation[a] (mg kg^{-1})	*Inorganic P (mg kg^{-1})*			*Available P (mg kg^{-1})*		
				Ca-P	Al-P	Fe-P	Olsen	Truog	Bray 2
Alfisol	6.9	0.05	1340	4	5	48	3.5	6.5	1.5
Vertisol	8.3	0.16	7380	58	20	55	1.5	49.2	18.1

[a] 20 mL of 2.5% ammonium phosphate was added to 10 g of soil. After 24 h, fixed P was measured from P content in the solution. (Source: Ae et al., 1991a).

Table 2. Response of grain yield (kg ha^{-1}) of pigeonpea and sorghum to phosphorus fertilizer application in an Alfisol and a Vertisol field. ICRISAT Center, rainy season, 1987

Crop	Soil	P fertilizer application (kg ha^{-1})					
		0	9	18	35	70	SE
Sorghum	Alfisol	87	673	2101	2621	3640	(727)
	Vertisol	3043	3364	3853	3697	3972	(570)
Pigeonpea	Alfisol	929	727	1113	629	678	(393)
	Vertisol	248	457	674	744	301	(128)

and sorghum 2.00 kg P ha^{-1} from the Alfisol. Phosphorus uptake by several crop species in potted Alfisol and Vertisol without P fertilizer added, under the restricted rooting volume that occurs in pots, is shown in Table 3. All of the crops, except pigeonpea, produced more dry matter and could extract more P up to the grain filling stage in the Vertisol than in the Alfisol. Without P addition, growth and P uptake of sorghum, soybean, pearl millet, and maize were severely limited on the Alfisol and these crops died as a result of P deficiency within 1 month after sowing. Phosphorus uptake by pigeonpea was the highest for all crop species tested in the Alfisol.

Mycorrhizal association

In order to determine whether VAM associations contribute to the different P response of pigeonpea compared to sorghum, in these soils, a pot experiment was conducted. Inoculation with VAM enhanced growth of pigeonpea in both soils whether P was applied or not, but it stimulated sorghum growth only on the Vertisol. On the Alfisol without P fertilizer, sorghum could not survive either with or without VAM inoculation (Table 4). VAM could enhance growth of sorghum on Alfisol when 8.7 mg kg^{-1} P fertilizer was added. Sorghum roots which had died on the Alfisol without P were colonised with VAM (infection rate: 18%) to the extent of half as much as on sorghum roots in the Vertisol (35%).

Better utilization of iron-associated P by pigeonpea

To confirm the ability of pigeonpea to solubilize Fe-P in the Alfisol with low-P fertility, a sand-vermiculite culture experiment in which P was

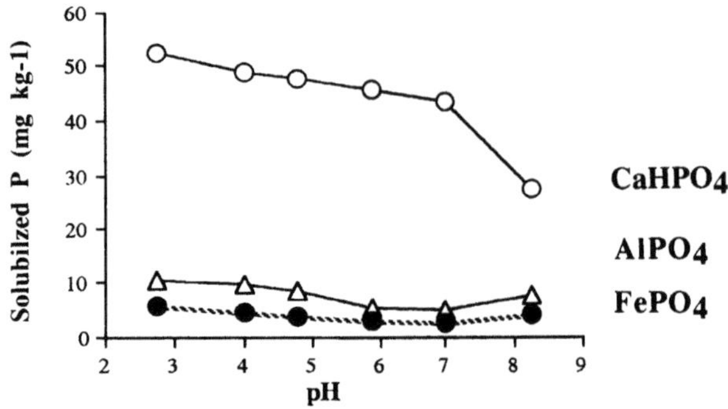

Fig. 1. Effect of pH on P released from $CaHPO_4$, $AlPO_4$ and $FePO_4$ applied in sand-vermiculite culture.

Table 3. Shoot phosphorus contents (mg per pot) of several crop species at the grain-filling stage after growth in potted Alfisol and Vertisol without phosphorus fertilizer in a greenhouse

Soil	Chickpea	Pigeonpea	Sorghum	Soybean	Pearl millet	Maize
Alfisol	4.73	5.72	0.59[a]	1.4[a]	0.64[a]	0.51[a]
Vertisol	7.79	2.34	3.91	6.53	5.38	6.13
(SE)	(0.77)	(0.82)	(0.39)	(0.20)	(0.34)	(0.25)

[a] Plants died one month after sowing.
(Source: Ae et al., 1991a).

Table 4. Effect of VAM fungi on phosphorus uptake (mg per pot) by sorghum and pigeonpea in sterilized Vertisol and Alfisol in pots. (Values are means ±SE. Three replications)

Crop	VAM inoculation	Vertisol		Alfisol	
		−P	+P[a]	−P	+P[a]
Sorghum	−	0.24 ± 0.04	14.08 ± 1.77	0.07 ± 0.03[b]	11.72 ± 4.59
	+	20.68 ± 4.52	44.84 ± 4.48	0.08 ± 0.02[b]	17.62 ± 6.39
Pigeonpea	−	0.32 ± 0.06	1.07 ± 0.08	0.26 ± 0.05	1.61 ± 0.44
	+	11.39 ± 0.54	17.50 ± 0.65	10.04 ± 1.43	15.23 ± 1.35

[a] P was applied at 8.7 mg P kg^{-1}.
[b] Plants died one month after sowing.

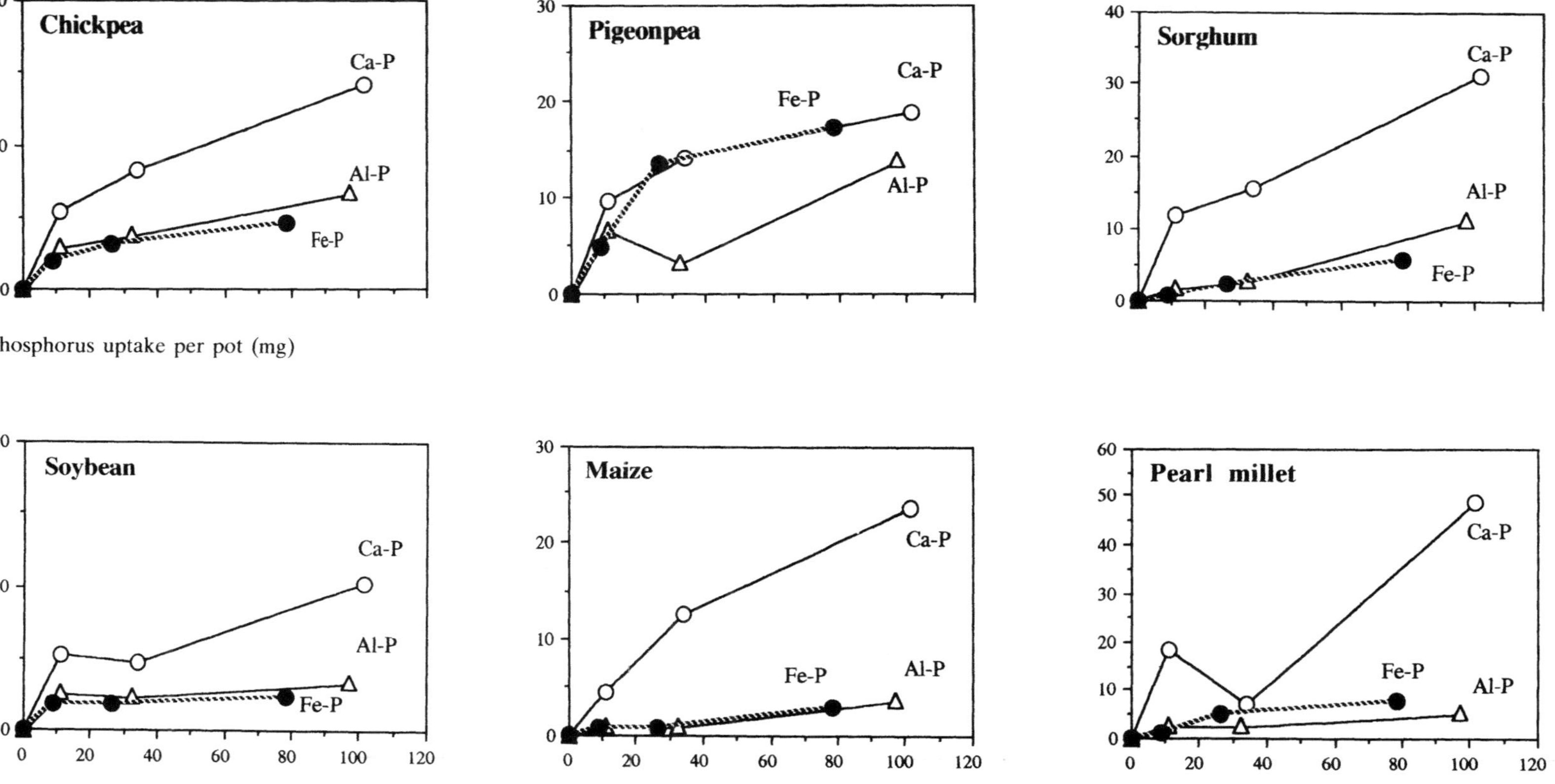

Fig. 2. Effect of P applied as different sources of phosphate ($CaHPO_4$, $AlPO_4$, and $FePO_4$) on P uptake by shoots of several crops (pigeonpea, sorghum, chickpea, maize, pearl millet, and soybean). SE = ±2.91, for comparing means (n = 3) for each crop at the same combination of P source and P level.

applied as different P sources of either $CaHPO_4$, $AlPO_4$ or $FePO_4$ was conducted. The solubility of these chemicals added to the sand-vermiculite mixture was tested at various pH levels before the pot experiment was carried out (Fig. 1). $CaHPO_4$ was the most soluble form over the pH range 3–8. For example at pH 7.0, values were 44 mg P kg^{-1} for $CaHPO_4$, 5.1 mg kg^{-1} for $AlPO_4$ and 2.9 mg kg^{-1} for $FePO_4$. All crops except pigeonpea absorbed much more P from $CaHPO_4$ than from $AlPO_4$ and P uptake from $FePO_4$ was less than from $AlPO_4$ (Fig. 2). On the other hand, pigeonpea could take up P from $FePO_4$ as well as it could from $CaHPO_4$ over the range of P levels used. Phosphorus uptake by pigeonpea from $AlPO_4$ was less than that from $FePO_4$ or $CaHPO_4$.

Root exudates

Root exudates were collected from pigeonpea grown on sand culture and separated into three fractions by ion-exchange resins. The acid fraction had twice as much ability to solubilize $FePO_4$ than the cationic fraction while the neutral fraction was inactive. The acid fraction from 2 month-old plants was collected and the major organic acids in each of the crops tested were analyzed (Table 5). Citric acid was the major organic acid found. Most was found in chickpea, followed by soybean, pigeonpea and sorghum in that order.

Ae et al. (1990) studied the acid fraction from the root exudates of four crop species, pigeonpea, sorghum, maize and soybean and found that (p-hydroxybenzyl) tartaric acid, named piscidic acid, and its p-methyl derivative, (p-methoxybenzyl) tartaric acid, were peculiar to pigeonpea. To test the ability of piscidic acid and related compounds (Fig. 3) to specifically release P from $FePO_4$, we prepared piscidic acid from *Narcissus poeticus* bulb (Smeby et al., 1954). Some derivatives of fukiic acid from *Petasites japonicus* (Sakamura et al., 1973) were also synthesized to investigate the chelating ability between Fe^{3+} and reactive groups, such as phenolic, alcoholic, and carboxylic groups in piscidic acid. Phosphorus released by these compounds was measured (Table 6). The P releasing ability of dimethyl fukiic acid was similar to that of piscidic acid. Trimethylfukiic acids, where an alcoholic group is replaced by a methoxyl group, are less able than piscidic acid to release P.

Table 5. Major organic acids of root exudates from sorghum, pigeonpea, soybean, and chickpea

Crop	Organic acids (mg/g-dry root)			
	Malonate	Succinate	Citrate	Malate
Sorghum	Trace	Trace	0.045	0.008
Pigeonpea	Trace	0.025	0.101	0.047
Soybean	0.324	0.046	0.481	0.078
Chickpea	Trace	0.054	1.292	0.025

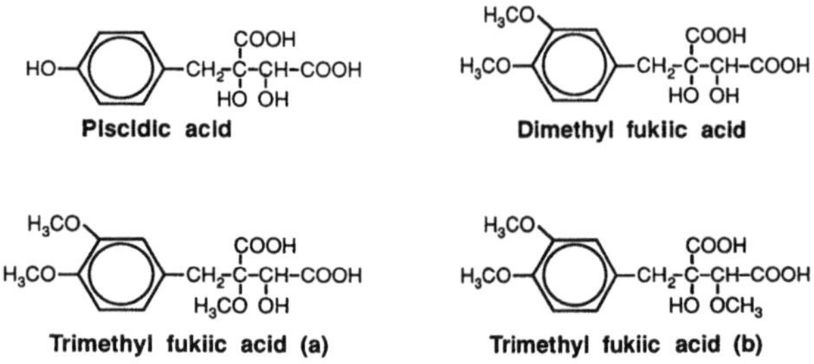

Piscidic acid — Dimethyl fukiic acid — Trimethyl fukiic acid (a) — Trimethyl fukiic acid (b)

Fig. 3. Chemical structure of piscidic acid and derivatives.

Table 6. The effects of piscidic acid and its derivatives on phosphorus release from $FePO_4$

Chemical	Released P ($\mu g\,mL^{-1}$)
Control (water)	1.48
Piscidic acid	4.37
Dimethyl fukiic acid	4.44
Trimethyl fukiic acid[a]	3.27
Trimethyl fukiic acid[b]	3.23
SE[c]	±0.395

[a,b] Refer to structural formulae (a) and (b) respectively in Figure 3.
[c] 3 replicates.

Discussion

Olsen's sodium bicarbonate extraction (Olsen et al., 1954) is generally recommended for evaluating P availability of soils, especially calcareous and/or alkaline soils (Banger et al., 1979). According to Olsen's method, P availability of the Alfisol (3.5 mg P kg^{-1}) is better than that of the Vertisol (1.5 mg P kg^{-1}). When we compared P fertility on both soils with low-P fertility, better growth and P uptake of sorghum was obtained on Vertisol than on Alfisol in both field and pot experiments (Tables 2 and 3). In the Alfisol most of the inorganic P is associated with iron (Fe-P), whereas in the Vertisol there is a large fraction of calcium bound P (Ca-P). The large Ca-P fraction in Vertisols can release soluble P as a result of acidification of the rhizosphere by secretion of organic acids and H^+ from roots (Ae et al., 1991a; Moghimi and Tate, 1978) but the Fe-P fraction cannot. Therefore, Olsen's method is not appropriate to compare the soil P fertility of Alfisols and Vertisols, and an acid extraction method, such as that of Truog (1930) or Bray No. 2 (Bray and Kurts, 1945), gives a better prediction of P fertility of soils (Ae et al., 1991a). Although P availability of the Alfisol is less than that of the Vertisol according to acid extraction methods, P uptake by pigeonpea on the Alfisol was much better than that of sorghum.

The better growth of pigeonpea than sorghum on the Alfisol (Table 2) may possibly be attributed to the deep-rooting characteristic of this crop. To eliminate the effect of rooting habit of pigeonpea, a pot experiment with the same soils, in which rooting volume was restricted compared to the field experiment, was conducted using several crop species (Table 3). In pots also, the growth and P uptake of chickpea, sorghum, soybean, pearl millet, and maize is higher on the Vertisol than on the Alfisol. Without P applied, sorghum, soybean, pearl millet and maize died on the Alfisol within 1 month after sowing. However, pigeonpea could not only survive and grow on the Alfisol but could also obtain more P from that soil than from the Vertisol. This shows that pigeonpea has an ability to take up P from the Alfisol irrespective of its deep rooting habit.

As it has been reported that pigeonpea has strong mycorrhizal associations (Manjunath and Bagyaraj, 1984), it is necessary to ascertain whether VAM associations contribute to the different P responses of pigeonpea and sorghum on these soils. Growth of pigeonpea and sorghum were markedly stimulated by VAM on the Vertisol. This means that inoculated VAM fungi on sterilized soils was effective for both crops. On the Alfisol, however, only the growth of pigeonpea could be stimulated, and sorghum failed to survive with or without VAM inoculation when no P fertilizer was added. However, when 8.7 mg kg^{-1} was added on the Alfisol, sorghum could grow well on the plot without inoculation and VAM fungi enhanced the growth of sorghum to a similar level as that of pigeonpea (Table 4). These results demonstrate that VAM acts not by dissolving relatively insoluble forms of P, but by allowing more efficient uptake of P that is already in a soluble form. This mode of action has been previously described (Mosse, 1981). Therefore, the ability to solubilize Fe-P in Alfisols appears to be an inherent characteristic of pigeonpea.

The fact that pigeonpea performed better on the Alfisol than on the Vertisol, and also better than the other crops (chickpea, maize, sorghum, and pearl millet) on the Alfisol (Table 3), suggests that it is better able to access the large Fe-P fraction in the Alfisol. Generally, Fe-P is more insoluble than Ca-P or Al-P. These results presented in Figure 2 demonstrate a unique ability of pigeonpea to solubilize Fe-P. Ability to solubilize P from Al-P or Fe-P has also been claimed for other plant species, such as *Eucalyptus* spp. (Mullette et al., 1974).

Gardner et al. (1983) proposed that citric acid exuded from the roots of lupin formed high molecular-weight complexes with Fe-P, causing P to be released on reduction of Fe^{3+} to Fe^{2+} on the root surface. Mullette et al. (1974) also proposed the hypothesis that citric acid and oxalic acid from *Eucalyptus gummifera* chelated with Al^{3+} and/or Fe^{3+} to release P in infertile soils. Of the 3 fractions (cationic, anionic and neutral fraction) collected from pigeonpea roots, the anionic fraction showed the highest activity to solubilize $FePO_4$. Citric acid was the major component of root exudates of all crop species tested (Table 5). Pigeonpea exuded much less citric acid than soybean or chickpea but more than sorghum. Pigeonpea also exuded less malonate, succinate and malate than soybean. Thus the citric acid mechanism proposed by Gardner and co-workers would not explain the particular advantage shown by pigeonpea in being able to solubilize Fe-P because soybean, which exuded more citric acid than pigeonpea, could not survive on the Alfisol (Table 3).

To identify the other organic acids which are

unique to pigeonpea, comparisons of gas chromatographs (GC) of the acid fraction from soybean, sorghum, maize and pigeonpea were made. The 2 peaks at 23–24 min retention time that were peculiar to pigeonpea were identified as piscidic acid ((p-hydroxybenzyl) tartaric acid) and its methyl derivative ((p-methoxybenzyl) tartaric acid) (Ae et al., 1990). Piscidic acid is one of the chemical constituents of hypnotic and narcotic drugs that have been extracted from the bark of Jamaica dogwood tree (*Piscidia erythrina* L.) (Freer and Clover, 1901; Bridge et al., 1948). However these substances have not been previously considered in relation to P acquisition ability from soils.

To test the relationship between ability to chelate with Fe^{3+} and chemical structure, piscidic acid and derivatives of fukiic acid (Yoshihara et al., 1974) were prepared (Table 6). The absolute configuration of fukiic acid is the same as that of piscidic acid (Yoshihara et al., 1974). Piscidic acid and dimethylfukiic acid have similar P releasing ability. Trimethylfukiic acids have less ability than the original piscidic acid. From the results obtained, alcoholic OH and carboxyl groups in the tartaric part of piscidic acid appear to be involved in chelation with Fe^{3+} (Ae et al., 1990).

In order to take up P from Fe-P in iron-rich soils like Alfisols, it is proposed that pigeonpea exudes piscidic acid. Therefore, it is necessary that piscidic acid-Fe^{3+} complexes be excluded from the rhizosphere because of the possibility of excess Fe uptake into roots which could result in precipitation of P in plant cells. Table 7 shows mineral element concentrations of several crop species which were grown on the Alfisol at various P levels (Ae et al., 1991a). Pigeonpea has the lowest concentration of Fe among the three legumes (chickpea, pigeonpea, and soybean) and it is also lower than in maize. The Fe/P ratio, which indicates the degree of exclusion of Fe in relation to P uptake, is also lowest for pigeonpea. As Mn is also expected to be chelated by piscidic acid, the Mn/P ratio may also be a good indicator of the exclusion ability of the piscidic acid-Mn complex. The Mn/P ratio of pigeonpea is also the lowest among the crop species tested (Table 7). Piscidic acid is a phenolic acid like p-hydroxybenzoic acid, ferulic acid, and p-coumaric acid, and these phenolic acids are considered to be toxic and contribute to 'soil sickness' (Börner 1955, 1956, 1958). It would be potentially detrimental for the piscidic acid-Fe complex to remain in the rhizosphere. The postulated mechanism involved is shown in Figure 4

Table 7. Concentrations of various elements, and Fe/P and Mn/P ratios, in shoots at the grain-filling stage of crop species grown on an Alfisol of low P status. Values are means for the P application rates of 0, 9, 22, 44, 87, mg P kg^{-1}. Standard errors are in parentheses. Date from an experiment described by Ae et al. (1991a); their Table 3

Crop	P	Ca	Mg	Zn	Fe	Cu	Mn	Ratios	
	(%)			(mg kg^{-1})				Fe/P	Mn/P
Sorghum	0.14 (0.02)	0.38 (0.07)	0.29 (0.02)	75 (13)	168 (17)	9 (1)	206 (37)	0.12	0.15
Pigeonpea	0.27 (0.03)	0.72 (0.06)	0.23 (0.01)	50 (1)	259 (15)	8 (1)	93 (6)	0.10	0.03
Chickpea	0.28 (0.04)	2.27 (0.22)	0.44 (0.02)	200 (135)	453 (35)	25 (1)	279 (28)	0.16	0.10
Soybean	0.19 (0.03)	1.09 (0.14)	0.44 (0.01)	82 (9)	410 (123)	43 (16)	269 (9)	0.22	0.14
Pearl millet	0.12 (0.03)	0.85 (0.10)	0.62 (0.05)	138 (27)	167 (23)	31 (1)	302 (31)	0.14	0.25
Maize	0.13 (0.02)	1.33 (0.25)	0.70 (0.07)	88 (16)	274 (65)	33 (7)	316 (40)	0.21	0.24

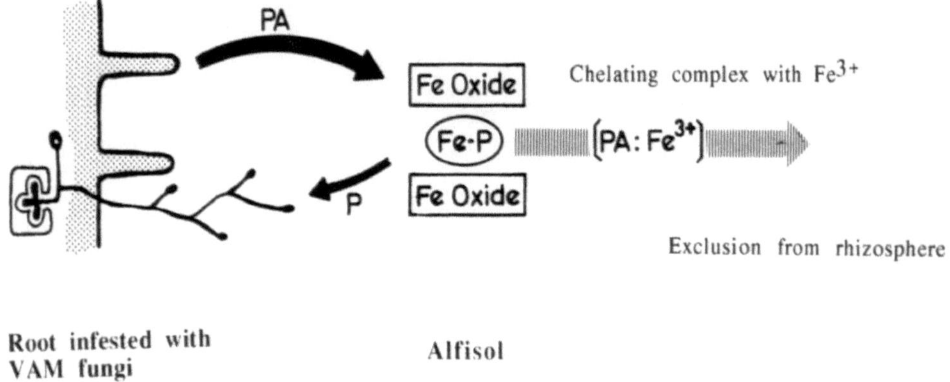

Fig. 4. Postulated P uptake mechanism in pigeonpea involving piscidic acid (PA).

and is similar to Mechanism III of Marschner (1986). It is suggested that piscidic acid, exuded from the roots of pigeonpea, releases phosphate from Fe-P compounds. This phosphate can then be taken up into the root surface or through VAM hyphae. The low Fe/P ratios in pigeonpea suggest that Fe^{3+} is excluded from the rhizosphere, we suggest as a complex with piscidic acid. However, further studies are needed to confirm this hypothesis.

References

Ae N, Arihara J and Okada K 1991a Phosphorus response of chickpea and evaluation of phosphorus availability in Indian Alfisols and Vertisols. *In* Phosphorus Nutrition of Grain Legumes in the Semi-arid Tropics. Eds. C Johansen, K K Lee and K L Sahrawat. pp 33–41. ICRISAT, Patancheru A P India.

Ae N, Arihara J and Okada K 1991b Phosphorus uptake mechanisms of pigeonpea grown in Alfisols and Vertisols. *In* Phosphorus Nutrition of Grain Legumes in the Semi-arid Tropics. Eds C Johansen, K K Lee and K L Sahrawat. pp 91–98. ICRISAT, Patanchera A P, India.

Ae N, Arihara J, Okada K, Yoshihara T and Johansen C 1990 Phosphorus uptake by pigeonpea and its role in cropping systems of Indian subcontinent. Science 248, 477–480.

Aiyer A K Y N 1949 Mixed cropping in India. Indian J. Agric. Sci. 19, 439–543.

Arnon D I 1938 Microelements in culture-solution experiments with higher plants. Am. J. Bot. 25, 322–325.

Banger A R, Mane V B and Zende G K 1979 Evaluation of soil tests at different depths and fertility levels of Vertisols. Indian J. Agric. Sci. 49, 542–549.

Börner H 1955 Untersuchungen über phenolische Verbindungen aus Getreidestroh und Getreiderückständen. Naturwissenschaften 42, 583–584.

Börner H 1956 Der papierchromatographische Nachweis von Ferulasäure in wässrigen Extrakten von Getreidestroh und Getreiderückständen. Naturwissenschaften 43, 129–130.

Börner H 1958 Untersuchungen über den Abbau von Phlorizin im Boden. Ein Beitrag zum Problem der Bodenmüdigkeit bei Obstgehölzen. Naturwissenschaften 45, 138–139.

Bray R H and Kurtz T 1945 Determination of total, organic and available forms of phosphorus in soils. Soil Sci. 59, 39–45.

Bridge W, Coleman F and Robertson A 1948 Constituents of 'Cortex pisciae erythrina'. Part 1. The structure of piscidic acid. J. Chem. Soc. 1948, 257–260.

Freer P C and Clover A M 1901 On the constituents of Jamaica dogwood. Am. Chem. J. 25, 390–413.

Gardner W K, Barber D A and Parbery D G 1983 The acquisition of phosphorus by *Lupinus albus* L. III. The probable mechanism by which phosphorus movement in the soil/root interface is enhanced. Plant and Soil 70, 107–124.

Itoh S 1987 Characteristics of phosphorus uptake of chickpea in comparison with pigeonpea, soybean, and maize. Soil Sci. Plant Nutr. 33, 417–422.

Johansen C 1990 Pigeonpea: mineral nutrition. *In* The Pigeonpea. Eds Y L Nene, Susan D Hall and V K Sheila. pp 209–231. C A B International, Wallingford, UK.

Manjunath A and Bagyaraj D J 1984 Response of pigeonpea and cowpea to phosphate and dual inoculation with vesicular-arbuscular mycorrhiza and *Rhizobium*. Trop. Agric. (Trinidad) 61, 48–52.

Marschner H 1986 Mineral Nutrition of Higher Plants. Academic Press, London, UK. 456 p.

Moghimi A and Tate M E 1978 Does 2-ketogluconate chelate calcium in the pH range 2.4 to 6.4? Soil Biol. Biochem. 10, 289–292.

Mosse B 1981 Vesicular-arbuscular mycorrhiza research for tropical agriculture. Research Bulletin, Hawaii Institute of Tropical Agriculture and Human Resources. University of Hawaii, Honolulu, HI. 82 p.

Mullette K J, Hannon N J and Elliott A G L 1974 Insoluble phosphorus usage by Eucalyptus. Plant and Soil 41, 199–205.

Murphy J and Riley J P 1962 A modified single solution method for determination of phosphate in natural waters. Anal. Chem. Acta 27, 31–36.

Olsen S R, Cole C V, Watanabe F S and Dean L A 1954 Estimation of available phosphorus in soils by extraction with sodium bicarbonate. USDA Circ. US Gov. Printing Office, Washington, DC.

Sakamura S, Yoshihara T and Toyoda K 1973 The constituents of *Petasites japonicus*: Structure of Fukiic acid and Fukinolic acid. Agric. Biol. Chem. 37, 1915–1921.

Sheldrake A R and Narayanan 1979 Growth, development and nutrient uptake in pigeonpeas (*Cajanus cajan*). J. Agric. Sci., Cambridge 92, 513–526.

Smeby R R, Zbinovsky V and Burris R H 1954 The organic acids of *Narcissius poeticua*. Am. Chem. Soc. 76, 6127–6130.

Truog E 1930 Determination of the readily available phosphorus of soils. J. Am. Soc. Agron. 22, 874–882.

Yoshihara T, Ichihara A, Nuibe H and Sakamura S 1974 The stereochemistry of fukiic acid and its correlation with piscidic acid. Agr. Biol. Chem. 38, 121–126.

P. J. Randall et al. (Eds.), Genetic aspects of plant mineral nutrition, 289–294.

PLSO SV61

Effect of mineral nutrients and combined nitrogen on the growth and nitrogen fixation of *Azolla-Anabaena* symbiosis

N. SHIOMI and S. KITOH
Research Institute for Advanced Science and Technology, University of Osaka Prefecture, Sakai, Osaka 593, Japan

Key words: *Azolla*, growth, mineral nutrients, nitrogen fixation, ammonia, nitrate, urea

Abstract

The largest difference in an optimum temperature of growth condition, i.e. 27°C for *Azolla japonica* and 31°C for *A. pinnata* var. *imbricata*, was found. The effects of mineral nutrients are combined nitrogen in the medium on the growth and N_2 fixation of *A. japonica* were studied. The optimum concentrations of K, Mg, Ca, P, and S in the medium for growth were about 0.5, 0.1–0.5, 0.5, 0.1 and 0.05–0.1 m*M* and those for N_2-fixing capacity were about 0.05–0.1, 0.05–0.1, 1.0, 0.5 and 0.05 m*M*, respectively. Ammonium in the medium inhibited strongly both growth and N_2 fixation. The total nitrogen budgets, as a consequence of the balance between nitrogen inputs from atmosphere and medium, was positive at concentrations of ammonium below 10 m*M*, whereas it was negative above this concentration. Nitrate was less effective than ammonium. Urea addition into the medium resulted in a substantial increase in growth rate and nitrogen content although N_2 fixation decreased.

Introduction

Azolla, a genus of aquatic fern, is able to grow in environments free from combined nitrogen since the nitrogen that it requires is supplied by symbiotic nitrogen-fixing blue-green algae, *Anabaena azollae*, within the leaf cavity (Peters and Mayne, 1974). The genus *Azolla* comprises seven species distributed all over the world. Hence it is likely that there are species or clones specially adapted to different climatic conditions.

In recent years, the interest in the use of the *Azolla-Anabaena* symbiosis for various purposes, e.g., green manure and feed for animals and fish, has increased, and there have been a number of basic and applied studies (IRRI, 1987). The authors studied the nutrient absorption by *Azolla* from water subjected to eutrophication, sewage and other waste water effluents with a view of utilizing this fern for the control of water pollution (Kitoh and Shiomi, 1984; Shiomi and Kitoh, 1987). Since, the levels of mineral nutrients vary widely with the types of waste water, information concerning the mineral requirements of *Azolla* is essential for successful propagation of the fern in these waste water effluents. Waste water usually contains combined nitrogen such as ammonium and nitrate, which are known to inhibit nitrogen fixation. The nitrogen fixation by the *Azolla-Anabaena* symbiosis is much more tolerant to combined nitrogen in the medium, which is one of the factors controlling the growth, than that of free-living green algae (Peters and Mayne, 1974) because the symbionts are not in direct contact with the medium.

In this study the effect of the mineral nutrient status in the medium on the growth and nitrogen-fixing activity of the *Azolla-Anabaena* symbiosis was investigated.

Materials and methods

Azolla plant

Two species, *Azolla japonica* (IRRI accession No. 1001) and *A. pinnata* var. *imbricata* (No.

44), which are widely distributed in Japan were used.

Culture media

The N-free medium contained the following macronutrients in mmole per liter: $CaCl_2$ $2H_2O$, 1.00; $MgSO_4$ $7H_2O$, 1.65; K_2SO_4, 0.50; and NaH_2PO_4 $2H_2O$, 0.65; and micronutrients in μmole per liter: $FeSO_4$ $7H_2O$, 27.0; $MnCl_2$ $4H_2O$, 1.13; $CuSO_4$ $5H_2O$, 0.08; $ZnSO_4$ $7H_2O$, 0.19, Na_2MoO_4 $2H_2O$, 0.05 and H_3BO_3, 5.77. Media deficient in K, Mg, Ca and P were prepared with lower levels of each component than the N-free media. To the P-deficient medium 2.5 m*M* MES buffer was added. In the S-deficient medium, K_2SO_4 and NaH_2PO_4 in the N-free medium were replaced with 0.35 m*M* KCl and 0.65 m*M* KH_2PO_4, and the $MgSO_4$ level was reduced to various levels.

Deficiency of Mg was compensated by the addition of $MgCl_2$. $FeSO_4$ was replaced with an equivalent concentration of EDTA-Fe. Media containing combined N as ammonium, nitrate of urea were prepared by the addition of $(NH_4)_2SO_4$, KNO_3 or urea to the N-free medium. In all cases, the initial pH of the media was 6.5.

Growth conditions

Azolla fronds were transferred to polypropylene trays (21.7 × 15.6 × 8.4 cm) containing 1 L of nutrient media and grown in a growth cabinet (Toshiba TGS-33LS). The temperature was 27/18°C and the light intensity was 15 klux with 16 h photoperiod. The fresh weight of the initial inoculum was approximately 0.7 g, with three replicates for each treatment. The media containing combined N were renewed every 2 days to prevent a rapid shift in the pH and the growth of green algae under nitrogen-rich conditions. At weekly intervals, the fronds were harvested, growth was determined and the nitrogenase activity assayed. A part of the harvested fronds, usually 0.7 to 1.5 g depending on the growth rate, was returned to new, freshly prepared nutrient media.

Growth rate and nitrogen fixation

Relative growth was calculated from the growth value (G.V. = wet weight of harvested plants/wet weight of initial explants). Doubling time (D.T.) was also calculated based on the equation D.T. = t log2/log G.V., where t is the duration (day) of the culture. Nitrogenase activity was assayed by the acetylene reduction assay (ARA) method of Tung and Shen (1981), and the result shown as μmoles C_2H_4 g^{-1} wet weight h^{-1}. Total N analysis of the harvested fronds was done with an elemental analyzer (Yanagimoto CHN Corder MT-1) after the digestion with HNO_3-$HClO_4$.

Results

Figure 1 shows the growth rate at different light intensities and temperatures varying between 23°C and 34°C. The optimum temperature for growth was 27°C for *Azolla japonica* and 31°C for *A. pinnata* var. *imbricata*. Both *Azolla* species showed the best growth at the initial pH of 6.5 which ranged from 3.5 to 8.5 and the pH was adjusted every day for 7 days. *A. japonica* was used in further experiments.

The *Azolla* plants were grown for 4 weeks on media containing various levels of each nutrient element, after which growth was observed and ARA determined. The results are shown in Table 1. The optimum concentration of P in the medium for growth was about 0.5 m*M*; at higher concentrations the growth rate decreased. For the other nutrient elements, no such decrease occurred within the limits of the concentrations tested. The ARA was also maximized when the concentrations tested. The ARA was also maximized when the concentrations of K, Mg, and P in the media were about 0.5, 0.1 and 1.0 m*M*, respectively.

If we assume that about 10% reduction of the growth rate or ARA is an index of nutrient deficiency, the optimum concentrations of K, Mg, Ca, P and S in the medium for growth were about 0.5, 0.1–0.5, 0.5, 0.1 and 0.05–0.1 m*M*, and for ARA about 0.05–0.1, 0.05–0.1, 1.0, 0.5 and 0.05 m*M*, respectively (Table 1).

As shown in Figure 2a, a concentration of

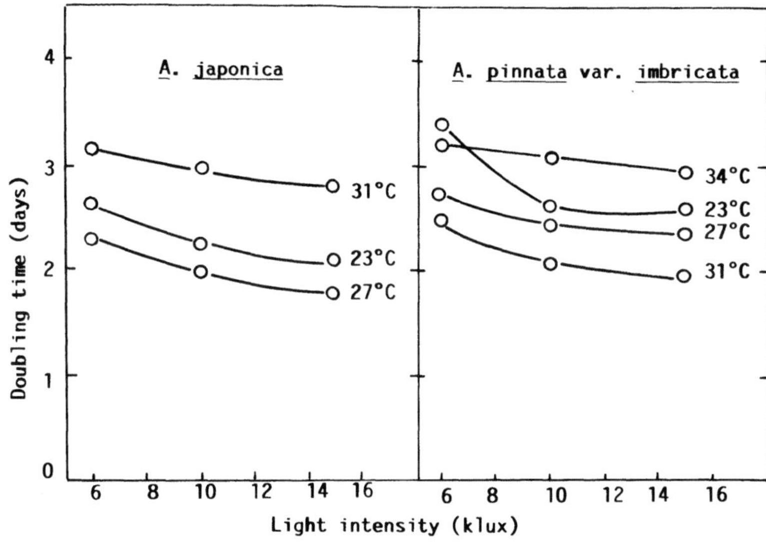

Fig. 1. Effect of temperature and light intensity on the growth of *Azolla*.

Table 1. Effect of the level of K, Mg, Ca, P and S in the medium on the growth rate and on the acetylene reduction activity of *Azolla japonica*

Concentration in the medium (m*M*)	K	Mg	Ca	P	S
	Growth rate as % of maximum observed				
0.05	44	43	4	65	83
0.1	73	84	21	93	94
0.5	98	92	93	98	97
1.0	97	98	99	87	99
	Acetylene reduction activity[a]				
0.01	ND[b]	ND[b]	ND[b]	29	61
0.05	82	83	2	52	93
0.1	95	94	10	72	94
0.5	97	94	64	93	95
1.0	90	81	90	94	94

[a]Maximum activities (100%) for K, Mg, Ca, P and S were 5.20, 5.07, 4.73, 4.52 and 4.16 μmole C_2H_4 g^{-1} wet weight h^{-1}. Each value is shown as % of maximum.
[b]Not determined.

ammonium up to 2.5 m*M* and nitrate up to 1.25 m*M* in the medium had no obvious effect on the growth of *Azolla*. About 90% growth of the control was maintained up to 5 m*M*. However, further increases of the concentrations of these combined nitrogen sources resulted in a decrease in the growth rate. On the other hand, urea up to 10 m*M* in the medium considerably accelerated the growth.

Combined nitrogen in the medium affected ARA more adversely than growth (Fig. 2b). Ammonia was most inhibitory, with 10 m*M* causing a 70% decrease in ARA relative to the control. The addition of nitrate at 10 m*M* and urea at 5 m*M* caused a decrease in ARA of about 45%.

The total nitrogen content in the tissue of *Azolla* was determined and shown as the ratio of the control (=100%) (Fig. 3). It increased with the increase of the concentration of urea in the

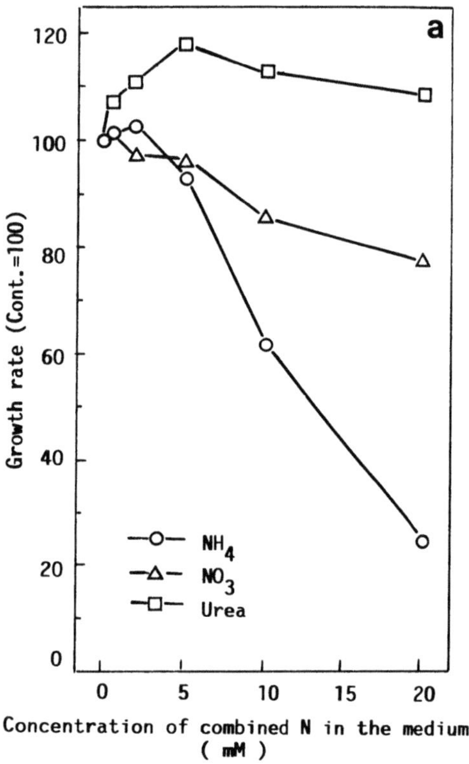

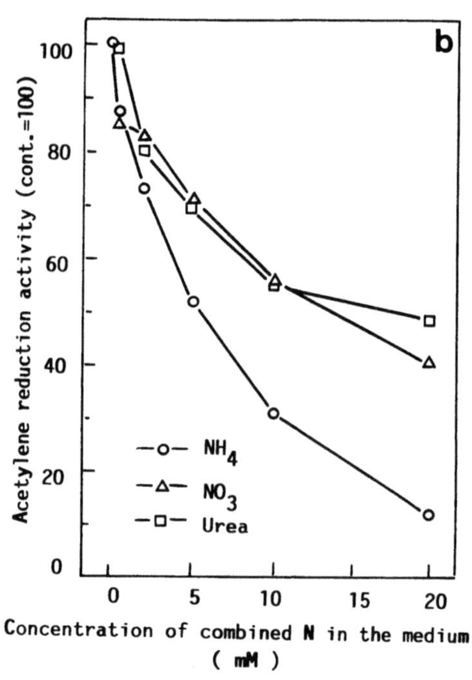

Fig. 2. **a.** Effect of combined nitrogen sources in the medium on the growth rate of *Azolla japonica. Azolla* was grown for 3 consecutive weeks and the measurements were performed every week before inoculation to the new medium. The average of 3 weekly determinations is shown here. **b.** Effect of combined nitrogen sources in the medium on the acetylene reduction activity of *Azolla japonica*.

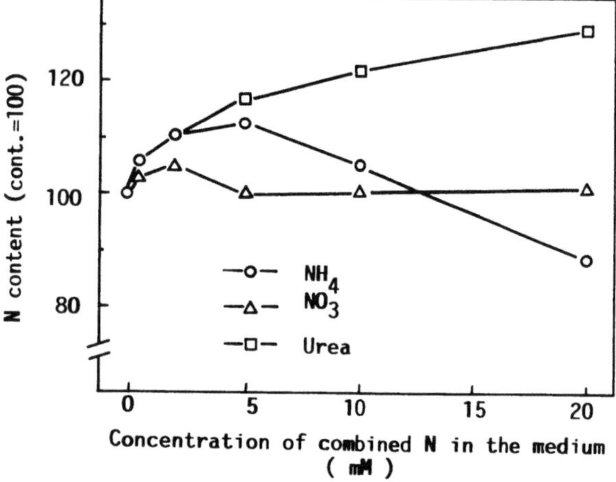

Fig. 3. Effect of combined nitrogen sources in the medium on the total nitrogen content of *Azolla japonica*. The measurements were performed after 3 weeks.

medium. It also increased in the medium containing ammonium up to 10 mM but decreased with a further increase of the ammonium concentration. The total nitrogen content was hardly affected by the increase of the nitrate concentration except for a moderate increase below 5 mM.

Discussion

The use of aquatic plants as agents for removing nutrients is not a new concept (Cornwell et al., 1977; Harvey and Fox, 1973; Hashimoto, 1983). However, the efficient removal of P would be expected even after N is consumed in the waste water because of the capability of N_2 fixation by *Azolla* use. However, in a diluted secondarily treated effluent which contained significantly higher concentrations of ammonium and Cl, both the growth and the N_2-fixing activity of

Azolla were less than in a synthetic nutrient solution (Shiomi and Kitoh, 1992).

The optimum concentrations of K, Mg and Ca in the medium for the growth of *A. japonica* obtained in this investigation agree with the values reported for *A. pinnata* var. *imbricata* by Yatazawa et al., (1980). Nevertheless, for ARA, the optimum concentrations of K and Mg were much lower in *A. japonica* than in *A. pinnata* var. *imbricata*. The reverse was observed in the case of Ca. However, it remains to be determined whether this discrepancy is due to the difference in the preculture method or to the difference in species of *Azolla*. The optimum concentrations of the medium P for growth and ARA of *A. japonica* are about 3 times and 10 times as high as those of *A. pinnata* var. *imbricata*, respectively. Subudhi and Watanabe (1981) classified *Azolla* species into two groups according to the P requirements: the *A. pinnata* species, which can efficiently utilize P at a low level, and the new World species such as *A. mexicana*, *A. caroliniana* and *A. filiculoides*, which cannot. *A. japonica* appears to belong to the latter group.

The inhibitory effect of combined nitrogen in the medium on the nitrogen-fixing activity in *Azolla* species has been confirmed by many investigators (Brotonegoro and Abdulkadir, 1976; Newton and Selke, 1981; Peters et al., 1981; Yatazawa et al., 1980). However, few reports have dealt with the effect of combined nitrogen sources in the medium on growth. Yatazawa et al. (1980) reported that ammonium, even at a concentration as low as 1 m*M*, caused a considerable decrease in the growth rate of *A. pinnata*. Peters et al. (1981) reported that the growth of *A. caroliniana* was comparable when this species was cultured on N_2 or on a medium containing ammonium up to 5 m*M*, nitrate up to 25 m*M* or urea up to 12.5 m*M*. The present results show that ammonium and nitrate in the medium exert an unfavorable effect on the growth of *A. japonica* except when their concentrations are very low, whereas urea is fully suitable up to 20 m*M*.

The decrease in the N_2-fixing capacity of free-living blue-green algae grown with combined nitrogen sources has been ascribed to the inhibition of *de* novo synthesis of nitrogenase (Mahl and Wilson, 1968), or to the inactivation of the nitrogenase system (Ohmori and Hattori, 1974). Kitoh et al. (1991) have shown that the *Azolla* fronds grown in media with a high concentration of ammonium are *Anabaena*-free.

Since *Azolla* plants can assimilate combined nitrogen in the medium (Ito and Watanabe, 1983; Meeks et al., 1987), two sources of nitrogen are available to them in the medium containing combined nitrogen. Combined nitrogen, however, depresses nitrogen fixation by the symbionts. Consequently, the total nitrogen budget, due to the balance between nitrogen inputs from the atmosphere and the medium, as well as the growth rate, is considered to indicate whether the combined nitrogen sources are beneficial to *Azolla*. When the fronds were grown on a medium containing ammonium or nitrate at concentrations less than 2.5 m*M*, the growth was comparable to that of fronds grown on N_2 alone and the total nitrogen budget was also positive. The current findings indicate that these combined nitrogen sources at low levels can entirely make up for the loss of nitrogen input from the atmosphere as a result of the reduction in the N_2-fixing activity. In this regard, it may be concluded that urea, which brought about a significant increase in both the growth rate and total nitrogen content up to about 10 m*M*, is a suitable nitrogen source for *A. japonica*.

References

Brotonegoro S and Abdulkadir S 1976 Growth and nitrogen-fixing activity of *Azolla pinnata*. Ann. Bogorienses 6, 69–77.

Cornwell A, Zoltek J Jr and Patrinely C D 1977 Nutrient removal by water hyacinths. J. Water Pollut. Control Fed. 49, 57–65.

Harvey R M and Fox J L 1973 Nutrient removal using *Lemna minor*. J. Water Pollut. Control Fed. 45, 1928–1938.

Hashimoto S 1983 Wastewater treatment by channel flow system and food production. Water Purification and Liquid Wastes Treatment 24, 17–24.

International Rice Research Institute 1987 *Azolla* Utilization. P.O. Box 933, Manila, Philippines. 295 p.

Ito O and Watanabe I 1983 The relationship between combined nitrogen uptake and nitrogen fixation in *Azolla-Anabaena* symbiosis. New Phytol. 95, 647–654.

Kitoh S and Shiomi N 1984 Nutrient removal by *Azolla* from the mineral nutrient solution and waste water. Water Purification and Liquid Wastes Treatment 25, 561–567.

Kitoh S, Shiomi N and Uheda E 1991 Disappearance of symbiotic algae in the *Azolla-Anabaena* association subjected to ammonium. Soil Sci. Plant Nutr. 37, 323–329.

Mahl M C and Wilson P W 1968 Nitrogen fixation by cell-free extracts of *Klebsiella pneumoniae*. Can. J. Microbiol. 14, 33–38.

Meeks J C, Steinberg N A, Enderlin C S, Joseph C M and Peters G A 1987 *Azolla-Anabaena* relationship. XIII. Fixation of ^{13}N–N_2. Plant Physiol. 84, 883–886.

Newton J C and Selke E S 1981 Assimilation of ammonia by the *Azolla-Anabaena* symbiosis. J. Plant Nutr. 3, 803–811.

Ohmori M and Hattori A 1974 Effect of ammonia on nitrogen fixation by the blue-green algae *Anabaena cyrindrica*. Plant Cell Physiol. 15, 131–142.

Peters G A and Mayne B C 1974 The *Azolla, Anabaena azollae* relationship. II. Localization of nitrogenase activity as assayed by acetylene reduction. Plant Physiol. 53, 820–824.

Peters G A, Ito O, Tyagi V V S and Kaplan D 1981 Physiological studies on N_2-fixing *Azolla*. *In* Genetic Engineering of Symbiotic Nitrogen fixation and Conservation of Fixed Nitrogen. Eds. J M Lyons et al. pp 343–362. Plenum, New York.

Shiomi N and Kitoh S 1987 Use of *Azolla* as decontaminant in sewage treatment. *In Azolla* Utilization. pp 169–176. IRRI, Los Baños, Lugna, Philippines.

Shiomi N and Kitoh S 1992 *Azolla* culture in a pond and its use as a biomass. Nippon Kagaku Kaisi No. 5, 512–516.

Subudhi B P R and Watanabe I 1981 Differential phosphorus requirements of *Azolla* species and strains in phosphorus-limited continuous culture. Soil Sci. Plant Nutr. 27, 237–247.

Tung H F and Shen T C 1981 Studies of the *Azolla pinnata - Anabaena azollae* symbiosis: Growth and nitrogen fixation. New Phytol. 87, 743–749.

Yatazawa M, Tomomatsu N, Hosoda N and Nunome K 1980 Nitrogen Fixation in *Azolla-Anabaena* symbiosis as affected by mineral nutrient status. Soil Sci. Plant Nutr. 26, 415–426.

P. J. Randall et al. (Eds.), Genetic aspects of plant mineral nutrition, 295–299.

PLSO SV63

Can maize cultivars with low mineral nutrient concentrations in the grains help to reduce the need for fertilizers in third world countries?

B. FEIL[1], R. THIRAPORN[2] and P. STAMP[1]
[1]*Institute of Plant Sciences, ETHZ, Universitätstrasse 2, CH-8092 Zurich, Switzerland and* [2]*Kasetsart University, Department of Agronomy, Bangkok, Thailand*

Key words: cultivars, grains, maize, nitrogen, phosphorus, potassium, tropical climate

Abstract

An earlier study revealed considerable genotypic variation in grain N, P and K concentrations (GNC, GPC and GKC, respectively) in tropical maize. The expression of varietal differences in GNC, GPC and GKC, however, may depend on environmental conditions such as the N status of the soil. Two tropical maize hybrids (*Suwan 2301* and *CP 1*) with comparable yielding capacity, but contrasting GNCs, GPCs and GKCs, were therefore grown at four levels of N in a field experiment at Farm Suwan (Thailand, latitude 14.5°N). *Suwan 2301* exhibited a higher GNC than did *CP 1* at all rates of N, but large differences in GPC and GKC were found only at high N fertilization. This was obviously due to individual grain yield responses of the cultivars to increasing rates of N fertilizer, demonstrating that grain nutrient concentrations are, at least in part, functions of the amount of grain carbohydrates which dilute a genetically and environmentally fixed amount of grain P and K. As compared to *Suwan 2301*, *CP 1* accumulated less N, P and K in the grains at almost all levels of N fertilization, confirming our hypothesis that the cultivation of maize genotypes with low grain mineral nutrient concentrations may help third-world cash-crop farmers to reduce the need for scarce and costly mineral fertilizers. This finding has to be verified at reduced availability of soil −P, −K, and water.

Introduction

A high percentage of protein and minerals in the grain is often considered to be an important criterion for the dietary quality of a cereal product. At least for monogastradians, grain protein and phosphorus of most cereals are of low nutritive value. In ordinary maize, for example, the protein is characterized by disproportionately low concentrations of the essential amino acids lysine and tryptophan, and by a non-optimal ratio of leucine and isoleucine (Alexander, 1988). The main storage form of cereal grain phosphorus is phytate (Michael et al., 1980). Phytate has been implicated in mineral-deficiency symptoms in man (Hambidge and Walraven, 1976) and animals (Forbes et al., 1984). Cultivation of maize mutants, in which phytic-acid synthesis is blocked (Raboy et al., 1990) may help to avoid such problems. Furthermore, phytate is a rather inefficient source of phosphorus for monogastradians because only about one-half of it is utilized, the remainder being lost by excretion. This may lead to severe environmental problems in the vicinity of farms which, for example, specialize in fattening pigs and poultry.

Consequently, farmers in developed countries, for whom alternative sources of high-quality protein and mineral nutrients are readily available, are almost exclusively interested in the energy content of maize. This complies with the interests of cash-crop farmers in third-world countries who produce maize as fodder for animals in developed countries. In these regions, mineral fertilizers are often scarce and costly as compared to the prices paid for agricultural products.

In an earlier study, we reported a considerable genotypic variation in nitrogen, phosphorus and potassium concentrations in the grains of high-yielding tropical maize (Feil et al., 1990). This variability could be exploited to save the nutrient resources of soils. The expression of varietal differences, however, may depend on the nutrient status of a site. Therefore, a further experiment was conducted to study the performance of two cultivars under different nitrogen supplies.

Material and methods

The trial was carried out at the National Corn and Sorghum Research Centre, Farm Suwan, Pakchong, Thailand (latitude 14.5°N) during a dry season.

Two maize hybrids were compared. One cultivar, *Suwan 2301*, originates from the maize-breeding program of the Kasetsart University. The other cultivar, *CP 1*, is a commercial product. The soil was fertilized, and the seeds sown on November 14, 1989. Atrazine was sprayed after sowing to control weeds. Additionally, weeds were removed manually 35 days after planting. Seeds were treated with Ridomil for downy mildew (*Sclerospora sorghi*) control. Furthermore, Azodrin was sprayed at 10, 25 and 40 days after planting, and Furadan 3% G was applied to the leaf whorl at 47 days after emergence for insect control. The stands were free of lodging and diseases. If necessary, plots were irrigated weekly by furrow irrigation. The stands were thinned to the final population density (53,000 plants ha^{-1}) on December 1.

The soil type is a reddish brown latosol. The analysis of 10 soil samples of the top-30 cm of the profile revealed the following average nutrient contents: 109 mg kg^{-1} P (Bray II), 99 mg kg^{-1} K (exchangable K by 1 *N* ammonium acetate at pH 7.0) and 97 mg kg^{-1} nitrate-N (extracted by 2 *N* KCl). The percentage of organic matter was 1.5 (wet oxidation with K_2CrO_7 and back-titration with $FeSO_4$). The pH (H_2O) was 7.5. All plots received 27 kg P ha^{-1} as TSP and 41 kg K ha^{-1} as KCl. N was supplied as $(NH_4)_2SO_4$ at 4 levels: 0, 40, 80 and 160 kg N ha^{-1}.

The plots had 7 rows, 5 m long and 75 cm apart. Two rows were harvested to measure grain yield. Grains of five plants were analyzed for N, P and K. Plant and soil samples were analyzed in the Central Laboratories of the Kasetsart University. There were 4 replications. Analyses of variance were peformed for all variables.

Results

Grain yields increased with increasing N fertilization. Cultivars, however, responded very differently to increasing rates of N (Fig. 1a): at 0 and 40 kg N ha^{-1} *Suwan 2301* was superior to *CP 1*, whereas at 160 kg ha^{-1} *CP 1* outyielded *Suwan 2301*.

N enhanced the nitrogen concentration in the grains (GNC) of both cultivars to a very similar degree (Fig. 1b). A large increase in GNC occurred between 0 and 40 kg N ha^{-1}. An additional application of 40 kg N ha^{-1} did not affect GNC consistently, but an additional 120 kg ha^{-1} caused a significant increase in GNC of both cultivars. *Suwan 2301* exhibited a higher GNC than did *CP 1* at all levels of N.

With phosphorus, the situation was more complicated (Fig. 1c). In *Suwan 2301*, increasing rates of N steadily enhanced the phosphorus concentration in the grains (GPC). *CP 1*, however, showed a different response: 40 kg N ha^{-1} increased the GPC but higher rates caused a decrease. Thus, averaged over the cultivars, there was no clear effect of N fertilization on GPC. The F test revealed an almost statistically significant ($p < 0.05$) cultivar × N fertilization interaction.

However, the cultivar × N rate interactions were statistically significant for the grain potassium concentration (GKC; Fig. 1d). At 0 and 40 kg N ha^{-1}, both cultivars showed about the same GKC. However, at the next higher level of N, *Suwan 2301* had a clearly higher GKC than did *CP 1*. Similar differences were observed at the highest rate of N.

The product of grain yield and grain nutrient concentration is the grain nutrient yield which corresponds to the export of nutrients from the plots if the straw is left on the field. At fertilizer rates of 0, 40 and 80 kg N ha^{-1}, *Suwan 2301*

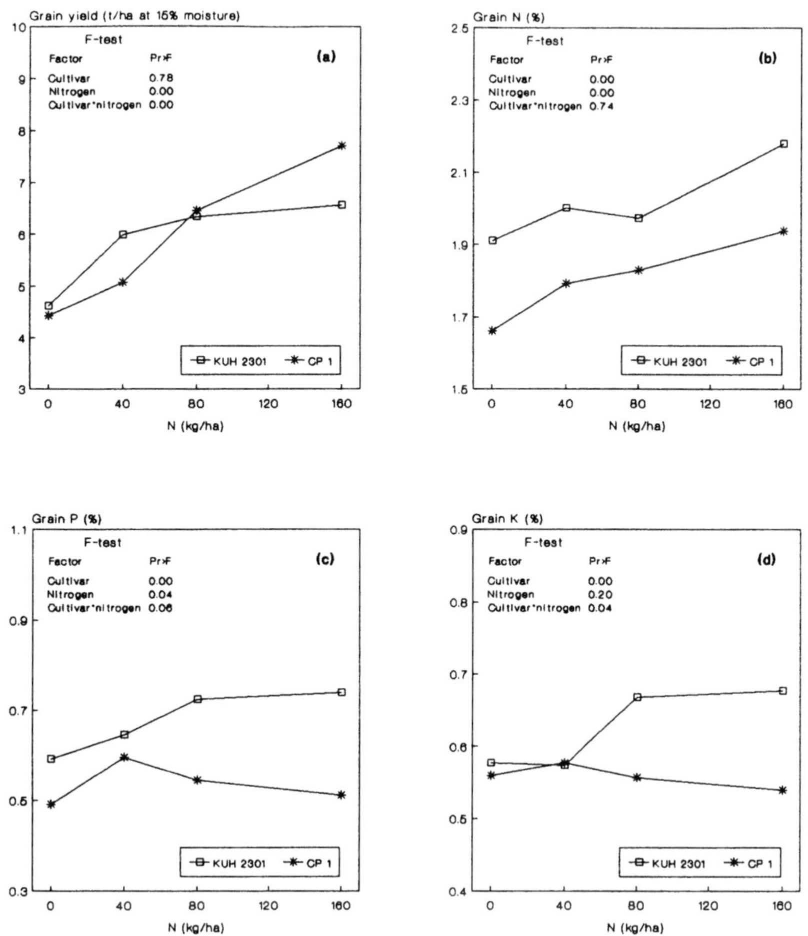

Fig. 1. Grain yield (**a**) and concentrations of N (**b**), P (**c**) and K (**d**) in the grains of two tropical maize hybrids grown at four levels of nitrogen.

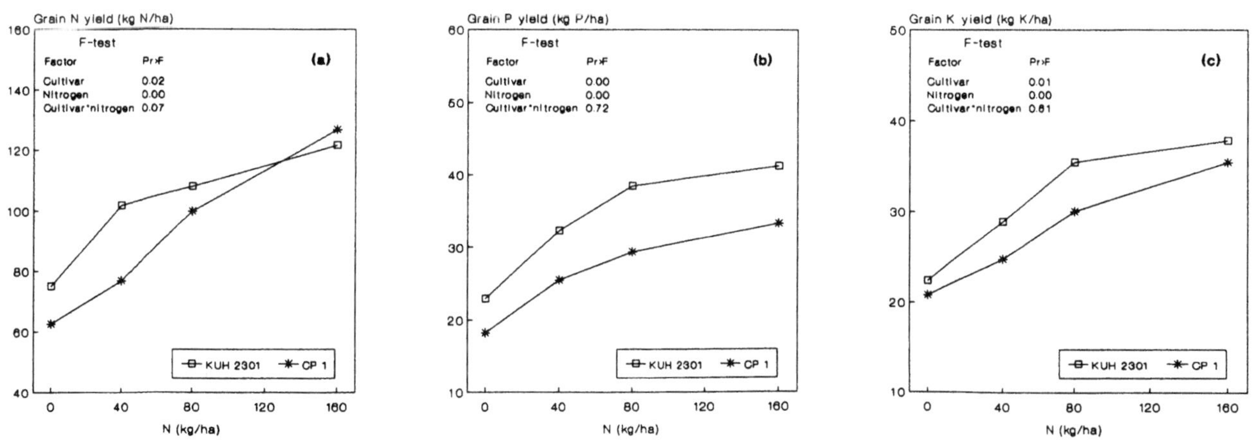

Fig. 2. Grain N (**a**), P (**b**) and K (**c**) yields of two tropical maize cultivars grown at four levels of nitrogen.

showed higher grain N yields (GNY) than *CP 1*. At the highest rate, however, *CP 1* accumulated more N in the grains than did *Suwan 2301*. The cultivar N interactions were not quite statistically significant at the 95% confidence level. In contrast, there was no evidence of any cultivar × N rate interactions in GPY and GKY, respectively. Both cultivars responded uniformly to increasing rates of N (Figs. 2b,c). At all levels of N, *Suwan 2301* exported more phosphorus and potassium from the field than did *CP 1*.

Discussion

Concern exists that breeding at high-fertility sites favors the selection of high-input types which, however, may fail in low-fertility environments. This concern appears to be justified, since this experiment as well as others (Muruli and Paulsen, 1981; Thiraporn et al., 1987; Tsai et al., 1984) revealed differential grain yield responses of maize varieties to N fertilization. On the other hand, modern (probably developed at high N fertilization) U.S. maize cultivars outperformed old ones (presumably selected at limited N supply) at all levels of N (Castleberry et al., 1984). Evaluations of grain yield responses of maize hybrids to varying supplies of N at different locations in the U.S. Corn Belt resulted in sporadic genotype N interactions which vanished when averaged across the environments (Gardner et al., 1990).

In the present experiment, cultivars *Suwan 2301* and *CP 1* showed a different response to increasing rates of N. A similar response was observed in two earlier experiments carried out at Farm Suwan during different cropping seasons. Further experiments at other locations and under different growth conditions are required to confirm these genotype × N interactions and to examine the environmental or cultural prerequisites which are responsible for the appearance of differential grain yield responses to increasing rates of N.

Genotype × N fertilization interactions in grain yield may cast more light on the complicated relationships between grain yield and grain nutrient concentrations. Reports on the inter-cultivar relationship between grain yield and GNC in maize are very contradictory (Dudley et al., 1977; Eberhard, 1977; Feil et al., 1990; Frey, 1977; Gupta et al., 1975; Kauffman and Dudley, 1979). The results of this investigation clearly confirm our earlier conclusion that genotypes of comparable yielding ability and of similar maturity can show considerable differences in GNC (Feil et al., 1990). Interestingly, these differences proved to be stable over a wide range of nitrogen availability, in spite of significant cultivar × N rate interactions in grain yield. Studies of wheat and other small grain cereals (Frey, 1977; Heitholt et al., 1990) resulted in consistent inverse inter-cultivar relationships between grain yield and GNC, probably reflecting the dilution of comparable amounts of kernel N by different amounts of grain carbohydrates. According to these findings, one would expect a reverse trend in grain yield and GNC response of maize cultivars to increasing availability of N. Our results, however, may indicate that there are fundamental differences in N utilization between small-grain cereals and maize whose identification might help to overcome the negative correlation between grain yield and GNC in small-grain cereals.

In contrast to GNY (Fig. 2a), increasing rates of N enhanced the GPY and the GKY of both cultivars to the same extent (Figs. 2b, c). Nevertheless, varietal differences in GPC and GKC increased with increasing N fertilization (Figs. 1c and d). Apparently, individual grain yield responses to increasing N fertilization slightly diluted (*CP 1*) or concentrated (*Suwan 2301*) genetically and environmentally fixed amounts of P and K in the grain.

Averaged over the levels of N, cultivation of *CP 1* instead of *Suwan 2301* would have saved about 21% of P and 11% of K, with grain yields being almost identical. The average GNY of *CP 1* was 10% lower, but, at the highest rate of N fertilizer, *CP 1* accumulated slightly more N in its kernels than did *Suwan 2301* which was obviously due to the farmer's higher grain yield. All values agree well with those reported in an earlier study (Feil et al., 1990).

Possibly, genotypic variation in GPC and GNC at comparable grain yield is expressed only at an ample supply of P, K and water. Therefore, further investigations at different levels of P, K and water in the soil are desired to confirm the differences between *CP 1* and *Suwan 2301* in GNC, GPC and GKC. Despite this restriction, the economically interesting results of the present study may encourage further efforts to identify high-yielding genotypes with low grain nutrient concentrations.

Acknowledgement

The authors wish to thank Miss Samran Phokham for technical assistance.

References

Alexander D E 1988 Breeding special nutritional and industrial types. *In* Corn and Corn Improvement. Ed. G F Sprague. pp 869–880. Agronomy Monograph no. 18, 3rd edition. ASA-CSSA-SSSA.

Castleberry R M, Crum C W and Crull C F 1984 Genetic improvement of U.S. maize cultivars under varying fertility and climatic environments. Crop Sci. 24, 33–36.

Dudley J W, Lambert R J and de la Roche I A 1977 Genetic analysis of crosses among corn strains divergently selected for percent oil and protein. Crop Sci. 17, 111–117.

Eberhard D 1977 Untersuchungen über den Einfluß von proteinreichen Inzuchtlinien auf die Ertragsgestaltung von Hybriden unter dem Aspekt der Körner- und Silonutzung von Mais (*Zea mays* L.). Ph.D. Thesis, Hohenheim, Germany.

Feil B, Thiraporn R, Geisler G and Stamp P 1990 Genotype variation in grain nutrient concentration in tropical maize grown during a rainy and a dry season. Agronomy 10, 717–725.

Forbes R M, Parker H M and Erdman Jr J W 1984 Effect of dietary phytate, calcium and magnesium levels on zinc bioavailability to rates. J. Nutr. 114, 1421–1425.

Frey K J 1977 Protein of oats. Z. Pflanzenzüchtg. 78, 185–215.

Gardner C A C, Bax P L, Bailey D J, Cavalieri A J, Clausen C R, Luce G A, Meece J M, Murphy P A, Piper T E Segebart R L, Smith O S, Tiffany C W, Trimble M W and Wilson B N 1990 Response of corn hybrids to nitrogen fertilizer. J. Prod. Agric. 3, 39–43.

Gupta D, Kovacs I and Gaspar L 1975 protein quality traits and their relationships with yield and yield components of opaque-2 and analogous normal hybrids and inbred lines. Theor. Appl. Genet. 45, 341–348.

Hambidge K M and Walravens P A 1976 Zinc deficiency in infants and preadolescent children. *In* Trace Elements in Human Health and Disease. Eds. A S Prasas and D Overleas. Vol. 1, Chapter 2, pp 21–32. Academic Press, New York.

Heitholt J J, Croy L I, Maness N O and Nguyen H T 1990 Nitrogen partitioning in genotypes of winter wheat differing in grain N concentration. Field Crops Res. 23, 133–144.

Kauffman K D and Dudley J W 1979 Selection indices for grain yield, percent protein, and kernel weight. Crop Sci. 19, 583–588.

Michael B, Zink F and Lantzsch H J 1980 Effect of phosphate application on phytin-phosphorus and other phosphate fractions in developing wheat grains. Z. Pflanzenernaehr. Bodenkd. 143, 377–384.

Muruli B I and Paulsen G M 1981 Improvement of nitrogen use efficiency and its relationship to other traits in maize. Maydica 26, 63–73.

Raboy V, Dickinson D B and Neffer M G 1990 A survey of maize kernel mutants for variation in phytic acid. Maydica 35, 383–390.

Thiraporn R, Geisler G and Stamp P 1987 Effects of nitrogen fertilization on yield and yield components of tropical maize cultivars. Z. Acker Pflanzenbau 159, 9–14.

Tsai C Y, Huber D M, Glover D V and Warren H L 1984 Relationship of N deposition to grain yield and N response of three maize hybrids. Crop Sci. 24, 277–281.

P. J. Randall et al. (Eds.), Genetic aspects of plant mineral nutrition, 301–309.

PLSO SV62

A physiological basis for genetic improvement to nitrogen harvest index in wheat

T.D. UGALDE
Institute for Sustainable Agriculture, Department of Agriculture, Tatura, Vic. 3616, Australia

Key words: amino acids, glutamine, nitrogen harvest index, nitrogen metabolism, proline, protein, wheat grain

Abstract

Both the rate and duration of protein deposition in the wheat grain are usually considered to be dominated by factors operating external to the grain, i.e. as source-limited events. However, limitations operating within (or close to) the grain become more obvious under conditions of increased nutrition, especially when the plants are able to take up extra nitrogen later in development. Nitrogen harvest index declines, decreasing the efficiency of nitrogen use.

This paper explores the opportunity for genetic manipulation of some of the processes involved in protein deposition in wheat, and in particular concentrates on some of the limitations that may be acting within the grain and endosperm.

The endosperm of wheat seems to contain abundant amounts of amino acids for protein deposition, yet substrate-driven responses are still observed. The explanation may be that the rate of protein deposition is limited by the supply of one or perhaps a few of the amino acids almost independent of the supply of amino acids as a whole. Concentrations of individual amino acids along the transport pathway, and turn-over rates within the endosperm point to metabolic limitations acting at key steps in amino acid synthesis within the endosperm itself; key steps that may be amenable to genetic manipulation.

Introduction

Grain protein percentage is one of the most important quality factors influencing the saleability of wheat on world markets. In particular, protein percentage needs to be at a level appropriate for the genetically established parameters of grain hardness and quality of the storage protein (Moss, 1973). This protein percentage value is determined mainly by the relative amounts of protein and starch that are deposited in the endosperm during the grain filling period; deposition events that appear to have largely independent regulation (Jenner et al., 1991).

In the simplest sense, there are only two ways to increase protein deposition in the grain. The first is to increase nitrogen nutrition; in other words, to increase the amount of substrate available in the plant for transfer to the grain. A problem here, however, is that nitrogen harvest index declines under increased nitrogen nutrition, and this decreases the efficiency of nitrogen use. The second option is to increase the values of nitrogen harvest index that are obtained possibly across the whole range of nutritional conditions; that is, to increase the proportion of plant nitrogen that becomes deposited in the grain as protein. Either way, nitrogen harvest index is an important part of the equation.

The purpose of this paper is to explore the possibility of increasing nitrogen harvest index in wheat, to explore where the limits to transfer of nitrogen from the plant to the grain may lie, and to outline the physiological basis for a possible genetic approach to increasing nitrogen harvest index.

Materials and methods

Wheat (*Triticum aestivum* L.) was sown (10 cm × 5 cm spacing) in stainless steel tanks (1.0 m × 0.8 m, in the horizontal plane, × 0.2 m) containing washed river sand. Nutrient solution (based on the Long Ashton recipe; all N as NO_3^-; Hewett, 1966) was flushed through the sand for 15 min six times daily. These solutions (60 l) were renewed weekly. The nitrate concentrations described below refer only to the concentrations at the start of each week.

Plants were grown in a glasshouse during the normal growing season (May to December) with no artificial light. The temperature was kept the same as outside, except with an upper limit of 30°C.

In one experiment, different levels of nitrate supply (0.5, 1.0, 2.0, 4.0, 8.0 m*M*) were held constant throughout crop development. Two cultivars were used: *Condor* is a high protein good quality bread wheat, and *Wyuna* is a lower protein good quality biscuit wheat. Nitrogen harvest index values were determined by growth analysis (including roots) at maturity.

In another experiment, cv. *Condor* only was used but with 3 different patterns of nitrate supply:

(1) 1.25 m*M* to 10 m*M* ten days after anthesis
This treatment will be referred to subsequently as *Current uptake*. Plants were supplied with low levels of nitrogen throughout most of the development such that phenological development appeared normal, but the plants contained low reserves of nitrogen within vegetative structures (see Zhen and Leigh, 1990). Nitrogen supply was increased coinciding with the rapid increase in the capacity for protein deposition in the endosperm (e.g. Barlow et al., 1974). In this treatment, therefore, most of the nitrogen arriving in the grains had only recently been taken up by the plant.

(2) 3 m*M* throughout (*Current + remobilised*)
This treatment provided moderate amounts of nitrogen continually throughout development such that the nitrogen delivered to the grains during grain filling contained a relatively even contribution from both remobilised and current sources.

(3) 10 m*M* to zero at anthesis (*Remobilised*)
Plants were provided with abundant amounts of nitrogen throughout development until anthesis. The nutrient solutions thereafter were made up free of nitrogen. Reserves of nitrogen in the plant were high, even during the grain filling period. In this treatment, all of the nitrogen delivered to the grains came from remobilised sources.

Amino acids were extracted and analysed from 4 regions along the transport pathway to the endosperm: (1) the rachis midway along the ear, (2) the pericarp tissue at the base of the crease of the grain containing the main vascular bundle that conducts the amino acids longitudinally (see Ugalde and Jenner, 1990c), (3) the fluid of the endosperm cavity that forms the apoplastic step between the maternal tissue and the new generation (see Ugalde and Jenner, 1990c), and (4) the endosperm itself.

Small pieces of tissue was sampled at 25 ± 1 days after anthesis, and extracted immediately in hot (75°C) 90% ethanol, as described elsewhere (Ugalde and Jenner, 1990a). Soluble amino acids were measured by PITC derivatisation and reverse-phase HPLC, and component amino acids in deposited protein were measured by a similar analysis of hydrolysates (Ugalde and Jenner, 1990b). Total nitrogen was measured by a semimicro version of AOAC Kjeldahl method 2.058 (1980).

Results

Influence of nitrogen nutrition on NHI

The wheat plants were grown in sand culture to minimise any environmental or nutritional constraints apart from those imposed by variations in nitrogen. Nitrogen harvest index (NHI) is the proportion of total plant nitrogen that is located in the seeds at maturity. The influence of nitrogen nutrition on NHI is shown in Figure 1.

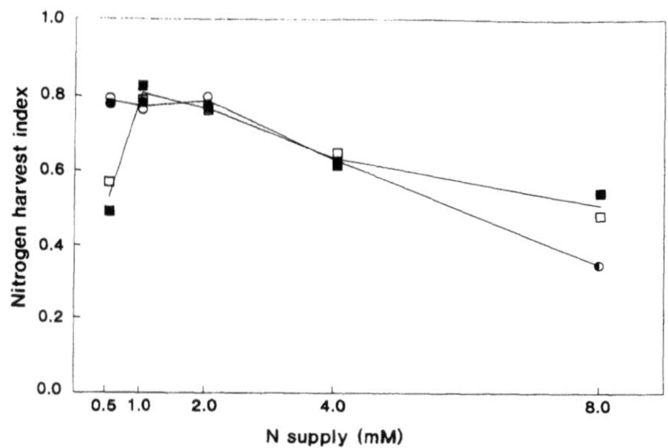

Fig. 1. Nitrogen harvest index of two cultivars of wheat (Condor, □, ■ and Wyuna, ○, ●) growing in sand culture as influenced by nitrogen supply in the nutrient solution. Open and closed symbols are replicates, and line is cultivar average.

NHI values decreased from about 0.8 to about 0.4 as nitrogen nutrition improved between the supply levels of 1 and 8 m*M*.

Amino acid supply to developing grains

An experiment was conducted which tested which group, or groups of amino acids in the supply to developing grains may be influenced by differences in the origin of nitrogen within the plant. Nitrogen supply was predominately from (1) current uptake, (2) a balance of current uptake and remobilised sources, or (3) remobilised sources.

Only the supply of total amino acids, glutamine and proline are reported now. These are shown in Table 1. A full description of the amino acid content along the transport pathway to the endosperm is being published elsewhere.

There was greater supply of total amino acids, and greater amounts of glutamine all along the transport pathway with increasing contribution from current uptake, except in the rachis. The rachis is not a good measure of amino acids in transporting tissues, however, because of storage capacity, and it is this storage capacity that is reflected by the high amino acid contents in the *remobilised* treatment. The proportional content of proline in the rachis is also much higher than in other regions, irrespective of treatment.

There are also differences in the glutamine: glutamate ratio across treatments and between tissues of the transporting pathway and the endosperm. The amount of glutamate, and the glutamine : glutamate ratio in the endosperm

Table 1. Soluble amino acids extracted from regions along the transport pathway to developing wheat endosperm 25 ± 1 days after anthesis (cv. *Condor*). Values for the rachis, vascular bundle, and endosperm are mean ±SEM of 8 replicates. Values for the endosperm cavity are the mean of 2 pooled extracts from 12 grains. Parentheses show percentage content (mol/mol)

	Nitrogen supply		
	Remobilised	Current + remobilised	Current uptake
Rachis ($\mu mol\ g^{-1}$ fwt)			
Total	12.6 ± 3.4	9.1 ± 1.3	13.5 ± 2.5
Glutamine	1.36 ± 0.49 (11.7 ± 1.9)	1.51 ± 0.10 (17.8 ± 1.8)	3.22 ± 1.00 (24.3 ± 2.8)
Proline	0.24 ± 0.04 (2.6 ± 0.4)	0.19 ± 0.02 (2.1 ± 0.2)	0.39 ± 0.08 (2.3 ± 0.2)
Pericarp containing vascular bundle of the grain ($\mu mol\ g^{-1}$ fwt)			
Total	25.3 ± 3.1	35.6 ± 4.8	44.1 ± 7.6
Glutamine	6.11 ± 1.29 (22.3 ± 2.9)	8.50 ± 1.78 (22.7 ± 2.3)	10.33 ± 2.09 (23.5 ± 3.2)
Proline	0.19 ± 0.03 (0.8 ± 0.1)	0.20 ± 0.06 (0.6 ± 0.1)	0.30 ± 0.09 (0.7 ± 0.1)
Endosperm cavity fluid ($\mu mol\ mL^{-1}$)			
Total	28.0	32.3	49.4
Glutamine	7.06 (25.3)	8.10 (24.7)	16.66 (33.8)
Proline	0.26 (0.9)	0.19 (0.6)	0.17 (0.3)
Endosperm ($\mu mol\ g^{-1}$ fwt)			
Total	20.0 ± 1.7	19.8 ± 1.5	31.0 ± 4.6
Glutamine	0.58 ± 0.08 (2.9 ± 0.2)	0.76 ± 0.12 (3.8 ± 0.5)	1.18 ± 0.19 (4.1 ± 0.6)
Proline	0.17 ± 0.01 (0.9 ± 0.1)	0.18 ± 0.01 (1.0 ± 0.1)	0.27 ± 0.07 (0.8 ± 0.1)

Table 2. Glutamate in the endosperm cavity fluid and endosperm 25 ± 1 days after anthesis (cv. *Condor*). Values for the endosperm cavity are the mean of 2 replicates of pooled extracts from 12 grains. Values for the endosperm are mean ±SEM of 8 replicates. Parentheses show percentage content (mol/mol)

	Nitrogen supply		
	Remobilised	Current + remobilised	Current uptake
Endosperm cavity fluid (μ*mol mL*$^{-1}$)			
Glutamate	1.38 (2.7)	1.20 (3.8)	1.34 (4.9)
Glutamine : Glutamate ratio	5.1	6.8	12.4
Endosperm (μ*mol g*$^{-1}$ *fw*)			
Glutamate	3.38 ± (0.38)	3.60 ± 0.35	4.99 ± 0.61
	(19.3 ± 1.3)	(19.6 ± 1.8)	(17.0 ± 1.2)
Glutamine : Glutamate ratio	0.15 ± 0.02	0.21 ± 0.03	0.24 ± 0.03

Table 3. Nitrogen increment in the grain between days 20 and 30 after anthesis, and percentage content of glutamine (including glutamate) and proline in the deposited protein (cv. *Condor*). Nitrogen values are mean ±SEM of grain samples from 8 plants, and percentage contents (mol/mol) are mean ±SEM of one endosperm from 8 plants at 25 ±1 days after anthesis

	Nitrogen supply		
	Remobilised	Current + remobilised	Current uptake
N content day 20 (mg/grain)	0.223 ± 0.009	0.299 ± 0.014	0.221 ± 0.015
N content day 30 (mg/grain)	0.378 ± 0.015	0.537 ± 0.030	0.575 ± 0.033
N increment (mg/grain)	0.155	0.238	0.354
Content of glutamine (%)	21.01 ± 1.37	23.20 ± 0.57	24.35 ± 0.65
Content of proline (%)	9.47 ± 0.44	10.16 ± 0.19	10.72 ± 0.20

cavity and the endosperm are shown in Table 2.

Table 3 shows features of protein deposition across the 3 treatments. First, there were differences in the amount of protein deposited consistent with the characteristics of total amino acid supply described in Table 1. Secondly, there were changes in the amino acid composition of the deposited protein. As with Table 1, glutamine (including glutamate) and proline only are shown. These were the *only* amino acids that *increased* in proportional content concurrent with increased protein deposition.

Discussion

Nitrogen harvest index

The proportion of total plant nitrogen located in the seeds at maturity (NHI) rarely exceeds 0.8 in bread wheats. Spiertz and de Vos (1983) reported NHI values in healthy crops of between 0.74 and 0.78, and Loffler and Busch (1981) reported values between about 0.55 and 0.74. Durum wheats possibly can return higher NHI values, exceeding 0.8 (Desai and Bhatia, 1978).

It is often reported that NHI values in wheat decrease with increasing nitrogen nutrition (Halloran, 1981; Morris and Paulsen, 1985; McMullen et al., 1988; Whitfield et al., 1989). In the experiments of Whitfield et al. (1989), NHI values decreased from a mean of 0.77 in treatments without added N fertiliser to 0.70 in treatments with 150 kg N ha^{-1} applied at sowing. An increasing imbalance between limiting water and increased fertilisation, reduced NHI to 0.64.

The decline in NHI with increasing nitrogen nutrition observed in the wheat grown in sand culture (Fig. 1) shows that this decline is a physiological response to nitrogen within the plant independent of other constraints, for instance, those imposed by decreasing availability of water.

Cultivar selection this century has led to improvements in NHI. Modern varieties of oats (Wych and Stuthman, 1983), wheat (Austin et al., 1980) and barley (Wych and Rasmusson, 1983) show higher NHI values than their earlier counterparts. NHI values for modern cultivars of maize (Muirhead et al., 1985), and barley (Wych and Rasmusson, 1983) are similar to those for wheat. Modern varieties of oats, like wheat, show a strong decrease in NHI with increasing nitrogen nutrition (Kairudin and Frey, 1988).

Rice appears to be an exception among cereals, with NHI values being quite low, ranging between 0.65 and 0.40. Even so, NHI values still show strong negative responses to increased fertility (Muirhead et al., 1985).

Grain legumes show entirely different NHI responses. Because of nitrogen fixation, they are nearly always well supplied with nitrogen, to levels rarely seen in cereals. Yet in contrast to what may be expected from experience with cereals, NHI values are high, indeed even higher than the maximum values for wheat. NHI values for field beans (*Vicia faba*) have been reported to be between 0.78 and 0.90 (Brunner and Zapata, 1984), and values for soybeans likewise are greater than 0.80 (Hammond et al., 1951). Furthermore, analysis of seed composition (Sinclair and de Wit, 1975) also leads to the conclusion that there are major differences between species in the regulation of transfer of nitrogen from the vegetative tissues to developing grains.

Factors that may limit NHI in wheat

Limits to protein deposition in the wheat endosperm may lie in one or more of the following areas: (1) provision of substrate within the plant, (2) activity of the systems that move substrate to the grain, and (3) activity of mechanisms in the grain converting substrate to endosperm protein.

There is abundant evidence to show that the amount of nitrogen available in the plant influences protein deposition in the grain. Most of the attempts to increase protein deposition either by agronomy (e.g. fertilizer use or crop rotations), or genetic means (e.g. selecting for N uptake or activity of nitrate reductase) aim to increase nitrogen supply. It is difficult, however, to attribute the inverse relation between NHI and nitrogen nutrition to events of supply, at least to supply in the simplest sense. Remobilisation would have to be determined by, or linked to, the amount of transfer that had occurred previously for this to be the case.

Long-distance transport to the ear appears adequate (Wardlaw and Moncur, 1976), and so it is unlikely that improving the capacity of the transport system would increase protein deposition in the grain. In contrast, the observed NHI response could be explained in terms of limitations that lie in the grain, or more particularly in the endosperm itself. The concentration of free amino acids in wheat endosperm increases substantially under conditions where the rate of protein deposition increases as a response to increased substrate supply. For instance, Barlow et al. (1983) reported that a substrate-induced doubling of protein deposition was accompanied by a 10-fold increase in amino acid concentration in the endosperm.

With regard to mechanisms within the endosperm, there are a number of anomalies when considering protein deposition as a response to substrate concentration. For instance, the endosperm seems normally to contain abundant amounts of amino acids for protein deposition; enough to sustain about $1\frac{1}{2}$ days of protein deposition (Ugalde and Jenner, 1990b). To put this in perspective, the endosperm contains enough sucrose for only about 3 hours of deposition of starch (Ugalde and Jenner, 1990a).

Yet with protein, additional supply of nitrogen to the ear and grain, especially in the form of glutamine, still enhances the rate of deposition, in contrast to the response of starch deposition to sucrose supply (Jenner et al., 1991). Why are there such high levels of amino acids in the endosperm, and why is there still such a pronounced substrate-driven response?

One explanation may be that one, or perhaps a few of the amino acids in the endosperm impose the limitations to the whole process, almost despite the level of total amino acids in the pool. For each amino acid, Ugalde and Jenner (1990b) calculated its rate of supply from external sources, its rate of incorporation into protein, and by difference, the rate of synthesis required in the endosperm. They also measured the amount in the soluble pool within the endosperm, and calculated turn-over times within this soluble pool.

There are a number of characteristics that may be expected of a specific amino acid if it were imposing a limit to protein deposition:

1. Inadequate supply from external sources, hence the need for synthesis in the endosperm.
2. Low levels of the amino acid in the soluble pool within the endosperm, with short turnover time, showing that as soon as it is synthesised it is rapidly incorporated into protein.
3. Possibly, an energetically expensive pathway, or for some other reason, difficult to make.
4. Possibly, high content in endosperm storage protein.

From data of Ugalde and Jenner (1990b), most of the amino acids do not fit with these expected characteristics, but a number do. And the ones that do, divide clearly into 4 groups (Table 4). Amino acids within each group are structurally related, and contain common steps in their synthetic pathway (see Wallsgrove, 1989).

As amino acid supply (including glutamine) was increased to the developing endosperm (*Remobilised* to *Current*, Table 1), the rate of protein deposition, and the proportional content of glutamine and proline in deposited protein increased also (Table 3). Proline was the one amino acid with characteristics suggesting a limiting role that increased in the deposited protein when a limit to protein deposition was alleviated.

The changes to proline deposition were only small, and there is probably no need at this stage to invoke action of gene control, as has been proposed, for instance, during sulphur deficiency (Spencer et al., 1990). It is possible that the changes came about simply through mass action.

A strategy for genetic manipulation

The few changes in nitrogen metabolism described above were brought about by substrate effects. It seems possible, however, that changes in protein deposition could equally come about through changes in the capacity of metabolism in the grain altering the source-sink balance independent of supply. A consequence of any increase in the capacity for protein deposition in the endosperm is likely to be reduced level of amino acids in the endosperm and increased (more steep) concentration gradients along the transport pathway between the source and the sink.

Turnover of protein in the plant is a continuous process. As older leaves senesce, their protein is mobilised and utilised for protein synthesis in developing tissues elsewhere around the plant (Leopold, 1980). Increased capacity for protein deposition in the endosperm may preferentially draw amino acids away from other

Table 4. Amino acids that could limit protein deposition in wheat endosperm. Taken from data of Ugalde and Jenner 1990b. All % values are mol/mol

Amino acid	Content in endosperm protein (%)	Content in endosperm soluble pool (%)	Synthesis in endosperm as % of amount for protein	Turn-over time in endosperm soluble pool (h)[a]
Proline	12.0	1.1	97	3.0
Branch chain amino acids				
Isoleucine	3.4	0.3	88	2.9
Leucine	6.5	0.4	94	2.0
Valine	4.6	0.9	72	6.4
Aromatic amino acids[b]				
Phenylalanine	3.7	0.7	81	8.9
Tyrosine	3.8	0.3	93	2.6
Sulphur-containing amino acids				
Cysteine	1.0	0.1	90	3.2
Methionine	1.1	<0.1	90	3.0

[a] Turn-over time for soluble amino acids as a whole within the endosperm is 30 h.
[b] No information on tryptophan

synthetic sites, or even enhance remobilisation of nitrogen reserves – in effect increase NHI.

Extracts from the main vascular bundle of the grain and endosperm cavity provide good measure of amino acid supply to the developing endosperm (Ugalde and Jenner, 1990c). The proline levels in these extracts are very low (Table 1; see also Ugalde and Jenner 1990b). Although substantial amounts of proline may exist elsewhere in the plant, the low levels of proline in the grain's vascular bundle and the endosperm cavity show that nearly all of the proline required for protein synthesis in the endosperm must be manufactured from other nitrogenous compounds in the endosperm, itself. Proline is the second most abundant amino acid in endosperm protein, so demand for synthesis is high.

The pathway of proline synthesis in wheat endosperm has not been described. But from studies on other tissues, it is most likely to come from glutamine via glutamate in a reaction that is energetically expensive and thermodynamically unfavourable, requiring 7 ATPs per molecule and mitochondrial location. Two pathways from glutamate are possible, via pyrroline-5-carboxylate, or via ornithine (Wallsgrove, 1989). Ornithine has been reported in wheat endosperm (Ugalde and Jenner, 1990b).

There is also very little glutamate in the fluid of the endosperm cavity, but changes in the glutamine : glutamate ratio from the endosperm cavity to the endosperm (Table 2) suggest a flow of nitrogen away from glutamine to glutamate within the endosperm. The explanation for the observation of possible parallel treatment effects on both free glutamine and proline in the endosperm (Table 1) may lie in a high Km value for the first step from glutamate to proline. The only measurements on the glutamate Km value for γ-glutamyl kinase are from rat intestinal mucosa (Wakabayshi and Jones, 1983) and *E. coli* (Baich, 1969). These values are high, 2.5 mM and 7–10 mM respectively.

All of the observations described above point to a metabolic block to protein deposition acting at the site of proline synthesis in the endosperm. There may be others. Nevertheless, one strategy to increase protein deposition in the developing wheat grain, and hence to increase NHI is to genetically augment the capacity of the endosperm to synthesise proline, and this process must be distinct from the capacity of the wheat plant to synthesise proline in vegetative tissues especially as a response to stress.

The milling industry has long realised that protein content within wheat endosperm is not uniform, as flours from different mill streams have different protein percentages. Indeed, the gradient in protein percentage increasing in an outward direction observed in mature dry endosperm (Hinton, 1947; Morris et al., 1945, 1946; Normand et al., 1965) is already being established midway through grain filling (Ugalde and Jenner, 1990b). The gradient in protein percentage across the endosperm is due almost entirely to a gradient of protein; starch is deposited uniformly. The outer layers of endosperm, even excluding the aleurone and subaleurone layers, contain nearly twice the amount of protein per unit of fresh volume than do the regions of endosperm adjacent to the endosperm cavity.

This gradient in protein deposition cannot be attributed to any parallel change in total amino acid supply. Indeed, protein deposition increases in the direction away from the source of substrate, and across an area near the endosperm cavity where the concentration of free amino acids decreases. Clearly the changes in the rate of protein deposition across the endosperm are due to changes in synthetic capacity.

Is there any evidence that the changes in synthetic capacity across the endosperm may be related to some factor associated with proline synthesis? If this were so, there may be detectable changes in proline content as a proportion of amino acid composition of the endosperm protein across the endosperm, in a similar way to that described above in whole endosperms, even though this technique is probably a most insensitive way to pick it up. The only information relating to this is in data of Ugalde and Jenner (1990b). The experiment was designed for another purpose, but nevertheless a trend can be seen with a positive correlation between the rate of protein deposition increasing in an outward direction and the proportional content of proline in the deposited protein (Fig. 2).

There are other factors that should be considered should an attempt be made to increase NHI

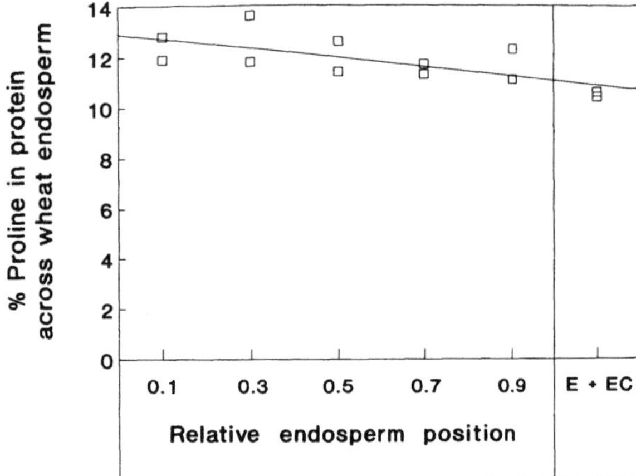

Fig. 2. Percentage proline in protein across the endosperm. E + EC = endosperm adjacent to the endosperm cavity, and relative position 0.1 is adjacent to the peripheral tissue on the dorsal surface. Two replicates for each position are shown, $R^2 = 0.49$.

through augmenting the capacity for proline synthesis in the endosperm. One concerns the type of protein that would be deposited. For instance, some ω-gliadins contain about 30% proline while some HMW glutenin subunits contain about 13% proline (Shewry and Tatham, 1991). Augmenting proline synthesis could alter the balance of the type of protein deposited. Indeed, in the field there appears a similar response. There is a shift in glutenin:gliadin ratios in favour of gliadins under increased nitrogen nutrition (R B Gupta and F MacRitchie, pers, comm.).

A second concern would be whether increased NHI may prematurely limit the capacity for photosynthesis in the vegetative tissue. Sinclair and de Wit (1975) have already classified wheat as a self-destructive crop on the basis of maintenance of photosynthetic capacity during grain filling, and alterations to the process of nitrogen remobilisation may affect yield.

Acknowledgements

The work was supported financially by the Wheat Committee of the Australian Grains Research and Development Corporation. I thank N E Nardella for excellent technical assistance and R M Wallsgrove for helpful discussions.

References

Association of Official Analytical Chemists 1980 Official Methods of Analysis, Thirteenth edition. Ed. W Horwitz. Association of Official Analytical Chemists, Washington, DC.

Austin R B, Bingham J, Blackwell R D, Evans L T, Ford M A, Morgan C L and Taylor M 1980 Genetic improvement in winter wheat yields since 1900 and associated physiological changes. J. Agric. Sci. 94, 675–689.

Baich A 1969 Proline synthesis in *Escherichia coli*. A proline-inhibitable glutamic acid kinase. Biochim. Biophys. Acta 192, 462–467.

Barlow K K, Lee J W and Vesk M 1974 Morphological development of storage protein bodies in wheat. *In* Mechanisms of Regulation of Plant Growth, Eds. R L Bieleski, A R Ferguson and M M Cresswell. pp 793–797. Royal Society of New Zealand, Wellington.

Barlow E W R, Donovan C R and Lee J W 1983 Water relations and composition of wheat ears grown in liquid culture: Effect of carbon and nitrogen. Aust. J. Plant Physiol. 10, 99–108.

Brunner H and Zapata F 1984 Quantitative assessment of symbiotic nitrogen fixation in diverse mutant lines of field bean (*Vicia faba minor*). Plant and Soil 82, 407–413.

Desai R M and Bhatia C R 1978 Nitrogen uptake and nitrogen harvest index in durum wheat cultivars varying in their grain protein concentration. Euphytica 27, 561–566.

Halloran G M 1981 Cultivar differences in nitrogen translocation in wheat. Aust. J. Agric. Res. 32, 535–544.

Hammond L C, Black C A and Norman A G 1951 Nutrient uptake by soybeans on two Iowa soils. Iowa Agricultural Experimental Station Research Bulletin. No. 384.

Hewett E J 1966 Sand and Water Culture Methods Used in the Study of Plant Nutrition. Commonwealth Agriculture Bureaux, Farnham Royal, Bucks., UK. pp 430–451.

Hinton J J C 1947 The distribution of vitamin B1 and nitrogen in the wheat grain. Proc. Royal Soc. (London) B134, 418.

Jenner C F, Ugalde T D and Aspinall D 1991 The physiology of starch and protein deposition in the endosperm of wheat. Aust. J. Plant Physiol. 18, 211–226.

Kairudin N MD and Frey K J 1988 Soil N availability and nitrogen harvest index of oats. J. Iowa Acad. Sci. 95, 73–78.

Loffler C M, and Busch R H 1982. Selection for grain protein, grain yield, and nitrogen partitioning efficiency in hard red spring wheat. Crop Sci. 22, 591–595.

McMullan P M, McVetty P B E and Urquhart A A 1988 Dry matter and nitrogen accumulation and redistribution and their relationship to grain yield and grain protein in wheat. Can. J. Plant Sci. 68, 311–322.

Morris C F and Paulsen G M 1985 Development of hard winter wheat after anthesis as affected by nitrogen nutrition. Crop Sci. 25, 1007–1010.

Morris V H, Alexander T L and Pascoe E D 1945 Studies on the composition of the wheat kernel. I. Distribution of ash and protein in the center sections. Cereal Chem. 22, 351–361.

Morris V H, Alexander T L and Pascoe E D 1946 Studies on the composition of the wheat kernel. III. Distribution of ash and protein in the central and peripheral zones of whole kernels. Cereal Chem. 23, 540–547.

Moss H J 1973 Quality standards for wheat varieties. J. Aust. Inst. Agric. Sci. 39, 109–115.

Muirhead W A, Blackwell J, Humphreys E and White R J G 1989 The growth and nitrogen economy of rice under sprinkler and food irrigation in South East Australia. Irrig. Sci. 10, 183–199.

Normand F L, Hogan J T and Deobald H J 1965 Protein content of successive peripheral layers milled from wheat, barley, grain sorghum and glutinous rice by tangential abrasions. Cereal Chem. 42, 359–367.

Shewry P R and Tatham A S 1990 The prolamine storage proteins of cereal seeds: Structure and evolution. Biochem. J. 267, 1–12.

Sinclair T R and de Wit 1975 Photosynthesis and nitrogen requirements for seed production by various crops. Science 189, 565–567.

Spencer D, Rerie W G, Randall P J, and Higgins T J V 1990 The regulation of pea seed storage protein genes by sulfur stress. Aust. J. Plant Physiol. 17, 355–363.

Spiertz J H J and De Vos N M 1983 Agronomy and physiological aspects of the role of nitrogen in yield formation of cereals. Plant and Soil 75, 379–391.

Ugalde T D and Jenner C F 1990a Substrate gradients and regional patterns of dry matter deposition within developing wheat endosperm. I. Carbohydrates. Aust J. Plant Physiol. 17, 377–394.

Ugalde T D and Jenner C F 1990b Substrate gradients and regional patterns of dry matter deposition within developing wheat endosperm. II. Amino acids and protein. Aust. J. Plant Physiol. 17, 395–406.

Ugalde T D and Jenner C F 1990c Route of substrate movement into wheat endosperm. II. Amino acids. Aust. J. Plant Physiol. 17, 705–714.

Wakabayashi Y and Jones M E 1983 Pyrroline-5-carboxylate synthesis from glutamate by rat intestinal mucosa. J. Biol. Chem. 258, 3865–3872.

Wallsgrove R M 1989 Amino acid biosynthesis inhibitors. *In* Prospects for Amino Acid Inhibitors in Crop Protection and Pharmaceutical Chemistry. Eds. L C Copping, J Dalziel and A D Dodge, pp 3–13. BCPC Monograph No. 42.

Wardlaw I F and Moncur L 1976 Source, sink and hormonal control of translocation in wheat. Planta 128, 93–100.

Whitfield D M, Smith C J, Gyles O A and Wright G C 1989 Effects of irrigation, nitrogen and gypsum on yield, nitrogen accumulation and water use by wheat. Field Crops Res. 20, 261–277.

Wych R D and Stuthman D D 1983 Genetic improvement in Minnesota-adapted oat cultivars released since 1923. Crop Sci. 23, 879–881.

Wych R D and Rasmusson D C 1983 Genetic improvement in malting barley cultivars since 1920. Crop Sci. 23, 1037–1040.

Zhen R G and Leigh R A 1990 Nitrate accumulation by wheat (*Triticum aestivum*) in relations to growth and tissue N concentrations. Plant and Soil 124, 157–160.

P. J. Randall et al. (Eds.), Genetic aspects of plant mineral nutrition, 311–319.

PLSO SV48

Post-transcriptional control of the expression of a plant gene by an environmental factor: Sulphur regulation of the expression of the Pea Albumin 1 gene

ROGER L. MORTON[1,2], MALCOLM WHITECROSS[1] and T.J.V. HIGGINS[2]
[1]*Division of Botany and Zoology, School of Life Sciences, Australian National Univeristy, Canberra ACT 2601 and* [2]*CSIRO Division of Plant Industry, Canberra ACT 2601, Australia*

Key words: pea, *Pisum sativum*, sulphur deficiency, tobacco, transgenic plants

Abstract

Pea albumin 1 (PA1) is a storage protein from the seeds of the pea, *Pisum sativum*. This protein contains an unusually large proportion of sulphur amino acids (11% methionine + cysteine). Earlier studies have indicated that the steady state levels of PA1 protein and PA1 mRNA are severely reduced when plants are grown under sulphur deficient conditions. Previous investigations, using in vitro transcription assays, indicated that the poor sulphur nutrition exerts its effect on PA1 gene expression by reducing the stability of the PA1 mRNA.

In this study we describe a PA1 gene engineered for expression in leaves and show that this chimeric gene is regulated by sulphur status in tobacco plants. A gene encoding a high sulphur protein from animals, engineered for expression in leaves, is not regulated by sulphur. Thus, the DNA sequences required for regulation of the PA1 gene by sulphur status are contained in the cloned portion of the gene and they function in the heterologous tobacco system. Possible mechanisms for this regulation are discussed.

Introduction

The seeds of all plants contain seed storage proteins whose function is to serve as a nutrient source for the germinating embryo (Higgins, 1984). Seed storage proteins from different species have been classified into a number of different classes based on similarities in their chemical properties and their sizes. (See Casey et al., (1986) for review.) Members of the same class of protein from different species also share sequence similarities and are evolutionarily related (Borroto and Dure, 1987; Kortt et al., 1991).

The seeds of peas contain large amounts of the two proteins legumin and vicilin and smaller amounts of four other proteins [pea albumins (PA) 1, 2 and 3, and pea seed lectin] Casey et al., 1986; Higgins, 1984; Schroeder, 1984). Under conditions of poor sulphur nutrition the relative amounts of these proteins is altered. The levels of legumin, PA1 and PA3 are dramatically reduced. A compensatory increase in the amount of vicilin is observed. The relative levels of PA2 and lectin appear unaffected (Chandler et al., 1983; Randall et al., 1979).

The sulphur amino acid content (Cys + Met) of the pea seed storage proteins vary considerably and are as follows: Legumin 1.7% (Lycett et al., 1984), PA1 11% (Higgins et al., 1986), PA2 2.6% (Higgins et al., 1987), Vicilin 0% (Spencer et al., 1983) and Lectin 0% (Higgins et al., 1983). Thus, under conditions of sulphur deficiency the pea plant reduces the amount of two sulphur containing seed proteins (legumin and PA1) and increases the level of a sulphur-free protein (vicilin).

This phenomenon appears to be widespread.

Sulphur deficiency has been shown to affect the balance between the sulphur-rich and the sulphur-poor seed proteins in most of the plants that were tested. These include lupins (Blagrove et al., 1976), wheat (Wrigley et al., 1980), barley (Shewry et al., 1983), soybean (Gayler and Sykes, 1985), cowpea (Evans et al., 1977), rape and sunflower (Spencer et al., 1990).

Previous studies of the sulphur dependant regulation of legumin and PA1 in peas have shown that the rate of synthesis of these proteins is reduced under sulphur deficient conditions and that the rates of degradation are unaffected (Chandler et al., 1983; Higgins et al., 1986). The steady state levels of the mRNAs encoding PA1 and legumin are reduced under S deficient conditions (Chandler et al., 1984), while the rates of transcription of these genes are not affected by sulphur deficiency (Beach et al., 1985; Higgins et al., 1986). These data imply that poor sulphur nutrition affects the level of PA1 and legumin mRNA by a post-transcriptional mechanism.

As part of an approach to understand the mechanism of this post-transcriptional regulation we have engineered the cloned PA1 gene for expression in leaves and examined the regulation of this modified gene in transgenic tobacco plants grown under sulphur deficient and control conditions. For comparison we have engineered a gene encoding the sulphur-rich animal protein, ovalbumin, for leaf expression using the same promoter.

Materials and methods

Plasmid construction

Leaf specific PA1 gene

A PA1 genomic clone (Genbank data base accession No. M81864) was modified to introduce a *NcoI* site at the translation initiation codon. A 2.2 kb *NcoI-BamHI* fragment from this clone was ligated into a *NcoI*, *BamHI* digested plasmid containing a ribulose-1,5-bisphosphate carboxylase small subunit (SSU) promoter (Dean et al., 1985). The resulting plasmid pJOO2 consists of approximately 1.4 kb of the 5′ flanking region of the ribulose-1,5-bisphosphate carboxylase small subunit gene of petunia (*SSU301*), 56 bp of *SSU301* 5′ untranslated region, 481 bp of PA1 coding region (including the intron) and approximately 1.7 kb of PA1 3′ flanking region (Fig. 1). A 3.6 kb *XbaI* fragment from pJOO2 was cloned into the *XbaI* site of the binary vector pGA492 (An, 1986) to produce pSJ21 (Fig. 1).

Leaf specific ovalbumin gene

The 3.6 kb *XbaI* fragment from pJOO2 was cloned into the *XbaI* site of the plasmid p118-NOS to produce the intermediate pRM5A (Fig. 1). pRM5A was digested partially with *XbaI* and completely with *NcoI* and the 5.6 kb fragment was ligated to a 1.3 kb partial-*NcoI* complete-*XbaI* fragment from an ovalbumin cDNA (McReynolds et al., 1978) to produce pRM5 (Fig. 1). pRM5 consists of 1.4 kb of SSU301 5′ flanking sequence, 56 bp of SSU301 5′ untranslated region, 1.28 kb of ovalbumin cDNA (from the *NcoI* site at the translation initiation codon to an *XbaI* site 122 bp 3′ to the translation stop codon) and a transcription termination signal (1.1 kb) from the nopaline synthase (NOS) gene of *Agrobacterium tumefaciens* (Bevan et al., 1983). pRM5 was linearised with *NarI* and the entire plasmid cloned into the *ClaI* site of pGA492 to produce pBRM5 (Fig. 1).

Plant transformation

The binary vector pSJ21 was transferred from *Escherichia coli* to *Agrobacterium tumefaciens* (LBA4404) by triparental mating with the plasmid pRK2013 supplying the transfer functions (Ditta et al., 1980). The binary vector pBRM5 was transferred to *Agrobacterium tumefacines* (LBA4404) by electroporation (Wen-jun and Forde, 1989). The structure of the plasmids in *Agrobacterium* was verified by Southern blotting. Tobacco leaf pieces (cv. Wisconsin 38) were transformed using *Agrobacterium* by the procedure of Horsch et al. (1985) as modified by Higgins et al. (1988).

Growth of sulphur deficient tobacco

For each of the two constructs a single transformed plant expressing high levels of mRNA from the transgene was selected. This plant was

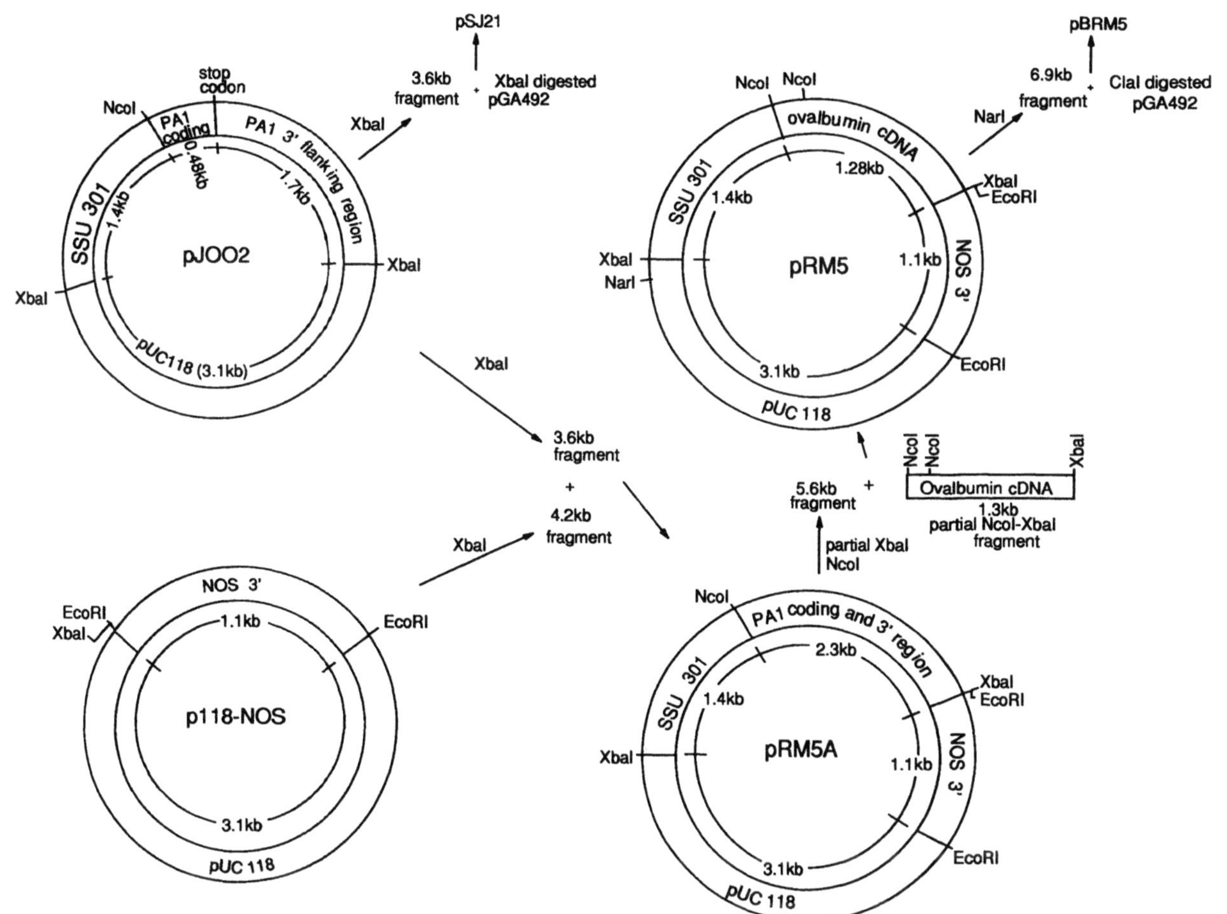

Fig. 1. Plasmid constructions. Circles represent plasmid DNA. The derivation of each segment of DNA is indicated. The size of each segment in kilobase pairs (kb) is indicated inside the circles. Only the relevant restriction sites are shown. See text for details of constructions.

multiplied up by vegetative propagation to produce four clonaly identical plantlets. These plantlets were transplanted into 20 cm pots containing washed river sand and grown in a glasshouse (26°C day/20° night) under natural light. The plants were watered each morning with 400 mL of nutrient solution containing various levels of sulphate.

The basal nutrient solution contained: 5 m*M* potassium nitrate (KNO_3), 4 m*M* calcium nitrate ($Ca(NO_3)_2$), 1 m*M* ammonium nitrate (NH_4NO_3), 1 m*M* potassium dihydrogen phosphate (KH_2PO_4), 0.1 m*M* ferric citrate, 23 μM boric acid (H_3BO_3), 4.5 μM manganese chloride ($MnCl_2.4H_2O$), 0.7 μM zinc chloride ($ZnCl_2$), 0.4 μM cupric chloride ($CuCl_2.2H_2O$) and 0.22 μM molybdic acid (H_2MoO_4). The magnesium sulphate ($MgSO_4$) was varied between 10 μM and 1 m*M* (0.32–32 mg S L^{-1}) and the magnesium concentration held constant by varying the level of magnesium chloride ($MgCl_2$) between 1.99 and 1 m*M*.

After transplanting, all plants were watered for 3 weeks with nutrient solution containing a 0.2 m*M* S. The plants were then divided into 2 groups. The control group was watered with nutrient solution containing a 1 m*M* S until leaves were harvested. The sulphur deficient group was watered with 0.1 m*M* S for one week and then with 10 μM S until leaves were harvested. When the sulphur deficient plants were starting to show the yellowing symptoms of sulphur deficiency (after 1–2 weeks on 10 μM S) leaves from both treatments were harvested by taking the 2nd or 3rd expanded leaf from the top and freezing in liquid nitrogen for RNA isolation.

RNA isolation and Northern analysis

Approximately 2 g of frozen leaves were ground in a mortar with 4 mL of extraction buffer (0.1 *M* NaCl, 10 m*M* Tris-HCl pH8, 1 m*M* EDTA, 1.0% SDS). The slurry was extracted with an equal volume of phenol:chloroform:isoamyl alcohol (25:24:1). After centrifugation the nucleic acids were precipitated from the aqueous phase with ethanol and resuspended in water. The RNA was precipitated from this solution by adding 0.25 volumes of 10 *M* LiCl and incubating on ice for >1 hr. The solution was centrifuged and the pellet was resuspended in water. The solution was then re-precipitated with ethanol before being washed in 70% ethanol. The pellet was finally resuspended in water and the RNA concentration determined by measuring absorbance at 260 nm.

RNA was fractionated on a 1.2% agarose gel in the presence of deionized formaldehyde as described previously (Higgins and Spencer, 1991). The RNA was capillary blotted onto nylon membrane (Hybond-N, Amersham). In the case of PA1 expressing plants the membranes were hybridized by the method of Khandjian (1987) using a probe prepared by random primed labelling of DNA fragment corresponding to the first 220 bp of the PA1 coding region. In the case of the ovalbumin expressing plants the filters were hybridized with an RNA probe complementary to the first 417 bp of the ovalbumin cDNA clone using 50% formamide, 2X SSPE[1], 1% SDS, 1% BSA, 10% dextran sulphate and 0.5 mg mL^{-1} sonicated herring sperm DNA as a hybridization solution. A one hour pre-hybridization and an overnight hybridization at 65°C were used. In both cases membranes were washed at 65°C in 2 × SSC[2], 0.1% sodium pyrophosphate and 0.1% SDS twice for 10 min each. The filters were then exposed to X-ray film at −70°C using an intensifying screen.

[1] 20 × SSPE = 3 *M* NaCl, 0.2 *M* NaH_2PO_4, 20 m*M* EDTA, pH 7.4 (Sambrook et al., 1989)

[2] 20 × SSC (standard saline citrate) = 3 *M* NaCl, 0.3 *M* sodium citrate.

Results

A PA1 gene responded to sulphur stress in tobacco leaves

The promoter and 5′ untranslated region of the PA1 gene was removed and replaced with the promoter and 5′ untranslated region from the leaf specific petunia gene *SSU301* (Dean et al., 1985) to produce pSJ21 (Fig. 1).

This chimeric gene was transferred to tobacco by *Agrobacterium* mediated plant transformation. A transformant that was expressing the gene at a high level was selected on the basis of Northern analysis (data not shown) and multiplied up by vegetative propagation. Four of these clonaly identical plants were planted in the glasshouse. Two of them were grown under sulphur deficient conditions and two under control conditions.

Equal amounts (5 μg) of total leaf RNA were analysed by Northern blotting to determine the relative amounts of PA1 mRNA in the leaves of the plants from the two treatments. The results from this analysis are shown in Figure 2. The level of PA1 mRNA in the sulphur deficient plants was reduced relative to the level in the control plants. It was estimated that the difference in RNA levels was approximately 25 fold based on densitometric scans of autoradiographs exposed for different lengths of time (data not shown).

An ovalbumin gene did not respond to sulphur stress in tobacco leaves

A leaf specific ovalbumin gene was constructed by linking the *SSU301* promoter and 5′ untranslated region to the coding region of an ovalbumin cDNA and by adding a plant transcription termination signal to the end of the ovalbumin coding region to produce pBRM5 (Fig. 1). This gene was transferred to tobacco and 10 transformants analysed by Northern blot using a random primed DNA probe. Four plants expressed an ovalbumin specific transcript of the

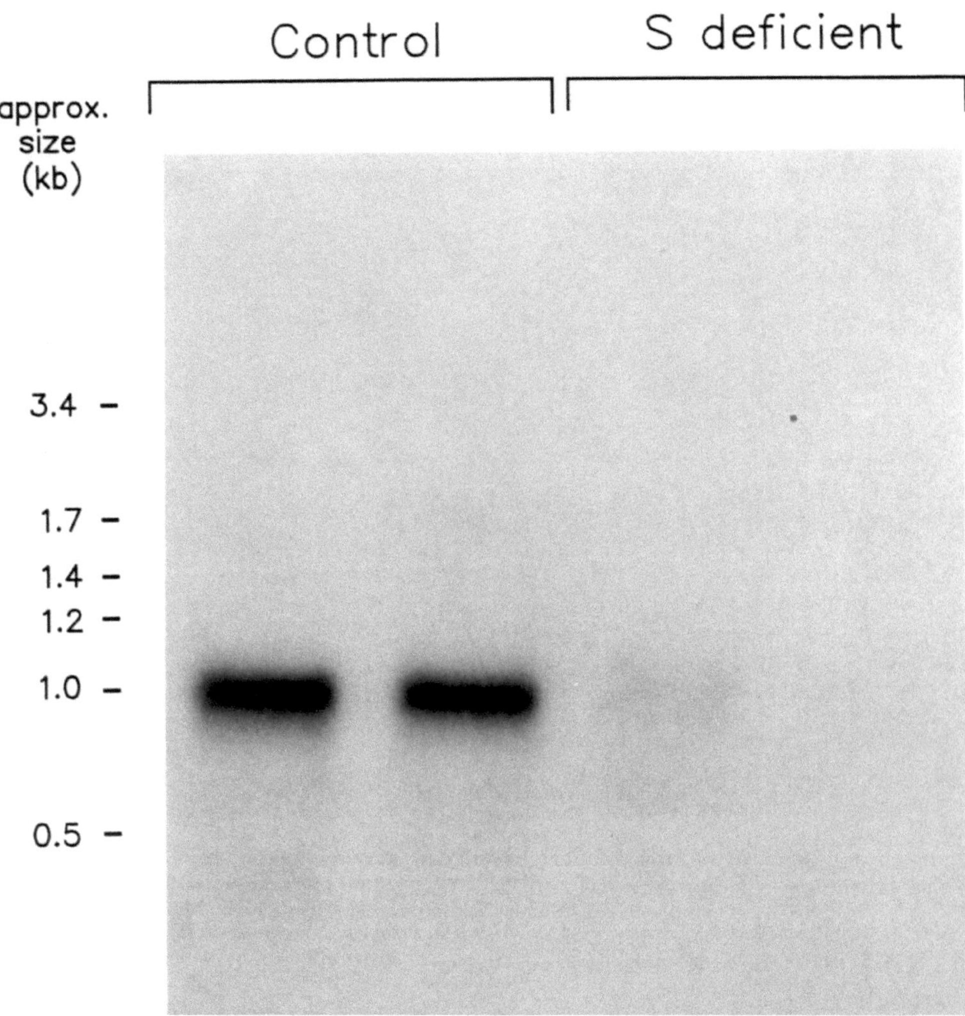

Fig. 2. Northern analysis of RNA extracted from four clonaly identical tobacco plants transformed with plasmid pSJ21 (SSU-PA1 chimeric gene). The first two lanes are the RNA from two plants grown under sulphur adequate conditions. The second two lanes are RNA from two plants grown under sulphur deficient conditions. 5 μg of total RNA was loaded. Filters were probed with random primed DNA probe complementary to part of the PA1 coding region. An approximate scale for the sizes of RNA (based on migration of the ribosomal RNAs) is given.

correct size (data not shown). The most strongly expressing plant from these four was used to determine the sulphur responsiveness of the ovalbumin gene as described for the PA1 gene above.

The results from this analysis appear in Figure 3. In order to optimize the detection of the low levels of ovalbumin transcript, 20 μg of total RNA was loaded and the filters were probed using a RNA probe which was labelled to a very high specific activity. The relative levels of ovalbumin mRNA did not differ significantly between the sulphur deficient and control treatments.

Discussion

There are numerous examples where heterologous genes are expressed in transgenic hosts in a

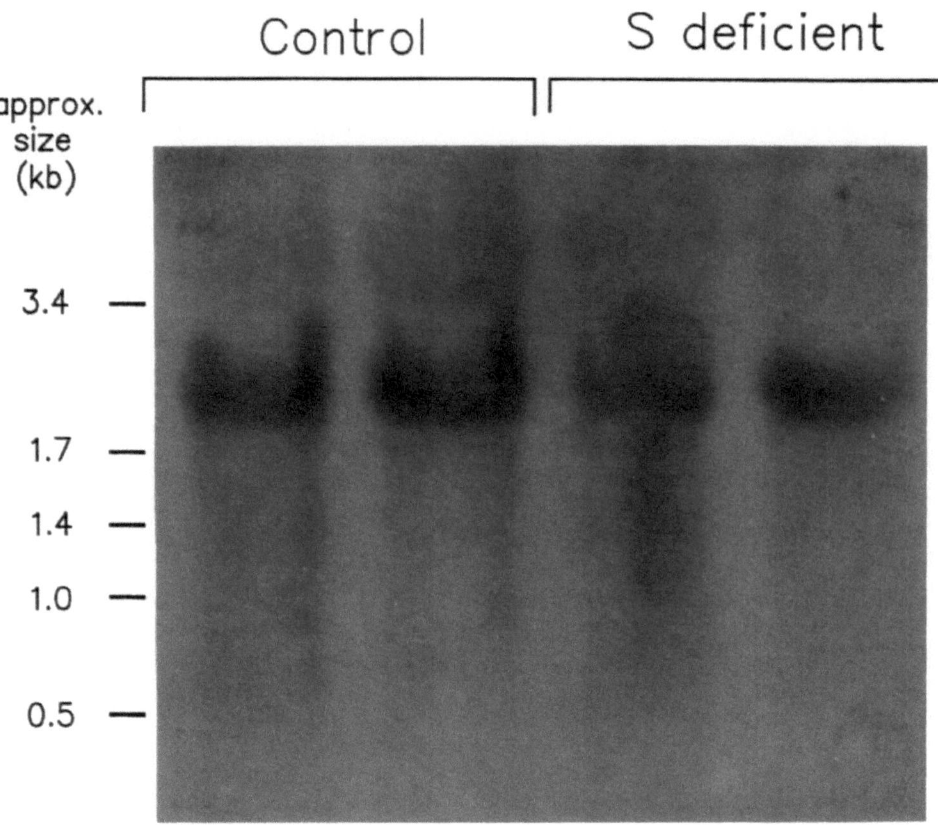

Fig. 3. Northern analysis of RNA extracted from four clonaly identical tobacco plants transformed with plasmid pRM5 (SSU-ovalbumin chimeric gene). The first two lanes are the RNA from two plants grown under sulphur adequate conditions. The second two lanes are RNA from two plants grown under sulphur deficient conditions. 20 μg of total RNA was loaded. Filters were probed with an RNA probe complementary to part of the ovalbumin cDNA. An approximate scale for the sizes of RNA (based on migration of the ribosomal RNAs) is given.

regulated manner (Beachy et al., 1985; Higgins et al., 1988; Sengupta-Gopalan et al., 1985). We are interested in characterising the mechanism by which sulphur nutrition regulates the expression of the sulphur-rich storage proteins. We have chosen the pea gene, pea albumin one (PA1) as a model and are studying its regulation in tobacco plants. Studies using transgenic plants to compare expression of different transgene constructs require the examination of many independent transformed plants from each construct so as to be able to separate any real differences in expression from the background variation present due to inter-transformant variability ('position effect'). The experimental design used here did not require large numbers of individual transformants to be examined since comparisons were only made between genetically identical plants containing the same transgene construct – inter-transformant variability has been eliminated from the experiment.

PA1 mRNA levels produced from a chimeric PA1 gene containing a light regulated, leaf and stem specific promoter were reduced by 25 fold under conditions of poor sulphur nutrition. This indicated that the 5′ flanking region of PA1 (which was omitted from this construct) was not necessary for sulphur regulation of the transgene. Since it was previously though that the PA1 gene was regulated by sulphur by a post-transcriptional mechanism (Higgins et al., 1986) we were interested in testing whether the ribulose-1,5-bisphosphate carboxylase small subunit (SSU) promoter used in this experiment was regulated transcriptionally by sulphur nutrition. We therefore constructed a chimeric gene containing the SSU promoter and an animal gene encoding a sulphur-rich protein (ovalbumin). We assumed that since ovalbumin is an animal gene it would not be regulated by sulphur nutrition by the post-transcriptional mechanism that occurs in plants. Hence, any reduction in the relative amounts of ovalbumin mRNA under sulphur deficient conditions could be regarded as being due to sulphur nutrition affecting the transcriptional activity of the SSU promoter. We found no difference in the levels of ovalbumin mRNA in the leaves of plants expressing this construct, indicating that the SSU promoter itself was not regulated by sulphur nutrition.

We conclude that the sequences responsible for conferring sulphur dependant regulation on the PA1 gene reside within the coding region, intron or 3′ flanking DNA. Since the transcription rate of the PA1 gene examined in this paper has not been measured in tobacco leaves there is a possibility that, in tobacco leaves, the regulation could be via a transcriptional mechanism acting through sequences contained in the coding region, intron or 3′ flanking region. However, since we know that the gene is regulated by a post-transcriptional mechanism in pea seeds (Higgins et al., 1986) it seems unlikely that the gene would be regulated by a completely different mechanisms in tobacco leaves. The data is, therefore, consistent with the previous conclusion that S regulation is via a post-transcriptional mechanism since any sequences responsible for such a mechanism must be contained within the transcribed portion of the gene.

It has previously been shown that legumin is regulated by sulphur in the seeds of tobacco (Rerie et al., 1991) leading to the conclusion that the *trans*-acting factors responsible for sulphur regulation have been conserved between the seeds of pea and tobacco. We conclude here that these factors are also conserved between the pea seed and tobacco leaves. This raises the following question. Why are factors, which regulate a seed specific gene, also found in the leaves? One possible explanation is that there are genes expressed in leaves which are also regulated by sulphur nutrition using regulatory factors functionally identical to those regulating the high-sulphur seed proteins. Leaf genes that could potentially be regulated in this fashion are the high-sulphur leaf thionins (Bohlmann and Apel, 1991).

Our working model for the mechanism for the regulation of the genes encoding sulphur-rich proteins is illustrated in Figure 4. We propose that there is a specific sequence of RNA contained in the sulphur regulated genes (a sulphur responsive element or SRE). The SRE takes a specific conformation which is recognised by a binding protein (the sulphur-responsive-element binding protein or SRE-BP). Under sulphur deficient conditions the SRE-BP takes up an active conformation, binds to the RNA at the SRE and cuts the RNA endonucleolitcally. The RNA is then exposed internally to the actions of non-specific exonucleases which degrade the RNA. Under sulphur adequate conditions the SRE-BP takes up an inactive conformation and can not bind the SRE. This means the RNA is not cut internally and it is protected from the action of the exonucleases by the poly (A) tail and the 5′ cap structures. Variations on this model can be envisioned but the major prediction from these models is the interaction between a protein and the SRE. Such models should be testable once the minimal SRE has been defined by making use of RNA binding assays such as used to identify factors interacting with iron regulatory elements in transferrin receptor and ferritin mRNA (reviewed in Theil (1990)) or with the AU destabilizing sequences found in some mammalian genes (Brewer, 1991).

Acknowledgements

We would like to acknowledge Peter Randall for his assistance with the sulphur nutrition experiments, Pam Dunsmuir for supplying the SSU promoter, Danny Lewellyn for supplying the plasmid 118-NOS and Megan Griffiths for constructing the plasmid pSJ21. R.L.M. is supported by a Wool Research and Development Corporation Postgraduate Research Fellowship.

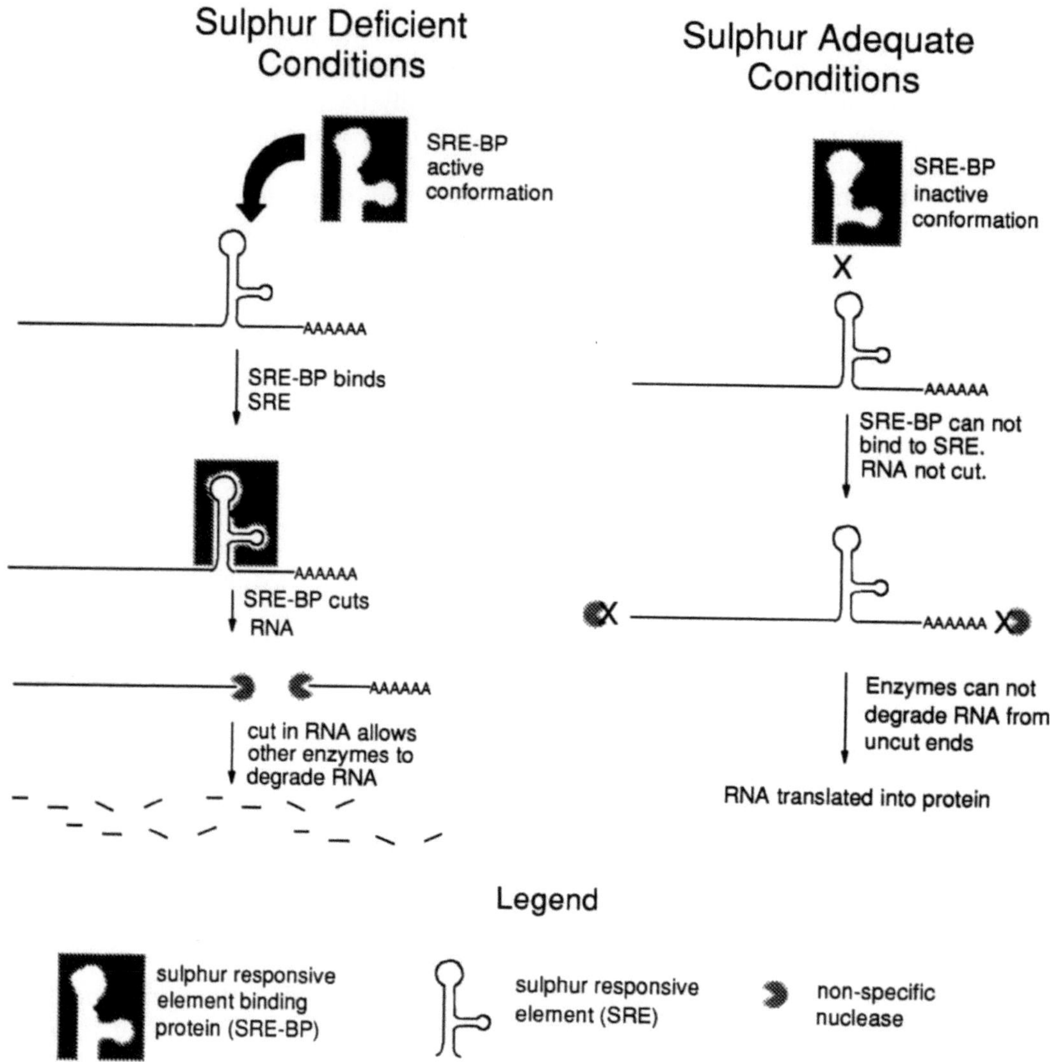

Fig. 4. Possible model for the sulphur dependant regulation of sulphur rich storage protein genes. For details see text.

References

An G 1986 Development of plant promoter expression vectors and their use for analysis of differential activity of nopaline synthase promoter in transgenic plants. Plant Physiol. 81, 86–91.

Beach L R, Spencer D, Randall P J and Higgins T J V 1985 Transcriptional and post-transcriptional regulation of storage protein gene expression in sulfur-deficient pea seeds. Nucleic Acids Res. 13, 999–1013.

Beachy R N, Chen Z-L, Horsch R B, Rogers S G, Hoffmann N J and Fraley R T 1985 Accumulation and assembly of soybean $-\alpha$ conglycinin in seeds of transformed petunia plants. EMBO J. 4, 3047–3053.

Bevan M W, Barnes W M and Chilton M-D 1983 Structure and transcription of the nopaline synthase gene region of T-DNA. Nucleic Acids Res. 11, 369–385.

Blagrove R J, Gillespie J M and Randall P J 1976 Effect of sulfur supply on the seed globulin composition of *Lupinus angustifolius*. Aust. J. Plant Physiol. 3, 173–184.

Bohlmann H and Apel K 1991 Thionins. Annu. Rev. Plant Physiol. Plant Mol. Biol. 42, 227–240.

Borroto K and Dure L 1987 The globulin seed storage proteins of flowering plants are derived from two ancestral genes. Plant Mol. Biol. 8, 113–131.

Brewer G 1991 An A + U-rich element RNA-binding factor regulates c-myc mRNA stability in vitro. Mol. Cell. Biol. 11, 2460–2466.

Casey R, Domoney C and Ellis N 1986 Legume storage proteins and their genes. Oxf. Surv. Plant Mol. Cell Biol. 3, 1–95.

Chandler P M, Higgins T J V, Randall P J and Spencer D 1983 Regulation of legumin levels in developing pea seeds under conditions of sulfur deficiency: Rates of legumin synthesis and levels of legumin mRNA. Plant Physiol. 71, 47–54.

Chandler P M, Spencer D, Randall P J and Higgins T J V 1984 Influence of sulfur nutrition on developmental patterns of some major pea seed proteins and their mRNAs. Plant Physiol. 75, 651–657.

Dean C, Van den Elzen P J M, Tamaki S, Dunsmuir P and Bedbrook J 1985 Linkage and homology analysis divides the eight genes for the small subunit of petuina ribulose-1,5-bisphosphate carboxylase into three gene families. Proc. Natl. Acad. Sci. USA 82, 4964–4968.

Ditta G, Stansfield S, Corbin D and Helinski D R 1980 Broad host range DNA cloning system for gram-negative bacteria: Construction of a gene bank of *Rhizobium meliloti*. Proc. Natl. Acad. Sci. USA 77, 7347–7351.

Evans I M, Boulter D, Fox R I and Kang B T 1977 The effects of sulfur fertilizers on the content of sulfo-amino acids in seeds of cowpeas (*Vigna unguicultata*). J. Sci. Food Agric. 28, 161–166.

Gayler K R and Sykes G E 1985 Effects of nutritional stress on the storage proteins of soybeans. Plant Physiol. 78, 582–585.

Higgins T J V, Chandler P M, Zurawski G, Button S C and Spencer D 1983 The biosynthesis and primary structure of pea seed lectin. J. Biol. Chem. 258, 9544–9549.

Higgins T J V 1984 Synthesis and regulation of major proteins in seeds. Annu. Rev. Plant Physiol. 35, 191–221.

Higgins T J V, Chandler P M, Randall P J, Spencer D, Beach L R, Blagrove R J, Kortt A A and Inglis A S 1986 Gene structure, protein structure, and regulation of the synthesis of a sulfur-rich protein in pea seeds. J. Biol. Chem. 261, 11124–11130.

Higgins T J V, Beach L R, Spencer D, Chandler P M, Randall P J, Blagrove R J, Kortt A A and Guthrie R E 1987 cDNA and protein seqeunce of a major pea seed albumin (PA2:M ~ 26,000). Plant Mol. Biol. 8, 37–45.

Higgins T J V, Newbigin E J, Spencer D, Llewellyn D J and Craig S 1988 The sequence of a pea vicilin gene and its expression in transgenic tobacco plants. Plant Mol. Biol. 11, 683–695.

Higgins T J V and Spencer D 1991 The expression of a chimeric cauliflower mosaic virus (CaMV- 35S) – Pea vicilin gene in tobacco. Plant Sci. 74, 89–98.

Horsch R B, Fry J E, Hoffmann N L, Eichholtz D, Rogers S G and Fraley R T 1985 A simple and general method for transferring genes into plants. Science 227, 1229–1231.

Khandjian E W 1987 Optimized hybridization of DNA blotted and fixed to nitrocellulose and nylon membranes. Biotech. 5, 165–167.

Kortt A A, Caldwell J B, Lilley G G and Higgins T J V 1991 Amino acid and cDNA sequences of a methionine-rich 2S protein from sunflower seed (*Helianthus annuus* L.). Eur. J. Biochem. 195, 329–334.

Lycett G W, Delauney A J, Zhao W, Gatehouse J A, Croy R R D and Boulter D 1984 Two cDNA clones coding for the legumin protein of *Pisum sativum* L. contain sequence repeats. Plant. Mol. Biol. 3, 91–96.

McReynolds L, O'Malley B W, Nisbet A D, Fothergill A D, Givol D, Fields S, Robertson M and Brownlee G G 1978 Sequence of chicken ovalbumin mRNA. Nature 273, 723–728.

Randall P J, Thomson J A and Schroeder H E 1979 Cotyledon storage proteins in *Pisum sativum*. IV. Effects of sulfur, phosphorus, potassium and magnesium deficiencies. Aust. J. Plant Physiol. 6, 11–24.

Rerie W G, Whitecross M and Higgins T J V 1991 Developmental and environmental regulation of pea legumin genes in transgenic tobacco. Mol. Gen. Genet. 225, 148–157.

Sambrook J, Fritsch E F and Maniatis T 1989 Molecular cloning: A laboratory manual. Cold Spring Harbor Laboratory Press, Cold Spring Harbor.

Schroeder H E 1984 Major albumins of *Pisum* cotyledons. J. Sci. Food Agric. 35, 191–198.

Sengupta-Gopalan C, Reichert N A, Barker R F, Hall T C and Kemp J D 1985 Developmentally regulated expression of the bean α-phaseolin gene in tobacco seed. Proc. Natl. Acad. Sci. USA 82, 3320–3324.

Shewry P R, Franklin J, Parmar S, Smith S J and Miflin B J 1983 The effects of sulfur starvation on the amino acid and protein compositions of barley grain. J. Cereal Sci. 1, 21–31.

Spencer D, Chandler P M, Higgins T J V, Inglis A S and Rubira M 1983 Sequence interrelationships of the subunits of vicilin from pea seeds. Plant Mol. Biol. 2, 259–267.

Spencer D, Rerie W G, Randall P J and Higgins T J V 1990 The regulation of pea seed storage protein genes by sulfur stress. Aust. J. Plant Physiol. 17, 355–363.

Theil E C 1990 Regulation of ferritin and transferrin receptor mRNAs. J. Biol. Chem. 265, 4771–4774.

Wen-jun S and Forde B G 1989 Efficient transformation of *Agrobacterium* spp. by high voltage electroporation. Nucleic Acids Res. 17.

Wrigley C W, Du Cros D L, Archer M J, Downie P G and Roxburgh C M 1980 The sulfur content of wheat endosperm proteins and its relevance to grain quality. Aust. J. Plant Physiol. 7, 755–766.

P. J. Randall et al. (Eds.), Genetic aspects of plant mineral nutrition, 321–326.

PLSO SV74

Strategies in population development for the improvement of Fe efficiency in soybean*

S.R. DE CIANZIO
Department of Agronomy, Iowa State University, Ames, IA 50011, USA

Key words: backcross, breeding methods, calcareous soil, chlorosis resistance, recurrent selection, single cross, three-way cross

Abstract

Iron (Fe) deficiency in some soybean [*Glycine max* (L.) Merr.] genotypes is observed when they are grown on calcareous soil. Leaves become chlorotic at early stages of development. Plants may recover later in the season; however, yield reduction occurs. The improvement of Fe-efficiency in soybean is an objective of the soybean breeding project at Iowa State University. For cultivar development, the usefulness of single-cross and backcross, and of single-cross, backcross, and three-way cross populations was evaluated in two studies. High-yielding genotypes were crossed to highly Fe-efficient lines to obtain single crosses. The F_2-derived lines of these crosses were evaluated for Fe efficiency on calcareous soil, and the most efficient lines were backcrossed to the high-yielding parent. For the three-way crosses, the selected lines were crossed to a third parent. Evaluation of lines for seed yield and Fe efficiency was conducted. Results indicated that high-yielding genotypes with improved Fe efficiency can be obtained from all population types. In both studies, the backcross populations had the highest yield and the lowest level of Fe efficiency. The single crosses had the highest level of Fe efficiency, but the lowest mean yield. The three-way crosses were intermediate to the single-cross and the backcrosses for seed yield and were similar in Fe efficiency to the backcrosses. For germplasm development, a recurrent selection program has been successful in improving Fe efficiency. Highly Fe-efficient genotypes have been released from this research.

Introduction

Soybean cultivars that are unable to utilize available iron (Fe) when grown on calcareous soil have leaves that become chloritic at early stages of plant development (Weiss, 1943). Plants may recover later in the season; however, yield is reduced (Froehlich and Fehr, 1981).

Genotypic differences for Fe efficiency in soybean were first reported by Weiss (1943). He crossed efficient and inefficient genotypes and evaluated the F_1, F_2, and F_2 progenies in nutrient solution media. He reported that Fe efficiency was controlled by a single gene, with Fe efficiency being dominant to Fe inefficiency.

In field plantings on calcareous soils, a continuous range in the expression of Fe efficiency is observed among genotypes. The segregation observed in crosses between Fe-efficient and Fe-inefficient genotypes when evaluated in field plantings on calcareous soil has indicated that Fe efficiency could be explained by a single major gene and modifying genes (de Cianzo and Fehr, 1980) or by multiple genes with additive gene action (de Cianzo and Fehr, 1982). The presence of a single major and modifying genes also has been reported in an interspecific cross between a Fe-efficient *G. soja* accession and an inefficient *G. max* experimental line (Diers et al., 1992).

* Journal Paper no. J-14748 of the Iowa Agric. and Home Econ. Exp. Stn., Ames, Iowa, Project 3107.

For breeding purposes, the trait is considered a quantitative character (de Cianzo and Fehr, 1980).

The improvement of Fe efficiency in soybean cultivars is one of the objectives of the soybean breeding project at Iowa State University. A second objective is to develop germplasm with improved Fe efficiency for use as parents in crosses. The purpose of this review is to discuss strategies in population development for the selection of high-yielding cultivars with improved Fe efficiency, and for the selection of germplasm with improved Fe efficiency. Results from research conducted in the soybean breeding program of Iowa State University will be discussed.

Strategies in population development

Development of high-yielding cultivars with improved Fe efficiency

Studies on breeding methodologies have been conducted to compare the relative efficiency of different strategies in population development for combining high Fe efficiency and high yield into new cultivars (Hintz et al., 1987; Voss et al., 1992). Populations obtained by single crosses (Hintz et al, 1987; Voss et al., 1992) were developed with high-yielding lines and highly Fe-efficient genotypes. The backcross (Hintz et al., 1987; Voss et al., 1992) populations were obtained by using as a recurrent parent the high-yielding line of the single cross. The three-way cross populations (Voss et al., 1992) were developed by mating the single crosses to a third high-yielding parent.

Experiment I

The research conducted by Hintz et al. (1987) provided the first direct comparison of the relative efficiency of single crosses and backcrosses. Four single-cross populations were developed by crossing three high-yielding experimental lines to one or two highly Fe-efficient genotypes at Ames, IA, during 1982 (Table 1). The Fe efficiency of 200 F_2-derived lines from each population was evaluated on calcareous soil, and the 20 most Fe-efficient lines were backcrossed to their respective high-yielding recurrent parent at Ames, IA, during 1983. Forth F_2-derived lines with the highest Fe efficiency and 400 BC_1F_2-derived lines from each cross were evaluated on calcareous soil in Iowa in 1984. An F_4-derived line or a BC_1F_3-derived line was obtained from the 20 F_2-derived lines and 20 BC_1F_2-derived lines with the highest Fe efficiency for each population. In 1985, these lines were tested for Fe efficiency on calcareous soil in Iowa in six replications at one location and for seed yield on noncalcareous soil in two replications at each of three locations.

The backcross populations had higher mean yields but lower average Fe efficiency than the single-cross populations (Table 1). Differences in mean yield of the two population types were significant for one of the four crosses, and differences in average Fe efficiency were significant for three of the four crosses.

The phenotypic correlations between the yield of lines on noncalcareous soil and their chlorosis score on calcareous soil were either not significantly different from zero or significantly negative. The negative correlations were considered desirable because they indicated that higher seed

Table 1. Mean seed yields and chlorosis ratings of two population types, parents, and a check cultivar (Hintz et al., 1987)

Population type and genotype	Seed yield[y] ($g\,m^{-2}$)	Chlorosis rating score[z]
Single cross	222 a[x]	2.5 a[x]
Backcross	234 a	3.0 b
High-yielding parents		
A79-135010	269 b	4.4 c
A81-157007	242 ab	3.4 b
A81-157024	237 b	4.2 b
Fe-efficient parents[x]		
AP9-82-176153		1.7d
AP9-82-176115		1.6 d
Check cultivar		
BSR 101	274 bc	3.0 ab

[z]Scores ranged from 1.0 = no yellowing to 5.0 = severe yellowing with some necrosis.
[y]Means followed by the same letter are not significantly differant at 0.05 probability level.
[x]Yield was not reported.

yields were associated with lower chlorosis scores and greater Fe efficiency. Therefore, the development of high-yielding, Fe-efficient cultivars should be possible.

Lines in the single-cross and backcross populations were compared for seed yield with the recurrent parent and for Fe efficiency with the donor parent and 'BSR 101', a cultivar that has an acceptable level of Fe efficiency on most calcareous soils (Table 2). An average of 36% of lines in the single-cross populations and 55% of lines in the backcross populations had seed yield not significantly less than that of the recurrent parent, and an average of 96% of lines in the single-cross populations and 80% of lines in the backcross populations had an Fe efficiency rating not significantly different from BSR 101. The average percentage of lines that were not significantly different from the recurrent parent in seed yield and from BSR 101 in Fe efficiency was 36% in the single-cross populations and 43% in the backcross populations. The results indicated that both population types would be suitable for combining high yield and high Fe efficiency.

Experiment II

In a study conducted by Voss et al. (1992) three-way cross populations were compared with single-cross and backcross populations for the development of high-yielding cultivars with Fe efficiency. Twelve populations were developed by using the three different strategies and four genotypes for crossing. Three high-yielding cultivars with poor Fe efficiency were each crossed to a highly Fe-efficient line to develop the single-cross populations (Table 3). The highly resistant line was obtained from cycle 3 of a recurrent selection program conducted to improve Fe efficiency (Fehr et al., 1984).

In the summer of 1985, 200 F_2-derived lines were evaluated for Fe efficiency in replicated field tests planted on calcareous soil in Iowa. From each cross, the 20 most Fe-efficient lines were backcrossed to their respective high-yielding parent to form the backcross populations and to a third parent to form the three-way cross populations. In 1986, 200 $BC_1F_{2:3}$-derived lines from each backcross population and 200 $F_{2:3}$-derived lines from each three-way cross population were evaluated for Fe efficiency on calcareous soil in Iowa. From each population, the 20 lines with the highest Fe efficiency were selected. In the summer of 1987, seed of the lines, parents, and checks was increased at Ames, IA.

The 20 lines selected from each of the 12 populations, the parents, and the check cultivar BSR 101 were evaluated for seed yield at one location in 1988 and two locations during 1989 on noncalcareous soil. Two replications were planted at each of the three environments. During the same years, Fe efficiency was evaluated on calcareous soil at each of two locations in Iowa. In 1988, two replications were planted at one location and four replications at the second location. In 1989, three replications were planted at each of the same two locations.

The mean seed yields and chlorosis scores of the three population types were similar for the backcrosses and three-way crosses (Table 3).

Table 2. Percentage of lines not significantly different from the recurrent parent for seed yield, not significantly less than BSR 101 for Fe efficiency, and desirable for both traits (Hintz et al., 1987)

Population type[a]	Cross	Seed yield	Fe efficiency	Yield and resistance
SC	A79-135010 × AP9-82-176153	9	91	9
BC	A79-135010^2 × AP9-82-176153	41	55	18
SC	A81-157007 × AP9-82-176153	55	100	55
BC	A81-157007^2 × AP9-82-176153	55	91	55
SC	A81-157024 × AP9-82-176153	45	100	45
BC	A81-157024^2 × AP9-82-176153	73	100	73
SC	A81-157024 × AP9-82-176115	36	95	36
BC	A81-157024^2 × AP9-82-17615	50	73	27
Average				
SC		36	96	36
BC		55	80	43

[a]SC = single cross; BC = backcross.

Table 3. Mean seed yields and chlorosis ratings of three population types, parents, and a check cultivar (Voss et al., 1992)

Population type and genotype	Seed yield ($g\ plot^{-1}$)	Chlorosis rating[a] score
Single cross	1281	2.1
Backcross	1377	2.6
Three-way cross	1347	2.6
$LSD_{(0.05)}$	32	0.2
High-yielding parents		
Profiseed 1138	1482	3.9
S42A	1607	3.3
Elgin	1581	3.8
Fe-efficient parent		
A7	1016	1.4
Check cultivar		
BSR 101	1476	2.8
$LSD_{(0.05)}$	332	0.5

[a]Scores ranged from 1.0 = no yellowing to 5.0 = severe yellowing with some necrosis.

The single-cross populations had significantly lower seed yields and significantly better chlorosis scores than the backcross and three-way crosses. For each population type, it was possible to identify high-yielding F_2-derived lines and lines that had acceptable Fe efficiency.

The relative effectiveness of the three population types for combining high yield and high Fe efficiency was evaluated by comparing seed yield and Fe efficiency of the lines with standards (Table 4). The standards used for seed yield were the highest-yielding parent used in the single cross or the cultivar BSR 101. For chlorosis scores, the standards were the resistant parent used in crossing or BSR 101. BSR 101 was selected as the check cultivar because it had the best combination of seed yield and chlorosis resistance available among publicly developed cultivars at the time of the study. The number of lines in each population similar to or better than the standards was expressed as a percentage.

For seed yield, the percentage of lines in the 12 populations that were similar or better than the highest-yielding parent of the single cross ranged from 0 to 15% (Table 4). The average percentage of superior lines for yield in each of the three populations types was similar. Compared with BSR 101, the percentage of superior lines in the individual populations ranged from 0 to 35%, and the backcross populations generally had a greater percentage of high-yielding lines than the other two population types. Averaged over populations, 23% of the lines in the backcrosses, 13% in the three-way crosses, and 5% in the single-crosses had seed yields that were similar or better than BSR 101.

Lines with chlorosis scores similar to A7 were observed in only two populations, and these

Table 4. Percentage of lines similar to or better than the highest-yielding parent of the single cross or BSR 101 for seed yield, similar to or better than BSR 101 for Fe efficiency, and desirable for both traits (Voss et al., 1992)

Population type[a]	Cross	Seed yield		Fe efficiency	Yield and resistance
		parent	BSR 101	BSR 101	BSR 101
SC	Profiseed 1138 × A7	15	15	95	10
BC	Profiseed 1138^2 × A7	15	15	70	15
TWC	(Profiseed 1138 × A7) × S42A	10	10	85	5
TWC	(Profiseed 1138 × A7) × Elgin	0	5	40	0
SC	S42A × A7	0	0	95	0
BC	$S42A^2$ × A7	5	35	80	20
TWC	(S42A × A7) × Profiseed 1138	0	5	75	5
TWC	(S42A × A7) × Elgin	5	35	60	25
SC	Elgin × A7	0	5	90	5
BC	$Elgin^2$ × A7	5	20	60	10
TWC	(Elgin × A7) × Profiseed 1138	0	10	60	10
TWC	(Elgin × A7) × S42A	10	10	60	10
Average					
SC		5	7	93	5
BC		8	23	70	15
TWC		4	13	63	9

[a]SC = single cross; BC = backcross; TWC = three-way cross.

were single-cross populations. Lines similar to or better in chlorosis scores than BSR 101 were observed in all populations (Table 4). The single crosses had an average of 93% of lines that were similar to or better than BSR 101, compared with 70% for the backcrosses and 63% for the three-way crosses.

The three population types were compared for the number of lines similar or better than BSR 101 for both seed yield and chlorosis score (Table 4). There were 15% of acceptable lines in the backcross populations, 9% in the three-way crosses, and 5% in the single crosses. These results indicated that the three population types may be used to combine high-yield with high Fe efficiency.

Development of germplasm with improved Fe efficiency

Recurrent selection has been used in the soybean breeding program of Iowa State University to develop Fe-efficient germplasm in the soybean population AP9 (Fehr and de Cianzo, 1980). From this population, six lines with improved Fe efficiency have been released: A7, selected from cycle 3 (Fehr et al., 1984), and A11, A12, A13, A14, and A15, selected from the cycle 7 (Jessen et al., 1988b) population.

In the recurrent selection program, each cycle consisted of the evaluation of the Fe efficiency of 100 S_0-derived lines from the population and the diallel mating of the 10 most Fe-efficient lines to form a new population for the next cycle of selection (Jessen et al., 1988a). Field evaluations on calcareous soil were utilized to identify the most Fe-efficient lines during the first four cycles of selection.

Substantial genetic improvement was made in the population during the first four cycles, which made it difficult to obtain adequate chlorosis expression to determine genetic differences among the lines in the cycle 5 population. The method used during cycles 5 and 6 to increase chlorosis severity was stem cutoff (Fehr et al., 1985; Piper et al., 1986). At stage V3, when the second trifoliolate leaf was fully developed, the main stem of each plant was cut off above the unifoliolate node. Two weeks later, plots were rated for chlorosis by evaluating the new leaf tissue that had developed after treatment. Because of genetic improvement obtained during cycles 5 and 6, the level of chlorosis expressed after plant cutoff was inadequate to discriminate among lines in the cycle 7 population. A nutrient solution system developed by Coulombe et al. (1984) with addition of HCO_3^- was adapted for evaluation of the Fe efficiency in the cycle 7 and 8 populations.

Discussion

The strategies for population development used by the soybean breeding project of Iowa State University have been successful in improving Fe efficiency of genotypes. For cultivar development, the studies of Hintz et al. (1987) and Voss et al. (1992) suggested that the choice of a population type to use in breeding high-yielding cultivars with improved Fe efficiency will depend on the performance of the parents and the resource allocation of the particular breeding program. Of the three population types, single-cross, backcross, and three-way crosses, the backcross stragegy seems superior when the Fe-efficient parent has relatively low yield. Single crosses involve less labor and time to develop populations than backcrosses and three-way crosses. Three-way crosses and backcrosses require similar labor and time to develop the populations.

For germplasm improvement, recurrent selection was effective for developing highly Fe-efficient lines. The seed yield of lines selected from the recurrent selection program was not adequate for commercial use. In the soybean breeding program of Iowa State University, research has begun to evaluate genetic improvement for yield and Fe efficiency by using recurrent selection. A population has been formed by crossing high-yielding parents that are moderately Fe efficient, with the highly Fe-efficient lines A11, A12, A13, A14, and A15. Lines from the populations will be selected simultaneously for yield and Fe efficiency. Field plantings on calcareous soils and nutrient solution screening will be used to select for Fe efficiency.

References

Cianzio S R de and Fehr W R 1980 Genetic control of iron deficiency chlorosis in soybeans. Iowa State J. Res. 54, 367–375.

Cianzio S R de and Fehr W R 1982 Variation in the inheritance of resistance to iron deficiency chlorosis in soybeans. Crop Sci. 22, 433–434.

Coulombe B A, Chaney R L and Wiebold W J 1984 Use of bicarbonate in screening soybeans for resistance to iron chlorosis. J. Plant Nutr. 7, 411–425.

Diers B, Cianzio S and Shoemaker R 1992 Possible identification of quantitative trait loci affecting iron efficiency in soybean. J. Plant Nutr. (in press).

Fehr W R and Cianzio S R de 1980 Registration of AP9(S1)C2 soybean germplasm. Crop Sci. 20, 677.

Fehr W R, Voss B K and Cianzio S R 1984 Registration of a germplasm line of soybean, A7. Crop Sci. 24, 390–391.

Fehr W R, Froehlich D M and Ertl D S 1985 Iron-deficiency chlorosis of soybean cultivars injured by plant cutoff and defoliation. Crop Sci. 25, 21–23.

Froehlich D M and Fehr W R 1981 Agronomic performance of soybeans with differing levels of iron deficiency chlorosis on calcareous soil. Crop Sci. 21, 438–441.

Hintz R W, Fehr W R and Cianzio S R 1987 Population development for the selection of high-yielding soybean cultivars with resistance to iron-deficiency chlorosis. Crop Sci. 27, 707–710.

Jessen H J, Dragonuk M B, Hintz R W and Fehr W R 1988a Alternative breeding strategies for the improvement of iron efficiency in soybean. J. Plant Nutr. 11, 717–726.

Jessen H J, Fehr W R and Cianzio S R 1988b Registration of germplasm lines of soybean A11, A12, A13, A14, and A15. Crop Sci. 28, 204.

Piper T E, Fehr W R and Voss B K 1986 Stem cutoff enhances selection for improved iron efficiecny of soybean. Crop Sci. 26, 751–752.

Voss B K, Cianzio S R de and Fehr W R 1992 Three crossing strategies for breeding high-yielding soybean cultivars with improved Fe efficiency. Crop Sci. (submitted).

Weiss M G 1943 Inheritance and physiology of efficiency in iron utilization in soybeans. Genetics 28, 253–268.

P. J. Randall et al. (Eds.), Genetic aspects of plant mineral nutrition, 327–333.

PLSO SV75

Genetics of tolerance to iron chlorosis in rice

N.T. HOAN[1], U. PRASADA RAO[2] and E.A. SIDDIQ[2]
[1]*Plant Breeding Department, Culong, Delta, Rice Research Institute, Vietnam and* [2]*Directorate of Rice Research, Hyderabad-500030, India*

Key words: inheritance, iron chlorosis, rice

Abstract

Genetics of tolerance to iron chlorosis was investigated in eight crosses involving parents distinctly different in their level of tolerance. The segregating populations with parents and F_1s were screened under actual stress conditions in the field. Also, selected crosses were studied for Fe^{3+} uptake capacity. Tolerance/moderate tolerance to Fe chlorosis was dominant over susceptibility and it was controlled by two sets of nonallelic genes with complementary interaction. Gene Ic_1 has been found to be basic and in complementation with Ic_3 it confers tolerance. Likewise, Ic_2 with Ic_4 confers tolerance. The basic genes Ic_1 and Ic_2 are nonallelic and, in the absence of their respective complementary genes Ic_3 and Ic_4, ineffective, which results in susceptibility. Of tolerant cultivars, ARC 10372 and Cauvery have been tentatively assigned the genotype of Ic_1, ic_2, Ic_3, ic_4, and moderately tolerant IET 7613, Prasanna and Akashi ic_1, Ic_2 Ic_3 Ic_4. The susceptible ARC 5723 has been assigned ic_1, ic_2 Ic_3, Ic_4, and IET 9829, ic_1, ic_2 Ic_3, ic_4. IET 7614 is susceptible, due to the presence of inhibitory genes I–Ic_1, I–Ic_2 together with ic_1, ic_2, Ic_3, Ic_4. Further, the gene *Pc* for purple coleoptile shows linkage with one of the complementary genes with a crossover value of 15.26%, while the gene(s) for seedling height *Ts* with Ic_1 with a crossover value of 1.7%. It is possible that the gene(s) for iron chlorosis tolerance might belong to the second linkage group, where genes for purple leaf were located.

Introduction

Experience with nutritional management shows that deficiency/toxicity of minor elements is as detrimental to growth and productivity of rice crop as deficiency of major nutrients. Deficiency of iron is one of the widely known disorders. Iron is indispensable for chlorophyll synthesis. The amount of Fe^{2+} fraction of total iron is believed to make the plant green or chlorotic. This assumption is in line with the evidence (De Kock, 1971) that the Fe^{2+} is involved in the condensation of succinic acid and glycine to form δ-amino livulnic acid (ALA), which condenses to form pyrrole groups. The pyrrole groups in turn condense to give rise to protoporphyrin IX, which by insertion of Mg^{2+} produces chlorophyll. Thus, nonavailability or availability of Fe at less than optimum level impairs chlorophyll development, leading to chlorosis of young rice seedlings and early death in extreme cases of deficiency. Iron deficiency occurs, in general, in high pH alkaline soils under submerged conditions, in high pH calcareous soils under upland conditions, and in high pH heavy clay soils and inadequate soil reduction under submerged conditions. Deficiency at high pH could be explained either by low solubility of iron in the rooting medium, by fast oxidation of ferrous iron and immobilization in the roots, or by various combinations of the above three (Tanaka and Yoshida, 1970). In India, calcareous and alkaline calcareous soils of Bihar, heavy clay soils of Maharashtra, Karnataka and Andhra Pradesh, alkaline soils in Indo-Gangetic alluvial of Uttar Pradesh and Bihar, and manganese rich soils of coastal Karnataka suffer from this malady. Depending upon the situation, pre-submergence or application of organic matter, sulphuric acid or ferric sulphate, are recommended methods to alleviate the problem of chlorosis. Except for cultivation of locally adapted varieties like Tuljapur-1, no systematic research effort has been made so far to breed varieties tolerant to iron deficiency. This has been largely due to inadequate basic information on the extent of genetic variation and genetics of tolerance to iron chlorosis. The present study was, therefore, undertaken to investigate the inheritance of tolerance to stress.

Materials and methods

Experimental material, consisting of F_1s, F_2s and F_3s and parents of eight crosses involving tolerant, moderately tolerant and susceptible parents (Table 1) was studied under actual stress conditions in the field using chlorosis per se as the index, as well as in the laboratory using ferric iron uptake capacity of roots as an index.

In respect of tolerance to chlorosis, the experimental material was raised under upland conditions at the Directorate's farm at Hyderabad. The soil of the experimental plot is highly aerobic, sandy loam, moderately alkaline (pH 8.3 and E.C. 1.59 m mhos/cm) with low organic carbon content, medium P level and a known history of iron chlorosis in rice.

Field screening for chlorosis

Seeds of parents, F_1s, F_2s and F_3s were directly sown on raised dry beds during wet season in 1989 and 1990 adopting 10 cm spacing between seeds and 10 cm between rows. The material was sown in three replications. At every 20th row, a resistant and a susceptible check were grown for comparison. Observations were recorded on seedlings at 10-day intervals up to 30 days. The seedlings were scored using the following 1–9 scale.

Tolerant (0–1)	Seedling growth normal, foliage green
Tolerant (2–3)	Seedling growth normal, yellowing in less than 5% of the leaf area
Moderately tolerant (4)	Seedling growth normal, yellowing in 5–25% of the leaf area
Susceptible (5)	Seedling growth stunted, yellowing in 50–70% of the leaf area
Susceptible (9)	Seedling growth severely stunted showing signs of death.

Table 1. Rice cultivars and selections in the study and their tolerance to iron chlorosis

Cultivar/ selection	Origin[a]	Parents	Iron chlorosis and plant habit
IET 9829	DRR, Hyderabad	Rasi/Chittaraikar	Susceptible, short
Cauvery	Tamil Nadu	T(N)1/TKM 6	Tolerant, short
IET 7614	DRR, Hyderabad	Rasi/Finegora	Susceptible, short
ARC 5723	Assam	Cultivar	Susceptible, tall
IET 7613	DRR, Hyderabad	M 63-83/Cauvery	Moderately tolerant, short
Akashi	DRR, Hyderabad	IR 8/N 22	Moderately tolerant, short
Prasanna	DRR, Hyderabad	IRAT 8/N 22	Moderately tolerant, short
Tuljapur-1	Maharashtra	Cultivar	Tolerant, tall
ARC 10372	Assam	Cultivar	Tolerant, tall
WBPH 25	Kerala	Cultivar	Moderately tolerant, tall

[a]From India

Laboratory screening for iron uptake capacity

Parents, F_1s and F_2s of two crosses were screened for Fe^{3+} uptake capacity of roots. The screening method adopted by Brown et al. (1961) to determine the reductive capacity of Fe^{3+} of soya bean roots, and later by Misal and Nerkar (1983) in rice to distinguish genotypes tolerant or susceptible to iron chlorosis, was employed in the present study. Roots of 7-day and 10-day-old seedlings raised in petri dishes were immersed in 7 mL of 10^{-4} *M* potassium ferric cyanide solution contained in test tubes for 24 hours at room temperature (32°C). Each treatment was replicated three times. Care was taken to see that seedlings with uniform root size were included. Fe^{3+} uptake capacity of the seedlings was indirectly estimated by measuring the amount of potassium ferric cyanide reduced in the solution in which the roots of the seedlings were immersed. This was done by measuring the optical density of the test solution with the aid of a spectrophotometer set at 400 nm. On the basis of relative reduction, the F_2 genotypes were grouped into tolerant, moderately tolerant and susceptible categories. In the case of moderately tolerant × tolerant crosses, seedlings which showed Fe^{3+} uptake less than the maximum of the range of a moderately tolerant parent were grouped as moderately tolerant; those with less than the minimum of the range of a moderately tolerant parent were grouped as susceptible; and those exceeding the minimum of the range of the tolerant parent as tolerant. In the case of a moderately tolerant to susceptible cross, seedlings which showed Fe^{3+} uptake capacity less than the maximum of the range of a susceptible parent were taken as susceptible and those which exceeded the minimum of the range of a moderately tolerant parent as moderately tolerant. The range and means of the parents, F_1s and F_2s varied with the age of seedlings studied.

Observed phenotypic variation in the field on F_1s, F_2s, F_3s and parents in all the crosses was interpreted on the basis of χ^2 tests for both the indices of tolerance to iron deficiency and other characteristics. Phenotypic segregation pattern of F_2 was confirmed by study of 96 F_3 families with 100 plants/family.

Joint segregation analysis was done to determine linkage relationships between genes governing iron chlorosis and those governing coleoptile pigment and seedling height. χ^2 analysis was done to test the significance of deviation at 5% level for three degrees of freedom. In the absence of independent assortment, crossover values were estimated by the minimum discrepancy method of Richharia et al. (1966).

Results and discussion

Mode of inheritance

Results from field studies indicated that tolerance to iron chlorosis followed a simple mode of inheritance, with tolerance and moderate tolerance showing dominance over susceptibility. One to three genes in varied interactions governed its expression (Table 2). Among the crosses representing tolerant (T) × susceptible (S), IET 9829 (S) × Cauvery (T) showed monogenic (3:1) inheritance, while IET 7614 (S) × ARC 10372 (T) showed digenic (3:13) inhibitory gene action. In moderately resistant (MT) × susceptible (S) crosses, IET 7614 (S) × WBPH 25 (MT) showed digenic (3:13) inhibitory gene action, while Akashi (MT) × ARC 5723 (S) and ARC 5723 (S) × Prasanna (MT) showed monogenic (3:1) inheritance. Of the three crosses representing tolerant × moderately tolerant, IET 7613 (MT) × Tuljapur 1 (T) and Prasanna (MT) × Tuljapur 1 (T) showed duplicate dominant gene action (15:1), while IET 7613 (MT) × ARC 10372 (T) followed a trigenic (48:9:7) mode of inheritance.

On the basis of the foregoing, IET 7613 (MT) is assumed to contain three genes Ic_2, Ic_3, Ic_4 in dominant state and ic_1 in recessive state, whereas ARC 10372 (T) is believed to contain the basic gene Ic_1 and its complementary gene (Ic_3) in dominant state. The other pair (ic_2 and ic_4) remained in recessive condition. F_2 plants with Ic_1, Ic_3 in homozygous state were tolerant, Ic_2, Ic_4 moderately tolerant and those lacking both the basic genes in dominant state were susceptible, thus segregating in a ratio of 48 tolerant: 9 moderately tolerant: 7 susceptible.

Seedling height inherited simply in the cross

Table 2. Behaviour of F_1s and segregation pattern of F_2 for iron chlorosis tolerance and other characteristics

Cross	F_1 reaction	Phenotypic Frequencies[a]			Ratio	X^2	P value
		T	MT	S			
Susceptible × Tolerant							
IET 9829/Cauvery	T	337		129	3T:1S	1.70	0.20–0.10
IET 7614/ARC10372	S	093		397	3T:13S	0.16	0.70–0.50
Susceptible × Moderately tolerant							
IET 7614/WBPH 25	S		123	620	3MT:13S	2.30	0.20–0.10
ARC 5723/Prasanna	MT		237	88	3MT:1S	0.74	0.80–0.70
Akashi/ARC 5723	MT		117	26	3MT:1S	3.54	0.10–0.05
Tolerant × Moderately tolerant							
IET 7613/Tuljapur-1	T	402		34	15T:1S	1.78	0.20–0.10
Prasanna/Tuljapur-1	T	277		10	15T:1S	3.73	0.10–0.05
IET 7613/ARC 10372	T	786	146	126	48T:9MT:7S	1.10	0.80–0.70
Purple coleoptile (Pc) vs Green coleoptile (pc)							
IET 7614/WBPH 25	Pc	Pc 407	pc 338		9Pc:7pc	1.10	0.30–0.20
Tall seedling (Ts) vs. short seedling (ts)							
IET 7613/Tuljapur-1	Ts	Ts 298		ts 94	3TS:1ts	0.21	0.70–0.50

[a]T = Tolerant to iron chlorosis. MT = Moderately tolerant to iron chlorosis. S = Susceptible to iron chlorosis.

IET 7613 (short-statured seedlings) and Tuljapur 1 (tall-statured seedlings) with talls remaining dominant over short ones. The gene symbol *Ts* was tentatively used for tall seedlings. Pigment was absent in the coleoptile of parents of the cross IET 7614 × WBPH 25 but appeared in the F_1 and F_2, segregated in the ratio of 9 purple coleoptile to 7 green coleoptile, indicating the interaction of two dominant complementary genes tentatively designated as Pc_1 and Pc_2.

The F_3 breeding behaviour (Table 3) confirmed the phenotypic ratio assumed in the F_2.

Of the crosses studied for the capacity of Fe^{3+} uptake, IET 7613 (MT) × Tuljapur 1 (T) segregated in the ratio of 15 high to 1 low Fe^{3+} uptake thereby showing duplicate dominant genes to govern high Fe^{3+} uptake capacity (Table 4). F_2 of IET 7614(S) × WBPH25(MT) segregated in a ratio of 3 high to 13 low Fe^{3+} uptake capacity. The digenic mode of inheritance here involved an inhibitory gene (Table 4) carried by the susceptible parent IET 7614.

Overall analysis of the inheritance of tolerance to iron deficiency as understood from the various cross combinations tends to suggest the existence of two basic genes, viz. Ic_1, Ic_2, which are not nonallelic and act in association with two complementary genes, viz. Ic_3, Ic_4. The basic gene Ic_1 in associated with its complementary gene Ic_3 confers tolerance, while Ic_2 with Ic_4 confers moderate tolerance. Susceptibility appears to be either due to an inhibitory gene or recessive state of both the basic genes (ic_1, ic_2). The complementary genes Ic_3 and Ic_4 also appear to be nonallelic as is evident from their lack of expression when either of the basic genes Ic_1 or Ic_2 is absent. More study involving tolerant/moderately tolerant and susceptible genotypes, however, is necessary to confirm the allelic relationships and genetic background.

On the basis of the mode of inheritance of the various genotypes, genotypic constitution of the tolerant, moderately tolerant and susceptible parents has been assumed and gene symbols are tentatively assigned as follows:

Tolerant

ARC 10372, Cauvery: $Ic_1\ Ic_1\ ic_2\ ic_2\ Ic_3\ Ic_3\ ic_4\ ic_4$

Tuljapur 1: $Ic_1\ Ic_1\ ic_2\ ic_2\ Ic_3\ Ic_3\ Ic_4\ Ic_4$

Table 3. F_3 breeding behaviour in 96 families for iron chlorosis tolerance and other characteristics

Cross	F_3 families segregating in ratio[a]										P value
	BTD T/MT	3:1 T/MT:S	3:13 T:S	1:3 T:S	BTR	15:1 T:S	12:3:1 T:MT:S	48:9:7 T:MT:S	9:7 T:S	χ^2	
Tolerance to iron chlorosis											
IET 9829/Cauvery (Exp. on 1:2:1)	20	53			23					1.22	0.70–0.50
IET 7614/ARC 10372 (Exp. on 7:2:4:2:1)	36	13	28	9	10					4.94	0.30–0.20
IET 7614/WBPH 25 (Exp. on 7:2:4:2:1)	41	13	20	18	4					4.44	0.50–0.30
ARC 5723/Prasanna (Exp. on 1:2:1)	19	60			17					5.40	0.10–0.05
Akashi/ARC 5723 (Exp. on 1:2:1)	21	49			26					0.48	0.80–0.70
IET 7613/Tuljapur-1 (Exp. on 7:4:1:4)	40	20			8	28				7.37	0.10–0.05
IET 7614/ARC 10372 (Exp. on 19:19:5:9:8:4)	28	29			9		14	10	6	3.76	0.70–0.50
Prasanna/Tuljapur-1 (Exp. on 7:4:1:4)	37	20			12	27				7.03	0.10–0.05
Purple pigmentation in the coleoptile											
IET 7614/WBPH 25 Exp. on 1:4:7:4)	3	20			37				36	6.82	0.10–0.05
Tall seedling											
IET 7613/Tujapur-1 (Exp. on 1:2:1)	30	46			20					2.24	0.50–0.30

[a]T: Tolerant to iron chlorosis; MT: Moderately tolerant; S: Susceptible; BTD: Breeding true for dominant character; BTR: Breeding true for recessive character.

Table 4. Inheritance of Fe^{3+} uptake capacity of parents, F_1 and F_2 seedlings

		Cross		F_1	No. of F_2 seedlings		Ratio	P value
					Tolerant	Susceptible		
		IET 7613/Tuljapur-1[a]		T	94	5	15:1	0.70–0.50
[c]Fe uptake capacity (mg) × 10^{-2}	Mean	3.12	5.12	8.11	6.83	1.27		
	Range	2.52–3.72	4.69–9.82	6.40–9.82	4.69–9.82	0.85–1.70		
		IET 7614/WBPH 25[b]		S	15	84	3:13	0.50–0.30
Fe uptake capacity(mg) × 10^{-2}	Mean	5.97	11.38	5.54	12.72	5.50		
	Range	5.12–6.83	9.39–13.66	5.14–6.87	9.30–13.69	5.10–5.98		

[a]7-day-old seedlings.
[b]10-day-old seedlings.
[c]Values/seedling (Formulae Jackson, 1967).

Moderately tolerant:
WBPH 25, Akashi
Prasanna and IET 7613: $ic_1\ ic_1\ Ic_2\ Ic_2\ Ic_3\ Ic_3\ Ic_4\ Ic_4$

Susceptible:
IET 9829: $ic_1\ ic_1\ ic_2\ ic_2\ Ic_3\ Ic_3\ ic_4\ ic_4$

ARC 5723: $ic_1\ ic_1\ ic_2\ ic_2\ Ic_3\ Ic_3\ Ic_4\ Ic_4$

IET 7614: $I\text{–}Ic_1\ I\text{–}Ic_1\ I\text{–}Ic_2\ I\text{–}Ic_2\ ic_1\ ic_1\ ic_2\ ic_2\ Ic_3\ Ic_3\ Ic_4\ Ic_4$

Identical modes of inheritance observed for

high Fe^{3+} uptake capacity and tolerance to iron chlorosis in the crosses studied indicate that the gene(s) governing the two indices could be one and the same. A highly significant negative correlation between Fe^{3+} uptake capacity and field score value for iron chlorosis reported by Hoan (unpublished, 1989) lends support to such an inference, but it warrants more intensive study, considering the occurrence of Fe deficiency-tolerant genotypes with low root Fe^{3+} uptake capacity and vice versa (Hoan, unpublished, 1989). It is quite possible that in addition to Fe uptake capacity, there could be other factors such as rooting depth, iron content in seed, Fe utilisation efficiency, etc., directly or indirectly conferring tolerance to the deficiency.

Römheld and Marschner (1986), working with barley, have reported that roots of graminacious species respond to iron deficiency through increased release of chelating substances (phytosiderophores) which are highly effective in solubilizing sparingly soluble inorganic Fe^{III} compounds and in the formation of Fe^{III} phytosiderophores. The role of phytosiderophores in iron uptake has been studied in a variety of graminacious crop species including maize, oats (Brown et al., 1991) and sorghum (Clark et al., 1988). Mori et al. (1991), after a detailed study, reason that high susceptibility of rice seedlings to iron deficiency was due to low secretion of deoxy-mugineic acid (MA), a phytosiderophore. They claim that when iron deficiency continues, root tips become chimeric and epidermal cells necrotic. As a result, section of deoxy-MA from the roots is diminished and activity of the transporter, which absorbs deoxy-MA Fe^{III} chelate, is reduced affecting finally the synthesis of deoxy-MA from methionine. Consequently, depletion of Fe^{II} in the shoot is induced, resulting in severe chlorosis in young rice plants. Present genetic study has relevance to the above findings, in that it is possible that secretion of phytosiderophores, level of Fe uptake and its transport to leaf area where it is utilised in the tolerant genotypes might be governed by different sets of dominant genes. This needs detailed investigation.

Alcantara et al. (1990) proposed different types of genetic control in sunflower depending on the cultivars, species and test conditions used to evaluate the characters, and indicated two pairs of genes controlling the trait. Weiss (1943) in soybean and Wann and Hills (1973) in tomato reported single recessive genes for iron chlorosis susceptibility, while Coyne et al. (1982) proposed two major genes for resistance to iron chlorosis in dry beans. In sorghum, inheritance of iron efficiency was reported to be a quantitative trait (Rodriguez de Cianzo, 1991).

Linkage relationship

With the objective of identifying key marker genes to distinguish tolerant/moderately tolerant from susceptible ones at the seedling phase,

Table 5. Joint segregation of F_2 phenotypes for iron chlorosis tolerance with morphological characters in two crosses

Cross		Purple coleoptile Fe chlorosis		Green coleoptile Fe chlorosis		χ^2	P	C.O.%
		M. Tolerant	Susceptible	M. Tolerant	Susceptible			
IET 7614 × WBPH 25								
Expected on	Observed	83	326	40	294			
27:117:21:91	Expected	78.36	339.57	60.95	264.11	11.40[a]	.02–.01	15.26
Cross		Tall seedlings Fe chlorosis		Dwarf seedlings Fe chlorosis		χ^2	P	C.O.%
		Tolerant	Susceptible	Tolerant	Susceptible			
IET 7613 × Tuljapur-1								
Expected on	Observed	336	1	66	33			
45:3:15:1	Expected	306.56	20.44	102.19	6.81	134.85[a]	0.001	1.75

[a]Significant at 5% level.

linkage relationships of two seedling characters, viz. coleoptile pigmentation and seedling height with iron chlorosis were studied in two of the crosses, viz. IET 7614 × WBPH 25 and IET 7613 × Tuljapur 1. Joint segregation analysis showed one of the two genes conferring purple colour to coleoptile (*Pc*) in 10-day-old seedlings to be linked with the gene for moderate tolerance (Ic_2) to iron chlorosis with a crossover value of 15.26% in the coupling phase. Similar analysis of seedling height and iron chlorosis done in 30-day-old seedlings of the other cross suggested gene *Ts* (tall seedling) to be very closely linked to the basic gene Ic_1, the recombination value being 1.7% in the coupling phase (Table 5). The genes may belong to linkage group II, since the genes for pigmentation in the leaf sheath are already located in this group (Kinoshita, 1985). Although seedling markers of this kind would be of help to breeders in rapidly screening segregating populations for tolerance to iron chlorosis, it is desirable to study the association in more diverse, and a greater number of crosses, before the markers are widely used as reliable indirect selection indices.

References

Alcantara E, Fernandoz M and De La Guardia M D 1990 Genetic studies on the acidification capacity of sun flower roots induced under iron stress. Plant and Soil 123, 239–241.

Brown J C, Holmes R S and Tiffin L O 1961 Iron chlorosis in soyabeans as related to the genotype of root stalk chlorosis susceptibility and reductive capacity at the root. Soil and Science 91, 127–132.

Brown J C, Jolley Von D and Lyle Mcl C 1991 Comparative evaluation of iron solubilizing substances (phytosiderophores) released by oats and corn: Iron-efficient and iron-inefficient plants. Plant and Soil 130, 157–163.

Clark R B, Römheld V and Marschner H 1988 Iron uptake and phytosiderophore release by roots of sorghum genotypes. J. Plant Nutr. 11, 663–676.

Coyne D P, Korban S S, Knudsen D and Clark R B 1982 Inheritance of iron deficiency in crosses of dry beans (*Phaseolous vulgaris* L.). J. Plant Nutr. 5, 575–585.

De Kock P C 1971 Fundamental aspects of iron nutrition of plants *In* Proceedings of Conf. of Trace Elements in Soils and Crops. London 1966. pp 41–44. H M 50, London.

Jackson M L 1967 Absorption spectrophotometry. *In* Soil Chemical Analysis. Published by Prentice Hall of India Pvt. Ltd., New Delhi. 49 p.

Kinoshita 1985 Standardisation of gene symbols and linkage maps in rice. *In* Rice Genetics Proceed. Int. Rice Genet. Symp. pp 215–227. Int. Rice Res. Inst. P.O. Box No. 933, Manila, Philippines.

Misal M B and Nerkar Y S 1983. Genotypic variation for iron reductive capacity of rice root. Cereal Res. Comm. 11, 291–292.

Mori S, Naoko Nishizawa, Hayashi H, Chino M, Yoshimura E and Ishihara J 1991 Why are young rice plants highly susceptible to iron deficiency? Plant and Soil 130, 143–156.

Richharia R H, Ghosh A K, Prakasa Rao S V S and Misro B 1966 Formulae for the estimation of linkage from F_2 data. Central Rice Res. Inst., Cuttack-753006, Orissa, India. Bull (2), 35 p.

Rodriguez de Cianzo S R 1991 Recent advances in breeding for improving iron utilization by plants. Plant and Soil 130, 63–68.

Römheld V and Marschner H 1986 Evidence for a specific uptake system for iron phytosiderophores in roots of grasses. Plant Physiol. 80, 175–180.

Tanaka A and Yoshida S 1970 Nutritional disorders of the rice plant in Asia. Int. Rice Res. Inst. Bull (10), 51 p.

Wann I V and Hills W A 1973 The genetics of boron and iron transport in the tomato. J. Heredity 64, 370–371.

Weiss M G 1943 Inheritance and physiology of efficiency in Iron utilization in soybeans. Genetics 28, 253–268.

P. J. Randall et al. (Eds.), Genetic aspects of plant mineral nutrition, 335–339.

PLSO SV79

Genotypic variation among Indian graminaceous species with respect to phytosiderophore secretion

KALYAN SINGH[1], M. CHINO[2], N.K. NISHIZAWA[2], T. OHATA[2] and S. MORI[2]
[1]*Department of Agronomy, Institute of Agricultural Sciences, Banaras Hindu University, Varanasi - 221 005, India and* [2]*Laboratory of Plant Nutrition and Fertilizer, The University of Tokyo, Bunkyo-Ku, Tokyo-113, Japan*

Key words: barley, gramineae, iron deficiency, maize, mugineic acid, oats, phytosiderophores, rye, sorghum, wheat

Abstract

Graminaceous plants can acquire Fe from sparingly soluble inorganic Fe (III) compounds through root secreted phytosiderophores by chelation.

Growth chamber experiments in nutrient solution were conducted with rye, maize, sorghum, barley, oat and two wheat cultivars. The root secretion of these Fe stressed cultivars were subjected to HPLC examination of mugineic (MA) acid and analogues (MAs). All the cultivars examined secreted deoxymugineic acid and this was an important component of the MAs pool in all species except oat where avenic acid predominated. These preliminary results also suggested that the species tested varied in quantity and composition of mugineic acid secretion.

Introduction

Chlorosis due to Fe stress is a widespread mineral deficiency condition in higher plants grown on calcareous soils (Vose, 1982). Poor availability of Fe in the soil, insufficient uptake and Fe inactivation within the plants are the main causes of Fe chlorosis in crops grown on such soils (Mengel and Geurtzen, 1986). The use of soil amendments and foliar sprays of Fe salts to overcome Fe deficiency are not economical (Hagstrom, 1984; Wallace and Wallace, 1982). Therefore, long term alternative methods to alleviate Fe deficiency chlorosis are desired.

Chlorosis resistance correlates closely with quantitative differences in Fe acquisition (Marschner et al., 1986; Mori, 1987; Römheld, 1987; Takagi, 1984). In Fe efficient dicotyledonous plants subjected to Fe deficiency, the stress response may act in a number of ways e.g. release of hydrogen ions or reducing compounds (phenolics) by roots which maintain and mobilize Fe (II); reduction of Fe (III) to Fe (II) and accumulation of organic acids particularly citric acid in the roots which chelate and transport Fe to the plant tops (Brown et al., 1991).

In contrast, graminaceous plants have a different mechanism for Fe acquisition, characterized by root-secreted Fe-mobilising mugineic acid (MA) and related compounds (MAs) in response to Fe deficiency chlorosis (Mino et al., 1983; Takagi et al., 1984). These are similar to siderophores derived from micro-organisms (Neilands, 1981). The system (Römheld, 1991) for ferrated phytosiderophores in highly specific and located in the apical root zone (Marschner, et al., 1986, 1987; Römheld, 1987). These Fe complexes are directly transported without a reduction step by a plasma membrane bound protein transporter (Mihashi and Mori, 1989). Fe acquisition is also influenced by the degree of Fe deficiency, growth rate and root morphology (Singh et al., 1991).

In general, the quantity and composition of

the phytosiderophore pool of MAs released by graminaceous species and their cultivars varies considerably (Kawai et al., 1988; Römheld, 1991). Chlorosis resistant species such as barley, wheat and rye are characterized by a high release rate of MAs in Fe deficient conditions (Kawai et al., 1988), whereas, chlorosis sensitive species such as upland rice (Mori et al., 1991; Takagi, 1976) and sorghum (Kawai et al., 1988), have a low rate. Maize is moderately chlorosis resistant and has an intermediate MAs release rate (Kawai et al., 1988; Mori et al., 1987). MAs release by oat cultivars generally follows the same pattern as maize (Brown et al., 1991; Jolly and Brown, 1989; Lytle et al., 1990). However little is known about the quantity and composition of MAs secretion by Indian graminaceous crops. With this objective, the present study was undertaken with Indian crop cultivars which are currently under cultivation.

Materials and methods

Plant cultivars used for experimentation were rye (*Secale cereale* L.) cv Russian rye, maize (*Zea mays* L.) cv. Deccan 103, sorghum (Sorghum bicolor L.) cv PKSS-6, barley (*Hordeum vulgare* L.) cv NDB 209, oat (*Avena sativa* L.) and two cultivars of wheat (*Triticum aestivum*) cv. KRL-1-4 and HD 2329. Seeds of rye were obtained from Vivekanand Research Institute (ICAR), Almorah; cultivars NBD 209 and KRL-1-4 were obtained from Central Soil Salinity Research Institute (ICAR), Karnal. Other crop cultivars were obtained from Indian Agricultural Research Institute (ICAR), New Delhi.

Plant culture condition

Seeds were germinated on a distilled water soaked paper towel in a tray covered with aluminium sheet at room temperature (20°C) for two days. After germination, these were transferred to a plastic net floating on tap water at pH 5.5. The growth cabinet temperature was maintained at 20°C/light and 15°C/dark with 75% humidity. One week old germinated seedlings were transplanted into a 5 L plastic culture box (18 × 12 × 15 cm in height) covered with a plastic lid. There were 6 plants/box supported with foam in each of the 3 holes in the plastic lid. The culture solution contained (mg L)$^{-1}$, N 10, P 5, K 10, Ca 10, Mg 10, B 0.5, Mn 0.05, Mo 0.05, Zn 0.05, Cu 0.02 and Fe 1.0 supplied as $NaNO_3$, KH_2PO_4, KCl, $Ca(NO_3)_2$, $MgSO_4$, H_3BO_3, $MnCl_2$, $NaMoO_4$, $ZnSO_4$, $CuSO_4$ and Fe citrate, respectively. The solution was made up in deionized redistilled water. The nutrient solution was initially half strength and was gradually increased to full strength. Solution pH was adjusted to 5.5 every day. The deficiency treatments were started 10 days after transplanting by changing to nutrient solution without Fe. After imposition of the deficiency treatment, pH was not adjusted but remained above 7 probably due to preferential absorption of NO_3. A high pH was also favourable to enhance chlorosis appearance. Water culture solutions were changed twice a week. Foam sheets were also changed when needed to avoid dust contamination without disturbing the roots.

Collection of root washings

Depending on species, root washing began 18–29 days after the start of the deficiency treatment when symptoms appeared on the plant tops. Root washings were collected just after daylight by soaking roots in deionized distilled water (31/6 plants for 3 hr at 25°C). The plastic container used was cleaned by soaking in 1*N* HCl (Takagi, 1976). After collection, 1 g thymol was mixed with each sample to prevent microbial degradation of MAs. Root washings were repeated thrice at intervals.

Column chromatography

Root washings were filtered through double layered Advantec 5A and 5C filter paper successively. The extracts were put on a column of 100 g Amberlite 1R-120(H^+) (60 cm length × 3 cm diameter), washed with deionized water and eluted with 2*N* NH_4OH to release MAs. The elute was evaporated under vacuum and the residue was dissolved in 1 ml deionized water, filtered through a 0.45 mμ filter (Millipore) and then 0.1 to 1.0 μLof the filtrate was injected into an HPLC as described by Mori and Nishizawa, (1987).

Results and discussion

Graminaceous plants under Fe deficiency chlorosis release various MAs (Fig. 1) for which methionine is a precursor of biosynthesis and nicotianamine is an intermediate (Mori et al., 1987, 1990; Shojima et al., 1989). It has been further observed that deoxymugineic acid (DMA) distinctly dominates in most of the gramineae genotypes (Kawai et al., 1988; Mori et al., 1990). In this present study, the various genotypes also varied markedly in the composition as well as quantity of MAs secreted (Table 1). DMA was a major contributor to the MAs pool in all species except oat where avenic acid (AA) predominated. In barley MA was the major form and in rye hydroxy mugineic acid (HMA) made a substantial contribution. The quantity of MAs can be ordered as barley = wheat > rye > oat > maize > sorghum. Sorghum

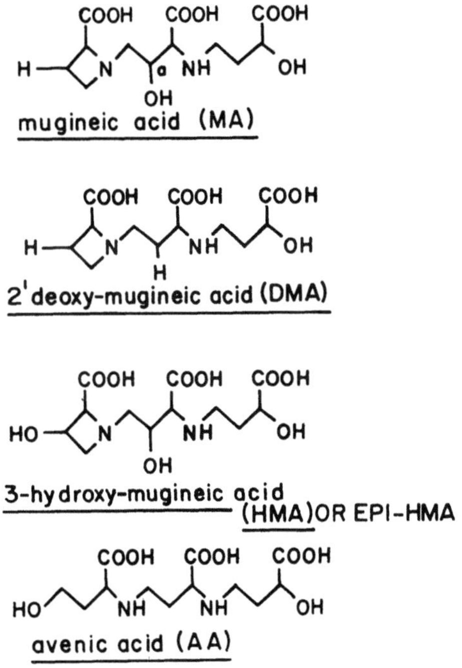

Molecular structure of MA,DMA,HMA,and AA

Fig. *1.* Molecular structure of mugineic acid, 2-deoxy-mugineic acid, 3-hydroxy-mugineic acid and avenic acid.

Table 1. Qualitative and quantitative analysis of mugineic acid family phytosiderophores (MAs) from Indian gramineae plants

Crop cultivar	Days after transplanting	Secreted MAs (μg/6 plants/3 hrs)				
		Epi-HMA	HMA	MA	DMA	A
Maize	28	–	–	–	51	1
Deccan 103	35	–	–	–	–	–
	40	–	–	–	–	–
Barley	30	–	–	1190	297	1
NBD-209	35	>1	–	1640	678	–
	40	–	–	1860	602	21
Oat	31	>1	–	–	–	306
PO-3	35	3	–	–	12	186
	45	–	–	–	–	815
Russian	36	154	257	110	641	–
rye	39	255	285	143	1160	–
	42	291	503	90	109	56
Sorghum	39	2	8	–	12	2
PKSS-6	42	3	2	–	3	–
	45	>1	13	1	13	1
Wheat	32	–	–	–	2830	2
KRL-1-4	39	–	–	20	1730	–
	42	–	–	–	3350	1
Wheat	32	–	–	–	2796	–
HD 2329	35	–	–	–	2349	–
	40	–	–	–	2420	–

– No MAs detected.

Table 2. Fresh top and root weight (g) of different crop genotypes at 45 days after transplanting

No.	Plant species	Fresh top weight in g/6 plants	Fresh root weight in g/6 plants	Total weight in g/6 plants
1	Maize (Deccan 103)	27.2	11.8	39.0
2	Barley (NDB-209)	51.2	48.0	99.2
3	Oat (PO-3)	39.5	19.1	58.6
4	Rye (Russian Rye)	37.1	29.1	66.2
5	Sorghum (PKSS-6)	10.0	6.0	16.0
6	Wheat (KRL-1-4)	41.5	31.1	72.6
7	Wheat (HD-2329)	39.5	29.1	68.6

is also the most susceptible species to Fe deficiency. Table 2 presents the top and root weights of plants. Clearly total MAs secretion is influenced by the root weight of the plants.

Genetic studies are now required on plant mugineic acid family phytosiderophores as they may prove valuable in breeding for more Fe tolerant crop genotypes. Selection for secretion of MAs could be more effective on Fe deficient calcareous soils because Fe acquisition through the Fe (III) MAs system is less affected by bicarbonate which results in high chlorosis on wet calcareous soils. For such conditions, selection based on other characteristics such as Fe deficiency induced increased in H^+ ion efflux (Kannan, 1980) and enhanced Fe (III) reduction at the root surface (McKenzie et al., 1985) would appear to be less promising. However, classification of graminaceous crops and crop cultivars should be interpreted with utmost care due to distinct diurnal patterns of MAs secretion (Takagi et al., 1984), increase (Kawai et al., 1984) or decrease (Clark et al., 1988) of MAs production with chlorosis index, enhanced biosynthesis of MAs in the roots (Mihashi and Mori, 1987), microbial degradation of MAs and Fe MAs by pseudomonas (Wada et al., 1987; Watanabe and Wada, 1989) and absorption demand of Fe (III) on the negatively charged groups of plasma membrane phospholipids (Mihashi et al., 1991).

Acknowledgement

We thank the Japanese Society for the Promotion of Science, Ministry of Education, Government of Japan, for supporting the research project by awarding a JSPS Fellowship to the senior author.

References

Brown J C, Jolly Von D and Lytle C 1991 Comparative evaluation of iron solubilizing substances (phytosiderophores) released by oats and corn: Iron efficient and iron-inefficient plants. Plant and Soil 130, 157–163.

Clark R B, Römheld V and Marschner H 1988 Iron uptake and phytosiderophore release by roots of sorghum genotypes. J. Plant Nutr. 11, 663–676.

Hagstrom G B 1984 Current management practices for correcting Fe deficiency in plants with emphasis on soil management. J. Plant Nutr. 7, 23–46.

Jolly V D and Brown J C 1989. Iron inefficient and efficient oat cultivars. II. Characterization of phytosiderophore released in response to Fe deficiency stress. J. Plant Nutr. 12, 923–937.

Kannan S 1980 differences in iron stress response and iron uptake in some sorghum varieties. J. Plant Nutr. 2, 347–358.

Kawai S, Takagi S and Sato Y 1988 Mugineic acid-family phytosiderophores in root secretions of barley, corn and sorghum varieties. J. Plant Nutr. II, 633–642.

Lytle C Mel, Jolly Von D and Brown J C 1991 Iron-efficient and iron-inefficient oats and corn respond differently to iron deficiency stress. Plant and Soil 130, 165–172.

Marschner H, Römheld V and Kissel M 1986 Different strategies in higher plants in mobilization and uptake of iron. J. Plant Nutr. 9, 665–673.

Marschner H, Römheld V and Kissel M 1987 Localization of phytosiderophore release and iron uptake along intact barley roots. Physiol. Plant. 71, 157–162.

McKenzie D B, Hossner L R and Newton R J 1985 Root reductant release as a measure of sorghum cultivar iron efficiency. J. Plant Nutr. 8, 847–857.

Mengel K and Geurtzen G 1986 Iron chlorosis on calcareous soils Alkaline nutritional condition as the cause for the chlorosis. J. Plant Nutr. 9, 161–173.

Mihashi S and Mori S 1989 Characterization of mugineic – Fe transporter in Fe-deficient barley roots using the multicompartment transport box method. Biol. Metol. 2, 146–154.

Mihashi, S, Mori S and Nishizawa N K 1991 Enhancement of ferric mugineic acid uptake by iron deficient barley roots in the presence of excess free mugineic acid in the medium. Plant Soil 130, 135–141.

Mino Y, Ishida T, Ota N, Inoue M, Nomoto K, Takemoto T, Tanaka H and Suguira Y 1983 Mugienic acid iron (III) complex and its structurally analogous cobalt (III) complex: Characterization and implication for absorption and transport of iron in graminaceous plants. J. Am Chem. Soc. 105, 4671–4676.

Mori S 1987 Physiology and biochemistry of muigenic acid. Nihon Nogeikagaku Kaishi 61, 389–394 (*In Japanese*).

Mori S and Nishizawa N K 1987 Methionine as a dominant precursor of phytosiderophores in gramineae plants. Plant Cell Physiol. 28, 1081–1092.

Mori S, Nishizawa N K and Fujigaki J 1990 Identification of rye chromosome 5R as a carrier of the genes for mugineic acid synthetase and 3. hydroxymugineic acid synthetase using wheat-rye addition lines. Jpn. J. Genet. 65, 343–352.

Mori, S, Nishizawa N K, Hayashi H, Chino M, Yoshimura E and Ishihara I 1991 Why are young rice plants highly susceptible to iron deficiency. Plant and Soil 130, 143–156.

Mori S, Nishizawa N K, Kawai S, Sato Y and Takagi S 1987 Dynamic state of mugineic acid and analogous phytosiderophores in Fe deficient barley. J. Plant Nutr. 10, 1003–1011.

Neilands J B 1981 Iron absorption and transport in microorganisms. Annu. Rev. Nutr. 1, 27–46.

Römheld V 1987 Different strategy for iron acquisition in high plants. Physiol Plant 70, 231–234.

Römheld V 1991 The role of phytosiderophores in acquisition of iron and other micronutrients in graminaceous species: An ecological approach. Plant and Soil 130, 127–134.

Römheld V and Marschner H 1987 Evidence for a specific uptake system for iron phytosiderophores in roots of grasses. Plant Physiol. 80, 175–180.

Shojima S, Nishizawa N K and Mori S 1989 Establishment of a cell free system for the biosynthesis of nicotianamine. Plant Cell Physiol. 30, 673–77.

Singh K, Chino M, Nishizawa N K, Goto S, Tomoko N, Takagi S and Mori S 1991 Iron extraction efficacy of plant borne mugineicacid family phytosiderophores in Indian calcareous soils. Paper presented in the VI Fe symposium held at Logan, Utah, USA from 14–19 July, 1991.

Takagi S 1976 Naturally occuring iron chelating compounds in oat and rice root washings. Soil Sci. Plant Nutr. 12, 423–433.

Takagi S 1984 Regulation mechanism of iron absorption in plant roots and its difference in plant species *In* Agriculture and Soil Science and Plant Nutrition in North Part of Japan Ed. Nippon Dojo Hiryo Gakkai. pp 190–195. (*In Japanese*).

Takagi S, Nomoto K and Takemoto T 1984 Physiological aspect of mugineic acid, a possible phytosiderophore of gramineous plants. J. Plant Nutr. 7, 469–477.

Vose P B 1982 Iron nutrition in plants: A world overview. J. Plant Nutr. 5, 233–49.

Wada H, Watanabe S and Kimura M 1987 Isolation of mugineic acid decomposing bacteria. Abstracts of the 1987 meeting. J. Soc. Soil Sci. Plant Nutr. 33, 282.

Wallace A and Wallace G A 1982 The enigma of synthetic metal chelates. *In* Plant Nutrition. Vol. 2. Ed. A. Scaife. pp 696–700. Agricultural Bureaux, Wallingford, Oxon, U K.

Watanabe S and Wada H 1989 Mugineic acid decomposing bacteria isolated from rhizosphere of iron deficient barley. Jpn. J. Soil. Sci. Plant Nutr. 60, 413–417. (*In Japanese*).

P. J. Randall et al. (Eds.), Genetic aspects of plant mineral nutrition, 341–348

PLSO SV84

Requirement and response of crop cultivars to micronutrients in India – a review

P. N. TAKKAR
Indian Institute of Soil Science, Bhopal-462 011, India

Key words: micronutrient deficiencies, Zn, Fe, Mn, B, cultivar differences, wheat, rice, maize, sorghum, finger millet, pearl millet, chickpea, lentil, pigeon pea, mustard, cotton, potato

Abstract

Micronutrient deficiencies are posing serious constraints to agricultural productivity on many soils in India. Research is being conducted to identify cultivars tolerant to micronutrient stresses as a means of overcoming these constraints. This review describes studies to identify and characterise such cultivars of wheat, rice, maize, sorghum, finger millet, pearl millet, chickpea, pigeon pea, lentil, mustard, cotton and potato with respect to their tolerance to Zn, Fe, Mn and B deficiencies.

Review

Micronutrient deficiencies have been emerging as a serious problem in many intensively cultivated soils of India and have become one of the serious constraints to productivity (Takkar et al. 1989, Takkar, 1991). To combat these constraints research has been undertaken in the State agricultural universities and research institutes of the ICAR under a coordinated scheme at 8 selected centres involving work with Zn, Fe, B and Mn in a range of crop species (Takkar et al. 1989). As correcting deficiencies through application of micronutrients in fertilizer is an expensive and not always effective technique, research is being conducted to select and/or develop cultivars tolerant to micronutrient stresses in order to increase production on deficient soils either without or with minimum micronutrient fertilization. Most of the studies have been conducted under conditions of micronutrient stress in the field or glass or screen house and the information obtained is reviewed in this paper.

Zinc

Zn deficiency is widespread in India. Nearly 46 per cent of the 113075 soil samples collected in the ICAR coordinated project, since its inception in 1967, were found to be low in Zn (Takkar, 1991). Most of the research in the coordinated scheme has been focussed on Zn. Cultivars of a large number of crops have been screened in sand culture, soil culture and under field conditions. In most of the individual studies, only a very limited number of cultivars of a crop has been screened. Their ratings into tolerant (non-responsive) or susceptible (responsible) groups have largely been done on the basis of: (1) degree of manifestation of visual deficiency symptoms of the element, (2) degree of response to applied micronutrient, and (3) depression in yield or absolute yield under micronutrient stress conditions. Using these criteria, cultivars of rice, wheat, maize, barley, sorghum, finger millet, pearl millet, chickpea, pigeon pea, lentil, peanut, raya (mustard), cotton and potato have been rated with respect to their tolerance to Zn deficiency. Other parameters such as low P/Zn ratio (Gupta and Hans Raj, 1983; Shukla and Hans Raj, 1976; Singh and Choudhary, 1987) and high root cation exchange capacity (CEC) and uptake of Zn (Verma, 1986) have been used in categorising the relative tolerance of cultivars to Zn deficiency (Table 1). The practical outcome of this work has been to identify tolerant

Table 1. Relative tolerance of cultivars of 16 crop species to zinc deficiency

	Cultivar/line[a]	Criteria[b]
	Rice	
Prasad et al. 1982	Sita (103), Jaya, Jayanti, IR8, IR20, Archana (3)**	A
Verma 1986	Sita (245), Rajendra Dhan (161), Pusa (83), Ratna (63), UPR 238 (30)**	B, C, G
Singh and Sakal, 1987	Pusa 33, Pusa 2-21, ES29-5-3, RAU 4004-127, IR 25890-82-5-3, NC 1626**, IET 2379**, IET 6148**, IET 7614**	A, C
Takkar et al. 1989	IR8 (59), Cul. 13493 (35), IR 62 (28), Bhawani (20), TNAU 658 (20), IR 60 (15), RP4 14 (10) CU 387 (10), IR 20 (9), ADT 36 (7), ASD 16 (6)*, CO 44 (4)*, Vatgai (4)*, Co 39 (3)*, TNAU 831520 (2)**, IR69 (2)**, Ponni (2)**, TNAU 801793 (−1)**, TNAU 801790 (−1)**, TNAU 831521 (−3)**	
Takkar et al. 1989	Jaya (113), Palman (25), Ratna (24), HM-95 (8)*, IR-8 (7)*, Padma (7)*, HM-484 (3)**, Barmati (3)**, HM-474 (2)**, Jhona 349**	A
Takkar et al. 1989	Jaya (15), C 24263 (7)**, C 13206 (6)**, C 7306 (6)**	
Takkar et al. 1989	Maduri (72), Kranti (19), Purua (12)*	
	Wheat	
Shukla and Hans Raj, 1974	UP 310, Sonalika, WG 357, WL 344, WG 377, HD 1977**, Kalyansona**	D
Brar et al. 1978	WL 334 (42), WG 377 (29), Sonalika (26), UP 301 (20), WG 357 (16)**, C 306 (5)**, Kalyansona (−1)**	
Sakal et al. 1984a	HP 1102 (25), UP 262, HP 1209, NP 852, HUW 12 (4)**, Sonalika (1)**	C
Gupta et al. 1984	WH147, HD 2009, WH 157, Kalyansona, MSML 7040, KSML 3**	C, E
Yadav and Vyas, 1987	Kalayansona, HD 2122, Raj 911 (26), Raj 821 (13)**	
Takkar et al. 1989	WG 357 (15), WG 377 (12), WL 711 (10)*, HD 2009 (9)*, PV 18 (9)*, Kalyansona (4)**	
Takkar et al. 1989	HD 1553 (32), WH 147 (22), TAWA (13)*, LOK-1 (8)*, HD 2236 (4)**	
	Maize	
Shukla and Hans Raj, 1976	Ganga 3, Ganga 5, HIM 123, KT 41, Vikram, Vijay, Amber, Ganga**, Safed-2**	A, B, E, F
Sinha and Singh, 1977	Ganga 4 (49), Sweet corn, J1 (W), Vijay, M2 Comp, Ganga 5, Hi-starch, M1 comp, GS 2, Pop corn (2)**	
Singh et al. 1983	Lakshmi (96), Suwan-1 (Yellow) (79), Keshari (65), Suwan-1 (White) (37), Obregaon (35)**	A, C
Takkar et al. 1989	Vijay (192), Ganga (154), HIM 123 (158), Ganga-5 (151), Kisan (140), Vikram (94), Talwandi local (72), JML 22 (64), Kalachal (37)**	
Takkar et al. 1989	Warangal local (90), Ganga 5 (59), Camp B VI (56), Syn B 21 (47), DMM 101 (43), Syn B 28 (41), Syn B 19 (16)**, H-3 (14)**	
	Barley	
Pathak et al. 1979	K 125-57, K 851-28, Ratna, Amber, Vijay, K 5451, Jyoti, K 24, K 853-28**, K 825-28**	C, E
Takkar et al. 1983	PL 101 (15), BG 25 (9), DL 120 (8), P 267 (4)** DL 70 (3)**, PL 56 (2)**	
	Sorghum	
Shukla et al. 1973	JS 263, CSH-1, 29/1, JS 20, 3677 × 6352, Son 127 × SL 245, Son 3 × SL209, CSH-3**, 2077 × 115**	B, C
Patel and Patel, 1988	AS 11, AS 15, AS 22, MP Chari, SSG 59-3, S 1049**	

Table 1 (*contd.*)

	Cultivar/line[a]	Criteria[b]
	Finger millet	
Sakal et al. 1985b	RAU-8 (54), BR-407 (26), RAU-3 (24), RAU-2 (22), BR-2 (7)**, RAU-7 (2)**	C
	Proso millet	
Sakal et al. 1985a	RAU-M1 (57), RAU M-6 (48), MS-4872 (38), RAU M-2 (31), RAU M-5 (27), BR-7 (15)**	
	Pearl millet	
Takkar et al. 1988	PHB-3 (25), 76/32 (26), PHB-12 (21), 76/2 (19), PHB-14 (18), PHB-10 (10)**	
	Pigeon pea	
Shukla and Hans Raj, 1980	Pant A-1 (200), H-72-44 (148), T-21 (117), Pant A-2 (97), H-73-207 (60), Prabhat (63), Pant A-3 (49)**	
Sakal et al. 1984b	Prabhat (1122), ICPL-1 (982), ICPL-289 (688), ICPL-161 (179), ICPL-186 (568), ICPL-181 (511), UPAS-120 (454), ICPL-187 (190)*	
Takkar et al. 1989	TT-6 (26), D-11 (22), Pusa-1 (12)*, Pusa-2 (−1)**, Bahar (2)**, DA-6 (−4)**	
	Chick pea	
Singh and Choudhary, 1987	DG-82-16 (49), DG-82-1 (39), DG-82-13 (21), Pant G 114 (16), DG 82-3 (13)*, DG-82-17 (7)**	A, F
Takkar et al. 1989	H-355 (42), C-214 (26), H-208 (17), G-24 (13), G-130 (7)*, H-355 (5)*, S-26 (3)**, C-214 (2)**, C-235 (1)**	
Singh et al. 1984b	DHG-82-10 (46), BG-2037 (31), Pant 6-114 (12)*, DHG-82-4 (4)**, C-235 (−2)**, DHG-8-12 (−4)**	
	Lentil	
Singh et al. 1987	BR-25 (21), Pant 639 (19), RAU 101 (15), L-9-12 (9)*, Pant 406 (3)**, DL-77-2 (-2)**	C
	Peanut	
Gupta and Hans Raj, 1983	GE 90, MH 1, DH 3-20, Sm 1, Rabout 33-1, MH2	F
Nayyar et al. 1990	C-501 (171), AK-12-24 (61), F-3-1-5 (44), M-13 (22)*, T-64 (15)**, M-145 (15), C 148 (-5)**, C 87 (-13)**	
	Soybean	
Rathore et al. 1986	Punjab-1 (46), Black Soybean (22), JS-72-280 (21), JS-72-44 (17)**	
	Raya (mustard)	
Takkar et al. 1989	LGL-1 (28), RLM-218 (20), RS-3 (12), RLM 202 (9)*, RLM 234 (4)**, RL-18 (1)**, RLM 198 (-2)**	
	Cotton	
Takkar et al. 1989	LSS (19), G-27 (16), J-34 (13), J-207 (8), LH-37 (5)*, RS-89 (4)*, BN (1)**, LH-38 (2)**, H-4 (-1)**, J-205 (-11)**, 320F (-12)**	
	Potato	
Trehan et al. 1988	Kufri, Alankar, Kufri Chandramukhi, Kufri Sheetman, Kufri Sindhuri, Kufri Chamlakar, Kufri Jyoti, SLB/Z 569, Kufri Badshah**, Kufri Bahar**	

[a] The per cent yield response to zinc is given in parentheses where the information is available. Ratings: **tolerant, *moderately tolerant, and no superscript denotes cultivar susceptible to Zn deficiency.

[b] The letters A to G indicate which criteria, other than response, were used for rating the tolerance. A = free from visual Zn deficiency symptoms; B = high Zn uptake under stress conditions; B = higher Zn concentration/uptake in plant; D = higher efficiency in Zn utilization under Zn stress; E = low yield reduction under Zn stress; F = low P/Zn ratio; G = higher root cation exchange capacity.

cultivars with high yield potential which can be cultivated without or with only minimal Zn fertilization on Zn-deficient soils in the respective agro-ecological zones to which they are adapted. At the other end of the scale it has identified susceptible cultivars which require Zn application to obtain optimum productivity on Zn deficient soils.

Some studies have shown that fertilizer Zn requirements differ for different cultivars depending upon their susceptibility. Takkar and Nayyar (1984) reported that wheat cultivars WG 357 and PV 18 had a higher Zn requirement i.e. 11.2 kg ha^{-1} to reach their maximum yield on a Zn deficient soil compared to 5.6 kg ha^{-1} for WG 377 and WI 711. Data for other cultivars of wheat (Rathore et al. 1986) and rice (Sakal et al. 1984b) are presented in Table 2. The optimum Zn requirement of pearl millet cultivars was: 11.2 kg ha^{-1} for PHB 37, 76/32 and 76/2; 5.6 kg ha^{-1} for PHB 12 and PHB 14; and 2.8 kg ha^{-1} for PHB 10 (Takkar et al. 1988).

Among four cultivars of mothbean (*Phaseolus aconitifolius*) tested over 3 years, T-18 was the most tolerant as it gave the highest grain yields when grown under Zn stress conditions. T-18 possessed the ability to absorb more Zn than the other cultivars and had the highest tissue concentrations at 22.5 mg kg^{-1} compared to 17.0, 15.5 and 11.2 mg kg^{-1} in Jadia, T-3 and a local cultivar, respectively (Saxena and Mathur, 1987).

The cultivars which exhibited greater efficiency for P, Fe and Mn were generally relatively more susceptible to Zn deficiency. Also, the values of K_m (dissociation constant) and K_3 (rate constant) were higher for the susceptible wheat cultivar WL 334 than for the tolerant C 306 (Brar et al. 1978). Randhawa and Takkar (1976) reported that wheat variety WG 377 was less susceptible to Zn deficiency and had higher biologically active Zn (680 μg leaf $protein^{-1}$) and higher carbonic anhydrase activity (13.7 mg CO_2 g^{-1} dry weight) as compared with the corresponding values of 360 and 7.6 for a highly susceptible variety WG 334.

Iron

Twenty-eight cultivars of chick pea were screened for Fe efficiency and divided into green and slightly chlorotic genotypes, which were non-responsive to Fe, and severely chlorotic genotypes which were responsive. It was observed that the higher Fe efficiency of the green genotypes as compared to the chlorotic ones was associated with the greater ability of the former to incorporate absorbed Fe into functional entities like chlorophyll and metallo-enzymes. The chlorophyll, Fe contents and catalase activities were higher in the green genotypes than in the severely chlorotic ones. The mean values were respectively: chlorophyll, 942 and 648 mg kg^{-1} fresh weight; Fe, 126 and 102 mg kg^{-1} dry weight and catalase 66.9 and 46.1 units (Kaur et al. 1984a and 1984b).

Based on per cent response in grain yield in the field, the relative tolerance of four sorghum cultivars varied in the order; 59-3 > SL 44 > JS 20 > and JS 263. The cultivar 59-3 was the least responsive to Fe and could be successfully grown on Fe deficient soils where other cultivars failed to produce comparable yields without Fe applications (Singh et al. 1984a).

Iron chlorosis is a severe problem in crops growing on calcareous soils low in organic matter. Singh et al. (1985b) observed that on such soils three genotypes of rice (IET 7972, IET 7973 and BR 34) out of the 20 under test were

Table 2. Effect of different rates of Zn application on grain yield (Mg ha^{-1}) of wheat and rice cultivars

Rice cultivar	Zn (kg ha^{-1})			Wheat cultivar	Zn (kg ha^{-1})		
	0	5.0	10.0		0	5.5	11.0
Ratna	2.8	4.3	4.6	Lok-I	2.5	2.9	3.0
UPR-238	3.4	4.2	4.4	HD 2236	2.5	2.7	3.2
Sita	1.4	4.6	4.8	HD 1553	2.8	2.8	3.3
Rajendra	2.0	4.6	5.3	WH 147	2.9	2.9	3.3
Pusa-S-21	2.5	3.9	4.5	Tawa	2.2	2.3	2.6
LSD (0.05)	0.4				0.15		

tolerant to Fe, when categorised on the basis of degree of visual deficiency symptoms and yield response to applied Fe. The Fe in leaves of these tolerant cultivars ranged from 42.0 to 52.0 mg kg^{-1} compared with 26.0 to 36.0 mg kg^{-1} for the group of highly susceptible cultivars Pusa 33, Jaya, Ratna, and Pusa 2-21.

Singh et al. (1984b) also observed that chickpea cultivar BR 78 displayed severe symptoms of Fe-chlorosis when grown under Fe stress conditions were H 208 and ST 4 showed few symptoms. BR 78 was highly responsive to applied Fe and H 208 and ST 4 were less responsive.

Singh et al. (1985a) observed that among the four lentil cultivars tested, No. 26 was highly susceptible and Pant 406 was tolerant to Fe deficiency in calcareous soil. On the basis of the per cent response in grain production to correction of deficiency, the tolerance of the cultivars was: No. 26 (68% response to Fe application) > Pant 639 (50%) > BR 25 (27%) > Pant 406 (−24%). Agarwala and Sharma (1979) reported that the poor efficiency of Fe absorption in susceptible cultivars could be attributed to low iron reductive capacity of the roots.

Rice cultivar Krishnasal, regarded as tolerant to Fe-chlorosis, was compared to Pusa 33 a susceptible cultivar under irrigated upland conditions (Kumbhar and Sonar, 1980). The tolerant cultivar produced markedly larger dry matter yield at tillering, panicle initiation and flowering stages. Grain and straw yields at maturity were higher. Tolerance was associated with lower retention of Fe in the roots and higher shoot/root ratio of Fe content (Table 3). Reddy and Shiv Prasad (1986) reported that Fe chlorotic rice plants had a high P/Fe ratio and that excess P might be responsible for inactivating Fe in these rice genotypes.

Shinde and Daftardar (1987) showed that absorption of ^{59}Fe in 14-day old intact rice seedlings was strikingly less in a susceptible cultivar than in 5 tolerant cultivars (Table 4). The data in Table 4 indicate that translocation of ^{59}Fe was significantly greater in two tall and tolerant cultivars Jalgaon 5 and Tuljapur 1 and one semi-dwarf PBN 1 than in the susceptible cultivar, Pusa 33. The organic acid content was greater in the root than in the shoots but the content in tolerant cultivars was greater as compared to the susceptible cultivar. The tolerant cultivars contains a higher content of caffeic, fumaric, oxalic and succinic acids than the susceptible cultivar but lower levels of citric and malic acids. Tartaric acid was not detected in the Fe-inefficient cultivar Pusa 33 and was present in appreciable amount in the Fe-efficient tall cultivars. Therefore, tartaric acid could serve as a useful parameter for identifying the trait of Fe tolerance.

Nerkar et al. (1987) isolated semi-dwarf and mid tall mutants of rice in the M_2 generation from 3 tall local cultivars tolerant to Fe-chlorosis viz., Ambemohar local, Jalgaon 5 and Tuljapur 1 after mutagenic treatment with EMS and gamma rays. One semi-dwarf mutant obtained from Ambemohar local by seed treatment with 0.2% EMS formed the basis of the cultivar Prabhavati. This was tolerant to Fe chlorosis and its roots had higher iron reductive capacity than sensitive genotypes. A mid-tall mutant, isolated following mutagenic treatment to an Fe sensitive semi-dwarf variety Pusa 33, was also found to be tolerant to Fe chlorosis and had good yielding ability. Segregation patterns, when Prabhavati

Table 3. Concentration and uptake of Fe in tolerant and susceptible rice cultivars

Parameter	Krishnasal, Tall[a]		Pusa 33, Dwarf[b]	
	Fe conc. (mg kg^{-1})	Fe uptake (g ha^{-1})	Fe conc. (mg kg^{-1})	Fe uptake (g ha^{-1})
Leaf	201	347	200	163
Sheath	113	261	289	312
Stem	763	1229	817	616
Ear	46	191	58	142
Root	137	127	448	207
Stem/root ratio	5.6	9.7	1.8	3.0

Grain yield of the cultivars under Fe stress [a]3.11 Mg ha^{-1} and [b]2.90 Mg ha^{-1}.

Table 4. Absorption and translocation of ^{59}Fe from 0.1 mM $FeSO_4$ by the rice cultivars after 6 hours

Cultivars		Tolerance to Fe chlorosis	^{59}Fe (μ Moles g^{-1} fresh weight)[a]		Organic acid ($\mu g\ g^{-1}$)		pH drop of Steenbergs solution[b]
			Absorption	Translocation	Shoots	Roots	
Pusa 33	Dwarf	Susceptible	9.6	0.058	947	993	0.3
IR 24	Semi dwarf	Tolerant	23.2	0.052	–	–	–
PBN 1	Semi dwarf	Tolerant	27.7	0.067	1115	1163	0.7
GAS 209	Tall	Tolerant	28.6	0.058	1117	1138	2.0
Jalgaon 5	Tall	Tolerant	30.2	0.064	1184	1214	0.9
Tuljapur 1	Tall	Tolerant	34.6	0.075	–	–	–
CD at 5%			2.2	0.003	–	–	–

Source: a = Shinde and Daftardar (1987); b = Daftardar and Shinde (1987).

was crossed with sensitive varieties, indicated tolerance to Fe deficiency in Prabhavati was controlled by dominant oligogenes.

Boron

Cotton cultivar Lohit and sunflower cultivar Sputnik were found to be more susceptible to B deficiency than cotton H 777 and sunflower EC 21991, when grown in sand culture at 0.0165 μg B mL^{-1}. The depression in tissue B concentration due to lack of B was greater in Sputnik than in EC 21991. The activities of catalase, peroxidase, acid phosphatase and aldolase were higher in the tolerant cultivar Sputnik than in the sensitive EC 21991 under B stress (Agarwala et al. 1984).

On the basis of the relative increase in B concentration of leaves and the grain yield response to applied B the relative susceptibility of maize genotypes to B was as follows: Suwan white = M 12 > UPB 742 > Suwan Yellow > M-9 > AB(W) white. Under B stress conditions AB (W) white contained two times more B in leaves than Suwan white (Sakal et al. 1989).

Sakal et al. (1991) showed that within sesamum and mustard, grain yield and boron uptake were highly correlated. Tolerant cultivars removed more B from soil than the susceptible cultivars under B deficiency. The order of tolerance of the cultivars was: (percent grain yield response to applied B given in brackets) Sesamum; RT54 (−8) > OMT-11-6-3(3) > RT49(13) > OMT-11-6-5(22) = Krishna (23) > TC25(38). Mustard; Pusa bold (7) > RH-30(11) = Kranti (13) > RAURD1002(26) = BR40(26) = Varuna (27).

Manganese

According to Kaur and Takkar (1987) Mn-efficient cultivars should have the ability to synthesize more organic matter by better utilization of the absorbed nutrient and also be able to better partition the organic matter into economically valuable parts e.g. grain in the case of cereals. Their results showed that the Mn-efficient cultivars produced a greater number of total as well as fertile tillers, longer spikes and more grains per spike resulting in higher harvest index and higher yields as compared to Mn-inefficient cultivars. Also the Mn-efficient cultivars WG 377, WG 357 and TL 419 were found to have significantly higher 'Hill' reaction activity and a higher content of active Mn i.e. the Mn fraction associated with chloroplasts as compared to the inefficient cultivars DWL 5023 and KSML 3. The activities of chloroplast peroxidase and cytoplasmic peroxidase were lower in the efficient as compared to the inefficient wheat cultivars (Kaur et al. 1989).

Takkar et al. (1990) screened 53 Indian wheat cultivars, on a loamy sandy soil containing 3.0 mg kg^{-1} DTPA-extractable Mn, for their tolerance to Mn deficiency under field conditions. Considering the absolute grain yield and the deficiency status of the plant, as expressed by Mn deficiency symptoms, the cultivars were rated into six categories of tolerance (Table 5). The magnitude of response to Mn application decreased successively as the rating of the tolerance increased and there were no significant responses in the most tolerant categories. The better harvest index in the most tolerant group of cultivars as compared to others revealed that

Table 5. Classification of wheat cultivars into nine groups and six categories of Mn tolerance based on absolute grain yield and development of Mn deficiency symptoms

Group No.	Yield without Mn (Mg ha^{-1})	Per cent yield response	Cultivars	Tolerance to Mn Stress	
				Rating	Category
1	<2.0	47.6	PBW 181*, J 24, C 306* CPAH 1676, Sonalika*]	(i)	Least
2	2.0–2.5	46.9	PBW 34*, HUW 213 (NWB)*, HD 1209*, UP 115*, VW 120*]		tolerant
3	2.5–3.0	41.1	PBW 179, HUW 213*, WH 147*, BW 11*]	(ii)	Slightly
4	3.0–3.5	40.3	HD 2189*, HD 2281*, HD 2285*, HD 2421, HUW 234*, Raj 2967*, Raj 3039*, WH 291, K 7410, J 17, Kalyansona]		tolerant
5	3.5–4.0	24.4	HD 2307, PBW 65, PBW 175, HUW 55, HUW 37*, Raj 1555* Raj 1972*, Raj 2593*, WH 283, WH 423*, WL 711, WL 2265, K 8020]	(iii)	Moderately tolerant
6	4.0–4.5	26.3	HD 2009, HD 2428, WH 332, WH 841	(iv)	Tolerant
7	4.5–5.0	20.8	HD 2270, HD 2429, PBW 159, WH 427	(v)	More tolerant
8	5.0–5.5	14.2	HD 4594, PBW 154, Raj 3038, WH 416]	(vi)	Most tolerant
9	>5.5	8.4	HD 2204, HD 2329, Raj 2535]		

* Cultivars developed Mn deficiency symptoms.

the tolerant cultivars possess genetic characteristics that withstand stress conditions of Mn deficiency. Thus, the tolerant cultivars HD 594, PBW 154, Raj 3038, WH 416, HD 2204, HD2329 and Raj 2535, particularly the latter three, can be recommended for use on Mn-deficient soils. The cultivar HD 2329 has been adopted on a large scale by the farmers on the Mn-deficient soils of Punjab, India. Nayyar et al. (1990) reported that the root weight as well as length and volume were high particularly in the later growth stages in the Mn-efficient HD 2059 as compared to the inefficient wheat cultivar DWL 5023.

The germplasm of different crop species tested in the collaborative co-ordinated project may be available from the workers referred to in this review or from the Heads of the Genetics or Plant Breeding Department of the respective organisations i.e. ICAR Research Institutes and State agriculture universities. Also, the germplasm of different crop species may be available from the Director, National Bureau of Plant Genetic Resources, IARI Campus, New Delhi 110012.

References

Agarwala S C and Sharma C P 1979 Recognizing micronutrient disorders of crop plants on the basis of visual symptoms and plant analysis. Department of Botany, Lucknow University, Lucknow, India.

Agarwala S C, Sharma C P, Chatterjee C and Nautiyal B D 1984 Annual progress report of all-India coordinated scheme of micronutrients in soils and plants. Department of Botany, University Lucknow, India. 127 p.

Brar S P S, Randhawa N S and David D J 1978 Causes of differential yield of wheat cultivars arising from zinc deficiency. Indian J. Agric. Sci. 48, 524–531.

Daftardar S Y and Shinde P H 1987 Effect of iron stress to different rice cultivars on organic acid contents of plants and reduction in pH of nutrient medium. Proc. Nat. Symp. Micronutrient Stresses in Crop Plants. MPKVV, Rahuri, pp 146–147.

Gupta V K and Hans Raj 1983 Response of groundnut genotypes to different levels of zinc. Crop Physiol. 1, 55–58.

Gupta V K, Hans Raj and Dabas D S 1984 Tolerance of wheat cultivars to zinc stress conditions. Haryana Agric. Univ. J. Res. 14, 170–175.

Kaur N P and Takkar P N 1987 Wheat Information Service, 64, 24–33.

Kaur N P, Takkar P N and Gupta A K 1989 Screening of wheat germplasms for Mn deficiency stress tolerance. *In* Genetics and Wheat Improvement. Eds. A K Gupta, S. Nagarajan and R S Rana. ICAR, New Delhi, pp 338–344.

Kaur N P, Takkar, P N and Nayyar V K 1984a Catalase, peroxidase and chlorophyll relationships with yield and iron deficiency chlorosis in *Cicer* genotypes. J. Plant Nutr. 7, 1213–1220.

Kaur N P, Takkar P N and Nayyar V K 1984b Relationship of total and ferrous iron with the incidence of chlorosis in some genetic lines of *Cicer arietinum*. Field Crops Res. 8, 273–280.

Kumbhar D D and Sonar K R 1980 Dry matter production and nutrient uptake pattern of two rice cultivars grown under upland condition. J. Indian Soc. Soil Sci. 28, 178–183.

Nayyar V K, Takkar P N, Bansal R L and Singh S P 1990 Micronutrients in soils and crops of Punjab. Res. Bull.

Dept. Soils. The Punjab Agricultural University, Ludhiana. 148 p.

Nerkar Y S, Misal M B and Reddy C D R 1987 Upland rice breeding for tolerance of iron chlorosis on vertisols and genetic control of iron chlorosis. Proc. Nat. Symp. Micronutrient Stresses in Crop Plants. MPKVV, Rahuri. 112 p.

Patel P C and Patel J R 1988 Effect of zinc with and without organic manures on growth and zinc nutrition of different genotypes of forage sorghum. J. Indian Soc. Soil Sci. 36. 820–822.

Pathak A N, Tiwari K N, Upadhyay R L and Jha A K 1979 Zinc reponse of barley as influenced by genetic variability. Indian J. Agric. Sci. 49, 25–29.

Prasad K, Sinha R B and Singh B P 1982 Response of rice varieties to zinc. Indian J. Agric. Sci. 52, 511–514.

Randhawa N S and Takkar P N 1976 Screening of crop varieties with respect to micronutrient stresses in India. Proc. Workshop Plant Adaptation to Mineral Stress on Problem Soils. Beltsville, MD. pp 393–400.

Rathore G S, Dubey S D, Kamparia R S and Sharma B L 1986 Annual progress report of all-India coordinated scheme of micronutrients in soils and plants. Department of Soil Science, JNKVV, Jabalpur, India. 95 p.

Reddy C K and Shiva Prasad G V S 1986 Varietal response to iron chlorosis in upland rice. Plant and Soil 94, 289–292.

Sakal R, Singh B P, Sinha R B and Singh A P 1984a Response of some wheat germplasm to zinc application in calcareous soil. Ann. Agric. Res. 5, 137–142.

Sakal R, Singh B P and Singh A P 1984b Annual progress report of all-India coordinated scheme of micronutrients in soils and plants. Department of Soil Science, RAU, Pusa, India. 42 p.

Sakal R, Singh A P, Singh R B and Bhogal N S 1991 Relative susceptibility of some important varieties of sesamum and mustard to boron deficiency in calcareous soils. Fert. News 36, 43–50.

Sakal R, Chaudhuri L B, Singh R B and Singh B P 1985a Differential response of some Cheena germplasms to Zn in calcareous soils. J. Indian Soc. Soil Sci. 33, 591–595.

Sakal R, Singh B P, Chaudhari L B and Sinha R B 1985b Screening of finger millet varieties for their relative response to zinc in calcareous soil. J. Indian Soc. Soil Sci. 33, 440–442.

Sakal R, Singh A P and Sinha R B 1989 Differential susceptibility of maize varieties to boron deficiency in a calcareous soil. J. Indian Soc. Soil Sci. 37, 582–584.

Saxena N B and Mathur S K 1987 Effect of available zinc on the productivity of *Phaseolus aconitifolius* cultivars. Proc. Nat. Symp. Micronutrient Stresses in Crop Plants. MPKVV, Rahuri. 124 p.

Shinde P H and Daftardar S Y 1987 Absorption and transport of ^{59}Fe in rice cultivars different in height and resistance to Fe chlorosis. Proc. Nat. Symp. Micronutrient Stresses in Crop Plants. MPKVV, Rahuri, pp 115–116.

Shukla U C, Arora S K, Singh Z, Prasad K G and Safaya N M 1973 Differential susceptibility in some sorghum genotypes to zinc deficiency in soil. Plant and Soil 39, 423–427.

Shukla U C and Hans Raj 1974 Influence of genetic variability on zinc response in wheat. Soil Sci. Soc. Am. Proc. 38, 477–479.

Shukla U C and Hans Raj 1976 Zinc response in corn as influenced by genetic variability. Agron. J. 68, 20–22.

Shukla G P and Hans Raj 1980 Zinc response in pigeonpea as influenced by genotypic variability. Plant and Soil 57, 323–333.

Singh B P and Choudhary J L 1987 Response of chickpea to zinc as influenced by genetic variability on calcareous soil. J. Indian Soc. Soil Sci. 35, 325–328.

Singh B N and Sakal R 1987 Screening for zinc deficiency tolerance in rice. IRRN 12, 15.

Singh A P, Sinha R B and Sakal R 1987 Screening of lentil varieties for their susceptibility to zinc deficiency in calcareous soil. Ann. Agric. Res. 8, 26–29.

Singh B P, Sakal R and Singh A P 1983 Differential susceptibility of some composite maize germplasms to zinc deficiency in calcareous soil. J. Indian Soc. Soil Sci. 31, 250–253.

Singh B P, Sakal R and Singh A P 1985a Response of lentil varieties to iron applicaiton on highly calcareous soils of Bihar. Indian J. Agric. Sci. 55, 56–58.

Singh B P, Singh A P and Sakal R 1984b Response of chickpea cultivars to iron application in calcareous soils. Int. Chickpea Newsl. 10, 11–12.

Singh B P, Singh R A, Sinha M K and Singh B N 1985b Evaluation of techniques for screening Fe-efficient genotypes of rice in calcareous soils. J. Agric. Sci. Camb. 105, 193–198.

Singh S P, Nayyar V K and Takkar P N 1984a Response of sorghum cultivars to iron sources and mode of its application. J. Agron. Crop Sci. 153, 342–347.

Sinha N P and Singh T P 1977 Responses of some maize germplasms to zinc in calcareous soils of north Bihar. J. Indian Soc. Soil Sci. 25, 422–425.

Takkar P N 1991 Zinc deficiency in Indian soils and crops. *In* Zinc in Crop Nutrition. Int. Lead Zinc Res. Org. Inc. and Indian Lead Zinc Information Center, New Delhi. pp 55–64.

Takkar P N, Bansal R L and Nayyar V K 1990 Response and tolerance mechanism of wheat cultivars to manganese deficiency 14th Int. Congr. Soil Sci. IV, 57–61.

Takkar P N, Chibba I M and Mehta S K 1989 Twenty years of coordinated research on micronutrients in soils and plants. Bull 1. Indian Institute of Soil Science, Bhopal, India. 314 p.

Takkar P N, Bansal R L, Singh S P and Nayyar V K 1988 Response of pearlmillet cultivars to zinc under field conditions. Int. J. Trop. Agric. 6, 247–251.

Takkar P N and Nayyar V K 1984 Response of crops to micronutrients. FAI-NR seminar on 'Balanced Fertilization Programme with Special Reference to Secondary and Micronutrients of Crops under Intensive Cropping'. Jaipur, India. pp 95–123.

Takkar P N, Singh S P, Bansal R L and Nayyar V K 1983 Tolerance of barley varieties to zinc deficiency. Indian J. Agric. Sci. 53, 971–972.

Trehan S P, Sharma U C and Grewal J S 1988 Differential susceptibility of potato cultivars to zinc stress and nutrient interaction. Indian J. Agric. Sci. 52, 922–927.

Verma M K 1986 Edaphic and plant factors influencing differential reaction of rice cultivars to zinc stress in calcareous soils. PhD thesis. Rajendra Agricultural University Pusa, Bihar, India.

Yadav B S and Vyas K K 1987 Response of wheat varieties to zinc. J. Indian Soc. Sil Sci. 35, 161–162.

P. J. Randall et al. (Eds.), Genetic aspects of plant mineral nutrition, 349–358.

PLSO SV76

Selecting zinc-efficient cereal genotypes for soils of low zinc status

ROBIN D. GRAHAM, JULIE S. ASCHER and SIMON C. HYNES
Waite Agricultural Research Institute, University of Adelaide, Glen Osmond, 5064, South Australia

Key words: *Avena sativa*, genotypic differences, grain zinc, *Hordeum vulgare*, *Triticum aestivum*, zinc concentration, zinc deficiency, zinc uptake

Abstract

Deficiencies of zinc are well known in all cereals and cereal-growing countries. From physiological evidence reported elsewhere, it would appear that a critical level for zinc is required in the soil before roots will either grow into it or function effectively; it is likely the requirement is frequently not met in deep sandy, infertile profiles widespread in southern Australia. Because fertilizing subsoils is impractical, this paper presents arguments for breeding cereal varieties with root systems better able to mobilise zinc from soil sources of low availability. Other agronomic arguments are presented in support of breeding for zinc efficiency.

Significant genetic variation for this character is described for wheat, barley and oats. Linkage to other efficiency traits (e.g., manganese) is poor suggesting independent mechanisms and genetic control not linked to gross root system geometry. Zinc efficiency traits for sandy and clayey soils appear to be genetically different. Zinc-efficient genotypes absorb more zinc from deficient soils, produce more dry matter and more grain yield but do not necessarily have the highest zinc concentrations in tissue or grain. Although high grain zinc concentration also appears to be under genetic control, it is not tightly linked to agronomic zinc efficiency traits and may have to be selected for independently. High grain zinc is considered a desirable quality factor which not only contributes to the seedling vigour of the next generation but could increase the nutritional value of the grain in areas where a high dependence on grains for food may result in zinc deficiency in humans.

Introduction

Genotypic variation for zinc efficiency in wheat has been recognized and reported by many: for example, by Solunke and Malewar (1987), by Randhawa, Takkar and colleagues over many years in the Punjab, India (Takkar et al., 1988), and by R.A. Hare in New South Wales (Table 3). In this paper we use the term zinc-efficient to apply to a genotype or phenotype which is better adapted to, or yields more in, a deficient soil than can an average cultivar of the species. It is an agronomic definition which can be measured in terms of grain yield in a field experiment, and does not imply a mechanism. Nevertheless, nearly all micronutrient efficiency traits so far studied arise from a superior ability to extract the limiting nutrient from soil, rather than a capacity to remobilize the nutrient within the plant or otherwise to survive on less of that nutrient. The thinking in this paper is therefore oriented towards factors which favour uptake.

Zinc deficiency is probably the most widespread micronutrient deficiency in cereals; many millions of hectares of wheat-growing soils are deficient in zinc since, unlike the other micronutrients, it is a common feature of both cold and warm climates, acid and alkaline soils, and both heavy and light soils. In particular, the subsoils of the drier regions are highly alkaline making zinc availability commonly lower than in the topsoils. Subsoils are of increasing research interest in the seasonally dry areas because they hold vital stored water, but are almost always highly infertile – and by virtue of their depth, untreatable. The potential for solving the zinc deficiency problems of top- and sub-soils by plant breeding is the subject of this paper.

Materials and methods

Field experiments with wheat, barley and oats were established for two years (1988–1989) at two sites, Lameroo and Yeelanna, in South Australia. The soil profiles are loamy sand over permeable medium clay (Lameroo) and loam over permeable medium clay (Yeelanna). DTPA extractable soil Zn was about 0.2 mg kg^{-1} at all sites. However, bicarbonate extractable P varied widely from 14 mg kg^{-1} at Lameroo, 1989 to 36 mg kg^{-1} at Yeelanna, 1988. Soil pH in water was 8.0–8.7 at all sites except the more sandy Lameroo 1989 site (6.1). A typical field experiment consists of 30 genotypes $\times$ $\pm$Zn application $\times$5 replications, with border plots, all plots being 0.8×4.5 m in size. A nearest neighbour (NN), balanced lattice design (Wilkinson et al., 1983), was generated by computer for a 30×5 replication design with borders as is done conventionally by breeders. The experiment was then sown with all plots (and borders) entered twice in paris, one of the pair (chosen at random) receiving zinc. For the oat variety trials a randomized complete-block, split-plot design was used with either nine (1988) or twelve varieties (1989). The Waite Institute's cereal breeding system (developed by A.J. Rathjen and D.H.B. Sparrow and colleagues) uses a magazine system for delivering seed to a cone seeder and zinc granules were delivered along with the seed via the magazine. The fertilizer box thus contained only basal nutrients (mono-ammonium phosphate + ammonium sulphate delivering 23 kg N, 20 kg P, 1 kg Cu and 4.5 kg Mn per ha) and did not need to be changed. The zinc granules were commercial zinc oxysulphate (~30% Zn) and were used at several times commercial rates (11–14 g/4.5 m^2 plot) because its effectiveness is less than when coated on macronutrient granules (ammonium phosphate), as is current commercial practice. The soil-applied zinc was supplemented with a foliar spray of $ZnSO_4$ at tillering (equivalent to 200 g Zn ha^{-1}).

Mid-season harvests were conducted where possible, to guard against loss of information due to end-of-season events and to give information on nutrient uptake. One metre of row was sampled from the middle two rows, tissue dried and weighed prior to grinding and digestion in nitric acid for analysis by ICP spectrometry. Grain yield was reaped by Wintersteiger small-plot harvester, and the results subjected to both NN (nearest neighbour) and ANOVA statistical analysis; not withstanding the greater efficiency of the NN analysis, we have generally given the more familiar LSDs from conventional ANOVA as a conservative guide to the results.

In addition to these analyses of grain yield data and the similar treatment of mid-season vegetative yields, a further index of considerable value is generated, after chemical analysis of the tissues and grain, by calculation of the total uptake of zinc into vegetative growth and/or grain. The most efficient genotypes are so because they absorb more zinc and often (but not always) they maintain higher concentrations of zinc in tissues and/or grain. Uptake, being the product

$$\text{Zn uptake} = \text{yield} \times \text{Zn concentration}\ ,$$

frequently shows greater variation among genotypes than yield. (By comparison with yield, variance in concentration data is small.)

For the pot experiment, there were three replicates of the five zinc and five cultivar treatments, 75 pots in all. Zn rates were 0, 0.25, 1.0, 4 and 16 mg Zn as $ZnSO_4$ added to 600 g of Laffer's sand, together with adequate amounts of all other essential elements. Plants (two per pot) were grown for 45 days at 15°C, 10 h light (500 $\mu E/m^2 s$), and harvested for top and root dry weights.

Results

Wheat

In South Australia, we have found considerable genetic diversity for zinc efficiency, an illustrative selection of which is shown in Table 1. Of these wheats, Excalibur, a recent Roseworthy variety, was the best and most consistently zinc-efficient line over two years and two sites (Table 2). RAC 629 also performed well. Gatcher (and Songlen – data not shown) are two zinc-inefficient bread wheats. Schomburgk is a rust-resistant backcross derivative of Aroona which appears to have lost the zinc efficiency of Aroona. It was consistently an average performer over sites and seasons. Aroona, an early maturing cultivar, was as outstanding for yield and efficiency as Excalibur in the drier 1988 season but well below average in 1989 when there was a long, cool finish to the season. Such seasonal interactions are quite common, and are further illustrated by the barley data discussed later.

Figure 1 is derived from a pot experiment in which the zinc response curves of five cultivars were compared under growth chamber conditions. Variability is greatest in the approach to maximum yield, but the overall trends indicate similar rankings at deficient levels of zinc supply, that is, the slopes are similar but the curves for cultivars identified from field testing as inefficient are displaced to the right and tend to reach

Table 1. Grain yields (t/ha) and calculated zinc efficiencies of seven wheats grown on zinc-deficient soils at two sites in South Australia in two years. Zinc efficiency (E) is calculated as $100 \cdot \frac{-Zn}{+Zn}$

		Lameroo			Yeelanna		
		−Zn	+Zn	E	−Zn	+Zn	E
Excalibur	1988	1.43	1.76	81	2.52	3.18	79
	1989	2.40	2.40	99	3.82	4.11	93
RAC629	1988	1.26	1.85	68	2.13	2.69	79
	1989	2.49	2.85	87	3.58	3.86	93
Aroona	1988	1.31	1.42	92	2.64	3.01	88
	1989	2.03	2.42	84	3.04	3.43	89
Schomburgk	1988	1.11	1.32	84	2.17	2.28	95
	1989	2.20	2.44	90	3.23	3.61	90
Warigal 5RL	1988	1.04	1.14	92	2.31	2.58	89
	1989	2.22	2.35	94	3.03	3.33	91
Warigal	1988	1.00	1.22	83	2.09	2.65	79
	1989	2.15	2.27	95	2.95	3.23	91
Durati	1988	0.45	1.12	40	1.73	2.14	81
	1989	1.36	1.57	87	2.01	2.32	87
LSD * (G × Zn)	1988	0.30			0.40		
	1989	0.45			0.79		

Table 2. Zinc efficiency rankings of nine selected wheats (out of 30 tested) grown on zinc-deficient soil at two sites in two seasons (based on nil Zn yield results). While some genotypes are stable in certain rankings, Aroona indicates a strong effect of season. Marion Bay manganese efficiency rankings (manganese-deficient site) included for comparison

	Lameroo (zinc) 1988	Yeelanna (zinc) 1988	Lameroo (zinc) 1989	Yeelanna (zinc) 1989	Marion Bay (Manganese) 1989
Excalibur	1	2	5	2	15
RAC 629	3	13	2	3	12
Aroona	2	1	20	19	1
Schomburgk	12	11	13	8	5
Warigal 5RL	14	6	12	20	4
Warigal	22	14	16	25	–
Takari	5	23	25	15	27
Gatcher	23	27	27	27	3
Durati	29	25	29	29	23

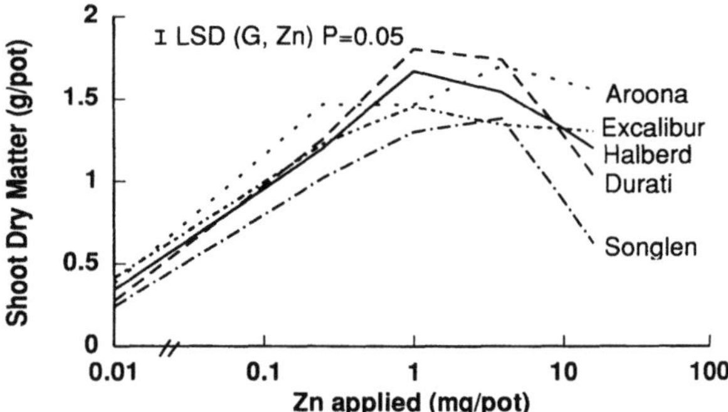

Fig. 1. Yield response of five wheat cultivars to five rates of zinc applied to zinc deficient Laffers sand in pots. LSD given for treatment main effects.

the yield plateau at a higher level of zinc supply. Notably, the most inefficient wheat, Durati produced the highest yields at adequate zinc, and (with Songlen) the lowest yield without added zinc. On the whole, these data suggest that two-point (±Zn) testing used in our field studies is adequate to identify the desirable zinc efficiency trait for South Australian conditions with reasonable probability.

Durati, a very sensitive durum wheat in the heavy black clay soils of New South Wales was also a poor performer in our light sandy soils. It is therefore a valuable indicator line. Kamilaroi is a derivative of Durati (Durati × Leeds) which not only incorporates yield, quality and disease resistance from Leeds but also zinc efficiency on the heavy black earths of New South Wales (Table 3). However, in South Australia on light sands, Kamilaroi appears worse than Durati for zinc efficiency (Tables 3, 5). Zinc deficiency in the black earths is a complex phenomenon involving very high levels of native soil phosphorus and manganese which appear to aggravate the low zinc status. Thus zinc efficiency on these soils may not be so much a 'foraging capacity of the roots' but a better discrimination for zinc over manganese and phosphate. In South Australia phosphorus and manganese are relatively low. We therefore recognize different types of zinc deficiency and corresponding efficiency traits.

Table 3. The wheat grain yield (GY) of Kamilaroi relative to that of its parent Durati, in contrasting soils in New South Wales and South Australia. Kamilaroi = Durati × Leeds. Data for N.S.W. kindly supplied by Dr Ray Hare, Agricultural Research Centre, Tamworth

	N.S.W. (black clay)		S.A. (light sand)	
	−Zn	+Zn	−Zn	+Zn
$100 \cdot \frac{\text{GY of Kamilaroi}}{\text{GY of Durati}}$	130	105	95	110

Rye and triticale are generally more zinc-efficient than wheat (Harry, 1982; K.V. Cooper, personal communication). Tables 1 and 2 show a slight advantage in zinc efficiency from the 5RL translocation which also confers marked copper efficiency to wheat (Graham et al., 1987). In particular, Warigal 5RL absorbs more zinc than Warigal and maintains higher concentrations in leaf tissues (Table 4). However, under zinc-deficient conditions. Warigal 5RL had lower zinc concentration in grain (Table 5).

Besides diversity for yielding ability on zinc-deficient soils, there may be genetic control over zinc concentrations in tissue and grain. From Table 4, Excalibur and Warigal 5RL are generally superior to the other lines (30 tested in all) in

Table 4. Effect of zinc fertilizer on concentrations of zinc and zinc uptake into shoots of six wheat genotypes at tillering stage on a marginally zinc-deficient soil at Yeelanna, S.A. in 1989

Genotype	Concentration (mg kg^{-1})		Uptake (g ha^{-1})	
	−Zn	+Zn	−Zn	+Zn
Durati	8.9	11.6	3.7	6.4
Machete	8.4	9.4	5.0	7.9
Millewa	9.9	9.6	6.1	7.0
Excalibur	9.3	12.5	5.5	10.2
Warigal	8.1	10.6	4.9	7.7
Warigal 5RL	10.8	11.2	7.1	9.6
LSD * (G)	1.5		1.8	

Table 5. Grain yields and grain zinc analyses of eight wheat genotypes (G) at Lameroo 1988, a low rainfall, zinc responsive sand-over-clay site. Zinc efficiency (E) is calculated as $100 \cdot \frac{-Zn}{+Zn}$

Genotype	Grain yield ($t\,h^{-1}$)			Grain zinc concentration ($mg\,kg^{-1}$)		Grain zinc content ($g\,ha^{-1}$)		
	−Zn	+Zn	E	−Zn	+Zn	−Zn	+Zn	E
Excalibur	1.43	1.76	81	7.6	25.3	10.8	45	24
Schomburgk	1.11	1.32	84	7.5	23.7	8.3	31	27
Warigal 5RL	1.04	1.14	91	10.5	29.2	10.9	33	33
Warigal	1.00	1.22	82	13.0	26.0	13.0	31	41
Kite	0.95	1.37	69	9.5	24.1	9.0	33	27
TJB * MKR	0.63	1.30	48	12.1	29.5	7.6	38	20
Durati	0.45	1.12	40	9.6	24.9	4.3	28	15
Kamilaroi	0.45	1.29	35	8.8	23.4	4.0	30	13
LSD * (G × Zn)	0.31			3.4		4.9		

TJB * MKR, a breeding line of A.J. Rathjen, Waite Institute.

zinc concentration in leaves and zinc uptake at tillering. However, Excalibur had low grain concentration, a condition apparently linked to its high yield since grain zinc content ($g\,ha^{-1}$) was high, especially after zinc treatment (Table 5). There is a distinct trend, as with grain nitrogen, for a lower grain zinc concentration with increasing yield (across genotypes). This is undesirable, both because of lower seedling vigour when low zinc seed is used for resowing in zinc deficient soil, and because wheat is generally considered to be too low in zinc for adequate human nutrition when it forms a high proportion of the diet (Welch and House, 1983). However, Warigal stands out as having the highest grain zinc concentration and content under zinc-deficient conditions. It is highly likely that grain zinc concentrations could be improved by breeding. It should be noted, however, that grain zinc concentration responds dramatically to fertilization under these conditions (Table 5), and therefore this quality factor could be readily improved, when markets require it, by soil or foliar application.

Barley

Table 6 shows a similar range of responses to zinc in barley cultivars as was shown for wheat in Table 1. Yields in nil zinc treatments varied from poorest to best by a factor of three and efficiencies varied by a factor of two. No one barley stands out as the most efficient. At Lameroo, the more zinc-deficient and drier site, three lines stand out as having a valuable degree of zinc efficiency: WI2597, WA73S276 and Forrest. Of the first two, which had almost equal top grain yields in both +Zn and −Zn treatments, WI2597 was outstanding in its early vigour. Forrest, however, had the highest calculated efficiency for grain yield, and although its yield potential is low, being an old, taller variety, it may be an excellent parent for introducing zinc efficiency because it maintained remarkably high concentrations of zinc in the grain on this deficient soil. High grain zinc is important in several respects. It indicates a high ability to absorb and translocate zinc to the grain; this in turn is an important aspect of grain quality for animal and human consumption (Welch and House, 1983); and if the grain is sown back into the zinc-deficient soil, as is commonly done by Australian farmers, we have observed that high seed zinc content under these conditions has a major effect on seedling vigour. It is a positive aspect of these data that the three most efficient cultivars for yield were also the three highest in zinc concentration in grain, indicating that for barley, unlike wheat, yield and concentration aspects of zinc efficiency may be linked, simplifying the breeding objectives.

At the higher yielding site, Yeelanna, deficiency was a little less severe, but the rankings are in any case considerably different, owing to late-season rains at this site which favoured later varieties and those capable of capitalizing on this

Table 6. Vegetative and grain yields of nine barleys grown on two zinc-deficient soils in 1988 and the concentrations of zinc in grain at final harvest. Lameroo was a more zinc-deficient, lower rainfall site than Yeelanna. Genotype (G); Zinc (Zn). Zinc efficiency (E) is calculated as $100 \cdot \frac{-Zn}{+Zn}$

Genotype	Vegetative harvest ($t\ ha^{-1}$)			Grain yield ($t\ ha^{-1}$)			Zn concentration ($mg\ kg^{-1}$)	
	−Zn	+Zn	E	−Zn	+Zn	E	−Zn	+Zn
Lameroo								
Prior	1.87	3.25	58	1.01	1.17	86	8.3	24.7
Galleon	1.85	3.04	61	1.17	1.70	69	11.1	23.1
CI3576	1.17	2.76	43	0.47	0.99	48	11.1	31.4
Schooner	1.72	2.88	60	1.03	1.51	69	11.5	23.2
Forrest	2.11	2.64	80	0.93	1.00	92	25.3	25.3
WI2645	2.03	2.80	72	1.03	1.40	74	10.9	20.8
WI2597	2.97	3.81	78	1.30	1.81	72	11.3	24.5
WI2585	2.07	3.59	58	1.02	1.46	70	10.5	24.4
WA73S276	2.13	3.38	63	1.28	1.73	74	12.7	25.6
LSD * (G,Zn,G × Zn)	0.50, 0.11, NS			0.22, 0.04, 0.31			NS, 1.0, 8.9	
Yeelanna								
Prior	2.10	2.86	73	1.20	1.17	102	7.4	11.1
Galleon	1.72	2.27	76	2.00	3.56	56	7.3	8.7
CI3576	2.08	2.92	71	2.73	2.82	97	6.3	10.5
Schooner	1.83	2.59	71	2.89	3.30	88	7.9	10.0
Forrest	1.88	2.59	72	2.14	2.69	79	7.7	12.3
WI2645	2.32	2.55	91	3.16	3.56	89	6.5	9.4
WI2597	2.04	2.24	91	2.71	2.87	95	7.1	10.6
WI2585	2.15	3.20	67	2.80	3.01	93	7.1	11.9
WA73S276	1.70	2.41	70	2.62	3.48	75	6.8	10.0
LSD * (G,Zn,G × Zn)	0.52, 0.09, NS			0.56, 0.09, 0.77			2.0, 0.04, 3.1	

event by producing more tillers, and ultimately more grain. Schooner is such a cultivar (D.H.B. Sparrow, pers. commun.), and its response, relative to Galleon, is shown in Table 6 by similar efficiencies for vegetative yields for both cultivars at both sites and for grain at Lameroo, but a considerably higher efficiency for grain yield in Schooner at Yeelanna. At this site, WI2645 outyielded the others at both harvests, although there was not much between the next five; and the low yielding, old and tall but well adapted variety, Prior, had the highest calculated efficiency. Prior has been an important parent to the modern barley breeding program at the Waite Institute and zinc efficiency may be one of its significant traits.

As in wheat (see Table 2), there appears to be no close linkage between zinc efficiency and manganese efficiency in barley (Fig. 2). The genetic independence of efficiency mechanisms argues strongly against the involvement of general mechanism such as H^+ extrusion or better root system geometry which would benefit the absorption of all micronutrients together.

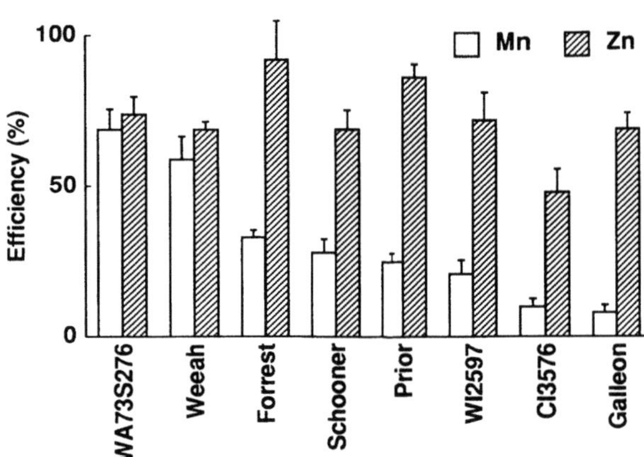

Fig. 2. Comparison of zinc and manganese efficiencies (for grain yield) of eight barley cultivars, based on results at Marion Bay, 1989 (Mn) and Lameroo, 1988 (Zn). Standard errors based on five replicates. Genotypes ranked according to manganese efficiency.

However, these two traits important to South Australia appear to be combinable as they are in the barley breeder's line WA73S276 (and also in certain wheats, ryes and triticales, Graham, 1988). Forrest and Prior are not notably manganese efficient in our field trials (data not published), while the acutely manganese inefficient barleys, Galleon, CI3576 and WI2585 are variable in expression of zinc efficiency. WI2597, overall the most zinc-efficient barley tested, is quite inefficient for manganese (Fig. 2).

Oats

Two sets of data representative of our results with oats are given in Table 7. The two sets of four cultivars have three oats in common which demonstrate a consistency in relative efficiency across the two environments, although again Yeelanna was less zinc-deficient. Mortlock was the most efficient in both years, but Bandicoot, a naked oat included in 1989, was even less efficient than Winjarde, the least efficient in 1988. It is clear that the Lameroo site was extremely deficient in zinc for oats as they not only had low zinc concentrations in shoots in the nil zinc treatments but in the +Zn as well, suggesting that the yield plateau may not have been reached in oats at this site without the foliar zinc (which was applied soon after sampling); at Yeelanna, the +Zn treated plants only approach the critical range for zinc in the shoot of 11–20 mg kg^{-1} (Reuter and Robinson, 1986), but a foliar application was also applied after sampling at this site. Like barley the more efficient oats appear to have higher concentrations of zinc in tissue than the inefficient, indicating more efficient uptake mechanisms are involved in underpinning the expression of agronomic efficiency.

Discussion

A case for a breeding program

Plant breeding is a numbers game, and any new objective such as in this case, zinc efficiency, represents a considerable escalation of the breeder's work or else a diversion of effort away from traditional targets such as quality and disease resistance. A strong case is therefore imperative; the lack of compelling arguments for doing so is the reason that little effort has until now been made to adapt crop plants to micronutrient-deficient soils. In earlier papers (Graham, 1984, 1988, 1988a), we demonstrated genetic diversity for micronutrient characters within wheat, and further argued that nearly all

Table 7. Vegetative and grain yields of four oat cultivars grown on zinc-deficient soils, and the zinc concentrations in shoots at the early harvest. (a) Lameroo, 1988. A low yielding, severely deficient site. (b) Yeelanna, 1989. A less deficient, more productive site in a higher rainfall year. Zinc efficiency (E) is calculated as $100 \cdot \frac{-Zn}{+Zn}$

Genotype (G)	Vegetative yield ($t\ ha^{-1}$)			Zn concentration ($mg\ kg^{-1}$)		Grain yield ($t\ ha^{-1}$)		
	−Zn	+Zn	E	−Zn	+Zn	−Zn	+Zn	E
(a) Lameroo 1988								
Mortlock	2.86	3.78	76	5.2	6.4	1.18	1.58	74
Winjarde	2.04	4.16	49	4.5	5.5	0.54	1.35	40
Marloo	2.92	3.30	89	5.8	7.0	0.99	1.41	70
Echidna	2.76	3.11	89	4.8	5.9	0.75	1.63	46
LSD * (G × Z)	0.54, 0.27						0.29, 0.08	
(b) Yeelanna 1989								
Mortlock	0.47	0.65	72	9.9	12.1	3.50	3.75	94
Winjarde	0.39	0.57	68	9.0	11.2	2.60	3.50	74
Bandicoot	0.27	0.48	57	7.6	10.8	1.46	2.31	63
Echidna	0.46	0.57	80	9.8	11.2	2.90	3.94	73
LSD * (G, Zn)	0.12, 0.04			NS, 1.25		0.56, 0.12		

soils no matter how poor, had sufficient content of micronutrients stored in the profile; the problem was usually one of availability, a problem which is half soil and half genotype. This brings us back to the agronomic arguments.

Six years ago, on ten soil types scattered across Sout Australia, various nutrient treatments were applied to the subsoils as they were returned to grave-size pits in their original layers. The topsoils were then replaced and the sites sown as part of the farmer's fields, receiving all the farmers' usual treatments and fertilizers, including in some cases micronutrients, placed in the topsoil.

Responses to the subsoil nutrients were immediate and often spectacular (Fig. 3), but generally there was little response to physical disturbance only or to gypsum, underling the nutritional nature of the problem. Importantly, these responses continue to the present (Fig. 3 shows, 100% yield increase due to nutrients after 3 seasons); and with the original nitrogen largely dissipated in the first year, the residual responses are principally to phosphorus and trace elements. (Analyses not presented show no difference in shoot nitrogen from subsoil treatments after three years). Indeed, at some sites the micronutrient treatment seems to have relatively greater residual value as time passes. In pot studies we have shown that wheat grows poorly in subsoil even when fertilized with nitrogen and phosphorus. Although we are experimenting

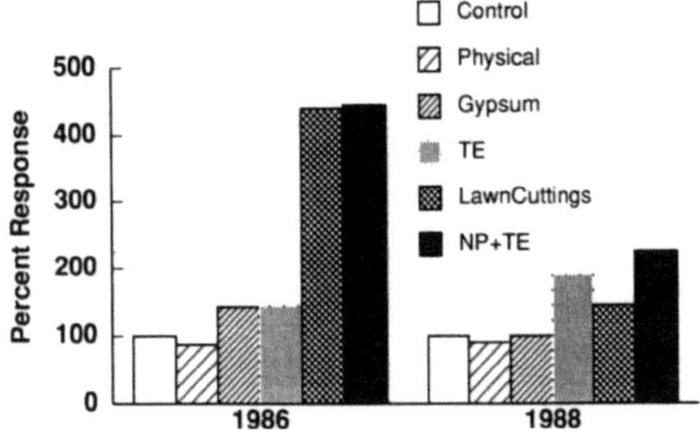

Fig. 3. Vegetative yield of barley, as a percent of control, at stem extension in 1986 and subsequently in 1988, as affected by subsoil treatments applied to a calcareous sand at Marion Bay in 1986. TE = micronutrients; NP + TE = nitrogen, phosphorus and micronutrients; Physical = physical disturbance with no nutrients added (Graham and Ascher, 1990).

with deep injection of micronutrients through tubes welded down the back of deep-ripping tynes, we believe the correct approach to this problem is to breed wheats with root systems which will penetrate subsoils of low phosphorus and micronutrient availability. In most respects, phosphorus and micronutrients are analogous in the arguments of this paper, but genetic progress for phosphorus efficiency has been particularly slow, as discussed elsewhere (Graham, 1984).

Immediately relevant to this argument is the picture emerging from physiological studies of roots spanning four decades. From the papers of Haynes and Robbins (1948), Epstein (1972), Pollard et al. (1977), Bowling et al. (1978), Graham et al. (1981), Welch et al. (1982), Nable and Loneragan (1984) and Loneragan et al. (1987), it appears that the elements phosphorus, zinc, boron, calcium and manganese are all required in the external environment of the root for membrane function and cell integrity. In particular, phosphorus and zinc deficiencies in the external environment promote leaking of cell contents such as sugars, amides and amino acids which are chemotaxic stimuli to pathogenic organisms. Although phosphorus is phloem-mobile, the other elements are not, or are poorly so; this means that the root tips may not be adequately supplied from elsewhere in the root system, such as for example, from those roots contacting a fertilizer band in the top soil. Moreover in the case of zinc, a high internal zinc content did not prevent leakiness due to a deficiency of zinc external to the membrane (Welch et al., 1982). It follows that the roots of those cereal genotypes which have a greater capacity to mobilize nutrients strongly bound to soil particles in the rhizosphere will probably be better able to penetrate an infertile, high pH subsoil. This view was confirmed by Nable and Webb (1992) in a pot experiment which compared wheat cultivars Excalibur and Gatcher in pots of zinc-deficient sand divided into upper and lower layers. Gatcher plants withdrew more water from the bottom layer if it were treated with zinc. The advantage of subsoil zinc treatment was less, as expected, in the zinc-efficient cultivar, Excalibur.

It follows too from the above that roots far from a fertilizer band, with leaky membranes, are at greater risk from pathogens. Recent studies have clearly linked trace element deficiencies with enhanced susceptibility to particular pathogens (Graham and Webb, 1991). For example, zinc deficiency decreased the resistance of wheat to *Fusarium graminearum*, the crown rot fungus (Sparrow and Graham, 1988); the implication is that nutrient-efficient genotypes in deficient soil, enjoying better nutrient status, should have greater resistance to such root pathogens. This hypothesis is supported by recent studies of take-all with wheats having a range of manganese efficiencies (Pedler et al., this symposium). Similarly, a strong association of susceptibility to crown rot (Burgess et al., 1984) with zinc inefficiency (our data) is apparent in the following:

Zinc efficiency: Cook = Takari > Kite > Gatcher > Songlen ⩾ Durati

Susceptibility: Cook < Kite < Takari < Gatcher < Songlen ⩽ Durati

Collectively, these results suggest causality in the concurrence in southern Australia of some of the world's most severe root disease and micronutrient deficiency problems.

Topsoil drying is another problem of wheat production on infertile soils of the seasonally humid zone. Most of the micronutrients (and phosphorus) is in the topsoil by virtue of fertilizer additions and nutrient cycling. Leaching of the heavy metals is negligible (Jones and Belling, 1967). When the topsoil dries as a result of a week or two of dry weather in spring, roots in the nutrient zone are largely deactivated and the plant must rely on deeper roots or retranslocation for further nutrition. With phloem-immobile micronutrients and inefficient genotypes, deficiency can result (Ascher and Graham, unpublished 1982); this has occured in the field (Grundon, 1980) where copper deficiency from topsoil drying at early boot stage caused severe sterility problems; these may, however, be overcome by copper-efficient genotypes such as triticale (Grundon and Best, 1981).

Two further advantages accrue to micronutrient efficient varieties if by virtue of their efficiency they also accumulate more of the limiting nutrient in the grain: firstly, better human nutrition if consumed (zinc, especially), and secondly, markedly better seedling vigour when resown on deficient soils. Finally, the degree of micronutrient efficiency currently available is generally adequate to overcome subclinical deficiency which is not usually treated because it is not usually recognized.

Concluding remarks

Breeding cereals with enhanced zinc efficiency is perceived as a worthwhile objective which can decrease fertilizer requirements, improve seedling vigour, overcome yield losses from unrecognized and subclinical deficiencies, increase resistance to pathogens and enhance the yield and quality of wheat for human consumption. More than that, zinc-efficient wheats may be necessary to permit full exploitation of stored water at depth in nutritionally inhospitable subsoils, a response that may not only permit a breakthrough in the yield potential of such semi-arid, high soil pH areas, but may be expected to help to limit the creep of dryland salinity.

Useful genetically-controlled zinc-efficiency factors exist in wheat, oats and barley and these may be further supplemented with efficiency genes from rye. However, their exploitation in a breeding program is limited by the high spatial variability of zinc in field sites, which remains a major experimental problem. For this reason, better selection criteria are needed to allow screening in early-generation segregating materials by laboratory-based procedures, but their development is restricted because the mechanisms of these efficiency factors are poorly understood. Although some progress has been made, much more work is needed in this area. Meanwhile selection may best be done in the field.

Acknowledgements

The authors wish to thank Mrs Teresa Fowles and Mr Nick Robinson for their dedicated analytical work for this paper, Drs. Bob Hannam and Nigel Wilhelm for soil analyses, Mr Trevor

Hancock and Dr Stan Eckert for statistical advice, Mr David Morris for expert field assistance, farmers Mr Jim Byrne, Mr Wayne Hayward, Mr David Smight, Mr Ian Glover, Mr John McEvoy and their families for the use of land and facilities on their properties, Top Australia Ltd and Hifert Pty Ltd for donations of fertilizers and the Grains Research and Development Corporation for financial support.

References

Bolwing D J F, Graham R D and Dunlop J 1978 The relationship between the cell electrical potential difference and salt uptake in the roots of *Helianthus annuus*. J. Exp. Bot. 29, 135–140.

Burgess L W, Klein T and Liddell C M 1984 Crown rot of wheat. *In* Research on Root and Crown Rots of Wheat in Australia. Proceedings of a Workshop held at the Waite Institute, Adelaide, 14–15 February 1984. Australian Wheat Industry Research Council, pp 62–65.

Epstein E 1972 Mineral Nutrition of Plants: Principles and perspectives. Wiley, New York.

Graham J H, Leonard R T and Menge J A 1981 Membrane-mediated decrease in root exudation responsible for phosphorus inhibition of vesicular-arbuscular mycorrhiza formation. Plant Physiol. 68, 548–552.

Graham R D 1984 Breeding for nutritional charactersitics in cereals. Adv. Plant Nutr. 1, 57–102.

Graham R D 1987 Triticale, a cereal for micronutrient-deficient soils. International Triticale Newsletter No. 1. University of New England, Armidale.

Graham R D, Ascher J S, Ellis P A E and Shepherd K W 1987 Transfer to wheat of the copper efficiency factor carried on rye chromosome arm 5RL. Plant and Soil 99, 107–114.

Graham R D 1988 Development of wheats with enhanced nutrient efficiency: Progress and potential. *In* Wheat Production Constraints in Tropical Environments. Ed. A R Klatt. pp 305–320. CIMMYT, Mexico.

Graham R D 1988a. Genotypic differences in tolerance to manganese deficiency. *In* Manganese in Soils and Plants. Eds. R D Graham, R J Hannam and N C Uren. pp 261–276. Kluwer Academic Publishers, Dordrecht.

Graham R D and Webb M J 1991 Micronutrients and resistance and tolerance to disease. *In* Micronutrients in Agriculture. 2nd ed. Eds. J.J. Mortvedt et al. pp. 329–370. Soil Science Society of America, Madison, WI.

Graham R D and Ascher J S 1990 Analysis of limitation to crop yields due to subsoil infertility. Biennial Report, Waite Agric. Res. Inst., Glen Osmond, S. Australia, 1988–89.

Grundon N J 1980 Effectiveness of soil dressings and foliar sprays of copper sulphate in correcting copper deficiency of wheat (*Triticum aestivum* L.) in Queensland. Aust. J. Exp. Agric. Anim. Husb. 20, 717–723.

Grundon N J and Best E K 1981 Tolerance of some winter and summer crops to copper deficiency. *In* Copper in Soils and Plants. Eds. J F Loneragan, A D Robson and R D Graham Academic Press, Sydney. p 360.

Harry S P 1982 Tolerance of wheat, rye and triticale to copper and zinc deficiency in soils of low and high pH. M. Ag. Sc. thesis, Unviersity of Adelaide.

Haynes J L and Robbins W R 1948 Calcium and boron as essential factors in the root environment. J. Amer. Soc. Agron. 40, 795–803.

Jones G B and Belling G B 1967 The movement of copper, molybdenum and selenium in soils as indicated by radioactive tracers. Aust. J. Agric. Res. 18, 733–740.

Loneragan J F, Kirk G J and Webb M J 1987 Translocation and function of zinc in roots. J. Plant Nutr. 10, 1247–1254.

Nable R O and Loneragan J F 1984 Translocation of manganese in subterranean clover (*Triolium subterraneum* L. cv. Seaton Park). II. Effects of leaf sensescence and of restricting supply of manganese to part of a split root system. Aust. J. Plant Physiol. 11, 113–118.

Nable R O and Webb M J 1992 Response of two wheat gentypes to low subsoil zinc supply. International Conference on the Genetic Aspects of Plant Mineral Nutrition, Canberra. p 86.

Pollard A S, Parr A J and Loughman B C 1977 Boron in relation to membrane function in higher plants. J. Exp. Bot. 28, 831–839.

Reuter D J and Robinson J B 1986 Plant Analysis: An Interpretation Manual. Inkata Press, Melbourne.

Solunke S N and Malewar G U 1987 Differential responses of wheat genotypes to zinc fertilization. J. Mahar. Agric. Univ. 12, 382–383.

Sparrow, D H and Graham R D 1988 Susceptibility of zinc-deficient wheat plants to colonization by *Fusarium graminearum* Schw. Group 1. Plant and Soil 112, 261–266.

Takkar P N, Chhibba I M and Mehta S K 1988 Twenty Years of Coordinated Research on Micronutrients in Soils and Plants, 1967–1987. Indian Institute of Soil Science, Bhopal, India. 312 p.

Welch R M and House W A 1983 Factors affecting the bioavailability of mineral nutrients in plant foods. *In* Crops as Sources of Nutrients for Humans. Eds R M Welch and W H Gabelman. pp. 37–54. American Society of Agronomy, Madison, WI.

Welch R M, Webb M J, and Loneragan J F 1982 Zinc in membrane function and its role in phosphorus toxicity. pp 710–715. *In* Plant Nutrition 1982. Proceedings of the Ninth Int. Plant Nutrition Colloquium. Ed. A Scaife. Commonwealth Agricultural Bureaux, Slough.

Wilkinson G N, Eckert S R, Hancock T W and Mayo O 1983 Nearest neighbour (NN) analysis of field experiments. J. Roy. Stat. Soc. B45, 151–211.

P. J. Randall et al. (Eds.), Genetic aspects of plant mineral nutrition, 359–361.

PLSO SV82

Combining ability of the response to boron deficiency in wheat

S. JAMJOD[1], C.E. MANN[2] and B. RERKASEM[1]
[1]*Agronomy Department, Faculty of Agriculture, Chiang Mai University, Chiang Mai, Thailand, 50002 and* [2]*CIMMYT, P.O. Box 9-188, Bangkok, Thailand, 10900*

Key words: boron, *Triticum aestivum* L., wheat

Abstract

Seven wheat lines (adapted to Thailand originated in Mexico) and their F1 hybrids from a diallel cross were grown in sand culture from November 1990 to February 1991. Plants were watered twice daily with a complete nutrient solution supplying a low level of boron at 0.2 μM. Plant response was measured as the average number of grains set in the two basal florets (F1 + 2) of 10 central spikelets of the wheat ears.

Grain set in low boron was significantly different among parents and offspring. Both general combining ability (GCA), a measure of additive gene effects, and specific combining ability (SCA), a measure of non-additive gene effects were highly significant ($p < 0.01$). The GCA effects were the major factor that controls this character. SCA effects were smaller, and found to be positive in six crosses between susceptible × tolerant and tolerant × tolerant lines. This indicates that it may be possible to transfer the B deficiency tolerance character to susceptible lines by a backcrossing program. The positive GCA effects of tolerant lines suggest that these lines can be useful in breeding for more tolerance in problem areas.

Introduction

Low boron (B) soils are distributed widely in agricultural areas of the world (Sillanpaa, 1982). B deficiency can severely depress grain yield in wheat through grain set failure (Li et al., 1978; Rerkasem et al., 1989; Silva and Andrade, 1983; Sthapit, 1988). In most of these reports, cultivars and lines differed widely in their responses. Wheat genotypes adapted to condition in Thailand, but originated in Mexico, were identified as tolerant or susceptible to B deficiency according to their ability to set grain in a low B soil at Chiang Mai, Northern Thailand (Rerkasem and Jamjod, 1989). Seven of these, five tolerant (Dt) and two susceptible (Ds) lines were used in this study to examine the genetic control of responses to B deficiency. General combining ability (GCA), a measure of additive genetic effects and specific combining ability (SCA), a measure of non-additive genetic effects, were calculated from the response of parents and their progeny when grown under low B conditions.

Materials and methods

The experiment was conduct at the Faculty of Agriculture, Chiang Mai University, Chiang Mai, Thailand, on seven wheat genotypes (Ds; SW41, SW23; Dt; SAMOENG1, INSEE1, SAMOENG2, CMU10 and CMU26) from November 1989 to February 1991. From a sowing in November 1989 a diallel cross was formed from all possible crosses excluding reciprocals. The parents and 21 F1 hybrids were harvested and grown in sand culture in 1990/1991 season. The seed were sown on November 9 1990 in 30 cm diameter earthenware pots containing washed river sand, and watered twice daily with a complete nutrient solution (Somasegaran and Hoben, 1985) with 9 mM N and a low B level at

0.2 μM. There were three replicates for each genotype, each pot containing 20 plants. At anthesis each ear was marked for the date of first pollen shedding; the first five ears to reach anthesis in each plant were sampled at maturity. Data on the average number of grains set in the two basal florets (F1 + 2) of 10 central spikelets of the wheat ear were recorded for the measurement of plant response to boron deficiency (Rerkasem et al., 1991). The combining ability was estimated according to Method 4 Model I of Griffing (1956).

Results and discussion

Grain set of the Dt and Ds wheat lines in the low B sand culture (Table 1) corresponded well with their relative performance in low B soils (Rerkasem and Jamjod 1989). The Ds parents, SW41 and SW23, set less than 0.1 grains/F1 + 2. This compared with more than 0.8 grains/F1 + 2 in the Dt parents: SAMOENG1, INSEE1, SAMOENG2, CMU10 and CMU26. With the exception of those involving SAMOENG1, grain set was high in crosses between the Dt lines, with >0.9 grains/F1 + 2, compared with 0.15 grains/F1 + 2 in the cross between the Ds lines. Crosses between a Dt and a Ds line all set more grains than the Ds parents.

Both general combining ability (GCA) and specific combining ability (SCA) for grains/F1 + 2 in 0.2 μM B were highly significant indicating that both additive and non-additive gene effects were involved in the inheritance of the response to B deficiency (Table 2). Furthermore, the GCA component of variance was more than seven times that of the SCA component of variance, indicating the predominance of additive gene effects in the inheritance of this character. Four Dt parents; INSEE1, SAMOENG2, CMU10 and CMU26 showed positive GCA effects suggesting that they are good general combiners. The negative GCA effects of the Ds

Table 1. Average number of grains per two basal florets of central spikelets (grains/F1 + 2) for 7 parents and 21 F1 hybrids from a diallel cross grown under 0.2 μM boron concentration

Genotypes	SW41	SW23	SAMOENG1	INSEE1	SAMOENG2	CMU10	CMU26
SW41	<u>0.02</u>	0.15	0.52	0.69	0.37	0.52	0.84
SW23		<u>0.08</u>	0.22	0.50	0.66	0.45	0.59
SAMOENG1			<u>0.84</u>	0.62	0.72	0.83	0.30
INSEE1				<u>0.82</u>	1.07	1.06	0.91
SAMOENG2					<u>1.04</u>	0.94	1.10
CMU10						<u>1.03</u>	1.21
CMU26							<u>1.22</u>

SE parents 0.17, SE hybrids 0.14.

Table 2. Estimates of general combining ability (GCA) and specific combining ability (SCA) effects for number of grains per two basal florets of central spikelets (grains/F1 + 2) of 7 parents and 21 F1 hybrids from a diallel cross grown under 0.2 μM boron concentration

Genotypes	GCA effects	SCA effects					
		SW23	SAMOENG1	INSEE12	SAMOENG2	CMU10	CMU26
SW41	−0.20**	−0.03	0.21**	0.05	−0.26**	−0.15**	0.18**
SW23	−0.30**		0.02	−0.03	0.12**	−0.11*	0.03
SAMOENG1	−0.17**			−0.03	0.12**	0.13**	−0.38**
INSEE1	0.15**				0.08	0.01	−0.10*
SAMOENG2	0.16**					−0.08	0.09
CMU10	0.18**						0.17**
CMU26	0.17**						

MS gca 0.02** SE gi 0.02 (* $p<0.05$; ** $p<0.01$)
MS sca 0.03** SE gi-gj 0.03
SE sij 0.05

parents, SW41 and SW23, and a Dt parent, SAMOENG1, showed these to be poor general combiners, suggesting that Dt lines should be examined for their combining ability before they are used for breeding purposes. However, among 21 crosses only 12 showed positive SCA effects, of which only six crosses were statistically significant.

The results of this study indicate that the genotypic variation in the response to B deficiency is mostly due to additive genetic effects. The variation in response among the Dt parents indicates that tolerance to low levels of B is expressed as a quantitative character. To increase the level of tolerance to B deficiency in breeding programs, this character should be responsive to direct selection. In this case, backcross method or early generation selection would be more effective. The estimate of the combining ability alone, however, does not present the complete picture. Along with this information, heritability and gene actions must be considered.

References

Griffing B 1956. Concept of general and specific combining ability in relation to diallel crossing systems. Aust. J. Biol. Sci. 9, 463–493.

Li W H, Chao M C, Jern N S, Li C R, Chu W J and Wang C L 1978 Studies on cause of sterility of wheat. Jour. Northern Agric. College 3, 1–19. (*In Chinese*).

Rerkasem B and Jamjod S 1989 Correcting boron deficiency induced ear sterility in wheat and barely. Thai Jour. Soils and Fertilizers 11, 200–209. (*In Thai*).

Rerkasem B, Jamjod S and Lodkaew S 1991 Assessment of grain set failure and diagnosis for boron deficiency in wheat. Proceedings of International Conference on Wheat for the Non-traditional Warm Areas. CIMMYT, Mexico.

Sillanpaa M 1982 Micronutrients and nutrient status of soils. FAO Soil Bull. 48.

Silva A R da and Andrade J M V da 1983 Influence of micronutrients on the male sterility on upland wheat and on rice and soybean yields in red-yellow latosol. Pesq. agropec. bras. Brasilia 18, 593–601. (*In Portuguese*).

Somaegaran P and Hoben H J 1985 Methods in Legume-Rhizobium Technology. NifTAL, University of Hawaii.

Sthapit B R 1988 Studies on wheat sterility problem in the Hills, Tar and Tarai of Nepal. Lumle Agricultural Centre. Technical Paper No 16/88.

P. J. Randall et al. (Eds.), Genetic aspects of plant mineral nutrition, 363–366.

PLSO SV97

Yield evaluation of a gene for boron tolerance using backcross-derived lines

D.B. MOODY[1], A.J. RATHJEN[1] and B. CARTWRIGHT[2]
[1]*Department of Plant Science, Waite Institute, Glen Osmond, South Australia 5064 and*
[2]*CSIRO Division of Soils, Glen Osmond, South Australia 5064*

Key words: boron, backcross, tolerance, toxicity, wheat, yield

Abstract

Wheat varieties show differential responses to high concentrations of soil boron and genetic studies have shown a single major gene, *Bo1*, is responsible for the higher level of boron tolerance found in varieties historically dominant in southern Australia. The yield advantage of this gene was evaluated over a range of soil types in southern Australia by comparing boron tolerant and boron intolerant derivatives from a backcrossing program. The advantage of boron tolerant lines ranged up to 11% with an average yield advantage of 3.3% in all trials conducted over a range of soil types. Other evidence suggests this is a conservative estimate of the average yield advantage of the *Bo1* gene in this region. Yields of the two groups of lines were not significantly different at sites with soil boron concentrations in the 'normal' range.

Introduction

Reduction in grain yield in South Australian cereal crops due to boron toxicity was first demonstrated in barley (Cartwright et al., 1984). Subsequent investigations have shown substantial genetic variation in response to boron toxicity in wheat both in glasshouse experiments (Moody et al., 1988; Paull et al., 1988) and field trials (Cartwright et al., 1987; Moody et al., 1989). In particular, the discovery that the widely grown varieties of the Federation-Currawa family are relatively boron tolerant (Paull et al., 1986) suggests that boron tolerance is an important feature in the adaptation of wheat varieties to southern Australia. These varieties have occupied over 70% of the Victorian, 50% of the South Australian and 20% of the Western Australian wheat acreages for extended periods of time (Paull, 1990).

The majority of the current Australian wheat varieties are moderately intolerant (MI) whereas Halberd and its ancestors are moderately tolerant (MT) (Moody et al., 1989). The transfer of a higher level of boron tolerance into otherwise adapted Australian cultivars is a major objective of the wheat breeding program at the Waite Institute. Inheritance of boron tolerance is under the control of a series of major genes (Paull et al., 1991) suggesting backcrossing may be employed to achieve this goal.

This paper reports on yield trials evaluating backcross derived lines. Comparative yields of closely related lines with varying levels of boron tolerance provide an estimate of the advantage of boron tolerance over a range of sites.

Materials and methods

The variety Halberd has been used as the MT donor parent and the MI variety Schomburgk as the recurrent parent in the backcrossing program. The segregation ratios during the backcrossing program and in the BC_3F_2 families were compatible with a single gene for boron tolerance being transferred from Halberd into Schomburgk.

The relatively tolerant BC_3F_1 plants were selected, allowed to self and selections from each of these families were multiplied through the BC_3F_3 and BC_3F_4. The BC_3F_4 lines were classified as either homozygous tolerant, heterogeneous or homozygous intolerant by visual assessment of 15 plants grown in boron enriched soil. MT, heterogeneous and MI BC_3F_4 lines were sown in trials at 3 sites during 1988 and at eight sites during both 1989 and 1990. these sites included Rudall, Minnipa, Snowtown, Two Wells, Roseworthy, Windsor, Yeelanna and the Waite Institute which are widely located across the South Australian cereal belt, and Walpeup in the Victorian Mallee.

Grain samples were taken at harvest from all lines included in the 1988 trial at Two Wells and from the Halberd and Schomburgk standards at all sites in both years. Whole shoots were collected at booting stage from the standards and six arbitrarily chosen MT or MI lines in the 1988 trial at Two Wells. Samples were digested in nitric acid and analysed using inductively coupled plasma (ICP) spectrometry to provide the concentrations of fourteen elements, including boron (Zarcinas et al., 1987).

Soil cores to a depth of 1 m were taken at all sites and the concentration of boron extractable in hot $CaCl_2$ (Aitken et al., 1987) determined at 10 cm intervals.

Results

Soil analysis

Soil boron concentrations showed considerable horizontal and vertical variability at all sites with the exception of the Waite Institute where the extractable soil boron levels were below 3 mg kg^{-1} throughout the profile. Concentrations in excess of 80 mg kg^{-1} were recorded below 30 cms depth at Two Wells while maximum concentrations at the other sites were in the range 13–40 mg kg^{-1}. The normal range for most soils is 0.5–5.0 mg B kg^{-1}.

Yield evaluation

While there were significant yield differences between the three groups (MT, heterogeneous and MI) of BC_3F_2 derivatives, in trials at Two Wells, Rudall, Walpeup, and Roseworthy (Table 1), the yield advantage of the boron tolerant genotypes was not uniform across all sites and the site * genotype interaction was highly significant ($p < 0.001$). The performance of the heterogeneous genotypes was generally intermediate between the MT and MI genotypes (Table 1). Over all sites the mean yield advantages of the MT genotypes and the heterogeneous genotypes in comparison with the MI genotypes were 3.3% and 2.6%, respectively.

Plant analysis

Mean grain boron concentration of all of the MT lines was similar to Halberd and significantly

Table 1. Grain yields for MT and heterogeneous backcross derived lines in replicated trials at eight locations during 1988, 1989 and 1990. Grain yields are expressed as a percentage of the MI lines

Site	1988		1989		1990	
	MT	Heterogeneous	MT	Heterogeneous	MT	Heterogeneous
Two Wells	110[a]	106	108[b]	106[a]	105[a]	97
Minnipa	106	100	101	102	100	104
Rudall			111[b]	111[b]	108[b]	110[b]
Roseworthy			107[a]	102	100	104
Walpeup			102	108[a]	105[a]	97
Snowtown			101	104	99	101
Windsor			101	100	97	100
Waite			100	97	101	100
Yeelanna	100	101				

[a]significantly different from MI types at $p < 0.05$. [b]significantly different from the MI types at $p < 0.01$.

Table 2. Boron concentrations (mg kg^{-1}) in grain and tissue samples taken from MT, heterogeneous and MI BC_3F_2 derivatives grown at Two Wells during 1988 compared with parental varieties Halberd and Schomburgk

Genotype	Grain B	Tissue B
MT BC_3F_2 derivatives	3.65	27.5
Heterogeneous BC_3F_2 derivatives	3.99	
MI BC_3F_2 derivatives	5.39	41.8
Schomburgk	5.50	39.0
Halberd	3.81	28.0
lsd (5%)	0.34	5.93

lower than those of the MI genotypes and Schomburgk (Table 2). Similarly, the tissue boron concentrations of three MT genotypes were similar to Halberd and significantly lower than Schomburgk and three MI genotypes. The mean grain boron concentrations of the heterogeneous lines were intermediate between the MI and MT genotypes.

Discussion

Although genetic variability in boron tolerance to wheat has been reported by other workers (Chhipa and Lal 1990, Mehrotra et al., 1980) the specific breeding of boron tolerant varieties has not been previously attempted. The results presented indicate that a simple glasshouse screening test can efficiently facilitate the transfer of boron tolerance in a backcrossing program. Boron tolerance in wheat has been reported to be under the control of a series of major genes (Paull et al., 1991) with a single gene difference (designated *Bo1*) occurring between the MT variety Halberd and the MI variety Warigal. In the experiments described in this paper the segregation ratios during the backcrossing were consistent with a single major gene for boron tolerance having been transferred from the donor variety Halberd to the recurrent parent, Schomburgk, which is closely related to Warigal. This gene is undoubtedly the same gene (*Bo1*) referred to by Paull et al., (1991).

Although yield benefits accruing from the incorporation of *Bo1* were variable, an advantage to the MT lines of 11% was achieved at Rudall. The average yield advantage conferred over all trials in this experiment was 3.3%. However, in the extensive Secondary Wheat Variety trial series, conducted in twenty two locations across South Australia by the South Australian Department of Agriculture, a MT BC_3F_2 selection has shown a yield advantage of 5% above the recurrent parent Schomburgk during 1990 (S P Jefferies, pers. comm.). The yield advantage of this selection was greatest in the Murray Mallee, the Yorke Peninsula and northern Eyre Peninsula where the importance of boron toxicity has been previously suggested by the historical dominance of varieties moderately tolerant to boron (Paull 1990; Rathjen et al., 1987).

Glasshouse studies (Nable et al., 1990) have indicated that barley genotypes with most tolerance to excess B supply were more susceptible to B deficiency. This notion was suggested by Rathjen et al. (1987) who showed disruptive selection in a wheat breeding program between sites with low and high soil B levels. However, in the present studies the MI lines did not show a significant yield advantage at any of the trial sites.

Although these studies were conducted over a range of sites considered representative of the cereal growing areas of South Australia, no attempt has been made to determine the maximum yield advantage conferred by this level of boron tolerance. Nor is there any data to show that is the optimal level of boron tolerance for particular regions in the South Australian cereal belt. Genotypes more tolerant than Halberd have been found by screening part of the Australian Wheat Collection (Moody et al., 1988) and the incorporation of these sources of boron tolerance into otherwise adapted genotypes is in progress. Trials are being conducted during 1991 using backcross derived lines with the boron tolerance of the donor parents exceeding Halberd.

Acknowledgements

The financial support of the Wheat Industry Research Council and the South Australian Wheat Industry Research Committee is gratefully acknowledge. The technical assistance of Jim

Lewis, Jim Chigwidden, Chris Stone, Michael Kroehn (Waite Institute) and Leonie Spouncer (CSIRO Soils Division) has been greatly appreciated.

References

Aitken R L, Jeffrey A J and Compton B L 1987 Evaluation of selected extractants for boron in some Queensland soils. Aust. J. Soil Res. 25, 263–273.

Cartwright B, Rathjen A J, Sparrow D H B, Paull J G and Zarcinas B A 1987 Boron tolerance in Australian varieties of wheat and barley. *In* Genetic Aspects of Plant Mineral Nutrition. Eds. H W Gabelman and B C Loughman. pp 139–151. Martinus Nijhoff, Dordrecht, The Netherlands.

Cartwright B, Zarcinas B A and Mayfield A H 1984 Toxic concentrations of boron in a red brown earth at Gladstone, South Australia. Aust. J. Soil Res. 22, 261–272.

Chhipa B R and Lal P 1990 A comparative study on the effect of soil B on yield, yield attributes and nutrient uptake by susceptible and tolerant varieties of wheat. An. Edafol. Agrobiol. 48, 489–498.

Mehrotra O N, Srivastava R D L and Mishra P H 1980 Some observations on the relative tolerance of wheat genotypes to boron. Indian Agric. 24, 223–238.

Moody D B, Rathjen A J, Cartwright B, Paull J G and Lewis J 1988, Genetic diversity and geographical distribution of tolerance to high levels of soil boron. Eds. T E Miller and R M D Koebner. pp 859–865. Proc. 7th Int. Wheat Genet. Symp, Cambridge, UK.

Moody D B, Rathjen A J and Cartwright B 1989 Boron tolerance of Australian wheat varieties. Ed. G P Ayling. p 566. Proc 5th Australian Agronomy Conference, Perth, Australia.

Paull J G, Rathjen A J and Cartwright B 1986 Boron tolerance in wheat varieties and advanced breeding lines. *In* Proc 5th Assembly Wheat Breeding Society of Australia, Perth/Merredin. Ed. R McLean. pp 245–247.

Paull J G, Cartwright B and Rathjen A J 1988 Responses of wheat and barley genotypes to toxic concentrations of soil boron. Euphytica 39, 137–144.

Paull J G 1990 Genetic studies on the tolerance of wheat to high concentrations of boron. Ph.D Thesis, University of Adelaide, Australia.

Paull J G, Rathjen A J and Cartwright B 1991 Major gene control of tolerance of bread wheat to high concentrations of soil boron. Euphytica 55, 217–228.

Pederson D G and Rathjen A J 1981 Choosing trial sites to maximise selection response for grain yield in spring wheat. Aust. J. Agric. Res. 32, 411–424.

Rathjen A J, Cartwright B, Paull J G, Moody D B and Lewis J 1987 Breeding for tolerance of mineral toxicities in Australian cereals with special reference to boron. *In* Perspectives and Priorities in Plant Production Research. Eds P G E Searle and B G Davey. pp 111–130. Sydney University Press, Australia.

Zarcinas B A, Cartwright B and Spouncer L R 1987 Nitric acid and multi-element analysis of plant material by inductively coupled plasma spectrometry. Commun. Soil Sci. Plant Anal. 18, 131–146.

P. J. Randall et al. (Eds.), Genetic aspects of plant mineral nutrition, 367–376.
 PLSO SV96

Physiological and genetic control of the tolerance of wheat to high concentrations of boron and implications for plant breeding

J.G. PAULL[1], R.O. NABLE[2] and A.J. RATHJEN[1]
[1]*Waite Agricultural Research Institute, PMB 1, Glen Osmond, South Australia, 5064 and* [2]*CSIRO Division of Soils, PMB 2, Glen Osmond, South Australia, 5064*

Key words: boron, genetic variation, major genes, uptake, wheat

Abstract

Physiological and genetic studies have been undertaken to further the understanding of genetic variation in response to high concentrations of B in the soil and so facilitate the breeding of tolerant varieties for cultivation in high B regions. Genetic variation in response to high concentrations of B has been identified for a number of crop and pasture species of southern Australia, including wheat, barley, oats, field peas and annual pasture medics. The wheat variety Halberd, which was the most widely grown variety in Australia during the 1970s and early 1980s, is the most tolerant of the current Australian wheat varieties. The mechanism of tolerance for all species studied is reduced accumulation of B by tolerant genotypes in both roots and shoots. Results from experiments of uptake kinetics indicate that control of B uptake is a non-metabolic process. The response of wheat to high B supply is under the control of several major additive genes, one of which has been located to chromosome 4A.

Introduction

High concentrations of B have recently been recorded in soil and plant samples obtained from widespread regions of the cereal growing districts of southern Australia (Cartwright et al.,1984, 1986). The concentration of B increases with depth and reaches a maximum, in the range 15 to greater than 100 mg kg^{-1} extractable B, usually between 30–100 cm of the surface. Soils with high concentrations of B are generally sodic and alkaline and include a number of soil types such as red-brown earths, calcareous earths, calcareous sands and grey clays (Cartwright et al., 1986). In a survey of the variability of distribution of B in agriculturally important land units of upper Eyre Penninsula, Holloway (1991) showed that some easily recognised subsoil features such as the presence of Blanchetown Clay or the reddish phase of the Wiabuna Formation were useful in indicating where high concentrations of B are likely to occur within a profile. In several land units, high B was associated with elevated portions of the landscape. A significant yield reduction due to high levels of B in soil and plants has been reported for barley (Cartwright et al., 1984) and the widespread distribution of high levels of B in plant samples (Cartwright et al., 1986) would indicate that B toxicity is a major factor affecting yield in southern Australia.

The extensive areas and low rainfall (275–500 mm) preclude amelioration of the high levels of B in soil, therefore the most feasible means of limiting loss of production due to B toxicity would be to identify and breed varieties tolerant to high concentrations of B. Selection for tolerance to B is now a priority of the South Australian breeding programs. Investigations into the genetic and physiological control of tolerance to B have been conducted to facilitate the breeding of tolerant varieties. This paper reviews the results of these studies, principally for wheat, although reference is made to other crops to demonstrate an apparently similar mechanism controlling tolerance to B for a number of species.

Genetic variation in response to B

The identification of genetic variation in response to B involved two stages, namely the screening of many lines followed by a more detailed study of a few contrasting lines. Initially, 150 locally adapted varieties and breeding lines were grown at a site with a high level (>100 mg kg^{-1} extractable) of B in the sub-soil, in an experiment of only two replicates to enable a maximum number of lines to be tested. Grain yield and the concentrations of B in grain were determined. (The procedure for determining the concentration of B in plant tissues for all experiments described, consisted of digestion of grain or tissue samples in nitric acid followed by analysis by inductively coupled plasma spectrometry according to the procedure of Zarcinas et al. (1987)). The concentration of B in grain of high yielding lines was significantly less than for the lower yielding lines (Cartwright et al., 1987). Six contrasting lines were selected and studied in detail at a range of B treatments in a pot experiment. A large range in response was observed and lines with the lowest concentrations of B in grain for the field experiment were the most tolerant with respect to dry matter and grain yield at high B treatments and contained the lowest concentration of B in shoots (Table 1) (Paull et al., 1988a).

Tolerant and sensitive lines were further tested for response to B, under controlled conditions, to identify parameters (e.g. symptom expression and plant development) which may be suitable for selecting tolerant lines when screening the large number of plants required in a breeding program (Paull et al., 1990). B is transported in the transpiration stream and accumulates at the site of evapotranspiration (Oertli and Kohl, 1961). As a consequence, symptoms consisting of tissue chlorosis and necrosis progress from the tip, or the margin, to the base of leaves and are most severe on the oldest leaves. Symptom expression differs markedly among genotypes and are least severe, and take longer to develop, for tolerant lines (Nable, 1988; Paull et al., 1990, 1991b). Other effects of B upon the growth of plants include reduced vigour, which is reflected in the height of plants, and delayed development (Paull et al., 1990). The delay in development is most pronounced with respect to tillering, rather than development of the primary culm, and the number of tillers produced is also reduced. These parameters, together with symptom expression, may therefore be used as criteria for selection for response to B (Paull et al., 1990).

The procedure adopted for screening wheat (including varieties, segregating breeding populations and germplasm collections) for response to B, at the Waite Institute, consists of applying a high concentration of B (typically 150 mg B kg^{-1} soil, although the concentration is varied according to the population being tested and is lower when a high proportion of sensitive lines is

Table 1. Yield and concentration of B in grain and shoots of six wheat genotypes when grown under high concentrations of B in the field and in a pot experiment. Adapted from Paull et al. (1988a)

Genotype	Field experiment[a]		Pot experiment[b]			
	Grain yield % site mean	Grain B (mg kg^{-1})	Dry matter production (g pot^{-1})			Shoot B (mg kg^{-1})
			0	50	100	
(Wq * KP) * WmH)[c]	133	4.9	32.2	33.5	32.1	131.1
Halberd	116	3.4	35.0	35.5	29.9	144.6
Bindawarra	113	8.7	32.4	34.0	4.2	270.3
(C8 * MM)	78	8.9	32.1	30.7	8.2	234.1
(Kins * CP)	60	7.7	30.0	29.6	8.0	191.1
(Wl * MMC)	58	9.1	32.3	28.6	1.0	276.3

[a]The field experiment comprised 150 breeding lines and cultivars of wheat and was conducted at Two Wells, a site with high concentrations of B in the sub-soil (Cartwright et al. 1987).
[b]The pot experiment comprised five B treatments. Yield results are presented for the 0, 50 and 100 mgB kg^{-1} soil treatments. Concentrations of B in tissue are those obtained for the 50 mgB kg^{-1} soil treatment.
[c]The full pedigrees of these lines are detailed by Paull et al. (1988a).

expected) to a fertile surface soil and placing the soil in a large box (2 m × 1 m × 0.25 m deep), in a glasshouse (Moody et al., 1988). Seeds are sown at a spacing of 3 cm × 5 cm and a grid consisting of either several varieties of known response, or the parents of a segregating population, is sown for reference. By this method in excess of 1500 accessions of wheat from the Australian Winter Cereals Collections, Tamworth, have been screened for tolerance to B. Approximately 7% (107) of the lines were more tolerant than Halberd, the most tolerant current Australian wheat variety (Moody et al., 1988). A number of the exotic tolerant lines are being used as donor parents in a backcrossing program to incorporate higher levels of B tolerance into Australian wheats. A similar method of screening has been applied to collections of barley (Lance, unpublished), medics and peas (Paull et al., 1991a) and oats (Barr and Tasker, unpublished) and in all instances lines more tolerant than all current Australian varieties have been identified.

Agricultural significance of B tolerance

A factor recognized early in the testing of Australian wheat varieties is that those which have been most widely grown in south-eastern Australia for the majority of the twentieth century are among the most tolerant of all Australian varieties. In particular, varieties which are descendents of the two varieties Federation (released 1901) and Currawa (1912), both of which are classified as moderately tolerant to B, have dominated wheat production in South Australia and Victoria, and also have low levels of B in shoots when grown under high B conditions (Table 2). Varieties of this family, which includes Halberd, have often accounted for greater than 40% of the production in each of the two States (Macindoe and Walkden-Brown, 1968).

The distribution of tolerant wheat varieties has been far from uniform, but rather they have been concentrated in regions where high levels of B prevail. For example, Halberd has accounted for greater than 70% of the total delivery of wheat to silos in areas of Eyre Penninsula and the Murray Mallee (Rathjen and Pederson, 1986), the regions of South Australia where high concentrations of B have been detected in barley grain (Cartwright and Hirsch, 1986). The dominance of tolerant varieties in specific regions indicates that high concentrations of B exert a significant selection pressure and influence the distribution of wheat varieties. Further evidence of the influence of B on agriculture in southern Australia is derived from the fact that the pea variety Early Dun, introduced to Australia during the 19th century, is not only among the most tolerant of Australian pea varieties (Paull et al., 1992), but is also still one of the most widely grown.

Physiological control of B tolerance

The initial pot and field experiments demonstrated a large range in the tolerance of wheat to high concentrations of B and tolerant genotypes contained low concentrations in shoots and grain (Cartwright et al., 1987; Paull et al., 1988a;

Table 2. Concentration of B (mg kg^{-1}) in shoots of wheat cultivars and lines of historical importance in Australia when grown under high B conditions at Two Wells. The named cultivars are listed by date of release

Federation/Currawa		Other tall varieties		Semi-dwarf varieties	
Federation	51.6	Bencubbin	58.2	*Mexican introductions*	
Currawa	39.6	Gabo	87.8	Mec-3	75.4
Ranee	60.0	Gamenya	89.4	Mexico 120	87.5
Dundee	49.7			WW-15	82.5
Quadrat	36.5			Siete Cerros	88.1
Insignia	40.2			*Australian varieties*	
Heron	52.5			Condor	70.6
Halberd	41.7			Kite	87.8
LSD $p < 0.05$, 15.0					

Rathjen et al., 1987). Low concentrations of B in plant tissues may arise through several mechanisms, namely –

(1) exclusion of B from the root system, either by a physical barrier to the entry of B, by a mechanism which pumps B from the root system, or by a chemical process which alters the pH of the rhizosphere thereby affecting the availability of B,

(2) reduced translocation of B from roots to shoots or

(3) avoidance of high concentrations of B in the sub-soil by virtue of a shallow root system.

Six wheat varieties and breeding lines selected as having contrasting response to B (Cartwright et al., 1987; Moody et al., 1988; Paull et al., 1988a,b) were cultured in solution at a range of B treatments from 'normal' to toxic and roots and shoots were analysed for concentrations of B to discriminate between the possible mechanisms (Nable, 1988). The dry matter response of the lines (Nable, 1988) concurred with previous results for plants grown in soil and the most sensitive variety, Kenya Farmer, died with the addition of 2000 μM B to the culture solution, a treatment which did not significantly affect the yield of G61450. Concentrations of B in tissues of the tolerant lines were significantly lower than for the more sensitive lines (Fig. 1) and B concentrations were much lower in roots than shoots for all lines. The lower concentrations of B in tissues of the tolerant lines were observed over the full range of treatments, including the lower treatments at which there was no effect of B upon yield.

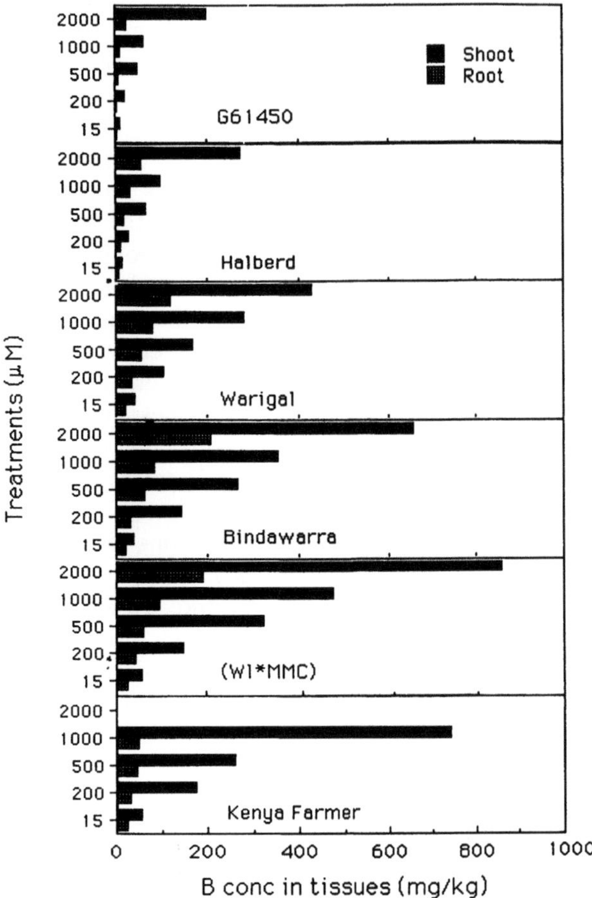

Fig. 1. Concentration of B (mg kg^{-1}) in roots and shoots of six wheat lines when grown in solution culture at five B treatments. Response of genotypes to B: G61450-tolerant, Halberd-moderately tolerant, Warigal-moderately sensitive, Bindawarra-moderately sensitive, (Wl * MMC)-sensitive, Kenya Farmer-very sensitive. Adapted from Nable (1988).

Further experiments conducted for barley (*Hordeum vulgare*) (Nable, 1988) and annual medics (*Medicago* spp.) and peas (*Pisum sativum*) (Paull et al., 1992) have identified an apparently similar mechanism controlling tolerance to B for these species. The low concentrations of B in roots indicate that the low concentrations measured in shoots of tolerant lines result from the lower uptake of B by tolerant lines rather than restricted translocation of B from roots to shoots. As the culture media for all experiments were maintained at near constant pH (5.5–6.0), were continuously mixed by aeration, and because lines of contrasting response were included in each pot, the difference in B accumulation between tolerant and susceptible lines could not be attributed to a plant mediated effect upon the availability of B in the rhizosphere.

There is a degree of controversy in the literature regarding the mechanism of uptake of B by plants. For example, Bingham et al. (1970) and Oertli and Grgurevic (1975) concluded that B was passively absorbed by excised barley roots over the range from normal to excessive concentrations, while Bowen (1972) and Bowen and Nissen (1976a,b) argued that B uptake is fundamentally an active process. In reviewing the short distance transport of B in plants, Raven (1980) concluded that although uptake of B is not simply related to mass flow of water, the high permeability of membranes to $B(OH)_3$ would greatly diminish the effect of any active regulation so that uptake would be principally regulated by passive transport processes.

Nable (1988) calculated that a seven-fold difference in water use efficiency would be required to account for the differences in B uptake between tolerant and sensitive genotypes of wheat and barley, a range far greater than that measured for the barley genotypes studied, namely 2.5–4.0 g dry matter kg^{-1} water (Walker and Lance, 1991). Although control of tolerance to B cannot be accounted for by differences in transport rates or water use efficiency, data have been obtained for several experiments which indicate that genetic variation in uptake of B is under passive control. As an example, the concentration of B in roots and shoots of three contrasting barley genotypes were determined at five root temperature regimes, from 5–25°C (Nable et al. 1990). Despite a large effect of temperature upon absolute yield, the relative yields and concentrations of B in roots and shoots were constant over the range of root temperatures, as were the differences in response among the genotypes (Table 3), suggesting that the differences in B uptake rates among the three genotypes results from non-active processes.

Differences in the response of wheat genotypes to high external concentrations of B are expressed at the organ and cellular level (Huang and Graham, 1990). Excised root tips of tolerant genotypes (e.g. G61450 and Halberd) continued to elongate and to produce lateral roots when cultured on agar medium containing a high (20 m*M*) level of B, whereas sensitive genotypes (e.g. (Wl * MMC)) ceased growth. Similarly, only tolerant genotypes were able to initiate callus in the presence of 25 m*M* B in the agar medium. Tolerance to B therefore appears to be regulated at the level of the cell wall or cell membrane, rather than being a function of root structure or mediated by the shoot.

The results of physiological studies have revealed an apparently similar mechanism controlling tolerance to B for barley, medics and peas with tolerant genotypes of these species also being able to restrict uptake of B by the root system. As the mechanism controlling tolerance of barley to B is unaffected by the temperature of the root system and tolerance of wheat is expressed at the cellular and organ level, it is probable that tolerance to B is a non-metabolic

Table 3. Effect of root temperature and B supply on the growth and B concentration of three barley genotypes. Adapted from Nable et al. (1990)

Genotype	Treatment (μM B)	Plant part	Nutrient solution temperature (°C)				
			5	10	15	20	25
			Whole-plant dry matter (mg $plant^{-1}$)				
Schooner	15		22 (5)	50 (7)	111 (4)	126 (20)	167 (10)
	1000		11 (2)	33 (4)	72 (17)	80 (10)	90 (9)
Ratio	1000/15		0.50	0.66	0.65	0.63	0.54
WI-276	15		31 (4)	121 (21)	164 (12)	181 (31)	206 (17)
	1000		20 (5)	79 (3)	117 (15)	141 (21)	152 (34)
Ratio	1000/15		0.68	0.65	0.71	0.78	0.73
Sahara 3771	15		21 (1)	71 (10)	85 (11)	94 (7)	111 (3)
	1000		21 (6)	65 (2)	75 (11)	98 (1)	110 (14)
Ratio	1000/15		1.00	0.92	0.89	1.04	0.99
			B concentration (mg kg^{-1})				
Schooner	1000	Roots	47 (15)	48 (12)	46 (9)	44 (9)	53 (10)
		Shoots	363 (45)	368 (27)	347 (64)	428 (51)	404 (43)
WI-276	1000	Roots	39 (3)	35 (4)	33 (4)	33 (6)	35 (16)
		Shoots	232 (33)	193 (18)	218 (95)	191 (30)	213 (15)
Sahara 3771	1000	Roots	20 (4)	16 (5)	15 (3)	17 (4)	20 (6)
		Shoots	80 (17)	62 (4)	79 (24)	87 (15)	90 (20)

Values are the means of three replicates with s.d. in parentheses.

process and may be related to the composition/structure of either the cell wall or cell membranes.

Genetic control of B tolerance

The genetic control of tolerance to B has been studied for wheat to allow the adoption of appropriate strategies for breeding tolerant varieties. Areas of investigation have included
(1) the identification of allelic variation of major genes,
(2) whether tolerance is expressed as a dominant character and
(3) the chromosomal location of genes conferring tolerance to high concentrations of B.
The results of first two aspects would determine the breeding strategy, for instance if major genes are identified, tolerance could be transferred to locally adapted sensitive varieties by backcrossing. Identifying chromosomes controlling B tolerance would assist in establishing linkage maps thereby allowing the identification of a closely linked marker which could be used for selection, or the elucidation of problems encountered during a backcrossing program (e.g. a gene for B tolerance in the donor parent may be linked in repulsion to a desired gene in the recurrent parent).

Five genotypes, which were also the subject of the physiological investigations, were crossed in all combinations and tested for response to B at the F_1, F_2 and F_3 generations. The response of an F_1 hybrid is intermediate to its parents for both dry matter production and concentration of B in tissues (Table 4) and comparisons between reciprocal hybrids have revealed no maternal influence upon tolerance to B (Paull et al., 1988; 1991b). As a consequence of the partial dominance it has been necessary to conduct segregation studies in the F_3 generation when it is possible to distinguish homozygous tolerant and sensitive families from segregating families.

The procedure for testing segregating generations, described by Paull et al. (1991b), consists of growing plants in plastic trays (400 mm × 285 mm × 120 mm) containing soil enriched with B and lower concentrations are used for combinations including a sensitive line as one of the parents. Approximately 120 F_2 derived families, each consisting of 10–12 F_3 plants, have been tested for eight of the 10 possible combinations of the five parents. The responses of individual F_3 plants were compared with a grid of the two parents and families were then classified as being either homozygous tolerant (all plants as tolerant as the tolerant parent), homozygous sensitive (all plants as sensitive as the sensitive parent) or heterozygous. By this means it has been possible to identify three combinations of lines which each segregate at a single, but different, gene with respect to tolerance to B and further combinations which segregate at two or more genes (Table 5). The combinations for which single gene segregation was identified were those between genotypes most similar for response to B (Fig. 1). The major genes act in an additive manner and have been named *Bo1*, *Bo2* and *Bo3* (Paull, 1990; Paull et al., 1991b). Transgressive

Table 4. Response of the F_1 hybrid of G61450 × Kenya Farmer to increasing levels of B supply in the soil. Adapted from Paull et al. (1991b)

Genotype	B treatment (mg B kg^{-1} soil)						
	0	25	50	75	100	125	150
	Dry matter (g pot^{-1})						
G61450	2.7	2.3	2.4	1.8	1.3	1.2	0.7
F_1 hydrid	2.9	2.7	2.2	1.2	1.0	0.5	0.4
Kenya Farmer	2.2	1.0	0.5	0.2	0.2	0.1	0.1
	B concentration of shoots (mg kg^{-1})						
G61450	16	154	392	697	1291	–	–
F_1 hydrid	17	199	680	1570	1983	–	–
Kenya Farmer	17	389	1192	–[a]	–	–	–

[a] Insufficient dry matter for analysis.

segregation occurred for G61450 × Halberd, with plants more tolerant than G61450 and more sensitive than Halberd among the progeny (Paull et al., 1991b). Transgressive segregation has also been observed for several combinations of exotic B tolerant genotypes (D B Moody, pers. comm.). Further research is required to determine the allelic relationships among these tolerant lines to identify those most suitable as donors in backcrossing programs.

A number of chromosomes of wheat have been implicated in the control of tolerance to B. The chromosome 4A intervarietal substitution line (post 7th International Wheat Genetics Symposium, 1988, designation) of Kenya Farmer (KF) into Chinese Spring (CS) is significantly more sensitive to B than CS and the 20 other CS/KF substitution lines (Paull et al., 1988b). CS(KF4A) develops a distinctive symptom in response to B, consisting of necrosis across the middle of the older leaves, in addition to the usual necrosis progressing from the leaf tip. This symptom, referred to as 'mid-leaf necrosis' is expressed only by sensitive genotypes and has been observed for Kenya Farmer and (Wl*MMC) (Paull et al., 1991b), while of a further eight chromosome 4A substitution lines tested only CS(Sapporo 4A) expressed this symptom. F_2 derived F_3 lines of CS × CS(KF4A) have been tested for segregation in response to B and results indicate a single gene located on the long arm of chromosome 4A (Paull, unpublished).

Further information regarding the location of genes conferring tolerance to B has been obtained by the method of monosomic analysis (Sears, 1953) and with alien addition lines. Although the results for monosomic analysis are somewhat equivocal, chromosomes 7B and 7D have been implicated as the most probable locations of genes for tolerance to B for the reciprocal monosomic analysis between Chinese Spring and Federation and for the analysis of the crosses between the Chinese Spring monosomics and G61450, respectively (Paull, 1990). The different critical chromosomes for the two combinations is consistent with transgressive segregation observed between G61450 and Halberd, a descendent of Federation.

Two amphiploids produced from hybridisations between diploid relatives of wheat and the wheat variety Chinese Spring, namely *Lophopyron elongatum* (syn. *Agropyron elongatum*)/Chinese Spring (AABBDDEE) (kindly provided by J. Dvorak) and *Thinopyrum bessarabicum* (syn *Ag. junceum*)/Chinese Spring (AABBDDE^{b}E^b) (kindly provided by T.E. Miller), are more tolerant to B than Chinese Spring (Paull et al., 1991a) and (Table 6), respectively, indicating that the tolerance of the related species is expressed in a wheat background. As the amphiploids appear to be no more tolerant

Table 5. Segregation of the F_3 generations among five wheat lines in response to high concentrations of B. Each F_3 family was derived from a single F_2 plant and families were classified as either homozygous sensitive, segregating or homozygous tolerant. Segregating and tolerant families were pooled for chi-square analysis. Adapted from Paull et al. (1991b)

Combination	Observed and expected frequencies			χ^2_1	P
		Tolerant and segregating families	Sensitive families		
(Wl*MMC)*KF	Obs	95	22		
	Exp 3:1	87.75	29.25	3.20	$0.05 < p < 0.10$
Warigal*KF	Obs	114	9		
	Exp 15:1	115.3	7.7	0.06	$0.70 < p < 0.90$
Warigal*(Wl*MMC)	Obs	95	27		
	Exp 3:1	91.5	30.5	0.40	$0.50 < p < 0.70$
Halberd*(Wl*MMC)	Obs	112	5		
	Exp 15:1	109.7	7.3	0.47	$0.30 < p < 0.50$
Halberd*Warigal	Obs	83	27		
	Exp 3:1	82.5	27.5	0.01	$0.90 < p < 0.95$
G61450*Halberd	Segregated beyond the range of the parents				

Table 6. Concentration of B (mg kg^{-1}) in shoots of Chinese Spring, G61450, the amphiploids of CX/*Lophopyron elongatum* and CS/*Thinopyrum bessarabicum* and chromosome 7E disomic and ditelosomic addition lines when grown in solution culture at two B treatments. (The two groups of lines were conducted as separate experiments, hence the differences in B concentration)

Genotype	B conc. (mg kg^{-1})		Genotype	B conc. (mg kg^{-1})	
	1500μM	3000μM		1500μM	3000μM
Chinese Spring	671	942	Chinese Spring	773	1013
G61450	302	474	CS/L. *elong*. amph	424	654
CS/*L. elong*. amph	413	647	Disomic 7E addition	519	891
CS/*Th. bess*. amph	454	754	Ditelo 7Eα addition	936	1166
			Ditelo 7Eβ addition	487	753
LSD $p < 0.05$, (interaction)	133		LSD $p < 0.05$, (interaction)	115	

than the most tolerant wheat lines identified, there would be no advantage from transferring genes from these species to what, but they may provide some information, through homologous relationships, on the location of genes conferring tolerance to B. A visual assessment of the response of the disomic addition lines of *L. elongatum* into Chinese Spring (Dvorak and Knott, 1974) indicated chromosome 7E to be critical in the control of tolerance to B (Paull, 1990). The two chromosome 7E ditelosomic addition lines have been compared with Chinese Spring and the disomic 7E addition line for concentration of B in shoots when grown in solution culture at high levels of B. The 7E disomic and the 7Eβ ditelosomic addition lines contained a significantly lower concentration of B in shoots than Chinese Spring and the 7Eα ditelosomic addition line (Table 6), which indicates that a gene(s) located n 7Eβ has a significant role in restricting the uptake of B. The results also confirm that chromosomes of homoeologous group seven play an important role in controlling tolerance to high concentrations of B.

Crop-plant breeding

Genetic variation in response to high concentrations of B has been identified for the major agricultural species of southern Australia, including lines more tolerant than the most tolerant Australian varieties. There is therefore the potential for improving the B tolerance of the crops which are cultivated in high B soils which occur over a large portion of this region. A program is under way at the Waite Institute to assess the transfer of the *Bo1* allele, by backcrossing, from Halberd to Schomburgk, a close relative to Warigal and also moderately sensitive to B. Results for BC_3 derived lines have shown that lines with the genotype *Bo1 Bo2 Bo3* produce significantly higher yields than *bo1 Bo2 Bo3* lines when grown under high B conditions at a number of sites in South Australia (Moody et al., 1992). These results indicate the selective advantage of B tolerance in southern Australia and support the theory that the success of Halberd and a number of its ancestors, all of which are tolerant to B, may be attributed in part to their tolerance of the high concentrations of B in the sub-soil.

In parallel with this genetic study, the *Bo1* derivatives of Schomburgk have been subject to normal selection in a commercial breeding program and one of the higher yielding segregants is now in the final stages of evaluation and seed multiplication with commercial release anticipated in 1991. Although its yield relative to Schomburgk varies widely from one location to another, presumably depending upon soil B and soil water status, its mean yield over two seasons and 27 locations in South Australian Department of Agriculture Primary and Secondary Wheat Variety Trials is greater than 5% above the recurrent parent Schomburgk (S Jefferies, pers. comm.).

An assessment of B tolerance is regularly listed in the Annual Report of the (Australian) Interstate Wheat Variety Trial, a co-operative series of experiments on the 20–25 most advanced lines from Australian breeding programs. While moderately tolerant entries (i.e. similar to

Halberd) are now common from southern Australian programs, intolerant entries are common from Queensland, possibly reflecting local adaptation to soils of low B status. Information on B tolerance is now reported in the registration of new commercial varieties and in 'Fact Sheets' advising farmers on the characteristics of alternative varieties, although their actual choice between varieties will be restricted until a greater range of tolerances become available from breeding programs over the next decade. Further, the potential for improvement of other crops, such as barley, oats and peas, where all Australian varieties are more sensitive than Halberd and chickpeas and lentils which are very sensitive, is very large. Lines more tolerant than the current Australian varieties have been identified in all species examined. Athough the results point to the conclusion that tolerance is attributable to variation in the uptake of B, the possibility that other mechanisms might also be operative cannot be discounted. As one mechanism of tolerance appears to be similar for all species, it is probable that the genetic control for other species is also similar to wheat, therefore the higher levels of B tolerance identified in germplasm collections should be readily incorporated into Australian varieties.

Acknowledgements

This research has been funded by grants from the Wheat and Barley Research Councils of Australia and the Wheat Industry Research Committee for South Australia.

References

Bingham FT, Elseewi A and Oertli J J 1970 Characteristics of boron absorption by excised barley roots. Soil Sci. Soc. Am. Proc. 34, 613–617.

Bowen J E 1972 Effect of environmental factors on water utilization and boron accumulation and translocation in sugarcane. Plant Cell Physiol. 13, 703–714.

Bowen J E and Nissen P 1976a Boron uptake by excised barley roots. I. Uptake into the free space. Plant Physiol. 57, 353–357.

Bowen J E and Nissen P 1976b Boron uptake by excised barley roots. II. Characteristics and kinetics of active uptake. Physiol. Plant. 41, 109–115.

Cartwright B and Hirsch M 1986 Boron toxicity in barley and wheat – a disorder resembling foliar disease. Dept. Agric. South Aust. Fact Sheet, FS 8/86.

Cartwright B, Zarcinas B A and Mayfield A H 1984 Toxic concentrations of boron in a red-brown earth at Gladstone, South Australia. Aust. J. Soil Res. 22, 261–272.

Cartwright B, Zarcinas B A and Spouncer L R 1986 Boron toxicity in South Australian barley crops. Aust. J. Agric. Res. 37, 351–359.

Cartwright B, Rathjen A J, Sparrow D H B, Paull J G and Zarcinas B A 1987 Boron tolerance in Australian varieties of wheat and barley. *In* Genetic Aspects of Plant Mineral Nutrition. Eds. H W Gabelman and B C Loughman. pp 139–151. Martinus Nijhoff Publishers, Dordrecht, The Netherlands.

Dvorak J and Knott D R 1974 Disomic and ditelosomic additions of dipolid Agropyron elongatum chromosomes to *Triticum aestivum*. Can. J. Genet. Cytol. 16, 399–417.

Holloway R E 1991 Factors affecting the growth of wheat roots in the subsoils of Upper Eyre Peninsula, South Australia. M Ag Sci Thesis, University of Adelaide.

Huang C and Graham R D 1990 Resistance of wheat genotypes to boron toxicity is expressed at the cellular level. Plant and Soil 126, 295–300.

Macindoe S L and Walkden Brown C 1968 Wheat breeding and varieties in Australia. N.S.W. Dept. Agric. Sci. Bull. 76.

Moody D B, Rathjen A J and Cartwright B 1992 Yield evaluation of a gene for boron tolerance. *In* Genetic Aspects of Plant Nutrition. Eds. P J Randall, E Delhaize, R A Richards and R Munns. pp 363–366. Kluwer Academic Publishers, Dordrecht, The Netherlands.

Moody D B, Rathjen A J, Cartwright B, Paull J G and Lewis J 1988 Genetic diversity and geographical distribution of tolerance to high levels of soil boron. *In* Proc. 7th International Wheat Genetics Symposium. Eds. T E Miller and R M D Koebner. pp 859–866. Institute Plant Science Research, Cambridge.

Nable R O 1988 Resistance to boron toxicity amongst several barley and wheat cultivars: A preliminary examination of the resistance mechanism. Plant and Soil 112, 45–52.

Nable R O, Lance R C M and Cartwright B 1990 Uptake of boron and silicon by barley genotypes with differing susceptibilities to boron toxicity. Ann. Bot. 66, 83–90.

Oertli J J and Grgurevic E 1975 Effect of pH on the absorption of boron by excised barley roots. Agron. J. 67, 278–280.

Oertli J J and Kohl H C 1961 Some considerations about the tolerance of various plant species to excessive supplies of boron. Soil Sci. 92, 243–247.

Paull J G 1990 Genetic studies on the tolerance of wheat to high concentrations of boron. Ph D Thesis, University of Adelaide.

Paull J G, Cartwright B and Rathjen A J 1988a Responses of wheat and barley genotypes to toxic concentrations of soil boron. Euphytica 39, 137–144.

Paull J G, Nable R O, Lake A W H, Materne M A and Rathjen A J 1992 Response of annual medics (*Medicago* spp.) and field peas (*Pisum sativum*) to high concentrations of boron: Genetic variation and the mechanism of tolerance. Aust. J. Agric. Res. 43, 203–213.

Paull J G, Rathjen A J and Cartwright B 1988b Genetic control of tolerance to high concentrations of soil boron for wheat. *In* Proc. 7th International Wheat Genetics Symposium. Eds. T E Miller and R M D Koebner. pp 871–878. Institute Plant Science Research, Cambridge.

Paull J G, Rathjen A J and Cartwright B 1991a Tolerance to high concentrations of boron for the amphiploid of *Triticum aestivum* × *Agropyron elongatum*. Plant and Soil 133, 297–299.

Paull J G, Rathjen A J and Cartwright B 1991b Major gene control of tolerance of bread wheat to high concentrations of soil boron. Euphytica 55, 217–228.

Paull J G, Rathjen A J, Cartwright B and Nable R O 1990 Selection parameters for assessing the tolerance of wheat to high concentrations of boron. *In* Genetic Aspects of Plant Mineral Nutrition. Eds. N. El Bassam, M Dambroth and B C Loughman. pp 361–369. Kluwer Academic Press, Dordrecht, The Netherlands.

Rathjen A J and Pederson D C 1986 Selecting for improved grain yield in variable environments. *In* Proceedings of Plant Breeding Symposium. Eds. T A Williams and G S Wratt pp 104–115. Agronomy Society of New Zealand Special Publication, Lincoln, New Zealand.

Rathjen A J, Cartwright B, Paull J G, Moody D B and Lewis J 1987 Breeding for tolerance of mineral toxicities in Australian cereals with special reference to boron. *In* Priorities in Soil/Plant Relations Research for Plant Production. Eds. P G E Searle and B G Davey. pp 111–130. School of Crop Sciences, The University of Sydney.

Raven J A 1980. Short- and long-distance transport of boric acid in plants. New Phytol. 84, 231–249.

Sears E R 1953 Nullisomic analysis in common wheat. Am. Nat. 87, 245–252.

Walker C and Lance R 1991 Silicon accumulation and ^{13}C composition as indices of water-use efficiency in barley cultivars. Aust. J. Plant Physiol. 18, 427–434.

Zarcinas B A, Cartwright B A and Spouncer L R 1987 Nitric acid digestion and multi-element analysis of plant material by inductively coupled plasma spectrometry. Commun. Soil Sci. Plant Anal. 18, 131–146.

P. J. Randall et al. (Eds.), Genetic aspects of plant mineral nutrition, 377–385.

PLSO SV99

Genetic variation in the response of pea (*Pisum sativum* L.) to high soil concentrations of boron

A. BAGHERI[1], J.G. PAULL[1], A.J. RATHJEN[1], S.M. ALI[2] and D.B. MOODY[1]
[1]*Waite Agricultural Research Institute, Glen Osmond 5064, South Australia and* [2]*South Australian Department of Agriculture, Northfield, South Australia*

Key words: boron, genetic variation, pea, toxicity

Abstract

In a greenhouse experiment nine current Australian cultivars of pea were grown to flowering time under five levels of soil boron (0, 10, 20, 30 and 40 mg kg^{-1}) applied to the soil. This study was conducted to identify the genetic range in tolerance to boron within the group and to identify specific responses which may be utilised as selection criteria in a breeding program. Significant differences in response to increasing levels of boron were found between cultivars for dry-weight yield, and boron concentrations were lowest in shoots of the most tolerant cultivars. Of the other parameters measured, emergence was not affected but plant height and the number of nodes were reduced and the severity of symptom expression increased at the higher boron treatments. Symptom expression was the most efficient observation for predicting the response of cultivars, as determined by dry-weight yield and concentration of boron in shoots, and it was found that the correlation coefficients between symptoms and the latter two measurements were $r = -0.78$ ($p < 0.01$) and $r = 0.81$ ($p < 0.01$), respectively. Early Dun, Dundale, Alma and Maitland were the more tolerant of the cultivars and these happen to be the most widely grown cultivars in southern Australia.

Introduction

The amount of boron required for normal plant growth is relatively close to the toxic level and therefore the range between deficiency and toxicity of boron in plant tissues is narrow compared to the range for other trace elements (Eaton, 1944; Nable, 1990).

Boron toxicity in crop plants in South Australia was first reported by Cartwright et al. (1984). They found that a 17% grain yield reduction in a barley crop could be attributed to a high concentration of boron in the soil. Soils with high concentrations of boron include red brown earths, calcareous earths and heavy grey clays (Rathjen et al., 1987) in which extractable boron in the subsoil is often above 20 mg kg^{-1} and in some cases may exceed 100 mg kg^{-1}.

There is considerable information on the genetic aspects of toxicity and tolerance to excess concentrations of some elements such as aluminium and manganese (Blum, 1988), but until recently only limited information was available in the case of boron. In recent years, breeding for tolerance to soil boron toxicity has been a major activity at Waite Agricultural Research Institute and C.S.I.R.O. in South Australia and genetic variation has been identified for cereals (Paull et al., 1988).

Although grain legumes are grown in rotation with cereals over wide areas in southern Australia where cereal crops show toxicity symptoms to boron, there is no published information on the response of the cultivars of grain legume crops to high concentrations of boron. The cultivation of grain legumes such as peas in areas where high concentrations of boron occur would suggest that boron toxicity of legumes may occur. Initial experiments (Materne, 1989) have shown that genetic variation in response to boron exists in Australian pea cultivars. The present investigation was undertaken to confirm the genetic variation of Australian pea cultivars, to study the mechanism of tolerance and to identify selection parameters which could be used for breeding cultivars tolerant to boron.

Materials and methods

Soil

The soil used was a bulk sample of silty clay loam texture from the surface (0–10 cm) of a red brown earth (Typic Haploxeralf) collected from the Glenthorne Research Farm, O'Halloran Hill, South Australia (Paull et al., 1988). The air-dried soil was sifted through a 1-cm grid. Boric acid (H_3BO_3) was applied to the soil by first dissolving in warm water and then mixing through the soil in a modified stock feed mixer (capacity 500 kg). Boron (as H_3BO_3) was applied in solution at the rate of 0, 10, 20, 30 and 40 mg B kg^{-1} soil; these treatments are designated as B0, B10, B20, B30 and B40, respectively. To ensure that there were no nutrient deficiencies and to improve the water-holding capacity and aeration of the soil, blood meal and peat were added at rates of 1.2 g kg^{-1} and 20 g kg^{-1}, respectively. Four kilograms of treated soil were added to each pot (200-mm diameter) lined with a water-tight polythene bag (38 micron × 305 × 455 mm).

Genotypes and experimental design

The pea cultivars Alma, Pennant, Dinkum, Buckley, Maitland, Derrimut, Early Dun, Dundale and Collegian were used. The pedigree, breeding institution and year of release of these cultivars are listed in Table 1. To ensure uniform germination, seeds were placed in plastic petri dishes containing moist filter paper, stored at 2–4°C for two days and then at room temperature for one day. Before seeding the pots were arranged in a randomized complete block as a split-plot design with three replicates. Four seeds were sown two cm below the soil surface in each pot. The pots were watered by adding distilled water as required.

Growth measurements and tissue analysis

Emergence was measured one and two weeks after sowing. After establishment (21 days after sowing) seedlings were thinned to two plants per pot and plant characteristics such as number of shoots, number of nodes, height of plants and visual symptoms of boron toxicity were documented for each pot. Harvested plants from the three replicates of each treatment were combined, oven-dried, ground and analysed for concentrations of boron in the plant tissue. As these values were measured on composite samples, the data were not statistically analysed. Plants were harvested when all cultivars had commenced flowering at the B0 treatment (45 days after

Table 1. Pedigree, breeding institution and year of release of Australian pea cultivars grown in this experiment

Cultivars	Breeding institution	Year of release	Pedigree
Early Dun		Pre-1900	Introduction, probably from UK
Collegian	RAC (SA)[a]	1939	White Brunswick × Early Dun
Derrimut	D.A. (Vic)[b]	1964	Collegian × M.U. 244 C
Buckley	D.A. (Vic)	1970	White Brunswick × M.U. 33
Dundale	D.A. (SA)[c]	1970	Selection from Early Dun
Pennant	WARI (SA)[d]	1977	White Brunswick × 15247
Alma	D.A. (SA)	1986	White Brunswick × P.I. 173052
Maitland	D.A. (SA)	1986	Early Dun × J.I. 143
Dinkum	D.A. (Vic)	1988	Complex cross involving Victorian Dippes Gelbe, Early Dun, Buckley and other introductions[e]

[a]Roseworthy Agricultural College, South Australia.
[b]Department of Agriculture, Victoria.
[c]Department of Agriculture, South Australia.
[d]Waite Agricultural Research Institute, South Australia.
[e]Daratech Pty. Ltd. (1988).

sowing) and plant characteristics (number of shoots, number of nodes, height of plants and visual symptoms of boron toxicity) were measured for each pot. All samples were dried at 80°C for 48 hours and weighed. Plants of Alma and Pennant were divided into leaves and stems and the separated tissues were weighed separately. The samples were then ground and digested in nitric acid at 140°C. The boron concentrations were determined by inductively coupled plasma spectrometry (Zarcinas et al., 1987).

Statistical analysis

Analysis of variance was based on a split-plot design. The significance between means was calculated by LSD test (Cochran and Cox, 1966). A paired t-test (Steel and Torrie, 1960) was performed to determine significant differences for the comparison between the two cultivars, Alma and Pennant, for leaf and stem dry matter.

Results

Emergence

The percentage emergence was measured one and two weeks after sowing. Boron treatments had no significant effect and approximately 70% and 90% of plants had emerged after one and two weeks, respectively (data are not presented). The emergence percentages of Pennant and Buckley were significantly lower than for all other cultivars and as this effect was observed at all treatments it could not be attributed to the effect of boron.

Visual symptoms

Cultivars were rated for the severity of symptoms of boron toxicity 21 and 45 days after sowing. They were scored for severity of symptoms on a scale of 0–8 (Table 2) adapted from Materne (1989). The initial symptoms were characterized by light brown specks near the margin of the distal half of the leaf. The development of boron toxicity symptoms followed a similar progression to that described by Salinas et al. (1981) for pea seedlings. As toxicity progressed, the leaf margins turned necrotic and symptoms developed progressively to the centre of the leaf and in severe cases resulted in the death of the leaf. Symptoms were first observed for the highest boron rate (40 mg B kg^{-1}) after 10 days. The severity of symptoms of boron toxicity increased with successively higher boron treatments and symptoms were more severe at the second stage of scoring (Table 3). The severity of the symptoms and the rapidity with which necrosis appeared differed among genotypes and the cultivar X treatment interaction for symptom expression was highly significant (Table 3). Early Dun, Dundale, Derrimut and Alma developed the least severe symptoms of toxicity while symptoms for Dinkum were most severe.

Table 2. Visual scoring system based on the severity and appearance of foliar symptoms of boron toxicity. Adapted from Materne (1989)

0. No apparent symptoms
1. Marginal necrosis on the bottom set of leaves
2. Marginal necrosis on ≤25% of leaves
3. Marginal necrosis on ≤50% of leaves
4. Marginal necrosis on ≤75% of leaves plus complete necrosis of bottom leaves
5. Marginal necrosis on ≥75% of leaves plus complete necrosis of second bottom leaves
6. Plants wilted
7. Only stem green
8. Dead

Score	*Classification*
≤ 3.0	Tolerant (T)
$3.0 < X < 4.0$	Moderately tolerant (MT)
$4.0 \leq X < 5.0$	Moderately sensitive (MS)
≥ 5.0	Sensitive (S)

Rate of plant development

The effect of boron treatments upon the development of secondary branches could not be assessed as plants generally did not develop these even at the control treatment, although Derrimut and Buckley produced a limited number of shoots at B10, B20 and B30. The experiment was undertaken in summer, so the low degree of branching was probably related to the growing season. Cultivars differed in their length of times to commencement of flowering. All cultivars had started flowering at the B0 treatment prior to harvest, while flowering was de-

Table 3. Mean visual score of pea cultivars[a] at two stages of scoring when grown at five levels of boron: an illustration of the use of LSD test for both factors

Cultivars	Boron treatments (mg B kg^{-1})											
	B0		B10		B20		B30		B40		Average	
	(21)[b]	(45)	(21)	(45)	(21)	(45)	(21)	(45)	(21)	(45)	(21)	(45)
Alma	0.0	0.0	1.8	2.2	3.2	3.3	3.5	3.5	3.8	4.3	2.4	2.7
Dundale	0.0	0.0	2.5	2.5	3.0	3.3	3.3	3.5	3.8	4.3	2.5	2.7
Early Dun	0.0	0.0	2.2	2.7	3.2	3.3	3.5	3.7	3.7	4.2	2.5	2.8
Maitland	0.0	0.0	2.5	3.2	2.8	3.7	4.0	4.0	4.2	4.8	2.7	3.1
Collegian	0.0	0.0	2.3	3.3	3.7	4.0	4.5	5.0	4.7	5.5	3.0	3.6
Buckley	0.0	0.0	2.2	2.7	3.3	3.3	4.0	4.2	5.0	5.3	2.9	3.1
Derrimut	0.0	0.0	2.5	2.7	3.2	3.0	3.5	3.7	4.2	5.0	2.7	2.8
Pennant	0.0	0.0	2.8	2.8	3.7	4.3	4.2	4.8	5.3	5.5	3.2	3.5
Dinkum	0.0	0.0	3.0	3.7	4.3	5.0	4.3	5.5	5.3	6.3	3.4	4.1
Average	0.0	0.0	2.4	2.8	3.4	3.7	3.9	4.2	4.4	4.6		

[a]Average of three replications. Interactions were $p < 0.01$ and $p < 0.05$ significant, respectively. To compare means of first scoring in a column LSD 0.01 = 0.18; in a row, LSD 0.01 = 0.37 and to compare means for second scoring in a column LSD 0.01 = 0.10; in a row, LSD 0.01 = 0.35.
[b](21) and (45) indicates days of first and second measurement.

layed by 3–4 days at B10 and B20 and by 5–7 days at B30 and B40 relative to B0. At the time the experiment was harvested, Buckley, Pennant and Dinkum had not commenced flowering at B40.

As shown in Table 4, the mean number of nodes decreased with increasing levels of boron, and particularly at B40 where plants developed 5% less than in the control treatment for the first stage of scoring. Differences were also observed among cultivars for the number of nodes (Table 4). The cultivars with the highest number of nodes included Collegian, Alma, Early Dun and Maitland, while the semi-dwarf cultivar Dinkum produced significantly fewer nodes than all other cultivars. The cultivar X treatment interaction at the first stage of scoring was also significant ($p < 0.05$; Table 5).

Boron treatments did not appreciably affect the height of plants, with the exception of B40 at which treatment heights were reduced 14% and 24% relative to B0 at the first and second stages, respectively (Table 5). Cultivars varied in their height and had a similar pattern at the first and second stages (Table 5). There was not significant variation among Alma, Collegian, Maitland, Early Dun and Dundale and these cultivars were taller than Derrimut, Pennant and Buckley

Table 4. Number of nodes 21 days after sowing for pea cultivars[a] when grown at five levels of boron at first stage of measurement: an illustration of the use of LSD test for both factors

Cultivars	Boron treatments (mg B kg^{-1})					
	B0	B10	B20	B30	B40	Average
Alma	11.0	11.3	11.6	10.7	11.2	11.2
Dundale	10.3	11.3	10.5	10.3	10.5	10.6
Early Dun	10.5	11.5	11.7	11.2	10.2	11.0
Maitland	10.3	11.5	10.7	11.5	12.0	11.2
Collegian	12.5	12.2	12.2	12.5	12.2	12.3
Buckley	10.7	10.5	7.2	9.0	7.8	9.0
Derrimut	10.5	10.7	10.5	11.2	10.6	10.7
Pennant	9.7	10.3	10.5	10.5	9.0	10.0
Dinkum	9.2	8.3	8.3	7.5	6.5	7.9
Average	10.5	10.8	10.3	10.4	10.0	

[a]Average of three replications. Interaction was significant ($p < 0.05$). To compare means in a column LSD 0.05 = 0.36; in a row, LSD 0.01 = 1.07.

Table 5. Mean height (cm) of pea cultivars[a] at two stages of measurement when grown at five levels of soil boron: an illustration of the use of LSD test for both factors

Cultivars	Boron treatments (mg B kg^{-1})											
	B0		B10		B20		B30		B40		Average	
	(21)[b]	(45)	(21)	(45)	(21)	(45)	(21)	(45)	(21)	(45)	(21)	(45)
Alma	64	117	63	108	67	128	58	128	58	107	62	118
Dundale	52	102	66	119	54	112	42	93	51	90	53	100
Early Dun	65	99	50	117	64	126	60	110	60	91	58	108
Maitland	63	123	62	117	53	111	65	119	57	100	60	114
Collegian	64	112	61	120	57	98	74	115	53	93	62	108
Buckley	46	101	57	107	34	77	39	89	25	52	40	85
Derrimut	44	94	44	82	47	93	45	84	50	74	46	86
Pennant	42	87	43	86	46	74	47	92	25	52	40	78
Dinkum	24	45	17	29	17	31	15	24	12	14	17	29
Average	50	98	52	98	49	95	49	95	43	75		

[a]Average of three replications. Interactions were significant ($p < 0.01$), for both measurements. To compare means of first measurement in a column LSD 0.01 = 31.1; in a row, LSD 0.01 = 54.4 and to compare means for second measurement in a column LSD 0.01 = 44.0; in a row, LSD 0.01 = 79.7.

[b](21) and (45) indicates days of first and second measurement.

which were not significantly different. Dinkum was significantly shorter than all other cultivars. The cultivar X treatment interaction was also highly significant ($p < 0.01$) at both the first and second stages, reflecting the differing responses of the cultivars to increasing levels of soil boron (Table 5).

Dry weight

The data in Table 6 indicate significant effects of both treatments and cultivars upon dry weight. The lowest dry weight was obtained with the application of 40 mg B kg^{-1} of soil. The boron treatments B10, B20, B30 and B40 resulted in mean decreases of 18, 31, 42, and 64 per cent relative to the control, respectively. Considerable variation was observed among cultivars for dry weight (Tables 6). Alma and Early Dun produced the highest and Dinkum the lowest yield. The cultivars X treatment interaction was not significant (Table 6).

The boron treatments affected the dry weight of leaves and stems of both Alma and Pennant (Table 7). There was no significant difference between the two cultivars for the reduction in yield of leaves, at B40 relative to the control, but the stems of Pennant were affected by boron treatments to a much greater extent than those of Alma. For Pennant, boron reduced the dry weight of stems considerably more than that of leaves, but this was not the case for Alma.

Table 6. Mean shoot dry weight[a] of nine pea cultivars (g) when grown at five levels of soil boron

Cultivars	Boron treatments (mg B kg^{-1} soil)					
	B0	B10	B20	B30	B40	Average
Alma	6.20	5.59	5.59	4.02	2.47	4.77
Dundale	5.11	5.09	4.78	3.57	2.70	4.25
Early Dun	5.59	5.53	5.11	3.89	2.82	4.59
Maitland	5.10	3.95	3.56	3.20	2.44	3.73
Collegian	6.34	5.37	3.45	3.52	1.98	4.13
Buckley	4.34	3.47	1.81	1.44	0.58	2.33
Derrimut	5.07	3.81	3.37	3.48	1.87	3.52
Pennant	4.99	4.19	3.53	3.23	1.33	3.45
Dinkum	3.21	1.21	0.93	0.64	0.47	1.29
Average	5.15	4.24	3.57	3.00	1.85	

[a]Averages of three replications and interactions not significant.

Table 7. Leaf and stem dry weight (g) of two pea cultivars grown at five levels of soil boron (mean of three replications)

Cultivars	Boron treatments (mg B kg^{-1})				
	B0	B10	B20	B30	B40
Leaf dry weight[a]					
Alma	2.83	2.66	2.67	1.98	1.06
Pennant	2.31	1.87	1.57	1.48	0.82
Stem dry weight					
Alma	3.36	2.92	2.04	2.04	1.38
Pennant	2.67	2.31	1.75	1.74	0.50

[a]t values for leaf and stem dry weight are 2.38 (n.s.) and 5.93 ($p < 0.05$), respectively.

Concentration of boron in shoots

The concentration of boron in shoots increased significantly with an increase in the boron concentration in the soil (Table 8) and there were also significant differences among the cultivars in tissue boron concentrations. The interaction between cultivars and treatments for concentration of boron in shoots was statistically significant ($p < 0.01$; Table 8). Boron concentrations were lowest for Alma, Dundale, Maitland and Early Dun. In general, concentrations of boron in shoots were highest for Dinkum, while Collegian was intermediate. The ranking of cultivars was similar between treatments although there were differences between treatments in distinguishing between cultivars. For example, at B20 the boron concentration of Dinkum was much higher than in Buckley, Derrimut and Pennant, but there was little difference between these four cultivars at B40.

Discussion

Substantial genetic differences were found among Australian pea cultivars in response to toxic levels of boron in the soil and in concentration of boron in shoots. The concentrations of boron in the shoots of cultivars which produced high relative yields at high boron treatments were significantly less than those of the other cultivars. Alma, Early Dun, Maitland and Dundale had low concentrations of boron in the tissue, while this value was very high for Dinkum (Table 8). The symptoms of boron toxicity were more severe at higher boron treatments and the maximum differences between cultivars occurred at the second time of scoring at the B40 treatment although significant differences between cultivars also resulted at the lower boron treatments and at the first stage. Alma developed the least severe symptoms of toxicity, while symptoms for Dinkum were very severe (Table 3). The significant genotype × treatment interaction ($p < 0.01$) for visual score could be attributed to a greater increase in severity of symptoms for the sensitive cultivars at the higher boron treatments.

Table 8. Concentration of boron (mg kg^{-1}) in shoots of nine pea cultivars[a] at two harvesting times when grown at five levels of soil boron: an illustration of the use of LSD test for both factors at second harvest

Cultivars	Boron treatments (mg B kg^{-1})											
	B0		B10		B20		B30		B40		Average	
	(21)[b]	(45)[b]	(21)	(45)	(21)	(45)	(21)	(45)	(21)	(45)	(21)	(45)
Alma	31.5	35.2	160	172	312	275	446	515	731	869	336	373
Dundale	22.7	36.0	132	163	275	297	387	503	700	717	303	343
Early Dun	24.1	41.5	127	156	270	291	544	524	787	966	350	395
Maitland	26.9	33.7	152	178	333	350	411	521	572	812	299	379
Collegian	34.4	35.2	179	174	471	384	564	676	911	965	432	447
Buckley	45.7	37.2	222	181	362	379	644	607	851	1193	425	479
Derrimut	44.3	39.0	225	196	414	394	628	613	862	1219	335	492
Pennant	33.2	40.9	211	235	431	345	645	683	1230	1242	510	511
Dinkum	36.9	53.4	242	275	526	629	754	1244	1409	1422	594	724
Average	33.3	39.1	183	192	377	372	558	654	895	1045		

[a]Average of three replications. Interaction was significant ($p < 0.01$). To compare means of second shoot analysis (45 days) in a column LSD 0.01 = 28; in a row, LSD 0.01 = 70.6.
[b](21) and (45) indicate days of first and second harvests.

The response of cultivars for shoot concentration of boron and visual score of toxicity symptoms for cultivars were consistent with results for dry weight. The shoot concentration of boron and the severity of symptoms were least for the cultivars rated as the most tolerant on the basis of dry weight. As summarized in Figure 1 the correlation coefficients that resulted between the three parameters shoot dry weight, concentration of boron in shoots and visual score of boron toxicity were: (1) relative dry weight and concentration of boron in shoots ($r = -0.78$, $p < 0.01$, Fig. 1a); (2) relative dry weight and visual score ($r = -0.74$, $p < 0.01$, Fig. 1b) and (3) concentration of boron in shoots and visual score ($r = 0.81$, $p < 0.01$, Fig. 1c). Based on the ranking of cultivars for these three parameters, cultivars can be grouped in four categories:

- Moderately tolerant: Alma, Early Dun, Dundale and Maitland,
- Moderately sensitive: Collegian and Derrimut,
- Sensitive: Buckley and Pennant,
- Very sensitive: Dinkum.

This experiment was conducted to identify the most reliable and efficient parameters for distinguishing between pea cultivars for response to high soil boron concentrations. The parameters emergence, height and number of nodes were less suitable than symptom expression, boron concentrations in the shoot and dry weight production for predicting the response of the cultivars to boron. There was a highly significant interaction between cultivars and boron treatments for plant height at both times of scoring with the relative heights of the sensitive cultivars Dinkum, Buckley and Pennant being affected to a much greater extent by high boron treatments than the more tolerant cultivars. There were, however, significant differences between the heights of cultivars at the control, and for this reason the height of plants would not be suitable as a selection criterion in a screening program where lines are grown at only a single treatment. The number of nodes also differed significantly between cultivars at the control with the tallest cultivars having the highest number of nodes. The response of cultivars to boron, with respect to number of nodes, was inconsistent between the two times of scoring, and the interaction was significant ($p < 0.05$) at only the first stage. Emergence of cultivars was not affected by boron treatments.

Previous investigators have misleadingly classified pea as a semi-tolerant (Eaton, 1944; Ayers and Westcot, 1976) or a sensitive (Salinas et al, 1981) plant species but these classifications were based on the response of single genotypes. There are several mechanisms that would enable plants to tolerate toxic concentrations of mineral elements in soils, namely avoidance (e.g. a shallow root system to avoid elements, such as boron, which accumulate in the subsoil), exclusion from the root system and internal tolerance (Rathjen et al., 1987). Several investigators have found the concentration of boron in shoots to be lower for tolerant genotypes of wheat and barley (Nable, 1988; Nable et al., 1990; Paull et al., 1988) and a similar mechanism appears to control the tolerance of peas to boron. Nable (1988) also measured low concentrations of boron in the roots of tolerant wheat and barley cultivars and suggested that the tolerance was governed by the ability of cultivars to exclude boron, but the mechanisms limiting uptake of boron are not yet understood.

A significant reduction in shoot dry weight occurred at each increase in level of boron, and yields were lowest at the B40 treatment (Table 6) but the interaction among genotypes and treatments was not significant, an unexpected finding in view of the other results. Among the genotypes studied Alma, Dundale, Early Dun and Collegian produced the highest and Dinkum the lowest dry weight (Table 6). The high mean dry weight yield of Collegian could be attributed to high yield at the control treatment rather than at high boron treatments. On the other hand, Maitland, which produced a low overall mean yield and a low yield at the control, was among the highest yielding cultivars at B40, and on this basis would be considered tolerant to boron.

The data in Table 6 indicate that the lowest boron treatment (B10) was toxic for pea and a 17.6% reduction in dry weight of shoot occurred in this treatment. Concentration of boron in shoots at this level over all cultivars was 192 mg kg^{-1}, although the value differed significantly between cultivars (Table 8). For example, Alma had the lowest concentration of boron in shoots (172 mg B kg^{-1}), while the concentration

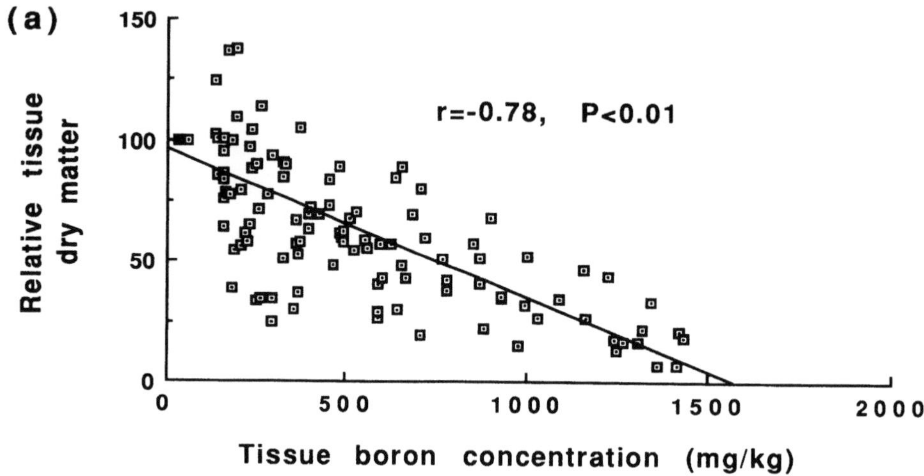

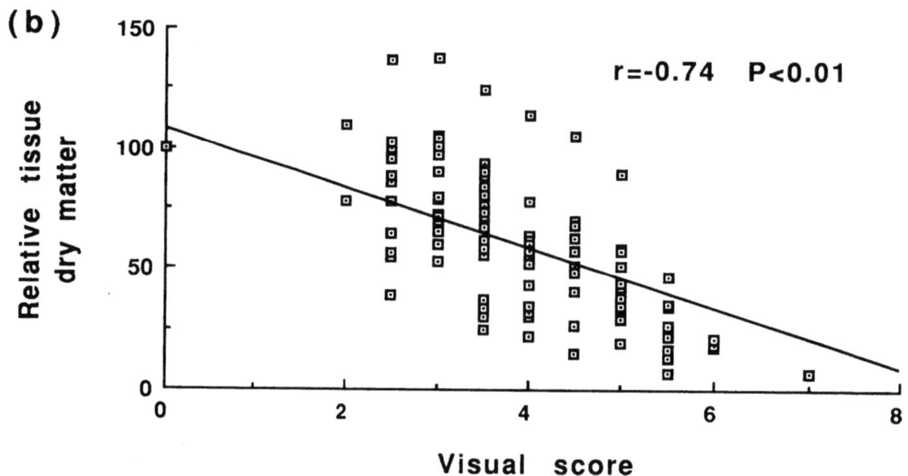

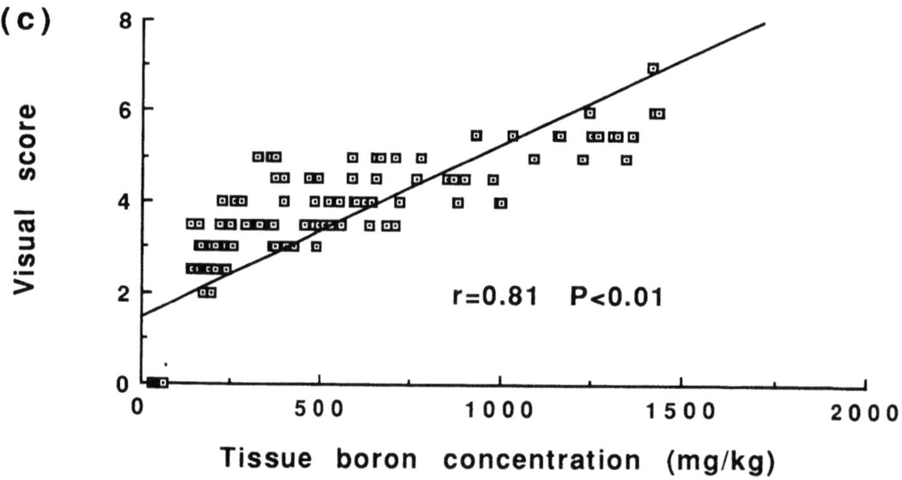

Fig. 1. Correlations between growth response parameters of peas to high levels of soil boron. (**a**) shoot dry weight (relative values) vs. shoot boron concentration, (**b**) shoot dry weight (relative values) vs. visual symptoms of boron toxicity and (**c**) shoot boron concentration vs. visual symptoms of boron toxicity.

was highest for Dinkum (275 mg B kg^{-1}). Salinas et al. (1981) suggested that a boron concentration of leaves in the range of 50–300 mg kg^{-1} at harvest time might be considered as sufficient for normal pea growth. They also reported that a leaf boron concentration of 350 mg kg^{-1}, when associated with 2 mg L^{-1} for solution culture, is within the toxicity range (Salinas et al., 1986). Toxicity to peas has also been reported when 4 mg kg^{-1} boron was supplied in irrigation water leading to 2 mg kg^{-1} soil solution boron and 213 mg kg^{-1} boron in plant tissues (Chauhan and Powar, 1978). Gupta and MacLeod (1981) reported the critical level for foliar toxicity symptoms of >61 mg B kg^{-1}. This critical value for toxicity may reflect their use of symptom development as the criterion for toxicity, whereas the other workers have measured dry weight reduction. Although there is considerable variation in toxic concentrations of boron reported in the literature, the concentrations of boron in plants associated with boron toxicity described in this paper are within the range of those reported by previous researchers.

This experiment has demonstrated that variation in boron tolerance exists among Australian commercial pea cultivars. A wide range of pea accessions should now be examined to determine the extent of genetic variation within the species *Pisum sativum* and to identify lines more tolerant than the Australian cultivars which may be used to introduce tolerance to the commercial pea cultivars by breeding.

Acknowledgement

The financial support of the Iranian Ministry of Culture and Higher Education is gratefully acknowledged.

References

Ayers R S and Westcot D W 1976 Water quality for irrigation. Irrigation and drainage paper 29, Food and Agricultural Organization of the United Nation, Rome Italy. *In* Toxicity and Deficiency: A Review. U C Gupta, Y M Jame, C A Campbell, A J Leyshon and W Nichlaichuk. Can. J. Soil Sci. 65, 381–409.

Blum A 1988 Plant Breeding for Stress Environments. CRC Press, Boca Raton, FL.

Cartwright B, Zarcinas B A and Mayfield A H 1984 Toxic concentrations of boron in a red-brown earth at Gladstone, South Australia. Aust. J. Soil Res. 22, 261–272.

Chauhan R P S and Powar S L 1978 Tolerance of wheat and pea to boron in irrigation water. Plant and Soil 50, 145–149.

Cochran W G and Cox G M 1966 Experimental Designs. Wiley, New York.

Daratech Pty. Ltd. 1988 Field pea (*Pisum sativum*) variety Dinkum. Plant Varieties J. 1, 19–22.

Eaton F M 1944 Deficiency, toxicity and accumulation of boron in plants. J. Agric. Res. 69, 237–277.

Gupta U C and MacLeod J A 1981 Plant and soil boron as influenced by soil pH and calcium sources on Podzol soils. Soil Sci. 131, 20–25.

Materne M A 1989 Genetic Variability in the Response of Field Pea Varieties to Soil Boron. B. Ag. Sc. (Hons) Thesis, University of Adelaide.

Nable R O, Cartwright B and Lance R C M 1990 Genotypic differences in boron accumulation in barley: Relative susceptibilities to boron deficiency and toxicity. *In* Genetic Aspects of Plant Nutrition. Eds. N El Bassam, M Dambroth and B C Loughman, pp 361–369. Kluwer Academic Publishers, Dordrecht, The Netherlands.

Nable R O 1988 Resistance to boron toxicity amongst several barley and wheat cultivars: A preliminary examination of the resistance mechanism. Plant and Soil 112, 45–52.

Paull J G, Cartwright B and Rathjen A J 1988 Response of wheat and barley genotypes to toxic concentrations of soil boron. Euphytica 39, 137–144.

Rathjen A J, Cartwright B, Paull J G, Moody D B and Lewis J 1987 Breeding for tolerance of mineral toxicities in Australian cereals with special reference to boron. *In* Perspectives and Priorities in Plant Production Research. Eds. P G E Searle and B G Davey. pp 111–130. Sydney University Press.

Salinas M R, Cerda A, Romero M. and Caro M 1981 Boron tolerance of pea (*Pisum sativum*). J. Plant Nutr. 4, 205–215.

Salinas R, Cerda A and Martinez V 1986 The interactive effect of boron and macronutrients (P, K, Ca and Mg) on pod yield and chemical composition of pea (*Pisum sativum*) Hortic. Sci. 61, 343–347.

Steel R G D and Torrie J H 1960 Principles and Procedures of Statistics. MacGraw-Hill, New York.

Zaricinas B A, Cartwright B and Spouncer L R 1987 Nitric acid digestion and multi-element analysis of plant material by inductively coupled plasma spectrometry. Commun. Soil Sci. Plant Anal. 18, 131–146.

P. J. Randall et al. (Eds.), Genetic aspects of plant mineral nutrition, 387–396.

PLSO SV93

Heavy-metal (Zn, Cd) tolerance in selected clones of duck weed (*Lemna minor*)

R.F.M. VAN STEVENINCK[1], M.E. VAN STEVENINCK[2] and D.R. FERNANDO[1]
[1]*School of Agriculture and* [2]*Department of Botany, La Trobe University, Bundoora, 3083, Australia*

Key words: cadmium, clones, heavy metal tolerance, *Lemna minor*, oxalate, phosphate, phytic acid, phytochelatin, sulphur, X-ray microanalysis, zinc

Abstract

Cryo-microprobe analysis of quench-frozen fronds of a Zn-tolerant clone of *Lemna minor* exposed to a high level of Zn (300 μM) showed the presence of cellular deposits consisting of Zn, Mg, K and P or Zn, K and P (Zn phytate). The same Zn-tolerant clone of *Lemna minor*, when exposed to a high level of Cd (30 μM), showed the presence of globular deposits consisting of Cd, K and P in mature fronds, but the immature cells of the enclosed daughter fronds contained relatively large deposits with Cd and S as the main components (Cd-phytochelatin?). Selection for Zn tolerance in a population of *Lemna minor* was easily achieved but selection for Cd tolerance has so far not been successful. The Zn-tolerant clone also tolerates high levels of phosphate.

Introduction

Certain plant species show a remarkable capacity for ecological adaptation to local enrichment of heavy metals in the environment. This acquired characteristic of heavy-metal tolerance is often highly metal-specific and, regardless of whether it is constitutive or induced upon exposure to heavy metals, it is usually a heritable phenomenon (Verkley and Schat, 1990). Multiple tolerance or cross-tolerance, i.e. the combined tolerance to several metals, is generally associated with co-occurrence of high levels of these metals in the environment. One might expect this particularly to be the case for Cd and Zn, two metals which, apart from co-occurrence in metalliferous soils, exhibit such a high degree of physical and chemical similarity that effective metallurgical separation is very difficult (Verkley and Schat, 1990).

Plant species which exhibit metal tolerance can be classified roughly into two major groups: 'includers' and 'excluders'. Although from an agricultural production point of view the latter type would provide distinct advantages in maintaining low levels of toxic elements in the food chain (e.g. Al in pasture species, Cd in tobacco), the former type occurs much more commonly (Verkley et al., 1990) and from a physiological point of view provides interesting alternatives in means of metal detoxification. These may range from mechanisms which rely largely on metal binding in the apoplast, on the formation of insoluble deposits in the cytoplasm (e.g. Zn, Cd), or on secretion into vacuoles (Zn, Cd).

Earlier work, especially with respect to Zn tolerance (Matthys, 1977), supported the sequestration of Zn by organic acids (malate, citrate, oxalate) in vacuoles (Godbold et al., 1984), a view which has recently received further support from laboratories principally involved in phytochelatin research (Krotz et al., 1989; Wagner and Krotz, 1989). It was proposed that in some cases the different organic acids could have a more specific role, such as the binding of Zn in the cytoplasm by malate followed by vacuolar sequestration of Zn citrate (Ernst, 1975; Godbold et al., 1984). Cd binding by oxalate as an alternative to the formation of Ca oxalate crystals in raphide cells of tomato was reported earlier by van Balen et al. (1980).

The binding of heavy metals, particularly Cd, by polypeptides through metal cysteinyl thiolate coordination was reported to occur in plants as early as 1975 by Casterline and Yip (1975). After a quiet interval, this was followed by a virtual explosion of information on a novel class of plant peptides: poly (γ-glutamyl cysteinyl) glycines [$(\gamma EC)_n$ G] during the mid 80's (Rauser, 1984; Robinson and Jackson, 1986; Wagner, 1984). Because these functional analogues of the well-known mammalian metallothioneins, now also known as phytochelatins (PC's; Grill et al., 1985; 1987), have characteristic γ-carboxyamide) linkages, it is implicit that they are not synthesized via mRNA as is the case with metallothioneins (Jackson et al., 1987). Nevertheless, it still appears that the production of Cd-binding proteins is transcriptionally regulated, perhaps via metal induction of transcription of the gene(s) for the enzyme(s) which synthesizes the metal chelatin peptides (Hirt et al., 1990).

A more recent development is the discovery that appreciable quantities of Zn can be sequestered by phytic acid (*myo*-inositol – *kis*-hexaphosphate) in the cortex of roots of Zn-tolerant *Deschampsia caespitosa* exposed to high levels of Zn (400 and 1000 μM; Van Steveninck et al., 1987). The presence of similar phytate globules was confirmed for *Lemna minor*. This aquatic macrophyte has been reported to be capable of producing and storing phytic acid in vegetative tissue (Roberts and Loewus, 1968) rather than in specific targeted tissues and cells which are involved in reproduction (pollen, aleurone layer and cotyledonary tissue in seeds) as in most higher plant species (Cosgrove, 1980).

Because we had no difficulty in selecting a Zn-tolerant clone (clone A) derived from a single frond of Lemna, we carried out comparative studies between this clone (A) and similarly derived Zn-sensitive clones (B and C) to further characterise their capacity to bind Zn and Cd. The dependence of this binding capacity on supply of phosphorus in the nutrient medium was also investigated. Most of these studies were carried out by means of electron probe X-ray microanalysis of freshly frozen specimen material and further checks were made using conventional analytical methods (e.g. atomic absorption spectroscopy).

Material and methods

Clones of *Lemna minor* L. derived from single fronds were grown in 250 mL beakers on 125 mL nutrient solution which was renewed twice per week and contained 3 mM KNO_3, 2 mM Ca $(NO_3)_2$, 1 mM $NH_4H_2PO_4$, 0.5 mM $MgSO_4$, 25 μM FeNa$(EDTA)_2$, 25 μM H_3BO_3, 1 μM $MnSO_4$, 1 μM $ZnSO_4$, 0.25 μM $CuSO_4$ and 0.25 μM H_2MoO_4. Experiments with elevated levels of Cd and Zn were carried out with Cd up to 100 μM and Zn up to 1000 μM. Plants were grown under controlled-environment conditions with 8/16 hrs light/dark, 25/22°C light/dark temperature and 220 μEinstein $m^{-2}s^{-1}$ light intensity.

For X-ray analysis, single mature fronds of Lemna were harvested after approximately 8 days of treatment, quench-frozen in melting nitrogen (−210°C), and fractured before insertion in the cold stage (−190°C) of a scanning electron microscope (JSM 840) by means of a Hexland CT 1000 cryo-transfer system. X-ray data were obtained using a TN 5500 X-ray analyser at an accelerating voltage of 15 kV with a probe current ranging from 0.3 to 0.6 nA. All spectra were obtained from a reduced scan area of 4 μm^2 at 10,000 × magnification or 0.45 μm^2 at 30,000 × magnification. To visualise detail of cellular features it was necessary to apply a certain degree of "etching" (differential removal of ice through sublimation) by temporarily raising the cold-stage temperature to −85°C. Unfortunately, this treatment invalidates direct quantitative analysis of X-ray data, although ratios of elements can still be estimated (Van Steveninck and Van Steveninck, 1991).

Atomic absorption analyses (Zn) were carried out on extracts of Lemna plants which were oven-dried at 80°C for 5 days, acid-digested in 2 mL 4:1 $HNO_3/HClO_4$, made up to 20 mL with 2.5% acetic acid and, for the high Zn-treated material, further diluted (1/50). Phosphorus (total P) analyses were carried out by the colorimetric method of Alexander and Robertson (1970).

Results

Excess Zn (300–1000 μM in the nutrient solution) results in the formation of globular deposits inside small vacuoles either singly (in mature cells of fully grown fronds) or in clusters (in immature cells of daughter fronds, buds or turions) (Van Steveninck et al., 1990). The X-ray spectra obtained from these globules show the presence of P, K, Mg and Zn as the principal mineral components (Figs. 1a and b) closely similar to the type of spectrum produced by small K Mg phytate crystals obtained commercially and used as a calibration standard (Fig. 1c). Globules inside cells of fully mature fronds tend to contain little or no detectable Mg and a relatively high proportion of K (Fig. 1a), while cells from immature tissue (daughter fronds or turions) contain significant amounts of Mg but less K than mature cells (Fig. 1b, Table 1). The ratio of Zn to P is much higher in mature cells than in immature cells and, in the Zn to P ratios of globules in mature tissue, there appears to be no significant difference between the Zn-tolerant clone (A) and the two Zn-sensitive clones (B and C) (Table 1). No Zn was detected by means of X-ray analysis in any of the Ca oxalate crystals of raphide cells (Fig. 1d).

When fronds were challenged with high concentrations of Cd (30–100 μM), globules containing Cd, K and P (Fig. 1e) were detected in parenchyma cells of mature fronds, but only Zn (no Cd) was detected in P-containing globules when fronds had been simultaneously exposed to high levels of Zn (300 μM) and Cd (30 μM). Cd was never detected as a component of Ca oxalate crystals in mature raphide cells. However, cells which may have just begun to function as raphide cells (Franceschi, 1989) and contained only 1 to 3 distorted crystals, seemed to have commenced Cd binding, since the spectra showed high levels of Cd as well as Ca (Fig. 1f).

A high level of Ca, in Cd-containing globules (Table 2) was a significant feature especially in clone B. Perhaps some of the high Ca levels were due to inclusion of data from globules in cells which were targeted to become raphide cells, or alternatively, the high Ca levels in globules could reflect a Cd-induced breakdown of cellular control of Ca levels in the cytoplasm. The ratio of Cd to P in globules from cells in clone A was significantly higher than in clone B (Table 2), yet growth and multiplication of fronds of both clones were equally inhibited and fronds became equally chlorotic in 30 μM Cd. All fronds of both clones became necrotic at 100 μM Cd.

Immature tissue of daughter fronds (buds, turions) contained Cd deposits consisting largely of Cd and S (Fig. 1g). X-ray maps showed these

Table 1. Elemental ratios in globular deposits obtained from "etched" frozen bulk samples of three clones, A (Zn-tolerant), B and C (Zn-sensitive) exposed to high concentrations of Zn (300–1000 μM)

Clone	μMZn	n[a]	Zn/P	Mg/P	Ca/P	K/P	ZnK_α/ZnL_α
A, mature cells	300	17	0.88 ± 0.06	0.003 ± 0.0003	0.001 ± 0.001	0.64 ± 0.06	0.75 ± 0.11
A, immature cells	300	10	0.31 ± 0.03	0.39 ± 0.03	0.005 ± 0.005	0.21 ± 0.02	1.01 ± 0.27
B, mature cells	300	6	0.79 ± 0.07	0.013 ± 0.007	0	0.51 ± 0.05	0.45 ± 0.13
B, immature cells	300	21	0.21 ± 0.02	0.43 ± 0.02	0	0.19 ± 0.01	1.37 ± 0.32
C, mature cells	1000	5	0.99 ± 0.06	0.003 ± 0.003	0.002 ± 0.002	0.49 ± 0.03	0.31 ± 0.07

[a]Number of analyses.

Table 2. Elemental ratios in globular deposits (Cd/P type) obtained from 'etched' frozen bulk samples of two clones A (Zn-tolerant) and B (Zn-sensitive) exposed to high concentrations of Cd (30–100 μM)

Clone	μMCd	n[a]	Cd/P	Mg/P	Ca/P	K/P
A	30	15	1.20 ± 0.07	0.09 ± 0.01	0.11 ± 0.02	0.37 ± 0.07
B	100	9	0.71 ± 0.08	0.17 ± 0.03	0.83 ± 0.08	0.49 ± 0.05

[a]Number of analyses.

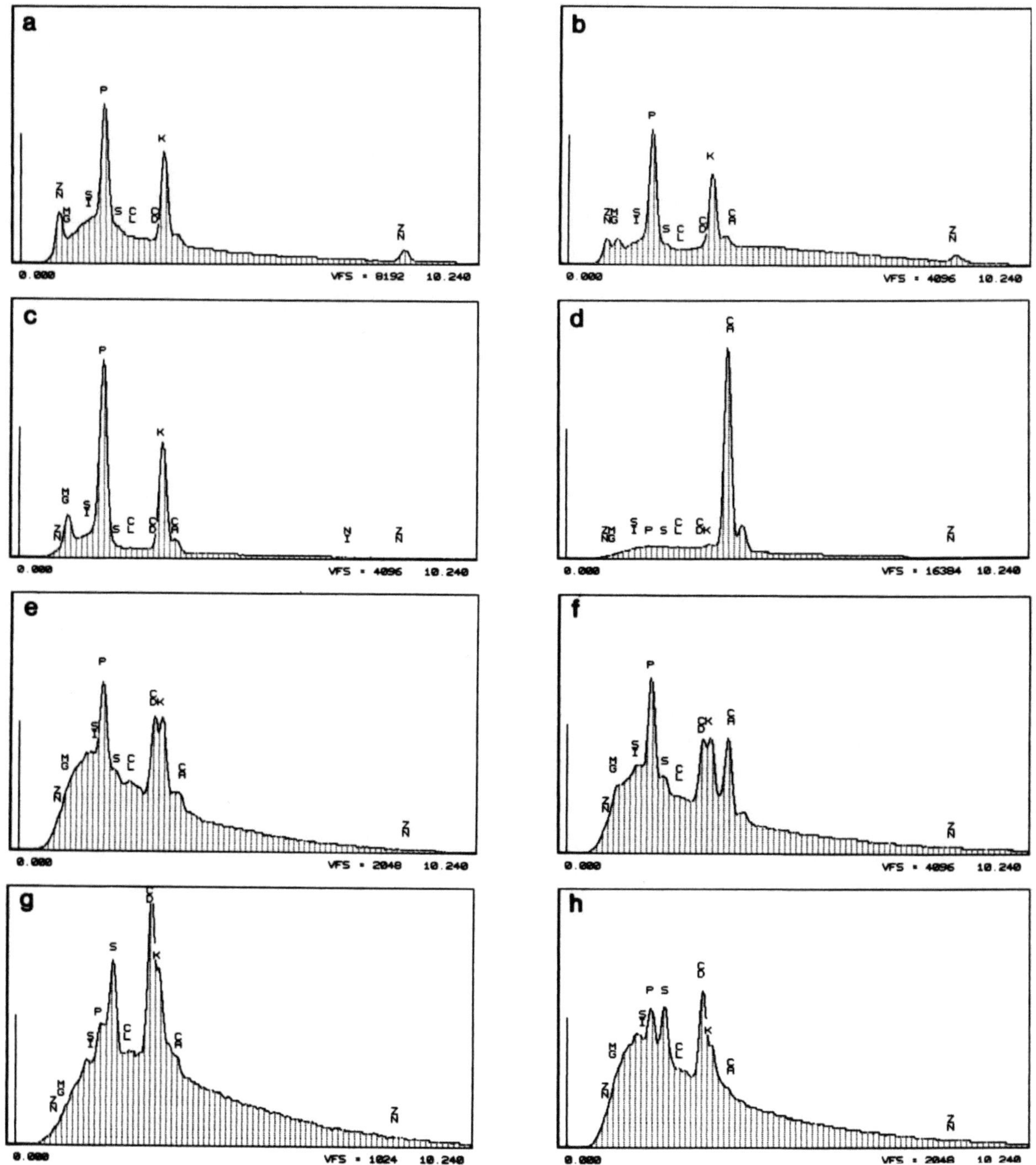

Fig. 1. Electron probe X-ray microanalysis spectra obtained at 10,000 × magnification and a reduced area scan of 4 μm^2 from:
a: a globule in a mature cell of clone A frond exposed to 300 μMZn
b: a globule in an immature cell of a daughter frond of clone A exposed to 300 μMZn
c: a small crystal of K Mg phytate (ex Sigma).
d: a Ca-oxalate crystal in a raphide cell of clone A exposed to 100 μMCd
e: approx. 1 m-diameter globule in a large parenchyma cell of clone A frond tissue exposed to 30 μMCd
f: a globule in an incipient raphide cell of clone A exposed to 100 μMCd.
g: a 3 × 4 μm deposit in immature daughter frond tissue (2nd turion) of clone A exposed to 30 μMCd
h: a 2 × 4 μm deposit in less immature daughter frond tissue (1st turion) of the same sample of clone A

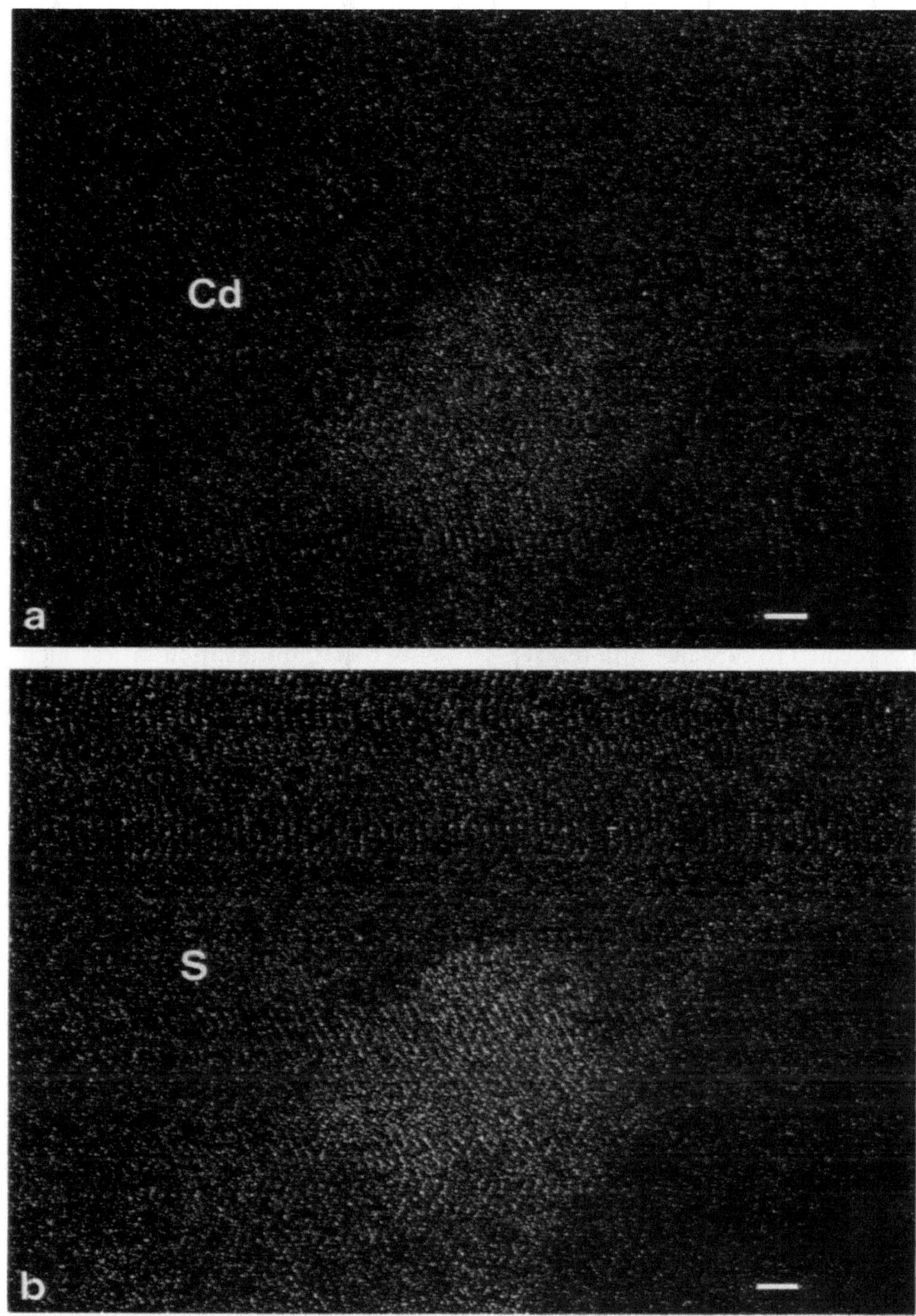

Fig. 2. X-ray maps of a particular cell region **a**, representing X-rays emitted by Cd, **b**, representing X-rays emitted by S showing the localisation of Cd and S to the same region in the cell. *Bar* represents 1 μm.

deposits to be about 4×10 μm in size (Figs. 2a and b). They had a sheet-like appearance, were relatively unstable under prolonged exposure to the electron beam, and probably consisted of proteinaceous or membranous material. If the fracture of a mature frond exposed several (up to 3) daughter fronds (turions), the youngest (smallest) exclusively contained Cd-S type de-

Table 3. Elemental ratios in sheet-like deposits (Cd/S type) obtained from "etched" frozen bulk samples of clones, A, B and C exposed to high concentrations of Cd (30–100 μM)

Clone	μMCd	n[a]	Cd/S	Mg/S	Ca/S	K/S	Particulars
A	30	9	1.60 ± 0.07	0.06 ± 0.01	0.04 ± 0.01	0.16 ± 0.015	low Ca, low Mg
B	30	7	1.14 ± 0.04	0.22 ± 0.05	0.09 ± 0.06	0.30 ± 0.09	low Ca, high Mg
B	30	4	0.65 ± 0.07	0.01 ± 0.01	0.32 ± 0.04	0.29 ± 0.01	med Ca, low Mg
C	100	4	0.66 ± 0.04	0.06 ± 0.05	0.25 ± 0.03	0.15 ± 0.04	med Ca, low Mg
C	30	4	0.48 ± 0.04	0.11 ± 0.01	0.66 ± 0.18	0.34 ± 0.11	high Ca, med Mg

[a]Number of analyses.

posits, while the others contained transition stages where Cd was present with both P and S as the main mineral elements (Fig. 1h).

Data obtained from these deposits in immature tissue gave elemental ratios (Table 3) showing a much higher value for the Cd to S ratio in clone A than in clones B and C. However, Cd/S varied considerably from cell to cell and between fronds and seemed to depend on the amounts of Ca, Mg and K in the deposits. Relatively high values of Cd/S seemed to correlate with relatively low levels of other elements, especially of Ca, but more material must be analysed to establish such a correlation.

Table 4. Effect of varying P_i (0.1 mM) and Zn (1 and 201 μM Zn) supply on mature frond size (maximum length × maximum breadth, mm^2) of clone A (Zn-tolerant) and clone B (Zn-sensitive). Values are means of 10 measurements

P_i concentration	Clone A		Clone B	
	Zn 1	Zn 201 (%)[a]	Zn 1	Zn 201 (%)[a]
0	2.9	2.2 (69)	8.9	4.3 (48)
10 μM	2.9	2.2 (69)	9.3	4.3 (46)
100 μM	3.6	2.6 (72)	7.8	4.9 (62)
1 mM	2.8	2.3 (82)	5.7	4.3 (75)

[a]Percentage of control (1 μM Zn) treatment.

An attempt was made to relate Zn tolerance to the supply of P in the nutrient solution. In these experiments Zn was present either at the normal nutrient solution concentration of 1 μM or at 201 μM. The latter concentration was chosen as the maximum possible for the Zn-sensitive clone (B) to survive during the experimental period (4 weeks). From the results shown in Table 4 it can be seen that high Zn levels inhibited frond growth of both clones but also that the inhibition is less at high P levels than at low P levels and less in clone A (Zn-tolerant) than in clone B (Zn-sensitive). In the controls (low-Zn) the largest average frond size occurs at 100 μM P in clone A and at 10 μM P in clone B. Frond growth is inhibited at high (1 mM) P levels, particularly in clone B, where growth was less than with no added P.

Root growth was particularly affected by the level of P supply. Low P promoted growth of roots, especially in clone B and at concentrations of 10 μM P and lower (Table 5). As expected, high Zn levels inhibited the growth of roots, but this inhibition became a slight promotion at high P levels (1 mM), especially in clone A (Table 5).

Table 6 summarises the contents of P and Zn for clone A and clone B. P deficiency was evident ($P_T \sim 0.25\%$ DWt) when no P was added to

Table 5. Effect of varying P_i (0 to 1 mM) and Zn (1 and 201 μM) supply on root length (mm) of clone A (Zn-tolerant) and clone B (Zn-sensitive)

P_i concentration	Clone A			Clone B		
	Zn 1	Zn 201	(%)[a]	Zn 1	Zn 201	(%)[a]
0	22.2 ± 1.7	2.6 ± 0.7	11.7	61.9 ± 4.2	1.4 ± 0.2	2.3
10 μM	20.0 ± 0.9	3.2 ± 0.4	16.0	54.2 ± 5.9	1.9 ± 0.1	3.5
100 μM	6.6 ± 1.3	6.5 ± 1.2	98.5	18.6 ± 3.4	3.9 ± 1.3	21.0
1 mM	1.3 ± 0.3	2.1 ± 0.5	161.5	1.6 ± 0.2	2.0 ± 0.2	125.0

[a]Percentage of control (1 μM Zn) treatment.

Table 6. Total phosphorus (P_T) and Zn contents of clone A and clone B plants exposed over a period of 4 weeks to a range of phosphate supply (0 to 1 m*M* P_i) in the presence of normal (1 μM) or toxic (201 μM) levels of zinc in the nutrient solution

	P_T %d.wt		Zn mg kg^{-1} d.wt		Zn/P by wt	
	Clone A	Clone B	Clone A	Clone B	Clone A	Clone B
0 P_i + 1 μMZn	0.22	0.14	79	53		
0 P_i + 201 μMZn	0.27	0.26	13400	8800	4.96	3.38
10 $\mu M P_i$ + 1 μMZn	0.50	0.30	118	67		
10 $\mu M P_i$ + 201 μMZn	0.56	0.44	17800	9100	3.18	2.07
100 $\mu M P_i$ + 1 μMZn	1.64	1.04	82	–		
100 $\mu M P_i$ + 201 μMZn	1.76	1.60	9690	13825	0.55	0.86
1 m$M P_i$ + 1 μMZn	2.08	2.05	84	50		
1 m$M P_i$ + 201 μMZn	2.16	2.06	6295	6985	0.29	0.34

the nutrient solution and it can be assumed that all P has originated from that present in the tissue at the beginning of the experiment. The P contents of clone A were generally higher than in clone B except when the clones were exposed to 1 m*M* P_i. P contents were also somewhat higher in the presence of high Zn levels compared to normal Zn supply. However, more replicated experiments are required to produce statistically reliable results. Zn contents of the tissue were also generally higher in clone A than in clone B with the exception of tissue exposed to high P_i as well as high Zn levels. The Zn/P ratio was higher in clone A than in clone B with the exception of tissue exposed to high P_i as well as high Zn levels. The Zn/P ratio was higher for clone A than clone B at low P_i levels and lower at high P_i levels (Table 6).

Discussion

Various mechanisms have been proposed to explain detoxification of potentially toxic heavy metals in plants. The role of a Cd-binding peptide (Cd-BP) or phytochelatin (PC) has received much attention in recent literature (Grill et al., 1987; Verkley et al., 1990). Although there is increasing evidence that the poly $(\gamma\text{-EC})_n$ G's or phytochelatins are major metal-complexing compounds in higher plants (Jackson et al., 1987; Reddy and Prasad, 1990), further experimentation, e.g. with buthionine sulfoximine (BSO) which inhibits PC synthesis, indicates that PC's may not be a primary agent in providing Cd tolerance in tomato cells (Huang et al., 1987). Zn tolerance in tobacco cells was also shown to be unrelated to PC synthesis (Reese and Wagner, 1987). Furthermore, S deficiency had no effect on the Cu tolerance of two clones of *Deschampsia caespitosa* adapted to Cu while it increased the susceptibility of two non-tolerant clones (Schultz and Hutchinson, 1988) possibly due to primary effects of Cu on membrane integrity (Wainwright and Woolhouse, 1977; De Vos et al., 1989). Because of close chemical similarity and co-occurrence of Cd and Zn in metalliferous soils, one might expect a high degree of cross-tolerance for these elements. However, recent work by Krotz et al. (1989) on responses to Cd and Zn in the presence or absence of Cd-BP in tobacco suspension cells seems to indicate that sequestration of Cd and Zn may involve entirely different mechanisms. In fact, no metal-binding peptides were observed in cells exposed to 300 μM Zn for up to 7 days (Krotz et al., 1989). Hence, it was suggested that sequestration of excess metal (especially Zn) was likely to depend on the formation of organic acid complexes stored in vacuoles (Wagner and Krotz, 1989). This, of course, is a complete return to the earlier theories of heavy-metal tolerance (Ernst, 1975; Matthys, 1977) which were often supported by the observation that the organic acid contents of tissues were elevated concurrent with high metal accumulation (Godbold et al., 1984).

In view of the above evidence it is interesting that we have found an alternative mode of Zn sequestration by means of phytic acid in the cortex of young roots of a Zn-tolerant clone of *Deschampsia caespitosa*, and in sufficient quantity to account for the observed heavy-metal accumulation (Van Steveninck et al., 1987). The same mechanism of Zn binding was shown to occur in *Lemna minor*, an aquatic macrophyte, which is known to have capacity for phytic acid synthesis in vegetative tissue (Roberts and Loewus, 1968). Although ubiquitous within the plant kingdom, its synthesis and storage are generally confined to specialised cells which are involved with some reproductive capacity such as pollen, cotyledonary tissue or aleurone layers of seeds (Cosgrove, 1980), but its presence in other storage organs (e.g. potatoes, carrots, culms) has also been reported (Dixon et al., 1983; Samotus and Schwimmer, 1962). Its evolutionary significance appears to be based upon providing a ready source of P and other minerals under conditions of rapid growth when acquisition of mineral resources is still limited (pollen tube growth, germination, clonal reproduction, cf Cosgrove, 1980).

High resolution X-ray microanalysis is ideal for the localisation of metals and determination of elemental ratios in characteristic features of cells (Van Steveninck and Van Steveninck, 1991). Using this method it was observed in *Lemna minor* that Zn levels were higher and Mg content lower in phytate globules of mature tissue compared to immature tissue (undifferentiated daughter fronds). However, the technique is unsuitable for an overall quantitative assessment of cellular or tissue contents of a particular element with a heterogeneous distribution such as phytate globules. Hence, in the absence of quantitative data on actual phytate content of the frond tissue it is at present not possible to relate the higher Zn tolerance of clone A to an increased capacity for phytate synthesis and/or storage.

After exposure to high levels of Cd, sequestration of excess Cd by means of phytate (Cd/P-type deposits) could only be observed in mature frond tissue. When high Cd (30 μM) and high Zn (300 μM) were supplied simultaneously only the Zn/P-type deposits, and no Cd/P type, could be detected. Thus it appears that sequestration by means of phytate may be more specific for Zn than for Cd, although it should be pointed out that Zn was supplied at 10 times higher concentration than Cd in order to produce a similar degree of toxicity.

Another significant point, especially in relation to recent observations (Wagner and Krotz, 1989), is that the Zn-tolerant clone of Lemna does not exhibit cross-tolerance to Cd. Thus, while normal development of daughter fronds is possible in clone A at high Zn levels, multiplication of fronds is totally inhibited at high Cd levels. In this respect, the total transition from a Cd/P-type sequestration into a Cd/S type in the youngest immature fronds inside the budding pouches of mature fronds seems highly significant. The general properties of the deposits, which are sheet-like, unstable, and probably a proteinaceous material, seem to indicate that excess Cd is being sequestered by a Cd-binding peptide. Nevertheless, operation of this type of sequestration in immature fronds does not seem to bestow Cd tolerance upon Lemna. In fact, although selection for Zn tolerance was easy, many months of mass-selection within populations exposed to moderately high levels of Cd (20 μM) has, so far, not resulted in establishing a clone which is more tolerant to Cd than the original population. This again may reflect the fact that Zn tolerance and Cd tolerance depend on entirely different mechanisms, one being based on phosphate co-nutrition while the other may depend on sulfate co-nutrition (Schultz and Hutchinson, 1988).

Preliminary results on the effect of phosphorus nutrition seem to indicate that Lemna grows optimally at fairly low levels of phosphate ($\sim$10 μM P_i). However, the Zn-tolerant clone (A) performs best at 100 μM P_i and still grows well at 1 mM P_i. The results of total-P analysis should take into consideration the fact that 0.5% P_i (of DWt) is the usual concentration of P_i in plant tissues (Marschner, 1986). Less than 0.2% is generally regarded as P-deficient while toxic levels usually occur at >1.0% (soybean, rice, oats, lupins), but in Protea and orchids, i.e. plant species which are adapted to very low P levels, toxic effects may occur at levels as low as 0.17% and 0.25%, respectively (Reuter and Robinson, 1986). For Zn 100 mg kg^{-1} is a normal level (<15 mg kg^{-1} is deficient, Marschner, 1986) but concentrations as high as 30,000 mg kg^{-1} (3.0% of DWt) have been recorded in a Zn-tolerant clone of *Deschampsia caespitosa* (Godbold et al., 1984). If all Zn was stored as Zn phytate then the highest possible ratio of Zn/P by weight would be 2. Because P must be present in many other forms (phospholipids, nucleic acids) a ratio of Zn/P of 2, or higher, indicates the presence of Zn in some form other than phytate. However, the much lower Zn/P ratios in plants grown on 100 μM or 1 mM P_i indicate that all Zn could be bound to phosphoryl groups or phosphate. It is at these external concentrations that luxury consumption of P takes place and excess P_i in the tissue can be stored as phytate (Michael, 1939). Thus the Zn tolerance exhibited by clone A may be by virtue of its tolerance to high external levels of P_i and its capacity to avoid toxic levels of P_i through storage of phytate, a property which may be less well developed in clone B.

Van Balen et al. (1980) concluded by means of autoradiography that Cd can be incorporated as oxalate in raphide cells of tomato. Using X-ray microanalysis we have found no evidence of Cd incorporation in any needle-like crystals in raphide cells of Lemna, whereas we had no difficulty in confirming the incorporation of Sr alongside Ca in mature raphide cells (Franceschi and Schueren, 1989; Van Steveninck, unpublished results). However, it should be noted that on several occasions incipient raphide cells with only a few distorted crystals were observed in fronds treated with high levels of Cd. In these cases it appeared that cells which had been targeted to become raphide cells had changed to an alternative mode involving the binding of Cd as an insoluble P complex with relatively large amounts of Ca present. In fact Cd complexing with phosphoryl groups included Ca in the complex to a much greater extent than with Zn. At present one can only speculate whether this qualitative change is due to the fact that Cd toxicity may cause a release of Ca in regions of phytate accumulation or a general breakdown of permeability barriers or control of Ca levels in the cytoplasmic compartment.

Finally, one might comment that neither the induction of metal tolerance by means of phytochelatin nor phytate synthesis involves a process which is under direct genetical control. In the case of PC the presence of peptide bonds through the γ-carboxyl group of glutamate, rather than the α-carboxyl group, suggests that these peptides are not encoded by structural genes and are the products of biosynthetic pathways (Robinson and Jackson, 1986). The same can be said for phytate with respect to synthesis of myo-inositol and the complete phosphorylation to phytic acid which in higher plants appears to occur in specially targeted cells involving reproduction in higher plants. However, it has recently been shown that phytic acid synthesis can be induced in cotyledons of already germinated bean seeds in response to exogenously supplied phosphate (Organ et al., 1988). It would be interesting to determine whether induction of phytic acid synthesis can be more generally achieved in tissues of higher plants which are not involved in reproduction.

Acknowledgement

Support from the Australian Research Council is gratefully acknowledged.

References

Alexander T G and Robertson J A 1970 Ascorbic acid as a reductant for inorganic phosphorus determination in Chang and Jackson fractionation procedure. Soil Science 110, 361–362.

Casterline J L and Yip G 1975 The distribution and binding of cadmium in oyster, soybean and rat liver and kidney. Arch. Environ. Contam. Toxicol. 3, 319–329.

Cosgrove D J 1980 Inositol Phosphates: Their Chemistry, Biochemistry and Physiology. Elsevier Scientific Publishing, Amsterdam, Oxford, New York. 191 p.

De Vos C H R, Schat H, Vooijs R and Ernst W H O 1977 Copper-induced damage to the permeability barrier in roots of *Silene cucubalis*. J. Plant Physiol. 135, 164–169.

Dixon K W, Kuo J and Pate J S 1983 Storage reservoir of the seed-like aestivating organs of geophytes inhabiting granite outcrops in Southwestern Australia. Aust. J. Bot. 31, 85–103.

Enrst W H O 1975 Physiology of heavy metal resistance in plants. *In* Proc. Int. Conf. on Heavy Metals in the Environment. Eds. T C Huchinson, S Epstein, A L Page, J Van Loon and T Davey. pp 121–136. CEP Consultants, Toronto, Canada.

Franceschi V R and Schueren A M 1989 Incorporation of strontium into plant calcium oxalate crystals. Protoplasma 130, 199–205.

Franceschi V R 1989 Calcium oxalate formation is a rapid and reversible process in *Lemna minor* L. Protoplasma 148, 130–137.

Godbold D L, Horst W J, Collins J C, Thurman D A and

Marschner H 1984 Accumulation of zinc and organic acids in roots of zinc tolerant and nontolerant ecotypes of *Deschampsia caespitosa*. J. Plant Physiol. 116, 59–69.

Grill E, Winnacker E-L and Zenk M H 1985 Phytochelatins: The principal heavy-metal complexing peptides of higher plants. Science 230, 674–676.

Grill E, Winnacker E-L and Zenk M H 1987 Phytochelatins, a class of heavy-metal binding peptides from plants, are functionally analogous to metallothioneins. Proc. Nat. Acad. Sci. 84, 439–443.

Hirt H, Sommergruber K and Barta A 1990 Effects of cadmium on tobacco: Synthesis and regulation of cadmium-binding peptides. Biochem. Physiol. Pflanzen 186, 153–163.

Huang B, Hatch E and Goldbrough P B 1987 Selection and characterization of cadmium tolerant cells in tomato. Plant Science 52, 211–221.

Jackson P J, Unkefer C J, Doolen J A, Watt K and Robinson N J 1987 Poly (γ-glutamyl cysteinyl) glycine: Its role in cadmium resistance in plant cells. Proc. Nat. Acad. Sci. 84, 6619–6623.

Krotz R M, Evangelou B P and Wagner G J 1989 Relationships between cadmium, zinc, Cd-peptide and organic acid in tobacco suspension cells. Plant Physiol. 91, 780–787.

Marschner H 1986 Mineral Nutrition of Higher Plants. Academic Press, London. 674 p.

Matthys W 1977 The role of malate, oxalate and mustard oil glycosides in the evolution of zinc-resistance in herbage plants. Physiol. Plant. 33, 161–165.

Michael G 1939 Phosphate fractions in oat grains and spinach related to a varied application of phosphorus. Bodenkd. Pflanzenernaehr. 14, 148–171.

Organ M G, Greenwood J S and Bewley J D 1988 Phytin is synthesized in the cotyledons of germinated castor-bean seeds in response to exogenously supplied phosphate. Planta 174, 513–517.

Rauser W E 1984 Isolation and partial purification of cadmium-binding protein from roots of the grass *Agrostis gigantea*. Plant Physiol. 74, 1025–1029.

Reddy G N and Prasad M N V 1990 Heavy metal-binding-proteins/peptides: Occurrence, structure, synthesis and functions – A review. Environ. Expt. Bot. 30, 251–264.

Reese R N and Wagner G J 1987 Effects of buthionine sulfoximine on Cd-binding peptide levels in suspension-cultured tobacco cells treated with Cd, Zn or Cu. Plant Physiol. 84, 574–577.

Reuter D R and Robinson 1986 Plant Analysis: An Interpretation Manual. Inkata Press, Melbourne, Sydney. 218 p.

Roberts R M and Loewus F 1968 Inositol metabolism in plants. VI. Conversion of myo-inositol to phytic acid in *Wolffiella floridana*. Plant Physiol. 43, 1710–1716.

Robinson N J and Jackson P J 1986 'Metallothionein-like' metal complexes in angiosperms: Their structure and function. Physiol. Plant. 67, 499–506.

Samotus B and Schwimmer S 1962 Phytic acid as a phosphorus reservoir in the developing potato tuber. Nature 194, 578–579.

Schultz C L and Hutchinson T C 1988 Evidence against a key role for methallothionein-like protein in the copper tolerance mechanism of *Deschampsia caespitosa* (L.). Beauv. New Phytol. 110, 163–171.

Van Balen E, Van de Geijn S C and Desomet G M 1980 Autoradiographic evidence for the incorporation of cadmium into calcium oxalate crystals. Z. Pflanzenphysiol. 97, 123–133.

Van Steveninck R F M, Van Steveninck M E, Fernando D R, Horst W J and Marschner H 1987 Deposition of zinc phytate in globular bodies in roots of *Deschampsia caespitosa* ecotypes: A detoxification mechanism? J. Plant Physiol. 131, 247–257.

Van Steveninck R F M, Van Steveninck M E, Wells A J and Fernando D R 1990 Zinc tolerance and the binding of zinc as zinc phytate in *Lemna minor*. X-ray microanalytical evidence. J. Plant Physiol. 137, 140–146.

Van Steveninck R F M and Van Steveninck M E 1991 Microanalysis. *In* Electron Microscopy of Plant Cells. Eds. J L Hall and C Hawes. pp 415–455. Academic Press, London.

Verkley J A C and Schat H 1990 Mechanisms of metal tolerance in higher plants. *In* Evolutionary Aspects of Heavy-Metal Tolerance in Plants. Ed. J Shaw. pp 179–193. CRC Press Boca Raton, FL.

Verkley J A C, Koevoets P, Van't Riet J, Bank R, Nydam Y and Ernst W H O 1990 Poly (g-glutamyl cysteinyl) glycines or phytochelatins and their role in cadmium tolerance of *Silene vulgaris*. Plant, Cell Environ. 13, 913–921.

Verkley J A C, Lolkema P C, De Neeling A L and Harmens H 1991 Heavy-metal resistance in higher plants: Biochemical and genetic aspects. *In* Ecological Responses to Environmental Stresses. Eds. J Rozema and J A C Verkley. pp 8–19. Kluwer Acadmeic Publishers, Dordrecht, The Netherlands.

Wainwright S J and Woolhouse H W 1977 Some physiological effects of copper and zinc tolerance in *Agrostis tenuis* Sibth.: Cell elongation and membrane damage. J. Exp. Bot. 28, 1029–1036.

Wagner G J 1984 Characterization of cadmium binding complex of cabbage leaves. Plant Physiol. 76, 797–805.

Wagner G J and Krotz R M 1989 Perspectives on Cd and Zn accumulations, accommodation and tolerance in plant cells: The role of Cd-binding peptides versus other mechanisms. UCLA Symp. Mol. Cell Biol. (New Series) 98, 325–336.

P. J. Randall et al. (Eds.), Genetic aspects of plant mineral nutrition, 397–405.

PLSO SV90

Biosynthesis and metabolic roles of cadystins $(\gamma\text{-EC})_nG$ and their precursors in *Datura innoxia*

PAUL J. JACKSON[1], EMMANUEL DELHAIZE[2] and CHERYL R. KUSKE[1]
[1]*Genomics and Structural Biology Group, Life Sciences Division, Los Alamos National Laboratory, Los Alamos, NM 87545, USA.* [2]*Present address: CSIRO, Division of Plant Industry, Canberra ACT 2601, Australia*

Key words: biosynthesis of cadystin, cadmium tolerance, Class III metallothionein, cysteine, *Datura innoxia*, glutathione, iron accumulation, phytochelatin, poly(γ-glutamylcysteinyl)-glycine

Abstract

Metal-tolerant *Datura innoxia* cells synthesize large amounts of cadystin, [poly(γ-glutamylcysteinyl) glycines, $(\gamma\text{-EC})_nG$, $n = 2 - 5$], a class of metal-binding polypeptides, when exposed to Cd. These polypeptides have a high affinity for Cd (II) and certain other metal ions and are thought to play a role in metal tolerance in higher plants. Cells rapidly synthesize these metal-binding polypeptides when exposed to Cd and cellular concentrations of glutathione and cysteine, precursors for the synthesis of these compounds, are initially depleted then rapidly replenished. The time-frame of de novo polypeptide, glutathione and cysteine biosynthesis suggests that this pathway is, at least initially, regulated at the enzyme level. Significant amounts of Fe are associated with Cd:polypeptide complexes isolated from *D. innoxia*. Exposure of cultures to Cd results in an increased Fe accumulation by the cells. All the additional Fe found in the soluble portion of cell extracts is associated with the Cd:polypeptide complexes. The physiological significance of the synthesis of these polypeptides and their precursors and its relevance to Cd tolerance and metal homeostasis are discussed.

Introduction

Numerous industrial activities result in the contamination of soil, ground water, and aquatic and marine environments with Cd (Natusch et al., 1974; Street et al., 1978; Takijima and Katsumi, 1973a, b; Van Bruwaene et al., 1984; Vlamis et al., 1985). This metal ion is toxic to plants and animals. Toxicity in plants is the result of interaction of Cd with several important physiological processes including respiratory carbohydrate metabolism, chlorophyll formation, the Calvin cycle, uptake of necessary nutrients, and DNA, RNA and protein biosynthesis and processing (Jackson et al., 1990; Lindberg and Wingstrand, 1985; Petolino and Collins, 1985; Reese and Roberts, 1985; Stobart et al., 1985; Weigel, 1985). Plants and plant cells can be selected for the ability to grow in normally toxic Cd concentrations (Jackson et al., 1984; Steffens et al., 1986). There are numerous mechanisms that might confer tolerance to Cd and other toxic ions. However, the large number of different physiological processes that are adversely affected by Cd suggests that mechanisms of tolerance must involve the exclusion of the metal ion from all critical processes negatively affected. Cd-tolerant plants and plant cells produce large amounts of poly(γ-glutamylcysteinyl)glycines, $(\gamma\text{-EC})_nG$, $n = 2 - 5$, a class of metal-binding polypeptides (Grill et al., 1987; Jackson et al., 1987; Steffens et al., 1986). These thiol-rich compounds were first discovered and characterized in the fission yeast *Schizosaccharomyces pombe* and given the trivial name cadystin (Kondo et al., 1983; 1984). They are also known as phytochelatins, γ-glutamyl metal-binding polypeptides, and Class III metallothioneins. $(\gamma\text{-EC})_n$Gs have a high affinity for Cd and Cu (Jackson et al., 1985). It has been suggested that they play an important role in Cd tolerance and, perhaps, metal ion homeostasis in higher plants, certain fungi and algae (Grill et al., 1987; Jackson et al., 1987; Robinson and Jackson, 1986; Robinson, 1989; Wikfors et al., 1991). These polypeptides accumulate rapidly in *Datura innoxia* cells exposed to Cd, even in the presence of protein synthesis inhibitors (Robinson et al., 1988). This rapid accumulation suggests that the enzymes required for polypeptide synthesis must be maintained constitutively in the cells. Synthesis of a large amount of these compounds has an effect on the total amount of gluthathione (GSH) and cysteine within the cells. The rapid depletion of $[^{35}S]$GSH and cysteine followed closely by the rapid increase in concentrations of radiolabeled $(\gamma\text{-EC})_n$Gs strongly suggests that these metal binding polypeptides are derived from GSH and that several steps of the biosynthetic pathway are regulated, at least initially, at the enzmye level. The observation that exposure to Cd results in an increased uptake of Fe and that polypeptide complexes contain Fe is also reported. Potential roles of these polypeptides in metal metabolism are suggested.

Materials and methods

Maintenance of cell cultures

Suspension cultures of Cd-tolerant and sensitive *D. innoxia* cells were maintained at 30°C as 50 or 100 cm^3 batch cultures in beveled Delong flasks at a ratio of flask volume to culture of 5 to 1 as described previously (Jackson et al., 1984). Under these conditions, cells divide approximately every 24 h. The cell cultures are predominantly diploid. Cd-tolerant cell cultures were derived from a single sensitive cell line by a stepwise selection protocol previously described (Jackson et al., 1984).

Radioisotope labeling of cells

Carrier-free ^{109}Cd (>37 GBq/m*M*) was prepared and provided by the Nuclear and Radiochemistry Group, Isotope and Nuclear Chemistry Division, Los Alamos National Laboratory. All other radioisotopes were purchased from duPont-New England Nuclear. Radiolabeling of polypeptides with ^{109}Cd was by addition of 5.55 kBq cm^{-3} $^{109}CdCl_2$ and unlabeled $CdCl_2$ to a final concentration of 250 μM. ^{109}Cd was added to cultures 24 h after transfer to fresh media, or as specified in the experiment. Radiolabeling with ^{59}Fe was accomplished in a similar manner except that $^{59}FeSO_4$ (>74 GBq/m*M*) was added to a final concentration of 11.1 kBq cm^{-3} to cells previously grown for at least 20 generations in media without $FeSO_4$ or Na_2EDTA. Addition of $^{59}FeSO_4$ was accompanied by unlabeled $FeSO_4$ and Na_2EDTA to a final concentration of 0.1 m*M*, the concentration normally found in the medium. Labeled Fe was added 24 h prior to the addition of Cd to assure that the cells were no longer starved for Fe. Labeling of polypeptides with ^{35}S was by addition of $[^{35}S]$cysteine (>11.1 TBq/m*M*) to 37 kBq cm^{-3} with no additional unlabeled cysteine.

Extraction of $(\gamma\text{-EC})_n G$

Analytical extraction

Cells were collected from media by centrifugation (800 *g* for 1 min) and washed twice in an ice-cold solution containing 10 m*M* Tris-HCl, pH 7.4, 10 m*M* KCl, and 1.5 m*M* $MgCl_2$. Pellets were aspirated dry, weighed and resuspended in an equal volume of the above solution containing 20 m*M* 2-mercaptoethanol. The cell suspension was transferred to a handheld homogenizer and ground. HCl (1.0 *N*) was added to the homogenate at a ratio of 1 part HCl to 3.6 parts cell weight, and the homogenate was subjected to further homogenization. The homogenate was centrifuged (15,000 *g* for 10 min) and the resulting supernatant passed through a Centricon-30 filter (Amicon; 30 kD exclusion limit).

Preparative extraction

Larger numbers of cells were collected, washed and extracted as above except that the extract was not acidified and the supernatant was applied directly to a 5.0 × 150 cm column of Sephadex G-50 (fine) (Pharmacia) equilibrated at 4°C with 50 m*M* Tris-HCl, pH 7.8. Gel filtration fractions containing metal-binding polypeptides, as determined by thiol and Cd content, were pooled, then further purified and concentrated by ion exchange chromatography on DEAE-CL-6B (Pharmacia). After loading, the ion exchange resin was treated successively with 50 m*M* Tris-HCl, pH 7.8 containing 100, 150, 200, 250, and 300 m*M* KCl. Metal-binding polypeptides were released from the column as a broad peak upon washing with the solutions containing 250 and 300 m*M* KCl. Purification of apopolypeptides was by covalent chromatography on thiolpropyl-Sepharose. Briefly, pooled fractions containing the polypeptides were assayed for thiol and metal content. Enough thiolpropyl Sepharose 6B (Sigma Chemical Co.) to make a bed 1-cm deep was equilibrated at 4°C with 50 m*M* Tris-HCl, pH 7.8. The polypeptide was loaded onto the column at a flow rate of 50 mL h^{-1}. Metal ions are released from the polypeptides and pass out of the column during loading. The column was then washed with 50 mL of a solution containing 50 m*M* Tris-HCl, pH 8.0, 1 *M* NaCl and 50 mL of a solution containing 50 m*M* Tris-HCl, pH 8.0. Polypeptides were eluted at 3 mL h^{-1} by washing the column with a solution containing 50 m*M* 2-mercaptoethanol in 50 m*M* Tris-HCl, pH 8.0. One mL fractions were collected and ultrafiltered to concentrate the polypeptides, transfer them to 50 m*M* Tris-HCl, pH 7.8 and remove 2-mercaptoethanol. Fractions were assayed for polypeptide content by HPLC.

Analytical separation and analysis of $(\gamma\text{-}EC)_nG$

The acidified supernatant described above was loaded onto a C-18 column (Bio-Rad) and eluted with a linear gradient of acetonitrile (0 to 20%) in 0.1% trifluoroacetic acid (TFA) (flow rate 1 cm^3 min^{-1}). A 0.2 cm^3 portion of each 1 cm^3 fraction was analyzed for radioactivity and the remaining 0.8 cm^3 mixed with an equal volume of Ellman's reagent (50 m*M* KH_2PO_4, pH 7.6, 75 μM 5,5'-dithiobis-2-nitrobenzoic acid). The change in absorbance at 410 nm was measured after 10 min to determine thiol content. GSH solutions were used to generate a standard curve for thiol determinations and amounts of $(\gamma\text{-EC})_n$G were expressed as GSH equivalents. HPLC yielded a 90 to 95% recovery of the material loaded on the column.

Analysis of GSH and cysteine

The above HPLC procedure was not adequate to separate GSH from cysteine. Therefore 0.2 cm^3 of extract was passed through two C-18 columns in series. Samples (0.25 cm^3 fractions) were eluted with 0.1% TFA in water (flow rate 1 cm^3 min^{-1}) and analyzed for radioactive or thiol content. The columns were washed between analyses with 20% acetonitrile containing 0.1% TFA to remove $(\gamma\text{-EC})_n$G loaded during the previous cycle. The reducing agent, 2-mercaptoethanol, was deleted from the extraction buffer because it interfered with thiol measurements and was not necessary in the highly acidic solution. No difference in results could be demonstrated in the presence or absence of this agent. Concentrations of GSH and cysteine were determined by comparison of results to those generated from solutions containing known concentrations of respective reagents (Gaitonde, 1967).

ICP analysis of polypeptide and protein samples

Non-radioactive Cd and Fe concentrations in purified samples of polypeptide : metal complexes were determined by analysis on a Perkin-Elmer Model 5500 Inductively Coupled Plasma atomic emission spectrometer. The instrument was calibrated with solutions of the Sephadex G-50 column buffer (in which the polpeptide : metal complexes were dissolved) contained known amounts of these metal ions.

Results

Synthesis of $(\gamma\text{-}EC)_nG$ *in response to Cd*

Figure 1a shows the synthesis of $(\gamma\text{-EC})_n$Gs (n = 2 – 5) in cells exposed to Cd. The very rapid production of the smaller (n = 2 and 3) polypeptides in Cd-tolerant *D. innoxia* cells is even more pronounced when compared to the larger polypeptides (n = 4 and 5) if one considers that these compounds contain only 2 and 3 GSH equivalents as compared to the 4 and 5 GSH

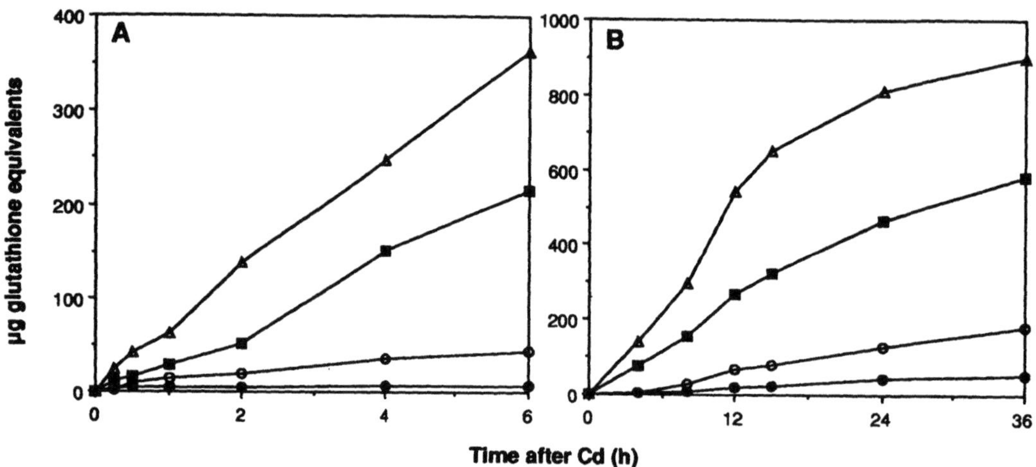

Fig. 1. The effect of Cd on the (**A**) short term and (**B**) long term synthesis of $(\gamma\text{-EC})_2$G(-△-); $(\gamma\text{-EC})_3$G (-■-); $(\gamma\text{-EC})_4$G (-○-); and $(\gamma\text{-EC})_5$G (-●-) in Cd-tolerant *D. innoxia* cells. Cells were grown for different times in media containing 250 μM Cd. Cultures (20 cm^3) were collected and extracted and the soluble portion of the extract analyzed for polypeptide content. The amount of $(\gamma\text{-EC})_n$G is expressed in GSH equivalents.

equivalents present in each molecule of the larger polypeptides. Virtually all of the polypeptides found in the soluble portion of cell extracts are found in Cd: polypeptide complexes (data not shown). If the same cells are grown for 2 h in media containing [^{35}S]cysteine then radiolabel is found in GSH and cysteine, prior to addition of Cd. If cells are then transferred to media containing Cd and unlabeled cysteine, the [^{35}S] rapidly accumulates in the newly synthesized small (n = 2 and 3) polypeptides (Fig. 2). Moreover, the ratio of [^{35}S] among the different polypeptides is the same as the ratio of the absolute amount of polypeptide present in the linear portion of the experiment suggesting that there is no obvious precursor-product relationship among the different polypeptides. After the initial increase in radioactive content, the amount of radiolabel in each polypeptide remains the same as the cells deplete the pool of [^{35}S] precursors. Experiments measuring the increase in polypeptide concentrations with longer exposure to Cd demonstrate a rapid increase in the accumulation of $(\gamma\text{-EC})_n$G for the first 12 h of exposure, followed by slower rates of increase that can be accounted for by the increase in cell number with time (Figure 1b). $(\gamma\text{-EC})_n$G is maintained at approximately 6% of the total cellular protein as long as Cd is present in the medium.

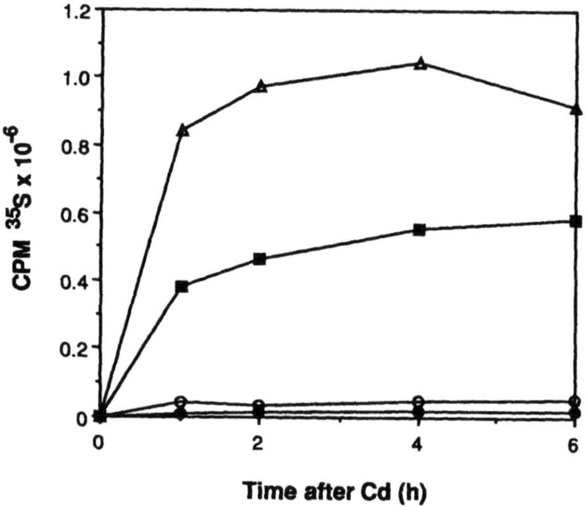

Fig. 2. The effect of Cd on the synthesis of [^{35}S] $(\gamma\text{-EC})_2$G (-△-); $(\gamma\text{-EC})_3$G (-■-); $(\gamma\text{-EC})_4$G (-○-); and $(\gamma\text{-EC})_5$G (-●-) in Cd-tolerant *D. innoxia* cells previously labeled with [^{35}S]-cysteine. Two-day old cultures (20 mL) were exposed to [^{35}S]-cysteine for 2 h prior to replacement of the medium with medium containing unlabeled cystein (1 mM) and 250 μM $CdCl_2$. Cell extracts were analyzed for radiolabeled polypeptide content.

Changes in GSH and cysteine concentrations in metal-tolerant D. innoxia *exposed to Cd*

In the experiment described above, radiolabeled and total cysteine and GSH concentrations were determined before, and at different times after addition of Cd (Fig. 3). Upon exposure to Cd, there is a rapid depletion of the [^{35}S]GSH pool that is accompanied by a rapid increase in the amount of [^{35}S](γ-EC)$_n$G synthesized. This suggests that there is a direct precursor-product relationship between GSH and the metal-binding polypeptides. There is a less pronounced depletion of the small [^{35}S]cysteine pool. A large increase in the cellular concentrations of unlabeled GSH and cysteine accompanies depletion of the radiolabeled pool. This is evident within 15 min following exposure to Cd, suggesting that, at least initially, new synthesis of GSH is regulated in part at the enzyme level, since this is not sufficient time to produce large amounts of new enzymes for this biosynthesis.

Changes in Fe uptake and binding of Fe to Cd: (γ-EC)$_n$G complexes

Cd-tolerant *D. innoxia* are normally grown in media containing 100 μM $FeSO_4$ and 100 μM Na_2EDTA. However, cells are capable of long-term growth (>50 generations) in media without Fe. If such cells are provided with carrier-free ^{59}Fe, they will rapidly accumulate this metal ion. The majority of the radiolabel is associated with insoluble fractions of the plant cell wall. However, more than 18 percent is found associated with soluble complexes within the cells. A measurement of total and soluble Fe in cultures grown for extended times in normal media containing Fe shows that the partitioning of the Fe to different fractions of the extracts is similar (data not shown). If metal-tolerant *D. innoxia* cells grown in media containing normal (100 μM) concentrations of Fe are exposed to Cd, the rapid uptake of Cd is accompanied by an increase in the uptake of Fe (Fig. 4). Similar results are obtained for cells growing in lower and higher concentrations of Fe. Virtually all of the additional Fe is found in fractions containing Cd:(γ-EC)$_n$G complexes (Fig. 5). Fe co-purifies with the Cd:polypeptide complexes and can not be removed by dialysis. ICP analysis demonstrates that purified Cd:(γ-EC)$_n$G complexes contain significant amounts of Fe (Table 1). However, if apopolypeptide is prepared from Fe:Cd:(γ-EC)$_n$G complexes by affinity chromatography on thiopropyl-Sepharose 6B, both the

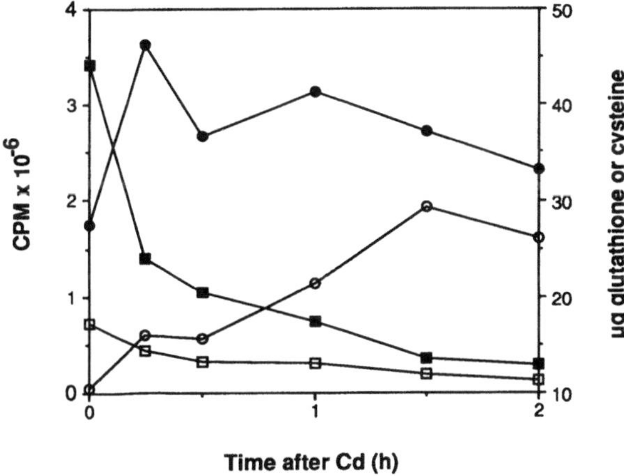

Fig. 3. The effect of Cd on cellular concentrations of total GSH (-●-), [^{35}S] GSH (-■-), total cysteine (-○-), and [^{35}S]-cysteine (-□-). Cells (20 mL) were labeled with [^{35}S]-cysteine for two h, then the medium was replaced with medium containing no radiolabeled cysteine. Cd was added at time 0. Cell extracts were analyzed for radiolabeled and total GSH and cysteine as described in Materials and methods.

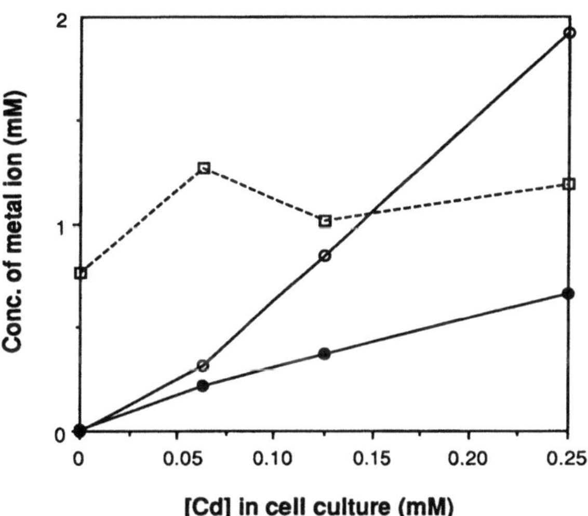

Fig. 4. The uptake of Fe by Cd-tolerant *D. innoxia* cells exposed to Cd. Cells were grown for at least 25 generations in the absence of Cd in media containing normal concentrations of $FeSO_4$ and Na_2EDTA, then were transferred to media containing different concentrations of Cd for 24 h. Cells were harvested and homogenized, and the Fe content was determined for the soluble portion of the extract (-□-) and purified polypeptide:metal complexes from the same extract (-●-). The Cd content of the polypeptide:metal complexes was also determined (-○-). Metal content was determined using ICP.

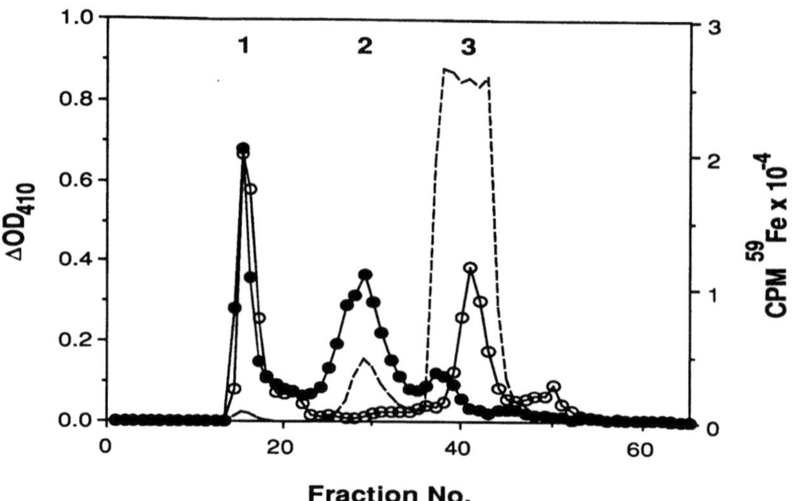

Fig. 5. Fe content of gel filtration fractions of extracts from Cd-tolerant *D. innoxia* cells grown with and without Cd. Cells grown in medium containing no additional Fe for 20 generations were transferred to medium containing ^{59}Fe plus 100 μM $FeSO_4$ and 100 μM Na_2EDTA for 24 h. 250 μM Cd was then added to one-half of the culture while the remainder was grown in media without Cd. Both cultures were grown an additional 24 h. Cells were harvested, homogenized and the soluble portion of the extract analyzed by G-50 Sephadex gel filtration. Fractions were then analyzed for radiolabeled Fe content. Fractions from cells grown in the absence of Cd (-○-); fractions from cells grown in the presence of Cd (-●-). Thiol content of fractions from cells grown in the presence of Cd was also determined by measurement of the change in OD_{410} following addition of Ellman's reagent (----). Peak 1 contains compounds which elute at the void volume of the column (>6000 D); Peak 2 contains metal:$(\gamma\text{-EC})_n$G complexes; and peak 3 contains 2-mercaptoethanol and other molecules <500 D.

Table 1. The association of Cd and Fe with purified preparations of $(\gamma\text{-EC})_n$G from metal-tolerant *D. innoxia* cells exposed to Cd

Conc. Cd in media (μM)	Amt. of Cd in $(\gamma\text{-EC})_n$G (μM)	Amt. of Fe in $(\gamma\text{-EC})_n$G (μM)	Ratio Cd:Fe
0	8	15	0.53
62.5	428	203	2.11
125	735	218	3.37
250	1698	459	3.70

Cd-tolerant *D. innoxia* cells grown many generations in media containing 100 μM Fe were grown for 48 hours in the same media containing different concentrations of cadmium. Cells were extracted and $(\gamma\text{-EC})_n$G purified, first by gel filtration chromatography, then by ion exchange chromatography. Samples were analyzed by ICP following dialysis for 24 hours at 4°C against 50 mM Tris-HCl, pH 7.8 to remove any unbound metal ions.

Cd and Fe are released from the polypeptides. Dialysis of the solutions containing the released Cd and Fe through membranes that exclude molecules 500 D and larger results in the removal of Cd and the retention of Fe. Concentration and analysis of the remaining Fe complex demonstrates the presence of one or more small, chemically uncharacterized molecules that contain no thiols but may contain two or more amino groups. This suggests that the Fe:Cd:$(\gamma\text{-EC})_n$G complexes contain one or more additional components. The study of Fe binding to specific components of the Cd:$(\gamma\text{-EC})_n$G complexes is in progress.

Discussion

The synthesis of $(\gamma\text{-EC})_n$G from GSH in response to Cd

The rapid depletion of [^{35}S]GSH accompanied by the accumulation of radiolabeled $(\gamma\text{-EC})_n$Gs implicates GSH as a precursor in the biosynthesis of these metal-binding polypeptides. The structure of the polypeptides and the lack of accumulation of [^{35}S] in any other potential precursors suggests that GSH is an immediate precursor. Grill et al. (1989) have reported the existence of a 96 kD Cd-activated transpeptidase that synthesizes $(\gamma\text{-EC})_n$Gs from GSH. The report that Cd increases this transpeptidase activity is supported by our biosynthesis data (Robinson et al., 1988 and Figure 1) which demonstrate that $(\gamma\text{-EC})_n$Gs are synthesized in vivo very rapidly following exposure to Cd, even in the presence of inhibitors which block de novo protein synthesis. However, results reported by Yoshimura et al. (1990) demonstrate that Cd is not required for transpeptidase activity in *S. pombe*. They suggest the possibility that metal regulation of polypeptide synthesis is by product removal. Moreover, Hayashi et al. (1991) have shown that there are at least two avenues of polypeptide synthesis. The first is γ-glutamylcysteine (γ-EC) dipeptidyl transfer from both GSH and $(\gamma\text{-EC})_n$Gs to GSH and $(\gamma\text{-EC})_n$G. The second is (γ-EC) polymerization from $(\gamma\text{-EC})_n$ and GSH to $(\gamma\text{-EC})_{n+1}$ followed by glycine addition with glutathione synthetase. The results presented here are also consistent with the existence of these pathways in *D. innoxia*. At least in *S. pombe*, neither pathway appears to be regulated directly by metal concentrations. Activation of the pathway must therefore be by some other means besides dipeptidyl transferase activation by metal ions.

The constitutive presence of biosynthetic enzymes in Cd-tolerant and sensitive plant cells and presence of more than one biosynthetic pathway in *S. pombe* suggest some other physiological role for the enzyme(s) responsible. Glutathione synthetase, implicated in polypeptide synthesis in *S. pombe*, clearly serves a dual function (Hayashi et al., 1991). It is possible that their role in $(\gamma\text{-EC})_n$G production may be important only in the presence of certain metal ions. Such differential enzyme activity in response to metal ions has been documented for some peptidases (Cheblowski and Coleman, 1976). GSH plays an important role in the defense against Cd toxicity in animals (Singhai et al., 1987) and this tripeptide is known to form complexes with Zn and Cd (Perrin and Watt, 1971 for an example). It is therefore likely that such complexes form in plants. If Cd functions to provide the enzyme with a required GSH:Cd substrate, then the availability of GSH, not $(\gamma\text{-EC})_n$Gs may be more important for Cd tolerance and these metal-binding polypeptides may have no other physiological activity and function solely to store toxic metal ions.

Potential roles of $(\gamma\text{-EC})_n G$ in metal metabolism

The constitutive presence of the enzymes response for $(\gamma\text{-EC})_n$G synthesis also suggests that the enzymes or $(\gamma\text{-EC})_n$Gs play some other role in metabolism besides their response to metals, since the ability to produce these metal-binding polypeptides appears to be almost universal among metal-tolerant and sensitive plants (Gekeler et al., 1989). It has been suggested that $(\gamma\text{-EC})_n$Gs might be involved in sulfur metabolism and some Cd:$(\gamma\text{-EC})_n$G complexes contain significant amounts of sulfide (Steffens et al., 1986; Verkleij et al., 1991). However, it is unlikely that Cd:$(\gamma\text{-EC})_n$G complexs are found in sufficient quantities to play such a role in plants or cells that have not been exposed to Cd, and Cu:$(\gamma\text{-EC})_n$G isolated from *D. innoxia* and several other sources do not contain significant amounts of sulfide (our own unpublished results). It has also been suggested that the transpeptidase or the metal-binding polypeptides are involved in metal ion homeostasis (Jackson et al., 1985; Robinson, 1989; Robinson and Jackson, 1986). Cu:$(\gamma\text{-EC})_n$G complexes are rapidly produced in metal-tolerant *D. innoxia* cells exposed to small or moderate amounts of Cu (Jackson et al., 1985). The high affinity of these polypeptides for Cu at physiological pH suggests that exchange of this metal ion between $(\gamma\text{-EC})_n$Gs and other cellular components must be facilitated by an intermediate that can transfer Cu from the polypeptide to an appropriate recipient. Such an intermediate has not yet been identified.

Results presented here demonstrate the presence of a large amount of Fe associated with Cd:$(\gamma\text{-EC})_n$G complexes. It is not known whether this association has some physiological significance. However, addition of Cd or Cu to metal-tolerant cells results in the rapid synthesis of metal-binding polypeptides that is accompanied by an increase in the amount of soluble Fe found in the cells. All of the additional soluble Fe is found associated with metal-:polypeptide complexes. Fe binding may be dependent upon the presence of Cd (or Cu) in the metal:polypeptide complex because removal of the group IIB metal ion from the complex results in a loss of affinity for Fe. A strong interaction with the sulfhydryl groups of the cysteine residues has been implicated in the binding of Cu and Cd (Jackson et al., 1987). This interaction may result in an increased negative charge on the free carboxyl groups of the adjacent glutamate residues. Carboxyl groups are often involved in Fe chelation (Nieboer et al., 1979). It is possible that the association of Fe with metal:$(\gamma\text{-EC})_n$G complexes is an artifact of the extraction or purification procedure. However, a recent report that an iron deficiency-specific cDNA from barley roots contains two cysteine-rich metallothionein-like domains similar to those found in cadystin (Okumura et al., 1991) suggests that cysteine-rich proteins or polypeptides might play a direct or indirect role in Fe metabolism. The association of Fe with metal:$(\gamma\text{-EC})_n$G complexes must therefore be further investigated and metal luminescence experiments are in progress to determine whether components of the polypeptides are directly involved in Fe chelation.

Conclusions

Metal-binding polypeptides, $(\gamma\text{-EC})_n$Gs, are produced in large amounts by Cd-tolerant *D. innoxia* suspension cultures exposed to Cd. These polypeptides are found associated with Cd in the tolerant cells. GSH is a direct precursor for $(\gamma\text{-EC})_n$Gs synthesis, *in vivo*. Cysteine and GSH concentrations drop rapidly when cells are exposed to Cd as the synthesis of $(\gamma\text{-EC})_n$Gs depletes cellular pools of these precursors. However, de novo synthesis of these two compounds rapidly replenishes the pools. The enzymes and intermediate metabolites responsible for synthesis of the metal-binding polypeptides produced by *D. innoxia* are present constitutively in the cells. These enzymes and the $(\gamma\text{-EC})_n$Gs may be implicated in some other aspect of micronutrient metabolism in addition to their involvement in the mechanism of tolerance to supra-optimal metal ion concentrations. However, it is also possible that the enzymes responsible for polypeptide synthesis play unrelated roles in GSH metabolism and the only function of these metal-binding polypeptides is to provide a sink for the chelation of excess metal ions within the cells. If this latter conclusion is correct, then it may be that GSH metabolism plays an important role in toxic trace metal tolerance in higher plants.

Acknowledgments

This research was supported by the US Department of Energy, grant RPI5004666/F482. The authors wish to thank Lt Krystal Guenther, USAF, Mr Amit Neeman, and Ms Mary E Rice for technical support.

References

Cheblowski J F and Coleman J E 1976 Zinc and its role in enzymes. *In* Metal Ions in Biological Systems. Vol. 6, Biological Action of Metal Ions. Ed. H Sigel. pp 1–140. Marcel Dekker, Inc., New York.

Gaitonde M K 1967 A spectroscopic method for the direct determination of cysteine in the presence of other naturally occurring amino acids. Biochem. J. 104, 627–633.

Gekeler W, Grill E, Winnacker E-L and Zenk M H 1989 Survey of the plant kingdom for the ability to bind heavy metals through phytochelatins. Z. Naturforsch. 44, 361–369.

Grill E, Löffler S, Winnacker E-L and Zenk M H 1989 Phytochelatins, the heavy-metal-binding peptides of plants, are synthesized from glutathione by a specific γ-glutamylcysteine dipeptidyl transpeptidase (phytochelatin synthase). Proc. Natl. Acad. Sci. USA 86, 6838–6842.

Grill E, Winnacker E-L and Zenk M H 1987 Phytochelatins, a class of heavy-metal-binding peptides from plants are functionally analogous to metallothioneins. Proc. Natl. Acad. Sci. USA. 84, 439–443.

Hayashi Y, Nakagawa C W, Mutoh N, Isobe M and Goto T 1991 Two pathways in the biosynthesis of cadystins $(\gamma\text{-EC})_n$G in the cell-free system of the fission yeast. Biochem. Cell Biol. 69, 115–121.

Jackson P J, Naranjo C M, McClure P R and Roth E J 1985 The molecular response of cadmium resistant *Datura innoxia* cells to heavy metal stress. *In* Cellular and Molecular Biology of Plant Stress. Eds. J L Key and T Kosuge. pp 145–160. Alan R. Liss, New York.

Jackson P J, Roth E J, McClure P R and Naranjo C M 1984 Selection, isolation, and characterization of cadmium-resistant *Datura innoxia* suspension cultures. Plant Physiol. 75, 914–918.

Jackson P J, Unkefer C J, Doolen J A, Watt K and Robinson N J 1987 Poly(γ-glutamylcysteinyl)glycine: Its role in cadmium resistance in plant cells. Proc. Natl. Acad. Sci. USA. 84, 6619–6623.

Jackson P J, Unkefer P J, Delhaize E and Robinson N J 1990 Mechanisms of trace metal tolerance in plants. *In* Environmental Injury to Plants. Ed. F Katterman. pp 231–255. Academic Press, New York.

Kondo N, Isobe M, Imai K, Goto T and Murasugi A. 1983 Structure of cadystin, the unit-peptide of cadmium-binding peptides induced in a fission yeast, *Schizosaccharomyces pombe*. Tetrahderon Lett. 24, 925–928.

Kondo N, Imai K, Isobe M and Goto T 1984 Cadystin A and B major unit peptides comprising cadmium binding peptides induced in a fission yeast – Separation, revision of structures and synthesis. Tetrahedron Lett. 25, 3869–3872.

Lindberg S and Wingstrand O 1985 Mechanism for Cd^{2+} inhibition of (K^+ + Mg^{2+}) ATPase activity and K^+($^{86}Rb^+$) uptake in roots of sugar beet (*Beta vulgaris*). Physiol. Plant. 63, 181–186.

Natusch D F S, Wallace J R and Evans C A 1974 Toxic trace elements: Preferential concentration in respirable particles. Science 183, 202–204.

Nieboer E, Richardson D H S, Lavoie P and Padovan D 1979 The role of metal-ion binding in modifying the toxic effects of sulfur dioxide on the lichen *Umbilicaria muhlenbergii*: 1. Potassium efflux studies. New Phytol. 82, 621–632.

Okumura N, Nishizawa N-K, Umehara Y and Mori S 1991 An iron deficiency-specific cDNA from barley roots having two homologous cysteine-rich MT domains. Plant Molec. Biol. 17, 531–533.

Perrin D D and Watt A E 1971 Complex formation of zinc and cadmium with glutathione. Biochim. Biophys. Acta 230, 96–104.

Petolino J F and Collins G B 1985 Manganese toxicity in tobacco (*Nicotiana tabacum* L.) callus and seedlings. J. Plant Physiol. 118, 139–144.

Reese N R and Roberts L W 1985 Effects of cadmium on whole cell and mitochondrial respiration in tobacco cell suspension cultures (*Nicotiana tabacum* L. var. *xanthi*). J. Plant Physiol. 120, 123–130.

Robinson N J 1989 Algal metallothioneins: Secondary metabolites and proteins. J. Appl. Phycol. 1, 5–18.

Robinson N J and Jackson P J 1986 'Metallothionein-like' metal complexes in angiosperms: Their structure and function. Physiol. Plant. 67, 499–506.

Robinson N J, Ratliff R L, Anderson P J, Delhaize E, Berger J M and Jackson P J 1988 Biosynthesis of poly(γ-glutamylcysteinyl)glycines in cadmium-tolerant *Datura innoxia* (Mill.) cells. Plant Science 56, 197–204.

Singhai R K, Anderson M E and Meister A 1987 Glutathione, a first line of defense against cadmium toxicity. FASEB J. 1, 220–223.

Steffens J C, Hunt D F and Williams B G 1986 Accumulation of non-protein metal-binding polypeptides (γ-glutamylcysteinyl)$_n$glycine in selected cadmium-resistant tomato cells. J. biol. Chem. 261, 13879–13882.

Stobart A K, Griffiths W T, Ameen-Bukhari I and Sherwood R P 1985 The effect of Cd^{2+} on the biosynthesis of chlorophyll in leaves of barley. Physiol. Plant. 63, 293–298.

Street J J, Sabey B R and Lindsay W L 1978 Influence of pH, phosphorus, cadmium, sewage sludge, and incubation time on the solubility and plant uptake of cadmium. J. Environ. Qual. 7, 286–290.

Takijima Y and Katsumi F 1973a Cadmium contamination of soils and rice plants caused by zinc mining: Production of high cadmium rice on the paddy fields in lower reaches of the mine station. Soil Sci. Plant Nutr. 19, 29–38.

Takijima Y and Katsumi F 1973b Cadmium contamination of soils and rice plants caused by zinc mining. IV. Use of Soil amendment materials for the control of Cd uptake by plants. Soil Sci. Plant Nutr. 19, 235–244.

Van Bruwaene R, Kirchmann R and Impens R 1984 Cadmium contamination in agriculture and zootechnology. Experientia 40, 43–52.

Verkleij J A C, Koevoets P, Van't Riet J, Bank R, Nijdam Y and Ernst W H O 1991 Poly(γ-glutamylcysteinyl)glycines or phytochelatins and their role in cadmium tolerance of *Silene vulgaris*. Plant Cell Environ. 13, 913–922.

Vlamis J, Williams D E, Corey J E, Page A L and Ganje T J 1985 Zinc and cadmium uptake by barley in field plots fertilized seven years with urban and suburban sludge. Soil Sci. 139, 81–87.

Weigel H J 1985 Inhibition of photosynthetic reactions of isolated intact chloroplasts by cadmium. J. Plant Physiol. 119, 179–189.

Wikfors G H, Neeman A and Jackson P J 1991 Cadmium-binding polypeptides in microalgal strains with laboratory-induced cadmium tolerance. Mar. Ecol. Prog. Ser. 79, 163–170.

Yoshimura E, Kabuyama Y, Yamazaki S and Toda S 1990 Activity of poly(γ-glutamylcysteinyl)-glycine synthesis in crude extract of fission yeast, *Schizosaccharomyces pombe*. Agric. Biol. Chem. 54, 3025–3026.

P. J. Randall et al. (Eds.), Genetic aspects of plant mineral nutrition, 407–415.

PLSO SV91

Genes with similarity to metallothionein genes and copper, zinc ligands in *Pisum sativum* L.

NIGEL J. ROBINSON, I. MARTA EVANS, JANET MULCRONE, JULIA BRYDEN and ANDREW M. TOMMEY
Department of Biological Sciences, University of Durham, Durham DH1 3LE, UK

Key words: copper metabolism, metallothionein, *Pisum sativum* L., *PsMT* genes, zinc metabolism

Abstract

The *PsMT* gene family of pea (*Pisum sativum* L.) encodes predicted proteins with sequence similarity to metallothioneins. However, PsMT proteins have not yet been characterised in planta and their functions remain obscure. *PsMT* transcripts were identified in the cortex tissue of pea roots using tissue squash-blotting techniques. Transcripts were not detected on northern blots of RNA isolated from the embryonic radicle, but *PsMT* transcript abundance in roots increased with age of germinating seedlings. The $PsMT_A$ gene was expressed in *E. coli* as a carboxyterminal extension of glutathione-S-transferase (GST). Fusion protein purified from crude cell lysates (500 mL cultures) bound an estimated amount of 5.99, 6.27 and 7.07 moles of Zn, Cu and Cd respectively per mole protein, compared to equivalent estimates of 0.37, 0.63 and 0.26 moles for GST alone. Similar estimates for Fe-binding were 0.28 moles for GST-$PsMT_A$ fusion protein and 0.1 moles for GST alone.

In summary, these data: 1, show that *PsMT* transcripts are abundant in roots of pea plants that have not been exposed to supra-optimal concentrations of trace metals and hence appear to be constitutively expressed and 2, indicate that $PsMT_A$ protein can bind certain trace metal ions. We have also identified and partially purified a Zn ligand (Zn-A) and two Cu ligands (Cu-A, Cu-B) from pea roots which have not been exposed to supra-optimal conditions of trace metal ions and are therefore defined as 'constitutive'. Whether or not these ligands include the products of *PsMT* genes remains to be established.

Introduction

Metallothioneins, MTs have been isolated from a wide range of vertebrates, invertebrates and micro-organisms. Proposed functions for these cysteine-rich proteins include roles in essential trace metal ion homoeostasis and the detoxification of certain non-essential metals (cited in Kägi and Schäffer, 1988). In animals, MT is most abundant in parenchymatous tissues (liver, kidney, pancreas and intestine) and its abundance varies with age, stage of development and diet. A Zn-binding protein has been isolated from wheat germ (Lane et al., 1987) and designated class II MT (Kägi and Schäffer, 1988) but equivalent proteins have not yet been purified from vegetative plant tissues. Therefore, any proposed role for MTs (class I or II) in essential trace metal nutrition or the detoxification of non-essential metals in vegetative plant tissue is hypothetical.

We have isolated cDNAs (pPR179, pPR705) and a gene, $PsMT_A$, from pea which encode proteins with sequence similarity to MTs (Evans et al., 1990). In addition, we have recently identified two other members of the *PsMT* gene family via PCR-mediated cloning. These genes have been designated $PsMT_B$ and $PsMT_C$. Related cDNAs have been independently isolated from cDNA libraries prepared from mRNA isolated from *Mimulus guttatus* (de Miranda et al., 1990). Homologous gene sequences have also been identified in several other higher plant species. The sequence of the Zn-binding protein from wheat germ (class II MT) does not closely align with the predicted products of these novel plant genes or with known class I MTs. It remains to be established whether or not the products of these novel genes bind metals in planta and their functions are therefore uncertain.

MT genes typically possess metal-regulatory elements. MT transcript abundance increases after exposure to elevated concentrations of certain metal ions in a range of eukaryotes (cited in Palmiter, 1987) and this has also been observed in a prokaryote (Robinson et al., 1990). However, in roots of the monkey flower plant, *M. guttatus*, no increase in the abundance of transcripts encoded by 'MT-like' genes was detected following exposure to elevated concentrations of Cu, Cd or Zn (de Miranda et al., 1990). In addition, a related cDNA (*ids-1*) has recently been isolated from barley, following specific hybridisation to a probe prepared from mRNA isolated from Fe-deficient roots (Okamura et al., 1991). This implies that the *ids-1* sequence is preferentially expressed in barley roots under Fe-limiting conditions. Upon exposure to elevated concentrations of trace metals, higher plants synthesize and accumulate increased amounts of the metal-binding polypeptide poly (γ-glutamylcysteinyl)glycine (Jackson et al., 1987), most commonly known as phytochelatin (Grill et al., 1985) but also referred to as cadystin (Kondo et al., 1985) and class III MT (Kägi and Schäffer, 1988). It has been proposed that phytochelatin and MT (class I) perform analogous functions in plants and animals exposed to elevated concentrations of certain trace metal ions (Grill et al., 1987). It is possible to rationalise that any MT gene present in organisms containing phytochelatins may not show a greatly elevated level of expression in response to supra-optimal concentrations of those metal ions that are efficiently detoxified by phytochelatins.

Animal MT genes additionally show programmed expression during embryogenesis and in different stages of fetal and perinatal development (cited in Kägi and Schäffer, 1988). For example, the hepatic concentration of MT, primarily Zn-thionein, is 20-fold greater in neonatal than in adult rats. The role of MT in these processes is uncertain, although it has been noted that changes in MT concentration coincide with changes in metabolic processes which require trace metals such as Zn. Moreover, there are reports of differential activation of MT isoforms during these processes. Accumulation of MT in nuclei of growing primary cultured-adult-rat hepatocytes has also been observed in early S phase, but not in G_0 or G_1 when the protein is found in the cytoplasm (Tsujikawa et al., 1991). In wheat, class II MT is present in dry embryos but its abundance declines immediately following germination (Hanley-Bowdoin and Lane, 1983). Lane et al. (1987) noted an analogy between wheat and animal MT production during embryogenesis and observed that deposition of a Zn MT during plant embryogenesis may be associated with a shift between proliferative and differentiating stages of development. A detailed description of the pattern of expression of $PsMT_A$, and its homologues, is clearly required.

We report here a preliminary examination of the temporal and spatial pattern of expression of *PsMT* genes in developing pea and provide evidence that the $PsMT_A$ protein is capable of binding certain metal ions. Three metal-complexes have also been identified in pea roots grown in the absence of supra-optimal concentrations of trace metal ions and these may include candidates for the products of *PsMT* genes.

Materials and methods

Plant material

Seeds of *P. sativum* L. (cv Feltham First) were surface-sterilized with 1% v/v Chloros, germinated in the dark for 4 days and grown as previously described by Evans et al. (1990). After a further 9 days, leaves were harvested directly into liquid nitrogen and stored at −80°C. Etiolated leaf material was obtained from plants grown for 4 days in the presence of light and then transferred into the dark for the final 5 days. Roots were harvested from pea plants grown in hydroponic cultures, as described previously (Evans et al. 1988). "Mid-development" cotyledons were obtained from plants grown in hydroponic culture and pods were harvested 15 days after flowering (the period from flowering to seed maturity under these conditions was ca. 22 days). Cotyledons, asceptically removed from the testa and embryonic axes, were frozen in liquid nitrogen and stored at −80°C. Radicles were dissected from seeds obtained from pods harvested 13 days after flowering.

Isolation of mRNA, northern and squash blotting

Total RNA and poly(A)$^+$RNA were isolated from different organs (radicle, 4-day-old root, 14-day-old root, developing cotyledon, etiolated leaf, green leaf) as previously described (Evans et al., 1988). Total RNA (5 or 10 μg/lane) and poly(A)$^+$RNA (2 μg/lane) was glyoxalated and electrophoresed through 1.5% agarose gels (McMasters and Carmichael, 1977).

Northern blot analyses were carried out as described previously (Evans et al., 1990). Northern blots were incubated with ^{32}P-labelled *PsMT*$_A$ cDNA excised from plasmid pPR179 and labelled by random oligonucleotide-priming (Feinberg and Vogelstein, 1983). Filter hybridisation and post-hybridisation washes were carried out under standard conditions.

Squash blotting involved gently pressing freshly excised root material onto a nitrocellulose filter that was previously equilibrated for 15 min with buffer containing 5 × SSC (1 × SSC = 0.15 *M* sodium chloride, 0.015 sodium citrate, pH 7.0), 0.1% w/v SDS (pH 7.5). The nitrocellulose was baked at 80°C under vacuum for 2 h and then incubated with buffer (10 m*M* Tris HCl, pH 7.8; 50 mM EDTA; 0.5% w/v SDS) containing 0.5 mg mL^{-1} proteinase K for 2 h at 37°C prior to hybridization with the ^{32}P-labelled cDNA probe (described above) in a solution containing 50% v/v formamide, 5 × SSC, 5 × Denhardts reagent, 100 μg mL^{-1} denatured herring sperm DNA, 0.1% SDS at 42°C for 48 h. Final post-hybridisation washes were in 1 × SSC, 0.1% SDS at 65°C.

Analysis of metal-binding to PsMT$_A$ protein expressed in E. coli

The *PsMT*$_A$ protein coding region was cloned into the vector pGEX3X to facilitate expression of the *PsMT*$_A$ protein in *E. coli* as a carboxyterminal extension of glutathione-S-transferase (GST) as described by Tommey et al. (1991). The fusion protein (M_r 34, 500) was purified from cell lysates using glutathione affinity chromatography (Smith and Johnson, 1988). GST (M_r 26, 500) was similarly purified from lysates of cells transformed with the pGEX3X vector alone. Samples (total volume 2.5 mL) containing either fusion protein or GST were purified from crude lysates of 500 mL cultures grown for 4 h in media supplemented with 500 μM of either Cd, Zn, Cu or Fe. Cd and Zn were added as metal chloride, Cu as metal sulphate. Fe was supplied as Fe citrate, which is available for *fec*-mediated uptake by *E. coli* (Pressler et al., 1988). These samples were passed through Sephadex G-25 (PD-10, Pharmacia) and protein content of the void fraction was estimated using a Coomassie blue based reagent (Bio-Rad) and bovine serum albumin as standard. Metal content was determined by atomic absorption spectrophotometry.

Identification of phytochelatins by reversed phase HPLC

Pea seedlings were grown hydroponically with a final 60 h of growth in either the presence or absence of 50 μM Cd. Extracts were prepared for reversed phase HPLC analysis of phytochelatins by methods previously used for extracts from plant cell suspension cultures (Delhaize et al., 1989) with the following modifications. Harvested roots (between 0.5 and 1.0 g) were washed three times in an excess of distilled water and once with an equal volume of extraction buffer before an equal volume of 1 *N* HCl was added. Tissue was homogenised (Polytron) and cell debris removed by centrifugation (10 000 *g* for 10 min) and also by passage through a 0.2 μm porosity polycarbonate membrane. Samples were then passed through Centricon filtration units (Amicon Corporation, >30,000 Da exclusion), fractionated by reversed phase HPLC and analysed for thiols as described previously (Delhaize et al., 1989).

Purification of constitutive Cu and Zn ligands from pea roots

Plants were grown hydroponically in a 0.05% w/v Phostrogen solution (initial concentrations of Cu and Zn were 0.6 and 0.2 μM respectively, which are assumed to be further depleted during growth) with no additional trace metal ions. Harvested roots (between 150 and 200 g) were washed three times in a large volume of distilled water and then mixed with an equal volume of buffer (0.1 M ammonium acetate, pH 5.5, 1% v/v β-mercaptoethanol) prior to homogenisation (Polytron). The homogenate was strained through two layers of muslin and the resulting solution centrifuged (2 500 g for 15 min). The supernantant was mixed with 4 volumes of acetone, incubated at −80°C for 3 h followed by centrifugation (2 500 g for 5 min). Pellets were resuspended in 50 mL of buffer (10 mM Tris-HCl, pH 7.2, 1% v/v β-mercaptoethanol) heated (60°C for 10 min; insensitivity to such thermal treatment has previously been used to purify MTs from other organisms) and heat denatured debris removed by further centrifugation (2 500 g for 5 min). Samples were passed through Sephadex G-25 (PD-10, equilibrated with 10 mM Tris-HCl pH 7.2, 1% v/v β-mercaptoethanol) and the void fractions loaded onto a column (10 mL) of DEAE Sephadex equilibrated in the same buffer. The matrix was washed with 200 mL of the same buffer followed sequentially by 200 mL of equivalent buffers containing 100 mM and 300 mM Tris-HCl. Fractions (10 mL) were analysed for Cu and Zn by atomic absorption spectrophotometry. Acetone precipitates of pooled fractions containing the major Cu and Zn peaks, prepared as described above, were resuspended in 5 mL of buffer (10 mM Tris-HCl, pH 7.2, 1% v/v β-mercaptoethanol) and fractionated on Sephadex G-50 equilibrated with the same buffer. Every third fraction (2.5 mL) was analysed for Cu, Zn and protein. The void and total volumes of the column were estimated by separation of samples containing blue dextran and free metal ions respectively.

Results and discussion

Figure 1 shows the predicted products of $PsMT_A$ and related plant genes. There is significant sequence similarity within N- and C-terminal cysteine-rich domains which have previously been aligned with know class I MTs (for examples of alignments see Evans et al., 1990; de Miranda et al., 1990). These MT-like domains are separated by a more divergent intervening region.

$PsMT_A$ probe hybridized to transcripts (ca. 640 bases) which were relatively abundant in

```
Tobacco NtMT[a]      --------------------------------------------------------------AAEGGN-GCKCGSNCTCDPCNC-
Alfalfa MsMTA[a]     ---------------------------------------TVIL--GVGPAKIHF-EGAEMGVAAEDG--GCKCGDSCTCDPCNCK
Pea PsMTA[b]         MS---G-CGCGSSCNCGDSCK-CNKRSSGLSYSEMETTETVIL--GVGPAKIQF-EGAEMSAASEDG--GCKCGDNCTCDPCNCK
Pea PsMTB[a]         MS---G-CGCGSSCNCGDSCK-CNKRSSGLSYSEMETTETVIL--GVGPAKIQF-NGAEMSVAAEDG--GCKCGDSCTCDPCNCK
Pea PsMTC[a]         MS---G-CGCGSSCNCGDSCK-CSKRSSGLSYSEMETTETVIL--GVGPAKIQF-NGAEMSVAAEDG--GCKCGDSCTCDPCNCK
M. guttatus[c]       MS--SG-CSCGSGCKCGDNCS-C-SMYPDME-TN--TTVTMIE--GVAPLKM-YSEGSEKSFGAEGGN-GCKCGSNCKCDPCNC-
Soybean[d]           MSCCGGNCGCGSSCKCGNGCGGC-KMYPDLSYTESTTTETLVM--GVAPVKAQF-EGAEMGVPAEND--GCKCGPNCSCNPCTCK
Maize MT-1[e]        MS-----CSCGSSCGCGSSC-KCGKKYPDLEETS--TAAQPTVVLGVAPEKKAAPEFVEAAAESGEAAHGCSCGSGCKCDPCNC-
Barley ids-1[f]      MS-----CSCGSSCGCGSNC-NCGKMYPDLEEKSGATMQVTVIVLGVGSAK----VQFEEAAEFGEAAHGCSCGANCKCNPCNC-

CONCENSUS            MS-----C-CGS-C-CG--C--C------------T--------GV---K-------E---------GC-CG--C-C-PC C
```

Fig. 1. Amino acid sequences of the products of plant genes with sequence similarity to metallothionein genes predicted from cDNA sequences, genomic DNA sequences or DNA sequences obtained from PCR mediated cloning. [a]Robinson and co-workers (unpublished data); [b]Evans et al. (1990); [c]de Miranda et al. (1990); [d]Kawashima et al. (1991); [e]de Framond (1991); [f]Okamura et al. (1991).

total RNA and poly(A^+)RNA isolated from roots of 14 day old plants (Fig. 2A, B). Transcripts were less abundant in poly(A^+)RNA from etiolated leaves and were barely detectable in isolates from leaves grown in the light. Weak hybridization to slightly smaller transcripts, which may be the product of a *PsMT* gene other than $PsMT_A$, was detected in developing cotyledons (Fig. 2B). *PsMT* transcripts were not detected in RNA isolated from embryonic pea radicles but increased in abundance in the roots during germination (Fig. 2C). ^{32}P-labelled $PsMT_A$ cDNA did not hybridize to sections of embryonic radicle using squash-blotting techniques (data not presented), but did hybridize to squash-blots of 14 day old pea roots (Fig. 3). These data were consistent with northern analyses (Fig. 2C). Hybridization was observed within the parenchymatous cortex, but not in the stele or the lateral root tips.

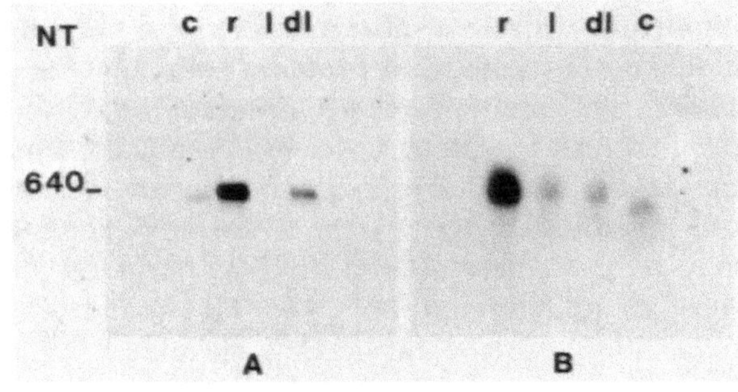

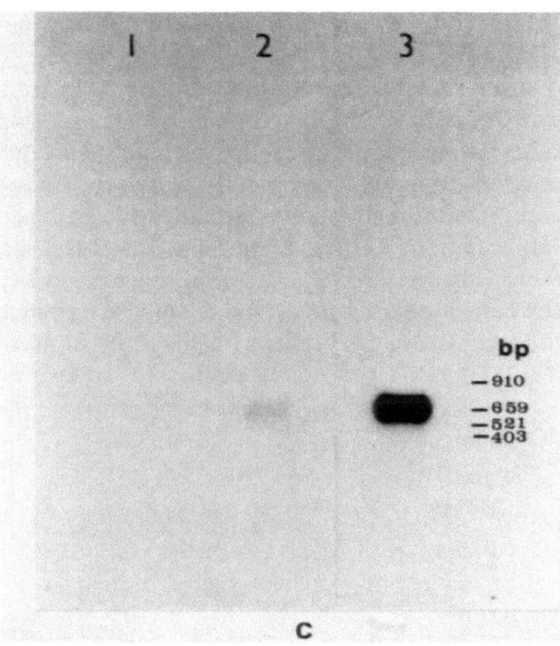

Fig. 2. Detection of *PsMT* mRNA in extracts from pea organs. Northern blot analysis was performed for samples of poly(A^+)RNA (panel **A**) and total RNA (panel **B**) isolated from different organs. Agarose gels (1.5%) were loaded with total RNA (5 μg/lane), or poly(A^+)RNA (2 μg/lane), from developing cotyledons 15 days after flowering (lane **c**), roots 14 days after germination (lane r), leaves (lane l), etiolated leaves (lane dl). Panel C shows the detection of *PsMT* mRNA in extracts from pea roots at different stages of development; radicle 13 days after flowering (lane 1), from roots 4 days after germination (lane 2) and roots 14 days after germination (lane 3).

Observed increases in the abundance of $PsMT_A$ transcripts in pea roots during the first 14 days after germination could suggest that this gene is subject to developmental control. Alternatively, analogy to the expression of *ids1* in barley, might indicate that the pea plants used in these experiments were subject to increasing Fe deficiency during the course of root development, causing activation of the $PsMT_A$ gene. This explanation seems feasible since the stringent growth conditions used by Grusak et al. (1990) to avoid activation of Fe efficiency mechanisms in *P. sativum* were not employed in these

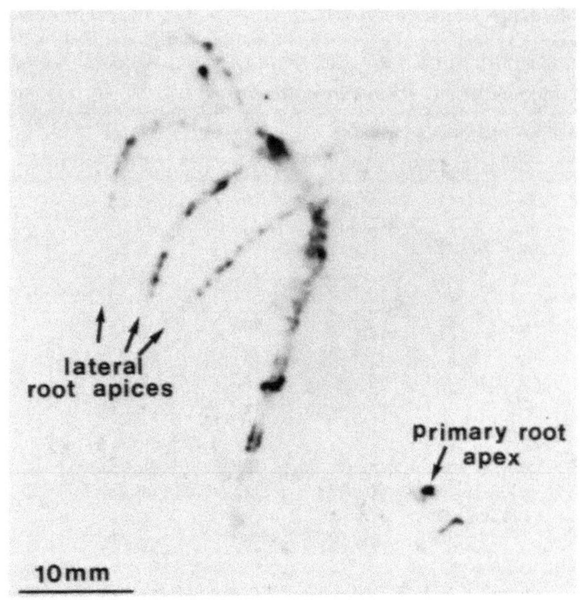

Fig. 3. Detection of *PsMT* mRNA on a tissue blot of pea roots 14 days after germination.

studies. It is also noted that preliminary results (data not shown) reveal that the Cu content of pea roots increased when plants were grown under conditions of Fe deficiency. It is possible that MT-like plant genes may be expressed under low Fe conditions in response to such elevated levels of Cu and function to detoxify excess of this metal ion. Clearly, further experiments are required to examine the pattern of expression of *PsMT* genes in pea roots following growth under different regimes of Fe availability.

In order to examine the metal binding characteristics of its product, the $PsMT_A$ gene has been expressed in *E. coli*. Table 1 shows the amount of individual trace metals associated with GST-$PsMT_A$ fusion protein, and with GST alone, following purification from lysates of 500 mL cultures grown in media supplemented with 500 μM Fe, Zn, Cu or Cd. A greater amount of metal associates with the fusion protein than with GST alone, suggesting that metal ions preferentially bind to the $PsMT_A$ portion of the protein. However, these ratios of metal to protein may not be reliable estimates of metal:protein stoichiometries, since differences could exist in the reactivities of GST, $PsMT_A$ and BSA with Coomassie blue, the reagent used to estimate protein concentrations. Furthermore, in subsequent experiments with replicated 50 mL cultures of *E. coli*, it was observed that isolates of GST-$PsMT_A$ from Cu exposed cells contained impurities which can also interfere with the estimated ratio of metal to protein (Tommey et al., 1991). The amount of Fe associated with the purified fusion protein was considerably less than the other metals. Explanations for this include the possibilities that 1) the Fe-protein binding may be extremely air sensitive, as observed for reconstituted animal Fe^{2+}-MT (Kägi and Kojima, 1987); 2) the intracellular levels of available Fe did not increase to allow enhanced Fe-binding; 3) the protein does not have a high affinity for Fe and preferentially binds other metal ions.

Table 1. Total amounts of metal and purified protein in the void fraction (3.5 mL) eluted from columns of Sephadex G-25 (PD-10, Pharmacia). Protein was purified by glutathione-affinity-chromatography from extracts of *E. coli* cells expressing either GST or GST-$PsMT_A$ fusion protein. Cells (500 mL) were grown in media supplemented with 500 μM of either Fe, Zn, Cu or Cd or without metal supplementation (figures in parentheses)

Purified protein		Metal (n moles)	Protein (n moles)	Ratio (metal:protein)
Fe	GST	16.49	167.7	0.1
		(0.0)	(111.9)	(0.0)
	GST-$PsMT_A$	19.75	71.2	0.28
		(0.0)	(72.1)	(0.0)
Zn	GST	28.28	76.4	0.37
		(1.35)	(111.9)	(0.01)
	GST-$PsMT_A$	397.8	66.4	5.99
		(55.98)	(72.1)	(0.78)
Cu	GST	92.6	146.4	0.63
		(18.07)	(111.9)	(0.16)
	GST-$PsMT_A$	414.1	66.0	6.27
		(54.22)	(72.1)	(0.75)
Cd	GST	37.6	144.0	0.26
		(7.58)	(111.9)	(0.07)
	GST-$PsMT_A$	299.9	42.4	7.07
		(11.36)	(72.1)	(0.16)

Many previous attempts to isolate MTs from plant roots have employed material excised from plants grown in the presence of highly elevated concentrations of trace metals. However, under these conditions plants accumulate phytochelatins (for reviews see Steffens, 1990; Rauser, 1990; Robinson, 1990), the presence of which may confound attempts to purify any MT protein which may also be present but possibly in much

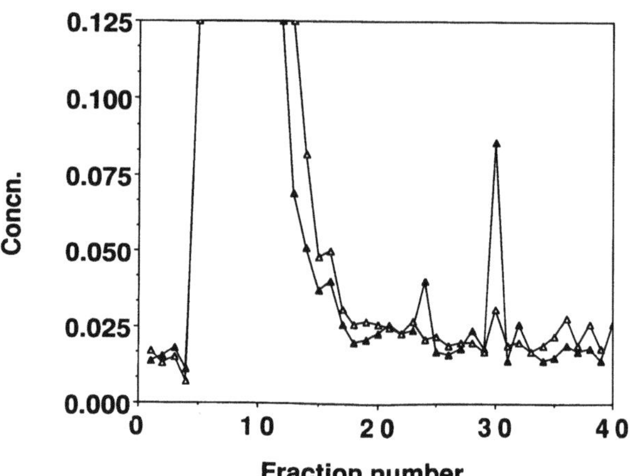

Fig. 4. Reversed phase HPLC analysis of phytochelatins in extracts from pea roots grown in the presence (▲) or absence of Cd (△). The thiol-content of fractions is shown (in μg glutathione equivalents mL^{-1}). The initial large peak (off-scale) is due to the presence of β-mercaptoethanol in the extraction buffer while the two lesser peaks (fractions 24 and 30) detected in extract from Cd-exposed material correspond to diagnostic retention times for phytochelatins.

lower quantities. Figure 4 shows the detection of phytochelatins in the roots of pea following growth in the presence of 50 μM Cd, consistent with previous observations that these polypeptides are present in pea. The assay was not sufficiently sensitive to clearly detect the lower levels of these polypeptides, which we assume to be present in pea roots grown in the absence of additional metal ions, although a small peak (putatively phytochelatin) is present in fraction 29 of the control extract (Fig. 4).

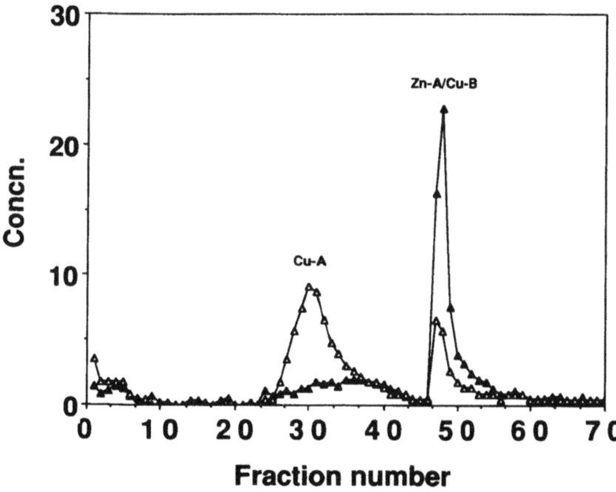

Fig. 5. Anion exchange chromatography on DEAE Sephadex of a 'constitutive' extract from pea roots. The Cu (△) and Zn (▲) content of fractions is shown in μM. The matrix was sequentially eluted with buffer containing 10 mM Tris (fractions 1 to 20), 100 mM Tris (fractions 21 to 40) and 300 mM Tris (fraction 41 onwards).

The data presented here show that $PsMT_A$ transcripts are abundant in roots grown without supra-optimal trace metal ion supplementation and indicate that the $PsMT_A$ protein will bind Cu and Zn following expression in *E. coli*. Purification of low M_r constitutive Cu and Zn complexes has therefore been initiated from pea roots with the eventual aim of establishing whether or not the ligands include the products of *PsMT* genes. Heat stable material from acetone precipitates of pea root extracts was crudely fractionated by stepwise elution from DEAE Sephadex (Fig. 5). Fractions eluted in buffer containing 100 mM Tris included a Cu complex which has been designated Cu-A, while fractions eluted in buffer containing 300 mM Tris included both Zn and Cu complexes, designated Zn-A and Cu-B, respectively. Figure 6 shows the subsequent frac-

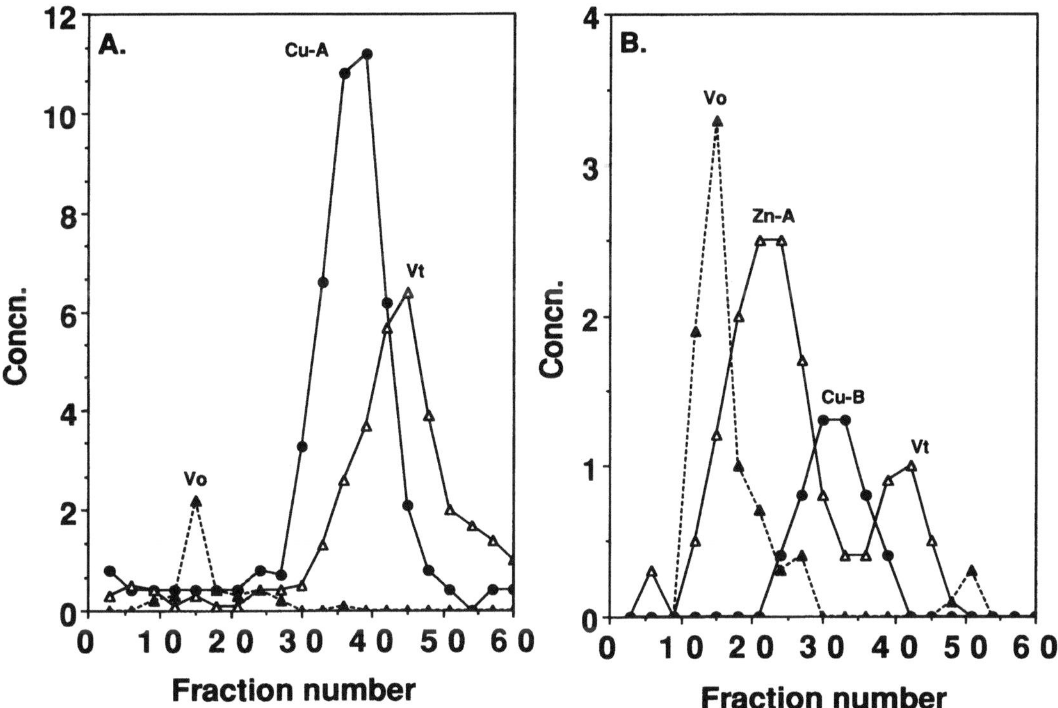

Fig. 6. Gel filtration on Sephadex G-50 of pooled fractions containing Cu-A (panel **A**) and Cu-B and Zn-A (panel **B**) after anion exchange chromatography. Every third fraction was analysed for protein (▲) in $\mu g\, mL^{-1}$, Zn (△) and Cu (●) in μM.

tionation of these complexes on Sephadex G-50. Cu-A and Cu-B have similar low M_r, while Zn-A is apparently larger than the Cu complexes. Antibodies are currently being raised to the expressed GST-PsMT$_A$ fusion protein. An initial aim will be to establish whether or not fractions containing Cu-A, Cu-B or Zn-A include antigens that cross-react with this antibody, indicative of metal-binding by the products of *PsMT* genes.

Acknowledgements

This work was supported by research grant PG12/519(PMB) from the Agricultural and Food Research Council. N. J. R. is a Royal Society University Research Fellow.

References

Delhaize E, Jackson P J, Lujan L D and Robinson N J 1989 Poly(γ-glutamylcysteinyl)glycine synthesis and aggregation with cadmium: Role in cadmium-tolerance. Plant Physiol. 89, 700–706.

Evans I M, Swinhoe R, Gatehouse L N, Gatehouse J A and Boulter D 1988 Distribution of root mRNA species in other vegetative organs of pea (*Pisum sativum* L.). Mol. Gen Genet. 214, 153–157.

Evans I M, Gatehouse L N, Gatehouse J A, Robinson N J and Croy R R D C 1990 A gene from pea (*Pisum sativum* L.) with homology to metallothionein genes. FEBS Lett. 262, 29–32.

Feinberg A P and Vogelstein B 1983 A technique for radiolabelling DNA restriction endonuclease fragments to high specific activity. Anal. Biochem. 132, 6–13.

Grill E, Winnaker E L and Zenk M H 1985 Phytochelatins: The principal heavy-metal complexing peptide of higher plants. Science 230, 574–576.

Grill E, Winnaker E-L, and Zenk M H 1987 Phytochelatins, a class of heavy-metal binding peptides from plants, are functionally analogous to metallothioneins. Proc. Natn. Acad. Sci. 84, 439–443.

Grusak M A, Welch R M and Kochian L V 1990 Physiological characterization of single-gene mutation of *Pisum sativum* exhibiting excess iron accumulation. Plant Physiol. 93, 976–981.

Hanley-Bowdoin L and Lane B G 1983 A novel protein programmed by the mRNA conserved in dry wheat embryos: The principal site of cysteine incorporation during early germination. Eur. J. Biochem 135, 9–15.

Jackson P J, Unkefer C J, Doolen J A, Watt K and Robinson N J 1987 Poly(γ-glutamylcysteinyl)glycine: Its role in cadmium resistance in plant cells. Proc. Natn. Acad. Sci. 84, 6619–6623.

Kägi J H R and Kojima Y 1987 Chemistry and Biochemistry of metallothionein. *In* Metallothionein II. Eds. J H R Kägi and Y Kojima. pp 63–80. Birkhäuser Verlag, Basel-Boston.

Kägi J H R and Schäffer A 1988 Biochemistry of metallothionein. Biochemistry 27, 8509–8515.

Kawashima I, Inokuchi Y, Chino M, Kimura M and Shimizu N 1991 Isolation of a gene for a metallothionein-like protein from soybean. Plant Cell Physiol. 32, 913–916.

Kondo N, Isobe M, Imai K and Goto T 1985 Synthesis of metallothionein-like peptides cadystin A and B occurring in fission yeast, and their isomers. Agric Biol. Chem. 49, 71–83.

Lane B, Kajioka R and Kennedy T 1987 The wheat germ E_c protein is a zinc-containing metallothionein Biochem. Cell Biol. 65, 1001–1005.

McMasters G K and Carmichael G G 1977 Analysis of single-stranded and double-stranded nucleic acids on polyacrylamide and agarose gels by using glyoxal and acridine orange. Proc. Natn. Acad. Sci. 74, 4835–4838.

de Framond A J 1991 A metallothionein-like gene from maize (*Zea mays*) Cloning and characterization. FEBS Lett. 290, 103–106.

de Miranda J R, Thomas M A, Thurman D A and Tomsett A B 1990 Class I metallothionein genes from the flowering plant *Mimulus guttatus*. FEBS Lett. 260, 277–280.

Okumura N, Nishizawa N-K, Umehara Y and Mori S 1991 An iron deficiency-specific cDNA from barley roots having two homologous cysteine-rich MT domains. Plant Molec. Biol. 17, 531–533.

Palmiter R D 1987 Molecular biology of metallothionein gene expression. *In* Metallothionein II. Eds. J H R Kägi and Y Kojima. pp 63–80. Birkhäuser Verlag, Basel-Boston.

Pressler U, Staudenmaier H, Zimmermann L and Braun V (1988) Genetics of the iron dicitrate transport system of *Escherichia coli*. J. Bacteriol. 170 2716–2724.

Rauser W E 1990 Phytochelatins. Annu. Rev. Biochem. 59, 61–86.

Robinson 1990 Metal-binding polypeptides in plants. *In* Heavy-metal tolerance in plants. Ed. A J Shaw. pp 195–214. CRC Press, Boca Raton FL.

Robinson N J, Gupta A, Fordham-Skelton A P, Croy R R D, Whitton B A and Huckle J W 1990 Prokaryotic metallothionein gene characterisation and expression: Chromosome crawling by ligation-mediated PCR. Proc. R. Soc. Lond. B. 242, 241–247.

Smith D B and Johnson K S 1988 Single-step purification of polypeptides expressed in *Escherichia coli* as fusions with glutathione-S-transferase. Gene 67, 31–40.

Steffens J C 1990 The heavy metal-binding peptides of plants. A. Rev. Pl. Physiol. Pl. Mol. Biol. 41, 553–575.

Tommey A M, Shi J, Lindsay W P, Urwin P E and Robinson N J 1991 Expression of the pea gene *PsMT*$_A$ in *E. coli*: Metal-binding properties of the expressed protein. FEBS Lett. 292, 48–52.

Tsujikawa K, Imai T, Katutani M, Kayamori Y, Mimura T, Otaki N, Kimura M, Fukuyama R and Shimizu N 1991 Localization of metallothionein in nuclei of growing primary cultured adult rat hepatocytes. FEBS Lett. 283, 239–242.

Developments in Plant and Soil Sciences

1. J. Monteith and C. Webb (eds.): *Soil Water and Nitrogen in Mediterranean-type Environments.* 1981 ISBN 90-247-2406-6
2. J. C. Brogan (ed.): *Nitrogen Losses and Surface Run-off from Landspreading of Manures.* 1981 ISBN 90-247-2471-6
3. J. D. Bewley (ed.): *Nitrogen and Carbon Metabolism.* 1981 ISBN 90-247-2472-4
4. R. Brouwer, I. Gašparíková, J. Kolek and B. C. Loughman (eds.): *Structure and Function of Plant Roots.* 1981 ISBN 90-247-2510-0
5. Y. R. Dommergues and H. G. Diem (eds.): *Microbiology of Tropical Soils and Plant Productivity.* 1982 ISBN 90-247-2624-7
6. G. P. Robertson, R. Herrara and T. Rosswall (eds.): *Nitrogen Cycling in Ecosystems of Latin America and the Caribbean.* 1982 ISBN 90-247-2719-7
7. D. Atkinson, K. K. S. Bhat, M. P. Coutts, P. A. Mason and D. J. Read (eds.): *Tree Root Systems and Their Mycorrhizas.* 1983 ISBN 90-247-2821-5
8. M. R. Sarić and B. C. Loughman (eds.): *Genetic Aspects of Plant Nutrition.* 1983 ISBN 90-247-2822-3
9. J. R. Freney and J. R. Simpson (eds.): *Gaseous Loss of Nitrogen from Plant-Soil Systems.* 1983 ISBN 90-247-2820-7
10. United Nations Economic Commission for Europe (ed.): *Efficient Use of Fertilizers in Agriculture.* 1983 ISBN 90-247-2866-5
11. J. Tinsley and J. F. Darbyshire (eds.): *Biological Processes and Soil Fertility.* 1984 ISBN 90-247-2902-5
12. A. D. L. Akkermans, D. Baker, K. Huss-Danell and J. D. Tjepkema (eds.): Frankia *Symbioses.* 1984 ISBN 90-247-2967-X
13. W. S. Silver and E. C. Schröder (eds.): *Practical Application of* Azolla *for Rice Production.* 1984 ISBN 90-247-3068-6
14. P. G. L. Vlek (ed.): *Micronutrients in Tropical Food Crop Production.* 1985 ISBN 90-247-3085-6
15. T. P. Hignett (ed.): *Fertilizer Manual.* 1985 ISBN 90-247-3122-4
16. D. Vaughan and R. E. Malcolm (eds.): *Soil Organic Matter and Biological Activity.* 1985 ISBN 90-247-3154-2
17. D. Pasternak and A. San Pietro (eds.): *Biosalinity in Action.* Bioproduction with Saline Water. 1985 ISBN 90-247-3159-3
18. M. Lalonde, C. Camiré and J. O. Dawson (eds.): Frankia *and Actinorhizal Plants.* 1985 ISBN 90-247-3214-X
19. H. Lambers, J. J. Neeteson and I. Stulen (eds.): *Fundamental, Ecological and Agricultural Aspects of Nitrogen Metabolism in Higher Plants.* 1986 ISBN 90-247-3258-1
20. M. B. Jackson (ed.): *New Root Formation in Plants and Cuttings.* 1986 ISBN 90-247-3260-3
21. F. A. Skinner and P. Uomala (eds.): *Nitrogen Fixation with Non-Legumes* (Proceedings of the 3rd Symposium, Helsinki, 1984). 1986 ISBN 90-247-3283-2
22. A. Alexander (ed.): *Foliar Fertilization.* 1986 ISBN 90-247-3288-3
23. H. G. v.d. Meer, J. C. Ryden and G. C. Ennik (eds.): *Nitrogen Fluxes in Intensive Grassland Systems.* 1986 ISBN 90-247-3309-X
24. A. U. Mokwunye and P. L. G. Vlek (eds.): *Management of Nitrogen and Phosphorus Fertilizers in Sub-Saharan Africa.* 1986 ISBN 90-247-3312-X
25. Y. Chen and Y. Avnimelech (eds.): *The Role of Organic Matter in Modern Agriculture.* 1986 ISBN 90-247-3360-X
26. S. K. De Datta and W. H. Patrick Jr. (eds.): *Nitrogen Economy of Flooded Rice Soils.* 1986 ISBN 90-247-3361-8
27. W. H. Gabelman and B. C. Loughman (eds.): *Genetic Aspects of Plant Mineral Nutrition.* 1987 ISBN 90-247-3494-0
28. A. van Diest (ed.): *Plant and Soil: Interfaces and Interactions.* 1987 ISBN 90-247-3535-1

Developments in Plant and Soil Sciences

29. United Nations Economic Commission for Europe and FAO (eds.): *The Utilization of Secondary and Trace Elements in Agriculture.* 1987 ISBN 90-247-3546-7
30. H. G. v.d. Meer, R. J. Unwin, T. A. van Dijk and G. C. Ennik (eds.): *Animal Manure on Grassland and Fodder Crops.* Fertilizer or Waste? 1987 ISBN 90-247-3568-8
31. N. J. Barrow: *Reactions with Variable-Charge Soils.* 1987 ISBN 90-247-3589-0
32. D. P. Beck and L. A. Materon (eds.): *Nitrogen Fixation by Legumes in Mediterranean Agriculture.* 1988 ISBN 90-247-3624-2
33. R. D. Graham, R. J. Hannam and N. C. Uren (eds.): *Manganese in Soils and Plants.* 1988 ISBN 90-247-3758-3
34. J. G. Torrey and J. L. Winship (eds.): *Applications of Continuous and Steady-State Methods to Root Biology.* 1989 ISBN 0-7923-0024-6
35. F. A. Skinner, R. M. Boddey and I. Fendrik (eds.): *Nitrogen Fixation with Non-Legumes* (Proceedings of the 4th Symposium, Rio de Janeiro, 1987). 1989 ISBN 0-7923-0059-9
36. B. C. Loughman, O. Gašparíková and J. Kolek (eds.): *Structural and Functional Aspects of Transport in Roots.* 1989 ISBN 0-7923-0060-2; Pb 0-7923-0061-0
37. P. Plancquaert and R. Haggar (eds.): *Legumes in Farming Systems.* 1990 ISBN 0-7923-0134-X
38. A. E. Osman, M. M. Ibrahim and M. A. Jones (eds.): *The Role of Legumes in the Farming Systems of the Mediterranean Areas.* 1990 ISBN 0-7923-0419-5
39. M. Clarholm and L. Bergström (eds.): *Ecology of Arable Land – Perspectives and Challenges.* 1989 ISBN 0-7923-0424-1
40. J. Vos, C. D. van Loon and G. J. Bollen (eds.): *Effects of Crop Rotation on Potato Production in the Temperate Zones.* 1989 ISBN 0-7923-0495-0
41. M. L. van Beusichem (ed.): *Plant Nutrition – Physiology and Applications.* 1990 ISBN 0-7923-0740-2
42. N. El Bassam, M. Dambroth and B.C. Loughman (eds.): *Genetic Aspects of Plant Mineral Nutrition.* 1990 ISBN 0-7923-0785-2
43. Y. Chen and Y. Hadar (eds.): *Iron Nutrition and Interactions in Plants.* 1991 ISBN 0-7923-1095-0
44. J. J. R. Groot, P. de Willigen and E. L. J. Verberne (eds.): *Nitrogen Turnover in the Soil-Crop System.* 1991 ISBN 0-7923-1107-8
45. R. J. Wright, V.C. Baligar and R. P. Murrmann (eds.): *Plant-Soil Interactions at Low pH.* 1991 ISBN 0-7923-1105-1
46. J. Kolek and V. Kozinka (eds.): *Physiology of the Plant Root System.* 1992 ISBN 0-7923-1205-8
47. A. U. Mokwunye (ed.): *Alleviating Soil Fertility Constraints to Increased Crop Production in West Africa.* 1991 ISBN 0-7923-1221-X; Pb 0-7923-1222-8
48. M. Polsinelli, R. Materassi and M. Vincenzini (eds.): *Nitrogen Fixation* (Proceedings of the 5th Symposium, Florence, 1990). 1991 ISBN 0-7923-1410-7
49. J. K. Ladha, T. George and B. B. Bohlool (eds.): *Biological Nitrogen Fixation for Sustainable Agriculture.* 1992 ISBN 0-7923-1774-2
50. P. J. Randall, E. Delhaze, R. A. Richards and R. Munns (eds.): *Genetic Aspects of Plant Mineral Nutrition.* 1993 ISBN 0-7923-2118-9

Kluwer Academic Publishers – Dordrecht / Boston / London

MIX
Papier aus verantwortungsvollen Quellen
Paper from responsible sources
FSC® C105338

If you have any concerns about our products,
you can contact us on
ProductSafety@springernature.com

In case Publisher is established outside the EU,
the EU authorized representative is:
Springer Nature Customer Service Center GmbH
Europaplatz 3, 69115 Heidelberg, Germany

Printed by Libri Plureos GmbH
in Hamburg, Germany